International Conference on Chemical and Bioprocess Engineering 2015

IOP Conference Series: Earth and Environmental Science
Volume 36

Kota Kinabalu, Malaysia
9 – 12 December 2015

Editors:

Awang Bono
Chu Chi Ming
Tham Heng Jin
Noor Maizura Ismail
Zykamilia Kamin

ISBN: 978-1-5108-2615-1
ISSN: 1755-1307

Printed from e-media with permission by:

Curran Associates, Inc.
57 Morehouse Lane
Red Hook, NY 12571

Some format issues inherent in the e-media version may also appear in this print version.

Printed by Curran Associates, Inc. (2016)

Institute of Physics
Dirac House, Temple Back
Bristol BS1 6BE UK

Phone: 44 1 17 929 7481
Fax: 44 1 17 920 0979

techtracking@iop.org

Additional copies of this publication are available from:

Curran Associates, Inc.
57 Morehouse Lane
Red Hook, NY 12571 USA
Phone: 845-758-0400
Fax: 845-758-2633
Email: curran@proceedings.com
Web: www.proceedings.com

Table of contents

Volume 36

International Conference on Chemical and Bioprocess Engineering
9–12 December 2015, Kota Kinabalu, Malaysia

Published online: 27 June 2016

Preface

Papers

Oil & Gas

Waste

Biomass & Biofuels

Environment

IOP Conf. Series: Earth and Environmental Science **36** (2016) 011001 doi:10.1088/1755-1315/36/1/011001

ICCBPE-2015

5th International Conference on Chemical and Bioprocess Engineering (ICCBPE)
9-12 December 2015

"From Earth for the Earth"

9th to 12th December 2015, The Pacific Sutera Hotel, Kota Kinabalu Sabah Malaysia

Preface

The 5th International Conference on Chemical and Bioprocess Engineering (ICCBPE) was organized by the Chemical Engineering Programme of the Faculty of Engineering, Universiti Malaysia Sabah. The primary purpose of ICCBPE 2015 was to discuss bridging technologies from laboratory and pilot scales to commercial scales for manufacturing advanced products by involving innovative processing of biomass and feedstock suitable for SME's.

This event serves as a platform for researchers, engineers, academicians as well as industrial professionals from countries such as the United State of America, Taiwan, Korea, Iran, Thailand and Malaysia to present their research and development activities in chemical engineering sciences and applications. During the conference, a total of 81 manuscripts under the Oil and Gas, Waste, Biomass and Biofuels, and Environment categories were presented and discussed in a series of parallel presentations.

We are honoured to have distinguished keynote speakers participating in this event. They were Prof. Dr. Ahmad Fauzi Ismail, the founding Director of Advanced Membrane Technology Research Center (AMTEC) (Universiti Teknologi Malaysia, Malaysia), Prof. Dr. Ghasem D. Najafpour (Chairman of the Biotechnology Research Center, Babol Noshirvani University of Technology, Iran) and Prof. Dr. Shizhong Li (Executive Director of MOST-USDA Joint Research Center for Biofuels, China).

Last but not least, we are grateful to all of the reviewers for maintaining the standards and quality of the manuscripts throughout the reviewing process.

Kota Kinabalu, Sabah April 2016

Chairman : Professor Ir. Dr. Rosalam Sarbatly

Chief Editor : Professor Dr. Awang Bono

Editorial Team : Associate Professor Dr. Chu Chi Ming
Dr. Tham Heng Jin
Dr. Noor Maizura Ismail
Ms. Zykamilia Kamin

Universiti Malaysia Sabah (UMS)

Peer review statement

All papers published in this volume of *IOP Conference Series: Earth and Environmental Science* have been peer reviewed through processes administered by the proceedings Editors. Reviews were conducted by expert referees to the professional and scientific standards expected of a proceedings journal published by IOP Publishing.

Characterization of Malaysian coals for carbon dioxide sequestration

M Abunowara[1], M A Bustam[1], S Sufian[1], U Eldemerdash[2]
[1]Chemical Engineering Department, *Universiti Teknologi PETRONAS*, 31750 Tronoh, Perak, Malaysia
[2]Chemical Engineering Department, Benha Faculty of Engineering, *Benha University*, Benha, Egypt

E-mail: [1]abunowara1980@gmail.com, [1]azmibustam@petronas.com.my, [1]suriati@petronas.com.my, [2]usama.nour@bhit.bu.edu.eg

Abstract. Coal samples from Mukah-Balingian and Merit-Pila coal mines were characterized with ultimate, approximate, petrographic analysis, FT-IR spectra patterns, FESEM images and BET measurements to obtain information on the chemical composition and chemical structure in the samples. Two coal samples were obtained from Merit-Pila coal mine namely sample1 (S1) and sample2 (S2). The other two coal samples were obtained from Mukah-Balingian coal mine namely sample3 (S3) and sample4 (S4), Sarawak, Malaysia. The results of ultimate analysis show that coal S1 has the highest carbon percentage by 54.47%, the highest hydrogen percentage by 10.56% and the lowest sulfur percentage by 0.19% and the coal S4 has the highest moisture content by 31.5%. The coal S1 has the highest fixed carbon percentage by 42.6%. The coal S4 has BET surface area by 2.39 m^2/g and Langmuir surface area by 3.0684 m^2/g respectively. Fourier-Transform Infrared (FT-IR) spectroscopy analysis of all coal samples shows a presence of oxygen containing functional groups which considered are as active sites on coal surface. The oxygen functional groups are mainly carboxyl (-COOH), hydroxyl (-OH), alkyl (–CH, –CH_2, –CH_3), aliphatic (C–O–C stretching associated with –OH), amino (–NH stretching vibrations), (–NH stretching vibrations), aromatic (C=C), vinylic (C=C) and clay minerals. In all FE-SEM images of coal samples matrix, it can be seen that there are luminous and as non luminous features which refer to the existence of various minerals types distributed in the coal organic matrix. The bright luminosity is due to the presence of sodium, potassium or aluminium. According to petrographic analysis, all coal sample samples are range in vitrinite reflectance from 0.38% to 56% (VRr) are sub-bituminous coals.

1. Introduction

Geological sequestration of carbon dioxide (CO_2) in deep unmineable coal bed seams considered as one of promising geologic reservoirs [1] because coal bed seams have large capacity of CO_2 and the worldwide unmineable coal bed seams capacity of CO_2 storage ranges between 140 and 3000 Gt [2] and it is calculated that one ton of coal could adsorb about 30 – 35 m^3 of CO_2 at pressure range 5 to 8

[1] To whom any correspondence should be addressed.

MPa [3]. In addition to that coal usually holds in its micropores methane (CH_4). Thus, unmineable coal bed seam could be utilized for enhanced coal bed methane recovery (CBM) and CO_2 storage. Since coal has high affinity to CO_2, the injected CO_2 displaces the preadsorbed CH_4 in the coal seam [4]. However, the laboratory experiments of CO_2 adsorption isotherms on coal has been considered as one of the basic steps for the evaluation of CO_2 geological storage as well as enhanced coal bed methane recovery (CBM) [5].

The fact that the structure of coal is a mixture of inorganic and organic materials in a complex, three-dimensional network, that may change during the adsorption or desorption process, can complicate matters considerably. However, coal being highly heterogeneous from basin to basin, seam to seam, and those within the seam, it is difficult to determine the properties of the complete seam. This makes it difficult to characterize and understand the coal properties for developing ideal models for optimization of sequestration process and generating useful correlations [6].

One of the main the coal physical properties is surface area. Coal surface area is originally measured by the adsorption of N_2 at -196 °C. The Brunauer-Emmett-Teller (BET) equation was generally used to determine the surface area. However N_2 surface areas of coals are varying in the range of 1-9 m^2/g and lower than expected. It was concluded that the surface area, as measured with N_2, is a measure of the macroporosity of coals [7] because N_2 molecules at -196 °C are not completely accessible to micropores in coals. However, the micropore in coal has large micropore surface area, and accounts for 97.3% of whole area, which is about 200 m^2/g calculated using Dubinin-Polanyi equation and such a big area determine the adsorption volume of the coal [8–11]. Fourier transform infrared spectroscopy (FTIR), a non-destructive technique, is one the most powerful techniques to characterize the chemical structure of coals. The FTIR technique focuses on determining information on the functional groups on coal [12].

The porous carbon materials such as coal usually include its structure of oxygen containing functional groups for example carboxyl and hydroxyl surface functional groups. This could lead to complex CO_2 functional group–surface interactions. Rather than interacting directly with a surface, CO_2 may react indirectly via a shared proton. These embedded functional groups will change the electrostatic properties of the adsorbent surface and thus are expected to play a significant role in the adsorption mechanisms associated with CO_2 in these systems depending on the temperature and pressure conditions [13]. Coal samples from different coal bed seams vary based on a number of factors, including coal rank and moisture and carbon content, organic and inorganic matters and oxygen containing functional groups. Hence, coal characterization analysis is necessary before and after any gas adsorption measurements.

In this work, the objectives were to characterize the structural features of the coal samples using ultimate analysis, approximate analysis, Brunauer-Emmett-Teller (BET) technique, Fourier-Transform Infrared (FT-IR) Spectroscopy, Powder X-ray diffraction (XRD), Field-Emission Scanning Electron Microscope (FE-SEM) and petrographic analysis.

2. Material and Methods

2.1 Coal specimens

The coal specimens were obtained from two locations. First location is Merit-Pila coal mine and second location is Mukah-Balingian coal mine, Sarawak. From Merit-Pila coal mine, we obtained coal sample1 (S1) and coal sample2 (S2). From Mukah-Balingian mine, we obtained sample3 (S3) and sample 4 (S4). Merit-Pila and Mukah-Balingian mines are at depth 16 to 20 m below existing ground and seam thickness is generally about 1.2 to 1.4 m.

2.2 Methods

Four coals, S1, S2, S3, and S4, were grinded and dried for the characterization. The techniques were used to characterize the micro-texture of the coal samples. The methods are ultimate analysis where characterized by CHNS-932 analyzer, approximate analysis where made by thermal gravimetric analysis (TGA), Brunauer-Emmett-Teller (BET) technique and Langmuir were performed with N_2(g) at 77 K (-196 °C), determining functional groups by Fourier-Transform Infrared (FT-IR) Spectroscopy, coal surface morphology by powder X-ray diffraction (XRD), Field-Emission Scanning Electron Microscope (FE-SEM) and coal maturity (rank) by petrographic analysis.

3. Characterization of coal samples

3.1 Ultimate analysis

Ultimate analysis tests produce more comprehensive results than the proximate analyses. The ultimate test was conducted using CHNS-932 elemental analyzer (ASTM5373, Truspec, Leco, USA). The analysis was performed to determine percentages of basic elements namely carbon, hydrogen, nitrogen and sulphur (CHNS) on dry coal samples basis. As shown in Table 1, the coal sample 2 has the highest carbon percentage by 54.47%, the highest hydrogen percentage by 10.56% and the lowest sulfur percentage by 0.19%.

Table 1. The ultimate analysis for S1, S2, S3 and S4 coal samples.

Coal sample	Carbon%	Hydrogen%	Nitrogen%	Sulfur%
Sample 1	32.81	4.25	0.0	0.25
Sample 2	54.47	10.56	0.832	0.19
Sample 3	44.00	5.85	1.52	0.34
Sample 4	34.07	6.66	0.83	0.28

3.2 *Proximate analysis*

The approximate analysis signifies the percentage by weight of the fixed carbon, volatiles, ash, and moisture content in coal. It was performed using thermal gravimetric analysis (TGA, PerkinElmer, Thermal Analysis, STA 6000, USA). Raw sample of crushed coal is weighed and placed in crucible of TGA and heated up to 900±15°C. The fixed carbon is a solid combustible matter (carbon in its free state) without the light and volatile compounds. It is a value obtained from by abstracting the sun of ash, moisture and volatile matter from 100 where all values are on the same moisture reference base. However, coal S4 has the highest moisture content by 21.5% and coal S2 has the highest fixed carbon percentage by 42.6%. Hence, it is essential to measure the moisture in coal samples to evaluate the potential moisture effect on CO_2 sorption on coal.

Table 2. The approximate analysis (%) of S1, S2, S3 and S4 coal samples.

Coal sample	Moisture%	Volatiles%	Fixed carbon%	Ash%
Sample 1	11.3	26.8	42.6	19.2
Sample 2	13.4	43.3	36.7	6.6
Sample 3	21.3	25.4	32.6	10.5
Sample 4	21.5	22.2	35.8	10.5

3.3 Brunauer-Emmett-Teller (BET) technique

The BET surface area measurement was proposed by Brunauer, Emmett and Taylor. It was built on subsequent assumptions. First, the adsorbate molecules physically adsorb on a adsorbent surface, the adsorbent surface is uniform and identical, the interaction between the adsorbent and adsorbate molecules are stronger than the interaction between the adsorbate molecules, the interactions of adsorbent molecules are considered only in the vertical to the adsorbent surface, and is accounted as condensation [14].

Micromeritics (3Flex 3.00) analyzer utilized measure BET and Langmuir surface area analysis (GAT scientific Sdn Bhd in Selangor, Malaysia). Nitrogen pressures were varies and introduced onto the coal samples and quantify the amount of N_2 gas adsorbed at each target pressures. The surface area was determined for relative pressure target range from 0.05-0.35 P/P_0. It is figured out that the surface area measured using N_2 gas is a measure of macroporosity range [7] because N_2 molecules at -196 °C could not manage completely to access micropores in coals. Micropore in coal has an enormous internal surface area is about 200 m^2/g, and it represents 97.3% of entire area and this huge surface area determine gas adsorption on coal [8].

The surface area of all coal samples were obtained by measuring BET and Langmuir surface areas of the coal samples as shown in Table 1, Table 2. The sample 4 has the highest BET surface area by 2.3988 m^2/g and Langmuir surface area by 3.0684 m^2/g respectively.

Table 3. BET and Langmuir surface areas of S1, S2, S3 and S4 coal samples.

Specific surface area	S1	S2	S3	S4
BET surface area (m^2/g)	0.1824	0.1661	2.3712	2.3988
Langmuir surface area (m^2/g)	0.2027	0.2393	2.9453	3.0684

3.4 Fourier transform infrared spectroscopy (FT-IR)

Fourier-Transform Infrared (FT-IR) spectroscopy was performed to investigate the variation in functional groups and their properties change and the FT-IR spectrum for every single sample was tested and recorded in FT-IR spectrophotometer (Perkin Elmer, Spectrum One/BX) with KBr pellet.\

The FT-IR spectra of functional groups in coals S1, S2, S3 and S4 are domonstrated in Figure 1. The distinct peak in all coal samples at 1078 cm^{-1}, 1031 cm^{-1},1008 cm^{-1} and 471 cm^{-1} are ascribed to clay minerals such as quarts, kaolinite, illite and momtmorillonite. The silicate minerals (Si–O–Si) stretching vibration causes absorption at 1031 cm^{-1} and 1008 cm^{-1}. The silicates (Si–O) bending vibration in coal S3attributes to strech at 533 and 471 cm^{-1} and both peaks can also be due to the presence of ionic sulphates. A small peak of absorption at 1149 cm^{-1} in S1, S2 and S4 can be attributed th the existance of mineral quartz.

There is small but visible band at 695 cm^{-1} in coal S3 may possibly be due to magnetite (Fe_2O_3). Meadium peak at 2375 cm^{-1} in the FT-IR peaks of sample1, S2 and S4 can be attributed to isocyanate functional group (–N=C=O). Strong peak at 2969 cm^{-1} in the FT-IR of S2 and sample4 coals can be attributed to C–H stretch (alkanes) or CH_3 (aliphatic). The peaks in coal sampl1, sample2 and sample 4 at 3202 cm^{-1}, 3305 cm^{-1}, 3466 cm^{-1} and 3530 cm^{-1} is assigned to alcohols (O–H stretch) and phenols (H–bonded).

Table 4 describes FTIR band assignments of all coal samples. The peak of 1624 cm^{-1} in coal sample3 and the tiny peak of 1627 cm^{-1} are assigned to amines functional group (N–H bend). The peaks of 1082 cm^{-1} and 1152 cm^{-1} in all coal samples assigned to C–O peak which represent alcohols, carboxylic acids, esters and ethers functional groups. These functional groups are considered as oxygen-containing functional groups.

Table 4. Band assignments for the FTIR spectra of coals [15].

Bands (cm^{-1})	Assignments
400-690	Clay and silicate minerals
530-580 (s)	C–C (cycloalkanes) and C–C–CN (nitriles)
1082-1150 (s)	C–O (alcohols, carboxylic, acids, esters and ethers)
1580-1650	C=C (aromatic) and C=O (esters) and NH_3 and NH_2 (primary amines)
2000-2500	–N=C=O (isocyanate functional group)
2350-2780	–OH strongly hydrogen-bonded and NH stretching mode
2850-2960 (s)	C–H stretch (alkanes) and CH_3 (aliphatic)
2500-3300 (s,b)	O–H stretch (carboxylic acids)
3200-3600	O–H stretch, H–bonded (alcohols and phenols)

s=strong, m=medium, b=broad.

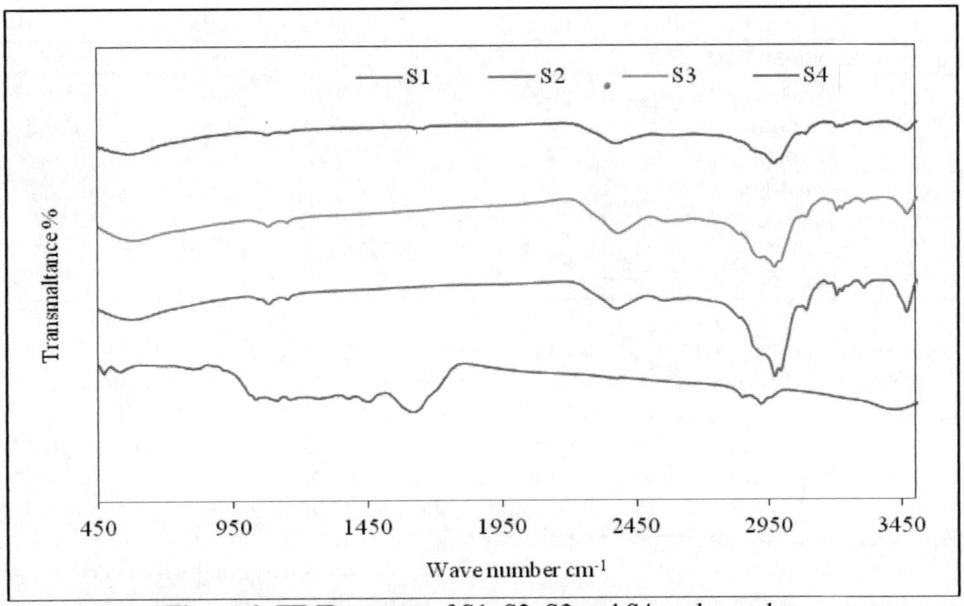

Figure 1. FT-IR spectra of S1, S2, S3 and S4 coal samples.

3.5 *Powder X-ray diffraction (XRD)*

X-ray diffraction analysis was conducted using Bruker D8 instrument. XRD is a qualitative and quantitive method conducted to identify crystal structure of various minerals in powder materials and provide information about mineralogical composition. Hence, the coal samples were subjected to XRD inspection. All coal samples were scanned 10 to 80° using Cukα radiation with step scan mode (step size 0.0173°).

The XRD patterns of all coal samples are presented in Table 5. There are two main major diffraction peaks of quartz are located around 21° and 26° in all the samples. As shown in Figure 2., XRD patterns of all coal samples show a dominance of quartz with a small percentage of kaolinite, calcite, priyite, hematite and aluminium silicate identified asandalusite (Al_2SiO_5).

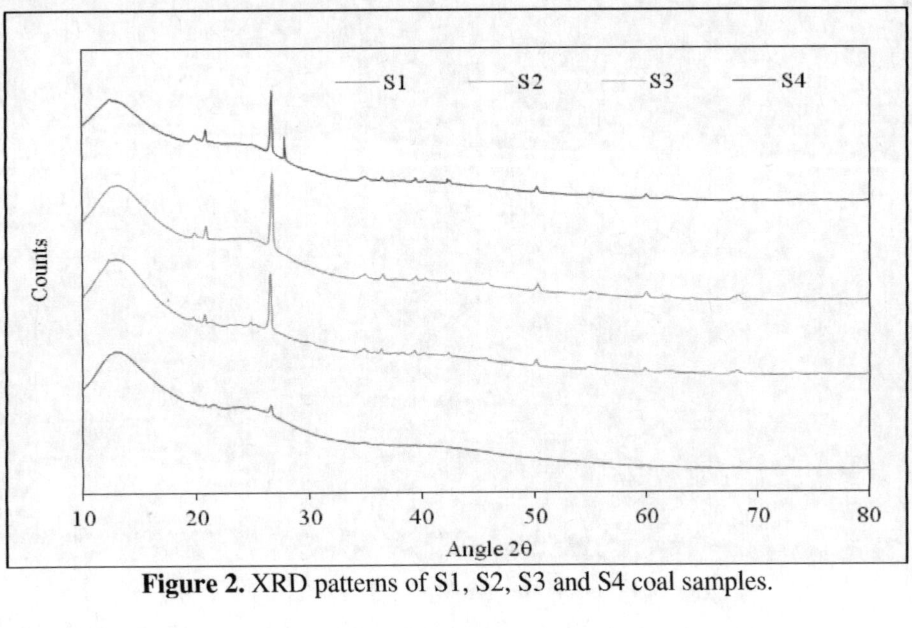

Figure 2. XRD patterns of S1, S2, S3 and S4 coal samples.

Table 5. Percentage of minerals in S1, S2, S3 and S4 coal samples.

Compound Name	S1	S2	S3	S4
Quartz	44.14%	13.90%	54.03%	19.85%
Kaolinite	15.64%	25.22%	11.35%	28.62%
Hematite	-	3.31%	2.57%	3.14%
Anatase	-	3.79%	1.37%	4.22%
Pyrite	1.65%	5.11%	-	8.65%
Illite	33.67%	39.98%	19.72%	21.64%
Dolomite	4.90%	8.68%	10.96%	13.88%

3.6 Field emission scanning electron microscopy (FE-SEM)

FE-SEM imaging was utilized to provide surface morphology as well as information about elements at magnifications of 10x to 300,000x with virtually at magnifications of 10x to 300,000x with virtually infinite depth of field. The intact and unpolished coal samples were investigated with FE-SEM instrument (SUPRA 55VP, Carl Zeiss AG, Germany). All coal specimens were scanned using high voltage with extra high tension (EHT) of 20 kV and magnification of 1.00 KX.

FE-SEM analysis was performed at effective distance of around 8 mm, utilizing derived electron imaging. The coal sample was placed on SEM stubs using double-sided conductive carbon tape. The bulk microstructure composed of homogeneously distributed network of small crystallites demonstrated existence of minerals. In all coal samples, it can be seen that there are luminous and as non luminous features as depicted in FE-SEM images in Figure 3. Those features signify the presence of minerals distributed in the organic coal matrix. The brightness sites are due to existence of aluminum, potassium or sodium. The dark sites are due to the presence of chalcophiles [1].

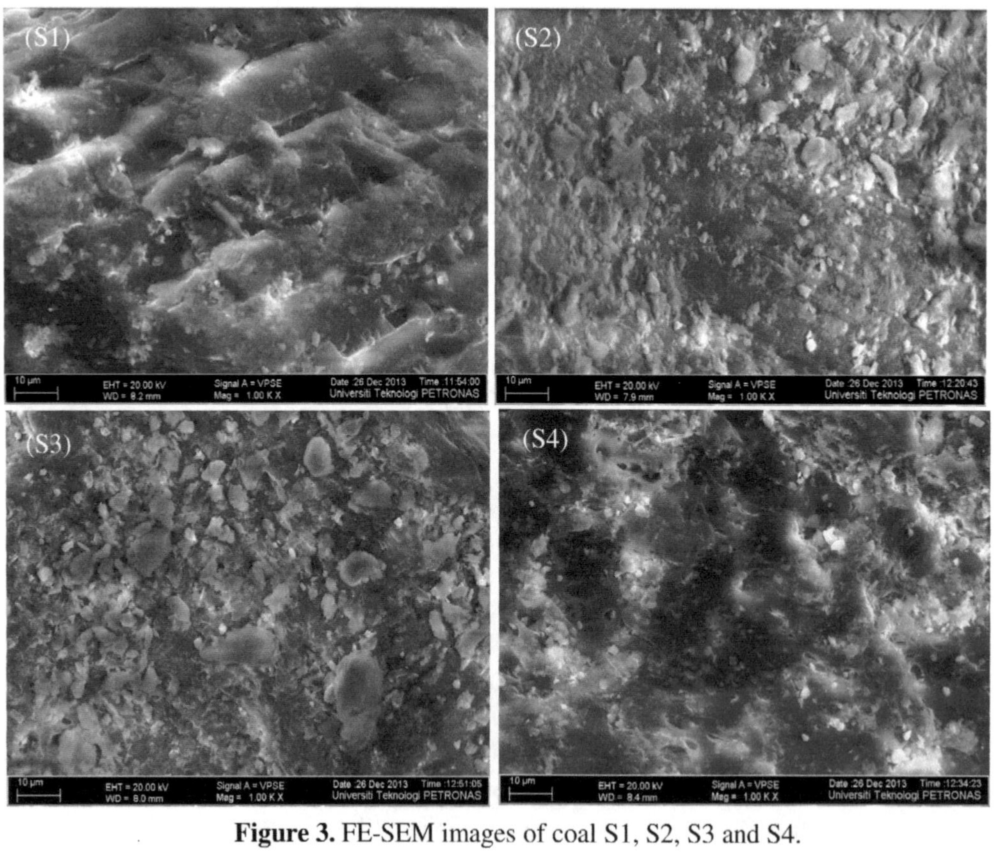

Figure 3. FE-SEM images of coal S1, S2, S3 and S4.

3.7 Petrographic analysis
Coal is often classified according to its rank, or thermal maturity. The three main classifications of coal rank from low-rank to high-rank are lignite, bituminous, and anthracite. Carbon content is often used as a measure of rank. As the carbon content of the coal increases, rank increases. Coal petrography is a microscopic method conducted to determine coal rank (degree of coalification) as well as amount and type of macerals on polished specimens of minus 20 mesh prepared coal.

Vitrinite reflectance (VR_o) usually utilized as an indicator of coal thermal maturity. Petrographic analysis measurements were conducted according to Australian standard AS 2856.3. The maximum reflectance (R_{max}) was measured in a green light with reflected light microscope-photometer using Leitz-Orthoplan. As shown in Table 6, the Vitrinite reflectance of coal samples range in vitrinite reflectance from 0.38% to 56% VRr. According to the classification of coal types based on vitrinite reflectance, sub-bituminous coals range between 0.3–0.7 of vitrinite reflectance [16,17]. These coal samples range in vitrinite reflectance from 0.38% to 56% VRr which are sub-bituminous coals.

Table 6. The Petrographic analysis of coal S1, S2, S3 and S4.

Petrographic analysis	**S1**	**S2**	**S3**	**S4**
Mean maximum reflectance[%]	0.38	0.38	0.56	0.51
Vitrinite	76.1	81.3	80.1	84.5
Liptinite	13.9	8.7	4.3	3.7
Inertintie	3.5	2.3	7.4	5.2
Minerals	6.5	7.8	8.2	6.6

4. Conclusions
The characterization study is a first step that requires to be performed before and after conducting CO_2 sorption capacity on coal. Thus, ultimate, approximate, petrographic analysis, FT-IR spectra patterns, FESEM images and BET measurements are the main basic characterization techniques. All coal samples are range in vitrinite reflectance from 0.38% to 56% (VRr) which are sub-bituminous coals. It is obtained that coal S2 has the highest carbon percentage by 54.47%, the highest hydrogen percentage by 10.56% and the lowest sulfur percentage by 0.19%. Coal S4 has the highest moisture content by 21.5% and coal sample1 has highest fixed carbon percentage by 42.6%. The FT-IR spectra for all coal specimens demonstrate the presence of alkanes (–CH, $–CH_2$ and $–CH_3$ groups), aliphatic (C–O–C stretching associated with –OH), amine (–NH stretching vibrations), aromatic (C=C), and clay minerals. In all coal samples morphology, there are luminous and as dark features. Those features signify the existence of organic minerals distributed in the organic matrix of coal. The bright luminosity is due to the presence of aluminium, potassium or sodium. The presence of moisture, minerals and oxygen functional groups do play a crucial on CO_2 sorption onto coal.

5. References

[1] Prusty B K 2008 *J. Nat. Gas Chem.* **17** 29–38
[2] Brown T D, Harrison D K, Jones J R and LaSota, K A 2007 *Int. J. Environ. Pollut.* **29** 474–483
[3] Cook P J, Rigg A and Bradshaw J 2000 *Aust. Pet. Prod. Explor. Assoc. J* **40** 654–666
[4] Ottiger S, Pini R, Storti G and Mazzotti M 2008 *Adsorption* **14** 539–556
[5] van Bergen F, Gale J, Damen K J and Wildenborg A F B 2004 *Energy* **29** 1611–1621
[6] Vishal V, Ranjith P G and Singh T N 2013 *Int. J. Coal Geol.* **105** 36–47
[7] Curt M White D H S 2005 *Energy Amp Fuels* **19** 559
[8] Qin Y, Wang Y, Yang X, Liu W and Luo W 2012 *Int. J. Min. Sci. Technol.* **22** 763–767
[9] Kutchko B G, Goodman A L, Rosenbaum E, Natesakhawat S and Wagner K 2013 *Fuel* **107** 777–786
[10] Han F, Busch A, Krooss B M, Liu Z and Yang J 2013 *Fuel* **108** 137–142
[11] Jessen K, Tang G-Q and Kovscek A R 2007 *Transp. Porous Media* **73** 141–159

[12] Wang S, Tang Y, Schobert H H, Guo Y, Gao W and Lu X 2013 *J. Anal. Appl. Pyrolysis* **100** 75–80
[13] Liu Y and Wilcox J 2012 *Environ. Sci. Technol.* **46** 1940–1947
[14] Abell A B, Willis K L and Lange D A 1999 *J. Colloid Interface Sci.* **211** 39–44
[15] Chen Y, Mastalerz M and Schimmelmann A 2012 *Int. J. Coal Geol.* **104** 22–33
[16] Wang K, Wang G, Ren T and Cheng Y 2014 *Int. J. Coal Geol.* **132** 60–80
[17] De Silva P N K and Ranjith P G 2014 *Fuel* **121** 250–259

Acknowledgments
The authors gratefully acknowledge and thank Ministry of Science, Technology and Innovation (MOSTI) for financial support provided for this research through E-science (0153AB-B41) and Centre of Postgraduate Studies (CGS), Universiti Teknologi PETRONAS (UTP), Centre of Postgraduate Studies (CGS), Universiti Teknologi PETRONAS (UTP), Malaysia.

Performance evaluation of carbon dioxide-alkanolamine-water system by equation of state/excess Gibbs energy models

H Suleman, A S Maulud[1] and Z Man
Department of Chemical Engineering, *Universiti Teknologi PETRONAS*
Bandar Seri Iskandar, 32610 Perak, Malaysia

E-mail: [1]halims@utp.edu.my

Abstract. Numerous thermodynamic techniques have been applied to correlate carbon dioxide-alkanolamine-water systems, with varying accuracy and complexity. With advent of high pressure carbon dioxide absorption in industry, the development of high pressure thermodynamic models have became an exigency. Equation of state/excess Gibbs energy models promises a substantial improvement in this field. Many researchers have shown application of these models to high pressure vapour liquid equilibria of said system with good correlation. However, no study shows the range of application of these models in presence of other competitive techniques. Therefore, this study quantitatively describes the range of application of equation of state/excess Gibbs energy models to carbon dioxide-alkanolamine systems. The model uses Linear Combination of Vidal and Michelsen mixing rule for correlation of carbon dioxide absorption in single aqueous monoethanolamine, diethanolamine and methyldiethanolamine mixtures. The results show that correlation of equation of state/excess Gibbs energy models show a transient change at carbon dioxide loadings of 0.8. Therefore, these models are applicable to the above mentioned system for carbon dioxide loadings beyond 0.8 mol/mol and higher. The observations are similar in behaviour for all tested alkanolamines and are therefore generalized for the system.

1. Introduction

Absorption of carbon dioxide in alkanolamines has become a technically and commercially proven technology. A number of alkanolamines have been tested on lab scale and classified for carbon dioxide absorption in industrial environments [1]. These include primary, secondary and tertiary alkanolamines. However, the biggest challenge is to transpire the laboratory based data into an effective industrial value and define the extent of the parametric performance. Therefore, a number of thermodynamic approaches have been proposed for description of vapour liquid equilibria (VLE) of carbon dioxide-alkanolamine-water system. These approaches can be generally classified into three types; namely, semi-empirical models, activity coefficient (G^E) models and equation of state/excess Gibbs energy (EoS/G^E) models. Semi-empirical models are easy in computation but lack precise theoretical agreement. On the other hand, G^E and EoS/G^E models are rigorous in nature but require complex simultaneous computations to retrieve specific results. Although similar in theoretical

[1] To whom any correspondence should be addressed.

development, EoS/G^E models are an extension of G^E models to high pressure non-ideal VLE. These models have been used in a variety of systems and carbon dioxide-alkanolamine-water system is no exception [2].

Vallée et al. [3] extended electrolyte equation of state [4] to carbon dioxide-diethanolamine (DEA)-water system with good correlation. The model showed low deviation at high loadings/pressure (<20%). However, the low pressure/loadings region was weakly correlated and high deviation from experimental values were observed. Chunxi and Fürst [5] extended the same model to aqueous methyldiethanolamine (MDEA) systems, with similar results. The work of Vrachnos et al. [6] used e-LCVM model to correlate absorption of acid gas in MDEA systems. Figure 1 and 2 present the correlation of their model on logarithmic pressure-solubility plot and linear deviation ratio-solubility plot. The results showed excellent correlation at high pressures/loadings but the correlation was considerably weak at low loading values.

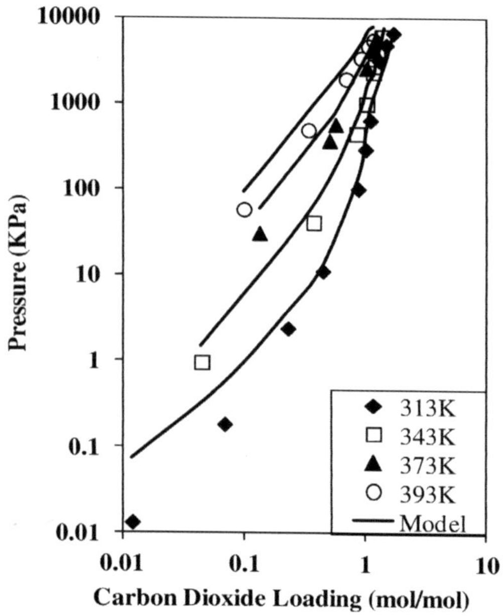

Figure 1. Pressure-Solubility plot for correlation of e-LCVM [6] against experimental data [7] for CO_2-MDEA-H_2O system

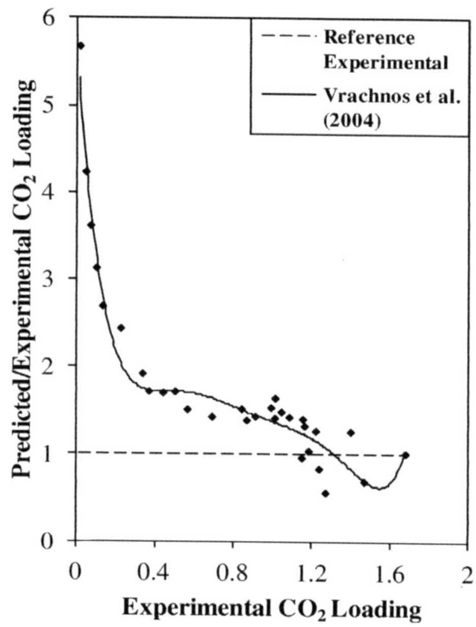

Figure 2. Ratio of Deviation against acid gas loading plot for e-LCVM [6] for CO_2-MDEA-H_2O system

Moreover, Vrachnos et al. [8] extended e-LCVM model to aqueous monoethanolamine (MEA) systems and blends of MEA-MDEA solutions with similar observations. Moreover, various researchers [9,10] have used EoS/G^E models with varying success for description of CO_2-alkanolamine-H_2O VLE. It is unanimously understood that EoS/G^E models are applicable to high pressure VLE of the said system due to pure EoS behaviour of EoS/G^E models at low loadings/pressures [11]. However, the range of application recommended in previous studies remains qualitative. Therefore, it is imperative to understand the model behaviour and develop a general quantitative range for application of EoS/G^E models to carbon dioxide-alkanolamine-water system.

In this study, Linear Combination of Vidal and Michelsen (LCVM) mixing rule [12] has been applied in conjunction with modified translated Peng Robinson EoS [13], original UNIFAC [14] and modified three characteristic parameter correlation (TCPC) electrolyte model [15] for correlation of carbon dioxide solubility in aqueous MEA, DEA and MDEA solutions. The results explain the correlative performance of EoS/G^E models and quantitatively define their range of application in the aforementioned system.

IOP Conf. Series: Earth and Environmental Science **36** (2016) 012002 doi:10.1088/1755-1315/36/1/012002

2. Thermodynamic Framework

LCVM mixing rule was selected for description of carbon dioxide-alkanolamine-water system, due to their superior correlation characteristics [16].

$$\alpha = \left(\frac{\lambda}{A_V} + \frac{1-\lambda}{A_M} \right) \cdot \frac{G^E}{RT} + \frac{1-\lambda}{A_M} \cdot \sum_i x_i . \ln\left(\frac{b}{b_i} \right) + \sum_i x_i . \alpha_i \tag{1}$$

where α defined as a/bRT for all components in the system (where a and b are defined below), λ is the LCVM constant, G^E is the excess Gibbs energy, x_i is the mole fraction of each component. Constants A_V and A_M represent individual Vidal and Michelsen mixing rules' constants.

The translated modified Peng Robinson equation of state was used for description of physical forces.

$$P = \frac{RT}{V+t-b} - \frac{a}{(V+t)(V+t+b)+b(V+t-b)} \tag{2}$$

where P is the pressure, T is the temperature, R is the universal gas constant, V is the volume of the system, a and b are the co-volume parameters and t is the translated volume.

The excess Gibbs energy function was taken as a summative function of short and long range forces.

$$G^E = G^E_{UNIFAC} + G^E_{TCPC} \tag{3}$$

The long range and columbic forces are described by modified Three Characteristic Parameter Correlation (TCPC) electrolytic model.

$$G^E_{\text{TCPC}} = \sum_{i=1}^{i} x_i \left(-z_i^2 A \left[\frac{I^{0.5}}{1+bI^{0.5}} + \frac{2}{b} \ln(1+bI^{0.5}) \right] + \frac{S}{T} \frac{I^{2n}}{v_i} \right) \tag{4}$$

where I is the ionic strength of the components, A is the Debye-Huckel constant, T is the temperature, whereas, b, S and n are fitted constants to the system.

Original UNIFAC was used to describe short range molecular forces (activity coefficients).

$$G^E_{UNIFAC} = x_i . \ln \gamma_i^C + x_i . \ln \gamma_i^R \tag{5}$$

where γ_i^C and γ_i^R represent combinatorial and residual forces, respectively. The binary interaction parameters of UNIFAC (BIP) were regressed to ternary experimental data, given in Table 1 for each alkanolamine. The selected experimental data was refined by using a recently developed statistical analysis technique [17].

Table 1. Sources of experimental data for CO_2-alkanolamine-H_2O system

Type of Alkanolamine	Researcher	No. of Data Points	Temperature, *T* (K)	Pressure, *P* (KPa)	MDEA Concentration (wt%)
MDEA	Sidi-Boumedine, et al. [18]	46	298.15 – 338.15	113 – 4560	25.73, 46.88
	Chakma and Meisen [19]	76	373.15 – 453.15	103 – 4930	19.8, 48.9
	Jou, et al. [7]	59	298.15 – 393.15	101 – 6630	23.34, 48.9
	Ma'mun, et al. [20]	34	328.15 – 358.15	118 – 813	50
DEA	Lee et al. [21]	160	273.15 – 413.15	217 – 6895	5.25 – 77.6
	Sidi-Boumedine, et al. [18]	15	298.15 – 348.15	110 – 4662	41.78
	Kennard and Meisen [22]	132	373.15 – 478.15	105 – 3747	10 – 30
	Lawson and Garst [23]	30	310.92 – 394.26	121 – 4373	25
MEA	Lee et al. [24]	112	298.15 – 393.15	100 – 7000	6.1 – 30.5
	Tong et al. [25]	48	313.15 – 393.15	101 – 984	30
	Jou, et al. [26]	34	273.15 – 393.15	103 – 6425	30
	Lawson and Garst [23]	14	333.15 – 393.15	127 – 2787	15.25

3. Results and Discussion

Figures 3 to 5 present the correlation of LCVM model in the prediction of carbon dioxide gas loadings in aqueous MDEA, MEA and DEA solutions. The model under-predicted for the values lower than 0.8 mol/mol, whereas the loadings were largely over-predicted for the loadings above 0.8 mol/mol, with a few high temperature values (353-393K) being under-predicted between loadings of 0.8 and 1.0. The over-prediction of the model is attributed to the weak description of physical absorption of carbon dioxide at high pressure.

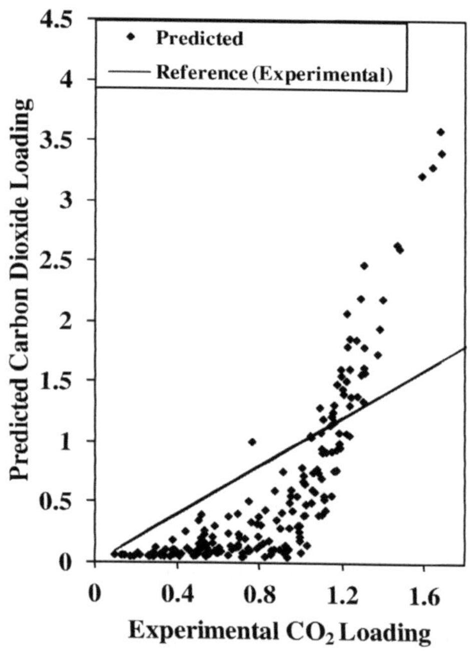

Figure 3. Correlation of CO_2 gas loadings in aqueous MDEA solutions.

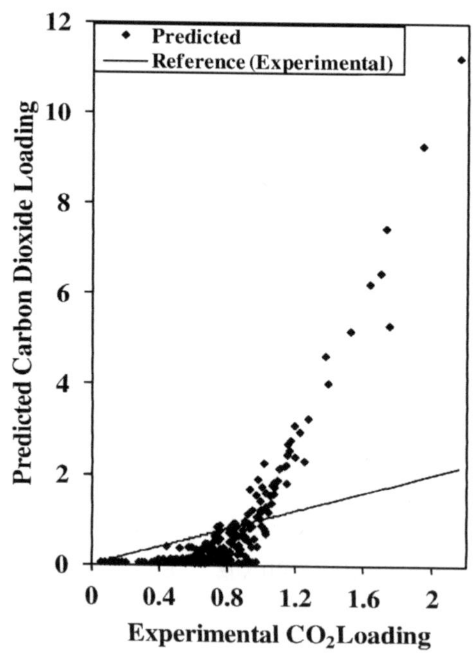

Figure 4. Correlation of CO_2 gas loadings in aqueous MEA solutions.

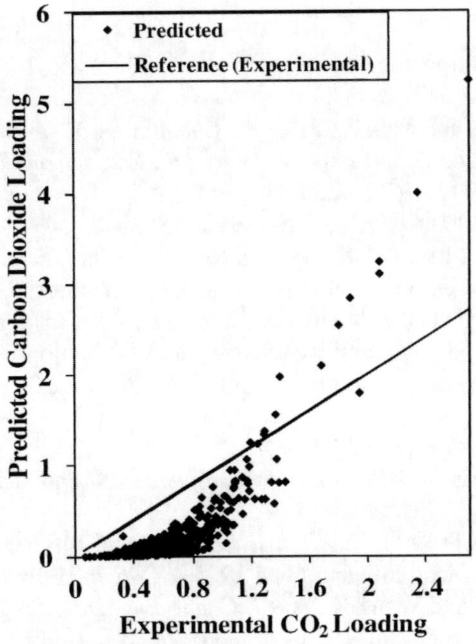

Figure 5. Correlation of CO_2 gas loadings in aqueous DEA solutions.

It is interesting to note that all figures show similar trends and the behaviour of the model is reverted around the loading values of 0.8. Thus, the loading value of 0.8 acts as a pinch point for the solubility correlation via EoS/G^E models. Theoretically, at higher loadings (at high pressures), the physical solubility of carbon dioxide in aqueous alkanolamines solutions is appreciable. However, these equilibrium absorption values are considerably lower than observed reactive dissolution at lower loadings (at low pressures) experimentally. Therefore, LCVM over-predicts the carbon dioxide loadings taking the reactive dissolution into consideration by its molecular and electrolytic equilibria. This effect is further worsened by large number of available experimental data points at low loadings (< 0.8), which heavily channels the regression of UNIFAC BIP to lower loadings. Therefore, the UNIFAC BIP should be regressed to either loadings lower than 0.8 or greater than 0.8 for adequate correlation. This would result in weak prediction of values on the other end. It is therefore, recommended that EoS/G^E model can be used for high pressure/loading data, as other comparative thermodynamic techniques are weak in correlation in this region [27], as done by previous researchers[3, 5, 6, 8] and shown in Figures 1 and 2. Moreover, it explains the qualitative statements by other researchers [27,28] that EoS/G^E models are applicable for higher CO_2 loadings. However, the current study quantitatively states that EoS/G^E models are applicable above carbon dioxide loadings of 0.8 in carbon dioxide loaded aqueous alkanolamine solutions.

4. Conclusions

The performance of EoS/G^E model has been evaluated for carbon dioxide absorption in aqueous MEA, DEA and MDEA solutions. The model showed similar performance in all tested alkanolamines and exhibited a transition in correlation characteristics at carbon dioxide loadings of 0.8 mol/mol. It defines the quantitative lower application limit for EoS/G^E models for the system, which have been previously reported by various researchers in qualitative terms. Thus, EoS/G^E models are suitably applicable to carbon dioxide loadings of 0.8 mol/mol and above for correlation of VLE in carbon dioxide solubility in aqueous MEA, DEA and MDEA solutions.

5. References

[1] Puxty G, Rowland R, Allport A, Yang Q, Bown M, Burns R, Maeder M and Attalla M 2009 *Environ. Sci. Technol.* **43** 6427–6433
[2] Suleman H, Maulud A S and Man Z 2015 *Rev. Chem. Eng.* **31** 599–639
[3] Vallée G, Mougin P, Jullian S and Fürst W 1999 *Ind. Eng. Chem. Res.* **38** 3473–3480
[4] Fürst W and Renon H 1993 *AIChE J.* **39** 335–343
[5] Chunxi L and Fürst W 2000 *Chem. Eng. Sci.* **55** 2975–2988
[6] Vrachnos A, Voutsas E, Magoulas K and Lygeros A 2004 *Ind. Eng. Chem. Res.* **43** 2798–2804
[7] Jou F Y, Mather A E and Otto F D 1982 *Ind. Eng. Chem. Process Des. Dev.* **21** 539–544
[8] Vrachnos A, Kontogeorgis G and Voutsas E 2006 *Ind. Eng. Chem. Res.* **45** 5148–5154
[9] Solbraa E 2002 (Norwegian University of Science and Technology)
[10] Huttenhuis P J G, Agrawal N J, Solbraa E and Versteeg G F 2008 *Fluid Phase Equilibria* **264** 99–112
[11] Georgios M. K and Georgios K. F 2010
[12] Boukouvalas C, Spiliotis N, Coutsikos P, Tzouvaras N and Tassios D 1994 *Fluid Phase Equilibria* **92** 75–106
[13] Magoulas K and Tassios D 1990 *Fluid Phase Equilibria* **56** 119–40
[14] Fredenslund A, Jones R L and Prausnitz J M 1975 *AIChE J.* **21** 1086–1099
[15] Ge X, Zhang M, Guo M and Wang X 2008 *J. Chem. Eng. Data* **53** 149–159
[16] Suleman H, Maulud A S and Man Z 2014 *Appl. Mech. Mater.* **625** 541–544
[17] Suleman H, Maulud A S and Man Z 2016 *Neural Comput. Appl.*
[18] Sidi-Boumedine R, Horstmann S, Fischer K, Provost E, Fürst W and Gmehling J 2004 *Fluid Phase Equilibria* **218** 85–94
[19] Chakma A and Meisen A 1987 *Ind. Eng. Chem. Res.* **26** 2461–2466
[20] Ma'mun S, Nilsen R, Svendsen H F and Juliussen O 2005 *J. Chem. Eng. Data* **50** 630–634
[21] Lee J I, Otto F D and Mather A E 1972 *J. Chem. Eng. Data* **17** 465–468
[22] Kennard M L and Meisen A 1984 *J. Chem. Eng. Data* **29** 309–312
[23] Lawson J D and Garst A W 1976 *J. Chem. Eng. Data* **21** 20–30
[24] Lee J I, Otto F D and Mather A E 2007 *J. Appl. Chem. Biotechnol.* **26** 541–549
[25] Tong D, Trusler J P M, Maitland G C, Gibbins J and Fennell P S 2012 *Int. J. Greenh. Gas Control* **6** 37–47
[26] Jou F-Y, Mather A E and Otto F D 1995 *Can. J. Chem. Eng.* **73** 140–147
[27] Zoghi A T, Feyzi F and Dehghani M R 2012 *Ind. Eng. Chem. Res.* **51** 9875–9885
[28] Kontogeorgis G M and Coutsikos P 2012 *Ind. Eng. Chem. Res.* **51** 4119–4142

Solubility prediction of carbon dioxide in water by an iterative equation of state/excess Gibbs energy model

H Suleman, A S Maulud[1] and Z Man
Department of Chemical Engineering, *Universiti Teknologi PETRONAS*
Bandar Seri Iskandar, 32610 Perak, Malaysia

E-mail: [1]halims@utp.edu.my

Abstract. The solubility of carbon dioxide in water has been predicted extensively by various models, owing to their vast applications in process industry. Henry's law has been widely utilized for solubility prediction with good results at low pressure. However, the law shows large deviations at high pressure, even when adjusted to pressure correction and improved conditions. Contrarily, equations of state/excess Gibbs energy models are a promising addition to thermodynamic models for prediction at high pressure non-ideal equilibria. These models can efficiently predict solubilities at high pressures, even when the experimental solubilities are not corroborated. Hence, these models work iteratively, utilizing the mathematical redundancy of local composition excess Gibbs energy models. In this study, an iterative form of Linear Combination of Vidal and Michelsen (LCVM) mixing rule has been used for prediction of carbon dioxide solubility in water, in conjunction with UNIFAC and translated modified Peng-Robinson equation of state. The proposed model, termed iterative LCVM (i-LCVM), predicts carbon dioxide solubility in water for a wide range of temperature (273 to 453 K) and pressure (101.3 to 7380 kPa). The i-LCVM shows good agreement with experimental values and predicts better than Henry's law (53% improvement).

1. Introduction

The solubility of carbon dioxide in water is of immense importance because of their extensive applications in chemical process industries such as beverages, enhanced oil recovery, carboxylic acids manufacture, etc [1]. Furthermore, the recent carbon capture systems for greenhouse gas control employ water based solvents, which demands better knowledge of vapour-liquid equilibrium (VLE) of the $CO_2 - H_2O$ system [2]. Carbon dioxide is a gas at ambient conditions and weakly soluble in water. The concentration of carbon dioxide in liquid phase $[CO_2]$ is usually measured by Henry's law [3], using the relationship given in equation (1). A number of relationships relate the Henry's constant given in equation (2) to temperature, both at low pressures [4] and high pressures [5].

[1] To whom any correspondence should be addressed.

$$[CO_2] = \frac{P_{CO2}}{H_{CO2}} \tag{1}$$

$$H_{CO2} = \exp\left(\frac{A}{T} + B\ln T + CT + D\right) \tag{2}$$

where T is the temperature and A-D are the regressed constants.
Carbon dioxide shows positive non-ideal behaviour upon reactive dissolution in water and correlates well with Henry's law at low pressures [4]. However, it deviates significantly from Henry's law at high pressures. The root cause of this deviation is the weak definition of constant of proportionality present in Henry's law, known as Henry's constant. Interestingly, the term 'constant' is quite misleading, as Henry's constant is a function of temperature [6].

$$H_{CO2} = \lim_{x \to 0} \frac{f_{CO2}}{x_{CO2}} \tag{3}$$

Equation (3) can be theoretically applied to any temperature and pressure in $CO_2 - H_2O$ system with limit of infinite dilution. However, real systems do not reach the limit of infinite dilution. Thus, for real systems, the infinite dilution limit has to be ignored and further assumed that f_{CO2}/x_{CO2} remains constant for non-zero values of x_{CO2}. This assumption is valid for very small values of x_{CO2}. This forms the main reason behind the large deviation of carbon dioxide from Henry's law at high pressures. The accuracy of fitting to experimental data for determination of Henry's constant also affects the end result [4] and the arbitrariness of units used in Henry's constant creates confusion over the use of Henry's law [7]. A number of previous works were evaluated for correlation of CO_2-H_2O system, using different Henry's Constant equations [4,5,8,9] similar to equation (2). The performance of these equations was either weak in correlation or limited in application terms. Therefore, a recently developed EoS/G^E model for determination of liquid phase equilibria in carbon dioxide-water system [10] has been modified for iterative calculations. The model uses an iterative procedure to determine the excess Gibbs energy values (by initial guess values of mole fraction of components), which are correspondingly used to determine new mole fraction of carbon dioxide in the binary system, by equalization of individual fugacities. The model has been compared to Henry's law prediction for carbon dioxide-water system.

2. Thermodynamic Framework

The translated modified Peng Robinson (tmPR) [11] was used as equation of state and given in equation (4).

$$P = \frac{RT}{V+t-b} - \frac{a}{(V+t)(V+t+b)+b(V+t-b)} \tag{4}$$

where P is the pressure, T is the temperature, R is the universal gas constant, V is the volume of the system, a and b are the co-volume parameters and t is the translated volume. The values of translation term and pure component values for water and carbon dioxide were taken from original work of Magoulas and Tassios [11].
Linear Combination of Vidal and Michelsen (LCVM) mixing rule [12] was used with the Universal Functional Activity Coefficients (UNIFAC) [13] to determine excess Gibbs energy and combined it with aforementioned EoS.

$$\alpha = \left(\frac{\lambda}{A_V} + \frac{1-\lambda}{A_M}\right).\frac{G^E}{RT} + \frac{1-\lambda}{A_M}.\sum_i x_i.\ln\left(\frac{b}{b_i}\right) + \sum_i x_i.\alpha_i \tag{5}$$

where α defined as a/bRT for all components in the system, λ is the LCVM constant, G^E is the excess Gibbs energy, x_i is the mole fraction of each component. Constants A_V and A_M represent individual Vidal and Michelsen mixing rules constants.

The input values to i-LCVM were vapour pressure of gas and equilibrium temperature (as in the original case of Henry's law). Initial guess values were zero and one for carbon dioxide mole fraction in liquid and vapour phase, respectively. These values were used in determination of excess Gibbs energy, which resultantly calculated the new mole fractions in the model. The new values were then used again for the determination of new excess Gibbs energy values and vice versa. The loop continued until the condition of equalization of individual fugacities was satisfied in both vapour and liquid phase. Thus, the excess Gibbs energy formed an iterative function of liquid mole fraction, hence the reason for the name. The work mode of i-LCVM is given in Figure 1. The fugacity correlation for i-LCVM was taken from Boukouvalas et al. [12].

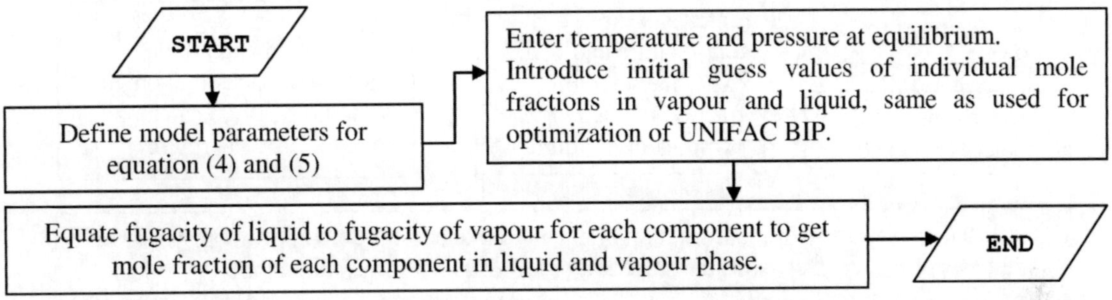

Figure 1. Work mode of i-LCVM

2.1. Optimization of UNIFAC Binary Interaction Parameters

The binary interaction parameters of UNIFAC (BIP) were fitted to experimental data given in Table 1. The selected experimental data was refined by using a recently developed statistical analysis technique [14]. Unconstrained minimization of the objective function, *OF* was carried out in MATLAB environment.

Table 1. Experimental data sources for carbon dioxide-water system

Data Source	No. of Data Points	Temp. (K)	Pressure (MPa)
Bamberger, et al. [15]	12	323 – 353	4.05 – 7.08
Campos, et al. [16]	17	303 – 323	0.10 – 0.54
Dalmolin, et al. [17]	33	288 – 308	0.11 – 0.47
Han, et al. [18]	16	313 – 333	0.37 – 2.00
Houghton, et al. [19]	103	273 – 373	0.10 – 3.65
Lucile, et al. [20]	29	298 – 373	0.54 – 5.14
Müller, et al. [21]	48	373 – 453	0.60 – 7.21
Steward and Munjal [22]	10	273 – 285	1.01 – 4.56
Valtz, et al. [23]	32	298 – 318	0.51 – 7.03
Zawisza and Malesinska [24]	27	323 – 453	0.15 – 4.62

$$OF = \frac{1}{N}\sum_{N=1}^{N}\left(\left(x_{pr}-x_{ex}\right)^2\right)^{0.5} \tag{6}$$

where, x_{pr} and x_{ex} are predicted and experimental carbon dioxide solubilities. *N* represents the number of data points. The optimized UNIFAC BIP, A_{nm} and B_{nm} for carbon dioxide and water system are given in Table 2. The root mean square error for liquid phase calculations was $7.84\text{x}10^{-4}$. The schematic for optimization of BIP of UNIFAC is given in Figure 2. The same schematic can be utilized for optimization of BIP for solubility of other gases in water/other solvents. Binary interaction parameters of UNIFAC for carbon dioxide and water system were optimized, taking zero and one as initial guess values of carbon dioxide mole fraction in liquid and vapour phase, respectively. These were based on the qualitative assumption (later quantitatively proven) that initially liquid phase completely consists of water (zero carbon dioxide) and vapour phase completely consists of carbon dioxide (no water), which is valid theoretically.

Table 2. Optimized UNIFAC BIP for carbon dioxide-water system

Component$_{(nm)}$	Binary Interaction Parameters		Initial Guess Value	
	A_{nm}	B_{nm}	Liquid$_{(nm)}$	Vapor$_{(nm)}$
Water – CO_2	283.121	2.717	1	0
CO_2 – Water	610.493	-2.929	0	1

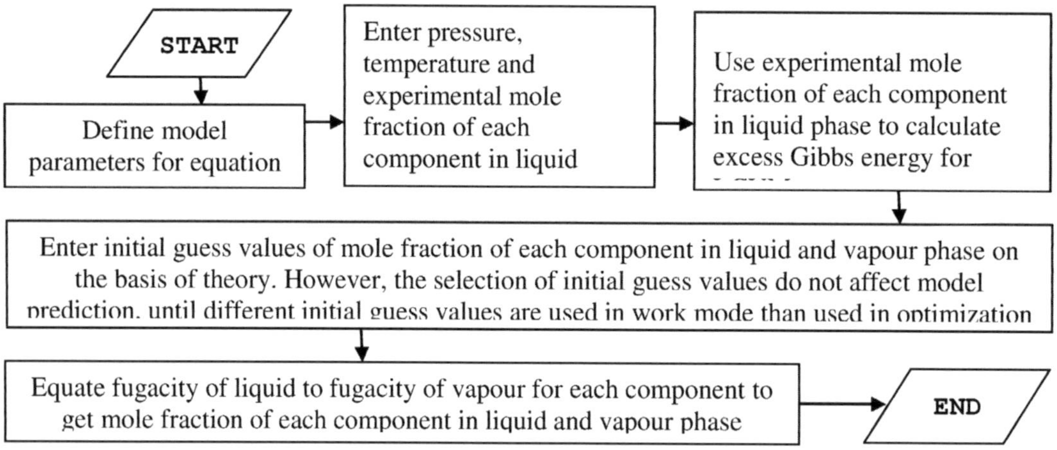

Figure 2. Algorithm for determination of UNIFAC BIP for i-LCVM

3. Results and Discussion

Figure 3 shows the prediction of Henry's law, basic LCVM and i-LCVM against experimental data.

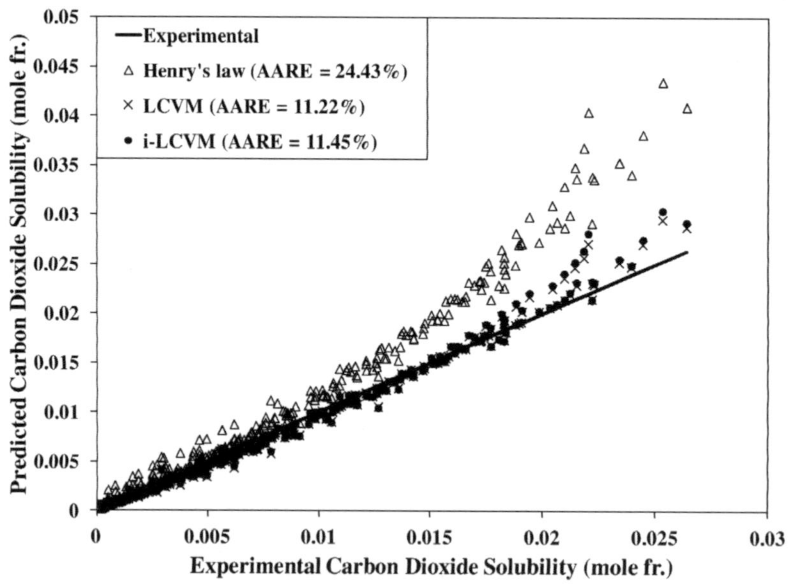

Figure 3. Comparison of i-LCVM (used predicted liquid mole fraction for excess Gibbs energy determination) with basic LCVM model (used experimental liquid mole fraction for excess Gibbs energy determination). Experimental data from Table 1.

The difference in absolute average relative percentage error (AARE) due to this iteration was found to be very small (0.23%) and deemed negligible. This proves that the effect of iteration on the prediction was insignificant and the assumption of initial conditions (liquid and vapour phase consist of water and carbon dioxide initially) was quantitatively valid. On the contrary, i-LCVM is better in prediction when compared to Henry's law for a wide range of temperature and pressure (53% improvement). However, all three models over predict at higher solubilities. This behaviour is attributed to near-critical fluid like behaviour of carbon dioxide in high pressure (≈ 7000kPa) and low temperature region (<313.15K). However, the deviation of i-LCVM is controlled as compared to Henry's law.
The presence of temperature independent UNIFAC BIP provides streamlined prediction behaviour for the i-LCVM model. Although, the prediction of Henry's law enhances with increase in temperature, the improvement is limited when compared to i-LCVM model. Similarly, the prediction of Henry's law weakens heavily beyond carbon dioxide pressures of 1000 kPa [25], the i-LCVM predictions are unaffected by increase the pressure, up to 6500 kPa at all temperatures.

4. Conclusions

An iterative EoS/G^E based model, termed as i-LCVM has been developed for prediction of carbon dioxide solubility in water. The model uses an iterative procedure to determine the activity coefficient values for both carbon dioxide and water by initial guess values of mole fraction of components. These determined values are correspondingly used to determine new mole fraction of carbon dioxide in the binary system, by equalization of individual fugacities. The algorithm is bound by the UNIFAC binary interaction parameters that results in termination of iterative procedure. The model shows good correlation for prediction of carbon dioxide solubility and results are heavily improved as compared to Henry's law. The proposed model can be used as an improved alternate to Henry's law in prediction of carbon dioxide solubilities. The solubility of other gases in different solvents can be predicted by using the proposed methodology.

5. References

[1] Kohl A L and Nielsen R B 1997 *Gas Purification, Fifth Edition* (Houston, Tex: Gulf Professional Publishing)
[2] Suleman H, Maulud A S and Man Z 2015 *Rev. Chem. Eng.* **31** 599–639
[3] Henry W 1803 *Philos. Trans. R. Soc. Lond.* **93** 29–276
[4] Carroll J J, Slupsky J D and Mather A E 1991 *J. Phys. Chem. Ref. Data* **20** 1201–1209
[5] Krichevsky I R and Kasarnovsky J S 1935 *J. Am. Chem. Soc.* **57** 2168–2171
[6] Smith F L and Harvey A H 2007 *Chem. Eng. Prog.* **103** 33–39
[7] Carroll J J 1992 *Chem. Eng. Prog.* **88** 53–58
[8] Visscher A D, Vanderdeelen J, Königsberger E, Churagulov B R, Ichikuni M and Tsurumi M 2012 *J. Phys. Chem. Ref. Data* **41** 013105–013105 – 67
[9] De V A and Conejo M S 2013 *Pure Appl. Chem.* **85** 2051–2058
[10] Humbul Suleman Q N 2015 *Chem. Eng. Trans.* **45** 589
[11] Magoulas K and Tassios D 1990 *Fluid Phase Equilibria* **56** 119–140
[12] Boukouvalas C, Spiliotis N, Coutsikos P, Tzouvaras N and Tassios D 1994 *Fluid Phase Equilibria* **92** 75–106
[13] Fredenslund A, Jones R L and Prausnitz J M 1975 *AIChE J.* **21** 1086–1099
[14] Suleman H, Maulud A S and Man Z 2016 *Neural Comput. Appl.*
[15] Bamberger A, Sieder G and Maurer G 2000 *J. Supercrit. Fluids* **17** 97–110
[16] Campos C E P S, Villardi H G D, Pessoa F L P and Uller A M C 2009 *J. Chem. Eng. Data* **54** 2881–2886
[17] Dalmolin I, Skovroinski E, Biasi A, Corazza M L, Dariva C and Oliveira J V 2006 *Fluid Phase Equilibria* **245** 193–200
[18] Han X, Yu Z, Qu J, Qi T, Guo W and Zhang G 2011 *J. Chem. Eng. Data* **56** 1213–1219
[19] Houghton G, McLean A M and Ritchie P D 1957 *Chem. Eng. Sci.* **6** 132–137

[20] Lucile F, Cézac P, Contamine F, Serin J-P, Houssin D and Arpentinier P 2012 *J. Chem. Eng. Data* **57** 784–789

[21] Müller G, Bender E and Maurer G 1988 *Berichte Bunsenges. Für Phys. Chem.* **92** 148–160

[22] Stewart P B and Munjal P K 1970 *J. Chem. Eng. Data* **15** 67–71

[23] Valtz A, Chapoy A, Coquelet C, Paricaud P and Richon D 2004 *Fluid Phase Equilibria* **226** 333–344

[24] Zawisza A and Malesinska B 1981 *J. Chem. Eng. Data* **26** 388–391

[25] Mondal B K, Bandyopadhyay S S and Samanta A N 2015 *Int. J. Greenh. Gas Control* **36** 153–160

IOP Conf. Series: Earth and Environmental Science **36** (2016) 012004 doi:10.1088/1755-1315/36/1/012004

Synthesis and characterization of supported sugar catalyst by dip coating method

J Janaun[1], N J Siambun, N N Safie
[1] Universiti Malaysia Sabah, Jalan UMS 88400 Kota Kinabalu, Sabah, Malaysia

E-mail: [1]jidon@ums.edu.my

Abstract
Sugar catalyst is a novel solid acid catalyst with reactivity comparable to that of sulphuric acid in biodiesel production. However, the fine powder form of sugar catalyst with the non-porous structure might cause large pressure drop in a packed bed reactor due to low bed porosity, affecting the reaction conversion especially in gas phase reaction. Furthermore, higher pressure drop requires higher electrical energy to drive the fluid through. Increasing the particle size is anticipated to be able to overcome the pressure drop matter. Hence, a deposition of sugar catalyst on larger particle materials was studied. Three types of materials were used for this investigation namely aluminum, silica and clay. The deposition was done via dip-coating method. The materials were characterized for their total acidity, thermal stability, functional groups, surface area, and element composition. The total acidity for SCDCAl, SCDCSi, and SCDCCl were 0.9 mmol/g, 0.2 mmol/g, and 0.4 mmol/g, respectively. The ratio of char deposited on SCDCAl, SCDCSi and SCDCCl were 0.9 g of support/g of carbon, 0.040 g of support/g of carbon, and 0.014 g of support/g of carbon respectively. FTIR and EDX analyses were carried out to determine the presence of active sites of the catalysis by identifying the functional groups such as –COOH, -OH, $-SO_3H$. The results showed that $-SO_3H$ was detected on the surface of synthesized catalysts, except for SCDCC1.The pore size of SCDCAl, SCDCSi and SCDCCl were classified as macropores because the average diameter were greater than 50nm.. The catalysts were stable up to 400 °C. The results showed that the dip-coating method could deposit sugar catalyst on aluminum, silica, and clay at low total acidity concentration.

1. Introduction

Recently, the research on biodiesel and related catalyst have attracted attention due to its potential as an alternative renewable energy and its ability to mitigate environmental pollution. Biodiesel, or known as fatty acid methyl ester (FAME), is an alternative diesel fuel that is derived from the renewable biomass that contains free fatty acids (FFA) or triglycerides, such as vegetable oils, animal fats, waste oils and waste greases [1]. Transesterifying such inexpensive feedstock with homogeneous acid catalysts is not susceptible to high FFA contents but require expensive equipment[2]. Basic homogeneous catalyst is however, not suitable as it yields bad side effect such as soap formation which require excessive water for its removal from the final products. Heterogeneous catalyzed process is widely favored compared to the other catalytic processes as it is easily separated and

[1] To whom any correspondence should be addressed.

purified. In addition to that, it reduces energy consumption, more environmentally benign and it is economically competitive due to the low total capital investment and manufacturing cost. West *et al.*,[3] showed through simulation that solid-acid catalyzed process is more economical compared to supercritical processes, homogeneous acid, and alkali catalysis processes.

A novel heterogeneous acid catalyst known as sugar catalyst or carbon-based solid acid catalyst was discovered by Toda et al.,[4] derived from the incomplete carbonization of D-glucose followed by sulfonating the pyrolyzed char with concentrated or fuming sulphuric acid. These types of catalyst were reported to show excellent catalytic performance in biodiesel synthesis especially for low quality feedstock due to the presence of polycyclic aromatic hydrocarbon containing $-SO_3H$ groups and phenolic –OH and –COOH groups [5]. Furthermore, the sugar catalyst afforded a high yield of above 90 % [6] that is comparable with sulphuric acid [7].

Fine powder of sugar catalyst is non-porous and has low specific surface area (<1 m^2g^{-1}) resulting in lack of usability in industrial operation which typically require large catalysts size. According to Melero et al.,[8], catalytic reactions are sensitive to the catalysts structure, surface atomic arrangement and coordination, which can be well controlled by tuning the specific composition, morphology and the size of catalyst. Hence, increasing the specific surface areas and its particle size are expected to overcome the limitation of sugar catalyst. The size increase can be achieved by depositing the powdered form of sugar catalyst onto supports which are industrially usable such as metal, ceramics, and glass. Chemical pretreatment known as wet-etching pretreatment is reported in this paper. It is proved to increase the adherence of the catalytic layer and forming an oxide layer by creating an anchor sites for deposition of catalyst to occur [9]. Further explanation by Reymond [10], the HCl treatment is often used to clean the metallic surfaces and helps forming a pseudo-layer accessible to chemisorption of small charged particles [11]. In this research, dip coating method with wet-etching pretreatment is reported for the coating of sugar catalyst on various commercially available supports namely aluminum, glass and clay to synthesize novel supported sugar catalyst. The aim is to evaluate the sugar catalyst deposition on the support, and subsequently chemically and physically characterized.

2. Methodology

2.1 Material

D-glucose (carbon precursor) (Systerm, D (+)-Glucose-monohydrate), Fuming Sulphuric acid (Across Organics; Extra Pure 20-30 % free SO_3), Fuming Hydrochloric acid 37 % (Merck), Fuming Nitric Acid, 65 % (Systerm), Sodium Hydroxide (Systerm pellets) were used as received. Commercial aluminum plate, low-cost microscope borosilicate glass slides (Sail Brand; typical composition: 80 % SiO_2, 14 % B_2O_3, 4 % Na_2O, and 2 % Al_2O_3)[12], and commercial clay (typical composition : 28.6 % Si, 15.1 % Al, 0.66 % Fe, 0.3 % Ti, 1.69 % K, 7.07 % loss of ignition) [13] were used as supports.

2.2 Preparation of supported sugar catalyst by dip coating on aluminium, silica and clay

The synthesis of supported sugar catalyst comprised of few important steps namely; (i) pretreatment of support, (ii) pretreatment of carbon precursor, (iii) pyrolysis of supported D-glucose on aluminum, glass and clay, and lastly (iv) sulphonation using fuming sulphuric acid. The steps are summarized in Figure 1.

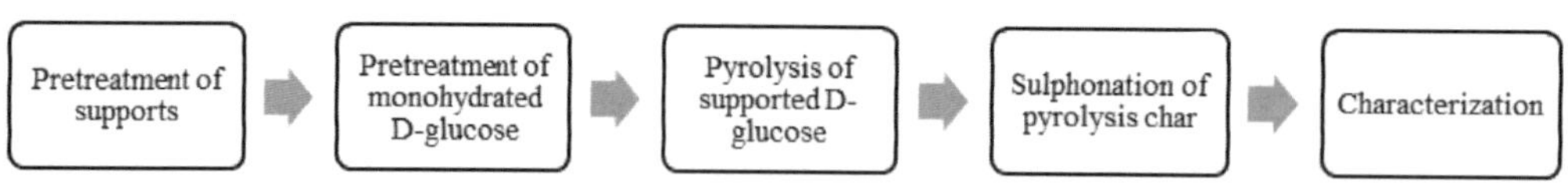

Figure 1. Schematic diagram of synthesis route of supported sugar catalyst on aluminium, silica and clay

Initially the supports were cut into smaller pieces. Aluminum supports were cut into rectangular shape size of 1.0 cm X 0.5 cm, silica supports were cut into size of 1.0 cm x 1.0 cm using glass cutter, clay support were crushed to irregular shapes and sizes larger than 4 mm. About 30 g of supports underwent a wet-acid etching pretreatments with 37 % fuming hydrochloric acid, HCl and 65 % fuming nitric acid, HNO_3. Supports were immersed in 20 mL of 37 % fuming HCl solutions at retention time of 2-3 minutes to improve superficial roughness, and subsequently in 20 mL of 65 % fuming HNO_3 solutions at 80 °C for 5 minutes. After drying in oven at temperature of 120 °C that is higher than boiling point of hydrochloric acid (83 °C) and nitric acid (110 °C), for 1 hour to remove the unreacted nitrate and chloride ions, the support was immersed in a solution contained 85 g of D-glucose and 25 g of deionized water. The supported D-glucose was dried in the oven at 104 °C to remove water. The pyrolysis was then conducted in a tube furnace at 400 °C for 4 hours under N_2 atmosphere at heating rate of 10 °C/min. Multiple vapor phase sulphonation was done accordingly [14]. In the multiple vapor phase sulfonation process, $-SO_3H$ group was introduced into the framework of carbon char through heating of concentrated sulfuric acid in the two neck round bottom flask at 150 °C for 6 hours, the sulfuric acid vapor produced is then diffused and absorbed onto the external surfaces and the pores of char particles. The sulphonation was repeated three times. The supported sugar catalyst was named as supported sugar catalyst on aluminum (SCDCAl), supported sugar catalyst glass (SCDCSi) and supported sugar catalyst clay (SCDCCl) accordingly.

2.3 Characterization of supported sugar catalyst

Total acidity was determined using a back-titration method. A mass of 0.1 g of sample was poured in a beaker containing 60 mL of 0.008 M NaOH, and stirred for 30 minutes at room temperature. The NaOH solution as the analyte was titrated using 0.02 M HCl as the titrant. Phenolphthalein was used as indicator to detect the end point at pH 7. The functional groups of synthesis catalysts were determined by using Fourier Transform Infrared Spectrometry (FTIR) with the scan range of 650 cm^{-1} to 4000 cm^{-1}. The pore size and surface area of catalysts were determined by using Automated Mercury Porosimeter, Pascal 440 series via mercury intrusion at density temperature of 25 °C and pressure at 200 MPa. The SEM-EDX analyses were carried out at Electro High Tension (EHT) = 5.00 kV, Working Distance (WD) = 5 - 7 mm, vacuum mode = variable pressure with the magnification power of 100x for support and 1000x for supported sugar catalyst, while signal VPSE was chosen for the non-conductive samples. The structure of carbon material was analyzed using X-Ray Diffraction (XRD) Philips Xpert Pro. The XRD patterns were analyzed under Cu Kα1 radiation (λ = 1.541 nm) and Kα2 radiation (λ = 1.541 nm). The determination of thermal stability of the catalysts was carried out using Thermal Gravimetric Analysis (Perkin-Elmer) in nitrogen atmosphere with flowrate of 20 mL/min and the temperature rate was set at 10 °C/min to 400 °C.

3. Results and discussion

3.1 Total char coated and total acidity of sugar catalyst deposited on different supports

As shown in Table 1, the ratio of char deposited on aluminum has the highest value of 0.9 (g_c/g_s), followed by clay with 0.4 (g_c/g_s), and silica with 0.2 (g_c/g_s). During the wet-etching treatment, only aluminum reacted vigorously which promoted its roughness compared to clay and silica. At the surface of the coating the chemical reactivity and the roughness must be considered in addition to the shear strength [15]. The roughness of the surface creates a better grip for the char, compared to the smooth surface; reducing the gliding of the carbon char from the surface. This suggested that the roughness of the aluminum support, increased its shear strength and resulted in high amount of char coated. As denoted in Table 1, SCDCAl, SCDCSi and SCDCCl possessed lower total acidity compared to the powdered sugar catalyst that has 3.05 $mmolg^{-1}$ [14]. This indicated that, the sulphonic ions were loosely bonded to the support that easily washed away during washing process, instead of forming a hydrogen bonding with the carbon char [16]. The SCDCAl has the highest total acidity, confirming higher char deposits gives higher total acidity.

Table 1. Amount of char successfully coated on three different supports with respective total acidity

Sample	Type of support	Char coating (g)	Ratio of char deposited on the support, weight of char, g_c /weight of support g_s, (g_c/g_s)	Total acidity, TA (mmol/g)
SCDCAl	Aluminum	49.74	4.97	0.9
SCDCSi	Silica	1.04	0.04	0.2
SCDCCl	Clay	0.43	0.01	0.4

3.2 FT-IR

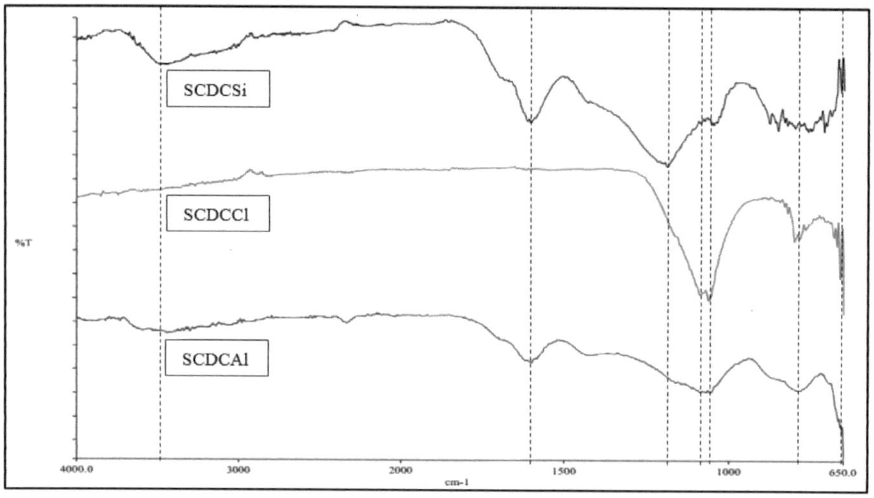

Figure 2. FT-IR spectra of supported sugar catalyst on aluminium, silica and clay

Figure 2 shows the FT-IR analysis of SCDCAl, SCDCSi and SCDCCl. SCDCCl and SCDCSi exhibit peak at 1050 cm^{-1} which indicated a presence of $–SO_3H$ groups [4]. The same peak observed in SCDCAl but with broader and weaker peak. SCDCAl and SCDCSi exhibits peak at 1600 cm^{-1} due to the C=C stretching [17]. The bands at 756 cm^{-1} observed for all supported catalyst were related to the out of plane blending modes of C-H bonds of aromatic and heterotopic compounds [18]. A strong stretching band at 1020 cm^{-1} to 1250 cm^{-1} indicated a presence of C-O bond. The band at 1251 cm^{-1} was related to Ar-OH stretching or aromatic acidic groups [20]. The FT-IR spectra coupled with EDX data which confirmed that SCDCAl and SCDCSi contained large amount of oxygen, the carbon deposited consist of polycyclic aromatic carbon sheets containing $–SO_3H$, –COOH, and –OH moieties which also in agreement Peng et al., [21] for the sugar catalyst.

3.3 Surface area

Based on Table 2, the surface area of aluminum and silica have increased by 63 % and 30 % respectively after catalyst deposition, whereas it has decreased 22 % for clay. The surface are in clay decreased due to the presence of sugar catalyst that cause blockage of pores of the clay thus reducing of surface SCDCAl, the pores mainly come from the surface area of the deposited sugar catalyst. The increase of surface area of SCDCAl and SCDCSi compared to their original support can be associated with the surface area of the deposited sugar catalyst. This is because Aluminum and Silica used were non-porous. The pore size of SCDCAl, SCDCSi and SCDCCl could be classified as macropores because the average diameters were greater than 50nm.

Table 2. Surface area and total porosity (%) of support and supported sugar catalyst obtained from Automated Mercury Porosimeter, Pascal 440 series, 200MPa, 25 °C

Parameters	Support			Supported Catalyst		
	Aluminum	Silica	Clay	SCDCAl	SCDCSi	SCDCCl
Total specific surface area (m^2/g)	1.24	0.71	6.29	3.38	1.02	4.91
Average pore diameter (nm)	587.70	4051.25	1612.01	3844.73	116.79	1442.22
Total porosity (%)	4.68	1.90	58.86	6.81	0.3339	15.57

3.4 EDX analysis

The EDX analysis showed that the weight percentage of sulphur was the highest for SCDCSi with 0.06 wt. %, followed by SCDCAl with 0.04 wt. %, sulphur was not detected in SCDCCl. SCDCAl had slightly lower composition of element C, S, and O compared to SCDCSi. Table 1 shows clay composed of the lowest ratio of char, that less sites available for sulphonic ions, $-SO_3H$ to attach during sulphonation. Despite of that, SCDCCl possessed the highest amount of oxygen. This was likely due to the nature of support that consisted of closely packed oxygen and hydroxyls in octahedral and tetrahedrons arrangements [23]. This data is in agreement with the total acidity analysis where SCDCCl composed of higher acidity compared to silica, due to presence of oxygen containing moieties such as hydroxyl and carboxylic acid groups.

Table 3. EDX analysis of supported sugar catalyst

Catalyst	Composition (wt. %)		
	C	S	O
SCDCAl	8.44	0.04	2.31
SCDCSi	8.90	0.06	3.03
SCDCCl	8.64	0.00	3.89

3.5 SEM analysis

Figs.4 (a) to (c) depicted the morphology of supports via SEM images. The supports had different morphologies, due to the variation in elements that present. Based on SEM images, the surface of glass is smooth compares to aluminum and clay.

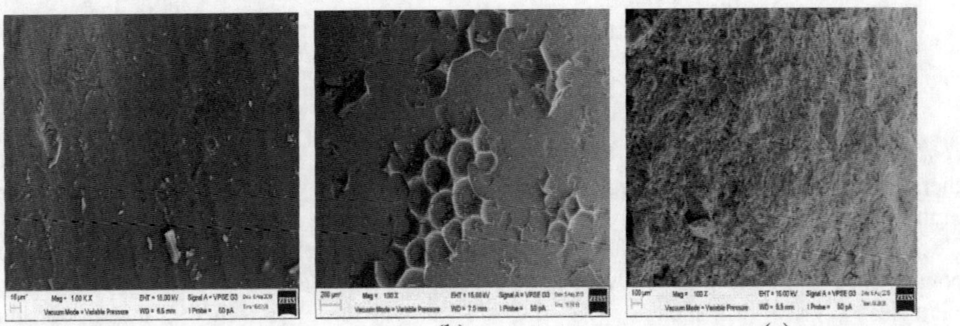

(a) (b) (c)

Figure 3.SEM Images (a) aluminum support (b) silica support (c) clay support at magnification 100x

As indicated in Table 1, the aluminum support attached most sugar catalyst. As shown in Fig 5(a), obvious layers of sugar catalyst with many pores is visible. On the other hand, only one layer of sugar catalyst was attached to the silica. Correspondingly, sugar catalyst attached on clay is less obvious due to its morphology explained by Nayak and Singh [23] ,that clay materials possess a layered structure. However, it can attach higher content of sugar catalyst compared to silica because according to Alkan et al., [24] clay sorption properties come from their high surface area and high porosity.

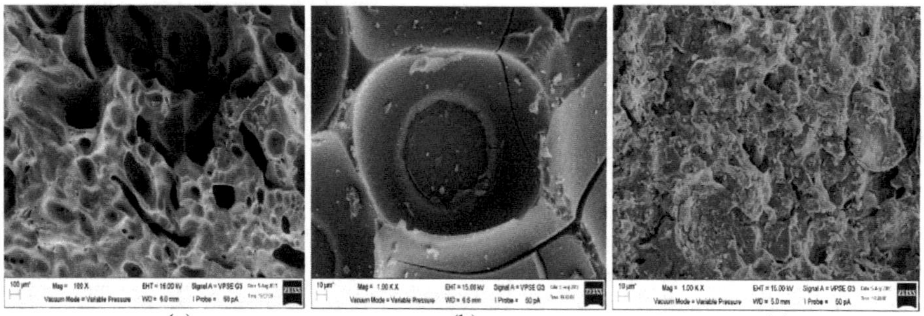

(a) (b) (c)

Figure 4. SEM Images (a) SCDCAl at magnification 100x (b) SCDCSi at magnification 1000x (c) SCDCCl at magnification 1000x.

3.6 XRD

As shown in Figure 6, X-ray diffraction (XRD) of SCDCAl and SCDCSi displayed a broad diffraction peak (2θ = 10 - 30 °) and a weak diffraction peak (2θ=35 -50 °) which can be assigned to amorphous carbon, composed of aromatic carbon sheets oriented in a considerably random fashion [16]. Meanwhile the SCDCCl has different diffraction patterns due to the least carbon contained. The weak diffraction peak (2θ = 70 ^{0}) could be associated with titanium [13].

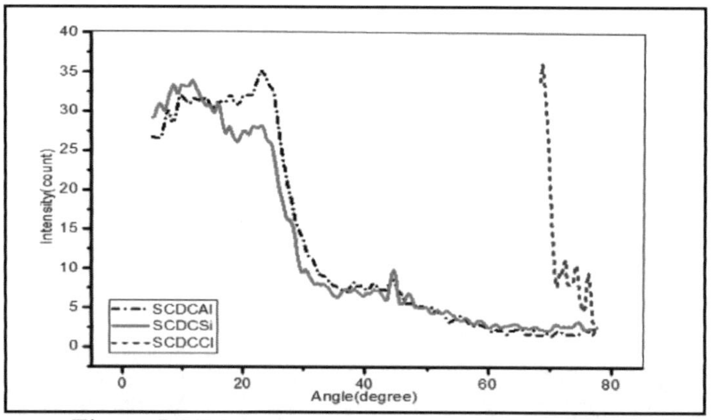

Figure 5. XRD patterns of supported sugar catalysts

3.7 TGA

The thermal stability of supported sugar shows that the supported sugar catalysts underwent two stages of weight loss from room temperature to 400 °C. The first weight loss occurred at 70 °C for SCDCAl, 50 °C for SCDCSi and 45 °C for SCDCCl. This is possibly due to the loss of low molecular weight compounds such as water. From 70 °C to 200 °C there is major decline of weight due to the hydration of moisture and compounds such as $-NO_3$ and –Cl that was loosely attached to the support during the wet-etching pre-treatment. The second stage weight loss was at 260°C for SCDCAl, 245 °C for SCDCSi and 195°C for SCDCCl. These are associated with the decomposition of -OH and -COOH compounds. The results showed that the temperature range was suitable to dehydrate most of the hydroxides, leaving only carbon char. The samples weight remained constant from 360 °C to 400 °C. The total weight loss was less than 10% from room temperature to 400 °C.

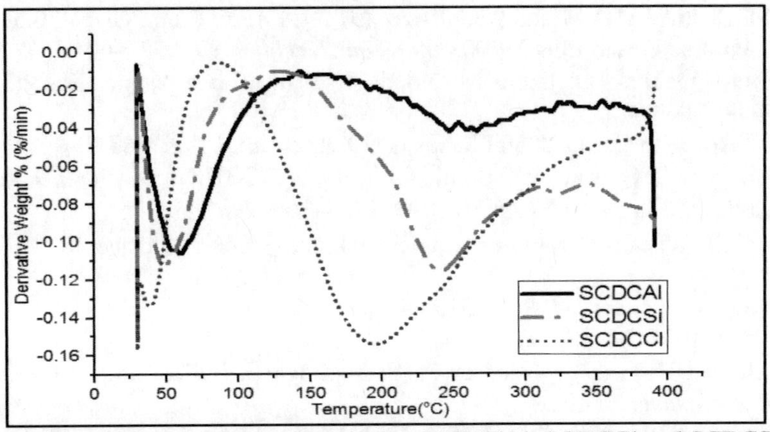

Figure 6. Derivative thermal analysis of SCDCAl, SCDCCl and SCDCSi

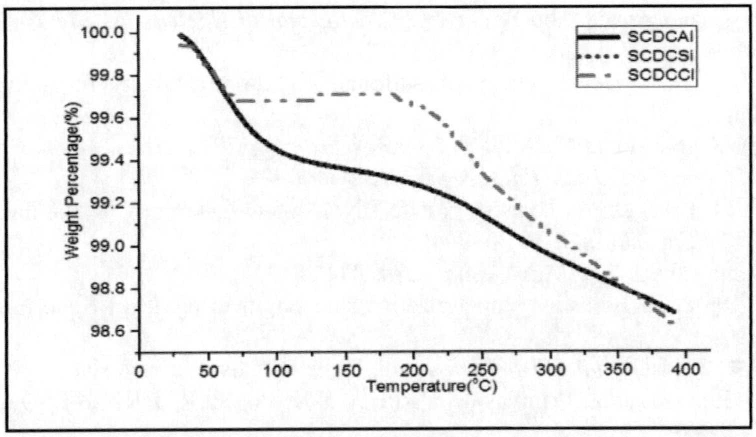

Figure 7. Thermo-gravimetric analysis of SCDCAl, SCDCSi, SCDCCl

4. Conclusions

Sugar catalyst was deposited on three different supports namely aluminum, silica and clay using dip coating method. The results showed that sugar catalyst could be deposited on these supports, but the amount of sugar catalyst is very low which influenced the total acidity. The total acidity of supported sugar catalyst at SCDCAl, SCDCSi, and SCDCCl were 0.9 mmol/g, 0.4 mmol/g and 0.2 mmol/g, respectively. Aluminum support was coated the highest ratio of char with 4.97 g/g, and has the highest total acidy. The higher amount of char deposited, the higher the total acidity could be achieved. The FT-IR spectra coupled with EDX data indicated that SCDCAl and SCDCSi contained high amount of oxygen, that the carbon sheets deposited consist of polycyclic aromatic carbon sheets containing –SO_3H, –COOH, and –OH moieties except SCDCCl. , which showed the presence of similar groups in FTIR, but EDX data showed the absence of sulphur element. The pore size of SCDCAl, SCDCSi and SCDCCl are classified as macropores. The sugar catalyst deposited was amorphous as shown by diffraction pattern in XRD analysis. The supported sugar catalysts underwent two stages of weight loss from room temperature to 400 °C, the total loss was less than 10 %.

Reference

[1] Janaun J and Ellis N 2010 *Renew. Sustain. Energy Rev.* **14** 1312–20

[2] Sani Y M, Daud W M A W and Abdul Aziz AR 2014 *Appl. Catal. Gen.* **470** pp 140–61

[3] West A, Posarac D, and Ellis N 2008 *Bioresour. Technol.* **99** 6587–601

[4] Toda M, Takagaki A, Okamura M, Kondo JN, Hayashii S Domen K, and Hara M 2005 *Nature*. **438** 178

[5] Mo X, Lotero E, Lu C, Liu Y, and James G G J 2008 *Catal. Lett* **123** 1–6

[6] Zong M H, Duan Z Q, Lou W Y, Smith T J, and Wu H 2007 Green Chem. **9** 434–7

[7] Nakajima K and Hara M 2007 *Am. Soc.* **90** 3725–34

[8] Melero J A, Bautista L F, Morales G, Iglesias J, and Sánchez-Vázquez R 2010 *Chem. Eng. J.* **161** 323–31

[9] Meille V 2006 *Elesevier Sci. Publ.* **315** 1–17

[10] Reymond J P 2001 *Catal. Today*. **69** 343–349

[11] Suknev A, van Veen A C, Toktarev A, Sadovskaya E, Bal'zhinimaev B, and Mirodatos C 2004 *Catal. Commun.* **5** 691–5.

[12] Naqshbandi M, Canning J, Gibson B C, Nash M M, and Crossley M J 2012 *Nat. Commun.* **3** 1188

[13] Saat A, Hamzah Z, and Abu Bakar Z 2009 *Journal of Nuclear and Related Technologies* **6** 230–6

[14] Janaun J A 2012 Development of sulfonated carbon catalysts for integrated biodiesel production

[15] Holmberg K and Matthews A 2009 *Coatings Tribology Properties, Mechanisms, Techniques and Applications in Surface Engineering*. Elsevier.

[16] Okamura M, Takagaki A, Toda M, Kondo J N, Domen K, Tatsumi T, Hara M, and Hayashi S 2006 Carbon *Chem. Mater.* **18** 3039–45

[17] Janaun J and Ellis N 2011 *Appl. Catal. Gen.* **394** 25–31

[18] Brewer C E 2012 Biochar characterization and engineering Iowa State University, Ames, Iowa 88-9

[19] Bruice P Y, 2011 *Organic Chemistry*, Sixth. United States: Prentice Hall

[20] Rafi J M, Rajashekar A, Srinivas M, Rao B V S K, Prasad R B N and Lingaiah N 2015 *RSC Adv.* **5** 44550–6

[21] Peng L, Philippaerts A, Ke X, Van Noyen J, De Clippel F, Van Tendeloo G, Jacobs P A, and Sels B F 2010 *Catal. Today*. **150** 140–6

[22] Nayak P S and Singh P K 2007 *Springer-Verl.* **30** 235–8

[23] Alkan M, Demirbas Ö, Çelikçapa S, and Doğan M 2004 *Elesevier Sci. Publ.*, **116** 135–45

[21] Peng L, Philippaerts A, Ke X, Van Noyen J, De Clippel F, Van Tendeloo G, Jacobs P A, and Sels B F 2010 *Catal. Today*. **150** 140–6

[22] Nayak P S and Singh P K 2007 *Springer-Verl.* **30** 235–8

[23] Alkan M, Demirbas Ö, Çelikçapa S, and Doğan M *Elesevier Sci. Publ.*, **116** 135–45

Physical explosion analysis in heat exchanger network design

M Pasha[1], **D Zaini**[1], **A M Shariff** [1]
Centre for Advance Process Safety (CAPS), Chemical Engineering Department, Universiti Teknologi PETRONAS, 32610 Seri Iskandar, Perak, Malaysia

E-mail: [1]mohsinpasha115@gmail.com , [2]dzulkarnain.zaini@petronas.com.my [3]azmish@petronas.com.my

Abstract. The failure of shell and tube heat exchangers is being extensively experienced by the chemical process industries. This failure can create a loss of production for long time duration. Moreover, loss of containment through heat exchanger could potentially lead to a credible event such as fire, explosion and toxic release. There is a need to analyse the possible worst case effect originated from the loss of containment of the heat exchanger at the early design stage. Physical explosion analysis during the heat exchanger network design is presented in this work. Baker and Prugh explosion models are deployed for assessing the explosion effect. Microsoft Excel integrated with process design simulator through object linking and embedded (OLE) automation for this analysis. Aspen HYSYS V (8.0) used as a simulation platform in this work. A typical heat exchanger network of steam reforming and shift conversion process was presented as a case study. It is investigated from this analysis that overpressure generated from the physical explosion of each heat exchanger can be estimated in a more precise manner by using Prugh model. The present work could potentially assist the design engineer to identify the critical heat exchanger in the network at the preliminary design stage.

1. Introduction

The explosion is defined as the sudden and violent release of energy. Violence of explosion depends upon the energy released rate [1]. Blast effect generated from the explosions can create transient change in the gas density, velocity and pressure of the surrounding air. Moreover, the loss of containment of pressurized gas, boiling liquid, flammable material and the reaction runaway could potentially lead to various types of explosions. Few well-known explosion mechanisms are presented in the Table 1.

Shell and Tube Heat Exchangers (STHE) are being extensively deployed in the chemical process industries. These heat exchangers serve as heater, cooler, partial or total condenser, evaporator, decomposer and boiler. Tubular Exchanger Manufacturing Association (TEMA) is responsible for issuing and updating its design guidelines on a regular frequency [2]. The latest version of these guidelines was updated by TEMA in 2007. The failure of STHE could potentially cause an enormous production loss and

[1-3] To whom any correspondence should be addressed.

have a potential of credible events such as explosion and fire. Occasionally, the failure of these heat exchangers has also been observed just after a few months of service [3, 4]. On 10th June, 2008, physical explosion observed through bursting of heat exchanger due to over pressurisation in rubber industry located in Houston, United State [5]. Recently, the catastrophic failure of STHE in Tesoro Anacortes Refinery, Washington caused seven fatalities [6].

Table 1 Various types of explosion

Explosion Type	Characteristics
Vapor cloud explosion (VCE)	Resulted from the combustion of flammable loss of containment
Boiling liquid expanding vapor explosion (BLEVE)	Resulted with the rapid released of pressurized liquid above from its atmospheric boiling point
Physical explosion	Caused by the catastrophic failure of pressurized system
Chemical explosion	Explosion resulted from the uncontrollable chemical reaction

An inadequately designed STHE could potentially have a high failure potential. The design errors frequently cause accidents in the chemical process industries [7, 8]. The failure frequency of STHE is relatively higher as compared to other process equipment [9]. Therefore, lack of consideration in performing safety analysis could possibly lead towards the failure of STHE. Hence, there is a need to properly explore the consequent impacts of explosions during the heat exchanger network design. Process design can easily be modified in the preliminary design stage. Although the level of process information is lower at this design stage, but process design can be changed smoothly and economically at this initial design stage [10]. The over pressure effect resulted from the physical explosions of heat exchangers is primarily discussed in this work.

2. Physical Explosion Models

2.1 Baker's Model

Baker et al. (1983) developed a model for estimating the burst pressure. In this method energy of explosion is evaluated by using Brode equation [11]. It is represented by below equation

$$E = \frac{(P_1 - P_0) \times V}{\gamma - 1} \tag{1}$$

P_1 and P_0 are the initial and final (ambient) pressure of expanding gas (Pa), V is total volume of gas (m^3), γ is heat capacity ratio and E is energy (J) of expanding gas. This expression indicates the required energy to raise the gas pressure from atmospheric pressure to initial pressure by keeping volume constant. However, the real gas explosions behave differently from this assumption. The expansion energy is partly dissipated in vessel deformation and rupture. The remaining explosion energy is utilized to generate blast over pressure. Peak overpressure (P') can be estimated from the graph as mentioned in Figure 1 by using Scaled distance (R').

$$R' = r \times \left(\frac{P_o}{E}\right)^{\frac{1}{3}} \quad (2)$$

Where r is absolute distance (m) and P_o is ambient pressure (Pa).

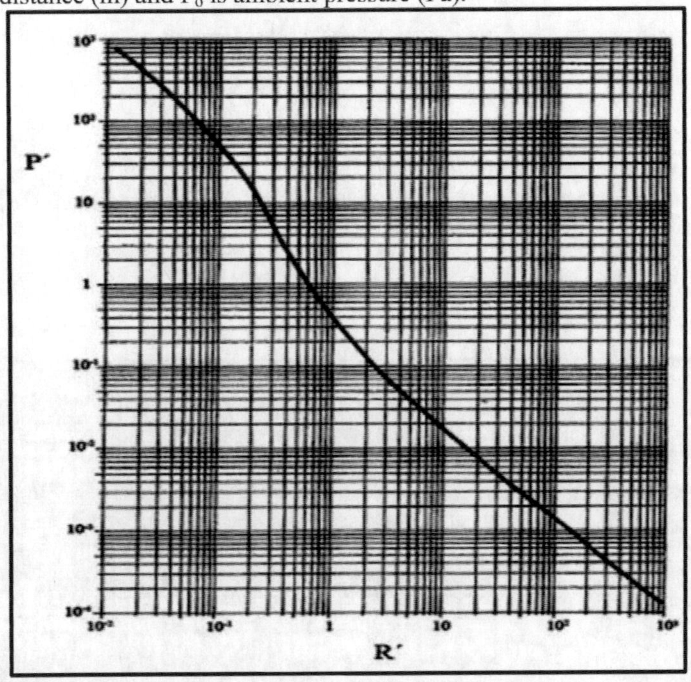

Figure 1. Estimation of overpressure from scaled distance in Baker Model [11]

2.2 Prugh's Model
Explosion energy can be estimated by using Brown equation in Prugh's model. It is based on isothermal gas expansion from atmospheric to initial conditions [11].

$$E = \frac{V}{V_R} \times \frac{P_1}{P_R} \times R \times T_1 \times \ln\left(\frac{P_1}{P_o}\right) \quad (3)$$

Where R is ideal gas constant, V is initial volume of compressed gas, V_R and P_R are volume and pressure at standard conditions. Explosion energy (E) is converted into equivalent mass of TNT and it can be used to estimate the blast effects.

$$W = 1.39 \times 10^{-6} V\left(\frac{P_1}{P_0}\right) RT_1 \ln\left(\frac{P_1}{P_2}\right) \quad (4)$$

Where W is equivalent mass of TNT, V is volume of compressed gas, P_1, P_2 and P_o are the initial, final and standard pressure and T_1 is the initial temperature of the expanding gas respectively. The analogy of the explosion of a pressurized gas container to a condensed phase point source of TNT is not adequate in the near field. As vessel is not a point source. Prugh suggested a modified method for estimating the virtual distance from explosion center. It can be estimated by subtracting the geometrical distance between center and external surface of the vessel. The actual distance is obtained by adding virtual distance to selected distance from the source. Maximum overpressure at the contact surface between the expanding gas and air can be evaluated by following equation [11].

$$P_b = P_a \left[1 - \frac{3.5(\gamma - 1)(P_a - 1)}{\sqrt{\left(\frac{\gamma T}{M}\right)}\left(1 + 5.9P_a\right)} \right]^{-2\frac{\gamma}{\gamma - 1}} \quad (5)$$

P_b is the burst pressure of the vessel and P_a is the pressure at the surface of the vessel in bar. Scaled distance (Z) would be required to evaluate scaled over pressure (P_s), and it can be evaluated by TNT equivalent mass (W) and source distance (r). Over pressure (P_s) can be estimated by using Figure 2.

$$Z = \frac{r}{(W)^{\frac{1}{3}}} \quad (6)$$

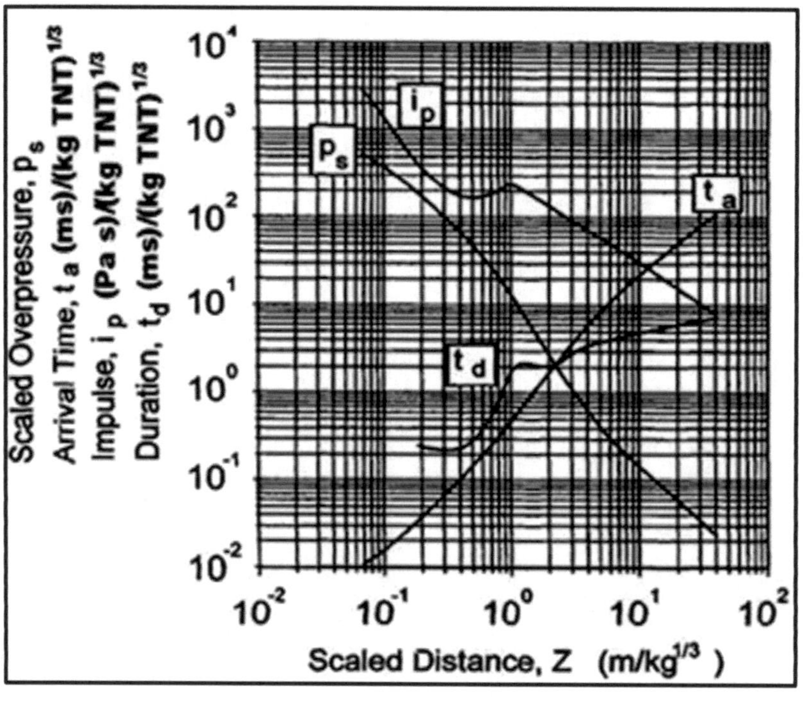

Figure 2. Estimation of overpressure from scaled distance in Prugh Model [11]

3. Case study of typical heat exchanger network

Steam reforming and shift conversion process are widely used for the production hydrogen (H_2) and carbon dioxide (CO_2) on the commercial scale. The simulated PFD of the process is given in the Figure 3.

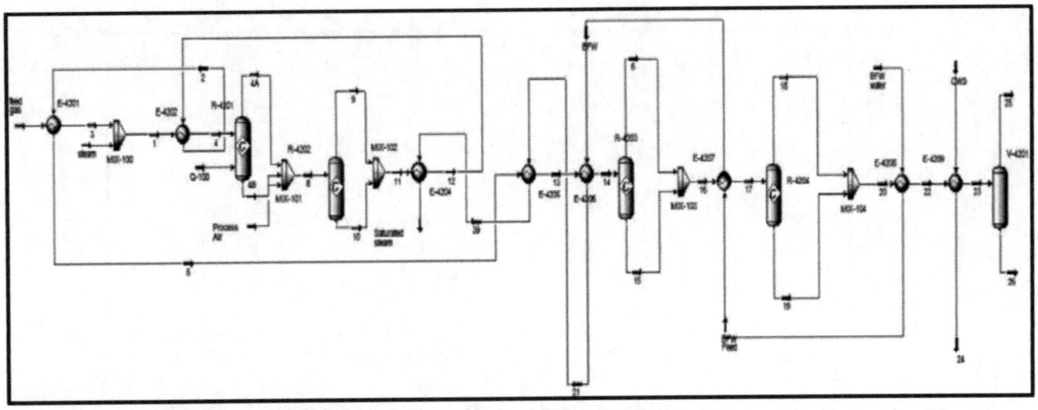

Figure 3. PFD of steam reforming and shift conversion process

The physical explosion effect is analysed for each heat exchanger by using both explosions models. The overpressure magnitude is evaluated at a constant distance from source and presented in the Table 2.

Table 2 Estimation of over pressure for the all heat exchangers in the network.

	Bakers Model			Prugh Model		
Heat exchanger	Explosion Energy(J)	Scaled distance(m)	Overpressure (Pa)	Explosion Energy(J)	Scaled distance (m)	Overpressure (Pa)
E-4201	1.02E+07	0.034	8.82	2.22E+10	1.22	1.824
E-4202	1.08E+07	0.035	8.92	1.84E+10	1.25	1.621
E-4204	1.15E+07	0.031	10.13	3.66E+10	1.10	2.533
E-4205	9.63E+06	0.036	9.02	1.71E+10	1.28	1.520
E-4206	9.00E+06	0.039	9.32	1.35E+10	1.38	1.419
E-4207	8.53E+06	0.038	9.63	1.33E+10	1.37	1.419
E-4208	9.78E+06	0.038	7.60	1.25E+10	1.37	1.013
E-4209	1.29E+07	0.037	9.63	1.34E+10	1.32	1.419

The over pressure value estimated from Bakers model is much higher than the Prugh model. Baker model is highly influenced by the physical properties of the working fluid and Prugh model is less influenced by the physical properties, but much sensitive to temperature and may be preferred for high temperature process equipment [12]. The trends of explosion energy and over pressure values are plotted to analyse the behaviour of both parameters in both explosion models. These trends are presented in Figure 4 and 5.

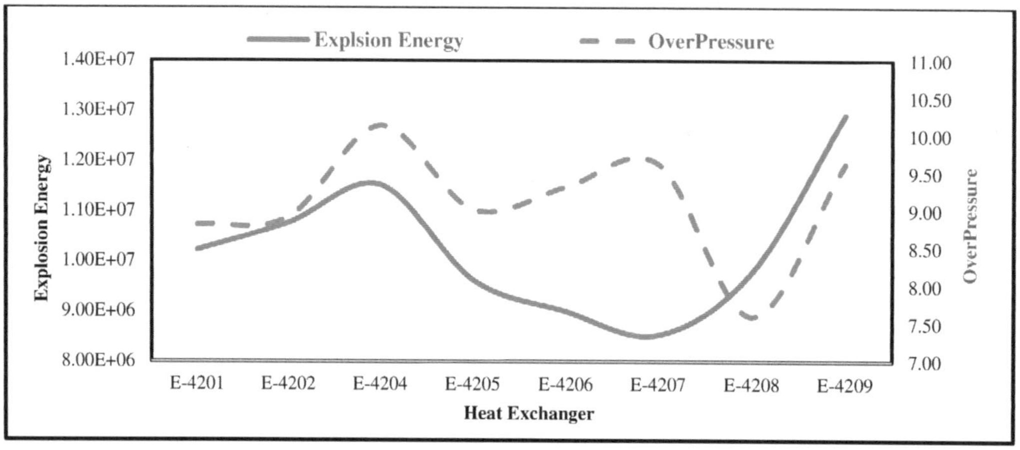

Figure 4 Comparison of explosion energy and over pressure in Baker Model

Wide variation is observed in the trends obtained from the Baker Model. However, similar configuration of the both parameter trends is observed in the Prugh model. The overpressure generated by the physical explosion of E-4204 is found maximum among the all the heat exchangers. Therefore, this heat exchanger is the critical one in the network and required a detailed safety review.

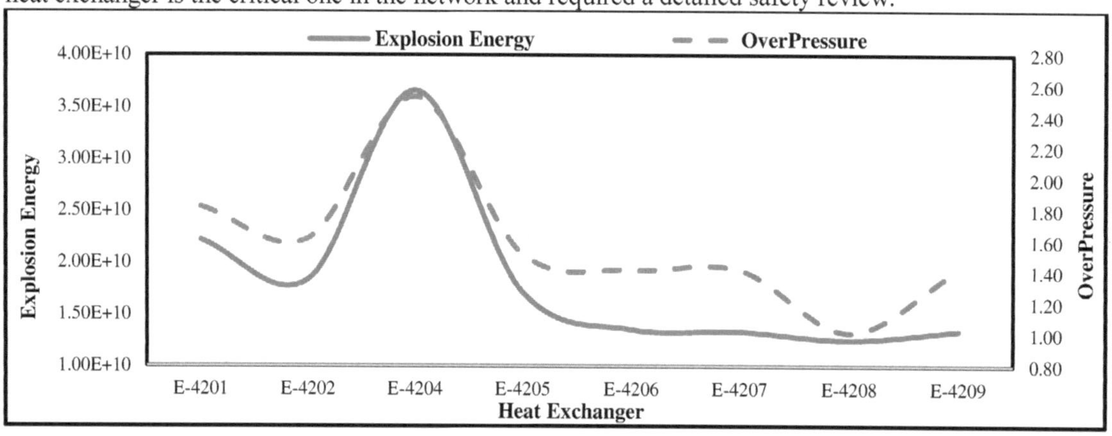

Figure 5 Comparison of explosion energy and over pressure in Prugh Model

4. Conclusion

Overpressure effect generated from physical explosion is analysed for the heat exchanger network by using two well-known explosion model. Prugh model can deliver a better approximation of overpressure as compared to the Baker model due to its consistent response for temperature sensitive equipment. Moreover, this analysis could potentially assist the design engineer to figure out the critical heat exchanger in the heat exchanger network.

5. References

[1] Mannan S 2012 Lees' Loss prevention in the process industries: Hazard identification, assessment and control: Butterworth-Heinemann.
[2] Ozden Ender T, Ilker 2010 *Energy Conv. and Manag.* .**51(5)** 1004-14.
[3] Allahkaram S, Zakersafaee P, Haghgoo S 2011 *Eng. Failure Ana.* **18(3)** 1108-14.
[4] Mousavian RT, Hajjari E, Ghasemi D, Manesh MK, Ranjbar K 2011 *Eng. Failure Ana.***18(1)** 202-11.
[5] CSB 2011 Available from: http://www.csb.gov/assets/1/19/Case_Study.pdf.
[6] CSB 2014 Available from: http://www.csb.gov/assets/1/19/tesoro_anacortes_2014-jan-29_draft_for_public_comment.pdf.
[7] Kidam K, Hurme M 2012 *J. of Loss Prev. in the Pro. Indus.***25(4)** 655-66.
[8] Kidam K, Hurme M 2012 *J. of Loss Prev. in the Pro. Indus.***25(6)** 937-49.
[9] Tugnoli A, Cozzani V, Landucci G 2007 *AIChE journal*. **53(12)** 3171-82.
[10] Hurme M, Rahman M 2005 *J. of Loss Prev. in the Pro. Indus.***18(4)** 238-44.
[11] CCPS 2000 *Safety and Relia.: Method. and App.* 2299-307.

Acknowledgement
The authors would like to thank to Universiti Teknologi PETRONAS, Malaysia for providing short term internal research fund (STIRF) for this project.

Preparation and characterization of Polyacrylonitrile/ Manganese Dioxides- based Carbon Nanofibers via electrospinning process

F E Che Othman[1,2], N Yusof[1,2], J Jaafar[1,2], A F Ismail[1,2], H Hasbullah[1,2], N Abdullah[1,2] , M S Ismail[1,2]
[1]Advanced Membrane Technology Research Centre (AMTEC), Universiti Teknologi Malaysia, 81310 Skudai, Johor Bahru, Malaysia.
[2]Faculty of Petroleum and Renewable Energy Engineering (FPREE), Universiti Teknologi Malaysia, 81310 Skudai, Johor Bahru, Malaysia.

E-mail: [1,2]norhaniza@petroleum.utm.my

Abstract. This research reports the production of precursor polyacrylonitrile (PAN)/ manganese dioxide (MnO_2) nanofibers (NFs) via electrospinning method followed by stabilization and carbonization processes. Nowadays, electrospinning has become a suitable method in manufacturing continuous NFs, thus it is employed to fabricate NFs in this study. The microstructural properties and adsorption competencies of the produced NFs were also studied. The NFs were prepared by electrospinning the polymer solution of Polyacrylonitrile (PAN) and Manganese Dioxide (MnO_2) in, N, N-Dimethylformamide (DMF) solvent. The factors considered in this study were various polymer PAN/MnO_2 concentrations which will significantly affect the specific surface area, fiber morphology and the diameter of the NFs prepared. Subsequently, heat treatment is applied by setting up the stabilization temperature at 275 °C and carbonization temperature at 800 °C with constant dwelling time (30 min). Nitrogen gas at constant rate 0.2 L/min was used for stabilization and carbonization with the stabilization rate (2 °C/min) and carbonization rate (5 °C/min). The carbon nanofibers (CNFs) produced were characterized using Scanning Electron Microscopy (SEM), Brunauer Emmett and Teller (BET) surface area and Fourier Transmission Infrared Spectroscopy (FTIR). It was found that the PAN/MnO_2 CNFs were successfully produced with the carbonization temperature of 800 °C. The prepared PAN/MnO_2 CNFs prepared showed an enhanced in specific surface area about two times compared to it precursor NFs.

1. Introduction

The environmental effect of continuously relying on using fossil fuel is already felt by the whole world in the form of greenhouse effects from the combustion of fossil fuels such as diesel, gasoline (petrol) and kerosene [1].

[1,2] To whom any correspondence should be addressed.

Due to this environmental crisis, researchers have proposed the use of less harmful fuels such as the natural gas in order to reduce dependable on other heavy fossil fuels. Natural gas (NG) has shown a growing interest because of its wide availability as it supplies 20 % of the world's commercial energy needs [2]. This is due to the fact that natural gas is still available in abundance and is the cleanest fossil fuels as it produces a cleaner combustion and a better consumption [1,3], with incredibly less non- carbon emissions [4,5,6,7].

In the case of natural gas, their uses as an automotive fuel present the advantage of their lower emissions relative to gasoline. However, the main limitation is the problem of their storage on a limited on-board tank volume. Advanced storage units with sufficiently high (volumetric and gravimetric) energy storage capacity need to be developed. Among the possible alternatives for storing NG such as liquefied natural gas (LNG) and compressed natural gas (CNG), adsorbed natural gas (ANG) happened to be the most suitable way as this technique can store NG at very low pressure (4 MPa)[8] by utilizing microporous solid as their adsorbents. Nevertheless, for on-board storage applications, additional constraints are still placed on developing further the carbon adsorbents to increase the storage capacity. Currently, the use of carbon adsorbents especially ACFs [9], has allowed the storage of considerable amounts of natural gas at room temperature and at pressures lower than 4 MPa, thus making it more attractive for practical use [10].

In order to implement the ANG method, the biggest drawbacks that need to be overcome by the researchers is the development and evaluation of adsorbents [11]. However, the fabrications of fibers also play a vital role in physical microstructure of the fibers itself. Based on valuable developments in the nano-era, wide potential applications for carbon fibers were expected and huge efforts were concentrated on their synthesis. This issue has caused an increasing in the quantity of correlated study articles published in the last two decades. It has been generally accepted that the quality of the high performance electrospun NFs greatly depend on the processing condition and composition of the precursor as well as the heat treatment. Therefore, electrospinning which is a simple and interesting method to make very fine carbon nanofibers (CNFs) in the diameter range about 10 μm to 10 nm with polyacrylonitrile (PAN) has been greatly developed. The development of the nanofibers contributed to significant improvement of the production of nanofibers offering high porosity and high surface area to volume ratio. To date, many researchers have developed CNFs embedded with metal and metal oxides to increase the gas adsorption activity. This is because metal oxides itself has large specific surface area that enhance the surface reactivity and consequently improved adsorptivity.

Therefore, in this study, the writers aim to fabricate CNFs by electrospinning method using a composite solution containing polyacrylonitrile (PAN) and different concentration of Manganese Dioxide (0, 1 and 2 wt.% MnO_2) in N,N-dimethylformamide (DMF). Then, PAN/ MnO_2 -based NFs are prepared by undergoing the heat treatment processes which are stabilization (275 °C) and carbonization (800 °C). The CNFs were characterized using Scanning Electron Microscopy (SEM), Fourier Transmission Infrared Spectroscopy (FTIR), and Brunauer Emmett and Teller (BET) surface area.

2. Experimental

2.1 Materials

Polyacrylonitrile (PAN, Mw = 150,000), which contained 91.4 wt.% acrylonitrile monomer (CH_2=CHCN) and 8.6 wt.% methylacrylate comonomer (CH_2=$CHCOOHCH_3$) was purchased from Sigma-Aldrich (USA). The average molecular weight of polymer was about 55,500 Da. Dimethylformamide (DMF; ~99.98 % purity) and manganese dioxides were also purchased from Sigma-Aldrich (USA).

2.2 Preparation of PAN nanofibers with N,N-dimethylformamide and metal oxides
The dope solution was prepared by adding 0, 1, and 2 wt.% of MnO_2 into 90, 89, and 88 wt.% of DMF, respectively. The mixture was stirred for at least 5 hours to ensure the MnO_2 dispersed well in the solution before adding up the PAN 10 wt.%. The dope was left to stir for 24 hours. The electrospinning process was carried out at optimum parameters as extracted from literature review; the voltage used is 12 kV, the infusion rate was 1 ml/min and the distance of the tip of needle to the collector was 20 cm, where aluminium foils were used as the collectors.

2.3 Carbonization of PAN-based nanofibers
All PAN nanofibers were stabilized at 275 °C under nitrogen gas flow at a heating rate of 2 °C/min with 30 minutes dwelling time. Then, the stabilized PAN nanofibers were carbonized by using nitrogen gas at 800 °C at the heating rate of 5 °C/min with 30 minutes dwelling time. The gas flow rate for both stabilization and carbonization were maintained at 0.2 L/min.

2.4 Characterization
The physical morphological features of both NFs and CNFs were observed using Scanning Electron Microscopy (SEM), whereas Brunauer Emmett and Teller (BET) method have been used to determine the specific surface area of the precursor NFs and CNFs, and Fourier Transmission Infrared Spectroscopy (FTIR) analysis depicts the presence of a variety of functional groups of both NFs and CNFs.

3. Results and Discussions

3.1 Scanning Electron Microscopy (SEM)
The SEM micrographs in Figure 3.1 (a, c and e) showed the electrospun nanofibers of various MnO_2 loading before carbonization. There was a variety of fibers' thread were produced such as straight line and bent fibers. Moreover, a few beads can be detected from the micrograph and this is due to the inconsistent or uncontrolled electrospinning condition such as surrounding environment temperature. Predictably, in Figure 3.1(e) 2 wt.% MnO_2 composition revealed more presence of white spot compared to 1 wt.% MnO_2 (Figure 3.1(c)). The electrospun precursor (0 wt.% of MnO_2) shows the average diameter ranging from 488 nm to 612 nm as can be seen in Figure 3.1 (a). From physical observations of SEM micrographs, the PAN nanofibers generated without MnO_2 loading (Figure 3.1(a)) have the most straight line fibers as compared to other precursors with 1 and 2 wt. % of MnO_2.

Figure 3.1 (c) showed the SEM image of fibers with 1 wt.% of MnO_2 composition which mostly has straight line fibers and a few bent fibers. In addition, the physical sign of white spot on the fibers indicated the presence of 1 wt.% of MnO_2 on the fibers. The fibrous nature of precursor nanofibers reveals a range of fiber diameters which from 792 nm to 1.45 μm. Figure 3.1 (e) shows SEM images of the precursor nanofibers with additives concentration of 2 wt.% MnO_2 revealed more presence of MnO_2 (white spot). The physical morphology of 2 wt.% MnO_2 precursor show mostly bent fibers and contain few beads and its average diameter ranging from 642 nm to 761 nm. Nevertheless, future study on solution and process parameters may give better understanding on how it affects the diameter of nanofibers. In theory, smaller diameter gives high specific surface area might due to the interaction of metal oxides with PAN precursor [12,13].

Figure 3.1 (b,d and f) demonstrated the SEM images of PAN/MnO_2- based NFs with each composition (0, 1 and 2 %) after carbonization. The rougher and wrinkled fibers portrayed the shrinkage of the diameter of nanofibers during stabilization and carbonization process. From the SEM observation of 800 °C carbonization temperature, it can be seen that 2 wt.% of MnO_2 composition produced smallest fibers diameter if compare to others composition and its average diameter around 335nm to 722nm.

This is due to the catalytic effects of MnO_2 with the PAN precursor such as graphitization of carbon as it helps bind the carbon element together which resulting nanofibers with smaller diameter that gives high specific surface area [12].

Figure 1 SEM images of 0 wt.% (a and b), 1 wt.% (c and d), and 2 wt.% (e and f) of MnO_2 before and after carbonization at 800 °C

3.2 Fourier Transmission Infrared Spectroscopy (FTIR)

PAN-based nanofibers (NFs) typically will show the existence of CH_2, C≡N, C=O, C–O and C–H bonds but after complete pyrolysis process most of these transition compounds were expected to evolve as volatilities and only carbon and hydrogen atoms remain. Figure 3.2 (a) shows the FTIR spectra of PAN NFs precursor of 0, 1, and 2 wt.% MnO_2 before carbonization. As can be seen, there are seven peaks at the wavenumber of 2914.44, 2241.28, 2160.27, 1618.28, 1450.47, 1350.17 and 1063.78 cm^{-1} that representing several functional groups appeared in both sole and modified PAN NFs.

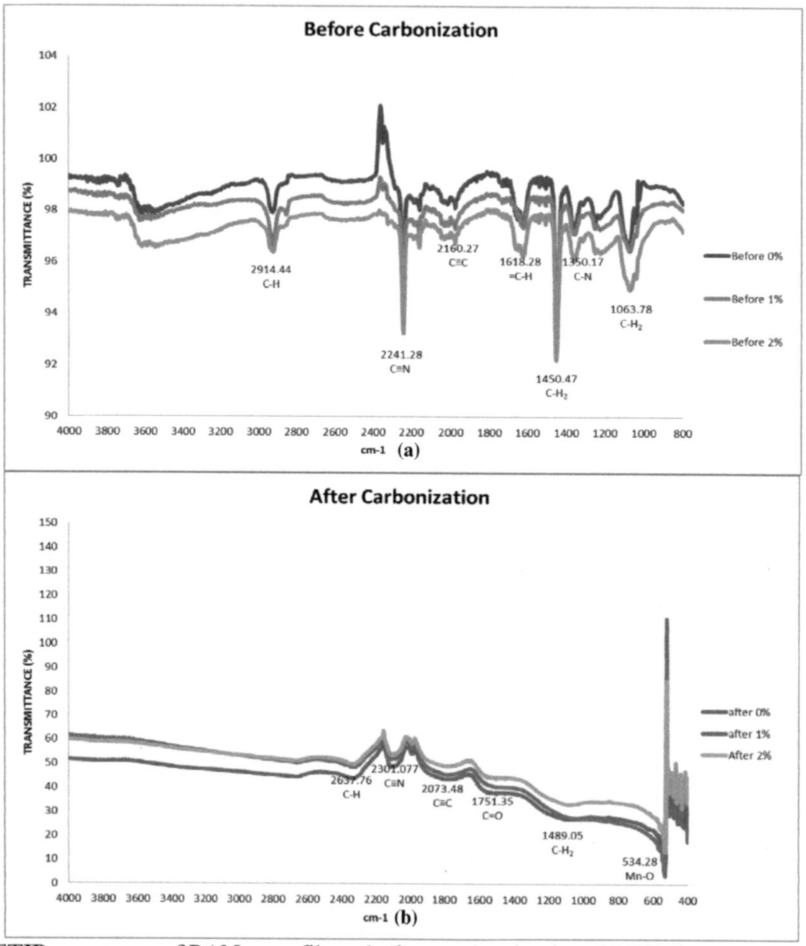

Figure 2 (a) FTIR spectrum of PAN nanofibers before carbonization (b) FTIR spectrum of PAN nanofibers after carbonization

Figure 3.2(b) show the spectrum of both carbonized sole and modified PAN NFs. After carbonization, most of the high peaks observed in Figure 3.2(a) either diminished or replaced with other peaks. This is because during oxidation, numerous new transition structures were formed such as ketones, aldehydes and carboxylic acid and all of these bonds were expected to be removed during pyrolysis process. However, carbonization conducted cannot completely converted the polymer into carbon as the existence of other groups (C≡N, C=O) as shown in the Figure 3.2(b). The presence of MnO_2 also can be observed in the FTIR spectrum of PAN. The wavenumbers of MnO_2 is approximately at 534.28 cm^{-1} corresponds to the Mn–O bond [14,15]. From the above result, it can be confirmed that the synthesized nanomaterial manganese oxide is added into the NFs. There is not much change of the FTIR spectrum between the NFs with the varied of MnO_2 loading because the compounds are still the same but only varies in quantity. It can be concluded that carbonization process seems not enough to remove all non-carbons atom (N, O) in all PAN NF samples and another pyrolysis steps (activation) are believed to convert almost 99% polymer into carbon can be done in future study.

3.3 Brunauer, Emmett and Teller (BET) surface area

In this study, the single point BET method have been used in order to determine the specific surface area of nanofibers with and without manganese dioxide (MnO2) loading before and after carbonization as summarized in Figure 3.3. The specific surface area of 0, 1, and 2 wt.% MnO2 before carbonization were 9.13 m2/g, 10.26 m2/g, and 12.45 m2/g, respectively.

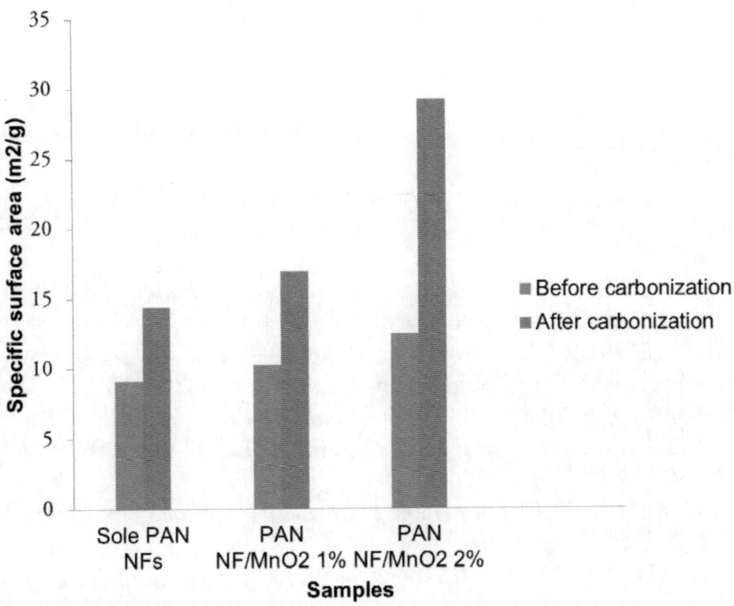

Figure 3. Specific surface area of nanofibers with different concentration of manganese dioxide loading before and after carbonization

As tabulated in Figure 3.3, the pure PAN NFs show the lowest specific surface area of 9.13 m^2/g and when the compositions of MnO_2 are increased from 1 wt.% to 2 wt.%, the surface area increased from 10.26 m^2/g to 12.45 m^2/g. Overall, the specific surface area values show 0<1<2 % MnO_2 while the diameter of the fibers show a reverse trend.

Bas and co-workers [16] showed that the BET surface areas measured before carbonization were in the range of 64-75 m^2/g. This study is slightly lower when comparing to the specific surface area of PAN NFs obtained which is from 9.13-12.45 m^2/g compared to Bas and co-workers. Carbonization of the nanofibers increased the surface areas of the fibers up to 29.20 m^2/g. The low SSA obtained in this study was due to the incomplete pyrolysis process as the temperature of the tube furnace does not increase regularly as set up in first hour and because of this problem, the burn-off of nanofibers was incomplete. This data indicates that the addition of MnO_2 in precursor solution resulting an increase in specific surface area of fiber materials, however the pyrolysis process also contribute great effects to the completeness combustion of the fibers.

The results concurred with theoretical ideas which stated that the amount of metal oxides will increase the specific surface area up to its optimum loading. Most of the journals revealed the results obtained are from CNFs. Thus, the experimental precursor might enhance the surface area once it's been carbonized. These results somehow provide an avenue for the enhancement of adsorption properties of CNFs by improving the surface area of its precursor NFs.

4. Conclusions

As a result, all precursors of PAN/ MnO_2-based CNFs were successfully fabricated by using electrospinning method. It is clearly demonstrates that, the PAN/ MnO_2 CNFs diameter and specific surface area were found to be affected by concentration of metal oxides. The average diameter of all precursors was approximately ranging from 980 to 287 nm. Moreover, the FTIR peaks at 534.28 cm^{-1} indicated the presence of MnO_2 inside the composite CNFs. The increment of the amount of metal oxide nanoparticles up to 2 wt. % embedded in precursor also increased its specific surface area about two times compared to its precursor NF. It can be concluded that CNFs embedded with higher MnO_2 (up to its maxiumum loading) possessed better characteristics in terms of fiber diameter or specific surface area.

5. References

[1] Vasiliev L L, Kanonchik L E, Kulakov A G and Babenko V A 2007 *Int. J. Therm. Sci.* **46** 914-25
[2] Azeman M 2009 Gas Processing & Liquefaction. SKG 5113.
[3] Zakaria Z, Mustafa A and Lim T P 2009 *International Journal of Oil, Gas and Coal Technology* 2 **4** 365–78
[4] Broom D P and Thomas K M 2013 *Materials Research Society Bulletin* **38** 412-21
[5] Yusof N and Ismail A F 2012 *J. Anal. Appl. Pyrolysis* **93** 1-13
[6] Yusof N and Ismail A F 2012 *Polym. Eng.& Sci. J.* **52** 360-66
[7] Yusof N, Ismail A F, Rana D and Matsuura T 2012 *Mater. Lett.* **82** 16-18
[8] Zubizarreta L, Arenillas A and Pis J J 2009 *Int. J. Hydrogen Energy* **34** 4575-78
[9] Lozano C D, Alcan˜iz M J, de la Casa L M A, Cazorla A D and Linares S A 2002 *Fuel* 81 **14** 1777-1803
[10] Linares S A and Cazorla A D 2013 Activated Carbon Fibers *Handbook Advanced Ceramics* (Amsterdam: Elsivier Inc.) Chapter 2.9
[11] Im J S, Park S J, Kim T and Lee Y S 2009 *International Journal of Hydrogen Energy* **34** 3382
[12] Cuenya B R 2010 *Thin Solid Films* **518** 3127
[13] Nowak J D and Ca C B 2009 *Journal of Materials Science* **44** 2408
[14] Kang L, Zhang M, Liu Z H and Ooi K 2007 *Spectrochim. Acta. A* **67** 864
[15] Li L, Pan Y, Chen L and Li G. 2007 *Solid State Chem.* **180** 2896
[16] Bas S, Hakil O, Dumanli A G and Yurum 2008 *Hacettepe J. Biol. & Chem.* 36 **3** 247-53

Acknowledgements
The authors would like to acknowledge the financial support from the Malaysian Ministry Education and Universiti Teknologi Malaysia under GUP grant (Q.J130000.2542.05H46) and Fundamental Research Grant (RJ130000.7842.4F279). The author would also like to acknowledge the technical and management support from Research Management Centre (RMC), Universiti Teknologi Malaysia.

International Conference on Chemical Engineering and Bioprocess Engineering IOP Publishing
IOP Conf. Series: Earth and Environmental Science **36** (2016) 012007 doi:10.1088/1755-1315/36/1/012007

The incorporation of graphene oxide into polysulfone mixed matrix membrane for CO_2/CH_4 separation

K Zahri[1], P S Goh[2], A F Ismail[3]
Advanced Membrane Technology Research Centre (AMTEC), *Universiti Teknologi Malaysia*, 81310 UTM Skudai, Johor Darul Ta'zim, Malaysia

E-mail: [1]zkhalisah@gmail.com, [2]peisean@petroleum.utm.my, [3]afauzi@petroleum.utm.my

Abstract. Carbon dioxide (CO_2) is often found as the main impurity in natural gas, where methane (CH_4) is the major component. The presence of CO_2 in natural gas leads to several problems such as reducing the energy content of natural gas and cause pipeline corrosion. Thus it must be removed to meet specifications ($CO_2 \leq 2$ mol%) before the gas can be delivered to the pipeline. In this work, hollow fiber mixed matrix membrane (MMM) were fabricated by embedding graphene oxide (GO) into a polysulfone (PSf) polymer matrix to improve membrane properties as well as its separation performance towards CO_2/CH_4 gas. The membrane properties were investigated for pristine membrane and mixed matrix membrane filled with filler loading of 0.25%. The synthesized GO and properties of fabricated membranes were characterized and studied using TEM, AFM, XRD, FTIR and SEM respectively. The permeance of pure gases and ideal selectivity of CO_2/CH_4 gas were determined using pure gas permeation experiment. GO has affinity towards CO_2 gas. The nanosheet structure creates path for small molecule gas and restricted large molecule gas to pass through the membrane. The incorporation of GO in PSf polymer enhanced the permeance of CO_2 and CO_2/CH_4 separation from 64.47 to 86.80 GPU and from 19 to 25 respectively.

1. Introduction

The key characteristics of membrane to be used in gas separation process are its permeability and selectivity. Powel *et al.* [1] has stated that other than permeability and selectivity, membrane for gas separation has to be chemically, thermally and physically stable. Furthermore, they should be resistant towards plasticization and aging effect which influences final cost and modularity. Polysulfone (PSf) based membrane has been an ideal choice as they possess the before listed characteristic [2-3]. However, neat polymeric membrane suffers the trade-off problem common with gas separation membrane, between permeability and selectivity [4]. The idea of incorporating inorganic particle into polymer matrix is expected to solve this problem.

Graphene oxide (GO) is a new material from the graphene nanomaterials family which has fascinating properties and characteristics that offers viable solutions in this industry. Shrier *et al.,* [5] has conducted a molecular simulation study to investigate the effect of graphene surface adsorption towards CH_4, CO_2, N_2, O_2, H_2S SO_2 and H_2O. The study proved that graphene was applicable for CO_2

[1-3] To whom any correspondence should be addressed.

capture and separation. GO is derived from graphene, owing the same carbon sheet structure as graphene but has oxygenated functional groups attached on its planar sheet. These active functional groups such as epoxy, hydroxyl and carboxyl groups help to enhance its dispersion in polymer substrate, which affects membrane gas separation performance [6]. GO filled in polymer nanocomposite membrane has shown promising improvement in its properties, thermally and mechanically [3]. Other than that, in terms of separation performance, GO act as barrier, which creates tortuous path in polymer matrix for gas with smaller molecules size to easily pass through the MMM, and hinder gas with larger molecule size from pass through it [7]. The incorporation of GO in MMM is expected to surpass the Robeson separation trade-off boundary by overcoming the polymeric membrane's drawbacks.

A literature survey reveals the gap of previous works on gas permeation studies for GO incorporated in PSf hollow fiber MMM. In this work, GO was embedded in glassy PSf polymer. The study aim was to prove that the incorporation of GO in PSf polymer improve gas separation performance as compared with pristine PSf. The properties of synthesized GO were investigated in terms of its structure and morphology. Further on, the membrane was characterized by SEM and a gas permeation study was conducted with pure CO_2 and CH_4 gas.

2. Experimental and characterization method

2.1 Materials

Polysulfone (PSf Udel-3500) was purchased from Amoco Chemicals. Dimethylacetamide (DMAC), tetrahydrofuran (THF) and ethanol were supplied by Acros, Merck and Fisher Scientific respectively. Natural graphite flake ($\leq$ 20 μm) and Potassium permanganate ($KMnO_4$) were purchased from Sigma Aldrich. Concentrated sulphuric acid (H_2SO_4, 95-97 wt%), hydrochloric acid (HCl) and hydrogen peroxide aqueous solution (H_2O_2, 30 wt%) were bought from Merck. All the reagents were purchased in analytical grade and used without further purification. PSf and graphite flake were dried in oven for 24 h at 60°C to completely remove moisture. Reverse osmosis water processed from Milipore system (Milli-Q) was used throughout experiments.

2.2 Preparation of GO nanosheets

GO was prepared by oxidation of graphite using modified Hummer's method [8-10]. Graphite powder (3.0 g) was dissolved in concentrated H_2SO_4 (100 ml) while the mixture was stirred in an ice bath to maintain temperature of 0 °C. Then, $KMnO_4$ (9.0 g) was added to the mixture gradually as the mixture was being stirred for 2 hrs at temperature below 10 °C. Next, the mixture temperature was increased to 35$\pm$2 °C and vigorous stirred for 30 mins. After that, 150 ml of RO water was added to the mixture slowly, producing a large exotherm up to 100 °C. The temperature was assured to be below than 100° C and was stirred under this condition for 20 mins. Additional water (500 ml) was added to the mixture, followed by 15mL of 30% H_2O_2. The mixture was filtered and further washed with 3% HCl and amount of RO water to remove metal ions and neutralize the filtrate respectively. The suspension was subjected to 1 hr sonication during the first RO water washing to remove the large and un-exfoliated GO. The resulting GO was dried in vacuum oven overnight and was ground to obtain GO powder.

2.3 Dope solution preparation

The preparation of pristine PSf membrane consists of PSf pellets, DMAC, THF and EtOH, with weight ratio of 30/35/30/5 respectively. The composition of dope solution was determined according to procedure described in elsewhere [11]. Initially, PSf pellets was added gradually to the solvent mixture of DMAC and THF and was left under medium speed of mechanical stirrer for 24 hrs. After ensuring the polymer pellets was dissolved, EtOH was added to the dope solution and was subsequently stirred for 2 hrs.

For the fabrication of MMM dope solution, 0.25% of GO nanosheets from the total solid of solution were added into DMAC whilst the operation of ultrasonic bath and was left under sonication for 30 min. THF was added to the suspension and stirred for 1 hr for wetting procedure. For optimum wetting procedure, 10 wt.% of total PSf was added into the suspension subsequently for 4 hours of priming. The remaining PSf pellets were added to the solution gradually and stirred overnight for complete dissolution of the polymer pellets. Finally, EtOH was added into the solution and further stirred for 2 hrs for completion of the formulation above. The solution was placed in an ultrasonic water bath for 1 hr and was left overnight at room temperature to remove trapped bubbles that may exist during dope solution preparation.

2.4 Asymmetric hollow fiber membrane fabrication

Asymmetric hollow fiber membranes were fabricated by spinning process via dry/wet phase inversion process. Details about spinning process described âre elsewhere [11]. The dope solution were extruded using spinneret dimension of 0.6/0.3 mm (OD/ID) at air gap distance of 7.0 cm, bore fluid flow rate and dope extrusion rate ratio of 1:2. The fabricated membranes were collected at rate of 15 m/min. Tap water was used as the coagulant during the whole fabrication process. Then, the fabricated membranes were immersed into water coagulation bath for 2 for complete solvent exchange. After that, the membranes were post-treated in methanol for 8 hours. Subsequently, the membranes were left to dry at room temperature for at least 2 days before ready for further characterization.

2.5 Characterization and gas permeation test

The surface morphology of GO was observed using a transmission electron microscope (TEM, Hitachi HT7700). The GO dimension was characterized using an atomic force microscope (AFM, XE-100 Park System, Korea).

The crystalline properties of GO and membranes were investigated using X-Ray diffractometer (XRD, Siemens D5000). The analysis was carried out using copper Kα radiation (λ= 1.541nm), in the scanning range of 2θ angle ranging from 5° to 60° with a step size of 2 degree.

The chemical structure of GO was analysed using Fourier transform infrared spectrophotometer (FT-IR, Perkin Elmer, Spectrum One). The morphology of prepared membranes was investigated using scanning electron microscopy (SEM, TM3000 Hitachi Japan).

The gas permeation test was conducted with a bubble soap flow meter to measure the membrane performance towards selected gases. Pure CO_2 and CH_4 gas were used and tested at room temperature with operating pressures of 5 bar. The pressure normalized flux or permeance (P_i/l) of hollow fiber membrane and its selectivity was calculated using equation described in elsewhere [11].

3. Result & Discussion

3.1 Characterization of GO nanosheets

3.1.1 TEM and AFM analysis

Various functional groups were introduced to graphite structure by oxidation process via Hummers' Method. During the oxidation process, the functional groups are introduced to the graphite's lamellar interlayer structure, making them suitable to be expanded into few layer or monolayer of GO after ultrasonication [12]. The morphologies of GO nanosheets were characterized using transmission electron microscopy (TEM) and atomic force microscopy (AFM). Figure 1(a) shows the typical structures of GO nanosheets from TEM analysis. The thickness of the nanosheets was measured via AFM analysis and the value was 0.716 nm which was same as reported by previous work [13]. Figure 1(b) shows the topography image of GO nanosheets while Figure 1(c) shows its line profile. The results indicated that the synthesized GO was successfully exfoliated into monolayer nanosheets.

International Conference on Chemical Engineering and Bioprocess Engineering IOP Publishing
IOP Conf. Series: Earth and Environmental Science **36** (2016) 012007 doi:10.1088/1755-1315/36/1/012007

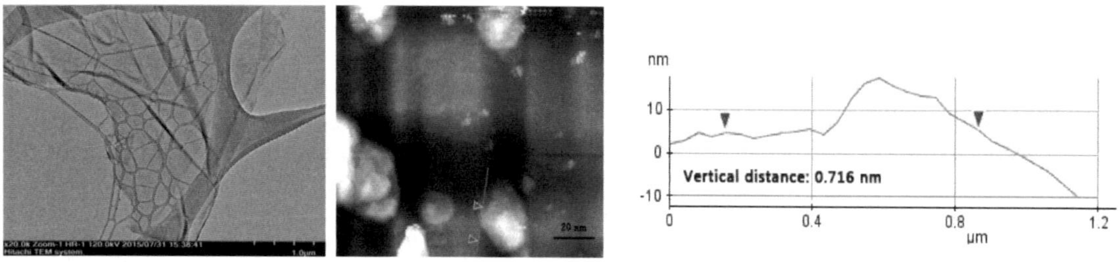

(a) (b) (c)

Figure 1. (a)TEM image, (b) AFM topography image, and (c) line profile of GO nanosheets

3.1.2 XRD analysis

The microstructure of GO was characterized using x-ray diffraction (XRD) analysis. Figure 2 shows the graphite powder exhibit a sharp diffraction peak at $2\theta = 26.41°$, corresponding to a d-spacing of 0.373 nm [14-15]. The peak of GO appeared at peak $2\theta = 10.5°$ with d-spacing 0.842 nm, which was consistent with value reported in literature [15] and also in line with AFM result. The GO d-spacing value was obviously larger than pristine graphite, which indicates the exfoliation of GO due to the attachment of functional groups during oxidation process and the absorbed water between GO interlayers [9,16].

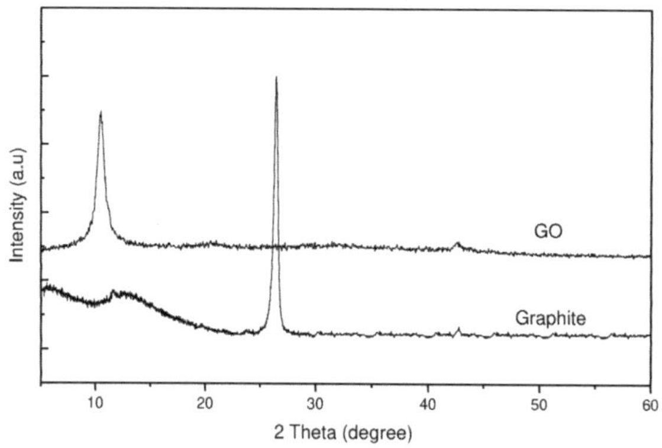

Figure 2. XRD patterns for graphite and GO

3.1.3 FTIR analysis

Figure 3 shows FTIR spectrum of pristine graphite and GO. The characteristic of pristine graphite FTIR spectrum shows vibration peak of C-C at wavelength 1600cm^{-1} [9] and two $-CH_2-$ peaks at wavelength 2920 cm^{-1}, 2850 cm^{-1} [10]. Compared with pristine graphite, more significant peaks of GO are observed on the GO spectrum after being oxidized. The spectrum shows broaden peak at 3450 cm^{-1}, attributed to water content adsorbed by polar functional groups on GO surface such as –OH [17]. Other vibration spectrum are observed at peak 1720 cm^{-1}, 1390 cm^{-1}, 1230 cm^{-1}, 1050 cm^{-1}, which corresponds to carboxyl (C=O), hydroxyl (C-OH), epoxy (C-O-C) and C-O respectively. Therefore, the existences of oxygenated functional groups are confirmed as well as in literature [9,18]. In addition, those oxygenated functional groups attached to GO are expected to improve the interaction between GO and polymer [19].

International Conference on Chemical Engineering and Bioprocess Engineering IOP Publishing
IOP Conf. Series: Earth and Environmental Science **36** (2016) 012007 doi:10.1088/1755-1315/36/1/012007

Figure 3. FTIR patterns for graphite and GO

3.2 Membrane characterization

3.2.1 SEM analysis of MMM

Figure 4 shows the SEM images of fabricated pristine PSf hollow fiber membranes as compared with PSf-GO hollow fiber MMM incorporated with 0.25% GO loading. From the SEM cross section images, a typical morphology of asymmetric membrane for gas separation consisting active dense skin layer structure supported by porous sub-layer comprising small sponge like structure sandwiched by large macrovoid structure were observed. It can be observed that the addition of GO to the membrane shorten the elongation of tear drop structure at the membrane lumen side and supress the formation of circular microvoid structure beneath the outer surface layer, which leads to a more dense and thicker active skin layer. The difference of dope polymer viscosity can explain this phenomena. This phenomenon was supported by previous study conducted by Ismail *et al.* [20] which suggested that the incorporation of filler influenced the membrane morphology structure due to change in polymer dope viscosity. The addition of GO increased the dope viscosity value and this affects the formation of asymmetric membrane structure. Higher viscosity delayed the solvent-nonsolvent exchange rate which restrained the formation of larger tear drops. Thicker and denser skin layer is expected for higher viscosity. However, the skin layer thickness cannot be measured via SEM analysis.

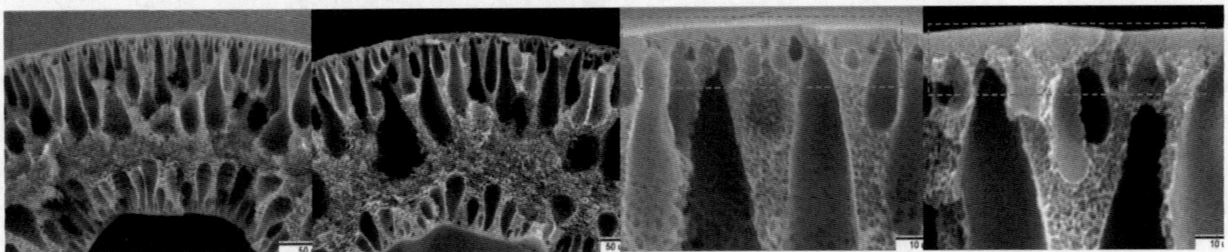

(a) (b) (c) (c)

Figure 4. SEM cross section images of (a) pristine PSf, (b) PSf-GO MMM at magnification x1.5k, (c) pristine PSf and (d) PSf-GO MMM at magnification at magnification x10k

International Conference on Chemical Engineering and Bioprocess Engineering IOP Publishing
IOP Conf. Series: Earth and Environmental Science **36** (2016) 012007 doi:10.1088/1755-1315/36/1/012007

3.2.2 The permeation and selectivity of MMM towards CO_2/CH_4 gas

Table 1. Gas permeation properties for CO_2 and CH_4 measured at 5 bar and 25±7 °C

Membrane	P_{CH4} (GPU)	P_{CO2} (GPU)	Selectivity
PSf	3.36	64.47	19.20
PSf-GO	3.34	86.80	25.98

Table 1 shows the effect of adding GO into PSf polymer on the gas permeation characteristic. Based on previous study, the incorporation of GO into rubbery polymer matrix enhanced the permeation of selective gas without major reduction in selectivity [7]. Li *et al.* [8] has reported that polyethyleneglycol-polyethylineimine GO nanosheets embedded into Pebax polymer matrix has improved MMM gas separation performance. However, the membrane performance deteriorated with the addition of filler at certain threshold amount. The present study shows that the addition of 0.25% GO nanosheets into PSf polymer does improve the MMM performance, especially for CO_2 gas permeation. CO_2 permeance significantly increased from 64.47 GPU to 86.80 GPU. It was observed that the permeation of CH_4 gas slightly reduce from 3.36 GPU to 3.34 GPU, which resulted improvement in CO_2/CH_4 selectivity by 35% with respect to the pristine PSf.

The increasing of CO_2 permeation can be contributed by the GO itself which possesses high absorption properties towards selective gas. GO own π-π stacking bond that has interaction with CO_2 gas and high affinity towards that polar gas [9]. Other than that, GO contained active functional groups which helped to improve its interaction with PSf molecular chain resulting in better GO dispersion in polymer matrix. These phenomenon's explains the significant increment of CO_2 permeation in MMM.

The low loading of GO can increase the nanoparticles aspect ratio by preventing GO sheets from agglomerating and restack back to multilayers GO. In addition, the oxygenated functional groups as mentioned above help GO to have good dispersion in polymer matrix and prevent interphase problem between GO and polymer matrix from occur. Other than that, GO structure in nanosheets or few layers sheet form creates tortuous diffusion path in polymer matrix which allows smaller gas molecule (CO_2: 3.3Å) to diffuse with less resistance and restricts the larger gas molecule (CH_4: 3.8 Å) from diffusing through the MMM [8]. This explains the slight reduction of CH_4 permeation and significant increment of CO_2/CH_4 selectivity.

4. Conclusion

PSf-GO hollow fiber MMM was fabricated. The characteristic of synthesized GO and membranes in term of its morphological structure as well as gas separation performance were analysed. From this work, the GO nanosheets have been successfully synthesized. The incorporation of GO nanosheets into PSf polymer gives significant improvement to gas separation properties even at low loading (0.25 wt.%). The special properties present in GO consisting of polar groups and π-π conjugate bond helps to improve CO_2 permeance in MMM by 34%, correspondingly increased the selectivity of CO_2/CH_4 gas.

5. References

[1] Powel C E and Qiao G G, 2006 *J. Membr. Sci.* **279** 1
[2] Ganesh B M, Isloor A M and Ismail A F 2013 *Desalination* **313** 199
[3] Ionita M, Vasile E, Crica L E, Voicu S I, Pandele A M, Dinescu S, Predoiu L, Galateanu B, Hermenean A and Costache 2015 *Composites: Part* B **72** 108
[4] Robeson L M 2008 *J. Membr. Sci.* **320** 390
[5] Schrier J 2011 *ACS Appl. Mater. Interfaces* **3** 4451
[6] Yang X, Wang X, Yang J, Li J and Wan L *Chem. Phys. Lett.* **570** 125

[7] Berean K J, Ou J Z, Nour M, Field M R, Alsaif M M Y A, Wang Y, Ramanathan R, Bansal, Kentish V S, Doherty C M, Hill A J, McSweeney C, Kaner R B and Kalantar-zadeh K 2015 *J. Phys. Chem.* C **119** (24) 13700
[8] Li X, Cheng Y, Zhang H, Wang S, Jiang Z, Guo R and Wu H 2015 *ACS Appl. Mater. Interfaces,* **7** 5528
[9] Zhao L, Cheng C, Chen Y F, Wang T, Du C H and Wu L G 2015 *Polym. Adv. Technol.* **26** 330
[10] Zhao Y, Ding H and Zhong Q 2012 *J. Appl. Surf. Sci.* **258** 4301
[11] Zulhairun A K, Ng B C, Ismail A F, Surya Murali S and Abdullah M S 2014 *Sep. Purif. Technol.* **137** 1
[12] Wang T, Zhao Li, Shen J N, Wu L G and Bruggen B V 2015 *Environ. Sci. Technol.* **49** 8004
[13] Li X, Ma L, Zhang H, Wang S, Jiang Z, Guod R, Wu H, Cao X Z, Yang J and Wang B 2015 *J. Membr. Sci.* **479** 1
[14] Cao K, Jiang Z, Zhao J, Zhao C, Gao C, Pan F, Wang B., Cao X and Yang J 2014 *J. Membr. Sci.* **469** 272
[15] Chen J, Li Y, Huang L, Li C and Shi G 2015 *Carbon* **81** 826
[16] Szabo T, Berkesi O and Dekany I 2005 *Carbon* **43** 3181
[17] Jin T X, Liu C, Zhou M, Chai S G, Chen F and Fu Q 2015 *Composites: Part* A, **68** 193
[18] Guo H L, Wang X F, Qian Q Y, Wang F B and Xia X H 2009 *ACS Nano.* **3** 2653
[19] Wang T, Zhao L, Shen J N, Wu L G and Bruggen B V 2015 *Environ. Sci. Technol.* **49** 8004
[20] Ismail N M, Ismail A F and Mustafa A 2015 *Procedia CIRP* **26** 461

IOP Conf. Series: Earth and Environmental Science **36** (2016) 012008 doi:10.1088/1755-1315/36/1/012008

Effect of evaporation time on cellulose acetate membrane for gas separation

W N R Jami'an[1,2], H Hasbullah[1,2], F Mohamed[1,2], N Yusof[1,2], N Ibrahim[1], R R Ali[3]

[1] Faculty of Petroleum and Renewable Energy Engineering, *Universiti Teknologi Malaysia*, 81310, Johor Bahru, Johor, Malaysia

[2] Advanced Membrane Technology Research Centre (AMTEC), Universiti Teknologi Malaysia, 81310, Johor Bahru, Johor, Malaysia

[3]Faculty of Chemical Engineering, *Universiti Teknologi Malaysia*, 81310 Johor Bahru, Johor, Malaysia

E-mail: [1]hasrinah@petroleum.utm.my

Abstract. Throughout this decades, membrane technology has been the desirable option among the others gas separation technologies. However, few issues have been raised regarding the membrane gas separation application including the trade-off between its permeability and selectivity and also its effects towards environment. Therefore, for this research, a biopolymer membrane for gas separation application will be developed with reasonably high on both permeability and selectivity. The main objective of this research is to study the effect of solvent evaporation time on the flat sheet asymmetric membrane morphology and gas separation performance. The membranes were produced by a simple dry/wet phase inversion technique using a pneumatically controlled casting system. The dope solution for the membrane casting was prepared by dissolving the cellulose acetate (CA) polymer in N-Methyl-2-pyrrolidone (NMP) and the solvent evaporation time was varied. Permeability and selectivity of the membrane was performed by using pure gases of carbon dioxide, CO_2 and methane, CH_4. The increase in solvent evaporation time had improved the membrane morphologies as the porosity of the membrane surface decrease and formation of a more mature skin layer. The gas permeation tests determined that increasing in solvent evaporation time had increased the selectivity of CO_2/CH_4 but reduce the permeability of both gases

1. Introduction

The ability of polymeric membranes to separate common gases including O_2/N_2; CO_2/CH_4; H_2/N_2; He/air; H_2O/air and H_2/CH_4 has been known for more than a century [1-2]. As the energy cost increases, membrane technology get a consideration to replace the conventional process in reducing cost of industrial process. Membrane separation offers a number of significant advantages in term of low energy and capital investment. The membrane device are compact and modular that can be operated with a simple equipment and mild condition and membrane system are energy efficient owing to large reduction in power electricity and fuel consumption [3]. In addition, membrane can also be used in the combination with traditional gas separation method to take advantages both technologies for certain occasions in which neither process can be achieve a proper result individually [4].

The goal of the membrane manufacturing is high selectivity and gas permeation flux with defect-free ultrathin dense layer of membrane structure, where most of membrane experts try to achieve it. According to Shilton *et al.,* rheological condition during manufacturing can affect the membrane performance as well as membrane material, dope preparation and fabrication technology [3]. Manipulation of solvent evaporation time prior to immersion in coagulation step strongly influences the surface skin-layer thickness. Dry-phase separation occurs at the initial stage of evaporation where the volatile solvent removed from casting solution. As a result, unstable top skin layer region of the nascent membrane will be formed due to the selective loss of the volatile solvent. It can be observed by the almost instantaneous onset turbidity in this top skin layer. Wet-phase separation occurs when the nascent membrane immersed in the nonsolvent coagulant. In this quench step, the bulk of the membrane structure is formed and the remaining solvent and nonsolvent are extracted [5-6].

There are three main types of membranes commercially available for CO_2 removal; cellulose acetate, polyimides and perfluoro-polymers and among of this three types, cellulose acetate (CA) had received the most attention for CO_2 separation from CH_4 [7]. CA is a biopolymer because its raw material which is cellulose, is made from a renewable resource and is readily degradable by a variety of mechanism such as biological degradation. Moreover, commercial availability, ease of processing, and favourable selectivity-permeability characteristics with acceptable mechanical and thermal properties make CA had received the most attention. In addition, CA consist higher thermal stability T_g of 198°C [8] compare to polysulfone (PSF) (T_g = 190°C).

Thus, the general aim for this research is to develop a biopolymer membrane using cellulose acetate polymer as membrane material for gas separation application. To achieve the aim of the research, the main objective is to study the effect of solvent evaporation time on the flat sheet asymmetric biopolymer membrane morphology and gas separation performance.

2. Experimental

2.1. Material

The base material used for the membrane material was cellulose acetate (CA) with 39.8% acetyl content produced by Aldrich Chemical Co. Inc. The solvent that was used for CA polymer is N-Methyl-2-pyrrolidone (NMP) supplied by Merck, Darmstadt, Germany. Tap water were used as a coagulation medium during wet phase inversion process and methanol and n-hexane was used in solvent exchange steps in order to produce membrane with the best permeation properties.

2.2. Method

2.2.1. Dope Solution Preparation. The method that was used in preparing 23 wt% the dope solution is by dissolving the cellulose acetate in the solvent used, which is N-Methyl-2-pyrrolidone. The solution was mixed in the round bottom reaction vessel, sealed with paraffin film and was stirred using a mechanical stirrer with temperature 56 °C, until a homogenous solution is produced. After that, the homogeneous solution was placed in the ultrasonic water bath to remove the gas bubbles that may exist in the solution prior to membrane casting process.

2.2.2. Flat Sheet Membrane Preparation. Dry/wet phase inversion technique was used to prepare the asymmetric flat sheet membranes. The polymer solution was casted on a clean glass plate using a pneumatically controlled membrane casting system [10]. The casting speed was fixed at $64s^{-1}$ at varied solvent evaporation time from 10s to 30s. After the evaporation step, the membranes were immersed in the water bath for about 30 minutes (wet phase inversion). Finally, the washed membrane was solvent exchanged with ethanol and n-hexane for about 20 minutes respectively (solvent exchange steps) and dry at atmospheric condition for another 24 h.

2.3. *Structural Characterization.*
Scanning electron microscopy (SEM) was used to examine the membrane dense skin layer, porous substructure and cross section. The dry membrane samples was mounted on a metal stub with double-sided carbon adhesive tape. The cross-section structure of the membrane sample was prepared by a liquid nitrogen freeze fracturing follow by gold sputtering before scanning on a SEM.

2.4. Membrane Testing
Gas permeation test was applied in this study to obtain the pressure-normalized flux of the membrane. Pure CO_2 and CH_4 was used as test gases, using upstream pressure of 1 to 5 bar and at ambient temperature. Volumetric gas permeation rates were determined with a local design soap-bubble flow meter.

3. Result and Discussion

3.1. Morphological and structural analysis of CA flat sheet asymmetric membrane
The solvent evaporation step was considered essential for the formation of the asymmetric membrane and was critically importance in determining the membrane structures. Different solvent evaporation time had revealed different membrane morphologies. Figure 1 and 2 shows the SEM image of top surface layer and cross-sections of asymmetric CA membrane prepared at various solvent evaporation time prior to immersion in the coagulation bath in the dry/wet phase inversion step. The membrane developed in this study achieved the asymmetric structure which consists of thin dense skin layer at the top and thick, porous substructure at the bottom that provides the mechanical strength for the membrane. However, different solvent evaporation time produced different membrane morphologies in terms of porosity, macro void substructure and the skin layer.

As shown in Figure 1, the increment of solvent evaporation time decreases the porosity of the

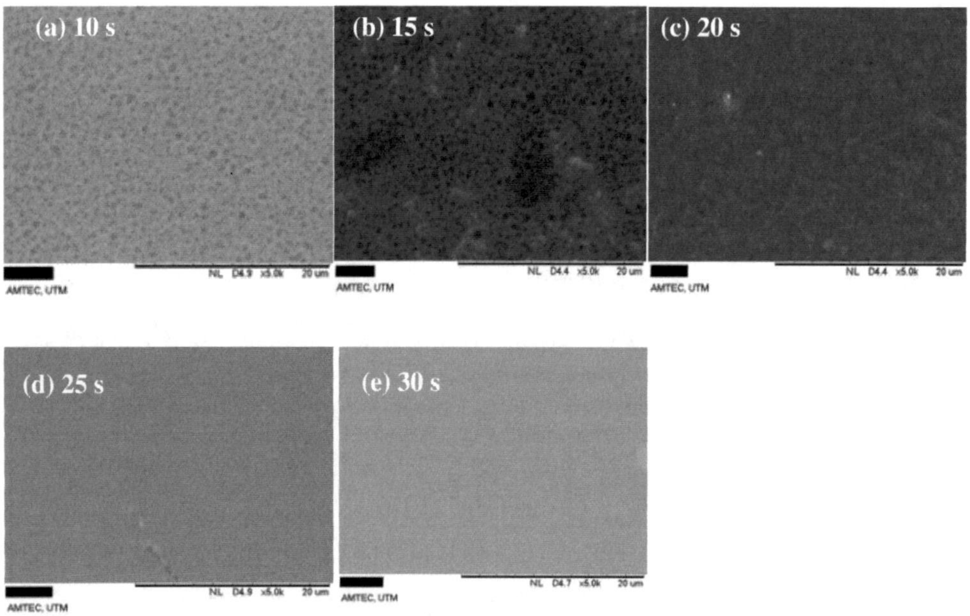

Figure 1. SEM pictures of 23 wt% cellulose acetate asymmetric membrane surface layer at different solvent evaporation time

membrane surface. This is because those membranes had enough time for the rearrangement of membrane molecular chains and causing the membrane surface to become closely pack thus decreasing the pore size on the membrane surface and reducing defects. Figure 2 shows the membrane cross-section. A distinct and dense skin layer was more visible on the longer evaporation time. Phase separation occurs initially at the surface of the casting membrane leading to a high concentration gradient of the polymer. A longer evaporation time would form a more concentrated polymer region at the open top of the casting membrane due to the loss of highly volatile solvents. The formation of the skin layer would affect the formation mechanism of asymmetric CA membrane substructure. A skin layer acts as a barrier for solvent–coagulant exchange during the immersion precipitation period. A longer evaporation step generally leads a to thicker skin layer [12]. A thicker skin slows down the solvent–coagulant exchange and thus reducing the membrane porosity. In short, by increasing the solvent evaporation time, the membrane substructure layer change from more porous to a less porous structure.

3.2. Effect of solvent evaporation time on gas separation performance

Table 1 and Figure 3 shows the gas permeability coefficients and selectivity of the CA asymmetric membrane respectively. Experimental result of this show that the optimum permeability of the CO_2 and CH_4 gases was at evaporation time of 15s with CO_2 permeance of 45.58 GPU and CH4 permeance of 14.62, while the selectivity of CO_2/CH_4 at 15s was 3.12.

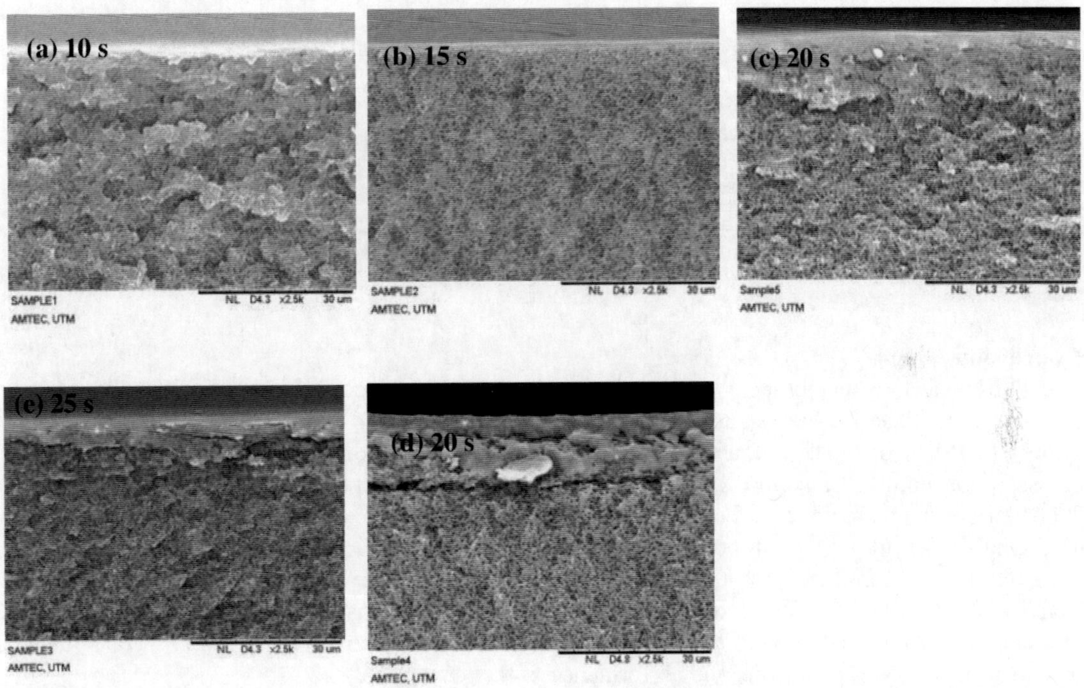

Figure 2. SEM pictures of 23 wt% cellulose acetate asymmetric membrane cross-section at different solvent evaporation time.

As we can see from the Table 1, the permeability for CO_2 and CH_4 decrease with increasing solvent evaporation time while the selectivity of CO_2/CH_4 increase with increasing of solvent evaporation time. The cause of this difference was due to the structural changes in the skin layer of asymmetric membrane or could be explained by referring to the morphologies of the membrane as shown by SEM images in Figure 1 and 2. Increasing the solvent evaporation time caused a decrease in pore size and surface porosity but an increase in skin thickness. This resulted in increasing selectivity but decreasing permeability of the membrane.

Table 1. CO_2 and CH_4 permeability coefficients at varied solvent evaporation time of CA flat sheet asymmetric membrane

Solvent Evaporation Time (s)	Permeability (GPU) CO_2	Permeability (GPU) CH_4
10	78.3378	69.5458
15	45.5758	14.6222
20	1.0083	0.1696
25	0.6444	0.0322
30	0.4490	0.0209

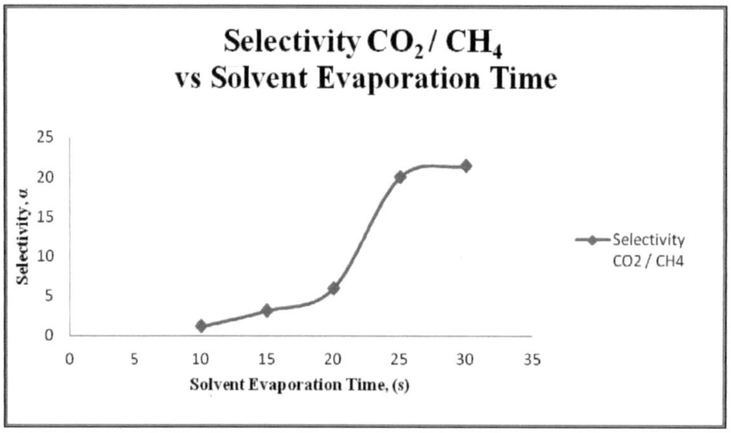

Figure 3. Selectivity of CO_2/CH_4 at varied solvent evaporation time of CA flat sheet asymmetric membrane

4. Conclusion

The cellulose acetate membrane was prepared by casting the dope solution at different solvent evaporation time. The polymer concentration at 23 wt% was used in the dope solution as a critical polymer concentration. The prepared membranes that formed via dry/wet phase inversion process possessed an asymmetric structure with thin skin layer at the top and porous structure at the bottom. The porosity of the membrane surface decreases with increasing evaporation time. As the evaporation time increase, the formation of bottom layer support structure become more compact. The gas permeation tests determined that the increasing in solvent evaporation time would increase the selectivity of CO_2/CH_4 gases but reduced the permeability of both gases through the CA asymmetric membrane. The membrane is recommended to be further modified by coating with silicone rubber coated material (PDMS) to enhance the performance of the membrane.

5. References

[1] Robeson L M 1999 *Solid State Mater. Sci.* **4** 549
[2] Robeson L, Burgoyne W, Langsam M, Savoca A C, and Tien C Fration 1994 *Polymer* **35** 4970
[3] Kusworo T D, Ismail A F, Mustafa A, and Matsuura T 2008 *Sep. Purif. Technol.* **61** 249
[4] Sadegi M, Mousavi S, Hashemi M, Pourafshari Chenar M and Roosta Azad R 2007 *J. Appl. Polym. Sceince* **105** 2683
[5] Kawakami H, Mikawa M, and Nagaoka S 1997 *J. Memb. Sci.* **137** 241
[6] Wan Jin L, Duek San K and Jin Hwan K 2000 *Korean J. Chem. Eng.* **17**, 143
[7] Scholes C A, Stevens G W and. Kentish S E 2012 *Fuel* **96** 15–28
[8] Puleo A and Paul D 1989 *J. Memb. Sci.* **47** 301

[9] Ismail A F and Lai P Y 2003 *Sep. Purif. Technol.* **33** 127
[10] Ismail A F, Ng B C and Abdul Rahman W A 2003 *Sep. Purif. Technol.* **33** 255
[11] Chung T, Teoh S, Hu X, and Shung T 1997 *J. Memb. Sci.* **133** 161
[12] Iqbal M, Man Z, Mukhtar H and Dutta B K 2008 *J. Memb. Sci.* **318** 167

Acknowledgement
We gratefully acknowledge the financial support from the Universiti Teknologi Malaysia and Ministry of Education Malaysia (Grant no: 4F652 and 4L116).

Effect of coagulant bath on the gas permeation properties of cellulose acetate asymmetric membrane

F. Mohamed[1,3], H. Hasbullah[1,3] , W. N. R. Jami'an[1,3], W. N. H. W. Salleh[1,3], N. Ibrahim[2,3], R. R. Ali[4]

[1]Advanced Membrane Technology Research Centre (AMTEC), FPREE, Universiti Teknologi Malaysia, 81310 Johor Bahru, Johor, Malaysia

[2]UTM-MPRC Institute for Oil and Gas, FPREE, Universiti Teknologi Malaysia, 81310 Johor Bahru, Johor, Malaysia

[3]Faculty of Petroleum & Renewable Energy Engineering (FPREE), Universiti Teknologi Malaysia, 81310 Johor Bahru, Johor, Malaysia

[4]Faculty Chemical Engineering, Universiti Teknologi Malaysia, 81310 Johor Bahru, Johor, Malaysia

E-mail: hasrinah@utm.my

Abstract. Membrane based gas separation process technology has been recognized as one of the most efficient and advanced unit operation for gas separation. One of the problems in membrane gas separation is membrane performance. This paper explores the application of cellulose acetate (CA) membrane for natural gas purification and separation by improving its permeability and selectivity. The main interest in this research is to study the effect of quench medium on the gas separation performance towards its physical characteristics and gas separation performance of CA membrane. Cellulose acetate polymer was dissolved in n-methyl-2-pyrrolidone solvent and casted onto a glass plate using a pneumatically controlled casting system with fixed shear rate and solvent evaporation times. The parameter varied was the non-solvent used as quench medium during membrane post treatment that were methanol and n-hexane. The different quench media as post treatment affected the O_2 and N_2 gas permeation and O_2/N_2 selectivity as well as the tensile strength of the flat sheet asymmetric membrane. Combination of methanol and n-hexane as quench media gave the best result than the other steps. This solvent exchange step influenced the morphology by producing thin skin layer and thus gives better gas separation performance than other steps

1. Introduction

Gas permeation system is the major breakthrough in the membrane technology's research and development. Approximately, more than 5000 membrane units has been installed for nitrogen-air separation, 500 membrane refinery units for hydrogen separation and 200 membrane units for carbon dioxide-natural gas separation [1]. Membrane becomes an industrial favourite option because the permeation rate of a chemical species through the membrane can be controlled as desired as well as offering many advantages such as low energy use, easy to operate and compact equipment compared to other technologies. Membrane gas separation is based on semi-selective barrier allowing one gases passing through while preventing the other under the influence of a driving force [2]. Performance of membrane-based gas separation process strongly depends on permeability and selectivity of membrane. Membrane with higher permeability gives higher productivity and lower capital costs whereas membrane with higher selectivity gives more efficient separations, higher recovery and lower power costs; indeed, membranes that simultaneously possess high values of selectivity and permeability would lead to the most economical gas separation processes. Solvent exchange is one of the important step in membrane fabrication that resulted in membrane of different average pore size and pore size distribution. Generally, solvent exchange can be done in two step: 1) using alcohol to replace water, 2) using non-polar volatile liquid to replace alcohol and then the volatile liquid is evaporated leaving the dry membrane.

Cellulose acetate is the acetate ester of cellulose produced by the acetylation of purified cellulose from wood pulp, which is a renewable resource. The commercial cellulose acetate membrane available in the industry consist of the combination of cellulose acetate, cellulose diacetate and/or cellulose triacetate [3]. The first developer of high flux cellulose acetate asymmetric membrane is Loeb and Sourirajan for reverse osmosis of water desalination [4]. Cellulose acetate membrane is still preferable due to its low cost, ease of handling, unproblematic availability and good strength [5]. The permeability of cellulose acetate depends on the function of the degree of acetylation cause by the replacement of polar hydroxyl group with bulky acetate groups , which affecting the polymer density and membrane free volume [6].

The main objective of this research is to study the effect different quench medium for solvent exchange step on morphology and gas separation performance of cellulose acetate membrane.

2. Experimental

2.1. Materials

Cellulose acetate (CA) with 39.8% acetyl content was used as the membrane material and obtained from Aldrich Chemical Co. Inc. N-methyl-2-pyrrolidone (NMP,>99%) supplied by Merck, Darmstadt, Germany. Tap water was used as a coagulation medium during wet phase inversion process. Methanol and n-hexane was used as quench medium in solvent exchange step.

2.2. Dope solution preparation

The dope solution of 23wt% concentration was prepared by dissolving the CA polymers with NMP solvent. The essential amount of CA was slowly added to avoid agglomeration. The solution was mixed in a capped bottle and stirred by mechanical motor until complete polymer dissolution. When the polymers were completely dissolved, as indicated by observing a clear solution, the solution was degassed for 1 hour and then stored 1 day in room condition in order for complete removal of air bubble inside the solution.

2.3. CA asymmetric membrane preparation

Asymmetric flat sheet membranes were prepared according to the dry/wet phase separation process. The polymer solution was casted on a clean glass plate using a pneumatically controlled membrane casting system as in figure 1. The casting time and solvent evaporation time was fixed at 15s and 15s respectively. After that, the membranes were immersed in water bath for 30 minutes for wet phase inversion process. Finally, the membrane was immersed the in quench medium as in table 1. Then, the membrane was dried for 24 h at atmospheric condition.

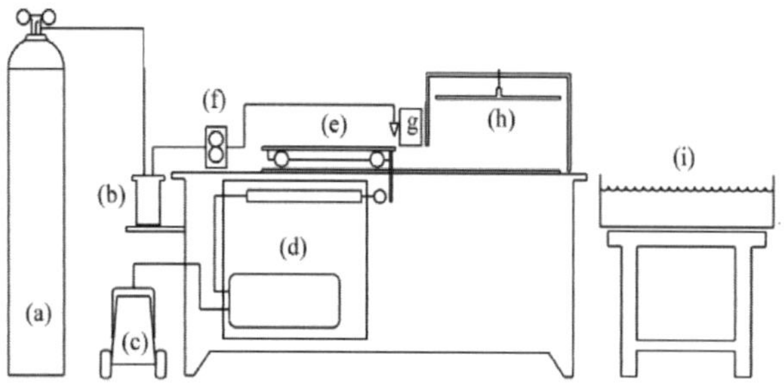

Figure 1 Pneumatically-controlled flat sheet membrane casting system, (a) Nitrogen gas tank; (b) solution dope reservoir; (c) air compressor; (d) pneumatic system; (e) trolley and glass plate; (f) gear pump; (g) casting knife; (h) evaporation chamber; (i) coagulant bath [7]

Table 1. Quench medium condition.

	Time (min)	
Sample	methanol	n-hexane
1	-	-
2	20	-
3	-	20
4	20	20

2.4. Membrane characterization

2.4.1. Scanning electron microscopy SEM. Membrane morphology was examined using scanning electron microscopy (SEM). The dried film samples were mounted on a metal stub with double-sided adhesive tape. The cross-section structure of the membrane sample was prepared by a liquid nitrogen freeze fracturing followed by gold sputtering before scanning on a scanning electron microscopy (SEM).

2.5. Membrane testing

2.5.1. Gas permeation test. Gas permeation tests were performed with a permeation membrane by using pure gases such as oxygen (O_2) and nitrogen (N_2). Circular membrane discs with an effective permeation area of 13.5 cm^2 were used. Feed pressure will be controlled with 10 bar as maximum pressure. Permeate side was opened to atmosphere. Experiments are carried out at ambient temperature (25°C). Gas permeation rates are measured by a soap bubble flow meter.

3. Results and discussions

3.1. Membrane morphology structure

Selection of the solvent and the non-solvent plays a vital role in controlling the membrane morphologies and properties. Figure 2 and figure 3 shows the SEM images of surface layer and cross-sections of asymmetric CA membrane respectively. It is found that asymmetric CA membranes could be produced using NMP and different non-solvents used. All these fabricated membranes consist of a skin layer supported above a closed-cell substructure. However, various non-solvents used in the study produced different membrane morphologies in terms macrovoid substructure and the skin layer.

It is observed that all four samples have similar membrane surface layer since it have similar solvent evaporation time which is 15 s as in figure 2. Solvent evaporation time plays important role for membrane surface condition because the longer time for solvent evaporation will affect the arrangement of membrane molecular orientation and causing the membrane surface to become closely pack. The closely pack membrane surface preventing the penetration of non-solvent thus decreasing the pore size on the membrane surface and reducing defects.

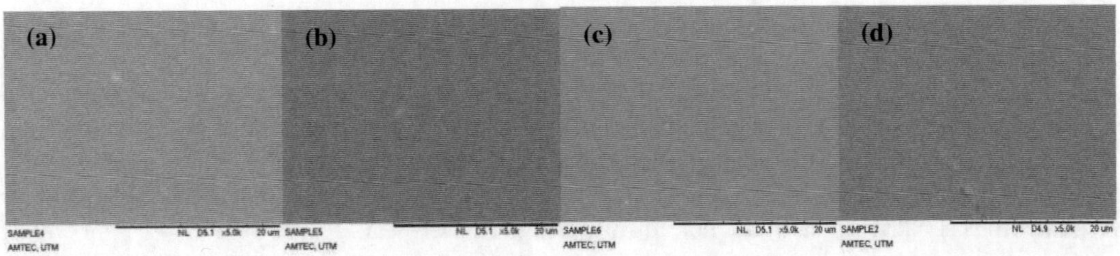

Figure 2. Micrograph of asymmetric membrane surface prepared by dry/wet phase casting method with different quench medium (a) no quench medium (b) quenched methanol (c) quenched n-hexane (d) quenched methanol followed by quenched n-hexane.

Figure 3(a-d) shows a distinct dense skin layer on top of porous substructure of CA membrane. The formation of membrane dense skin layer is due to the evaporation of solvent on membrane surface causing the membrane molecular chain rearrangement forming closely pack membrane skin layer. The penetration of water during wet phase inversion process gives rise to the sponge-like microvoids due to the fast solvent exchange and quick precipitation of polymer [8].

Based on figure 3, the order of decreasing of membrane substructure porosity is (d) > (b) > (c) > (a). No quench medium membrane has less macrovoids in the substructure as in figure 3(a) compared to other membranes due to the collapsed of membrane substructure cause by high surface tension water leaving the membrane consist of strong capillary forces within the finely porous substructure during the drying process [9]. Figure 3(c) shows that n-hexane quench membrane has more porous substructure than no quench medium membrane due to the n-hexane that helps water leaving the membrane better and preventing the membrane structure from collapsing during drying process [9]. The methanol quench membrane as in figure 3(b) has more porous substructure due to the swelling of polymer matrix that enhance the membrane free volume as well as complete removal of residual solvent [10]. The quench methanol followed by quench n-hexane membrane shown in figure 3(d) has the greatest membrane structure porosity compared to the other membrane. The immersions of membrane in methanol quench medium causing the membrane polymer matrix to swell and replacing the water. The further immersions in n-hexane quench medium cause the substructure of swell membrane polymer matrix become stronger and prevented from being collapsed during membrane drying process.

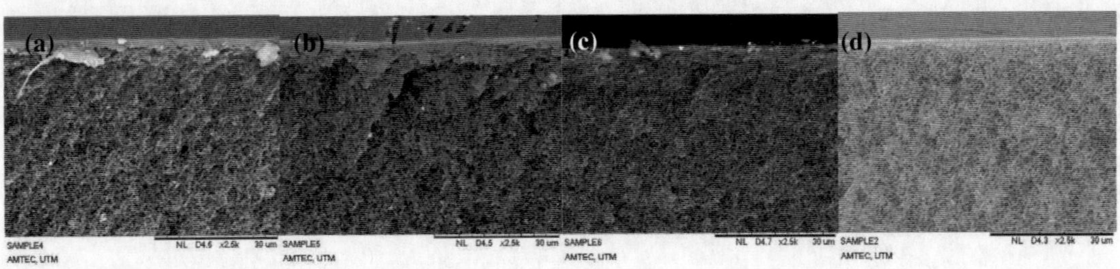

Figure 3. Micrograph of asymmetric membrane cross section prepared by dry/wet phase casting method with different quench medium (a) no quench medium (b) quenched methanol (c) quenched n-hexane (d) quenched methanol followed by quenched n-hexane.

3.2. Gas permeation measurement

The gas separation characteristic is determined by plotting the permeance of O_2, N_2 and O_2/N_2 ideal selectivity of each membrane against the membrane with different type of quench medium. The permeate side was maintained at essentially atmospheric pressure. The permeance of O_2 and N_2 of membrane with various quench medium are presented in table 2.

It is clearly shown that combination quench medium methanol and n-hexane gives the best permeability of O_2 and N_2 gas. This means that this type of solvent exchange steps produces high permeability with good pore size of membrane. The significant differences of gas permeance among membranes prepared from various solvent–non-solvent pairs could be explained by referring to their morphologies as shown by SEM images (figure 3). Since high value of permeability can be achieved in asymmetric membrane with very thin skin layer [11], it can be concluded that the rest of membranes have less porous substructure that did not allow gas to permeate well.

Table 2. Gas O_2, N_2 permeance with different quench medium

Permeability	O_2	N_2
No quench medium	0.1354	0.1059
Methanol	0.3811	0.3626
N-Hexane	0.1502	0.1338
Methanol and N-Hexane	4.4187	1.6856

Figure 4 shows selectivity of O_2/N_2 gas towards CA membrane that prepared with different quench medium. Membrane with water, methanol and n-hexane as quench medium gives the best selectivity of gas O_2/N_2 than other membrane that prepared with different quench medium. It indicated that membrane prepared with water, methanol and n-hexane as quench medium was able to do the gas separation very well. While the other membranes have low selectivity of membrane which indicate that the membrane did not be able to do gas separation very well and it may be because of CA membranes prepared have less pore size that allowed O_2 and N_2 passes through the membrane. As discuss earlier, since combination methanol and n-hexane as quench medium succeed to prepared a membrane with thin skin layer it will gives better selectivity than other membranes that have less porous substructure.

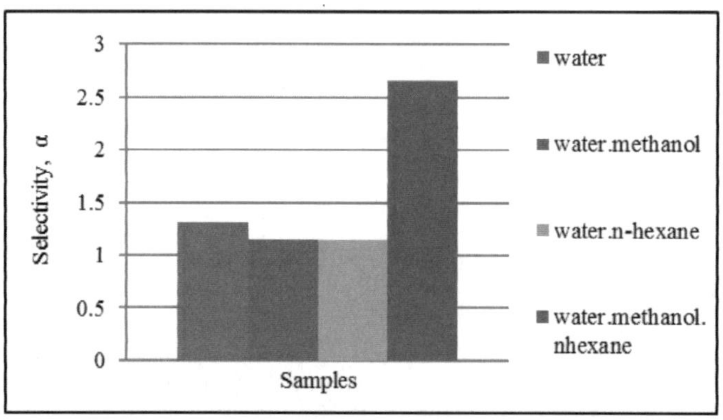

Figure 4. Graph of selectivity of O_2/N_2 against CA membrane with different quench medium

4. Conclusion

The effects of solvent–non-solvent pairs on the morphology of asymmetric CA membrane and its relation to the O_2/N_2 separation performance have been investigated. The prepared membranes that formed via dry/wet phase inversion process possessed an asymmetric structure with thin skin layer at the top and porous substructure at the bottom. The porosity of the membrane surface is almost the same for all the membranes prepared but differ in the compactness of substructure. This difference occurred because of different quench medium that affect the solvent exchange process. Quench medium using methanol and followed by n-hexane possessed the best CA membrane since it shows double performance of gas permeability and selectivity compared to other type of quench medium thus indicates that the membrane have very thin skin layer that allowed gas to pass through the membrane very well.

5. References

[1] Baker R W, Wijmans J G, and Kaschemekat J H 1998 The design of membrane vapor–gas separation systems *J. Memb. Sci.* **151** 55–62

[2] Scholes C A, Anderson C J, Stevens G W, and Kentish S E 2013 Membrane gas separation – physical solvent absorption combined plant simulations for pre-combustion capture *Energy Procedia* **37** 1039–49

[3] Scholes C A, Stevens G W and Kentish S E 2012 Membrane gas separation applications in natural gas processing *Fuel* **96** 15–28

[4] Donohue M D, Minhas B S, and Lee S Y 1989 Permeation behavior of carbon dioxide-methane mixtures in cellulose acetate membranes *J. Memb. Sci.* **42** 197–214

[5] Farrukh S, Javed S, Hussain A and Mujahid M 2014 Blending of TiO2 nanoparticles with cellulose acetate polymer: to study the effect on morphology and gas permeation of blended membranes *Asia-Pac. J. Chem. Eng.* **9** 543–51

[6] Sanders D F, Smith Z P, Guo R, Robeson L M, McGrath J E, Paul D R and Freeman B D 2013 Energy-efficient polymeric gas separation membranes for a sustainable future: A review *Poly.* **54** 4729–61

[7] Ismail A F, Ng B C and Abdul Rahman W A W 2003 Effects of shear rate and forced convection residence time on asymmetric polysulfone membranes structure and gas separation performance *Sep. Pur. Tech.* **33** 255–72

[8] Iqbal M, Man Z, Mukhtar H and Dutta B K 2008 Solvent effect on morphology and CO2/CH4 separation performance of asymmetric polycarbonate membranes *J. Memb. Sci.* **318** 167–75

[9] Wang L K, Hung Y T and Shammas N K 2007 *Handbook of Environmental engineering: Advanced Physicochemical Treatment Processes* vol. 4 New Jersey, Springer Science & Business Media

[10] Seong J G, Zhuang Y, Kim S, Do Y S, Lee W H, Guiver M D and Lee Y M 2015 Effect of methanol treatment on gas sorption and transport behavior of intrinsically microporous polyimide membranes incorporating Tröger's base *J. Memb. Sci.* **480** 104–14

[11] Ismail A F and Yean L P 2003 Review on the development of defect-free and ultrathin-skinned asymmetric membranes for gas separation through manipulation of phase inversion and rheological factors *J. App. Poly. Sci.* **88** 442–51

Acknowledgments
We gratefully acknowledge the financial support from the Universiti Teknologi Malaysia and Ministry of Education (Grant No: 4F652 and 4L116).

The impedance characterization of Carbon Nanotubes - Fumed Silica Poly (vinyl alcohol) Composites

R N Othman[1, 2], **A N Wilkinson**[1]
[1] Materials Science Centre, School of Materials, *The University of Manchester, Grosvenor Street*, Manchester, M13 9PL, United Kingdom.
[2] Department of Mechanical Engineering, *National Defence University Malaysia*, Kem Sungai Besi, 57000 Kuala Lumpur, Malaysia.

E-mail: [1]izawati@upnm.edu.my , [2]Arthur.Wilkinson@manchester.ac.uk

Abstract. Carbon Nanotube (CNT) was grown on the surface of fumed silica via chemical vapor deposition (CVD) method. In this work, silica acted as a site that holds CNT together, which prevents further agglomeration during composite processing. Iron catalyst at different loading (7.5 wt. % up to 25 wt. %) was introduced via impregnation method to synthesize CNT at 1000°C, under methane flow. Floating catalyst method was used where ferrocene (2.5 wt. % and 5 wt. %) was used as starting reactants together with toluene at 760°C. The reaction time was set at 1 hour for both methods. It was later confirmed via SEM images that the floating catalyst method is more suitable to produce a large amount of CNTs. The sample synthesized via floating catalyst method at both 2.5 wt.% and 5 wt. % ferrocene was later used to prepare composites. Composite films of the particles in poly (vinyl alcohol) (PVOH) were cast and their TEM images show that the dispersion is indeed uniform. From impedance measurement, it was found that the particles synthesized via floating catalyst method were found to form an electrically-conductive percolated network with percolation threshold of 1 wt. %, obtained via percolation equation.

1. Introduction

Carbon nanotubes (CNTs) are multifunctional fillers that impart enhanced mechanical [1], electrical [2], and thermal [3] properties to polymer composites. Electrically-conducting composites have found wide application in areas ranging from aerospace to electronics. However, the majority of nanotubes available on the bulk scale are entangled, making them difficult to disperse into a polymer matrix [4]. Among the approaches used to improve the dispersion of CNTs in polymer matrices include chemical functionalization, in situ polymerization, and enhanced physical blending [4].

Grafting of CNTs onto substrates such as fibers [5-7] and micro particles [8-10] offers an alternative approach to tackle the issues associated with dispersion in a composite matrix, as well as additional benefits (hybrid effects) provided by these dual-filler systems. One approach to obtain such hybrid systems is the direct growth of nanotubes on the supporting fibre or particle. Studies have shown that fibers of carbon [5], silica [6], and alumina [7] acted not only as sites for CNT growth, but also as a substrate carrier to disperse CNTs into polymer matrices. Direct growth of CNTs on silica particulate substrates has been reported [8, 11-13]. The growth of aligned CNT on alumina particles

(μAl_2O_3) has also been studied and a low percolation threshold and a high thermal conductivity were achieved in an epoxy matrix [9]. CNT grown on spherical stainless steel particles were reported to give improvements in the tensile strength and Young's modulus of polyurethane composites [10].

Polyvinyl alcohol (PVOH) has been widely used as a media to study the dispersion of CNT [14-16]. Shafer and Windle assessed the effectiveness of chemically treated CNTs as conductive filler in PVOH and reported a percolation threshold of 5-10 wt. % [14]. On the other hand, Kilbride *et al.* [15] achieved a lower percolation threshold value of 0.005 wt. %, in their non-treated arc discharge CNT – PVOH composite. Although Kilbride *et al.* [15] did not contrast their findings with Shaffer and Windle [14], this difference in percolation is thought to be due to the types of CNTs used. Without treatment, both CVD and arc discharge produced CNTs are tangled in nature. In contrast to that, the treated, hydrophilic CNTs form an individual dispersed fibre, which is not favourable for the formation of conductive pathway in PVOH.

In this study, CNTs were grown on fumed silica nanoparticles by Chemical Vapor Deposition (CVD) method. Two methods were performed to introduce catalyst into the system which is via iron impregnation and also via floating catalyst where the ferrocene was co-flown together with toluene during reaction. In the floating reaction, the ferrocene formed in-situ within the iron catalyst particles.

2. Experimental

2.1 Synthesis and the characterization of the CNT-Fumed Silica

In this section, the procedures taken to grow the CNTs on the fume silica substrates via Chemical Vapor Deposition method were described. As shown in figure 1, the morphology of the fume silica can be described as highly agglomerated nanostructured particles. There were two approaches in introducing catalyst into the system was taken, which were impregnated catalyst and floating catalyst.

Figure 1. The typical FESEM images of the fumed silica before the growth of CNT

In the case of catalyst introduced via impregnation, the fume silica was first coated with Iron (III) nitrate $Fe(NO_3)_3$ in acetone solution to achieve 7.5 wt.%, 10 wt.%, 15 wt.%, 20 wt.%, and 25 wt.% catalyst loading. The catalyst was dispersed throughout the fumed silica via the assistance of sonic tip with 10% amplitude, held for 10 minutes. The particles were then left to dry overnight in fume cupboard. The pre-coated particles were then placed with the heating zone of the furnace catalysts. The CNTs were grown under the flow of methane, hydrogen and argon for 1 hour at 1000°C.

The CNTs were also grown on the surface of fumed silica via floating catalyst method. In this approach, the mixture of ferrocene–toluene was injected into the furnace during reaction at a rate of 0.4 ml/min. The loading of ferrocene was kept at 2.5 wt. % and 5 wt. %. The reaction temperature and time was kept at 760°C and 1 hour, respectively.

The Scanning Electron Microscope (SEM) imaging was performed for all samples in determining the effects of reaction conditions on the yield of CNTs. An adhesive carbon disc on a pin stub was lightly pressed in the nanoparticles. The sample was then coated with gold using sputter coater to enhance the sample imaging. Philips XL 30 FEG SEM was used to view the morphology. Based on the results obtained, the sample obtained that yield the highest density of CNT was selected for further impedance investigation.

2.2 Synthesis and characterization of PVOH – CNT-Fume Silica

The preparation of the thin films can be described as follow. 80 mg of the hybrid particles from the selected sample was added into 20 ml of 2 wt. % PVOH aqueous solution ($PVOH_{(aq)}$). The particles were dispersed in the solution by magnetic stirring for 20 minutes at 80°C. 10 ml of this solution was then transferred to a plastic mould and left to dry to yield a thin PVOH film with 20 wt.% hybrid particle loading. Another 10 ml of (PVOH(aq)) was then added to the remaining solution, in which the process was repeated further to yield solution containing 10 wt. % dispersed hybrid particle loading. This process was stopped until the final dispersion containing 0.04 wt. % hybrid particles loading. All the thin films were left to dry in fume cupboard up to 5 days. To prevent water uptake, the dried samples were kept in dessicator for 2 days before performing impedance measurement.

The impedance characterization was performed using an Impedance Analysis Interface/NumetriQ PSM 1735, with a frequency ranging from 1 Hz to 1 MHz. The rectangular thin films (~ 30 mm x 10 mm) were cut and placed on glass slides. Silver paint was applied to both ends of the films to act as electrodes which were connected to the instrument by crocodile clips. At least five measurements were taken for each film sample to ensure the results reproducibility.

3. Results and Discussion

Figure 2 shows the typical SEM images of the particles synthesized via catalyst impregnation, CVD method. From the images, it can be seen that CNTs started to grow with catalyst loading as low as 7.5 wt. % Fe (figure 2a), with the amount seemed to increase as the catalyst loading was further increased. The diameter of the CNTs also seemed to be increased as the amount of the catalyst increased, implying a bigger initial catalyst nucleation before growth took place [17-19], based on the growth mechanism as suggested by Baker [20]. Although more catalysts were introduced (up to 25 wt. %), this attempt failed to result in the higher density of CNT formation. It can be concluded from figure 2 that catalyst impregnation fails to results in the formation of high amount of CNTs under the specified reaction conditions.

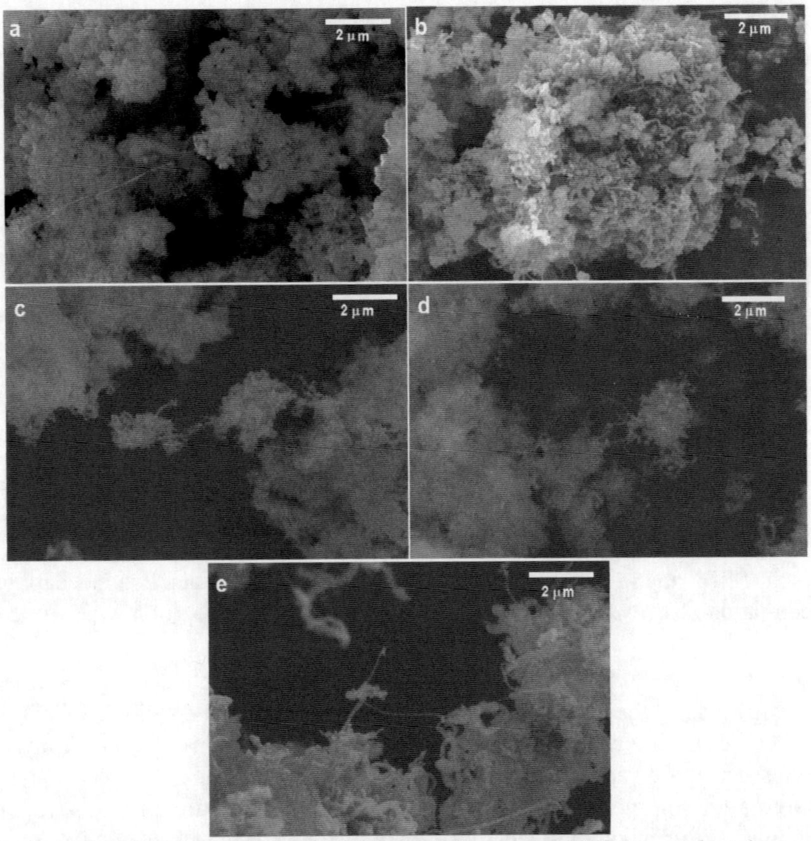

Figure 2. Typical SEM images of the hybrid particles via $Fe(NO_3)_3$ catalyst impregnation CVD method; with the following Fe loading; (a) 7.5 wt. %, (b) 10 wt. %, (c) 15 wt. %, (d) 20 wt. %, and (e) 25 wt. %.

As the pre-coated catalyst method did not seem to work, further effort in synthesizing CNT on the fumed silica substrate was performed via floating catalyst CVD method. Figure 3 shows the typical SEM images of the hybrid particles obtained. It can be seen that high density of CNTs were obtained via this method although the ferrocene loading employed was as low as 2.5 wt. %. As the loading was further doubled to 5 wt. %, tremendous improvement in the CNT yield could be observed. The efficiency of this floating catalyst method in enhancing the growth of CNTs was also reported previously on spherical micron silica gel [2], spherical alumina substrate [21], silica fibers [6], just to name a few.

IOP Conf. Series: Earth and Environmental Science **36** (2016) 012010 doi:10.1088/1755-1315/36/1/012010

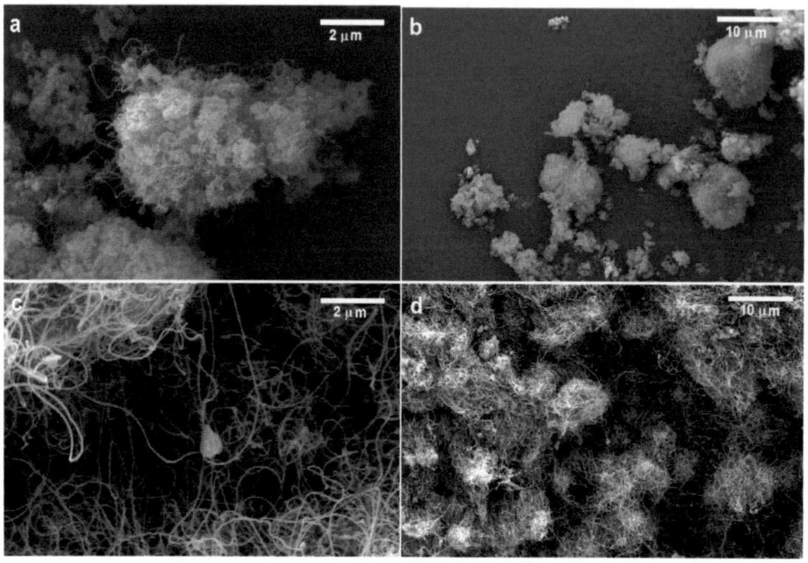

Figure 3. Typical SEM images of the hybrid particles synthesized via floating catalyst CVD with ferrocene concentration of 2.5 wt. % (a & b) and ferrocene concentration of 5 wt. % (c & d).

The particles synthesized via floating catalyst CVD were later selected for further electrical impedance investigation, as the condition with floating catalyst method yielded CNT with the high density (figure 3). The specific conductivity of the particles in poly (vinyl alcohol) films was measured using impedance spectroscopy and the results were summarized in figure 4 and 5.

Figure 4 shows a log-log plot of the film's conductivity as a function of frequency. In the case of the films containing particles synthesized from 2.5 wt. % ferrocene, the conductivity values of the films increased more than two orders of magnitude when loading is doubled from 2.5 wt. % to 5 wt. % (figure 4a). These values were recorded at frequency less than 10 Hz, as it is close to the dc condition [22, 23]. It can also be observed that for samples less than 5 wt. %, the resistance was constant until 10Hz after which it raised by three orders of magnitude as frequency increased up to 10^4 Hz, indicative of an increase in the capacitive component, compared to higher loadings. Figure 4b shows the impedance pattern for films containing higher CNT density (produced from 5 wt. % ferrocene). It can be seen that the conductivity values of the films for all loadings above 0.625 wt. % are generally higher than those containing particles synthesized from 2.5 wt. % ferrocene (Figure 4a). This simply implied that higher CNT density contributes to the higher values of the composite's conductivity.

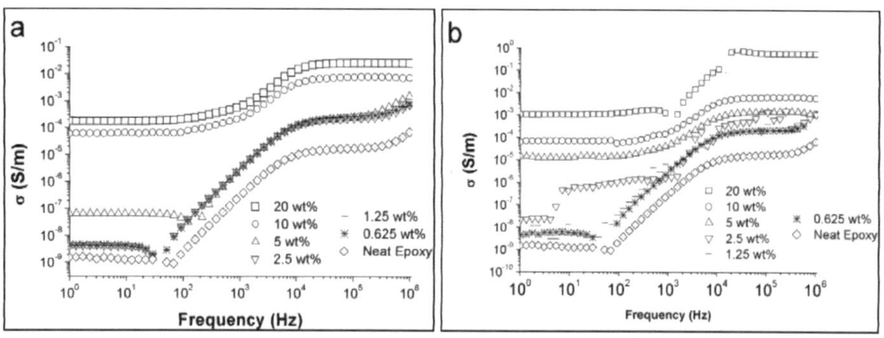

Figure 4. Log-log plot of the film's conductivity as a function of frequency for films containing hybrid particles synthesized with 2.5 wt. % ferrocene (a) and 5 wt. % ferrocene (b).

Figure 5 shows the relation between the film's conductivity as a function of particle loading for both cases. Samples were also made with 0.04, 0.08, 0.16, 0.32 wt % loading and all these were found to be insulating within the limits of the analyzer. As such, the values were not shown on these figures. In the case of films containing particles synthesized with 2.5 wt. % ferrocene, the significant increase in conductivity is overserved upon increasing the particle loading from 5 wt. % to 10 wt. % (figure 5b). This indicated the formation of a conductive network of particles above 5 wt. % loading. In the case of films containing particles synthesized with 5 wt. % ferrocene, the advancement in conductivity was observed to occur above 1.25 wt. %, implying the formation of percolated conductive network.

The critical percolation threshold can be determined from the percolation scaling law; $\sigma = \sigma_0 (p - p_c)^t$; where σ is the conductivity, σ_0 is a constant, p is the particle loading, p_c is the critical loading, and t is a constant. The transition from insulator to conductor occurred at p_c. The constant t defines the dimensionality of the filler forming percolated network; i.e. $t \sim 1.33$ for 2D and $t \sim 2.0$ for 3D [24]. The values of p_c equals to 4.9 and 1.2 are calculated from the films synthesized from 2.5 wt. % ferrocene amd 5 wt. % ferrocene, respectively. As the 5 wt. % sample containing higher CNT, the critical loading needed to achieve percolated network is definitely lower. Both films yielded the t values approximately equals to 2, indicating formation of 3-dimensional conductive network, as confirmed by the SEM images (figure 3).

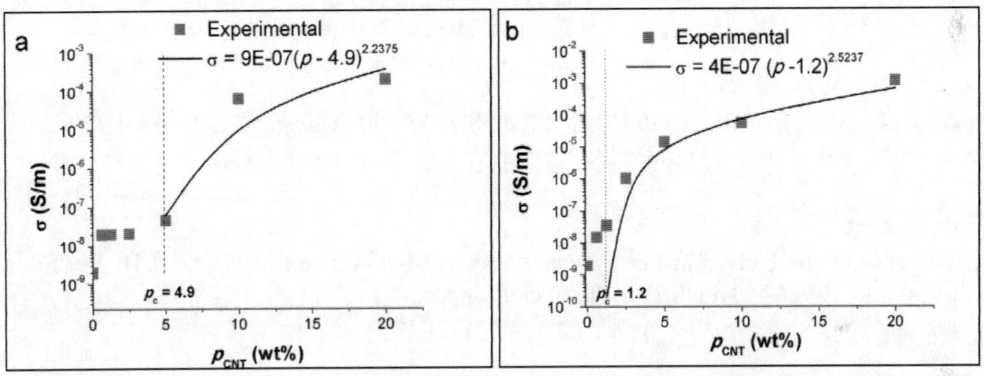

Figure 5. Semi log conductivity of the films as a function of weight loading for particles synthesized with 2.5 wt. % ferrocene (a) and 5 wt. % ferrocene (b). The equations showed the corresponding fitted percolation equations for each particle.

4. Conclusions

In conclusion, CNTs were successfully grown on the surface of fumed silica. The catalysts were introduced via both impregnation and floating catalyst method. The silica acted as to hold the CNTs and to prevent them from forming bundles during composites processing. This study presented a simple approach in controlling bundle formation during CNT-polymer processing. Its ability in forming a conductive network at low p_c demonstrated the efficiency of grafting CNT in fumed silica. The values of t obtained from percolation scaling law indicated that the network formed was in 3-dimensional.

5. References

[1] Feng L, Li K, Zhao Z, Li H, Zhang L, Lu J, *et al.* 2016 *Mater Des.* **92** 120
[2] Othman RN, Kinloch IA, and Wilkinson AN 2013 *Carbon* **60** 461
[3] Hong SK, Kim D, Lee S, Kim BW, Theilmann P and Park SH 2015 *Composites Part A:Applied Science and Manufacturing* **77** 142
[4] Xie XL, Mai YW, and Zhou X 2015 *Materials Science & Engineering R-Reports* **49** 89-112

[5] Qian H, Bismarck A, Greenhalgh ES, Kalinka G, and Shaffer MSP 2008 *Chem Mat.* **20** 1862
[6] Qian H, Bismarck A, Greenhalgh ES, and Shaffer MSP 2010 *Carbon* **48** 277
[7] Yamamoto N, Hart AJ, Garcia EJ, Wicks SS, Duong HM, Slocum AH, *et al.* 2009 *Carbon* **47** 551
[8] Agrawal S, Kumar A, Frederick MJ, and Ramanath G 2005 *Small* **1** 823
[9] Bozlar M, He DL, Bai JB, Chalopin Y, Mingo N, and Volz S 2010 **Adv Mater**. **22** 1654
[10] Nguyen XH, Lee YB, Lee CH, and Lim DS 2010 *Carbon* **48** 2910
[11] Huang SM 2003 *Carbon* **41** 2347
[12] Xiang R, Luo GH, Yang Z, Zhang Q, Qjan WZ, and Wei F 2009 *Mater Lett.* **63** 84
[13] Singh C, Shaffer MS, and Windle AH 2003 *Carbon* **41** 359
[14] Shaffer MSP and Windle AH 1999 *Adv Mater.* **11** 937
[15] Kilbride BE, Coleman JN, Fraysse J, Fournet P, Cadek M, Drury A, *et al.* 2002 *J Appl Phys.* **92** 4024
[16] Zhang J, Mine M, Zhu D, and Matsuo M 2009 *Carbon* **47** 1311
[17] Huang L, Wind SJ, and O'Brien SP 2003 *Nano Lett.* **3** 299
[18] Li WZ, Xie SS, Qian LX, Chang BH, Zou BS, Zhou WY, *et al.* 1996 *Science* **274** 1701
[19] Ramesh P, Okazaki T, Taniguchi R, Kimura J, Sugai T, Sato K, *et al.* 2005 *J Phys Chem B.* **109** 1141
[20] Baker RTK, Barber MA, Waite RJ, Harris PS, and Feates FS 1972 *J Catal.* **26** 51
[21] He DL, Bozlar M, Genestoux M, and Bai JB 2010 *Carbon* **48** 1159.
[22] Martin CA, Sandler JKW, Shaffer MSP, Schwarz MK, Bauhofer W, Schulte K, et al. 2004. *Compos Sci Technol.* **64** 2309
[23] Kovacs JZ, Velagala BS, Schulte K, and Bauhofer W 2007 *Compos Sci Technol.* **67** 922
[24] Stauffer D and Aharony An Introduction to percolation theory. 1994

Acknowledgment

RNO acknowledges the Universiti Pertahanan Nasional Malaysia and Ministry of Higher Education, Malaysia and RACE/F2/TK/UPNM/1 for financial support.

Prediction aluminum corrosion inhibitor efficiency using artificial neural network (ANN)

Sh Ebrahimi[1], E G Kalhor[2], S R Nabavi[3], L Alamiparvin[2], R Pogaku[1*]
[1] Faculty of Engineering, Universiti Malaysia Sabah, Kota Kinabalu, Sabah, Malaysia
[2] Department of Chemistry, Faculty of Science, Tabriz branch, Islamic Azad University, Tabriz, Iran
[3] Department of Applied Chemistry, University of Mazandaran, Babolsar, Iran

[1]E-mail: [1]ebrahimi_shamsi63@yahoo.com, [1]ravindra@ums.edu.my

Abstract. In this study, activity of some Schiff bases as aluminum corrosion inhibitor was investigated using artificial neural network (ANN). Hence, corrosion inhibition efficiency of Schiff bases (in any type) were gathered from different references. Then these molecules were drawn and optimized in Hyperchem software. Molecular descriptors generating and descriptors selection were fulfilled by Dragon software and principal component analysis (PCA) method, respectively. These structural descriptors along with environmental descriptors (ambient temperature, time of exposure, pH and the concentration of inhibitor) were used as input variables. Furthermore, aluminum corrosion inhibition efficiency was used as output variable. Experimental data were split into three sets: training set (for model building) and test set (for model validation) and simulation (for general model). Modeling was performed by Multiple linear regression (MLR) methods and artificial neural network (ANN). The results obtained in linear models showed poor correlation between experimental and theoretical data. However nonlinear model presented satisfactory results. Higher correlation coefficient of ANN (R>0.9) revealed that ANN can be successfully applied for prediction of aluminum corrosion inhibitor efficiency of Schiff bases in different environmental conditions.

1. Introduction

Corrosion is devastating effect of environment on metal. In another word, corrosion is destruction of metal due to chemical or electrochemical reaction between the metal and its surroundings. Corrosion may be the most common electrochemical phenomenon experienced in everyday life. Common examples of corrosion include rusting of iron and steel, darkening of silver and copper, breaking out and blowing of chromium coating and paint on the body of automobile [1]. Corrosion is a serious problem, since spends natural resources, weakens steel structures, increases the cost of maintenance, deprive human comfort and lead to human mortality [2]. Aluminum is soft and light but strong metal, with a silver – gray appearance, opaque thin oxidation layer that is formed on the surface due to exposure to air, that prevents further rusting. The weight of aluminum is almost one-third of the weight of steel or copper. It is very durable and resistant to rust. Due to the high technological importance of aluminum, methods are provided to prevent aluminum corrosion. One of these methods

[1] To whom any correspondence should be addressed.

is application of the organic inhibitors of Schiff Bases. Schiff bases are organic inhibitors with considerable importance in organometallic chemistry. They are usually used as ligands because of easy connection of their nitrogen to transition metals. Schiff bases are compound with a functional group that contains carbon-nitrogen double bond with the nitrogen atom connected to an aryl or alkyl group. Schiff base is formed from the condensation reaction of an amine group with the carbonyl group of an aldehyde or ketone. For the first time in 1864 it was reported by Professor Schiff, and named in his honor [3]. The inhibitory effect of some Schiff bases on the corrosion of aluminum has been studied by polarization, electrochemical impedance spectroscopy (EIS) and weight loss measurements. According to the results of all three methods, Schiff bases are excellent inhibitors and their efficiency increases with temperature reduction and increase of their concentration [4].

Quantitative Structure Activity Relationship (QSAR) is one of the major application fields of chemometrics in the studies linking molecule properties to their structural characteristics. QSAR is the logical relationship between the properties of compounds (active) and their chemical structure. QSAR is a theoretical method beneficial in relating structural based parameters to corrosion inhibition efficiencies of a group of compounds. Reviewing the relationship between chemical structure and activity of compounds enables predicting the activity of new compounds based on information about their chemical structure. The main objective of the present study is application of QSAR to predict inhibition rate of Schiff Bases for aluminum corrosion [5-6]. There is variety of models in QSAR studies; in this study the focus is on multiple linear regression models (MLR) and Artificial Neural Network (ANN).The regression model is linear, while artificial neural network is nonlinear. In this study, artificial neural network modeling indicates the best correlation between the actual data of corrosion inhibition of Schiff bases for aluminum corrosion and data from the model.

2. Methodology

The aim of the present study is establishment of quantitative structure-activity relationship to acquire the inhibitor efficiency of Schiff bases for aluminum corrosion based on theoretical method.

Application of this method in estimation of the inhibitor efficiency of Schiff bases for aluminum corrosion includes:

2.1. Selection of Schiff base molecule from resources and collection of data for the inhibitory efficiency

In this study the Schiff base molecules and required data, including corrosion inhibitory activity of different molecules for aluminum element, were extracted from scientific resources where the inhibitory efficiency of them is examined in different conditions such as temperature, time, pH, and concentration of inhibitors. Temperature, time, pH, concentration are independent variables, and the inhibitory efficiency is considered as the dependent variable. Number of Schiff base molecules is 42 extracted from references [3-4] and [7-24].

2.2. Drawing the form of molecules by Hyperchem

After extracting Schiff base molecules from resources, the structure of organic compounds (Schiff bases) is drawn by Hyperchem software.

2.3. Optimization of molecules by Hyperchem

To optimize Schiff base molecules the Hyperchem is used. Inhibitory molecules are in the most volatile state regarding energy and spatial formulation. In order to the inhibitory molecules to reach the most appropriate conformation and stable energy state, they can be optimized by the software.

2.4. Calculation of descriptors by Dragon3

Molecular descriptors are the final result of mathematical reasoning and methods that codify chemical information and indicate them as symbols providing the molecule as a valuable number. Descriptors are numerical values representing the properties of molecules. Each of the descriptors provides specific information about the molecule. For example, descriptors may be physical and chemical properties of a molecule such as molecular weight or the number of hydrogen bonds, the number of non-hydrogen bonds, the number of connected halogen, etc.

After determination of the most appropriate conformation, molecule descriptors are derived and selected by the software; here the Dragon3 has been used. Dragon3 Software is used to calculate descriptors simultaneously. The software can calculate 1497 descriptors for the molecules with up to 150 atoms. If the inhibitory molecule (Schiff bases) is anionic or cationic, after calculation of anion and cation descriptors, to calculate Schiff base descriptors, anion and action descriptors are collected and general molecule descriptor obtained.

2.5. Reduction of descriptors using PCA

As mentioned above, 1497 descriptors were calculated for each Schiff base by Dragon3 software. The number of independent variables to create QSAR model is very high and not all of these descriptors contain information. Because of high number of descriptors, to put them in neural network meaningful descriptors should be used. In this study, to further reduce the descriptors to meaningful descriptors, the principal component analysis (PCA) is used.

To perform the PCA on a series of data, the coordination axes are rotated; most of data changes are in the first new axis and from remaining changes, most of the changes are on the second new axis and from the remaining changes, most of the changes are on the third axis, etc. Data coordination in the new axis is called score. Applying the PCA, primary data converted to score. To reduce the dimensions of data, scores with the maximum changes (information) are selected and scores with changes around the system noise are removed [25].

2.6. Transferring descriptors (structural parameters) to Microsoft Excel and addition of environmental parameters (temperature, concentration, pH)

2.7. Data Matrix

Since the total number of molecules is 42, 5 descriptors for each molecule were obtained from PCA method. Therefore, molecular descriptors, as well as T (temperature), C (inhibitory concentration), t (time), and pH are independent and dependent variables of corrosion inhibitory activity of Schiff base for aluminum, respectively. Adding these environmental variables to structural descriptors, the number of independent variables will be 9. The total number of data is 241, hence, dimension of independent data matrix will be 241 × 9 and dependent data matrix will be 241 × 9. Then, for randomization of data, they were codified as 0, 0, 1, 2 and classified in three sets: training data matrix with code (0), test set with Code (1) and simulation set with code (2); training set is used to create a model between dependent and independent variables, test set to predict and evaluate the performance of the model in education, and simulation set is used for general training model. Training series cover most of the data because the probability for this series is double other series. From the total 241 data, 140 data belong to training set; 49 data for test set, and 52 data for the simulation set, regarding that training data matrix is 140 × 9, test data matrix is 49 × 9 and simulation data matrix is 52 × 9.

2.8. Multi-stage linear regression modeling and MLR model

In the simple linear regression, a linear relationship is obtained between the independent and the dependent variables. In fact, in the multiple linear regressions there is a relationship between several independent variables and the dependent variable. The goal is to study first-order linear model as Eq. (1).

$$y=\beta 0+\beta 1X1+\beta 2X2+\beta 3X3+\beta 4X4+\beta 5X5+\beta 6T+\beta 7C+\beta 8t+\beta 9pH \tag{1}$$

The confidence level has selected 95%. Therefore, independent variables with meaningfulness (p-value) of more than 0.05 will be removed in the final stage equation [26].

2.9. Modeling with artificial neural network (ANN)

Since the relationship between dependent and independent variables may be non-linear, it's normal for linear regression models to show high error for non-linear functions. Therefore, it is better to use a method for approximation of nonlinear functions. One of these methods is modeling using artificial neural networks. Artificial neural networks are nonlinear mapping structures based on the performance of the human brain. Interest to neural networks is due to the ability of the network to mimic the human brain, in addition to his ability to learn and respond. Therefore, neural networks, with its remarkable ability to derive meaning from complicated or imprecise data, can be used to extract patterns and identify the ways very complex and difficult to humans and other computer techniques. A trained neural network can be considered a specialist in analyzing the given information [27]. Modeling by artificial neural network, parameters should be optimized.

2.9.1. Optimization of artificial neural network parameters and modeling. At this stage, neural network is designed and optimized using Neural Power software to predict corrosion inhibition of Schiff bases for aluminum corrosion. To create a model by neural network, network parameters such as the number of neurons, epoch, learning rate and momentum, learning function and learning algorithms, are optimized. Optimization is done one at a time.

2.10. Calculation of error to assess the accuracy of the model using the following formula:

$$MS = \left(\frac{\sum_{s=1}^{ns}\sum_{i=1}^{ni}(y_{si}-\mathrm{out}_{si})^2}{n_i n_s}\right)^{1/2} \tag{2}$$

$$MPE = \frac{1}{N}\sum_{i=1}^{N}|r_i - y_i| \tag{3}$$

$$.S.E(\%) = \left[\frac{\sum_{i=1}^{N}(r_i-y_i)^2}{\sum_{i=1}^{N}(r_i)^2}\right]^{1/2} \times 100 \tag{4}$$

Where, N, ri and yi are the number of data, the real response and the network response or output to the desired parameters, respectively. Ni and n are the number of inputs and outputs of variables. Ysi, is the i th component of expected response ys and outsi, i- th output component produced by the network for the s- th input vector.

3. Results and discussion

3.1. Multiple linear regression model and its results

After creating the appropriate descriptors by PCA and Dragon3, Schiff base divided into three sets of training and testing, and simulation. Training set moved to MINITAB software to create multiple linear regression models. After two linear steps, the equation calculated by the training set. To check if the resulting equation is acceptable, it is studied by test and simulation set. Equation 5 shows first-order linear equation fitted in the first phase for training set.

$$y=89.5-0.000011x1-0.000124x2+0.000382x3+0.00044x4+0.000203x5-0.780T+33.5C-0.00683t-13.4 \tag{5}$$

Coefficient of multiple linear regression model is presented in table 1.

Table 1. P- Values for the independent variables in linear equation (first stage of) with multiple linear regression method

Independent variables	X_1	X_2	X_3	X_4	X_5	T	C	t	pH
Values P	0.210	0.412	0.269	0.26	0.769	0.000	0.022	0.001	0.001

As indicated in Table 1, regarding the selected level of confidence (95%) and in some descriptors P-value is greater than 0.05, these coefficients are not meaningful and should be removed. After removal of these coefficients training equation will be fit and change to Eq. 6 in the second level.

$$y=91.5-0.757T+27.9C-0.00784t-12.3pH \quad (6)$$

3.2. Modeling with multi-layer neural network
Data of neural network design, like multiple linear regressions, is randomly divided into three groups of training, test, and simulation. Training set creates model, test set predicts network performance and simulation generalizes the created network. In desired network design, initial weight values were randomly from [1, -1] and before the training input values were between [1, 1], and output regulated from [2.0 to 8.0]. The number of entries in the input layer is selected equal to the number of variables such as temperature, concentration, pH, time, and the number of descriptors (Scores) selected by the PCA method. Latent layer neurons are determined during optimization and the output layer neurons are equal to the number output variables, with the same inhibition percentage of corrosion. Neural Power is used in modeling the artificial neural network and normalization of the data is done by the software.

3.2.1. Optimization of neural network parameters in Neural Power software. At this stage parameters of the neural network, including the number of latent layer neurons, momentum, learning rate, repetition, learning algorithms and functions, are optimized one at a time. The optimization results of neural network parameters are shown in Table 2.

Table 2. shows the results from the optimization of neural network parameters

Optimized parameters	**Results**
The number of hidden layer neurons	20
Learning rate	0.32
Momentum coefficient	0.68
The number of repetitions	64
Hidden layer and output layer transfer function	Sigmoid function
Learning algorithm	Levenberg Marquardt
The number of hidden layers	1

3.2.2. Calculated errors for optimized neural network. The neural network model was evaluated after optimization. According to the results of neural network model and correlation coefficients of training, test and simulation, the output values of the model were compared and evaluated to experimental values. The error between the predicted and actual values for the training and test of designed neural network is calculated in three forms of RMSE, MPE, RSE and the correlation coefficient (R). Error and correlation coefficients are given in Table 3. According to table 3, the errors of ANN are less, and correlation coefficient is higher than regression models.

Table 3. The correlation coefficient values and errors for neural network optimized.

set	**R**	**RMSE**	**MPE**	**RSE**
training	0.908	8.2799	6.2674	12.8557
test	0.906	9.6604	7.2995	13.2178
simulation	0.878	14.9288	12.0745	19.8114

Results of comparing errors of test set, for three models of multiple linear regression, and ANN (nonlinear method) are indicated in table 4.

Table 4. Comparison of the errors for linear regression models and artificial neural network for test set

models	RMSE	MPE	RSE	R
multiple linear regression	16.9626	14.1420	23.2089	**0.5193**
ANN	9.66049	7.2995	13.2178	**0.9069**

Correlation coefficient between output of neural network and experimental amount of corrosion inhibition for evaluation is reviewed according to figure1.

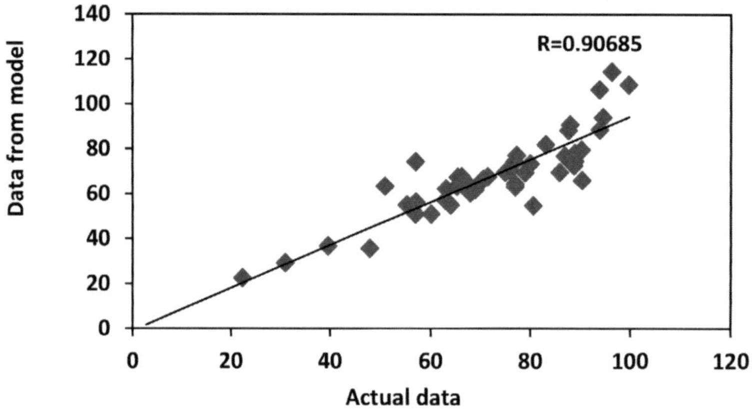

Figure 1. The correlation coefficient (R) show good correlation for evaluation (test) of data between the real data and the data from artificial neural network model.

4. Conclusion

RSE error of test set is high in the multiple linear regression models (23.2089), while the value of R is low that it is the reason for the rejection of this model for corrosion inhibition aluminum by Schiff Bases. R (0.5193) does not indicate good correlation between the model and the real value.

Due to the reduction of RSE error rate for test set in multilayer neural network model, and increase of R (~0.9069), it was concluded that this model, as a non-linear model, can predict the corrosion inhibition of aluminum by Schiff bases. Due to the increase of R in this model, there is good correlation between the actual value and the value of model.

Modeling in this study is through variable descriptors along with the high number of environmental variables, including temperature, concentration, time and pH; high R value (0.9069) for test set of multilayer neural networks model shows the importance of this modeling. It worth to note that, so far in no studies the modeling has been through these environmental variables (Figure 1). The figure shows the correlation coefficient between the actual data of inhibitory Schiff bases for aluminum corrosion and the data of artificial neural network for test set.

5. References

[1] Faran Hofar 2012 *Principals of Corrosion* Jahad Daneshgahi, Sharif Science and Technology University.

[2] Razi S M 1997 *Corrosion Control in Industries* vol. **1** Corrosion Association of Iran

[3] Gomma G K and Wahdan M H 1995 *Mater. Chem. Phys.*, **39** 209-13.
[4] Ashassi-Sorkhabi H b B S, Aligholipour B and Seifzadeh D 2006 Appl. Surf .Sci **252**, 4039-47.
[5] Sayed El, Ashry E and Senior S 2011 *Corros.Sci.,* **53**, 1025-34.
[6] Mousavi M, Mohamadalizadeh M and Khosravan A 2011 *Corros,Sci.,* **53**, 3086-91.
[7] Khaled K F 2010 *Corros.Sci.,* **52**, 2905-16.
[8] Obot1 I B, Obi-Egbedi N O and Umoren S A 2009 *Int. J. Electrochem. Sci*., **4**, 863-77.
[9] Foda A S, Elewady G Y, EL-Askalany A and Shalabi K 2010 *ZAŠT. MAT.,* **51**.
[10] Fakrudeen S P, B L H, C A M H and B B R V 2012 *I.J.E.R.A*, **5**, 2049-61.
[11] Naik U J, Panchal V A, Patel A S and Shah N K 2012 *J. Mater. Environ,* **3**, 935-46.
[12] Halambek J, Jukić M, Berković K and Vorkapić-Furač J 2012 *Int. J. Electrochem. Sci*, **7**, 1580-1601.
[13] Yurt A and Aykın Ö 2011 *Corros.Sci.,* **53**, 3725-32.
[14] A Y, S U and Dal H 2006 *Appl. Surf.Sci.,* **253**, 919- 25.
[15] Safak S, Duran B, Yurt A and Türkog˘lu G 2012 *Corros.Sci.,* **54**, 251-59.
[16] Pinto G M, Nayak J and Shetty A N 2011 *Nano. Metal. Chem.*
[17] Noor E A 2009 *Mater.Chem.Phys.,* **114**, 533-541.
[18] Li X, Deng S and Fu H 2011 *Corros.Sci.,* **53**, 1529-36.
[19] Bhat J I and Alva V D P 2012 *Nano. Metal. Chem.,* **42**, 951-57.
[20] Oguzie E E, Akalezi C O, Enenebeaku C K and Aneke A J N 2011 *Chem. Eng. Comm,* **198**, 46-60.
[21] Aytac A, zmen U O and Kabasakalog˘lu M 2005 *Mater.Chem.Phys.,* **89**, 176-81.
[22] Obot, I. B. and Obi-Egbedi, N. O 2008 *Phy. chem. Eng.,* **330**, 207-12.
[23] Shabab R S A 1999 Retrieved from www.alazhar.edu.ps/Library/aattachedFile.asp?id_no=0044099
[24] Patel A S, Panchal V A and Shah N K 2012 *Bull. Mater.Sci,* **35**, 283-90.
[25] Esbensen K and Geladi P 1987 *Chemometr. Intell. Lab*., **2**, 36-52.
[26] Soleimanian M H 2009 *M.A. thesis*, Islamic Azad University, Tabriz branch, Tabriz, Iran
[27] ilouchian A, and Gamshidi M O 2001 *CRC. press.,* **65** pp. 42-49

IOP Conf. Series: Earth and Environmental Science **36** (2016) 012012 doi:10.1088/1755-1315/36/1/012012

Reliability analysis on a shell and tube heat exchanger

S Lingeswara, R Omar[1] and T I Mohd Ghazi
[1]Department of Chemical and Environmental Engineering, Faculty of Engineering, *Universiti Putra Malaysia* 43400 Serdang, Selangor, Malaysia

E-mail: [1]rozitaom@upm.edu.my

Abstract. A shell and tube heat exchanger reliability was done in this study using past history data from a carbon black manufacturing plant. The heat exchanger reliability study is vital in all related industries as inappropriate maintenance and operation of the heat exchanger will lead to major Process Safety Events (PSE) and loss of production. The overall heat exchanger coefficient/effectiveness (U_0) and Mean Time between Failures (MTBF) were analyzed and calculated. The Aspen and down time data was taken from a typical carbon black shell and tube heat exchanger manufacturing plant. As a result of the U_0 calculated and analyzed, it was observed that the U_0 declined over a period caused by severe fouling and heat exchanger limitation. This limitation also requires further burn out period which leads to loss of production. The MTBF calculated is 649.35 hours which is very low compared to the standard 6000 hours for the good operation of shell and tube heat exchanger. The guidelines on heat exchanger repair, preventive and predictive maintenance was identified and highlighted for better heat exchanger inspection and repair in the future. The fouling of heat exchanger and the production loss will be continuous if proper heat exchanger operation and repair using standard operating procedure is not followed.

1. Introduction

A shell and tube heat exchanger is used to transfer heat source from one stream to another to optimize energy and cost in a plant. It is the latent heat from a condensing vapor stream which is transferred to a colder stream in order to heat it. The unit may also function as heat exchange between the two streams without necessary phase change. In this case, the hot stream has to be connected to the shell side of the heat exchanger [1]. A shell and tube heat exchanger can also be used to transfer the heat from an evaporating liquid stream, which is typically water to the tube side stream.

As far as fouling is concern, a typical shell and tube heat exchanger is more likely to foul or corrode on the tube side when higher pressure is exerted. The deposition of unwanted particles or substance on heat transfer surfaces is called fouling. Fouling significantly impacts the mechanical and thermal performance of heat exchangers. Fouling is also known to be a dynamic phenomenon which changes over a period of time. Cleaning inside a tube is easier than cleaning the outside [2].

When a vapor or gas is used as a heat exchange stream, it is typically introduced on the shell side to allow the highest temperature difference (calculated as log mean temperature difference)

[1] To whom any correspondence should be addressed.

between the hot and cold streams. As for liquids with high viscosity, the pressure drop of the liquid flow through the tubes will be prohibitively large. Thus, liquids with high viscosity can be channeled on the shell side [3].

The most common material of construction of a typical shell and tube heat exchanger is carbon steel. Other materials such as stainless steel or copper are used when needed, and the choice is made by corrosion concerns as well as the mechanical strength requirements. Expansion joints are used to accommodate differential thermal expansion of dissimilar materials used. Fouling changes expansion rate of expansion joints by reducing total heat transferred [4].

In a carbon black petrochemical plant, the shell and tube heat exchanger (air preheater in Figure 1) is used to preheat air (in shell side) to temperature between 650 and 700°C for the combustion to produce carbon black which runs 24 hours daily. The combustion air heat exchanger is the primary means of direct energy recycle in carbon black production accounting for about 5% of the total energy needed. Carbon black heat exchangers are designed to use both radiation and convection to transfer heat. The heat exchanger is one of the critical equipment in the carbon black production operation as it is high in costs. Of late, the heat exchanger was found to perform below design capacity causing loss of product yield. The problem is that air cannot be heated to the set temperature, giving indication of severe fouling on the tube side (product from furnace in gas form and solid particles) thus requiring a lot forced and unplanned shut down to clear the fouling. Furthermore, there were incidents of temperature excursions due to heat exchanger fire (in tube side) because of the presence of excess oxygen from the carbon black production furnace. If the fire was significant, it could cause heat exchanger damage and worst if ruptured.

Therefore, the objective of the present work is to evaluate the performance parameters of a shell and tube heat exchanger through Aspen Plus® data history analysis by analyzing overall heat transfer coefficient and mean time before failure (MTBF). From this evaluation guidelines for heat exchanger inspections and repair were proposed for future use.

2. Methodology

2.1. Unit operation

The study will be based on a shell and tube heat exchanger in a Carbon Black Manufacturing Plant as in Figure 2. Cold Low Pressurized Air flow in and out the shell side named Cold LPA in and Cold LPA out, respectively. The set temperatures for the air were 100 and 650°C for inlet and outlet, respectively. Meanwhile, the products from the reactor for carbon black production at 850°C named Hot LPA In was fed into the tube side to heat the air, and flowed out at 648°C named as Hot LPA Out.

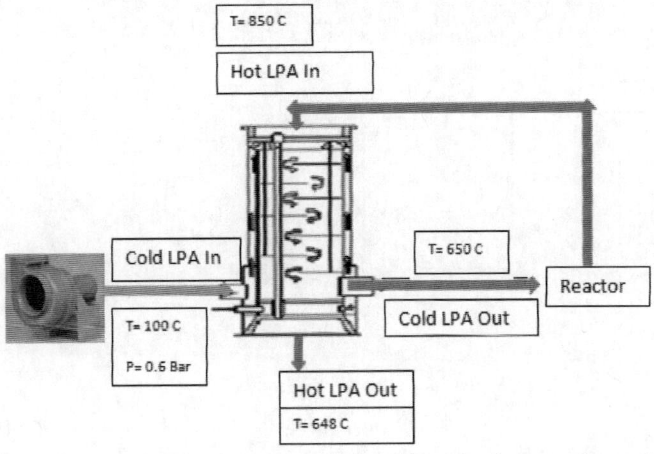

Figure 2: A typical carbon Black Plant Heat Exchanger Operation

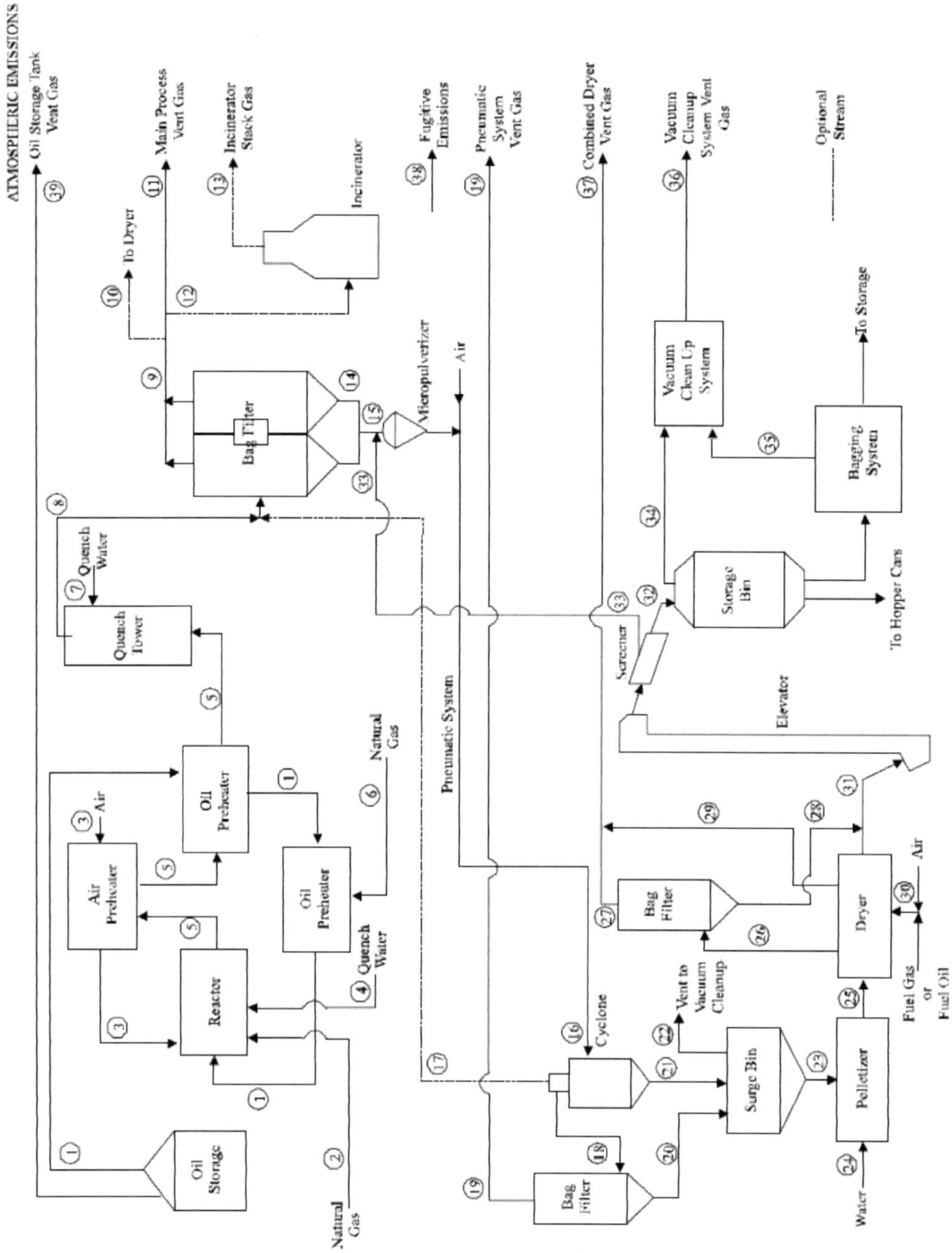

Figure 1: Typical process flow diagram of a carbon black production using oil furnace method

2.2. Calculation of overall heat transfer coefficient (U_o)

Figure 3 shows the flow of U_o determination using the basis of Equation 1 and data from Aspen (Figure 4). It should be noted that this method only accounts for conduction within materials and does not take into account heat transfer through methods such as radiation (Perry, 1973).

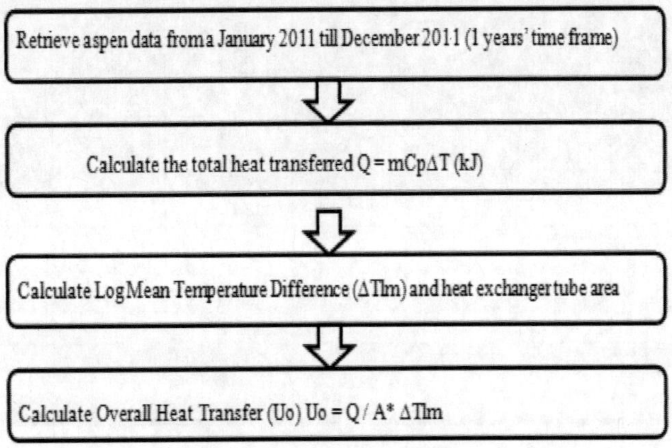

Figure 3: Determination of U_o

$$Q = U_o A \Delta T_{lm} \tag{1}$$

Where Q is total heat transferred (kJ), U_o is overall heat transfer coefficient (kJ/m^{2o}C), A is total surface area (m^2) and ΔT_{lm} is log mean temp difference. The ΔT_{lm} is calculated using Equation 2.

$$\Delta T_{lm} = \frac{(\Delta T_1 - \Delta T_2)}{\log\left(\frac{\Delta T_1}{\Delta T_2}\right)} \tag{2}$$

where $\Delta T_1 = T_1 - t_2$ and $\Delta T_2 = T_2 - t_1$. T_1 is inlet temperature of the shell-side (or hot) fluid, T_2 is exit temperature of the shell-side (or hot) fluid, t_1 is inlet temperature of the tube-side (or cold) fluid and t_2 is the exit temperature of the tube-side (or cold) fluid.

2.3. Mean Time between Failure (MTBF) Calculations

Reliability is the probability of the equipment or process functioning without failure when operated as prescribed for a given interval of time under stated conditions. Long failure free periods result in increased productivity. Thus, fewer spare parts are needed to be stocked and less manpower employed in maintenance activities hence lowering the costs. Increased availability, decreased down time, smaller maintenance costs and lower secondary expenditure result in a bigger profit. Setting reliability requirements is a corner stone of any reliability strategy. The reliability requirements have a quantitative part and that is mean time between failures (MTBF). MTBF is specified for a component and systems whose failures are characterized by a constant hazard rate [5].

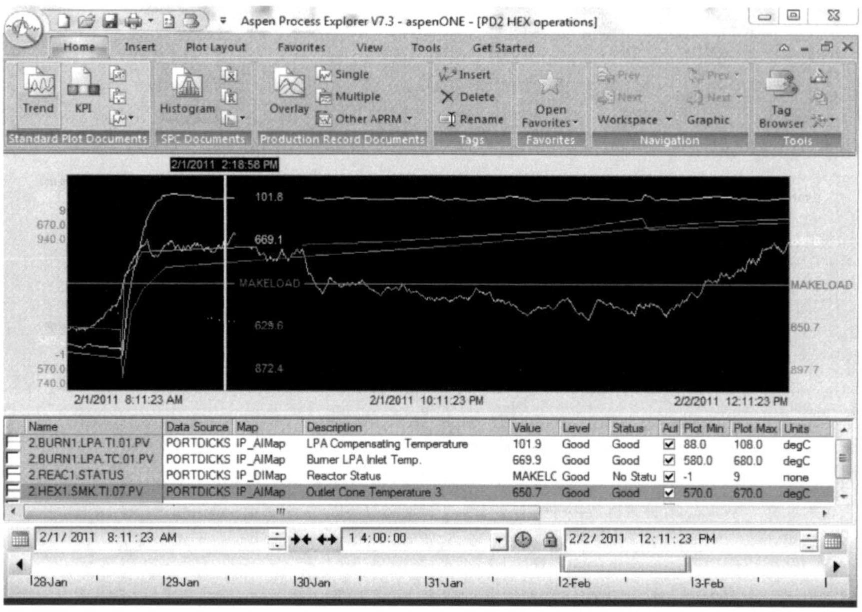

Figure 4: A Typical Heat Exchanger Operations Aspen Trending (Source: A carbon black plant)

Ideally the failure rate (FR) parameters should be determined on the basis of observations for the specific component types. Consequently, the failure rate parameters are determined on the basis of general experiences, both with the precursors of the relevant type and with the component types in general industrial applications. The parts count method is a technique for developing an estimate or prediction of the average life of the Mean Time between Failures (MTBF) of an assembly. It is a prediction process whereby a numerical estimate is made on the ability with respect to failure of a design to perform its intended function. Once the failure rate is determined, MTBF is easily calculated as the inverse of the failure rate as in Equation 3 [6]. The failure rate was analyzed for the 1 year time frame from January 2011 to December 2011.

$$MTBF = \frac{1}{FR1 + FR2 + FR3 + ...FRn} \quad (3)$$

2.4. Highlights of Temperature Excursions to Determine A Heat Exchanger Fire

Figure 5 shows live Aspen data and a possible heat exchanger fire can be seen. The sudden temperature excursion in heat exchanger outlet temperature (colored pink) indicates fire in heat exchanger tubes which could lead to significant catastrophic process safety event. The sample data below shows the burner inlet temperature shot from 530 to 597°C during start up.

3. Results and Discussions

3.1. Overall Heat Transfer Coefficient (U_o) Analysis

Overall heat transfer coefficient represents how fast energy can be transferred between two points. It is related to the total thermal resistance between the two points. The resistance to heat transfer between the two fluids takes place in three steps: (1) resistance between the hot fluid and the tube wall, (2) resistance within the tube wall, and (3) resistance between the tube wall and the cold fluid. This causes loss in yield and the volume of the carbon black produced.

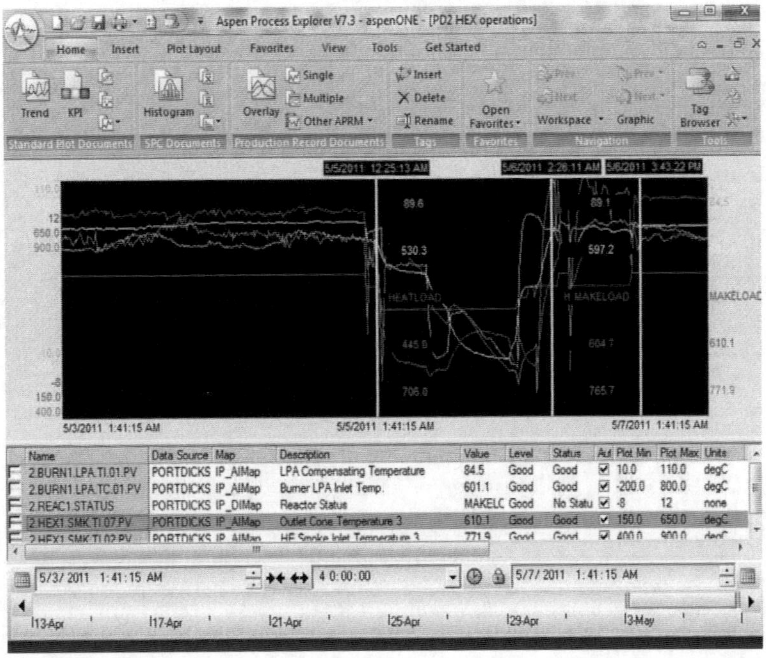

Figure 5: Sample live Aspen data for possible heat exchanger fire

Figure 6 shows the overall heat exchanger coefficient (U_o) for the year 2011 starting from month of February until December. There was no data for the month of March, June, October, November and December due to maintenance activity and planned curtailment. On the 3rd February, the U_o was calculated to be 99 kJ/m^{2o}C. It was found that on 6th February, the U_o had dropped to 95 kJ/m^{2o}C. For example, on February 3rd, the log mean temperature was 384°C with corresponding U_o of 99 kJ/m^2 C. Meanwhile, on 6th February, the log mean temperature increased to 397°C gradually which reduced the U_o of the shell and tube heat exchanger to 95 kJ/m^2 C.

Over a period of 3 days, the heat exchanger U_o dropped 4 kJ/m^2 C gradually according to the data observed from Aspen. The probable major contributor of this is caused by fouling as there were no upsets in the process and no changes in the operating parameters. The U_o in April, May, July August and September also shows the decrease in U_o gradually upon start-up of production and after a few days. Increase in fouling gradually coats the tubes with carbon black thus decrease the preheated air temperature for the production of carbon black.

Fouling is the resultant effect of deposition and removal of carbon black on a heat exchanger tube surface. The processes occurred simultaneously and depended on the operating conditions. Usually, the removal rates increased with increasing amounts of deposit whereas deposition rates were independent of the amount of deposit. In the case of a shell and tube heat exchanger, the changes are caused by deposits such as increase in flow velocity and surface roughness. In the application of constant combustion air temperature or constant heat transfer coefficient boundary conditions, the interface temperature decreases as deposits build up. This in return forces optimum desired temperature for combustion of carbon black production to be reduced [5].

Coking and polymerization is a major cause of fouling in the shell and tube of heat exchanger of a typical carbon black application which directly causes the log mean temperature to increase, thus reducing the heat transfer. The decomposition of organic products (in the products after reaction in reactor) can lead to the formation of very viscous tar or solid coke particles at high temperatures and polymerization involves the formation of undesirable organic sediments or polymers in the tube side. The coke particles and polymers formed in the heat exchanger grows to such a large size that they drop out of solution and deposit on the process equipment. Such deposits can be extremely tenacious and may require burning off the deposit in a timely manner to return the heat exchanger to satisfactory operation [3].

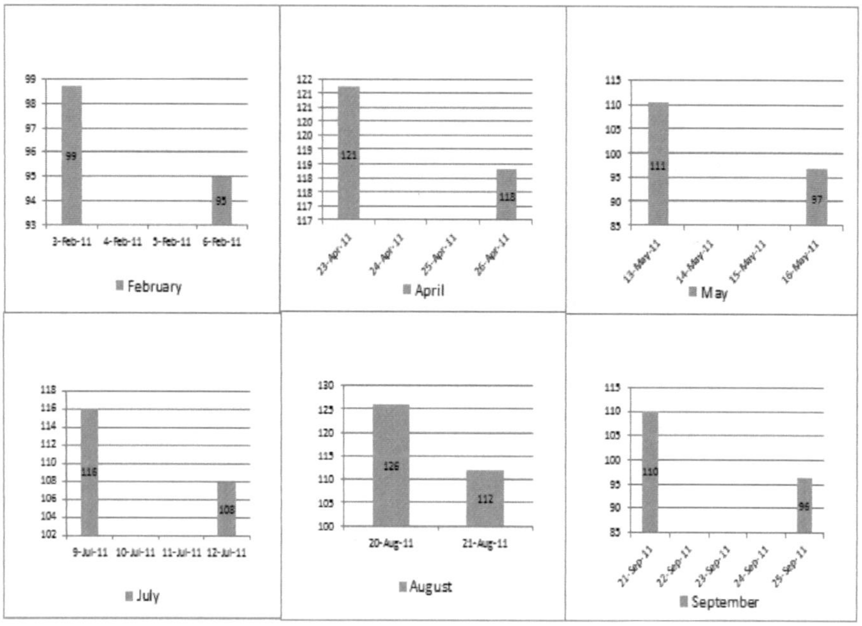

Figure 6: U_o value changes during production of carbon black

Figure 7 below shows bar graph on the impact of the hot stream outlet temperature on the U_o of a heat exchanger. From the bar graph below, it is deduced that at a constant cold stream outlet temperature (air), the increase in the hot stream outlet temperature indicates the initial stage of fouling. Surface is conditioned in the initiation period. Rough surfaces encourage particulate deposition and provide a good chance for deposit sticking. After the initiation of fouling, the persistence of the roughness effects will be more a function of the deposit itself. Even smooth surfaces may become rough in due course due to scale formation, formation of corrosion products, or erosion [7].

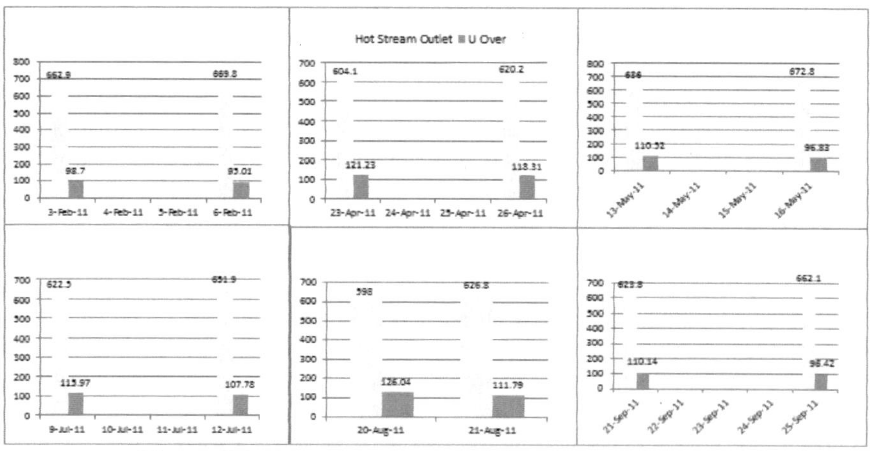

Figure 7: Effect of hot stream outlet temperature on U_o.

A simple way to monitor a heat transfer system is to plot the graph of outlet temperature versus time. Loss of heat transfer and subsequent charge outlet temperature decrease as a result of the low thermal conductivity of the fouling layer or layers which is generally lower than the thermal conductivity of the fluids or conduction wall. As a result of this lower thermal conductivity, the overall thermal resistance to heat transfer is increased and the effectiveness and thermal efficiency of heat exchangers are reduced [8].

In one unit at an oil refinery, in Homs, Syria, fouling led to a feed temperature decrease from 210°C to 170°C in a crude distillation unit. In order to bring the feed to the required temperature, the heat duty of the furnace may have to be increased with additional fuel required and resulting increased fuel cost [9]. Alternatively, the heat exchanger surface area may have to be increased with consequent additional installation and maintenance costs. The required excess surface area may vary between 10-50%, with an average around 35%, and the additional extra costs involved may add up to a staggering 2.5 to 3.0 times the initial purchase price of the heat exchangers [9].

With the onset of fouling and the consequent buildup of fouling layer or layers, the cross sectional area of tubes or flow channels is also reduced. In addition, increased surface roughness due to fouling will increase frictional resistance to flow. Such effects inevitably lead to an increase in the pressure drop across the heat exchanger, which is required to maintain the flow rate through the exchanger.

Figure 8 shows the declining of the U_o after a period of time and the burn out required to clear the fouling on the heat exchanger. Burn out means removal of fouling on the tube side of the shell and tube heat exchanger by introducing oxygen and heat source to burn off the fouled particles. During this burn out period, production has to be stopped thus a shutdown is required. Figure 8 reveals that after every burn out event, there is inclination of the U_o.

This burn out is required approximately once in 3-4 weeks to maintain a high U_o. However, it is noticed that after every burn out event, the U_o is higher but differs from one burn out event to another. This is due to the incomplete burn off period or interval. An incomplete burn off heat exchanger will have residues of fouled material on the tubes which will cause the lower U_o.

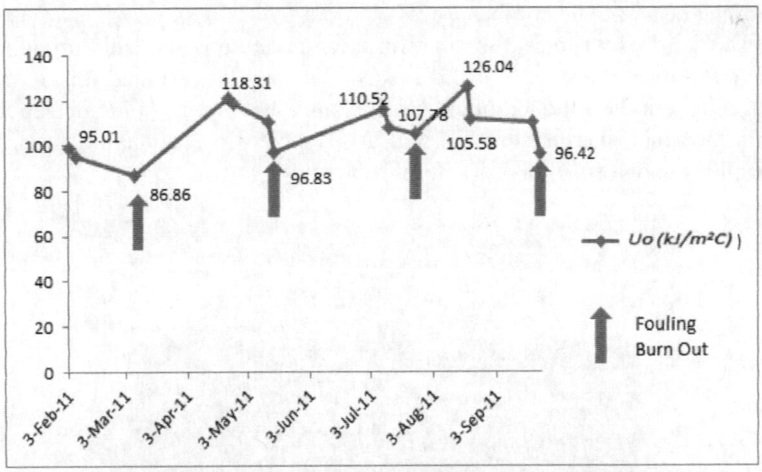

Figure 8: Trend on U_o and the fouling burn out interval

3.2. Analysis on the Mean Time between Failures

The calculated MTBF of a shell and tube heat exchanger is 649.35 hours (0.9 month). This value is very much lower than the standard MTBF of a typical shell and tube heat exchanger in the industry.

IOP Conf. Series: Earth and Environmental Science **36** (2016) 012012 doi:10.1088/1755-1315/36/1/012012

The standard MTBF is 6000 hours interval. The MTBF is an important system parameter in systems where failure rate needs to be managed, in particular for safety reasons.

The MTBF appears frequently in the engineering design requirements, and governs frequency of required system maintenance and inspections. In special processes called renewal processes, where the time to recover from failure can be neglected and the likelihood of failure remains constant with respect to time, the failure rate is simply the multiplicative inverse of the MTBF [10].

MTBF and reliability are really all about failure. Every product ever made will eventually fail, but ideally, not within its service life. Reliability and MTBF calculations are used to determine the likelihood of failure. Many people will look at the MTBF of a product and make assumptions. For example, if an MTBF is 100000 hours, one may think they have a long time before having to replace their product [10].

Service life is the time that a product is in service out in the field. The useful life of a product is the time after the device is broken in/burned-in and before the device starts to wear out. In order to have the maximum service life, a customer would want to ensure that the service life matches the useful life. In other words, it is a good idea to have the product burned in before put into service. This can prevent failures due to defective parts [11].

3.3. Determining Heat Exchanger Fires on Shell and Tube Exchanger

Figure 9 highlights the significant change in the heat exchanger hot stream outlet temperature. The figure shows the significant increase of the carbon black hot stream outlet temperature from 445 to 604.7°C. This sudden spike in the temperature indicates there was a fire in the heat exchanger tubes. It important to identify these heat exchanger fires and overcome before it becomes a process safety event.

Aspen trending can be used to troubleshoot this but it takes time. As the first line of barrier for preventing such heat exchanger fires will be to configure layers of alarm such as advisory high temperature, critical high temperature. Besides that, an emergency heat exchanger spray has to be installed in case of a heat exchanger fire. Unfortunately by introducing water during a heat exchanger fire can cause significant damage to the heat exchanger tubes. Thermal shock will happen which can cause distortion of the heat exchanger tubes.

Thermal shock arises when a solid at uniform temperature is suddenly brought in contact with a fluid at a different temperature. If the solid is small, it may be assumed that its temperature will remain uniform throughout the solid although it will change with time. In the presence of temperature differences within the solid, thermal stresses will arise, increasing in magnitude as the temperature gradient increases. Plastic deformation or fracture may result.

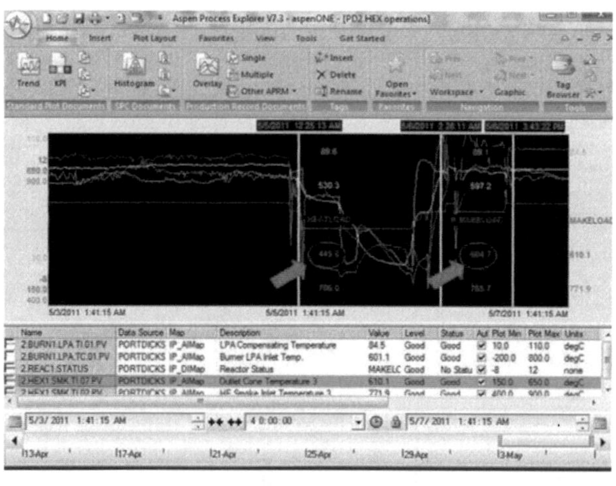

Figure 9: Heat exchanger Hot Stream Outlet temperature

4. Proposed Guidelines for Heat Exchanger Maintenance

Exceptional efforts should be made to save tubes since this will, in general terms, ultimately prolong the useful, rated service life of the vessel. A quick fix should be avoided in virtually all cases. The apparent time saved, in short term, will be unlikely to compensate for the degradation of the equipment so provoked. Even a quick fix needs to be correctly executed and can be optimized as a temporary repair such that later restoration can be carried out at a future, but not too distant, planned unit shutdown. It is always best to have a source of spare parts available for immediate repair of the problem. Since this is not always possible, it is the most likely reason to execute what has been called above the temporary repair.

4.1. Routine maintenance on heat exchanger pressure vessels

Air pre-heaters, which are pressure vessels in their own right, also form part of a pressure system. Their design and construction is made in accordance with pressure vessel codes and first installation is often required to be verified by the local, national authorities. This verification is frequently made on the basis of the entire system: changes to which could invalidate the original approval notwithstanding any name plate rating.

The standard pressures which are normally requested by clients usually mean that the more onerous requirements of pressure vessel inspections and repair procedures are not compulsory or mandatory though they may apply to some of the higher-pressure exceptions which exist at some facilities. It is good and recommended practice that the repair procedures and maintenance records be carried out and kept as though the vessels were required to meet more stringent statutory requirements. There is a legal requirement to implement an appropriate maintenance regime on all plant and keep sufficient records.

4.2. Preventive and predictive maintenance

Like all major items of plant, heat exchangers are best subject to a schedule of preventive maintenance tasks. Some of these are properly part of operator care philosophies. These include all operational personnel. Many data are collected in the Distributed Control System (DCS) and Aspen systems. It is good practice to walk the site once a shift at least. Look for anything which is damaged as it may just be loose insulation. A maintenance work order is usually issued to execute a repair.

A pressure test with air is probably the easiest and most reliable test which can be carried out on a heat exchanger. It is recommended that the vessel be cleaned first. It is often useful to test the hot air duct at the same time so a blind flange mounted at the burner position is usually sufficient. There is no need to specify a flange on the nozzle of the heat exchanger simply for pressure testing. The easiest, quickest and surest way to not to over-pressure the system is usually to use the standby blower. It is not necessary to reach the operating pressure, if this is impossible, since this is not a certification test. As for inspection, verification and evaluation work it must be done as early as possible in the turnaround. It is no use carrying out the cleaning and pressure test in the closing days of the turnaround.

5. Conclusion

U_o declines after a few days of operations and continues to decline further till a shutdown takes place to burn off the fouling coke on the tube side of the heat exchanger. This happens in a cycle of 2 to 3 weeks depending on the severity of the fouled heat exchanger. The coke particles and polymers formed in the heat exchanger grow to such a large size that they drop out of solution and deposit on the process equipment. Such deposits can be extremely tenacious and may require burning off the deposit in a timely manner to return the heat exchanger to satisfactory operation. The calculated MTBF of a heat exchanger was 649.35 hours (0.9 month) which is significantly low than the standard MTBF of 6000 hours interval led to major loss of yield and production volume to the carbon black industry. The heat exchanger repair guidelines and robust standard operating procedures are vital in day to day plant operation. These guidelines spell out the critical operating envelope and detailed repair methods technically to ensure the reliability and operability of the shell and tube heat exchanger.

References

[1] Suzuki K, Hiral K. and Miyake E 1998 *Int. J. Heat Mass Tran* **28** (4) pp 823-36.
[2] Manglik R and Bergles A 2008 *Exp. Fluid Sci.* **10** pp 171-80.
[3] Serth R. and Lestina T 2014 *Process Heat Transfer, Second Edition: Principles, Applications and Rules of Thumb.* (Oxford: Academic Press)
[4] Patiño D, Crespo B, Míguez JL 2016 *Appl. Therm. Eng.* **100** pp 849-860
[5] Jiang W, Gong J, Chen H and Tu S 2008 *Int. J. Pres. Ves. Pip.* **85** pp 569- 74.
[6] Arsenyeva O, Tovazhnyansky L, Kapustenko P and Khavin G 2013 *Chem. Eng. Trans.* **18** pp 791-96.
[7] Zubair S and Shah R 2001 Fouling in Plate-and-Frame Heat Exchangers and Cleaning Strategies *Proc. Compact Heat Exchangers and Enhancement Technology for the Process Industries* eds R K Shah, A Deakin, H Honda and T M Rudy (New York: Begell House) pp 553–65.
[8] Bryan Research and Engineering Inc. 1997 *Cause of Edmeston Heat Exchanger Failure: Unit 2* Altamira. 1998.
[9] Awad M, Abd El-Samad S, Gad H and Asfour F 2007 *Mansoura Eng. J.* **32** (1) pp 27-37.
[10] Thulukkanam, K 2013 *Heat Exchanger Design Handbook, Second Edition (Mechanical Engineering)* (Boca Raton: CRC Press)
[11] Lin S, Fan L, Azer N 1978 Augmentation of Single Phase Convective Heat Transfer with In-Line Static Mixers *Proc. Of the 1978 Heat Transfer and Fluid Mechanic Institute* ed C T Crowe, W L Grosshandler (Pulmann, Standford University Press).

IOP Conf. Series: Earth and Environmental Science **36** (2016) 012013 doi:10.1088/1755-1315/36/1/012013

Flipping organic chemistry course: Possibilities and challenges

J Cha[1], H B Kim
Division of Science Education, Daegu University, Gyeonsan, Gyeongbuk 38453, Republic of Korea

E-mail: [1]chajh@daegu.ac.kr

Abstract. The flipped classroom approach was applied to an introductory organic chemistry course. A total of 76 video clips (15 hours of running time) were developed and delivered to 41 sophomores (21 females and 20 males) through Youtube in addition to the university's learning management system. The students were asked to preview the lecture contents before each class by watching a pre-class video. For in-class activities, exercise problems were presented to groups of 3-5 students. An instructor and a teaching assistant guided each group to solve problems cooperatively, monitored the students' group activity and answered their questions. At the end of every chapter, the students were asked to evaluate their group work and personal preparedness for the class and also to write a short reflective journal. The muddiest point of each chapter, i.e., the topic posing the most difficulty to students' understanding, was surveyed through *Google Forms®*. The students liked watching the videos before each class and performing student-centered, in-class group activities but a few limitations were also found and reported.

1. Introduction

Since a documentary entitled "Search for future classroom" was broadcasted in March 2014, the flipped classroom has attracted a lot of attention in the Korean educational society. Teachers who are practicing and want to adopt flipped classrooms have built a teachers' network called the "future classroom network", and some regional educational offices have adopted it for the innovation of public education [1]. One year later in March 2015, four broadcasted episodes of the documentary showed various educational effects of flipped classrooms based on the experiments with more than a thousand schools. This further increased interest that has been expressed through professional development programs, local teacher networks, and research papers.

Korean students' attitude towards math and science are quite low despite their high academic achievement from international comparison studies such as TIMMS and PISA [2]. Many Korean educators attributed this to teacher-centered instruction focusing on the transfer of knowledge. Although the national curriculum aimed participatory instruction and most teachers agree that the student-centered approach is better, teachers have to spend most of their teaching time explaining knowledge. This is related to the Korean educational system, which aims almost exclusively at the university entrance exam. Within this educational environment, teachers have suffered from the discrepancy between theory and practice [3].

[1] To whom any correspondence should be addressed.

In order to reinforce the student-centered environment and decrease teacher-centered instruction, various teaching strategies based on constructivism have been suggested including the flipped classroom. However, this simple way of delivering a flipped classroom with "video lectures viewed at home along with problem-solving group work completed in-class" [4] is not a completely new approach. It was begun by two high school chemistry teachers, Jonathan Bergman and Aaron Sams, who were trying to solve a problem of frequent student absences in 2007 [5]. The simple flipped classroom is described as replacing "after-lecture homework with the expectation that students study course material prior to class" [6]. As an extension to this concept, before-class study was recently reinforced with the help of online videos, such as Khan Academy which have become widely spread worldwide at all school levels.

At the university level, extensive research has been conducted [6-8] with a limited numbers of papers in the field of engineering education [9;10]. Organic chemistry has been considered the most difficult subject for students [11;12]. Attempts have been made to adopt a flipped classroom to organic chemistry courses, with positive influences on grades and attitudes being reported [7;11]. However, its application still remains limited in the field of science at the university level, and no research on the flipped classroom has been conducted in university chemistry courses in Korea. Therefore, the present study explores the effect of flipped classrooms in the Korean context.

2. Method

2.1. Subjects

A flipped organic chemistry course worth 3 credits and entitled *Organic Chemistry 1* was run with 41 sophomores (21 female and 20 male) majoring in chemistry education at a private university in South Korea.

2.2. Lecture details

Before each lecture, the students were advised to watch a lecture video through the Learning Management System (LMS). All contents were uploaded at least two to three days before. Students could watch the movies with their smartphones and desktop computers. During the class, exercise problems were provided. These problems were mainly application problems sourced from the textbook, and were supplemented with additional problems for upgrading problem-solving abilities. The students solved the problems within small groups of 4-5 members. The groups were formed according to the students' friendship. After the midterm exam, however, the small group members were changed to heterogeneous groups based on the midterm exam result.

At the end of every chapter, various types of quiz were implemented such as (a) individual quiz and group reward, (b) relay group quiz in which students solved one problem in a sheet on the wall one by one, and (c) solving together and group reward. The quiz scores were not added to final grade, but only used as a reference to check the students' understanding. As the students solved problems in groups, the lecturer interacted with the students in the lecture room. He answered students' questions, but would not give direct answers. A teaching assistant, one of the authors, also helped students as the lecturer did.

2.3. Lecture Videos

Pre-class video consisted of three main parts: introduction, concept explanation, and exercise problem and summary. Figure 1 presents some screenshots of these parts. The videos began with an introductory talk recalling the previous lecture and providing motivation to study the topic. The lecturer's face was shown in this slide to familiarize the students. Then the learning goals were presented briefly. Total time of introduction was less than a minute.

The lecture slides for concept explanation using figures and schemes were based on the textbook. The lecturer attached some memos during the explanatory talk. Sometimes, the lecturer was shown during concept explanation. For example, in chapter 4, the lecturer demonstrated a molecular model showing the conformation of 1-substituted cyclohexane.

At the end of the lecture, the lecturer explained how to solve the exercise problem. Usually, the problem answers were not given directly in order to let the students solve the problems themselves. General directions were given for other application problems. Lastly, the learning goal was re-presented, and the main topics and concepts were summarized in terms of the goals. Problem numbers which needed to be solved were also presented for further study. According to the general rule for making pre-class videos, the lecture tried to make each clip shorter than 20 *min*. As a result, a total of 76 video clips averaging 11.9 *min* and totalling 905 *min* (15*h* 05*m*) were produced for the semester.

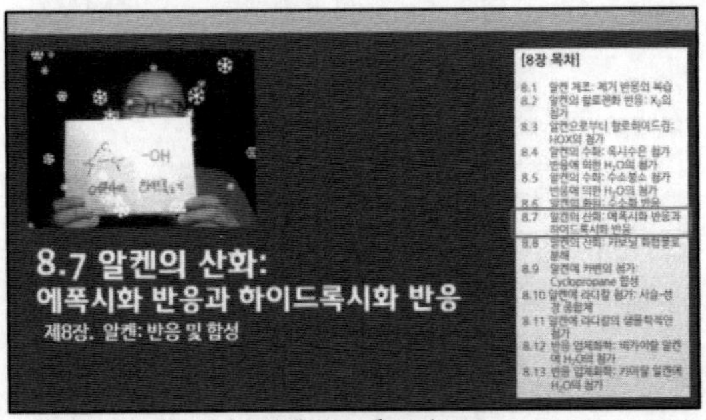

(a) Introduction

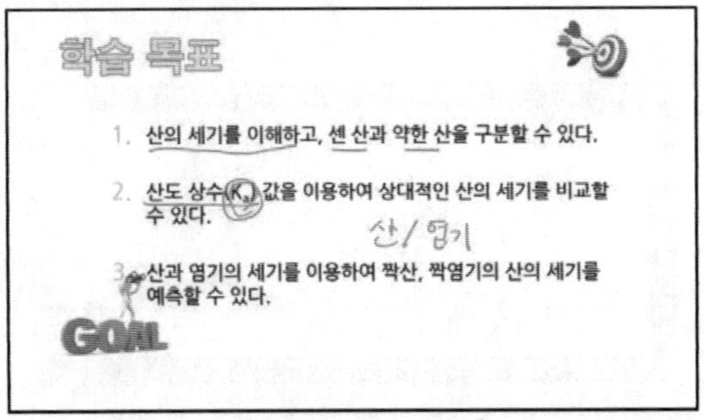

(b) Learning goal

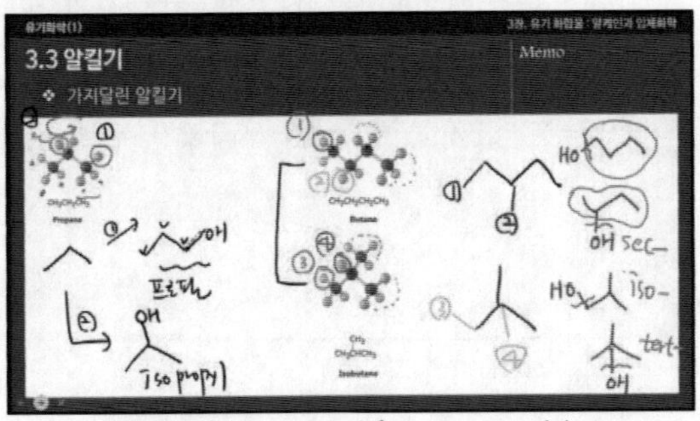

(c) Explanation with screen writing

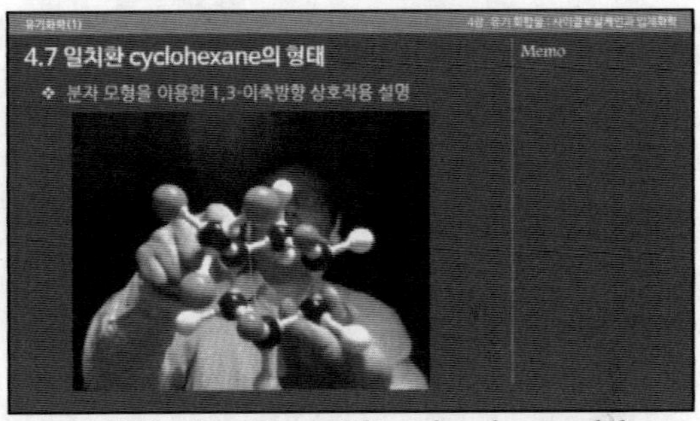

(d) Explanation with molecular model

Figure 1. Selected screenshots of the pre-class videos

2.4. Tools

The lecture contents were prepared by *Microsoft PowerPoint®* with *Office Mix,* which is a free add-in for PowerPoint available at http://mix.office.com. The lecture slides were prepared first, then *Wacom®* pen tablet was used for screen writing during the recording. The voice was recorded with movement of mouse writing together.

2.5. Survey Questionnaire

The authors developed a feedback questionnaire on the flipped learning that was administered at every chapter from chapter 3 to chapter 8. The questionnaire consisted of questions on the students' satisfaction with the pre-class lecture videos, in-class activity, self-evaluation of watching video and in-class participation with a scale from 1 to 3 (1 = dissatisfied; 2 = neither satisfied nor dissatisfied; 3 = satisfied). The muddiest point of each chapter and reflective writing on their performance were also asked. The questionnaire was delivered to students through *Google Forms®* and they answered through their smartphones.

3. Results and Discussion

3.1. Students' perception of the flipped classroom

Students' perception of the flipped classroom including satisfaction with the pre-class videos and in-class activities, and self-evaluation on their group work, preparedness, and participation were monitored during chapter 3 to 8, as shown in table 1, figure 2 and 3. Reflective journals by students were also collected and analyzed.

Table 1. Students' feedback on the flipped classroom

	Chapter 3		Chapter 4		Chapter 5		Chapter 6		Chapter 7		Chapter 8		Total	
	M	*SD*	*M*	*SD*	*M*	*SD*	*M*	*SD*	*M*	*SD*	*M*	*SD*	*M*	*SD*
Satisfaction														
Pre-class video	2.68	0.47	2.79	0.41	2.65	0.54	2.72	0.45	2.66	0.48	2.64	0.55	2.69	0.48
In-class activity	2.42	0.55	2.50	0.56	2.54	0.61	2.72	0.45	2.63	0.49	2.61	0.56	2.57	0.54
Self-evaluation														
Group work	2.53	0.6	2.61	0.55	2.73	0.45	2.75	0.44	2.63	0.49	2.64	0.49	2.64	0.51
Preparedness	2.05	0.61	2.34	0.63	2.16	0.73	2.08	0.69	2.06	0.72	1.97	0.85	2.12	0.71
Participation	2.37	0.59	2.47	0.56	2.43	0.55	2.53	0.51	2.38	0.61	2.36	0.60	2.43	0.57

M=mean score; *SD*=standard deviation

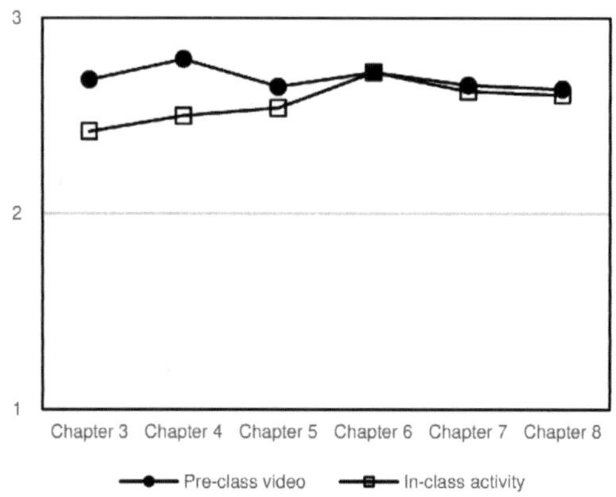

Figure 2. Students' feedback on the satisfaction with the pre-class videos and class activities

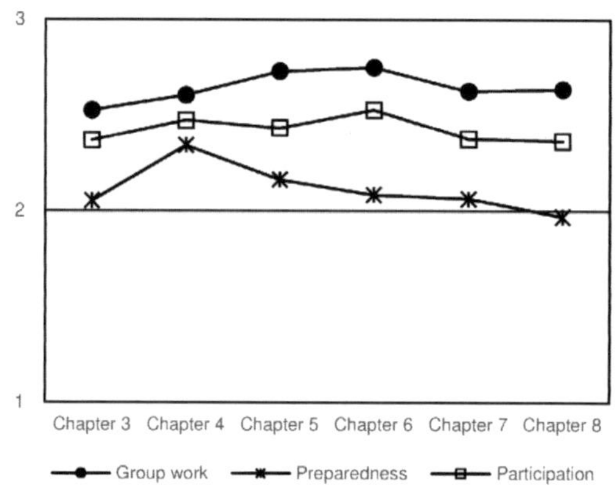

Figure 3. Students' self-evaluation of their group work, preparedness, and participation

3.1.1. Feedback on the Pre-class videos and In-class activities

The mean scores of satisfaction with the pre-class videos and in-class activities were 2.69 and 2.57, respectively, which indicated that the students were generally satisfied (figure 2). Students' positive perceptions of the pre-class videos and in-class activities were also found in their journals below. The most frequent feedback comment was the usefulness of the lecture videos. They liked the online videos since the videos could be watched freely and repeatedly at their convenience. Some students mentioned the helpfulness of using a molecular model in the video.

- *We've already reached to chapter 3, and I just felt that didn't review them well. So, I'm planning to watch the videos again. I think this is a big benefit of the flipped classroom!*
- *I love the way the professor explained the concepts with the molecular model. Well, it is not so easy to understand organic chemistry only based on 2D structure. But I could understand more with the model in the video and also with a tangible model in class.*

The satisfaction level of the in-class activities was relatively low during chapters 3 to 5 (mean score range: 2.50~2.54), but it increased from chapter 6, which implied that the students had some trouble with the flipped classroom initially, but became familiar with the new approach which demanded that they study by themselves as the course proceeded. The students' feedback on group work in their journal was also positive. Many students commented that they got help from group members to understand a concept in organic chemistry. Contrary to the lecture-centered classroom, the students could ask a question freely and obtain and offer some help from others. They also liked to solve more problems than before.

- *I concentrated more than before since I had realized the importance of watching the pre-class videos. I hated group activities before, but now I loved solving problems with group members.*
- *In chapter 3, just as in chapter 2, it was good to ask questions whenever I had something to ask without investing any additional time after the lecture. Actually the time for group presentation was decreased and the time for problem solving was increased. Although both have strengths and weaknesses, but with the way we are trying now, lectures can be speedy.*

However, there were some technical limits of the pre-class videos. One student reported a problem that the video would not play on her smartphone. A slight noise was also reported in another student's journal. Many students mentioned that the length of the video was too long to concentrate. Much attention needs to be paid to these technical issues.

A few students reported the fundamental constraint of learning in the flipped classroom. When they studied topics with the pre-class videos, sometimes they had some trouble in understanding the difficult concepts by themselves. However, they could not ask a question to the lecturer before the lecture. This problem could be solved by organizing study groups for in-class and out-class activities. Additionally, the conceptual difficulty of organic chemistry cannot be overlooked, although the key concepts can be explained briefly and questions can be asked at the beginning of each class. There were also problems in group works such as the free-rider effect. In this study, the lecturer stressed that all group members study together in cooperation, although there was no obligation or group reward. But one or two negligent students became a burden to the other group members.

- *As I watched the video, I was sorry that I could not ask a question immediately. This might be a limit of studying with a video at home.*
- *There was a certain limit of solving problems just after watching the video. So I felt the need to review concepts with the textbook after watching the video. I will study the textbook before watching the video. Unless watching video first, then study the textbook.*
- *I was upset that one of our group members fell asleep during group activities. I felt that I fell behind since I couldn't solve all the given problems due to spending some time to awaken and teach him.*

3.1.2. Self-evaluation of group work, preparedness, and participation

The students' self-evaluation of their group work and participation were positive with mean scores of 2.64 and 2.43, respectively (Figure 3). The students liked and tried their best in their group. They were satisfied with their group work which made them participate more in the group activities. The students' journals also showed their positive perception of group work.

- *It was difficult for me to identify the R, S configuration of a certain molecule and I was a little bit confused. However, I got to understand with the help of my colleagues. I loved to study together more than to do alone.*

- *Chapter 3 has many concepts and contents, but attracted me more than previous chapters. The ring flip of cyclohexane and calculating the energy from the interactions was so attractive to me that I studied pleasantly without any difficulty.*

However, the self-evaluation on their preparedness (M=2.12) was slightly lower than the other factors. Many students confessed that they did not watch the pre-class videos, and increasingly so after the midterm exam, possibly because of the increasing difficulty of the concepts after the introduction of stereochemistry and organic reactions. A heavy homework load from other lectures also reduced their available time to watch the pre-class videos. As they came to the class without watching the pre-class videos, they had difficulty in following the lecture topics. As one of the students noted, they were lazy in watching the videos since this was not obligatory. So, this issue needs to be taken into account carefully before starting a flipped classroom.

- *I came to the class without watching all the videos of chapter 4. So I couldn't follow the other member's pace, which held back the other members. However, they helped me, so I didn't give up participating. Next time, I will watch the pre-class video.*
- *It was difficult for me to follow the lectures without watching the pre-class videos at first. Watching the video in this chapter, however, made easier for me to understand the topics and concepts.*

3.2. Muddiest point of the lecture topic
The muddiest points [13], i.e., the topic that students found the most difficult to fully understand, were surveyed at the end of chapters 3 to 8. The student choices of the muddiest point are shown in Table 2. One of the most difficult topics among the muddiest points from the overall chapters for students was the oxidative cleavage of alkenes in chapter 8 (55%), followed by assigning R or S configuration in chapter 5 (54%). The muddiest point selections naturally tended to be concentrated in the later chapters. For example, in chapters 3 and 4, only one topic was selected as a muddiest point by more than 40%, compared to two to three topics in chapters 5-7. This tendency may have been influenced by the reduction in studying of the pre-class videos, and the students' decreasing preparedness.

Table 2. Major muddiest points of each chapter

Topic	Muddiest point	Percentage
Ch3. Alkanes and stereochemistry	Gauche and anti conformation	42%
Ch4. Cycloalkanes and stereochemistry	1,3-Diaxial interaction	32%
Ch5. Stereochemistry at Tetrahedral Centers	Assigning R or S configuration	54%
Ch6. Overview of organic reactions	Describing transition state	50%
Ch7. Alkenes: Structures and reactions	The Hammond postulate	53%
Ch8. Alkenes: Reactions and synthesis	Oxidative cleavage of alkenes	55%

4. Conclusion
In this study, college students' perception of the flipped classroom was investigated after this approach was adopted in an organic chemistry course for 15 weeks. The students' attitudes towards the flipped organic chemistry course were generally positive. The use of pre-class videos assisted the students in understanding the basic concepts in organic chemistry in advance, which laid a foundation for solving many problems in class. The lecture video could be watched anywhere that was convenient to the students, and watch repeatedly until they understood it fully. In class time, the students had the benefits of solving many problems, such as developing their understanding and upgrading their

problem solving ability. They also benefited in terms of their social relationships with their peers through the small group activities. The lecturer had enough time to interact with the students individually in the classroom and many students liked this personal relationship.
However, the flipped classroom approach needs to be carefully implemented to prevent any decrement in effectiveness. As watching the video before the lecture was not mandated, the students' degree of preparedness may have been lowered, which reduced the effectiveness of the in-class activities. As requested by some students, a certain range of mandatory homework for watching the videos seems to be needed. Another issue was choosing the most effective method for grouping the students among heterogeneous, homogeneous, and friendship-based. Active learning environments like think-pair-share [14] or uncritical inference test [12] can be alternative in-class activities to increase the level of student engagement.

Finally the flipped classroom is not a method for making, nor is it about the technology. Rather it is a method for organizing the class structures. Considerable time is needed to make and upload the videos before the course is commenced. Designing the in-class activities is crucial in ensuring an effective flipped classroom. There also have been some argues on the advantages and disadvantages of flipped classroom in engineering education [10]. Therefore, more trials of adopting flipped classrooms are needed in the various science and engineering fields in order to fully explore the possibilities and challenges of this new approach.

5. References

[1] Park S J 2015 *Research in Social Studies Education* **22** 1
[2] OECD 2014 *PISA 2012 Results: What Students Know and Can Do: Student Performance in Mathematics, Reading and Science* (PISA, OECD Publishing)
[3] Lee M K 2014 *Korean Journal of Sociology of Education* **24** 181
[4] Blair E, Maharaj C, Primus S 2015 *Education and Information Technologies* 1
[5] Bergmann J, Sams A 2012 *A Flip your classroom: Reach every student in every class every day*
(Eugene, OR: International Society for Technology in Education)
[6] Davies R S, Dean D L, Ball N 2013 *Educational Technology Research and Development* **61** 563
[7] Flynn A B 2015 *Chem. Educ. Res. Pract.* **16** 198
[8] Teo T W, Tan K C D, Yan Y K, Teo Y C, Yeo L W 2014 *Chem. Educ. Res. Pract.* **15** 550
[9] Mason G S, Shuman, T R, Cook K E *IEEE Trans on Education* **56** 430
[10] Kerr B International 2015 *Proc. 2015 Conf. on Interactive Collaborative Learning (Florence: IEEE)* p 815.
[11] Fautch J M 2015 *Chem. Educ. Res. Pract.* **15** 179
[12] Kan S-Y, Cha J, Chia P W 2015 *J. Korean Chem. Soc.* **59** pp 156
[13] King D B 2011 *J. Chem. Edu.* **88** pp 1485
[14] Butler A, Phillmann K-B, Smart L 2001 *Teaching of Psychology* **28** 257

Acknowledgement
This research was supported by the Daegu University Research Grant, 2014.

Development and performance evaluation of Polydimethyl siloxane/Polysulfone (PDMS/PSF) composite membrane for CO_2/CH_4 separation

M S Suleman[1], K K Lau[1*], Y F Yeong[1]
[1]Department of Chemical Engineering, *Universiti Teknologi PETRONAS*, Bandar Seri Iskandar, 32610, Perak, Malaysia.

E-mail: [1]laukokkeong@petronas.com.my, [1] shoaibmalik1@live.com

Abstract. Asymmetric polysulfone (PSF) membrane was developed by phase inversion in this study. Polysulfone (PSF) membrane was modified to develop a composite polymeric membrane. Polydimethyl siloxane (PDMS) was used to modify the PSF membrane. PDMS/PSF composite membrane was developed by dip coating of PDMS over PSF. Developed membranes were characterized in terms of membrane morphology by scanning electron microscope (SEM). Micro structure of polysulfone (PSF) membrane confirmed that developed membrane was asymmetric. Dense PDMS coating in microstructure of composite membrane was observed. Membrane swelling experiments were performed by immersing developed membranes in water for specific time period. PDMS/PSF composite membrane resisted water swelling as compared to PSF membrane. Performance of the membrane was evaluated before and after swelling within the pressure range of 2-10 bar. Reasonable decrement in permeance and selectivity was observed after membrane swelling. Thus, membrane swelling affected the separation performance of both PSF and PDMS/PSF composite membrane by decreasing the permeance and selectivity values.

1. Introduction

Membrane technology is one of the most suitable options in removal of carbon dioxide from natural gas [1, 2]. Conventional separation processes used for natural gas processing are absorption, adsorption and cryogenic distillation. All these processes have limitations of high cost and high energy consumption. The advantages in membrane based gas separation are low cost, simple process and high separation efficiency [3]. Polymeric membranes have gained importance because of their excellent separation efficiency in CO_2 removal from natural gas or methane [4]. Due to the excellent thermal and mechanical stability, polymeric membranes can be used in harsh operating conditions or offshore locations [5]. Performance of the polymeric membrane is affected by the problem of swelling in membrane. Swelling can be caused by solvent during membrane synthesis or it can be caused by humidity in feed during the separation process [6]. Swelling affects the separation performance of polymeric membrane by changing its molecular structure [7]. Cross linking and thermal treatment is one of the techniques used to overcome membrane swelling [8]. In the present work, effect of

[1] To whom any correspondence should be addressed.

membrane swelling on separation performance of composite polymeric membrane was studied. Polydimethyl siloxane/Polysulfone (PDMS/PSF) composite membrane was developed. Polysulfone (PSF) is a glassy polymer with excellent thermal stability and high plasticization resistance [9]. Polydimethyl siloxane (PDMS) is a rubbery polymer having excellent permeability in CO_2/CH_4 separation [10]. Composite polymeric membrane exhibits good separation performance in CO_2/CH_4 separation [11]. Madaeni et al. [12] developed a PDMS/PES composite polymeric membrane by dip coating and manual casting of PDMS over PES. Mansoori et al. [13] analyzed the gas permeation properties of PDMS/PSF composite membrane in terms of CO_2 permeability and CO_2/CH_4 selectivity. CO_2/CH_4 selectivity was reported as 39.81 for PDMS/PSF composite membrane.

2. Experimental

In this work, PDMS/PSF composite membrane was developed by using dip coating method. Membrane morphology was studied by scanning electron microscopy (SEM). Separation performance of the developed membrane was evaluated for pure carbon dioxide and methane at the pressure of 2-10 bar. Permeation experiments were performed under dry and wet conditions. In dry conditions, membrane samples were tested before swelling in membrane. In wet conditions, membrane samples were soaked in water for 5 minutes prior to permeation analysis. Water was used as a swelling agent. Permeation results of both dry and wet conditions were compared to study the effect of water swelling on separation performance of developed composite membrane.

2.1 Chemicals

Commercial grade polysulfone (PSF) Udel ® 1800 was purchased from Solvay Advanced Polymers U.S. Liquid polydimethyl siloxane (PDMS) was purchased from Acros organics, Belgium. Solvents *N*-methyl-2-pyrrolidone (NMP) and *n*-hexane were supplied by Merck.

2.2 Membrane Fabrication

Polysulfone (PSF) was in powder form and dried for 24 hours before use. To develop asymmetric polysulfone membrane, a homogeneous casting solution was obtained by adding 20 wt. % of polysulfone (PSF) in NMP solvent, followed by continuous stirring of 24 hours at room temperature. After that, the casting solution was degassed for 2 hours using ultra sonication bath to remove any bubbles formed during stirring. Homogeneous dope solution of PSF was obtained with complete dissolution of polymer in solvent. Membranes were cast on glass plate using casting knife by adjusting the gap opening as 30 microns between knife and glass plate. Cast membranes were placed in non-solvent bath (deionized water) for 24 hours to complete the solvent precipitation. After 24 hours, membranes were removed from glass plates and placed for atmospheric drying of 12 hours. As a last step, after atmospheric drying membranes were dried overnight in vacuum oven at 80 ^{0}C.

2.3 Modification to Composite membrane

Composite membrane was developed by dip coating of pure polysulfone membrane. 30 wt. % of PDMS solution was prepared by adding it to n-hexane solvent followed by gentle stirring at room temperature. A homogeneous PDMS solution was obtained. Developed PSF membrane was dip coated for 10 minutes in PDMS solution by immersing the membrane in PDMS solution for the required time period. After that membrane was taken out from PDMS solution and kept for atmospheric drying for four hours. After that, membrane was thermally annealed at temperature of 80 ^{0}C for 45 minutes. After thermal treatment, PDMS/PSF composite membrane was cured with complete crosslinking of PDMS and PSF.

2.4 Characterization and performance evaluation

Micro structures of the developed membranes were studied by scanning electron microscopy (SEM). Performance of the developed membrane was evaluated before and after membrane swelling within the pressure ranges of 2-10 bar. Permeance and selectivity results of both the conditions were compared to study the effect of membrane swelling on separation performance of developed membrane. Performance of the membrane was evaluated by using permeation equipment described by Jusoh et al.[14] in another study. Permeance values for carbon dioxide and methane and CO_2/CH_4 selectivity was estimated by method reported by Rafiq et al. [15] .

2.5 Swelling Experiments

Water swelling behavior of developed membrane was studied by soaking the developed membranes in distilled water for 15 minutes. Membrane sample was cut according to the size (1.98 cm) of permeation test cell. Weight of the sample before and after swelling was measured. Swelling degree defines the total water uptake in membrane for that specific time period. Swelling degree was calculated by the following equation reported in [16].

$$\text{Swelling degree (SD) \%} = \frac{w_2 - w_1}{w_1} \times 100 \quad (1)$$

w_1 is the weight of dry membrane, while w_2 is the weight of swollen wet membrane.

3. Results and Discussion

PDMS/PSF composite membrane was developed by dip coating of PDMS on PSF. Membrane morphology was studied by scanning electron microscopy (SEM).

3.1 Scanning Electron Microscopy (SEM)

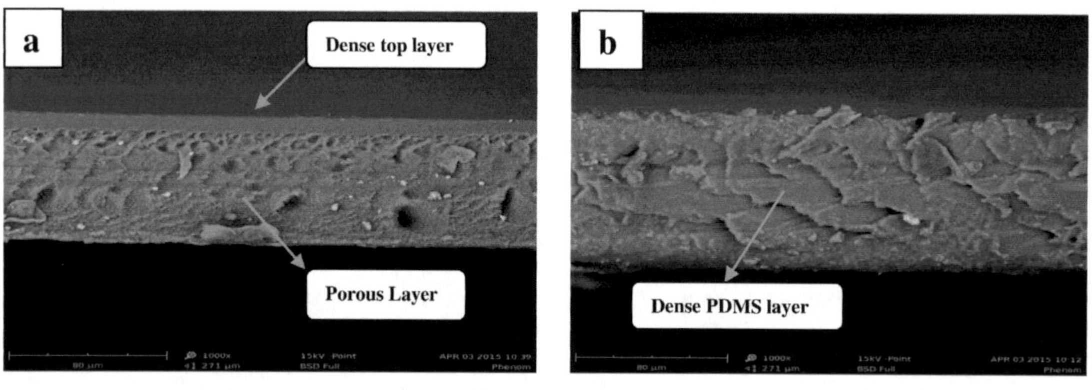

Figure 1. Cross sectional SEM images of (a) PSF membrane (b) PDMS/PSF composite membrane

Figure 1 shows SEM micrographs of developed membrane samples. Cross sectional SEM image (Figure 1-a) of pure polysulfone membrane shows the two layers in membrane morphology. Upper layer is dense layer with porous support layer. This shows that the developed membrane is asymmetric with two different morphologies (dense and porous) [14]. Porous layer shows different pore sizes in membrane morphology, while the uniformity in upper dense layer was observed with no defects. Figure 1(b) shows the cross sectional SEM image of PDMS/PSF composite membrane. Clear difference between the morphology of pure PSF and PDMS/PSF composite membrane can be observed from the SEM micrograph. Composite membrane shows dense morphology. Porous layer completely disappeared due to PDMS coating and pore filling phenomena was observed in case of composite membrane.

3.2 Performance evaluation before swelling

Figure 2 shows the CO_2 and CH_4 permeance of the developed PSF and PDMS/PSF composite membrane. With increase in pressure decrease in both CO_2 and CH_4 permeance was observed. This attributes to the characteristic behaviour of glassy polymers [17].

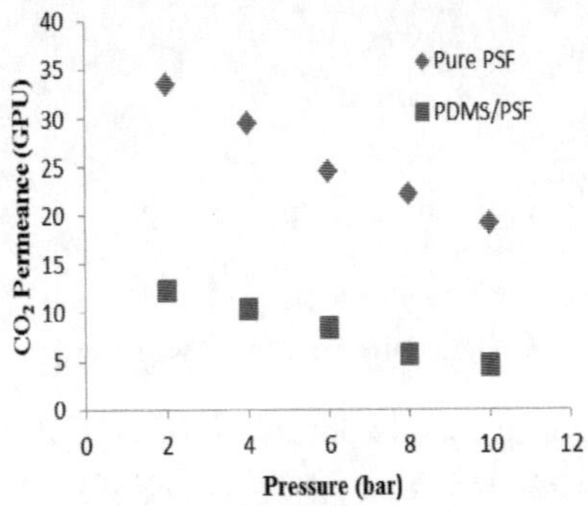

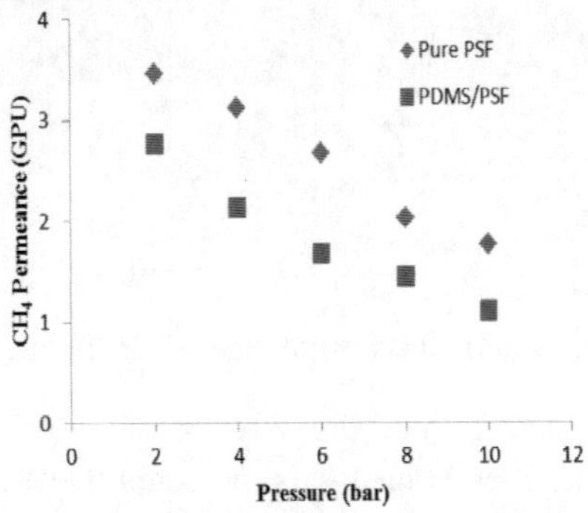

Figure 2. CO_2 and CH_4 permeance across pressure for PSF and PDMS/PSF composite membrane.

For pure polysulfone (PSF) membrane, CO_2 permeance at 2 bar was calculated as 33.4 GPU but at the pressure of 10 bar permeance decreased to 19.1 GPU. Similar behavior was observed for PDMS/PSF composite membrane that permeance decreases uniformly with increase in pressure. This indicates that there is no plasticization in developed membrane samples [18]. CH_4 permeance also decreases with increase in pressure in case of both pure and composite membrane. Both CO_2 and CH_4 permeance in case of composite membrane is less than the permeance in case of pure polysulfone membrane. This shows that PDMS coating affected the permeance of pure polysulfone membrane. Denser PDMS layer over PSF might be the possible reason for decrease in CO_2 and CH_4 permeance values.

3.3 Degree of swelling

Swelling degree for pure polysulfone membrane was calculated as 157.14%. High swelling degree in PSF membrane indicates that the membrane was completely swollen after immersion in water. However, swelling degree in PDMS/PSF composite membrane was very low (74.5%) as compared to PSF membrane. This indicates that the developed PDMS/PSF composite membrane resisted water and prevented water swelling in membrane. The possible reason of lower swelling degree in PDMS/PSF composite membrane was the water resistant nature of PDMS. Based on results of swelling degree, it can be concluded that PDMS/PSF composite membrane resisted water swelling at atmospheric conditions.

3.4 Performance evaluation after swelling

After swelling experiments, permeation analysis was carried out to check the effect of swelling on separation performance of developed membranes. Figure 3 shows the CO_2 and CH_4 permeance of pure and composite membrane after swelling.

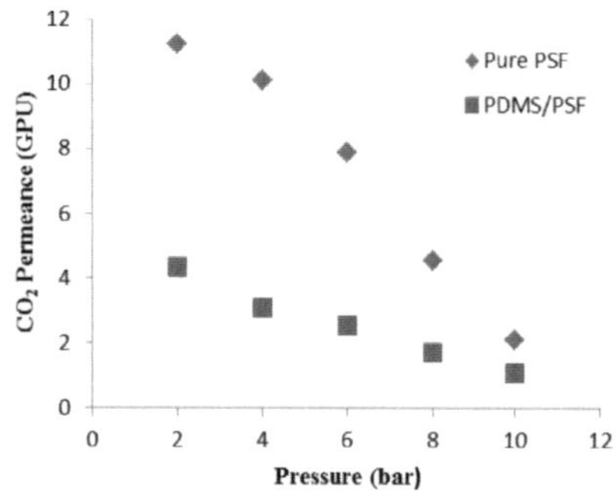

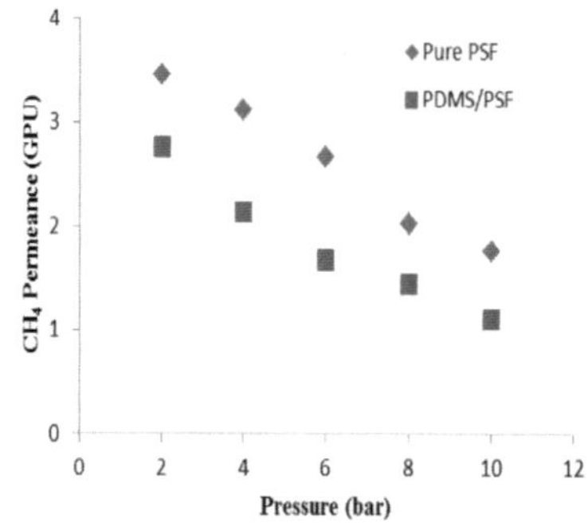

Figure 3. CO_2 and CH_4 permeance for PSF and PDMS/PSF composite membrane after swelling

With the increase in feed pressure, decrease in CO_2 and CH_4 permeance was observed for both pure and composite membrane. This indicates that there is no plasticization in membrane at any pressure [18]. If the permeance results for both CO_2 and CH_4 were compared with the results before swelling, then after swelling a remarkable decrease in CO_2 and CH_4 permeance was observed. For example, before swelling the CO_2 permeance in PSF membrane at 10 bar was 19.1 GPU. But after swelling permeance value for same membrane at same pressure decreased to 2.11 GPU. Similarly in case of CH_4 permeance, decrease in permeance was observed as compared to the CH_4 permeance before swelling. Thus, swelling affected the performance of the pure and composite membrane by decreasing CO_2 and CH_4 permeance. Water swelling in membrane caused contraction in membrane pores[20], resulting a remarkable decrease in CO_2 and CH_4 permeance .

3.5 CO_2/CH_4 selectivity before & after swelling

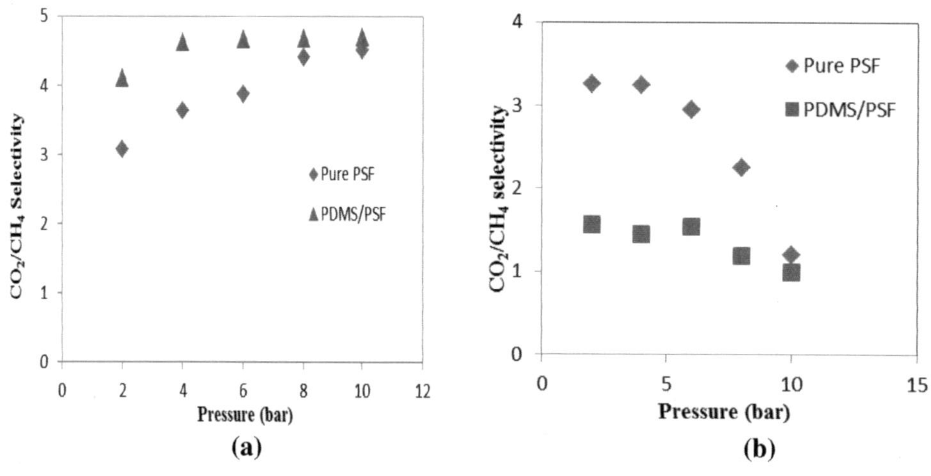

Figure 4. CO_2/CH_4 Selectivity across pressure (a) Before Swelling (b) after swelling.

Figure 4 (a) shows the CO_2/CH_4 selectivity of pure and composite membrane before swelling. Selectivity was observed to increase with increasing feed pressure. In case of pure polysulfone membrane, maximum selectivity at 10 bar was observed as 4.50. For PDMS/PSF composite membrane at 10 bar selectivity was recorded as 4.69. Permeation results for both PSF and PDMS/PSF composite membrane are in accordance with the dual sorption model for gas separation [19]. Figure 4 (b) shows CO_2/CH_4 selectivity of developed membranes after swelling. Decrease in selectivity with increasing pressure was observed for PSF and PDMS/PSF composite membrane. This behavior is opposite as compared to selectivity before swelling. Thus, water swelling in membrane also decreased the separation factor in both PSF and PDMS/PSF composite membrane. Before swelling, selectivity of PSF membrane at 10 bar was 4.50, but after swelling selectivity at same pressure decreased to 1.19. Similarly, for PDMS/PSF composite membrane selectivity before swelling was 4.69 at 10 bar, but after swelling selectivity decreased to 1.00. Based on these results, it can be concluded that water swelling affects the separation performance of membrane adversely by causing a remarkable decrease in separation factor of membrane.

4. Conclusions

PDMS/PSF composite membrane was developed by dip coating of PSF membrane with PDMS. Membrane morphology was studied by scanning electron microscopy (SEM). SEM micrographs reveal that developed membrane is asymmetric having dense and porous layer, while dense morphology for composite membrane was observed. Water swelling in developed membrane was studied by immersing the developed membranes in water for stipulated time period. Swelling degree value indicated that PDMS/PSF composite membrane resisted water swelling up to some extent. Permeation analysis shows that water swelling in membrane affected the separation performance of both PSF and PDMS/PSF composite membrane by decreasing permeance and selectivity.

5. References

[1] Scholes, C.A., G.W. Stevens, and S.E. Kentish, 2012 *Fuel* **96** pp 15-28
[2] Nasir, R., H. Mukhtar, and Z. Man, 2014 *J. App. Sci.* **14**(11) pp 1186-1191
[3] Zhang, Y., Sunarso. J, Liu. S, Wang. R, 2013 *Int. J. Greenhouse Gas Control* **12** pp 84-107
[4] Ghosal K. and B.D. Freeman. 1994 *Polym. Adv. Technol* **5**(11) pp 673-697
[5] Cui. L., Qiu. W, Paul D. R. and Koros, W. J. 2011 *Polymer* **52**(15) pp 3374-3380
[6] Catalano J Myezwa , D. Angelis, Sarti G. C. 2012 *Int. J.Hydrogen Energy* **37**(7) pp 6308-6316
[7] Izák. P. Hovorka S Bartovsky T Bartovska L Crespo J G. 2007 *J. Memb. Sci.* **296**(1–2) pp 131-138
[8] Aziz, F. and A.F. Ismail. 2010 *Sep. Purif. Technol.* **73**(3) pp 421-428
[9] H.A. Mannan, Mukhtar H, and T. Murugesan. 2014. *Appl. Mech. Mater.* **625** pp 172-175
[10] Banihashemi, F., M. Pakizeh, and A. Ahmadpour. 2011 *Sep. Purif. Technol.***79**(3) pp 293-302.
[11] Sadrzadeh M, Amirilargini M, Shahidi K, and Mohammadi T. 2009 *J. Memb. Sci.*, **342**(1–2) pp 236-250
[12] Madaeni, S.S., M.M.S. Badieh, and V. Vatanpour. 2013 *Polym. Eng. Sci.* **53**(9) pp 1878-1885
[13] Mansoori SAA, Pakizeh M, and J. A. 2011 *J. Memb. Sci. Technol.*, **1**(106)
[14] Jusoh , Lau. K.K, Shariff. A. M. Yeong. Y. F. 2014 *Int. J. Greenhouse Gas Control*, **22** pp 213-222
[15] Rafiq. S., Man. Z., Maitra. S., Maulud. A., Ahmad. F 2011 *Korean J. Chem. Eng*, **28**(10) pp 2050-2056
[16] Chen. J. Li. J., Lin. Y., and Chen. C. 2009 *J. App. Polym. Sci.* **112**(4) pp 2425-2433
[17] Mannan. H.A., H. Mukhtar, and T. Murugesan. 2015 *App. Mech. Mat.* **699** pp 325-330
[18] Nasir. R., Mukhtar , Man. Z, Dutta , Shaharun Abu Bakar. 2015 *J. Memb. Sci.* **483** pp 84-93
[19] Nasir. R., Mukhtar, Man. Z., Shahrun Abu Bakar 2015 *RSC Advances*, **5**(75) pp 60814-60822
[20] Farid, O.M. 2010. *Ph.D. Thesis*. University of Nottingham, United Kingdom.

Acknowledgements

Financial support provided by Universiti Teknologi PETRONAS is duly acknowledged by the authors.

Morphological analysis of Polyethersulfone/polyvinyl Acetate blend membrane synthesized at various polymer concentrations

S H A A Hadi[1], H A Mannan[1], H Mukhtar[1*], M S Shaharun[2], T Murugesan[1]
[1] Department of Chemical Engineering, *Universiti Teknologi PETRONAS*, Bandar Seri Iskandar 32610, Perak, Malaysia
[2] Department of Fundamental and Applied Sciences, *Universiti Teknologi PETRONAS*, Bandar Seri Iskandar 32610, Perak, Malaysia

E-mail: [1]hilmi_mukhtar@petronas.com.my

Abstract. This paper reports the effect of varying polymer concentration i.e. solvent/polymer ratio on the morphology and gas transport behaviour of polyethersulfone/polyvinyl acetate blend membrane. The solvent used was dimethylformamide, while the manipulated variable was polymer concentration. The concentrations were varied from 75-90 weight % solvent. A homogeneous surface with dense cross-section structure membranes were successfully developed as deduced from FESEM images. The permeability of CO_2 and CH_4 gases increased with increasing polymer concentration and an improved CO_2/CH_4 selectivity was observed. These observation made from the characterization justified the applicability of the blend to be synthesized as membrane for gas separation.

1. Introduction

The current specification of carbon dioxide (CO_2) content in natural gas composition as per US pipeline grid is less than 2%, which urges the advancement of natural gas treatment technology [1]. As compared to other conventional method, membranes are becoming a reliable alternative method equipped with attractive features such as its simple operation, low energy and maintenance requirement, environmental friendly as well as high separation performance [2]. Developments of membrane material usually consist of polymeric and inorganic membrane, where native polymeric membranes have been recognized as a competent membrane base material since 1980s. However, the empirical trade-off between the intra-segmental mobility and inter-segmental packing as demonstrated by Robeson through the upper bound plot in 1991 and 2008 [3] where their performance are still below the par, becomes the challenge to be implemented in industrial scale. The separation performance of a membrane is justified by calculating the permeability and selectivity of the membrane towards the test gases (i.e. CO_2, N_2 and CH_4). However, this performance key feature is directly proportional to its morphology and physico-chemical properties, thus modification needs to be done to improve the performance of pristine polymer [4].

[1] To whom any correspondence should be addressed.

Various enhancement methods have been carried out to improve the performance of pure polymeric material as a membrane base material, such as incorporation of polar groups (i.e. amines), coating with poly(dimethyl siloxane), crosslinking, production of fixed carrier membranes, increasing the free volume by incorporation of nanoparticles and polymer blending [5]. Above all, polymer blending is an attractive alternative in terms of economical and energy-efficiency point view whilst not only retaining the attributes from desired individual polymer, yet producing a new material with synergistic properties.

The existence of polar acetate group in PVAc polymer chain structure increases its competency as a potential membrane base material, together with easy film forming attributes [6]. However, fabrication limitation is the major issue in its individual applicability that urges the enhancement of its mechanical properties. Thus, blending of PVAc with a polymer of high chemical resistance as well as good mechanical properties, more likely from glassy polymer group is preferable. In addition to PES good mechanical properties and chemical resistance, its ether-oxygen which provides a binding mode for CO_2, became the prominence benefit [7]. The combination of these two materials will provide three binding modes for CO_2 which is preferable in solution-diffusion gas separation mechanism and more likely being able to separate CO_2 from other gases [6].

Herein, we investigated the effect of polymer concentration on the polymeric blend membrane, taking PES and PVAc as the component polymers. The developed membranes were prepared by solvent evaporation method. The morphology, inter-polymer interaction, and miscibility of the developed membranes were studied by FESEM, FTIR, and DSC respectively.

2. Experimental

2.1. Materials

Polyethersulfone (PES ULTRASON® E-6020 P) with MW of 50,000 g/mol was purchased from BASF. The PES flakes were pre-treated prior to processing at 105°C for 24 hours to remove moisture. Polyvinyl acetate (PVAc) beads with MW of 100,000 g/mol supplied by Sigma Aldrich were used as received. The solvent used to prepare the dope solution was Dimethyl formamide (DMF) EMPLURA®, supplied from Merck, Germany.

2.2. Membrane preparation

The method of solution casting and solvent evaporation were used to prepare the membrane. The blends were prepared according to optimum (90/10) polymer ratio with manipulation of the solvent/polymer ratio [8]. The blend with composition of 75-90 wt. % solvent/polymer ratios were labelled as Blend 1, Blend 2, Blend 3, and Blend 4, respectively. As a comparison, pure glassy and rubbery membranes were developed as well as shown in Table 1.

Table 1. The composition and thickness of the developed membrane

Membrane	Composition (wt %)	Thickness (μm)	Glass Transition Temperature (°C)
Pure PES	19.6	216.0	210.0
Pure PVAc	19.6	243.8	31.95
Blend 1	10.0	88.0	182.5
Blend 2	15.0	67.7	182.5
Blend 3	20.0	97.4	188.5
Blend 4	25.0	46.0	186.3

The dope solutions were prepared by dissolving the component polymers in a closely-tight blue cap bottle using DMF as solvent for 24 hours. The dope solutions were degassed for an hour to remove the bubbles formed during mixing process. For homogenization process, the dope solutions were left to stand at room temperature for 2 hours. The pure PVAc polymer was casted on a Teflon plate due to its sticky nature, while others were casted on a glass plate using an adjustable knife casting machine at 100 μm. The casted flat sheet membranes with half A4 size were dried at room atmosphere for three hours in an aluminium foil-closed tray prior to drying in vacuum oven at 90°C for 24 hours. Finally, the developed membranes were left at room temperature to dry completely. The thickness of the developed membranes was measured using FESEM analysis and summarized in Table 1.

2.3. Membrane Characterization and Gas Permeation Studies

Fourier transform infrared (FTIR) spectroscopy study was used to determine the functional groups exhibited in the developed blend membranes. The infrared was collected under Nicolet iS10 FTIR Spectrometer in the frequency range of 500-4000 cm^{-1}. The morphology of the membranes was observed under field emission scanning electron microscope FESEM (ZEISS SUPRATM 55VP). The membranes were fractured in liquid Nitrogen prior to imaging step. In order to justify the miscibility of the developed blend membrane, differential scanning calorimetry analysis (DSC) analysis had been carried out. The glass transition temperature T_g of pure and polymer blend membranes were observed by Metler Toledo model DSC-1 equipment. The detailed aspect of gas permeability unit and procedure for gas permeability analysis were described in our previous work [9]. The gas permeation experiments were performed at 25°C and 4 bar pressure.

3. Results and Discussion

3.1. Fourier Transform Infrared Spectroscopy

The effect of polymer concentration at fixed blend ratio towards the stabilization of inter polymer bonding, foremost the hydrogen bonding was analysed under the FTIR spectrum as presented in Figure 1.

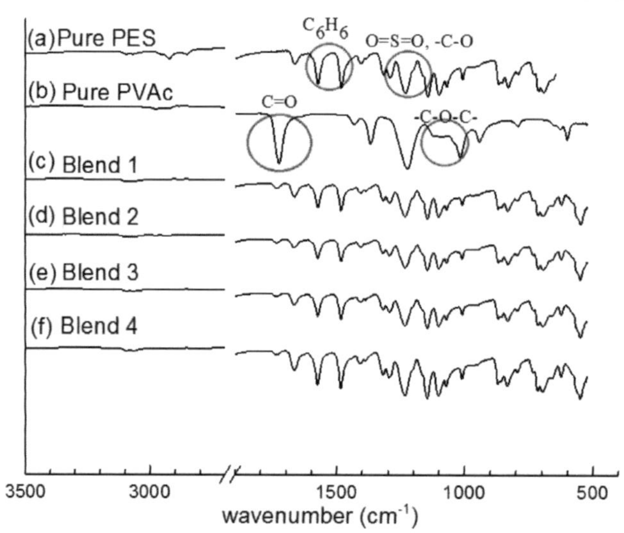

Figure 1. FTIR spectra of (a) Pure PES, (b) Pure PVAc, (c) Blend 1, (d) Blend 2, (e) Blend 3, and (f) Blend 4

The FTIR spectra of PES/PVAc blend membranes (Figure 1.c-f) have similar trends as their component polymers. However, the characteristic peaks are shifted. The effect of blending on the inter polymer interaction has been studied previously, which shown that the component polymers interact at the molecular level, apparently justified by the shift of the wavenumber of the potential functional group to the lower scale, where the potential functional groups in each polymer are the one who tends to interact via hydrogen bonding [10]. The sulfone groups (O=S=O) has the most significant shift from the pure PES polymer, where they move from 1297.43 cm^{-1} to lower wavenumber around 1296 cm^{-1} (the discrepancy of the wave numbers depended on the polymer solution concentration). A shift of the ether oxygen of pure PES was shifted from 1273.79 cm^{-1} to around 1232 cm^{-1}. Besides, the broad band of carbonyl group of pure PVAc shifted from 1728.00 cm^{-1} to around 1667 cm^{-1}, while the ester oxygen of this compound shifted from 1019.24 cm^{-1} to around 1010 cm^{-1}. These movements of the component polymers functional groups portrayed the existence of hydrogen bonding in the system [2]. The shifting of the aromatic benzene group is not significant as the group is already stable which is impossible to create hydrogen bonding.

3.2. Morphology of Blend Membranes

The morphological study of the cross section of the developed membranes was imaged under FESEM analysis, as shown in Figure 2. The microstructure images exhibit homogeneity of the polymers pair with dense structure, however, the cross-sectional views of the developed membranes showed the presence of micro pores. However, the blend membrane apparently showed no phase separation. Besides, it was observed that the number of pores reduces with increasing solvent/polymer ratio. This observation is in line with the reported studies on the effect of polymer concentration in hindering the micro pores formation. However, Blend 4 shown a big pore defect which correspond from the membrane-glass plate detachment effect which affecting its thin membrane thickness (46.0 μm). Bakeri et al. [11] and Rafiq et al. [12] observed similar suppression effect of the micro pores formation as they studied the effect of critical polymer concentration in the development of membrane for gas separation. The FESEM images of the pure component polymers can be found in the previous work [8]. These dense structures without any defect i.e. pinholes or cracks observed from the cross-section views of the developed blend membranes with respect to the pure component polymers suggest the miscibility is attained between the component polymers [2]. Both polymers are well dispersed in each other, where at varied polymer concentration, no phase separation is observed.

3.3. Differential Scanning Calorimetry (DSC) Analysis

The miscibility of the polymers upon blending can be justified with DSC analysis with the existence of single glass transition temperature in the blend for the pairs with significant difference in T_g [13]. Table 1 summarizes the T_g's of the developed membranes. It is observed that each blend displayed a single glass transition temperature, intermediate between the pure polymeric membranes and slightly varies with varying solvent/polymer ratios. This suggests the miscibility is attained by the blend. From the table, the T_g's of the developed blend membrane varied with different polymer concentration, however, the difference is not really significant. The lower the polymer concentration, the more reduction from the T_g of the pure PES membrane.

3.4. Gas Permeability Studies

Table 2 shows the preliminary investigation of gas permeation properties of synthesized membranes at 25°C and 4 bar pressure. The performance of blend membranes can be compared with pure PES and PVAc membrane [8]. By increasing the polymer ratio in the dope solution, the CO_2 permeability is reduced from 19.97 barrer to 0.42 barrer and then shows a gradual increase and reaches to 1.00 barrer and finally the permeability of Blend-4 membrane is recorded as 11.59 barrer. A similar behaviour is observed for CH_4 permeability and ideal selectivity. This is due to the fact that by increasing the polymer ratio in dope solution the cross section of membrane represents a defect free dense structure which caused a decrease in permeability and an increase in ideal selectivity.

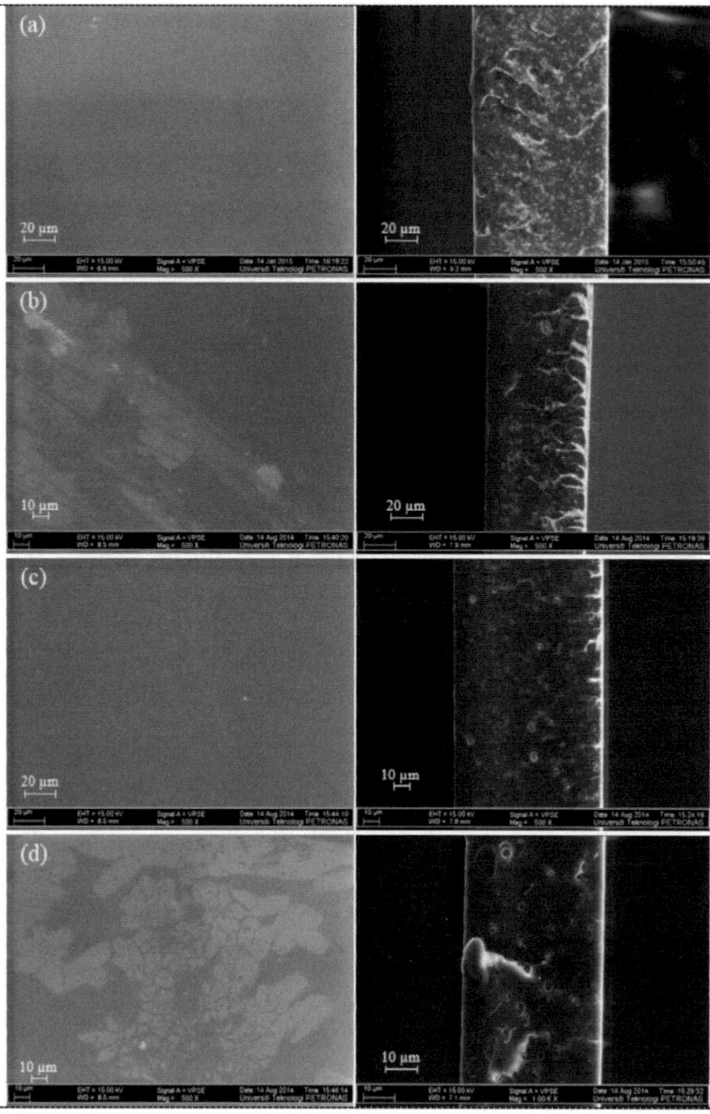

Figure 2. FESEM images of surface and cross section of the blend membranes (a) Blend 1 (b) Blend 2 (c) Blend 3 (d) Blend 4; Surface (left), Cross section (right)

Table 2. Gas permeation performance of blend membranes at 25°C and 4 bar pressure

Sample	CO_2 Permeability (barrer)	CH_4 Permeability (barrer)	Ideal Selectivity
Blend 1	19.97	11.79	1.69
Blend 2	0.42	0.39	1.08
Blend 3	1.00	0.79	1.26
Blend 4	11.59	4.94	2.34

4. Conclusions

PES/PVAc blend membranes with various polymer/solvent ratios have been developed and characterized via FESEM, FTIR, DSC as well as TGA analysis. From micro-scale, the FESEM cross-section images justified that the developed membranes were dense with varying number of micro pores. The higher polymer/solvent ratio (viscous dope solution) tends to inhibit the micro pores formation producing dependable membranes for gas separation. The interaction and miscibility between the components polymers were attainable and deduced via FTIR and DSC analysis. From TGA scan, the thermal stability of the developed blend membrane apparently lies between the parent polymers, where the addition of the more stable polymer (PES) stabilized the thermal stability of the weaker polymer (PVAc). The results suggest the applicability of the PES/PVAc blend to be developed as a membrane to push the performance of native polymeric membrane for gas separation.

5. References

[1] Baker R W 2002 *Ind. Eng. Chem. Res.,* **41** pp 1393-1411
[2] Mannan H A, Mukhtar H and Murugesan T 2015 *Applied Mechanics & Materials* **699** pp 325-330
[3] Robeson L M 2008 *J. Membr. Sci.,* **320** pp 390-400
[4] Hunger K, Schmeling N, Jeazet H B T, Janiak C, Staudt C and Kleinermanns K 2012 *Membranes* **2** pp 727-763
[5] Mannan H A, Mukhtar H and Murugesan T 2014 *Applied Mechanics & Materials* **625** pp 172-175
[6] Sanaeepur H, Ebadi A A, Moghadassi A, Kargari A, Moradi S and Ghanbari D 2012 *Polym. Adv. Technol,* **23** pp1207-1218
[7] Kapantaidakis G and Koops G 2002 *J. Membr. Sci.* **204** pp 153-171
[8] Hadi S H A A, Mukhtar H, Mannan H A and Murugesan T 2015 *Applied Mechanics and Materials* **754** pp 44-48
[9] Mannan H A, Mukhtar H, Othman M R, Shahrun M S and Murugesan T 2015 *J. Appl. Polym. Sci.,* **133** 42946
[10] Ahmad J and Hågg M B 2013 *J. Membr. Sci.,* **445** pp 200-210
[11] Bakeri G, Ismail A F, Shariaty-Niassar M and Matsuura T 2010 *J. Membr. Sci.,* **363** pp 103-111
[12] Rafiq S, Man Z, Maitra S, Maulud A, Ahmad F and Muhammad N 2011 *Korean J. Chem. Eng.,* **28** pp 2050-2056
[13] Mohshim D F, Mukhtar H and Man Z 2014 *Sep. Purif. Technol.,* **135** pp 252-258

Acknowledgments
The authors gratefully acknowledge the Ministry of Education (MOE) Malaysia for providing the fund (FRGS 0153AB-I68).

Polymer blend membranes for CO_2 separation from natural gas

H Mukhtar[1], H A Mannan, D Minh, R Nasir, D F Moshshim, T Murugesan

[1] Department of Chemical Engineering, *Universiti Teknologi PETRONAS*, 32610 Bandar Seri Iskandar, Perak Darul Ridzuan, Malaysia

Email: [1]hilmi_mukhtar@petronas.com.my

Abstract: Polymeric membranes are dominantly used in industrial gas separation membrane processes. Enhancement in membranes permeability and/or selectivity is a key challenge faced by membrane researchers. The current work represents the effect of poyetherimide blending on separation performance of polysulfone membranes. Polysulfone/poyetherimide (PSF/PEI) blend flat sheet dense membranes were synthesized and tested for permeation analysis of CO_2 and CH_4 gases at 6, 8 and 10 bar pressure and 25°C temperature. Morphology and thermal properties of membranes were characterized by field emission scanning electron microscope (FESEM) and thermo gravimetric analysis (TGA) respectively. Blend membranes were dense and homogeneous as deduced from FESEM analysis. Thermal stability of synthesized blend membranes was maintained by blending with PEI as characterized by TGA results. Decrease in permeability of both gases was observed by the addition of PEI due to rigidity of PEI chains. Additionally, selectivity of synthesized blend membranes was enhanced by blending PEI and blend membranes show improved selectivity over pure PSF membrane. This new material has the capability to be used as gas separation membrane material.

1. Introduction

Polymer blends are attractive alternative materials for gas separation membranes as compared to inorganic or mixed matrix membranes due to ease of processing, simplicity and synergetic properties of polymer blends [1]. It is a general requirement for gas separation membranes to have high permeability and high selectivity [2]. However, the typical trade off line known as Robeson upper bound line suggests that increase in permeability is at the expense of selectivity and vice versa [3]. According to solution diffusion model, the permeability of a gas across the membrane is a function of diffusivity and solubility [2]. Thus, by molecular engineering of membrane materials, diffusivity and/or solubility of a membrane for a target gas can be enhanced. Polymer blending of a base material with high glass transition temperature (T_g) rigid polymers provide diffusion selectivity to polymer blend matrix as reported by [1, 4].

Polysulfone is a commodity polymer used for membrane synthesis. It has moderate gas separation performance, good film forming ability and high plasticization resistance [4]. Thermal and chemical

[1] To whom any correspondence should be addressed.

stability, adequate permeability and selectivity of polyetherimide made it a suitable choice for gas separation membranes. Its reported selectivity for CO_2/CH_4 gas pair is 30-40 [5]. Therefore, blending small amount of PEI in PSF matrix will enhance the selectivity of the resultant blend membrane. The miscibility of PSF/PEI blend was reported in literature [6]. But no study is available in literature for application of PSF/PEI blend as gas separation membranes.

Therefore, the present study reports the blending effect of PEI in base PSF matrix on morphology and gas transport properties of PSF membrane by varying the PEI concentration from 0-3%. The synthesised membranes were characterized and tested for permeation of CO_2 and CH_4 gases at room temperature and 6, 8, 10 bar pressure.

2. Experimental

2.1. Chemicals

Polysulfone (PSF) Udel® P-1800 taken as a base polymer was acquired from Solvay Advanced Polymers, L.L.C, U.S. Polyetherimide (PEI) was obtained from Sigma Aldrich, Germany. 1-Methyl-2-pyrrolidone EMPLURA® (NMP) was used a solvent and it was obtained from Merck. CO_2 and CH_4 gases with 99.99% purity were purchased from MOX-Linde Gases Sdn. Both polymers were pre dried at 110°C for 24 hours to remove the moisture.

2.2. Membrane Synthesis

Pre-dried polymers (20% w/w) were dissolved in NMP solvent to prepare casting solution in a closed container. The casting solution was stirred for 24 hour at room temperature to completely dissolve the polymers followed by degassing for 30 minutes to remove the air bubbles formed during the mixing process. The casting knife was adjusted at 180 μm opening and the casting solution was poured onto a clean glass plate to cast the membrane films. The casted membranes were isolated in the vacuum oven at 25°C for the first 48 hours. For the next 24 hours, the temperature was raised 20°C for each 2 hours until the temperature reached 65°C which was maintained for 20 hours. After 72 hours of drying process, the membranes were secured.

2.3. Membrane Characterization

The synthesized membranes were characterized by FESEM and TGA in order to investigate the morphology and thermal properties of synthesized membranes.

2.4. Gas Permeation Analysis

The permeability of pure CO_2 and CH_4 gas across the membrane was recorded in the pressure range of 6-10 bar and room temperature. The detailed aspects of membrane permeation unit can be found in our previous work [7]. The CO_2 permeability (P_{CO_2}) was calculated according to the following formula:

$$P_{CO_2} = \frac{J_{CO_2} l}{\Delta p_{CO_2}} \tag{1}$$

where, J_{CO_2} is flux of CO_2, Δp_{CO_2} is differential partial pressure of CO_2 across the membrane and l is thickness of membrane. Similarly, permeability of CH_4 can also be calculated using the same formula. Then, the ideal selectivity (α) of membrane was obtained by dividing the permeability of CO_2 by permeability of CH_4.

$$\alpha_{CO_2/CH_4} = \frac{P_{CO_2}}{P_{CH_4}} \tag{2}$$

3. Results and Discussion

3.1.1. Membrane Morphology

Morphology of synthesized membranes is shown in Figure 1. PSF membrane has smooth, dense and homogeneous structure. The blend membranes have a similar dense structure with no pore formation. At a magnification of 1000, no cracks or defects are observed in membrane structure at macroscopic level [8]. However, with the increase in PEI content the resultant blend membrane has some irregularities in the cross-section. At the same time, the structure is not phase separated [9]. Thus, higher concentration of PEI would result in partially miscible blend. It is worth mentioning that membrane samples with higher weight percent of PEI were synthesised and it was noticed that the blend was not miscible at concentrations higher than 5 weight percent PEI in PSF matrix and the synthesised membrane samples were not stable. Thus, PEI in lower concentrations (< 5 %) makes miscible or partially miscible blend.

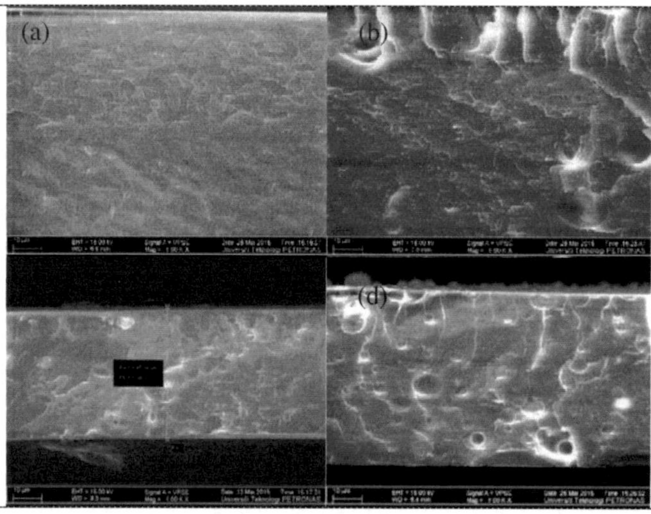

Figure 1. Cross sectional images of (a) PSF, (b) PSF-PEI (99-1), (c) PSF-PEI (98-2) and (d) PSF-PEI (97-3) blend membrane.

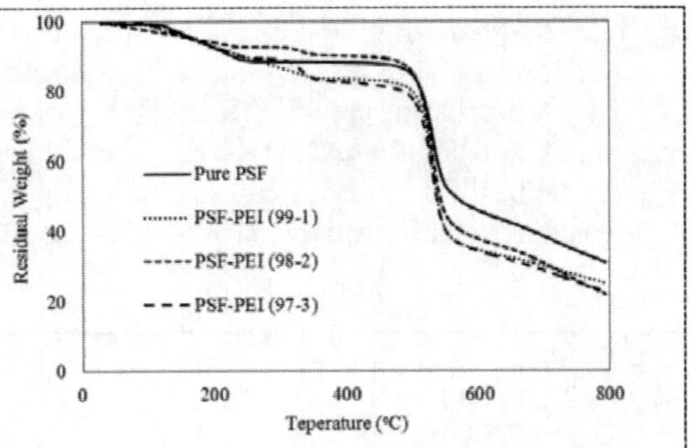

Figure 2. TGA curves of synthesized membranes.

3.2. Thermal properties of Membranes

Thermal properties of PSF/PEI blend membranes are portrayed in Figure 2. The first weight loss is observed at 200°C which might be due to trapped solvent. The second weight loss represents the degradation of polymer backbone at 490°C [7]. Blending of PEI does not affect the thermal properties of PSF membranes. Thermal properties of PSF membrane and blend membranes are in agreement with literature [10].

3.3. Gas Permeation Analysis

3.3.1. Effect of blend composition on gas permeability

The blend membranes were tested for gas permeation properties of CO_2 and CH_4 gases for the effect of PEI concentration on gas transport properties of membranes. The experiments were run at 6 bar pressure and 25°C. The permeability of both gases decreases with increase in PEI concentration as can be seen in Figure 3. The decrease in permeability is due to higher glass transition temperature (T_g) of PEI polymer which is reported in literature as 217°C [11]. Moreover, PEI has stiff, rigid and compact structure as compared to PSF [1]. Figure 4 shows an opposite trend to membrane permeability and selectivity of blend membranes is improved by increasing PEI concentration. Again, stiffness and rigidity of PEI chains caused the enhanced selectivity.

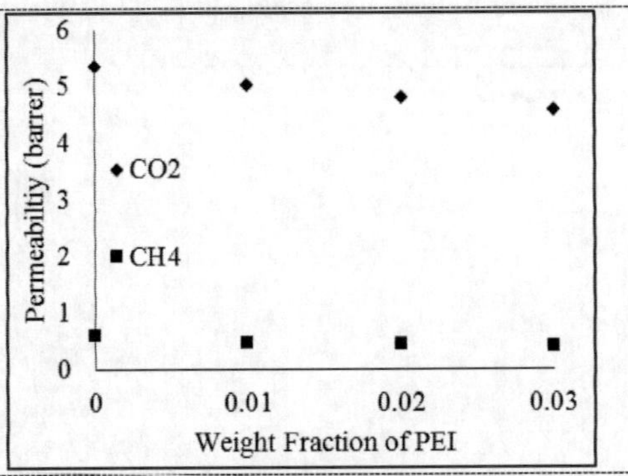

Figure 3. Effect of PEI weight fraction on permeation of CO_2 and CH_4 gases at 6 bar and 25°C.

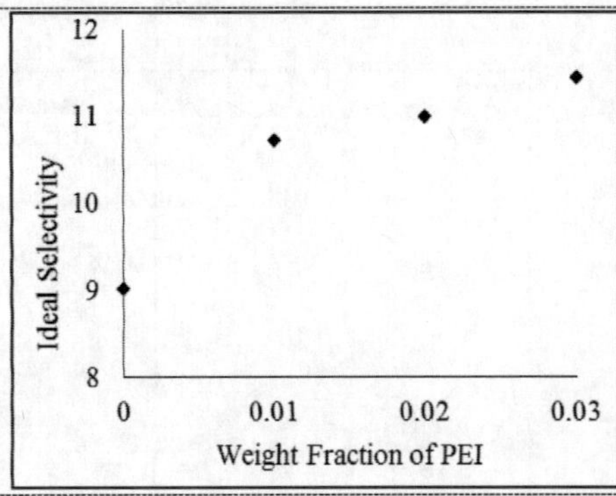

Figure 4. Effect of PEI weight fraction on ideal selectivity of blend membranes at 6 bar and 25°C

IOP Conf. Series: Earth and Environmental Science **36** (2016) 012016 doi:10.1088/1755-1315/36/1/012016

3.3.2. Effect of feed pressure on gas permeability

Figure 5 and 6 shows the effect of feed pressure on CO_2 and CH_4 permeability of blend membranes. The pressure was varied from 6-10 bar. A decreasing trend in permeability of both gases is observed which is known trend for glassy polymers [7]. For both gases, native PSF membrane has the highest permeability. PSF-PEI (97-3) shows the lowest permeability of both gases. The CH_4 permeability of blend membranes is significantly lower than CO_2 gas due to larger relative size of CH_4 gas (3.8 Å) as compared to CO_2 gas (3.3 Å).

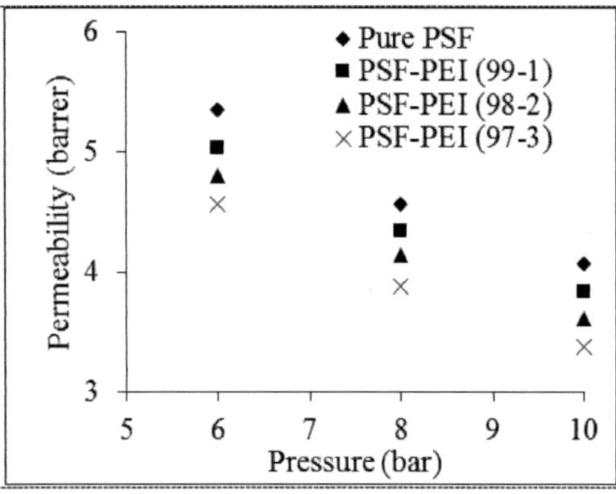

Figure 5. Effect of feed pressure on CO_2 permeability of blend membranes.

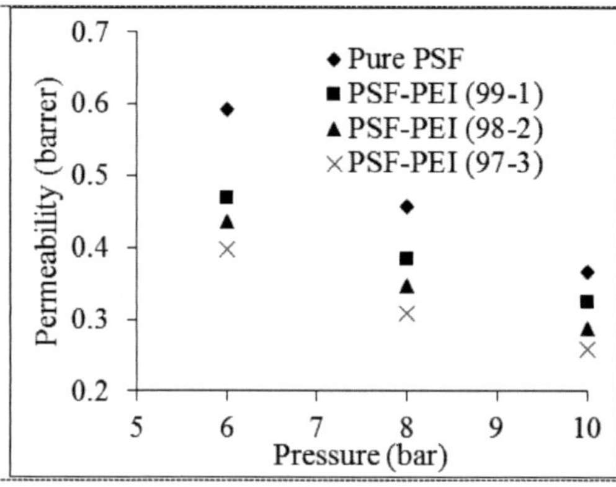

Figure 6. Effect of feed pressure on CH_4 permeability of blend membranes.

Shown in Figure 7 is the ideal selectivity of synthesized membranes. The ideal selectivity of blend membranes is increasing with increase in feed pressure. Within this pressure range, no plasticization effect was observed. Unblended PSF membrane has the lowest selectivity. However, by blending with PEI, the resultant selectivity is significantly improved. The stiffness and rigidity of PEI chains impart the diffusion selectivity and sieving ability to blend membrane. The selectivity of native PSF membrane at 10 bar pressure is 11.06 whereas for PSF-PEI (97-3) membrane it is 13.06 which accounts for 10.43 % improvement in ideal selectivity. Thus, the selectivity of PSF membrane can be improved by blending rigid glassy polymers [7]. A similar trend of effect of pressure was observed by [12]. The selectivity of PSF-PEI (97-3) blend membrane falls near attractive region for natural gas processing [13].

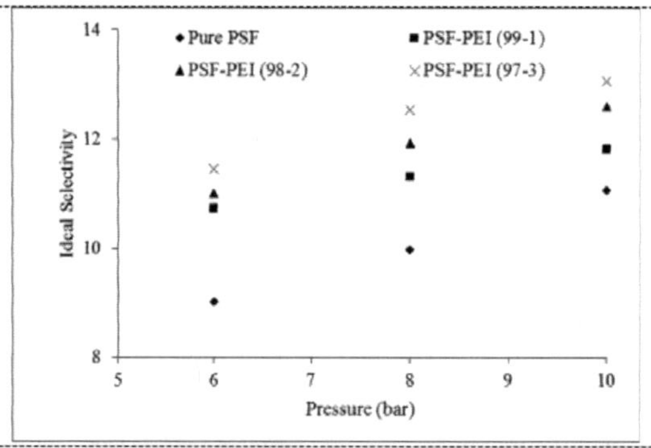

Figure 7. Effect of feed pressure on ideal selectivity of blend membranes.

4. Conclusion

This study presented the synthesis, characterization and gas permeation properties of polysulfone/poyetherimide (PSF/PEI) blend membranes. The polymer blends were miscible or partially miscible in low PEI concentration (< 5 % PEI). The thermal properties of PSF/PEI blend membranes followed the general degradation trend of PSF membranes. Higher concentration of PEI led to decrease in CO_2 and CH_4 permeability and increase in CO_2/CH_4 selectivity due to rigidity of PEI chains. Significant improvement is observed in selectivity of PSF/PEI blend membranes. The blend membranes followed the typical behaviour of glassy polymers with respect to effect of feed pressure. This new material has the capability to be used as gas separation membrane material in CO_2 separation from natural gas.

5. References

[1] Dorosti F, Omidkhah M R, Pedram M Z and Moghadam F 2011 *Chem. Eng. J.* **171** pp 1469-1476
[2] Chung T S, Jiang L Y, Li Y and Kulprathipanja S 2007 *Prog. Polym. Sci.* **32** pp 483-507
[3] Mannan H A, Mukhtar H, Murugesan T, Nasir R, Mohshim D F and Mushtaq A 2013 *Chem. Eng. Technol.* **36** pp 1838-1846
[4] Rafiq S, Man Z, Maitra S, Maulud A, Ahmad F and Muhammad N 2011 *Korean J. Chem. Eng.* **28** pp 2050-2056
[5] Arabi Shamsabadi A, Kargari A and Bahrami Babaheidari M 2014 *Int. J. Hydrogen Energy* **39** pp 1410-1419
[6] White D M, Giles H F, Matthews R O 1983 Google Patents
[7] Mannan H A, Mukhtar H, Othman M R, Shahrun M S and Murugesan T 2015 *J. Appl. Polym. Sci.,* **133** 42946
[8] Feng S, Ren J, Li H, Hua K, Li X and Deng M 2013 *Journal of Energy Chemistry* **22** pp 837-844
[9] Anil Kumar P V, Anilkumar S, Varughese K T and Thomas S 2011 *J. Polym. Res.* **19** pp 1-12
[10] Rafiq S, Man Z, Maitra S, Muhammad N and Ahmad F 2012 *J. Appl. Polym. Sci.* **123** pp 3755-3763
[11] Chung T S and Xu Z L 1998 *J. Membr. Sci.,* **147** pp 35-47
[12] Basu S, Cano-Odena A and Vankelecom I F J 2010 *Sep. Purif. Technol.* **75** pp 15-21
[13] Baker R W and Lokhandwala K 2008 *Ind. Eng. Chem. Res.* **47** pp 2109-2121

Acknowledgement
This work was funded by Fundamental Research Grant Scheme (FRGS) via cost centre number (0153AB-68). The authors gratefully acknowledge the technical support provided by Universiti Teknologi PETRONAS.

Anaerobic digestion of wastewater screenings for resource recovery and waste reduction

N Wid[1], **N J Horan**[2]
[1]Faculty of Science and Natural Resources, *Universiti Malaysia Sabah*, Jalan UMS, 88400, Kota Kinabalu, Sabah, Malaysia
[2]School of Civil Engineering, Faculty of Engineering, University of Leeds, Leeds, LS2 9JT, United Kingdom

E-mail: [1]newati@ums.edu.my, [2]n.j.horan@leeds.ac.uk

Abstract. Wastewater screenings are produced during the first stage of the wastewater treatment process and at present are disposed of to landfill. This material may not only cause operational failure to the treatment system, but also lead to environmental problems. In view of the high organic content of screenings, anaerobic digestion method may not only offer the potential for energy recovery, but also nutrient. In this study the, anaerobic batch digestion was performed at different dry solids concentrations of screenings to study the potential of biogas and phosphorus recovery. The tests demonstrated wastewater screenings were amenable to anaerobic digestion with methane yield was 355 m^3/kg VS, which are comparable to the previous results. The digestate was high in P content and can be recovered up to 41%. This study also shows that anaerobic digestion was not only to turn this waste into useful resources, but also has a potential in reducing the organic content up to 31% for safe disposal. In this way the amount of wastewater screenings going to landfill is not only can be reduced, but also valuable products such as methane and phosphorus can also be recovered.

1. Introduction

The recovery of resources from organic waste has received a lot of attention over the past decade and the theoretical basis for resource recovery is well understood. However one such waste that has largely been ignored is the wastewater screenings (WS) produced during the first stage of the sewage treatment process, including rags, paper, plastic, wood and other untreatable solid material. With a growing stress on global resources, wastewater is no longer viewed as a waste stream, but as a resource from which to generate many resources. Currently, WS are disposed of to landfill which results in greenhouse gases (GHGs) emissions, with lost energy, nutrients, with other recoverable resources and also other problems such as odour problems as well as limited space remaining for landfilling.

There are different methods are employed for the treatment of solid wastes, which one of them is anaerobic digestion (AD). AD is a complex biological process in degrading organic matter in the absence of oxygen, with the major stages including hydrolysis, acidogenesis, acetogenesis and methanogenesis. It is a sustainable approach that combines wastewater treatment with the recovery of valuable by-products, including renewable energy and nutrient [1]. Therefore, AD treatment is not only alleviate environmental problems, by stabilising sludge and reducing the mass of waste sludge for final disposal, but also eases the stress on depleted resources and growing energy insecurity, by converting waste to biogas. The digestate liquor has a high phosphorus concentration that can be recovered [2–5], with the potential to be used in agricultural activities.

[1]To whom any correspondence should be addressed.

The long term oil crisis and consequent price rises have spawned considerable interest in the exploration of renewable energy sources. Bioenergy will be the most significant renewable energy source in the next few decades until solar or wind power production offers an economically attractive large-scale alternative. In AD study, special attention has been paid to the final stage of the process i.e. methanogenesis to produce methane. Besides producing a significant amount of versatile energy in the form of methane, anaerobic digestion also solubilises the nutrients phosphorus and nitrogen, making them easy to recover. Meanwhile, phosphorus (P) is a limited resource which is anticipated will be exhausted in the 21st century [6–8]. While natural sources like oil can be substituted with other sources when its reserves peak (such as wind, solar, or thermal energy), P has no substitute and cannot be manufactured or synthesized. The main source of P is extracted from the phosphate rock in the earth's crust. Due to the depleting P resource and higher extraction cost, there are increasingly studies have been undertaken to recover phosphorus from waste streams [5,9–13]. The best way to recover phosphorus from solid waste is to solubilise it in the form of liquid. The use of AD technique to extract and solubilise phosphorus from solid waste has been extensively reported for the past 20 years. There are a number of studies have been reported in using anaerobic digestate liquor from various type of solid wastes to recover phosphorus [14–17], but as yet non from screenings has been reported. As this material has received little attention, research on treatment of screenings by AD is rare. Therefore, this study aims to examine the potential of wastewater screenings for methane production and phosphorus recovery following its mesophilic AD. Prior to AD, WS will be characterised to allow for proper examination of the feasibility of the WS for conversion into recoverable resources. The biodegradability of the material will also be investigated to evaluate its potential in reducing the volume of the waste for final disposal.

2. Materials and methods

2.1. Collection and preparation of wastewater screenings (WS) and primary sludge (PS)

The WS used in this study were collected from the Knostrop Wastewater Treatment Plant (KWWTP), located in West Yorkshire, Leeds. The sample was simply collected when needed. Four major sampling regimes were undertaken, labelled as 1, 2, 3 and 4. Primary sludge (PS) was also collected from the same treatment plant to be used as seed inoculum (SI). The PS supplied the seed bacteria that needed to boost the AD process. After removing plastics, the fresh WS were ground, to achieve homogenized sample as representative specimens. The sample was then kept in a fridge at 4°C prior to use.

2.2. Characterisation of WS

Characterisation involved analysis of component composition and physical character. This was done using the fresh WS. The composition was determined by manually segregating different type of materials in the sample such as paper, rags, plastic and kitchen waste. The percentage was determined by taking the wet weight of each material over the wet weight of the total sample. The physical analysis, i.e. pH, total solids (TS) and volatile solids (VS) were carried out according to Standard Methods [18]. For TS determination, the sample waste dried in an oven for 24h at 105°C, subsequently ignited in a furnace for 4h at 550°C for VS determination.

2.3. Anaerobic digestion (AD)

Anaerobic batch experiments were performed at a laboratory scale based on the method described by [19] and revised by others [20,21]. The reactors were developed using serum bottles with effective volume of 400mL. The reactors were sealed and tightened with neoprene bungs. There were four reactors were used with different dry solids concentrations, i.e. 3, 6, 9 and 12%, represented by R1, R2, R3 and R4, respectively. While the control represented by R5. Each reactor was filled with WS and seed inoculum at 1.5:1 ratio. Nutrient was also added at 1mL/L containing all the micro and micro nutrients and vitamins essential to the growth of the anaerobic microorganisms. While, R5 contained only seed inoculum and nutrient, as a control, to correct the biogas production of the inoculum. To perform the batch experiments, pH was measured and nitrogen gas was sparged to create anaerobic conditions in the reactors before starting the experiments. The experiment was operated in a shaker incubator at 37°C and 100 rpm for 30 days. The pH was maintained in the range of 6.8-7.2 using 0.1M HCl and 0.1M NaOH.

2.4. Biogas measurement

Biogas produced from the reactors was measured by a liquid displacement method using 5% NaOH to scrub CO_2. The measured biogas volume was corrected to standard temperature and pressure. The biogas production was determined by subtracting the background values for the incubation period of 30 days which ensures that complete biodegradation of all the organics.

2.5. Phosphorus (P) determination in the digested liquor

To study the potential of P recovery, the digested effluent produced on the final day was centrifuged, filtered and the resultant supernatant analysed for its P concentration according to the procedures outlined in [18]. The concentration of P was determined using Agilent Cary 60 UV-Vis Spectrophotometer at wavelength 880 nm. The calibration curve was plotted using standard solutions with different concentrations prepared from 100 ppm of P stock solution.

3. Result and discussion

2.6. Wastewater screenings characterization

Ideal operation of the AD process requires that the organic and inorganic composition of the WS is known prior to feeding into the digester, to permit modifications to the key parameters. The physical characteristics of WS sampled at different collection times are presented in Table 1. Despite the wide gaps between collection periods, the composition of each sample was very consistent. The average dry solids content was 29.56%, which is considered to be quite high. According to [22], treatment involving compaction of the screenings increases the dry solids content by dewatering the waste. This explains the high dry solids as the WS collected were dried by compaction, during preliminary treatment at the plant. Dry solid content of dewatered screenings was also reported as 24 – 30% [23], and 30% [24]. Clay et al. [25] suggested the dryness of screenings should be at least 25% solids to avoid spillage during transportation and for reducing the costs of disposal. Despite the dryness, WS were still high in moisture and unsuitable for the modern landfill. The VS were always higher than 90%, suggesting a high organic matter content. This agrees with the observations of [22], who reported the VS of WS was about 90%. High volatile solids illustrate the potential of methane production as well as nutrient recovery. The average pH was 6.2 suggesting that the material will digest well without the need for external pH correction.

Table 1. Physical characteristics of wastewater screenings.

	Characterisations at different collection times				Average (std. dev.)
	1	2	3	4	
TS (% dry)	31.75	29.91	32.02	24.55	29.56 (3.00)
VS (% TS)	93.74	91.42	93.05	95.03	93.31 (1.30)
pH	6.08	5.83	6.40	6.55	6.22 (0.28)

The composition of fresh screenings in terms of their gross composition was also similar and the predominant fractions were papers and rags, followed by vegetable and organic wastes, plastics and grits (Table 2). Both collected WS showed similar composition, albeit differing slightly in percentage. The high proportion of paper suggests common practice in using paper products, such as cardboards and paper bags as packaging materials in the local area. The predominant fraction of paper may results in higher VS, biodegradability and thus produce high methane yield from digestion of WS. This finding also opens the possibility for the material to be recycled from waste stream, especially paper recycling. It has been noted that paper and paperboard are always the largest component of the municipal solid waste (MSW) stream, compromising more than one-third of total generation [26]. According to [27], a high paper composition results in higher VS, i.e. above 82%, consequently higher biodegradability and resource recovery.

The high proportion of rags, underlined the increasing use of disposable wipes [22], which are mainly made of cotton comprise a large fraction of the rags in WS. This large fraction of sanitary material is another reason why WS are categorised as a difficult waste, which may lead to difficulties in storing and disposal [25]. While the low proportion of plastics, vegetable and organic wastes as well as grits, indicates low composition of these materials arrived at the treatment plant.

Table 2. Composition of WS collected from the KWwTP, Leeds.

	Composition (% of wet screenings)		
Sampling time	1	3	Average (%)
Papers	67.02	61.65	64.34
Rags	22.83	31.57	27.20
Plastics	2.25	2.16	2.21
Vegetable and organic wastes	6.01	4.59	5.30
Grits	1.89	0.02	0.96

3.2 Biogas production and P recovery

The profile of cumulative gas production (GP) at standard temperature and pressure (STP) was measured and over the first 20 days there was negligible difference in gas production between feed solids of 3 and 6%. (Figure 1). However the higher dry solids (also represented by total solids, TS) continued to produce gas after this period whereas there was a plateau for the 3% solids. By contrast gas production for both 9 and 12% dry solids was poor. Thus the cumulative GP for 6% and 3% reached 4,466 mL and 3,097 mL of biogas, respectively, whereas gas production from 9% and 12% was low with a cumulative GP of 835 and 756 mL, respectively. This suggests that 6% is the optimum feed concentration of TS for anaerobic digestion, and 3% is the second place. By contrast 9% and 12% suggested that high concentrations of dry solids were inhibitory to methanogenesis possibly as a result of overloaded with high ammonia or volatile fatty acids (VFA) concentrations.

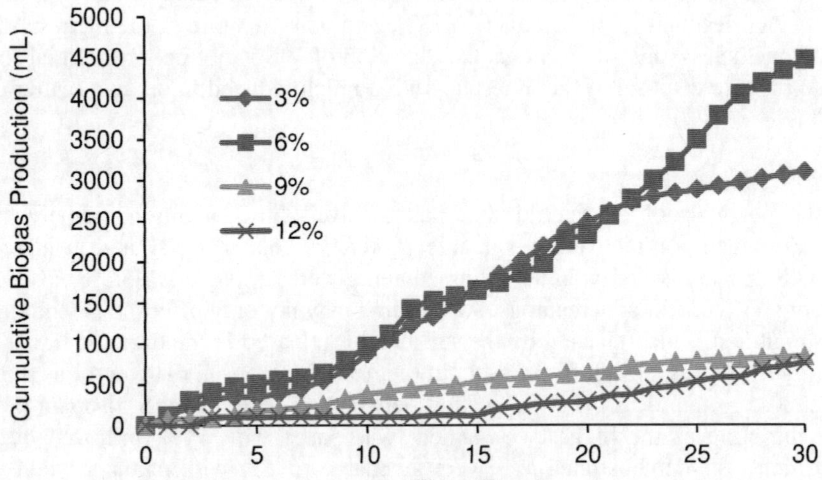

Figure 1. Cumulative GP at different dry solids concentrations

During the AD process, the first stage is the hydrolysis of large organic material (generally measured as volatile solids) to smaller, more readily biodegradable monomers. For complex wastes, this process is slow and indeed it is generally the rate-limiting step in the anaerobic digestion process.

Although the methanogenesis stage is also slow, for such complex wastes the rate of methane production is able to keep pace with the hydrolysis of organic material. Consequently the gas yield is expressed in terms of volume of gas produced/ amount of volatile solids applied (Table 3). This is clearly seen for the digestion of WS at 6%, where hydrolysis proceeds rapidly over the first 8 days and a rapid increase after 12 days indicates more complex material is destroyed, thus produced 0.355 m^3CH_4/kg VS methane yield. Previous studies also found the yield obtained from WS is typical of the value of other readily biodegradable wastes, such as organic fraction municipal solid waste (OFMSW) with 0.349 m^3CH_4/kg VS [28], food waste with 0.385 m^3CH_4/kg VS [29] and complex organic substrate with 0.243 m^3CH_4/kg VS methane yields [30]. The potential of P recovery was evaluated based on the percentage of P produced in the digestate from 1g dry solids of WS during the AD tests. The results show not much variation on the percentage when studied at different dry solids concentrations. It ranged between 37%-41%, with 6% shows the highest. These indicate that WS provides an ideal feedstock for resource recovery.

Table 3. The effect of dry solids on the anaerobic digestion performance.

Dry Solids (%)	VS destruction (%)	Total GP (mL)	Yield (m^3 CH_4/kg VS)	P recovery (%)
3	31.28	3097.11	0.298	37
6	31.44	4465.74	0.355	41
9	26.68	835.21	0.45	39
12	30.03	756.57	0.28	38

Note: the gas is assumed to be in the form of methane

3.3 Volatile solids (VS) reduction

In this study, VS reduction defined as the amount of organic content degraded from initial concentrations of dry solids by digestion process. The results show a constant value at different concentrations of substrate feed which ranged between 26.68%-31.44% (Table 3), with the highest reduction was at 6% dry solids. These values were less than some of the previous studies, which achieved up to 88.5% VS reduction [31]. This may due to the digestion system used, where previous studies used two-stage system. The results however, are similar to those found for digestion of terrestrial weeds which resulted in 46% reduction [32], and dairy manure at the highest VS removal about 35% [33]. Nonetheless, this study shows the digestion of WS is not only beneficial for resource recovery, but also for safe disposal of the digestate with a much reduced in organic matter content up to 31%.

4. Conclusion

In this study, the potential of a material that has received little attention, namely wastewater screenings, was examined for recovering valuable products. This material is considered high in organic matter with dry solids and volatile solids contents were 29.6% and 93.3%, respectively. In view of the high organic content of screenings, AD method may not only offer the potential for energy recovery, but also nutrient. This indicated by the sufficient P released in the digested liquor with 41% P recovery. The digestion was well-buffered without the need for external pH control as the pH of the material was 6.2 close to the ideal pH for digestion between 6.8 and 7.2. WS also can be used as a source of paper recycling as the highest component was paper with 64%, followed by rags with 27.2%. The digestion tests demonstrated WS were amenable to AD with methane yield was 0.355 m^3/kg VS at initial 6% of dry solids. Dry solids concentrations higher than 6% were inhibitory to methanogenesis possibly as a result of high ammonia or VFA concentrations. The organic content (in the form of volatile solids content) can be reduced up to 31%, which is environmentally beneficial and can reduce sludge volume for final disposal. The low VS reduction compared to previous studies implying WS is a difficult solid waste to digest, which may due to the high proportion of rags. Therefore, for future work it is suggested to perform pre-treatment and use two-stage digestion in treating difficult waste like WS. Overall, it can be concluded that AD treatment is not only alleviate environmental problems, by stabilising sludge and reducing the mass of waste sludge for final disposal, but also eases the stress on depleted resources and growing energy insecurity by converting waste to methane and phosphorus.

5. References

[1] Khanal S K 2008 *Wiley: Anaerobic Biotechnology for Bioenergy Production: Principles and Applications* (USA: Wiley and Sons Ltd)
[2] Marti N, Bouzas A, Seco A and Ferrer J 2008 Struvite precipitation assessment in anaerobic digestion processes *Chem. Eng. J.* **141** 67–74
[3] Münch E V and Barr K 2001 *Water Res.* **35** 151–159
[4] Jaffer J and Pearce P 2004 *Phosphorus recovery via struvite production at Slough sewage treatment works, UK - a case study* (IWA Publishing)
[5] Jaffer Y, Clark T A, Pearce P and Parsons S A 2002 *Water Res.* **36** 1834–42
[6] Cordell D, Drangert J-O and White S 2009 *Glob. Environ. Change* **19** 292–305
[7] Smil V 2000 *Annu. Rev. Energy Environ.* **25** 53–88
[8] Herring J R and Fantel R J 1993 *Nonrenewable Resour.* **2** 226–246
[9] Huang H, Xu C and Zhang W 2011 *Bioresour. Technol.* **102** 2523–28
[10] Jordaan E M, Ackerman J and Cicek N 2010 *Water Sci. Technol. J. Int. Assoc. Water Pollut. Res.* **61** 3228–34
[11] Tünay O, Kabdasli I, Orhon D and Kolçak S 1997 *Water Sci. Technol.* **36** 225–228
[12] Zang G-L, Sheng G-P, Li W-W, Tong Z-H, Zeng R J, Shi C and Yu H-Q 2012 *Phys. Chem. Chem. Phys.* **14** 1978–84
[13] Wilsenach J A, Schuurbiers C A H and van Loosdrecht M C M 2007 *Water Res.* **41** 458–466
[14] Celen I and Türker M 2001 *Environ. Technol.* **22** 1263–72
[15] Yoshino M, Yao M, Tsuno H and Somiya I 2003 *Water Sci. Technol. J. Int. Assoc. Water Pollut. Res.* **48** pp 171–178
[16] Pastor L, Mangin D, Ferrer J and Seco A 2010 *Bioresour. Technol.* **101** 118–125
[17] Song Y-H, Qiu G-L, Yuan P, Cui X-Y, Peng J-F, Zeng P, Duan L, Xiang L-C and Qian F 2011 *J. Hazard. Mater.* **190** 140–149
[18] APHA 2005 *Standard methods for the examination of water and wastewater.* (Washington D.C, USA: American Public Health Association)
[19] Owen W F, Stuckey D C, Healy Jr. J B, Young L Y and McCarty P L 1979 *Water Res.* **13** 485–492
[20] Chynoweth D P, Turick C E, Owens J M, Jerger D E and Peck M W 1993 *Biomass Bioenergy* **5** 95–111
[21] Hansen T L, Schmidt J E, Angelidaki I, Marca E, Jansen J la C, Mosbæk H and Christensen T H 2004 *Waste Manag.* **24** 393–400
[22] Le Hyaric R, Canler J-P, Barillon B, Naquin P and Gourdon R 2009 *Water Sci. Technol. J. Int. Assoc. Water Pollut. Res.* **60** 525–531
[23] Huber H, Tanik A B and Gerçek M 1995 *Water Sci. Technol.* **32** 265–271
[24] Bode H and Imhoff K R 1996 **33** 219–228
[25] Clay S, Hodgkinson A, Upton J and Green M 1996 *Water Sci. Technol.* **33** 229–234
[26] Tchobanoglous G and Kreith F 2002 *Handbook of Solid Waste Management* (McGraw-Hill , NewYork)
[27] Mata-Alvarez J, Cecchi F, Pavan P and Llabres P 1990 *Biol. Wastes* **33** 181–199
[28] Zhang D, Chen Y, Jilani G, Wu W, Liu W and Han Z 2012 *Bioresour. Technol.* **116** 386–395
[29] El-Mashad H M and Zhang R 2010 *Bioresour. Technol.* **101** 4021–28
[30] Labatut R A, Angenent L T and Scott N R 2011 *Bioresour. Technol.* **102** 2255–64

[31] Fongsatitkul P, Elefsiniotis P and Wareham D G 2012 *Waste Manag. Res. J. Int. Solid Wastes Public Clean. Assoc. ISWA* **30** 720–726
[32] Gunaseelan V N 1994 *Biomass Bioenergy* **6** 391–398
[33] Demirer G and Chen S 2004 E *J. Chem. Technol. Biotechnol.* **79** 1381–87

Enhancement of activated sludge disintegration and dewaterability by Fenton process

G C Heng[1] and M H Isa
Department of Civil and Environmental Engineering,
Universiti Teknologi PETRONAS, 32610 Bandar Seri Iskandar, Perak, Malaysia

E-mail: chgan26@gmail.com, hasnain_isa@petronas.com.my

Abstract. Municipal and industrial wastewater treatment plants produce large amounts of sludge. This excess sludge is an inevitable drawback inherent to the activated sludge process. In this study, the waste activated sludge was obtained from the campus wastewater treatment plant at Universiti Teknologi PETRONAS (UTP), Malaysia. Fenton pretreatment was optimized by using the response surface methodology (RSM) to study the effects of three operating conditions including the dosage of H_2O_2 (g H_2O_2/kg TS), the molar ratio of H_2O_2/Fe^{2+} and reaction time. The optimum operating variables to achieve MLVSS removal 65%, CST reduction 28%, sCOD 11000 mg/L and EPS 500 mg/L were: 1000 g H_2O_2/kg TS, H_2O_2/Fe^{2+} molar ratio 70 and reaction time 45 min. Fenton process was proved to be able to enhance the sludge disintegration and dewaterability.

1. Introduction

Sludge production in the industrialized world has steadily been increasing during the last decades due to biological processes for treating wastewater [1]. Though these processes are effective in treating wastewater to ensure minimum residual impact on the aquatic environment, they have the serious drawback of producing huge amounts of excess sludge. With more stringent disposal regulations, this has caused a demand for more efficient sludge conditioning and dewatering techniques to reduce the volume of sludge in order to lower cost of processing as well as transport and disposal of sludge.

Pretreatment methods are attracting much attention for their suitability to alter the structure and composition of the biomass and hence enhance sludge stabilization. These methods disrupt cell walls which result in lysis or disintegration of sludge cells. Organic materials are converted to low molecular weight and readily biodegradable compounds, thus bypassing the rate-limiting hydrolysis stage. Different sludge disintegration methods have been tested such as acid/alkaline treatment [2–4], chemical oxidation [5,6], sonication [4], thermal treatment [4,7] and electrical treatment [8]. Fenton process is an advanced oxidation process (AOP) that utilizes activation of ferrous (Fe^{2+}) ion and hydrogen peroxide (H_2O_2) to catalyze the formation of hydroxyl radicals (•OH) in acidic condition according to Reaction [9]. Past studies showed pH 3 to be optimum pH for Fenton treatment [10,11].

$$Fe^{2+} + H_2O_2 \longrightarrow Fe^{3+} + \bullet OH + OH^- \quad (1)$$

---

* To whom any correspondence should be addressed.

Fenton process has been applied on activated sludge [11–14]. In Fenton process, the microbial cells in sludge burst and the bound water is released; hence improving sludge dewaterability. A common characteristic of different sludges is the very high water content. Consequently, a reduction of sludge volume is generally required before handling and disposal. Due to their relatively low energy consumption, mechanical processes such as filter presses or centrifuges are often used for dewatering, albeit with low efficiency. Sludge dewatering is difficult mainly because of the colloidal nature of particles and the gel-like structure of the flocculated system known as extracellular polymeric substances (EPS) [15]. Proteins and polysaccharides are two of the predominant organic components of EPS; the total mass has been found to represent up to 80% of the mass of activated sludge [16]. Considering the high water binding properties of EPS, it is concluded that the dewatering efficiency of activated sludge can be increased by degrading the EPS. The objective of this paper was to investigate the effectiveness of Fenton process in enhancement of sludge disintegration and dewaterability.

2. Materials and methods

In this study, waste activated sludge (WAS) was obtained at the return pipe from secondary clarifier sludge hopper of a municipal wastewater treatment plant in Universiti Teknologi PETRONAS. The sludge samples were thickened to the required solids concentration and stored in the cold room prior to use. The characteristics (range from triplicate samples) of the raw samples are presented in Table 1.

Table 1. Sludge characteristics

Parameter	Unit	Value (mean)
pH	-	6.27
Total solids (TS)	mg/L	29470
Volatile solids (VS)	mg/L	18270
Mixed liquor suspended solids (MLSS)	mg/L	25391.3
Mixed liquor volatile suspended solids (MLVSS)	mg/L	16000
Soluble chemical oxygen demand (sCOD)	mg/L	1154.4
Capillary suction time (CST)	s	120
Extracellular polymeric substances (EPS)	mg/L	395.2

2.1. Experimental procedures and data analysis

The Fenton process was performed in a 500-mL Pyrex reactor using 250 mL of the activated sludge at pH 3. Ferrous sulfate ($FeSO_4 \cdot 7H_2O$) and hydrogen peroxide (H_2O_2) were added according to the selected dosage of H_2O_2 (g H_2O_2/kg TS) and the molar ratio of H_2O_2/Fe^{2+} simultaneously. The mixture was stirred continuously to ensure complete homogeneity during reaction. Aliquots were taken at selected reaction time, adjusted to pH above 10 with sodium hydroxide and mixed for 10 min to avoid interference by H_2O_2 in COD determination [17]. Design expert software (version 6.0.7) was used for statistical design of experiment and data analysis. Central composite design (CCD) and response surface methodology (RSM) were applied to optimize the three important operating variables of the Fenton process: dosage of H_2O_2 (g H_2O_2/kg TS), molar ratio of H_2O_2/Fe^{2+} and reaction time. A total of 20 experimental runs were set and the experimental data were fitted to the empirical second order polynomial model of a suitable degree for the optimum conditions of sludge treatment by the Fenton process. The study ranges were chosen as 464-1136 g H_2O_2/kg TS, 26-94 molar ratio of H_2O_2/Fe^{2+} and reaction time 34.8-85.2 min based on the preliminary tests. In order to obtain the optimum operating conditions, four parameters were analyzed as responses: MLVSS removal (%), CST reduction (%), sCOD concentration and EPS (total protein and polysaccharides) concentration. Regression analyses, graphical analyses and analyses of variance (ANOVA) were conducted using the design expert software. The optimum region was identified based on the main parameters in the overlay plot.

IOP Conf. Series: Earth and Environmental Science **36** (2016) 012018 doi:10.1088/1755-1315/36/1/012018

2.2. Analytical methods

Solids analysis was made according to Standard Methods [18]. sCOD was measured by the Reactor Digestion HACH Method No. 8000 [19]. Capillary suction time (CST) was measured using Triton type 319 Multi-CST (Triton Electronics Ltd.). Protein and polysaccharide concentrations were measured by Bicinchoninic acid (BCA) method [20] and sulfuric acid-UV method [21], respectively.

3. Results and Discussion

The effect of Fenton treatment was studied to evaluate the optimum operating conditions in terms of organic matter solubilization and hence to improve the disintegration and dewaterability of sludge.

3.1. Statistical analysis

ANOVA results for the response surface quadratic model are shown in Table 2. Adequate precision (AP) compares the range of the predicted values at the design points with the average prediction error. Ratios greater than 4 indicate adequate model discrimination and justify navigation of the design space defined by the CCD. AP for all the responses was greater than 4 in this study. The coefficient of variance (CV) is the ratio of the standard error of estimate to the mean value of the observed response and defines reproducibility of the model. A model normally can be considered reproducible if its CV is not greater than 10% [22]. All of the responses are less than 10% except CST reduction. The R^2 value gives the proportion of the total variation in the response predicted by the model. The R^2 value was found to be close to 1 indicating that the regression model explained the prediction well [23].

3.2. Process analysis

The response surface plots for MLVSS removal, CST reduction, sCOD and EPS are shown as two-dimensional contour plots on the dosage of H_2O_2/kg TS and H_2O_2/Fe^{2+} molar ratio at reaction time 1 h (Figure 1). The maximum MLVSS removal, maximum CST reduction, maximum sCOD and minimum EPS were 70.2%, 34.7%, 11627.4 mg/L and 423.2 mg/L at about 600-1000 g H_2O_2/kg TS and H_2O_2/Fe^{2+} molar ratio 60-80 at 1 h reaction time. Figure 1(a) shows MLVSS removal was increasing with the increase in H_2O_2 dose and H_2O_2/Fe^{2+} molar ratio. This was presumably due to the fact that increased H_2O_2 dose produced more hydroxyl radicals leading to higher substrate degradation [24]. On the other hand, higher molar ratio of H_2O_2/Fe^{2+} consumes lesser ferrous salts and hence contributes lesser iron sludge. Figure 1(b) shows the CST reduction reached the maximum value of about 35% at 800-900 g H_2O_2/kg TS and H_2O_2/Fe^{2+} molar ratio 60-70 and reaction time 1h. Further increase of H_2O_2 dose or Fe^{2+} did not improve the removal efficiency. This was due to scavenging of •OH radicals (Reaction 2) [17]. This leads to the production of hydroperoxyl radical, a species with much weaker oxidizing power compared to hydroxyl radical [25]. On the other hand, excess Fe^{2+} and organic compounds compete to react with hydroxyl radicals and thus affect the efficiency in terms of MLVSS removal and sCOD concentration as in Reaction 3 [26].

$$\bullet OH + H_2O_2 \longrightarrow HO\bullet_2 + H_2O \qquad (2)$$

$$Fe^{2+} + \bullet OH \longrightarrow Fe^{3+} + OH^- \qquad (3)$$

Table 2. ANOVA for response surface quadratic model

Response	AP	CV	R^2
Response	11.529	6.65	0.8854
MLVSS removal (%)	10.757	34.86	0.9194
CST reduction (%)	9.658	5.67	0.8864
sCOD	10.903	8.87	0.9162
EPS	11.529	6.65	0.8854

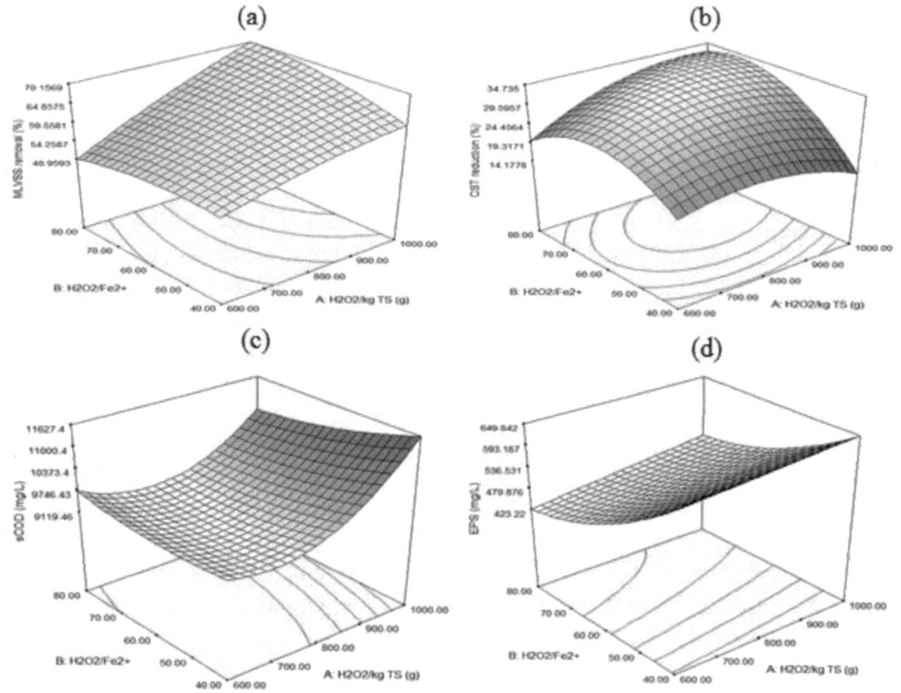

Figure 1. Response surface plot of (a) MLVSS removal, (b) CST reduction, (c) sCOD (d) EPS as a function of H_2O_2/kg TS and H_2O_2/Fe^{2+} molar ratio at reaction time 1 h

Figures 1(c) and 1(d) show that sCOD and EPS concentrations increased with increase of H_2O_2 dose due to release of cellular substances and EPS from activated sludge into the aqueous phase. Sludge dewaterability deteriorates with the increase of the EPS concentration. Higher dosages not only disrupt floc structures and cell membranes but also reduce floc size and increase the concentration of EPS [8]. While lower dosages would only destroy the sludge surface and some organic substance would be degraded. The effect of reaction time (results not shown) shows that the MLVSS removal and EPS concentration increased when the reaction time increased. However, CST reduction and sCOD decreased with the increase of reaction time. Soluble organic matters (mainly protein and polysaccharides) in the inner fraction of sludge flocs could become soluble and released to the slime owing to the pretreatment, resulting in deterioration of sludge dewaterability [8]. It could be that soluble EPS had strong water binding capability and affect sludge dewaterabilty. Increase of reaction time above 1 h did not reduce sCOD significantly due to rapid breakdown of organics within 1 h.

3.3. Process optimization

With multiple responses, the optimum operating variables where all parameters simultaneously meet the desirable response criteria could be visualized graphically by superimposing the contours of the response surfaces in an overlay plot. Graphical optimization displays the area of feasible response value in the factor space and the regions that do fit the optimization criteria would be shaded. In order to obtain a moderately precise optimum zone, response limits as the minimum permissible values were chosen for each parameter close to their acquired levels – MLVSS removal 65%, CST reduction 28%, sCOD 11000 mg/L and EPS 500 mg/L (Figure 2). The shaded region shows the optimum parameters – 1000 g H_2O_2/kg TS, H_2O_2/Fe^{2+} molar ratio 70 and reaction time 45 min and constitute the optimum operating variables (conditions). The results agreed well with literature; optimum dosage of H_2O_2/kg

TS in the range 400-1000 [11,12,14] and optimum H_2O_2/Fe^{2+} molar ratio at 40 [14] reported for Fenton treatment of activated sludge.

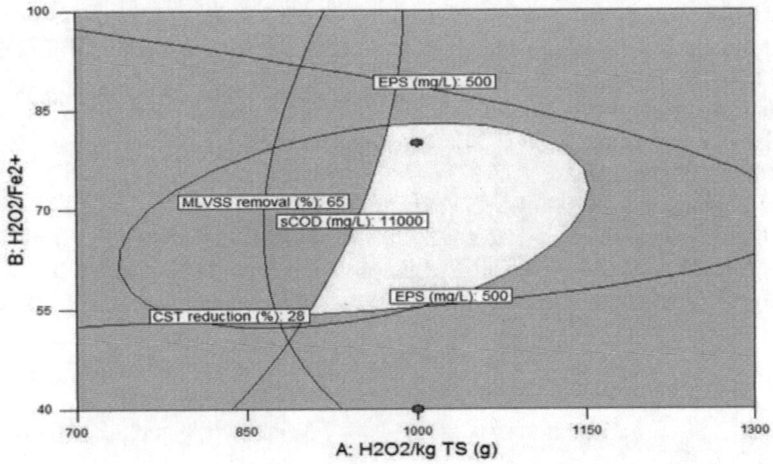

Figure 2. Overlay plot for optimal region at reaction time 45 min

4. Conclusions

CCD and RSM were used with three operating variables viz. dosage of H_2O_2 (g H_2O_2/kg TS), molar ratio of H_2O_2/Fe^{2+} and reaction time to investigate their interactive effects on the MLVSS removal, CST reduction, sCOD and EPS concentration from activated sludge samples by the Fenton process. The optimum operating variables to achieve MLVSS removal 65%, CST reduction 28%, sCOD 11000 mg/L and EPS 500 mg/L were: 1000 g H_2O_2/kg TS, H_2O_2/Fe^{2+} molar ratio 70 and reaction time 45 min. The study has revealed that RSM is a useful tool to optimize the treatment process and Fenton is an effective pretreatment of activated sludge for sludge disintegration and dewaterability.

References

[1] Feng X, Deng J, Lei H, Bai T, Fan Q and Li Z 2009 *Bioresour. Technol.* **100** 1074–81
[2] Liao B Q, Allen D G, Leppard G G, Droppo I G and Liss S N 2002 *J. Colloid Interface Sci.* **249** 372–380
[3] Chen Y, Yang H and Gu G 2001 *Water Res.* **35** 2615–20
[4] Apul O G, Atalar I, Zorba G T and Sanin F D 2010 *Dry. Technol.* **28** 901–909
[5] Neyens E, Baeyens J, Dewil R and De heyder B 2004 *J. Hazard. Mater.* **106** 83–92
[6] Kwon J H, Ryu S H, Park K-Y, Yeom I-T and Ahn K-H 2001 *J. Chin. Inst. Chem. Eng.* **32** 555–558
[7] Neyens E and Baeyens J 2003 *J. Hazard. Mater.* **98** 51–67
[8] Yuan H, Zhu N and Song F 2011 *Bioresour. Technol.* **102** 2308–15
[9] Yoon J, Lee Y and Kim S 2001 *Water Sci. Technol. J. Int. Assoc. Water Pollut. Res.* **44** 15–21
[10] Mohajeri S, Aziz H A, Isa M H, Zahed M A and Adlan M N 2010 *J. Hazard. Mater.* **176** 749–758
[11] Şahinkaya S, Kalıpcı E and Aras S 2015 *Process Saf. Environ. Prot.* **93** 274–281
[12] Dewil R, Baeyens J and Neyens E 2005 *J. Hazard. Mater.* **117** 161–170
[13] Erden G and Filibeli A 2010 *Desalination* **251** 58–63
[14] He M and Wei C 2010 P *J. Hazard. Mater.* **176** 597–601
[15] Dursun D and Dentel S K 2009 *Water Sci. Technol. J. Int. Assoc. Water Pollut. Res.* **59** 1679–85
[16] Frølund B, Palmgren R, Keiding K and Nielsen P H 1996 *Water Res.* **30** 1749–58

IOP Conf. Series: Earth and Environmental Science **36** (2016) 012018 doi:10.1088/1755-1315/36/1/012018

[17] Andreozzi R, Canterino M, Marotta R and Paxeus N 2005 *J. Hazard. Mater.* **122** 243–250
[18] APHA 2005 *Standard methods for the examination of water and wastewater.* (Washington D.C, USA: American Public Health Association)
[19] HACH 2003 *Water analysis handbook* (Loveland, CO, USA: Hach Company)
[20] Zuriaga-Agustí E, Bes-Piá A, Mendoza-Roca J A and Alonso-Molina J L 2013 *Sep. Purif. Technol.* **112** 1–10
[21] Albalasmeh A A, Berhe A A and Ghezzehei T A 2013 *Carbohydr. Polym.* **97** 253–261
[22] Beg Q K, Sahai V and Gupta R 2003 *Process Biochem.* **39** 203–209
[23] Ölmez T 2009 *J. Hazard. Mater.* **162** 1371–78
[24] Deng Y and Englehardt J D 2006 *Water Res.* **40** 3683–94
[25] Ting W-P, Lu M-C and Huang Y-H 2009 *J. Hazard. Mater.* **161** 1484–90
[26] Mohajeri S, Aziz H A, Isa M H, Bashir M J K, Mohajeri L and Adlan M N 2010 *J. Environ. Sci. Health Part A Tox. Hazard. Subst. Environ. Eng.* **45** 692–698

Synthesis, Characterization, and Catalytic Activity of Sulfonated Carbon-Based Catalysts Derived From Rubber Tree Leaves and Pulp and Paper Mill Waste

J Janaun[1], E Sinin, S F Hiew, A M T Kong, F A Lahin
Catalysis, Reaction Engineering & Drying Technology Research Group (CREDYT), *Universiti Malaysia Sabah*, Jalan UMS, 88400, Kota Kinabalu, Sabah, Malaysia

E-mail: [1]jidon@ums.edu.my

Abstract. Sulfonated carbon-based catalysts derived from rubber tree leaves, and pulp and paper mill waste were synthesized and characterized. Three types of catalyst synthesized were sulfonated rubber tree leaves (S-RTL), pyrolysed sludge char (P-SC) and sulfonated sludge char (S-SC). Sulfonated rubber tree leaves (S-RTL) and sulfonated sludge char (S-SC) were prepared through pyrolysis followed by functionalization via sulfonation process whereas, P-SC was only pyrolyzed without sulfonation. The characterization results indicated sulfonic acids, hydroxyl, and carboxyl moieties were detected in S-RTL and S-SC, but no sulfonic acid was detected in P-SC. Total acidity test showed S-RTL had the highest value followed by S-SC and P-SC. The thermal stability of S-RTL and S-SC were up to 230°C as the loss was associated with the decomposition of sulfonic acid group, whereas, P-SC showed higher stability than the S-RTL and S-SC. Morphology analysis showed that S-RTL consisted of an amorphous carbon structure, and a crystalline structure for P-SC and S-SC. Furthermore, traces of metal components were also detected on all of the catalysts. The catalyst catalytic activity was tested through esterification of oleic acid with methanol. The results showed that the reaction using S-RTL catalyst produced the highest conversion (99.9%) followed by P-SC (88.4%) and lastly S-SC (82.7%). The synthesized catalysts showed high potential to be used in biodiesel production.

Introduction

Biodiesel or fatty acid methyl ester (FAME) that derived from vegetables oils or animal fats is a great alternative to fossil fuel. Due to the increasing price of petroleum and environmental concerns such as greenhouse effect and emission of particulate matter [1], alternative renewable fuels such as biodiesel have gained much attention.

In biodiesel synthesis, the reaction is facilitated with a suitable catalyst. Generally, the reaction on the production of biodiesel can be done with or without the presence of catalyst [2]. There are different types of catalyst that can be used in biodiesel production, which are homogeneous and

1 To whom any correspondence should be addressed.

heterogeneous catalyst. Heterogeneous catalyst such as sulfonated carbon-based solid acid catalyst has recently reported as promising catalyst for biodiesel production [3].

Typically, base catalyst is more reactive than acid catalyst for biodiesel production. Nevertheless, base catalyst is suitable only for high quality feedstock (low FFA and water content) as opposed to acid catalyst where it can be used for low grade feedstock. This attributes promote the use of acid catalyst for large scale biodiesel production. It is more economical to produce biodiesel at a lower cost and this can be achieved by using low quality feedstock with higher catalytic activity. Low quality feedstocks which primarily contains high free fatty acids and water content, hence, producing a solid acid catalyst is ideal since it has the ability to carry out both the esterification of free fatty acid and transesterification of triglycerides simultaneously [2].

Solid acid catalyst, mainly sulfonated carbon-based catalyst can be prepared at a lower cost, environmental friendly and renewable therefore making catalysts such as sulfonated carbon-based catalyst, with a high conversion and low cost of production from renewable sources is needed for the production of biodiesel.

In this study, the readily available resources in large quantity, rubber tree leaves and pulp and paper mill waste, were utilized as the carbon precursors to synthesize sulfonated carbon-based catalysts. Their catalytic activity was tested on the esterification of oleic acid with methanol. The catalysts were characterized using modern analytical tools in order to understand their catalytic activity.

Preparation of Catalyst and Reactivity Test

1.1. Preparation of Carbon Supports

The samples were initially prepared according to an established method by [4] to prepare amorphous structured carbon support. All samples were pyrolysed in a Thermolyne 79300 Tube Furnace using Eurotherm 2116 Controller under 100 mL/min nitrogen atmosphere at 400°C for 4 h. After that, the resultants were grounded and sieved.

1.2. Functionalization of Chars via Sulfonation

The carbon char was then functionalized by heating in fuming sulfuric acid (20 wt.% free SO_3; 40 mmol loading for 20.0 g of char) at 150°C for 12 hours under nitrogen atmosphere at the flow rate of 100 mL/min. The resultants were then washed with hot distilled water until it reached constant pH, and there was no sulfate ion detected in the washed water. The samples were then dried at 100°C overnight to remove moistures from the samples using Memmert Universal Oven Model 400. The samples were denoted as Sulfonated Rubber Tree Leaf (S-RTL), Sulfonated Suldge Char (S-SC) and Pyrolyzed Sludge Char (P-SC). P-SC sample had not undergone the functionalization process.

1.3. Catalysts Characterization

The catalyst samples were characterized for their chemical and physical properties. The catalyst total acidity was measured using acid (0.02M HCl)-base (0.01M NaOH) back-titration method. The functional groups of the char were determined by using Fourier Transform Infrared Spectrometry (FTIR). The catalyst was scanned at a range from 650 cm^{-1} to 4000cm^{-1} on Perkin Elmer spectrum 100. Energy-dispersive X-ray microanalysis (EDX) (Bruker Quantax Esprit 1.0) used to determine the composition of the features in the SEM image with the target composition configuration setting of Carbon, Sulphur, and Oxygen. The analysis molecular structure was determined using Philip Expert pro X-ray diffraction (XRD) (Philips Expert Pro pw3040) analysis using Cu-Kα radiation source scanned from angle 5 to 80 (2Θ). Besides that, Z-5000 Polarized Zeeman Atomic Absorbance Spectrophotometer (AAS) was used for the analysis of metal elements for the catalyst as there were five standard solutions selected metal elements prepared which includes calcium, potassium, magnesium, zinc and sodium. The porosimetry analysis were carried out using Automated Mercury Porosimeter by Thermo Pascal 440 Series as the samples tested from atmospheric to 200Mpa pressure using medium gradual pressure increase and decrease. Determination of thermal stability was studied using Thermal Gravimetric Analysis (TGA) as each catalysts sample was analysed at a temperature rate of 10°C/min until it reached 400°C under nitrogen atmosphere.

1.4. Esterification Reaction of Oleic Acid with Methanol

In this study, oleic acid was used as FFA model feedstock for the reactivity test. The molar ratio of oleic acid to methanol used was 1:10 with catalyst loading of 0.3wt% with respect to the oleic acid, stirred at 800 rpm for 5 hours. The reaction temperature was conducted at 60°C. Samples were analysed using titration method based on its relation to acid value.

1.5. Acid Value

The concentration of the free fatty acid can be determined using titrimetric method by determining its corresponding acid value to the real concentration of oleic acid. In this study, 0.1M Potassium Hydroxide (KOH) was used as titrant while diethyl-ether and ethanol was used as the solvent with a mixture of diether-ether:ethanol (v:v) (1:1) using phenolphthalein as the end-point indicator. Result of titration is expressed in form of acid value. The equation of acid value is shown below:

$$\text{Acid Value} = (\text{A} - \text{B}) \times \frac{56.1\text{g}}{\text{mol}} \times \frac{\text{M}}{\text{m}} \tag{1}$$

(A-B) is the volume of the titrant used (ml), **M** is the concentration of KOH solution (mol/L) while **m** is the weight of sample (g). To calculate the conversion of oleic acid to biodiesel, the result can be further expressed in term of conversion percentage. The conversion equation used shown below:

$$\text{Conversion}\,(\%) = \frac{\text{Initial acid number} \times \text{Final acid number}}{\text{Initial acid number}} \times 100\% \tag{2}$$

Results and Discussion

1.6. Characterization of The Catalysts

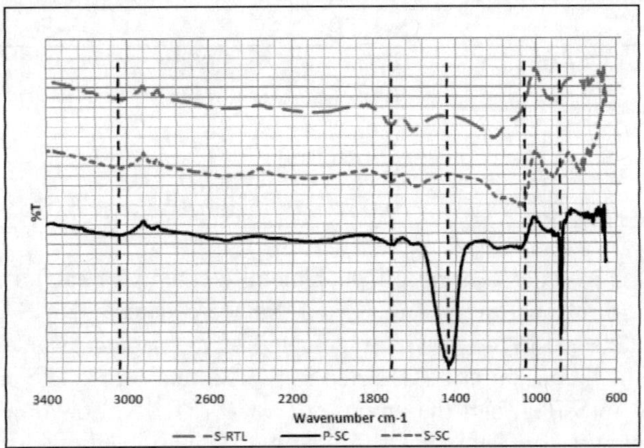

Figure 1. FTIR Spectra of S-RTL, P-SC and S-SC

Figure 1 shows the overall result of the FTIR for S- RTL, P-SC and S-SC. All samples showed bands at around ~1600 cm^{-1} which can be assigned to C=C stretching [5]. Moreover, a broad band was observed in the region of 3400 cm^{-1} which were assigned to the –OH stretching modes of the –COOH and -OH groups. According to Méndez [6], the band between 3200 cm^{-1} to 3600 cm^{-1} could be attributed to the presence of hydroxyl groups of cellulose. This result suggested the presence of OH group in the catalyst. Besides that, the band at around ~1700 cm^{-1} in S-RTL and S-SC was assigned to the C=O starching mode of the COOH group [7], but it can be observed it is weak band at around ~1700 cm^{-1} for P-SC. Bands at 1010 cm^{-1} and 1004 cm^{-1} for S-RTL and S-SC can be attributed to the presence of S=O stretching [8]. Both samples showed bands at 1716 cm^{-1} attributed to the presence of SO_3H stretching. Therefore, FT-IR analysis showed that, there were SO_3H, -COOH, and –OH present in S-RTL and S-SC which was in agreement with literature [9] and [7]. However, SO_3H was not detected in P-SC only -COOH, and –OH moieties.

Table 1. Summary of Total Acidity of S-RTL, P-SC and S-SC

Catalyst	Total Acidity (mmol/g)
S-RTL	3.80 ± 0.05
P-SC	0.50 ± 0.02
S-SC	2.40 ± 0.04

Table 1 shows the summary of total acidity analysis by back-titration method. Based on the results in Table 1, S-RTL has the highest total acidity followed by S-SC, which is 3.80mmol/g and 2.40mmol/g, and P-SC with only 0.50mmol/g. According to Lee [10], it was mentioned that the higher value of total acidity than sulfonic group can be interpreted as due to the presence of many acid groups which include phenolic or carboxyl groups in addition to sulfonic groups in the catalysts. The high amount of total acidity in S-RTL is mainly related to the high amount of carbon content in the biomass.

On the other hand, the low amount of total acidity for P-SC can be related to the absence of sulfonic group. This was attributed to the presence of weak acid groups, which is –OH and –COOH, by the incomplete carbonization of pulp and paper mill waste (sludge). However, S-SC shows an increase in amount of total acidity after sulfonation suggesting that the presence of sulfonic group boosts the acidity of the sample.

Table 2. Average Concentration (ppm) of Metal Elements in Each Sample

Metal Elements	**Average Concentration (ppm)**		
	S-RTL	**P-SC**	**S-SC**
Ca	0.180	3.095	1.943
K	0.056	0.690	0.160
Na	0.243	6.672	1.475
Mg	0.013	0.909	0.000
Zn	0.005	0.000	0.015

Table 2 shows the average concentration of metal elements in each sample in terms of ppm analyzed using AAS. From the results, Ca, K and Na were detected in S-RTL, P-SC and S-SC. However, Mg was not detected in S-SC, which indicated by the negative value. Likewise, Zn was not detected in P-SC, but small concentration detected in S-RTL and S-SC. The composition of Ca in P-SC is higher compared to S-RTL and the amount decreases after sulfonation (S-SC). Metal species concentration reductions were obvious for Ca, K and Na. This probably due to the neutralization via sulfonation thus reduces the concentration of those particular elements.

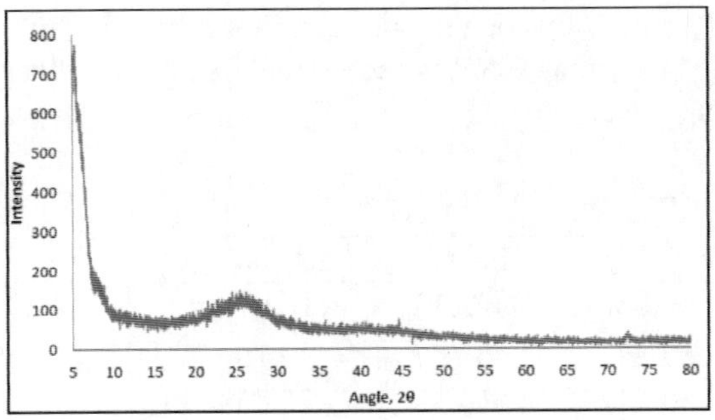

Figure 2. XRD Spectra of S-RTL

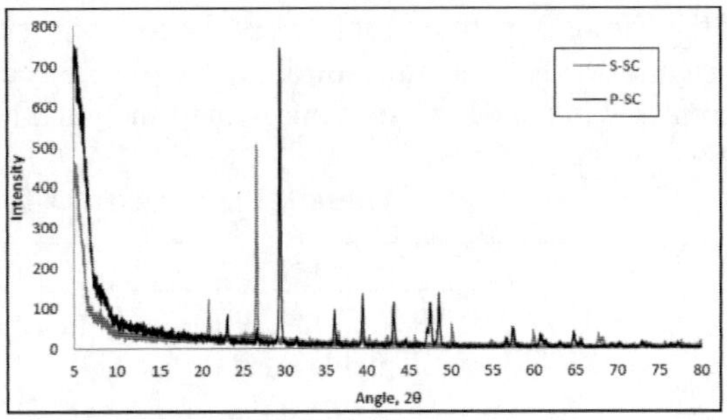

Figure 3. Superimposed XRD spectra of S-SC and P-SC

Figures 2 and 3 shows the results of X-Ray Diffraction plot analysis of S-RTL, and P-SC and S-SC, respectively. The analysis from XRD is important to determine the structure of carbon support as well as element identification. The XRD pattern exhibited diffraction peaks at 20 – 30° (2θ) for sample S-RTL in Figure 2. This pattern is highly attributed to the non-graphitic carbon and indicated that this sample comprised of high non- graphitic carbon content [3]. Likewise, upon sulfonation this value is more consistent with the presence of amorphous carbon [11]. In addition, Mar et. al, [12] also recorded similar XRD pattern with the diffraction peaks which was assigned to amorphous carbon.

On the other hand, for both P-SC and S-SC show a strongest peak at 26.59° and 29.39° as seen in Figure 3 due to origination from the cellulose crystalline plane. The intensity of the peak for the bands of crystalline cellulose at this region after P-SC has been sulfonated (into S-SC) decreases as observed in Figure 3. This is probably due to its less cellulose content and the degradation took place after functionalization process [6]. Furthermore, P-SC shows peaks at 2 = 36.01°, 39.40°, 43.21°, 47.49° and 48.53° which indicates that they were predominantly composed by calcite ($CaCO_3$) and small quantities of kaolinite as this data is in agreement with literature [6]. Likewise, the intensity of the peaks for P-SC at these regions decreases after sulfonation hence it can be inferred that the amount of $CaCO_3$ decrease after P-SC has been sulfonated (S-SC).

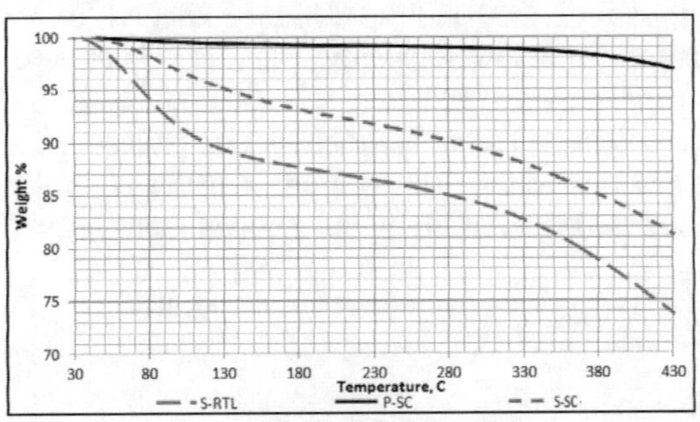

Figure 4. (a)

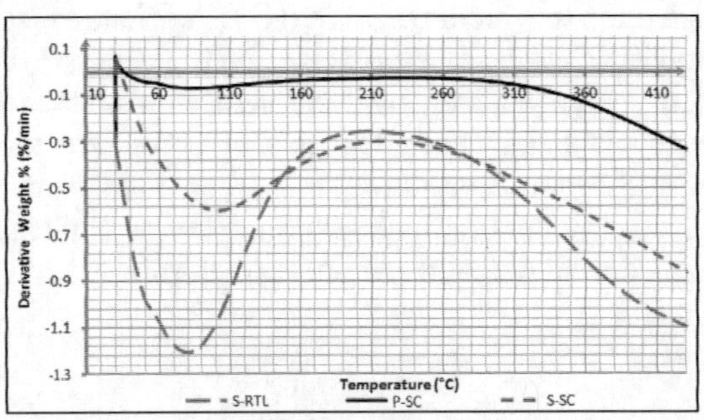

Figure 4. (b)

Figure 4. (a) Thermo-gravimetric Analysis (TGA) and (b) Derivative Thermal Analysis (DTA) of S-RTL, P-SC and S-SC under Nitrogen

Figures 4 (a) & (b) show the stability analysis graph derived on its TGA data. According to the results in Figure 4(a) it is clearly seen that, S-SC gradually starts to lose weight at 40°C followed by S-SC at 50°C and P-SC at 60°C while in Figure 4 (b), the decrease in weight of all the samples at around 90°C can be attributed to the loss of moisture [4]. Furthermore, both of the S-RTL and S-SC recorded weak second stage weight loss around 230°C though this trend was undetectable in sample P-SC. The second stage weight loss marks the first decomposition of the functional group on the surface of the carbon support interpreted as the loss of the sulfonic groups in agreement to literature [4] at that particular range of temperature. And so, it can be deduced that P-SC is more stable since the other samples starts to decompose important functional group faster.

Table 3. Porosimetry Data for S-RTL, P-SC and S-SC

Samples	S-RTL	P-SC	S-SC
Total Specific Surface Area	19.86	39.41	16.07
Total Porosity (%)	35.36	66.33	20.16
Average Pore Diameter (nm)	5104.42	2710.33	3375.53
Total Cumulative Volume (mm^3/g)	437.84	828.79	269.94

Table 3 shows the BET analysis using Automated Mercury Porosimetry in determining the total specific surface area and total porosity of S- RTL, P-SC and S-SC. P-SC sample shows the highest surface area followed by S-RTL and finally S-SC sample. The surface area of sample P-SC sample decreased after sulfonation (sample S-SC) and likewise, the same decrement trend was also recorded for the total porosity (%) of P-SC and S-SC sample as much as 46.17%. The reduction of pores and the surface area might be due to the penetration of acid groups into the surface of the porous pyrolysed sludge char, and the impregnation of -SO_3H groups into the pores of the char [13] caused the internal pores of sample P-SC to collapse [4]. On the other hand, S-RTL was observed to have highest average pore diameter, while P-SC sample had the smallest average pore diameter compared to all of the samples. Due to these reasons; high total porosity as well as smaller average pore diameter, caused the P-SC sample to have higher total specific surface area as well as high total cumulative volume in contrast to S-RTL and S-SC.

1.7. Catalytic Activity of The Catalysts

Biodiesels are mainly comprised of many Fatty Acid Methyl Esters (FAME) components such as Methyl Palmitate, Methyl Oleate, Methyl Myristate, and many more. These components can be derived from their respective triglycerides or fatty acids through transesterification or esterification process respectively. In this study, Oleic Acid was used as a model compound to represent the production of biodiesel through esterification reaction with methanol to produce methyl oleate (biodiesel) catalysed by S-RTL, P-SC and S-SC. Equation (3) below shows the chemical equation of esterification reaction by oleic acid with methanol in the presence of a catalyst to produce methyl oleate and water as adapted from [4]. From equation (3), it can be seen that the reaction pathways for esterification is reversible. To overcome and push the reaction to the desired product pathway, which is the right side, Le Chatelier principle was applied by increasing the molarity of methanol ten times of the oleic acid during the reactivity analysis.

R - $CH_3(CH_2)_7CH{=}CH(CH_2)_7$ (3)

The esterification reaction of oleic acid with methanol requires 1 molar of oleic acid to produce 1 molar methyl oleate and hence through this information correlated with equation (1) and (2), the conversion of methyl oleate can be associated with the formation of oleic acid during the reactivity test. Figure 5 shows the conversion percentage (%) of all the catalysts throughout the 5h reaction timespan. All of the conversions data were mathematically fitted using several mathematical models through computational means. The software used for the model fitting was OriginPro version 8.6. All of the lines were fitted with nonlinear curve fit method where S-RTL line was fitted growth/sigmoidal settings using Hill1 function model while samples P-SC and S-SC lines were fitted with exponential settings with ExpAssoc as its function.

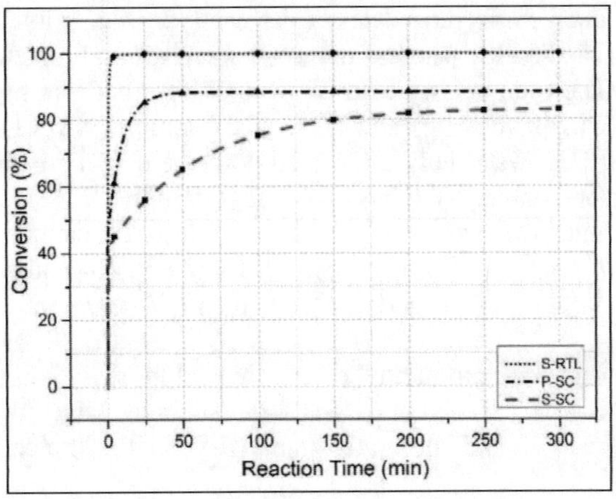

Figure 5. Graph of Conversion of Oleic Acid using S-RTL, P-SC and S-SC Catalysts

Table 4. Mathematical Fitting for Sample S-RTL, P-SC and S-SC

Catalyst	Mathematical Model Equation	Fitting (R^2)
S-RTL	$y = -0.46 + (100.36)\dfrac{x^{2.52}}{1.90 + x^{2.52}}$	0.999
P-SC	$y = -3.26\times10^{-17} + 45.83(1-e^{-x/9.35}) + 42.63(1-e^{-x/3.39\times10^{-4}})$	0.997
S-SC	$y = 5.31\times10^{-4} + 41.85(1-e^{-x/0.51}) + 41.45(1-e^{-x/0.56})$	0.990

Table 4 shows the mathematical fitting for sample S-RTL, P-SC and S-SC. The R^2 fitting value, which corresponds to the statistical measure of how close the data are to the fitted regression line, are all capped at 0.99 values thus displays high mathematical fitting for all the samples conversion lines.

By comparing all of these catalysts from the graph that was demonstrated in Figure 5, S-SC had the lowest formation rate compared to the others. All of the samples shows logarithmic conversion growth trend and reached maximum approximately at the 25th, 50th and 250th minutes of the reaction for catalyst S-RTL, P-SC and S-SC samples, respectively. The reactions were stopped after 5 hours. Among all of the catalysts, the reaction by using S-RTL contribute the highest conversion of oleic acid at the end of the reaction which are 99.86%, then P-SC (88.39 %) and lastly S-SC (82.65%).

S-RTL had the highest conversion may be associated to its highest total acidity among all other catalysts. The amorphous structure of S-RTL could have also contributed to the reactivity as amorphous structure is reported to be more flexible as opposed to the rigid structure of crystalline (P-SC and S-SC). Crystalline structure was arranged orderly compared to amorphous, hence it is very hard for the active site to attach into the catalyst. Similar result was reported by that of Toda et. Al [14].

In the FTIR analysis, the presence of sulfonic acid was detected in S-RTL and S-SC, however it was missing in P-SC. P-SC had the lowest total acidity compared to S-SC though it achieved higher conversion from the latter. This trend might be due to P-SC possessing higher total surface than S-SC.

In addition, the amount of metals on P-SC was higher than on S-SC. The presence of metal in the catalyst might have contributed to the higher catalytic reaction of P-SC.

Conclusions

Sulfonated carbon-based catalysts derived from rubber tree leaves, and pulp and paper mill waste were successfully synthesized, characterized, and tested its catalytic reactivity. Sulfonic acid moiety was detected in samples S-RTL and S-SC, whereas carboxyl and hydroxyl moieties were detected in all the samples including P-SC. The total acidity tests showed S-RTL had the highest total acidity of 3.80 mmol/g followed by S-SC (2.40 mmol/g) and P-SC (0.50 mmol/g). S-RTL and S-SC samples were thermally stable up to 230°C, it was likely associated with the decomposition of sulfonic acid group. P-SC showed thermally more stable. XRD analysis suggested that S-RTL was an amorphous carbon structure, whereas P- SC and S-SC were a crystalline structure. Porosimetry analysis showed that P-SC had highest porosity of 66.33 %, followed by S-RTL (35.36 %) and S-SC (20.16 %), while the specific surface area of S-RTL, P-SC and S-SC were 19.86 m^2/g, 39.41 m^2/g and 16.07 m^2/g, respectively.

The conversion of oleic acid into methyl ester achieved more than 80% within 5 hours using all catalysts. In conclusion, S-RTL, P-SC, and S-SC are suitable catalysts for biodiesel production because they are highly reactive and thermally stable beyond the typical temperature of biodiesel reaction.

References

[1] Atadashi I M, Aroua M K, Abdul Aziz A R and Sulaiman N M N 2012 *Renew. Sustain. Energy Rev.* **16** pp 3456–3470
[2] Janaun J and Ellis N 2010 *Renew. Sustain. Energy Rev.* **14** pp 1312–1320
[3] Shu Q, Nawaz Z, Gao J, Liao Y, Zhang Q, Wang D and Wang J 2010 *Bioresour. Technol.* **101** pp 5374–5384
[4] Janaun J 2012 *Development of sulfonated carbon catalysts for integrated biodiesel production* Doctoral dissertation (Vancouver: The University of British Columbia)
[5] Jidon Janaun N E Preparation of Mesoporous Sugar Catalyst
[6] Méndez A, Fidalgo J M, Guerrero F and Gascó G 2009 *J. Anal. Appl. Pyrolysis* **86** pp 66–73
[7] Chen G and Fang B 2011 *Bioresour. Technol.* **102** pp 2635–2640
[8] Geng L, Wang Y, Yu G and Zhu Y 2011 *Catal. Commun.* **13** pp 26–30
[9] Janaun J and Ellis N 2011 *Appl. Catal. Gen.* **394** pp 25–31
[10] Lee D 2013 *Mol. Basel Switz.* **18** pp 8168–8180
[11] Deshmane C A, Wright M W, Lachgar A, Rohlfing M, Liu Z, Le J and Hanson B E 2013 *Bioresour. Technol.* **147** pp 597–604
[12] Mar W W and Somsook E 2012 *Procedia Eng.* **32** pp 212–218
[13] Pua F, Fang Z, Zakaria S, Guo F and Chia C 2011 *Biotechnol. Biofuels* **4** pp 56
[14] Toda M, Takagaki A, Okamura M, Kondo J N and al et 2005 *Nature* **438** pp 178

Adsorption study of Ammonia Nitrogen by watermelon rind

A Ibrahim[1], L Yusof[1], N S Beddu[1], N Galasin[1], P Y Lee[1], R N S Lee[1], A Y Zahrim[1]

[1]Faculty of Engineering, Universiti Malaysia Sabah, Jalan UMS, 88400, Kota Kinabalu, Sabah, Malaysia.

E-mail: [1]azreen@ums.edu.my

Abstract. The utilization of fruit waste for low-cost adsorbents as a replacement for costly conventional methods of removing ammonia nitrogen from wastewater has been reviewed. The adsorption studies were conducted as a function of contact time and adsorbent dosage and it were carried out on four different adsorbents; fresh watermelon rind and modified watermelon rind with sodium hydroxide (NaOH), potassium hydroxide (KOH) and sulphuric acid (H_2SO_4). Adsorbents were tested for characterization by using zeta potential test and all samples shows negative values thus makes it favourable for the adsorption process. The batch experimental result showed that adsorption process is rapid and equilibrium was established within 40 minutes of contact time. The ammonia nitrogen removal rate amounted in range of 96% to 99%, and the adsorption capacities were in range of 1.21 to 1.24 mg/g for all four different types of adsorbents used.

1. Introduction

High levels of ammonia nitrogen (NH_3-N) in water can be toxic to some aquatic organisms and leads to eutrophication [1]. Hence, it is important to remove the NH_3-N from waterways and the traditional processes of it are nitrification de-nitrification [2], air stripping [3] and chemical precipitation [4]. However, extensive studies have been made to identify new economically feasible and environmentally friendly treatment alternatives, of which adsorption have received significant attention.

Adsorption is inexpensive, universal nature and ease of operation. Adsorption can remove soluble and insoluble organic pollutants due to its removal capacity may be up to 99.9% [5]. The use of fruit peels based adsorbent for removing various pollutants from wastewater offers many attractive features such as the outstanding adsorption capacity and the fact that these materials are low-cost, non-toxic and biocompatible[6]. Watermelon biomass can be categorised into three main components; the flesh, seed and rind. The flesh constitutes approximately 68% of the total weight, the rind approximately 30%, and the seeds approximately 2% [7]. In this study, watermelon rind was used as potential low-cost adsorbent. The skin of fully ripened watermelon contain approximately 20% cellulose, 23% hemicellulose, 10% lignin, 13% pectin, 7 mg/g silica and 12% silica free minerals[8].

Many studies have been conducted to investigate the ability of watermelon rind (WR) as an adsorbent and the results shows great rate of removal of various pollutants such as dye [9] and heavy

[1] To whom any correspondence should be addressed.

metals [10]. Therefore, this study was intended to evaluate the effect of contact time and adsorbent dosage on the adsorption of NH_3-N by using four different types of WR adsorbents; fresh WR, modified WR with sodium hydroxide (NaOH), potassium hydroxide (KOH) and sulphuric acid (H_2SO_4).

2. Materials and Methods

2.1 Preparation of the adsorbent

Watermelon rinds were collected from Kota Kinabalu area. The collected biomaterials were washed thoroughly using tap water and distilled water to remove any impurities such as soil or fertilizers that may exist on the surface of it. Next, the rind was cut into small pieces (around 1 to 2 mm^2) and dried in an oven at 40°C for 48 hours [11].

Watermelon rind was used as the raw material for the preparation of the chemically modified adsorbent. WR was chemically modified by pre-treatment with NaOH, KOH and H_2SO_4 solution in order to improve the capacity of ammonium (NH_4^+) ions uptake. WR with mass of 20g was treated in each of 1 L of NaOH solution (20 mmol/L), KOH solution (20 mmol/L) and of H_2SO_4 solution (20 mmol/L) respectively for 24 h [12]. Finally, the modified WR was dried in an oven at 40°C for 48 h.

2.2 Preparation of ammonia nitrogen solution

Ammonium chloride salt (NH_4Cl) with calculated weight and distilled water were used for the preparation of a stock NH_3-N solution of 50 mg/L. The initial pH was adjusted to 7.0 using dilute solutions of NaOH (0.1 N) [13]. The pH of the NH_3-N was constant throughout the experiments.

2.3 Characterization of adsorbent

Zeta potentials of the adsorbent were performed in three runs by using Malven model zetasizer 2010 zeta potential analyser.

2.4 Batch equilibrium studies

Batch equilibrium studies were carried out by adding 20g of watermelon rind adsorbent into 1000mL of ammonium nitrogen solution at room temperature for 40 minutes. Each experiment consisted of two runs and the average value from both experiments was taken. The effects of contact time and adsorbent dosage on the adsorption process for all four types of adsorbents were studied. All samples were adjusted to the optimum pH prior to the addition of the adsorbent. Sample solutions were withdrawn at certain intervals in order to determine the residual concentration by using UV-vis spectrophotometer with HACH Spectrophotometer DR2010 brand at maximum wavelength of 425 nm[14]. The amount of NH_3-N adsorbed at equilibrium, qe (mg/g) was calculated as followed:

$$t = \frac{(C_0 - C_t)V}{W} \quad (1)$$

where C_0 and C_e (mg/L) are the liquid-phase concentrations of initial adsorbate and equilibrium, respectively. V is the volume of the solution (L) and W is the weight of the adsorbent (g).

2.5 Ammonia nitrogen (NH_3-N) analysis

The analysis was carried out using Nessler Method of HACH Spectrophotometer DR2010 Procedures Manual. A calibration graph was plotted in order to calculate the concentration of the residual.

3. Result and discussions

3.1 Characterization of adsorbent

3.1.1 Zeta potential of the adsorbent

The characterization of WR adsorbent was evaluated by using zeta potential and the results obtained at different pH conditions are shown in Figure 1. The unmodified and modified WR adsorbent exhibits negative zeta potential over the pH range from 4.9 to 7.0. The zeta potential values decrease significantly with increase of pH. Negative zeta potential values recorded by all four types of adsorbent should be favourable to the attraction between active sites and NH_4^+ ions [15]. Zeta potentials of unmodified or fresh WR were higher than those of modified WR with NaOH, KOH, and H_2SO_4 at different pH values. This indicates that fresh WR adsorbent has more active sites and able to adsorb more NH_4^+ ions than modified WR adsorbents do.

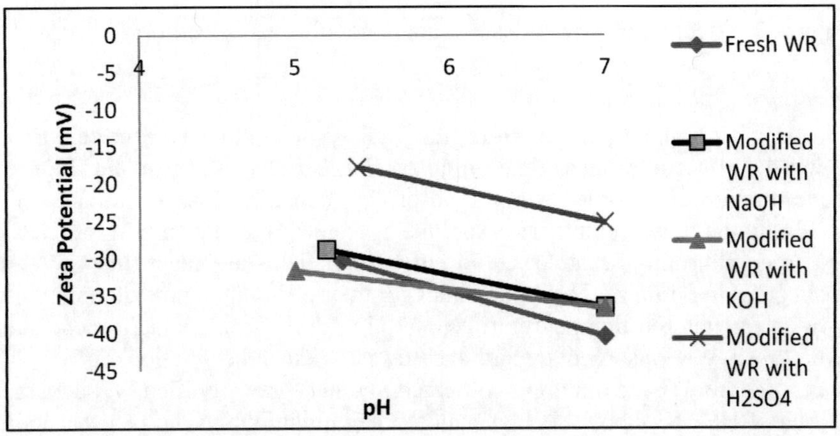

Figure 1. Zeta potentials of adsorbents

3.2 Effect of adsorbent dosage

The dependence of NH_3-N adsorption on adsorbent dose was studied by varying the amount of adsorbent used (10g, 20g, 30g, 40g) while keeping initial concentration (50 mg/L) and pH value constant. Based on Figure 2, final concentration of NH_3-N left in solution decreased from 1.2 mg/L to 0.4 mg/L as the mass of WR adsorbent increased from 10g to 20g. This indicates that the surface area and availability of adsorption sites increased as the adsorbent dose increased. However, for mass of adsorbent greater than 20 g, the NH_4^+ ions uptake decreased and final concentration left in solution were in range of 0.9 to 1.2 mg/L. This reduction may related to overlapping of adsorbent as the mass of adsorbent increased on which prevent good contact between the NH_4^+ ions and sorption sites[16] and thus decreasing the adsorption capacity of the adsorbent. It can be concluded that 20g of WR adsorbent is the optimum dosage for the adsorption process of NH_3-N in this study.

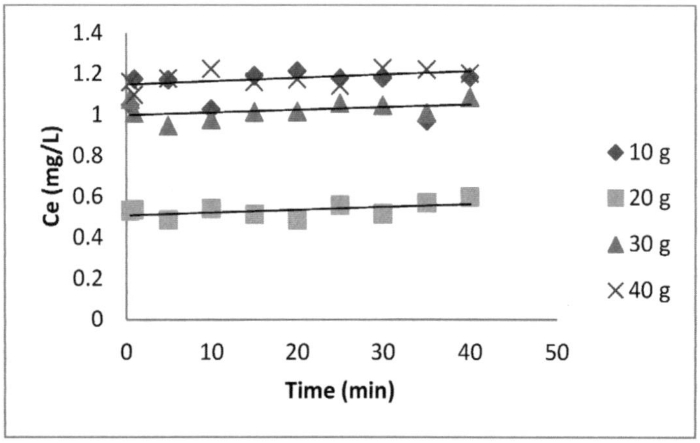

Figure 2. Effect of adsorbent dose on ammonia nitrogen adsorption

3.3 Effect of contact time

The evolution of the transfer ammonium ions from liquid phase to solid phase at certain time can be observed by studying the effect of contact time on adsorption capacity of adsorbent. Figure 3and 4 shows the final concentration of adsorbate and the adsorption capacity along 40 minutes of contact time respectively. Based on the preliminary experiment results, the adsorption of NH_3-N onto adsorbent is rapid and equilibrium is established within 40 min. Thus a contact time of 40 min was fixed for this study in order to optimize the contact times for the maximum uptake of ammonium ions.

Figure 3 shows a fast rate of ammonia nitrogen sorption during the contact time of adsorbent-adsorbate. However, fresh WR adsorbent record the lowest final concentration of NH_3-N left in solution with values range from 0.4 to 0.6 mg/L compared to other three modified WR adsorbents (0.9 to 1.5 mg/L). The rate of NH_3-N removal is higher in the beginning due to the fact that more active sites of adsorbent are available and gradual occupancies of active sites in later stage resulting slower rate of adsorption [10].

The adsorption capacity of fresh WR adsorbent was slightly higher compared to other three modified WR adsorbents as shown in Figure 4. The adsorption capacity of fresh WR was in range of 1.235-1.237 mg/g. As for modified adsorbents, the adsorption capacities were in range of 1.210-1.225 mg/g. The adsorption capacity of adsorbent decreased after 40 minutes of contact time and this may due to the presence of other co-cation and complex forming anions in the solution [17].

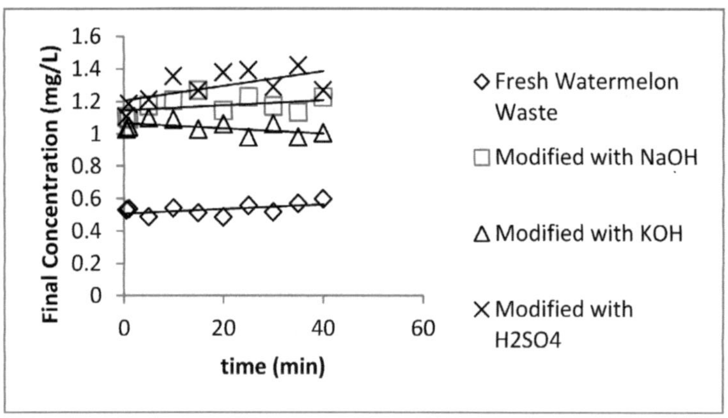

Figure 3. Effect of time on ammonia nitrogen adsorption

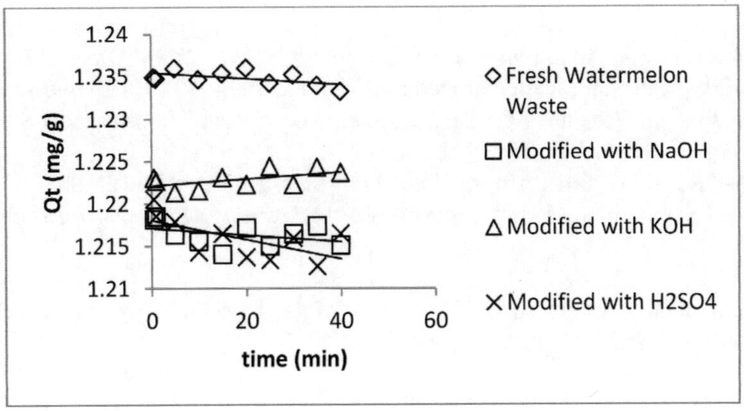

Figure 4. Adsorption capacity against contact time

4. Conclusion

The adsorption behaviour of NH_3-N onto fresh WR, modified WR with NaOH, KOH, and H_2SO_4 adsorbent were evaluated as a function of contact time and adsorbent dosage. The NH_3-N showed a fast rate of sorption and adsorbents were able to remove the pollutant along 40 minutes of contact time. 20g of adsorbent is the optimum dosage for removal process of NH_3-N in this study. The adsorption capacity shows not much difference for all four types of adsorbent. The adsorption capacity obtained shows not much difference between all four types of adsorbents used where all lie in range of 1.2 mg/g. the adsorption capacity of fresh WR (1.230-1.235 mg/g) is slightly higher compared to other three modified WR adsorbent (1.210-1.237 mg/g). Thus, it can be assured that WR have great potential to be used as low-cost adsorbent in ammonia nitrogen removal process.

5. References

[1] Hasan H, Abdullah S R S, Kamarudin S K and Kofli N T 2011 *World Applied Science Journal* **12** 1890-6
[2] Roshvanlo R B, Rezaee A, Hossini H and Shiri M 2014 *Health Scope* **3** 1-6
[3] He C, Wang K, Yang Y, Amaniampong P N and Wang J-Y 2015 *Environmental Science & Technology* **49** 6872-80
[4] Yigit A and Akbal F 2014 *Journal of Selcuk University Natural and Applied Science* 391-6
[5] Ali I, Asim M and Khan T A 2012 *Journal of Environmental Management* **113** 170-83
[6] Bhatnagar A, SillanpAA M and Witek-Krowiak A 2015 *Chemical Engineering Journal* **270** 244-71
[7] Hani M N F, Zahidah W W N, Saniah K and Irwani H M 2014 *J. Trop. Agric. and Fd. Sc* **42** 115-23
[8] Singh R, Kumar J and Nandpuri K 1975 *Indian Journal of Horticulture* **32** 98-101
[9] Alexander J, Jayanthi G, Lakshmipathy R, Kulasekaran A and Andal V Colour removal studies on treatment of textile dyeing effluent by Chitosan modified watermelon rind composite (CWR). In: *National conference on Nanomaterials for Environmental [NCNER-2015]*, pp 10-5
[10] Othman N and Asharuddin S M Cucumis melo Rind as Biosorbent to Remove Fe (II) and Mn (II) from Synthetic Groundwater Solution. In: *Advanced Materials Research*: Trans Tech Publ) pp 266-71
[11] Wahab M A, Boubakri H, Jellali S and Jedidi N 2012 *Journal of Hazardous Materials* **241** 101-9

[12] Feng N, Guo X and Liang S 2010 *Transactions of Nonferrous Metals Society of China* **20** s146-s52
[13] Jellali S, Wahab M A, Anane M, Riahi K and Jedidi N 2011 *Desalination* **270** 40-9
[14] Hach D 2000 Spectrophotometer Procedures Manual. 1994 *HACH Company: Loveland, CO*
[15] Boonamnuayvitaya V, Chaiya C, Tanthapanichakoon W and Jarudilokkul S 2004 *Separation and Purification Technology* **35** 11-22
[16] Gonen F and Serin D S 2012 *African Journal of Biotechnology* **11** 1250-8
[17] Reddy N A, Lakshmipathy R and Sarada N 2014 *Alexandria Engineering Journal* **53** 969-75

Acknowledgement
This project was financially supported by University Malaysia Sabah under research Project Grant FRG0420-TK-1/2015 and SGI0013.

Adsorption isotherm of non-azeotropic solution onto porous adsorbents

A Bono, N A Ramlan, S M Anisuzzaman[1], C M Chu, Y Y Farm
Chemical Engineering Programme, Faculty of Engineering,
Universiti Malaysia Sabah (UMS), Jalan UMS, 88400
Kota Kinabalu, Sabah, Malaysia

E-mail: [1]anis_zaman@ums.edu.my, [1]dr.anis.ums@gmail.com

Abstract. Adsorption isotherm is essential component in the understanding of the adsorption process. Several methods of the measurements, analysis and interpretation of adsorption from solution have been reported in the literature. Most of the measurements of adsorption isotherm from solution were involved the measurement of excess isotherm conducted at low region of sorbates concentration. Direct interpretation of excess adsorption isotherm as adsorption isotherm is always been practice. Therefore, in this work a study on the measurement of the adsorption isotherm from solution of non-azeotropic organic solvent mixture onto porous adsorbents for whole range of liquid concentration was conducted. The study included the measurement of excess adsorption isotherm using conventional technique. Theoretical analysis and interpretation of adsorption isotherm from the excess isotherm were conducted using Pseudo Ideal Adsorption, Gibbs Dividing Plane Model and Langmuir-Fruendlich binary isotherm model. For organic solvents, acetone and propanol were chosen as the adsorbates due to the non-azeotropic properties in the mixture. Activated carbon and silicalite were chosen as adsorbents due to the different in their porosity such as macro porous and micro porous structure. The result of the study has revealed that the adsorption isotherm of non-azeotropic mixture onto activated carbon and silicalite can be interpreted as monolayer type of adsorption.

1. Introduction

Adsorption process is a well-established technology in the separation process due to its relative simplicity of design, insensitivity to toxic substances, ease of regeneration and low cost [1]. Adsorption process have been widely employed in the industrial applications such as removal of unwanted component from a solution [2], removing dye from colored textile wastewater [4], removing color and taste [5], [6] and also phenol removal from water and wastewater [7]. In order to utilize the technology of adsorption process, knowledge in adsorption isotherm and adsorption kinetic is essential. The information of adsorption isotherm is always required for the basis of analysis, design and prediction of adsorption processes[8]. For example, adsorption isotherm, plays an important role in predictive modeling for analysis and design of adsorption systems [3,9], and gives useful information, such as determination of surface area, volume of pores, size distribution of adsorbent [10], description of the adsorbates interacts with adsorbent and thus optimizing the adsorption of

[1] To whom any correspondence should be addressed.

adsorbent [3,11] and determination information of relative adsorptivity of a gas or a vapor on a given adsorbent with respect to chosen standards [12].

Therefore precise measurement and accurate analysis of adsorption isotherm is vital key of success in the prediction of realistic industrial scale adsorption processes. Many measurement and models of adsorption isotherm are reported in the literatures [8,11], [13–15]. Most of measurements were conducted in lower liquid concentration [13, 16-17] and the adsorption isotherms were directly interpreted from measured excess isotherm [18]. The models of adsorption isotherm are also well documented in the standard text book [19]. The measurement of adsorption across whole range of mixture concentration will provide a lot of information on the behavior of the adsorption of the adsorbates onto the adsorbent itself [19–21]. There are few analysis of adsorption isotherms based on the combination of measured excess isotherm across whole range mixture concentration and the theoretical model and are available in the literature [15], [20–22].

Thus, in this work a study on the determination of adsorption isotherm from across whole range of liquid mixture concentration was conducted. Non-azeotropic binary mixture, acetone-propanol was chosen as adsorbates and activated carbon and silicalite were chosen as the adsorbents. The selection of solvent was based on the similarity of molecular size. Nevertheless, the selection of the adsorbent was based on pore structure. Activated carbons (ACs) are usually having macropore structure, whereas silicalite contain micropore structure. The molecular size of solvents and adsorbent pore size properties were selected due to the capability to form monolayer adsorption behavior. Three independent theoretical models which were Pseudo-Ideal monolayer theory, Gibbs Dividing Plane monolayer model and Langmuir-Freundlich isotherm model were chosen for the analysis and interpretation of adsorption isotherm along with the measured excess isotherm.

2. Experimental methods

a. Materials

Two types of adsorbents, AC and silicalite were used in this work. AC was purchased from Fluka, whereas silicalite from Zeolyst International. Both solvents used (acetone and propanol) were High-performance liquid chromatography (HPLC) grade and purchased form Fisher Scientific.

b. Excess adsorption isotherm

A conventional technique of excess adsorption isotherm measurement was employed in this study [21]. The quantities of adsorbent and liquid solvent mixture were set to a weight ratio of 1:10. To achieve equilibrium, the mixture of adsorbent and liquid sorbates were kept in a 250 mL sealed conical flask and placed in orbital shaker at 30°C and 100 rpm for 4 h. The system of the experimental work was designed as shown Table 1.

Table 1. Adsorption system of experimental work

System	Component	Adsorbent	Preference Adsorbate
1	Acetone-propanol	commercial powdered AC	Acetone
2	Acetone-propanol	Silicalite (alumina free ZSM-5, MFI framework)	Acetone

The liquid phase concentration of the adsorption system was determined using Gas chromatography mass spectrometer (GCMS). The method was involved, the withdrawal of liquid sample using 0.45μm Whatman syringe filter. Samples were then injected into GCMS using auto sampling device. The compositions of initial mixture of component 1, x_{10} , equilibrated mixture, x_1 , weight of liquid mixture, W_0 and the weight of adsorbent, W_s were then used for the determination of the numerical value of excess adsorption isotherm of component 1, Γ_1 as in (1), [20, 23-25].

$$\Gamma_1 = \frac{W_0(x_{10}-x_1)}{W_s} \quad (1)$$

c. Analysis procedure of mixture composition

The mixture liquid phase concentration was determined using Agilent GCMS Model 7890. Capillary column, 24079 Supelcowax types of column was used with helium as carried gas at 35 kPa and 1.6 mL/min. The injection volume was 1 μL and the initial temperature was kept at 40°C for 2 min before it was ramped at 25°C/min to 65°C with holding time for 1 minute. Finally, the GCMS oven as ramped at 25°C/min to 100°C and held for another 1 minute and injector temperature was set at 150°C with split ratio value of 100:1.

2.4 Adsorption isotherm analysis and interpretation

For the purpose of analysis and interpretation of adsorption isotherm, Gibbs dividing plane isotherm model [15,26], Pseudo-ideal monolayer adsorption theory [20,23-24] and Langmuir-Freundlich isotherm model [22] were used. Following section will discuss and explain briefly of the theory and model used here.

2.4.1 Gibbs dividing plane model

Based on excess isotherm of adsorption in whole range of sorbates concentration , the volume of sorbates adsorbed can be identified using Gibbs dividing plane monolayer adsorption model [15]. The volume of adsorbed layer $a_{tot}(C_e)$ was calculated by the addition of experimentally measured excess adsorption numerical value and the product of the equilibrium adsorbates concentration on the specific volume of adsorbed layer as shown in equation 3. Whereas, the specific adsorbed volume of sorbates, V_a was calculated using the information of the negative of gradient of excess isotherm as shown in equation 2.

Specific volume of adsorbed layer:

$$\frac{d\Gamma(C_e)}{d(C_e)} = -V_a \quad (2)$$

Volume adsorbed layer:

$$a_{tot}(C_e) = \Gamma(C_e) + C_e.V_a \quad (3)$$

In getting the value of excess isotherm gradient, the excess isotherm itself was plotted in mole/m^2 unit and the equilibrium concentration (C_e) in mole fraction.

2.4.2 Pseudo-ideal monolayer adsorption theory

Pseudo-ideal monolayer adsorption was derived from the combination of the materials balance of adsorption system and idealized equilibrium adsorption equation [20]. Below are the important equations that derived from the theory and used in the analysis in this work;
conformity of Pseudo-ideal equation:

$$\frac{x_1 x_2}{\Gamma_1} = \frac{1}{N_S}\left[x_1 + \frac{1}{K-1}\right] \quad (4)$$

Concentration of adsorbed phase:

$$x_1^S = \frac{K x_1}{1+(K-1)x_1} \quad (5)$$

Number of moles of adsorbed phase:

$$n_1^s = N_s \frac{Kx_1}{1+(K-1)x_1} \quad (6)$$

Here, K is the equilibrium constant, N_s is the total mole of adsorbed phase (adsorption saturation value), equilibrium concentration, $x_2 = (1 - x_1)$, x_1^s is the concentration of component 1in adsorbed phase and n_1^s is the mole of component 1 adsorbed.

2.4.3 Langmuir-Freundlich isotherm model

For the curve fitting purposes, isotherm model of the Langmuir-Freundlich of binary adsorption [22] as shown in equation 7 was used. The value of adsorbed concentration, x_1^s was calculated using equation 8, this equation was derived from the material balance of adsorption system as explained elsewhere [20]. In order to use equation 8, the value of excess isotherm, Γ_1 was obtained experimentally and the values of total adsorption, N_s were obtained from the intercept of the plot of equation 4. The values of parameters $\bar{K}_{12}$and m, were estimated using Solver of Microsoft Excel. The solver is devised to minimize a nonlinear sum of squares (SSQ) function subject to theconstraints given.

$$x_1^s = \frac{(\bar{K}_{12}x_{12}^l)^m}{1+(\bar{K}_{12}x_{12}^l)^m} \quad (7)$$

$$x_1^s = \frac{\Gamma_1}{N^s} + x_1 \quad (8)$$

3. Results and discussions

The excess adsorption isotherms for both binary mixtures of non azeotropic solution onto activated carbon and silicalite were determined experimentally as shown in figure 1. The excess isotherm for both systems (acetone-propanol/AC and acetone-propanol/silicalite) obtained confirmed to mono-layer type of U shape. The U-shape excess isotherm indicates a preference for one component from the mixture/solution over the entire concentration range. It can be seen clearly that the excess adsorption of silicalite is higher than activated carbon in comparison with the same binary mixture which is acetone-propanol. It shows that silicalite were able to be more selective toward smaller sorbates molecule compare to AC. This behavior is also can be related to the porous structure of the adsorbents used. The U shape ofexcess adsorption of organic solvent on porous adsorbent are also reported in previous work of Farhadpour and Bono [20,21] and Swiatkowski et. al [22].

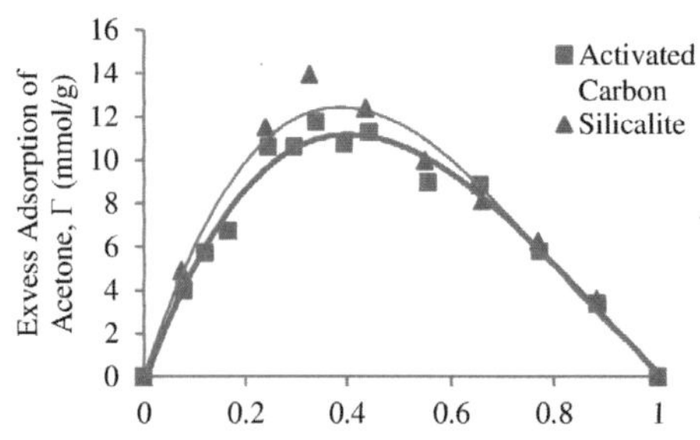

Figure 1. Excess adsorption isotherm for binary organics solvent non-azeotropic mixture onto AC and silicalite at 30°C

Surface specific adsorption isotherm using Gibbs dividing plane model shown in figure 2. It can be seen that the total adsorption of acetone onto ACs were slightly higher than onto silicalite which is in line with the pore volume of AC is higher than silicalite used. The adsorption of acetone for both

IOP Conf. Series: Earth and Environmental Science **36** (2016) 012021 doi:10.1088/1755-1315/36/1/012021

adsorbents was found to be predominantly monolayer fashion. This is essentially consistent with the same behavior observed previously for the adsorption organic effluent components on the surface of reversed-phase phenyl modified adsorbents [15].

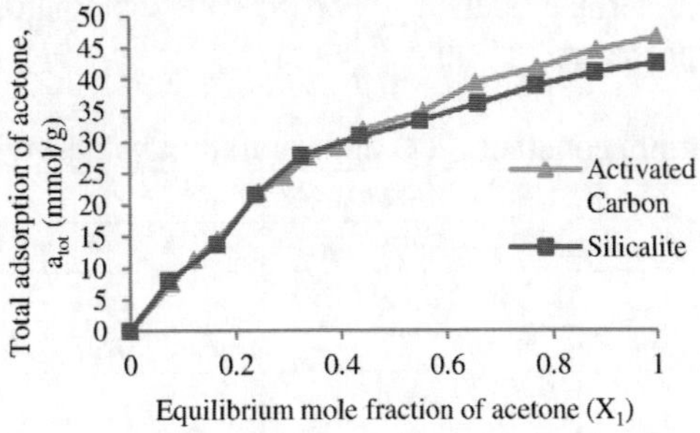

Figure 2. Analysis of total of acetone adsorbed from binary mixture onto AC and silicalite using Gibbs dividing plane model

The analysis of both adsorption systems using equation 4 has shown conformity to Pseudo-ideal monolayer adsorption theory. The slopes and intercepts of the plots in equation 4 were used for the determination of the equilibrium constant, *K* and saturation adsorption capacity, N_s as presented in Table 2.

Table 2. Equilibrium constant *K* and saturation adsorption capacity, N_s of adsorption acetone-propanol mixture

System	Parameters	
	K	Ns (mmol/g)
AC	2.00	62.50
Silicalite	3.00	45.45

The total adsorption sorbates onto AC shows higher compared to the adsorption onto silicalite as expected due to the porosity volume of AC is higher than silicalite. The adsorption plot based on Pseudo-ideal monolayer adsorption theory is presented in figure 3, it also show that the adsorption of acetone in AC is higher than silicalite. Both of adsorption isotherms are also showing the monolayer behavior across sorbates concentration range.

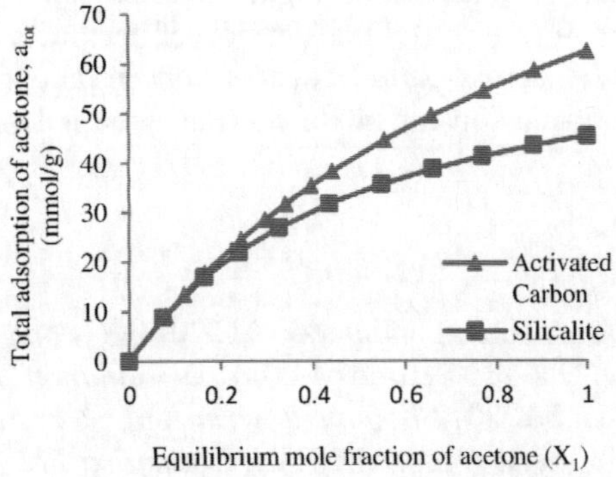

Figure 3. Analysis of total of acetone adsorbed from binary mixture onto AC and silicalite using Pseudo-ideal monolayer adsorption theory

In the curve fit exercise of adsorption isotherm of non-azeotropic adsorption using Langmuir-Fruendlich isotherm model as in equation 7, the adsorbed phase concentration were calculated using equation 8 with the value of excess isotherm obtained experimentally. The values of parameters of Langmuir-Fruendlich binary isotherm model, $\overline{K}_{12}$ and m were obtained by fitting equation 7 using Solver of Microsoft Excel and presented in Table 3.

Table 3. Equilibrium constant, $\overline{K}_{12}$ and saturation value, m of adsroption

System	Parameters	
	$\overline{K}_{12}$	m
AC	1.22	1.05
Silicalite	1.72	0.96

The plots of Langmuir-Freundlich binary isotherm shown in figure 4. Both adsorption isotherms of acetone onto AC and silicalite are fitted well with Langmuir-Freundlich isotherm model and also showed that their well explainable as mono-layer adsorption type.

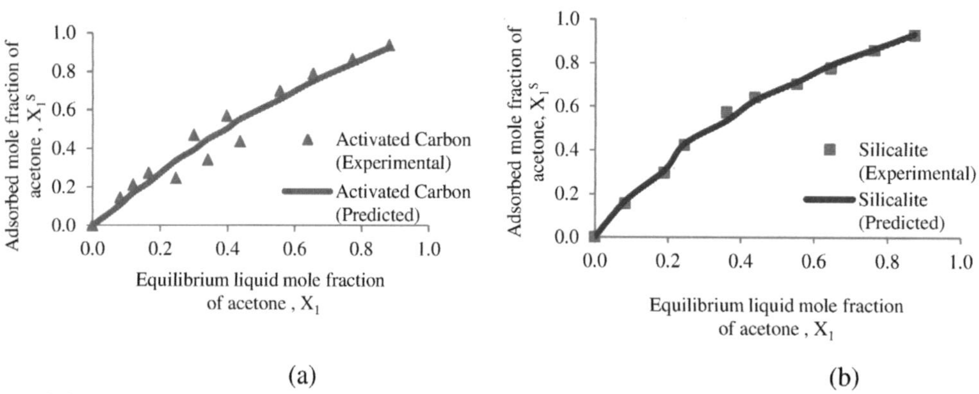

Figure 4. Experimentally measured and predicted of acetone adsorbed onto (a) AC and (b) silicalite based on Langmuir-Freundlich isotherm model.

4. Conclusion

The excess adsorption isotherm of acetone from non azeotropic mixture of acetone and propanol onto AC and silicalite was confirmed to U-shaped of preference adsorption isotherm. Silicalite shows more on the preference adsorption of acetone over AC due to the micropore structure. The total adsorption of sorbates for adsorption carbon found to be higher than silicalite which is in line with the pore volume of AC is higher than silicalite. Both adsorption of acetone from non-azeotropic mixture are explainable as mono-layer type across entire range of concentration. Comparison of the adsorption analysis of non-azeotropic organic solvent mixture using three independent adsorption model show very small variation.

5. References

[1] Soto M L, Moure A, Domínguez H and Parajó J C 2011 *J. Food Eng.* **105** pp 1–27
[2] Rahchamani J, Mousavi H Z and Behzad M 2011 *Desalination* **267** pp 256–260
[3] Ahmed M J and Theydan S K 2012 *Ecotox. Environ. Safe.* **84** pp 39–45
[4] Mahmoodi N M., Salehi R and Arami M 2011 *Desalination* **272** pp 187–195
[5] Juang R S, Lin S H and Cheng C H 2006 *Ultrason. Sonochem.* **13** pp 251–260
[6] Hanzlík J, Jehlicka J, Sebek O O, Weishauptová Z and Machovic V V 2004 *Water Res.* **38** pp 2178–2184

[7] Rengaraj S., Moon S H, Sivabalan R, Arabindoo B and Murugesan V 2002 *Waste Manage*. **22**, pp 543–548
[8] Allen S J, Gan Q, Matthews R and Johnson P A 2003 *Bioresource Technol.* **88** pp 143–152
[9] Hameed B H, Ahmad A A and Aziz N 2007 *Chem. Eng. J.* **133** pp 195–203
[10] Qi S. and Schideman L C 2008. *Water Res.* **42** pp 3353–3360
[11] Abdullah M A, Chiang L and Nadeem M 2009 *Chem. Eng. J.* **146** pp 370–376
[12] Bansal R C and Goyal M 2005 *Activated Carbon Adsorption*. CRC Press Taylor and Frances Group, pp 1-167
[13] Ahmed M J, Halim A, Mohammed A K, Amir A and Kadhum H 2012 *Fluid Phase Equilibr.* **313** pp 165–170
[14] Al-Degs Y, Khraisheh M A M., Allen S J, Ahmad M N and Walker G M 2007 *Chem. Eng. J.* **128** pp 163–167
[15] Chan F, Yeung L S, LoBrutto R and Kazakevich Y V 2005 *J. Chromatogr. A* **1082** pp 158–165
[16] Abdelwahab O 2008 *Desalination* **222** pp 357–367
[17] Chan L S, Cheung W H and McKay G 2008 *Desalination* **218** pp 304–312
[18] Hamdaoui O, Naffrechoux E, Tifouti L and Pétrier C 2003 *Ultrason. Sonochem.* **10**, pp 109–14
[19] Wei P C, May C Y, Ngan M A and Hock C C 2005 *University of Malaya* **1** pp 264–269
[20] Farhadpour F A and Bono A 1988 *J. Colloid Interf. Sci.* **124** pp 209-227
[21] Bono A, Sarbatly R, Krishnaiah D, San P M and Yan F Y 2008 *Catal. Today* **131** pp 472–476
[22] Świątkowski A, Deryło-Marczewska A, Goworek J and Błażewicz S 2004 *Appl. Surf. Sci.* **236**, pp 313–320
[23] Farhadpour F A and Bono A 1996 *Chem. Eng. Process.: Process Intensification* **35** pp 157–168
[24] Farhadpour F A and Bono A 1996 *Chem. Eng. Process.: Process Intensification* **35**, pp 141–155
[25] Anisuzzaman S M, Joseph C G, Daud W M A W, Krishnaiah D and Ho S Y 2015 *Int. J. Ind. Chem.* **6** pp 9-21
[26] Kazakevich Y V, LoBrutto R , Chan F and Patel T 2001 *J. Chromatogr. A* **913**, no. 1–2, pp 75–87

Acknowledgement
The authors are grateful to Ministry of Higher Education, Malaysia for the financial support through the research grant number SLB0007-TK-1/2011.

Development of carbon dioxide adsorbent from rice husk char

S Abang, J Janaun, S M Anisuzzaman[1], F S Ikhwan
Chemical Engineering Programme, Faculty of Engineering,
Universiti Malaysia Sabah (UMS), Jalan UMS, 88400
Kota Kinabalu, Sabah, Malaysia

E-mail: [1]anis_zaman@ums.edu.my, [1]dr.anis.ums@gmail.com

Abstract. This study was mainly concerned about the development of carbon dioxide (CO_2) adsorbent from rice husk (RH). Several chemical treatments were used to produce activated rice husk char (RHAC) from RH. Initially the RH was refluxed with 3M of sodium hydroxide (NaOH) solution, activation followed by using 0.5M of zinc chloride ($ZnCl_2$) solution and finally acidic treatment by using 0.1M of hydrochloric acid (HCl). Then, the RHAC was functionalized by using 3-chloropropylamine hydrochloride (3-CPA) and noted as RHN. RHN samples were characterized with scanning electron microscopy (SEM), mercury intrusion porosimetry (MIP), fourier transform infrared spectroscopy (FTIR). Based on the SEM, the RHN sample had a large pore diameter compared to RH sample after being treated. Based on MIP data, the average pore diameter between RH and RHAC samples were increased significantly from 0.928 microns to 1.017 microns. The RHN sample also had higher total porosity (%) compared to RHAC and RH (58.45%, 47.82% and 45.57% respectively). The total specific surface area of the sample was much increasing from RHO to RHAC (29.17 m^2/g and 62.94 m^2/g respectively) and slightly being decreasing from RHAC to RHN (58.88 m^2/g). FTIR result showed the present of weak band at 1587 cm^{-1} which demonstrating of the amine group present on the sample. The CO_2 capture result showed that the decreasing of operating temperature can increase the breakthrough time of CO_2 capture. On the contrary decreasing of CO_2 gas flow rate can increase the breakthrough time of CO_2 capture. The highest total amount of CO_2 adsorbed was 25338.57 mg of CO_2/g of RHN sample by using 100 mL/min of gas flow rate at 30°C. Based on adsorption isotherm analysis, the Freundlich isotherm was the best isotherm to describe the CO_2 adsorption on the sample.

1. Introduction

Carbon dioxide (CO_2) gas is one of the largest contributors for the increasing of the greenhouse level in the atmosphere about 60% of global warming effects. The international panel on climate change (IPCC) predicts that, by the year 2100, the atmosphere may contain up to 570 ppmv CO_2 which can increase the global temperature of around 1.9°C and can increase of mean sea level about 38 m [1].

[1] To whom any correspondence should be addressed.

There are several methods which have been suggested to reduce the CO_2 emission into the atmosphere, to reduce energy intensity, to reduce carbon intensity and to enhance the sequestration of CO_2. The first method would require efficient use of energy, the second one requires switching to using of non-fossils fuels such as hydrogen and renewable energy and the third one would be the development of technologies to capture and sequester more CO_2 gas from the surrounding environment [1-4]. Based on the previous researchers, the carbon which has a pore size less than 1nm is very suitable for CO_2 capture under ambient conditions [5]. It has been also found that the activated carbons (AC) with impregnation of sterically hindered amines had enhanced the CO_2 adsorption performance [6].

According to Liou and Wu [7], there are two major methods to produce rice husk (RH) activated carbon (RHAC), the physical activation and chemical activation. The physical activation is getting involved the carbonization process on the sample by using a gaseous activating agent such as steam or CO_2. While, the chemical activation is getting involved the chemical agent such as phosphoric acid (H_3PO_4), Zinc chloride ($ZnCl_2$) or alkali hydroxides (NaOH and KOH) plus with heating treatment to activate the RHAC. The chemical activation is provided lower reaction temperature than the physical activation method [4, 8, 9]. The surface area of RHAC can be increased by removing the ash contents inside the rice husk (RH). It is important to remove the ash from RH because the previous researchers have been mentioned that the higher the ash content, the lower the N_2 surface area [7]. One of the methods to remove ash from RH is basic leaching process by using a high concentration of base such as NaOH and KOH to leach out the silica from the sample.

Generally, the developing countries around the world produce 500 million tons of rice annually. The rice milling produces about 20% of rice husk for the total weight of rice plants [4]. Its contains an abundance of lignocellulosic biomass and ash. RH also can provide good pore structure for a functionalization process to get better CO_2 adsorbent [10-13]. However, the proper method and pre-treatment are needed to get the better pore structure. By using RH as adsorbent in industry is hopefully can reduce the production of carbon dioxide from the mill; by integrating the RH adsorbent for CO_2 from the flue gas.

Thus, the aim of this study was to produce of adsorbent from rice husk char. The char was undergone three phases of treatment with different types of chemicals before it was continued with next functionalization method by using amine functional group.

2. Materials and methods

2.1 Materials

RH was collected from Seri Dusun Rice Mill, Kota kinabalu, Sabah, Malaysia. All the chemicals were supplied by Sigma Sdn Bhd., Malaysia. The RH undergoes several chemical treatments before being functionalized by using 3-chloropropylamine hydrochloride (3-CPA).

2.2 *Preparation of functionalised AC*

2.2.1Base treatment

The RH was refluxed with 3M of sodium hydroxide (NaOH) solution at a ratio of 10ml of NaOH over 1 gm of RH char for 2 h. Then, the refluxed RH was being heated, stirred and washed by using distilled water till the washing solution was changed into a range of pH 8 to 9. After that, the washed RH was dried in oven at 105°C for 24 h.

2.2.2 Activation of RH

The activation of RH was done by using 0.5 M of zinc chloride ($ZnCl_2$) solution. Firstly, the RH and $ZnCl_2$ were mixed in a beaker in ratio of 1g of treated NaOH rice husk with 8 mL of 0.5 M of $ZnCl_2$ in the form of mixture. Then, this mixture was mixed and stirred for 2 h in a beaker with a temperature range between 50°C to 60°C. Finally, the mixture was impregnated in oven at 105°C for 24 h.

2.2.3 Acidic treatment
The NaOH treated rice husk was refluxed with 0.1M hydrochloric acid (HCl) in ratio of 10 ml/g for 2h at a temperature of 100 °C. Then, the RH were stirred, heated and washed with distilled water until the pH of mixture solution change to pH 5 to 6. Lastly, the mixture solution was drained and the AC was dried in an oven at 105 °C for 24h.

2.2.4 Functionalization of sample
0.003M 3-chloropropylamine hydrochloride (3-CPA) and 0.02M of potassium hydroxide (KOH) was mixed in a ratio of 1:1 volume. A weight of 1g of the rice husk activated carbon (RHAC) was refluxed with 25 mL mixture of 3-chloropropylamine hydrochloride (3-CPA) and KOH for 6h at 40°C. Then, the solution was dried in an oven for 24h by using the impregnation method to functionalize the AC at temperature 105 °C. This funtionalised AC was labelled as RHN.

2.3 Characterisations of functionalised AC

2.3.1 Scanning electron microscopy (SEM)
The surface morphology of the sample was determined by using SEM. The setting was chosen as follows; working distance (WD) was 6.0 mm, 6.5 mm and 7.0 mm and electron high tension (EHT) was 10 kV.

2.3.2 Mercury intrusion porosimetry (MIP)
The sample porosity and surface area were determined by using automated MIP model Pascal 440 series. The external pressure was applied to the sample to force the liquid into a pore against the opposing force of the liquid's surface tension. The pressure was created by the intrusion of a non-wetting liquid such as mercury. Then, the chosen parameter value was inserted into Washburn's equation to find the pore diameter ("Porosimetry," 2013).

2.3.3 Fourier transform infrared spectroscopy (FTIR)
The determination of the sample's functional group was determined by using FTIR and the model which was used is "Perkin Elmer Spectrum". Before starting the analysis, the sample analysis was cleaned with acetone to ensure that no others impurities present during the analysis run. The scan ranges were from 650 cm^{-1} to 4000 cm^{-1} with a resolution of 4cm^{-1}. The result data was saved as an ASCII file and was plotted again by using Microsoft Excel.

2.4 CO_2 capture experimental setup
For CO_2 capture experiment, 7 gm of RHN sample were used for every set of experiment. So, there are 9 sets of CO_2 capture experiment. The parameters were used to capture of CO_2 shown in Table 1. Initially, the surrounding temperature was measured by using a thermometer which was used for CO_2 meter setting. After that, the N_2 gas was purged into setup until CO_2 meter showed a 300 ppm reading and started the CO_2 capture experiment. The CO_2 capture experiment was stopped immediately once the CO_2 concentration was achieved their saturated concentration based on the calibration data. Then, The CO_2 concentration value from the CO_2 meter was recorded. Figure 1 shows the schematic diagram for CO_2 capture experiment

Table 1. The parameters for CO_2 capture experiment

Gas flow rate (mL/min)	Temperature (°C)	Pressure (atm)
	30	
100	60	1
	90	
	30	
150	60	1
	90	
	30	
170	60	1
	90	

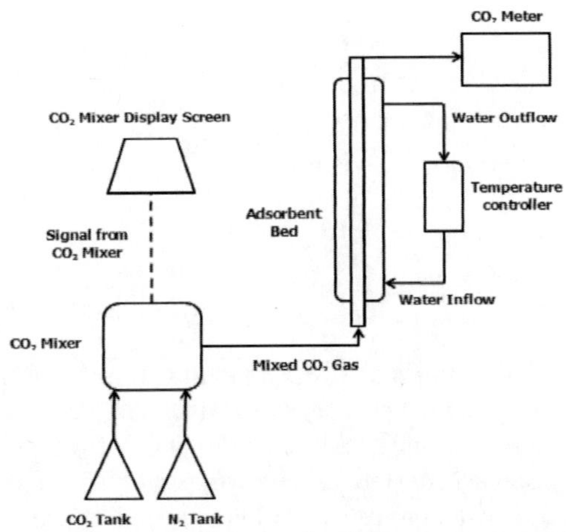

Figure 2. Schematic diagram for CO_2 capture experiment

2.5 Adsorption isotherms

Langmuir model is based on the assumption that adsorption energy is constant and independent of surface coverage where the adsorption occurs on localized sites with no interaction between adsorbate molecules and that maximum adsorption occurs when the surface is covered by a monolayer of adsorbate. The mathematical expressions of Langmuir equation can be written as follows,

$$\frac{C_e}{q_e} = \frac{1}{Q_0 K_L} + \frac{1}{Q_0} C_e \qquad (1)$$

where, C_e = Equilibrium concentration (mg/L); q_e = Amount adsorbed per unit mass (mg/g); Q_0 = Langmuir constant related to adsorption capacity (mg/g); K_L = Langmuir constant related to rate of adsorption (L/mg)

In order to find the value of R^2, Q_0 and K_L for the Langmuir isotherm for the adsorption of CO_2 on RHN samples, the graph of $\frac{C_e}{q_e}$ against C_e was plotted.

The Freundlich equation was calculated by using equation below,

$$q_e = K_f \, C^{1/n} \qquad (2)$$

$$logq_e = logK_f + \frac{1}{n} logC_e \quad (3)$$

where, K_f and n are the Freundlich constants with n as a measure of the deviation of the model from linearity of the adsorption and K_f indicates the adsorption capacity of the adsorbent. In order to find the value of R^2, n and K_f for the Freundlich isotherm for the adsorption of CO_2 on the RHN samples, the graph of $logq_e$ against $logC_e$ was needed was plotted.

Temkin isotherm assumes that the heat of adsorption of all the molecules in the layer would decrease linearly with coverage due to adsorbent-adsorbate interaction and the adsorption is characterized by a uniform distribution of binding energies. Temkin model is expressed as,

$$q_e = \left(\frac{RT}{b}\right) \ln(A_T C_e) \quad (4)$$

where $\frac{RT}{b} = B$ (J/mol), is the Temkin cons`ant related to heat of sorption whereas A_T (1/g) is the equilibrium binding constant corresponding to the maximum binding energy. In order to find the value of R^2, A_T and B for the Temkin isotherm for the adsorption of CO_2 on RHN samples, the graph of q_e against $\ln C_e$ was plotted.

3. Results and discussions

3.1 Surface morphology

Figure 2 shows the SEM images of samples at 1000x magnification: (a) RH (b) RHAC (c) RHN. Based on the manual calculation, the average pore diameter of RH sample was 9.1 microns, RHAC sample was 11.7 microns and RHN sample was 14.6 microns. The pore diameter of RH sample looks very small and likely has been clogging with some materials on the sample's pore. The clogging material might be caused by ash and metal impurities which were also being reported by previous researchers. This problem can reduce the cumulative volume sample. The RH sample also consists of large particles which contribute to the decreasing surface area value on the MIP result. The RHAC sample consists of small particles after being treated with three chemical treatments. The increasing of surface area value has been proven by using MIP characterization method which had been done to support this result data. Based on manual calculation, the average pore diameter of RHAC sample is 11.7 microns. This result value is only applicable for the sample's pore diameter, which can be seen through 1000x magnification. The largest pore diameter is 15 microns and the smallest is 1 microns. Based on the manual calculation, the average pore diameter of RHN sample is 14.6 microns. This result value is only applicable for the sample's pore diameter, which can be seen through 1000x magnification. Based on the SEM image result, the largest pore diameter for RHN was 20 microns and the smallest was 1 micron.

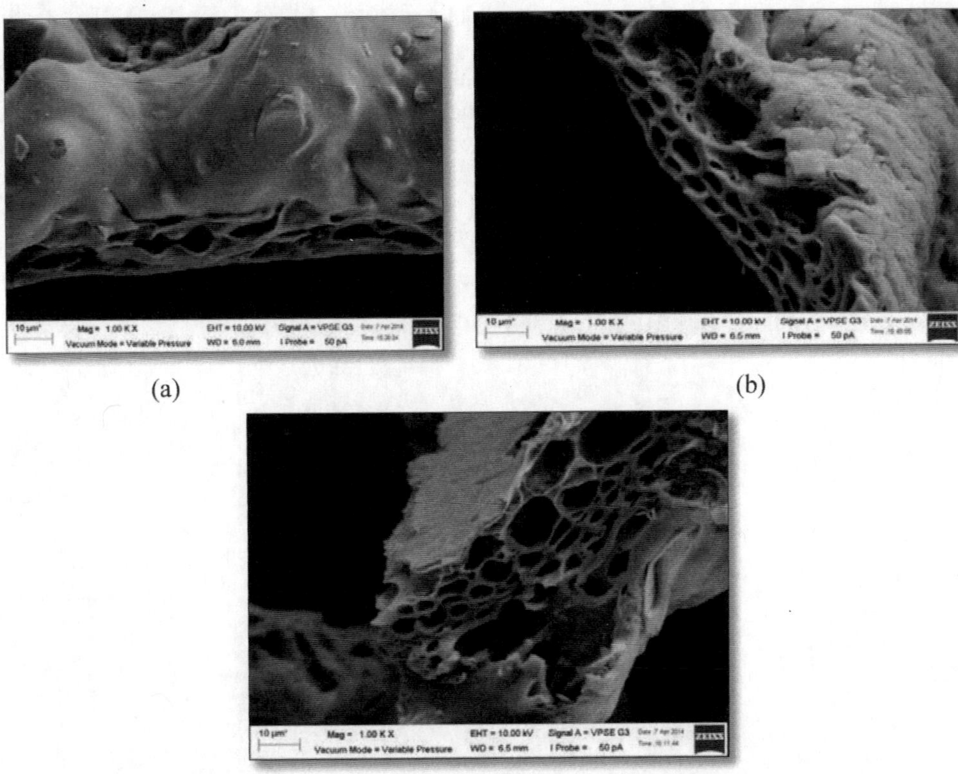

(a) (b) (c)

Figure 2. SEM images of samples at 1000x magnification: (a) RH (b) RHAC (c) RHN

3.2 Surface area and porosity

Table 2 shows the surface area, porosity and pore diameter of RH, RHAC and RHN samples.

Table 2. The surface area, porosity and pore diameter of RH, RHAC and RHN

Sample	Total specific surface area (m^2/g)	Average pore diameter (µm)	Total cumulative volume (mm^3/g)	Total porosity (%)
RH	29.17	0.928	410.50	45.57
RHAC	62.94	1.017	983.00	47.82
RHN	58.88	1.225	988.50	58.45

Based on Table 2, the total specific surface area of the sample was much increasing from RHO to RHAC and slightly being decreasing from RHAC to RHN. The decreasing of total surface area between

RHAC and RHN was a natural phenomenon which had been proven by Bhagiyalakshmi et al. [9]. The average pore diameter between RH and RHAC samples were increased significantly from 0.928 microns to 1.017 microns after being treated with several chemical treatments. The treatments could remove the metal and ash impurities on the sample especially on the basic and acidic chemical treatments. The result also shows the increasing of total cumulative volume of samples which corresponding to the increasing of sample's pore diameter after being treated with several chemical treatments. The total porosity of samples also becomes increasing from RH to RHN. The functionalization of 3-CPA on the RH sample can increase its total porosity value which is much better than using basic and acidic treatment.

3.3 Functional groups

Figure 3 shows the FTIR spectra of RH, RHAC and RHN.

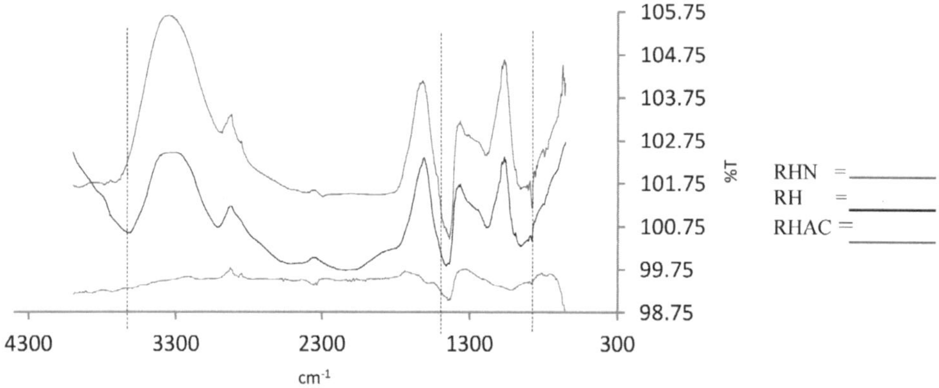

Figure 3. The FTIR spectra of RH, RHAC and RHN

Figure 3 shows the FTIR data between RH, RHAC and RHN. This data was used to determine of the presenting of the amine functional group on the RH sample by comparing the present of weak band between RHAC and RHN. The result shows the RHN's demonstrate a weak band at 1587 cm^{-1} to represent of the presence of the amine functional group in RHN sample. The large band between 3300 and 3600 cm^{-1} was represented to the -OH stretching frequency of the silanol group in the inorganic framework due to formation of hydroxyl groups from the adsorbed water. The spectrum of RHN also showed the weak band at 3700 cm^{-1} which indicating as the presents of –OH group in on the sample. Then, the two stretches weak bands of RHN's spectra at 2840 cm^{-1} and 2880 cm^{-1} was represented as the present of N-H stretching vibration. The peaks at 2370 cm^{-1} may be because of the existence of ammonium ions on the RHN sample. Probably, the amine group was attached to C instead of Si due to the presence of vibration peak at 1020 cm^{-1}.

3.4 Amount of CO_2 adsorbed

Table 4 shows the amount of CO_2 adsorbed by using 100, 150 and 170 mL/min of flow rate

Table 4. Amount of CO_2 adsorbed by using 100, 150 and 170 mL/min of flow rate

Gas flow rate (mL/min)	Temperature (°C)	Amount of CO_2 adsorbed (mg of CO_2/g of sample)	Breakthrough time to achieve saturated CO_2 conc. (min)
	30	25338.57	110

100	60	2854.86	48
	90	220.14	14
	30	21740.72	68
150	60	6351.79	36
	90	2087.36	26
	30	13166.74	34
170	60	6299.96	24
	90	14.16	16

Based on Table 4, the amount of CO_2 adsorbed was increased with the decreasing of the operating temperature of CO_2 capture. The breakthrough time of CO_2 capture also was increased with the decreasing of temperature by using 100, 150 and170 mL/min of gas flow rate.

3.5 Adsorption isotherms

On the basis of the R^2 value, Freundlich isotherm seemed to represent the equilibrium adsorption data with better fit as compared to other isotherms. The reason is because the R^2 value for the Freundlich isotherm is higher than that of Langmuir and Temkin isotherms except for the CO_2 capture data which used 150 mL/min at 30°C operating temperature; the adsorption data was fit to Temkin isotherm model which has higher regression correlation value than others isotherm model. Most of the results show that the $\frac{1}{n}$ values of adsorption data were higher than one. These values represent a concave, curved upward, solvent affinity type isotherm, where the marginal sorption energy increases with increasing surface concentration. It can be interpreted also with strong adsorption of the solvent, strong intermolecular attraction within the adsorbent layers, penetration of the solute in the adsorbent, and mono- functional nature of the adsorbate. The value of $\frac{1}{n}$ represents a joint measure of both the relative magnitude and diversity of energies associated with a particular sorption process. A $\frac{1}{n}$ = 1 indicates linear adsorption and, therefore, equal adsorption energies for all sites. Linear adsorption generally occurs at very low solute concentrations and low loading of the sorbent.

4. Conclusion

The RH with amine functionalized can adsorb high capacity of CO_2 with a suitable parameter chosen. The CO_2 adsorption capacity could be increased by the decreasing of the operating temperature in CO_2 capture experimental setup. On contrary, the decreasing of operating temperature can increase the total amount of CO_2 adsorbed. The highest total amount of CO_2 adsorbed was 25338.57 mg of CO_2/g of RHN sample by using 100 mL/min of gas flow rate at 30°C operating temperature. The breakthrough time of CO_2 capture also was increased with the decreasing of temperature. Based on SEM result obtained and being supported by BET result it could be concluded that the chemical treatments on the RH increases the average pore diameter, total specific surface area and porosity of sample. The Freundlich adsorption isotherm was the best isotherm to describe CO_2 adsorption on the adsorbent.

References

[1] Yang H, Xu Z, Fan M, Gupta R, Slimane R B, Bland A E and Wright I 2008 *J. Environ. Sci.* **20** pp 14–27
[2] Berstad D, Anantharaman R and Nekså P 2013 *Inter. J. Refrig.* **36** pp 1403–1416
[3] Chen C, Kim J and Ahn W S 2012 *Fuel*, *95* pp 360–364

[4] Dibenedetto A, Aresta M and Narracci M 2002 *Fuel Chemistry Division Preprints*, University of Bari and METEA Research Center, Italy, **47** pp 53-54
[5] Cai J, Lv X, Xing Y and Zhao X 2014 *Mater. Lett.* **114** pp 37–39
[6] Lee C S, Ong Y L, Aroua M K and Daud W M A W 2013 *Chem. Eng. J.* **219** pp 558–564
[7] Liou T H and Wu S J 2009 *J. Hazard. Mater.* **171** pp. 693–703
[8] Anisuzzaman S M, Joseph C G, Taufiq-Yap Y H, Krishnaiah D and Tay V V 2015 *J. King Saud Univ. Sci.* (in press)
[9] Anisuzzaman S M, Joseph C G, Daud W M A W, Krishnaiah D and Ho S , 2015 *Int. J. Ind. Chem.* **6** pp 9-21
[10] Putri A K, Ding W H and Kuo H W 2012 *J Pure Appl. Chem. Res.* **1** pp 1–10
[11] Bhagiyalakshmi M, Yun L J, Anuradha R and Jang H T 2009 *J. Porous Mat.* **17** pp 475–484
[12] Nagrale S D, Hajare H and Modak P R 2012 *Int. J. Eng. Res. Applications*, **2** pp 1-5
[13] Kartini K 2010 Pusat Penerbitan Universiti, *Universiti Teknologi Mara*, Malaysia

Acknowledgements
Authors would like to thank Universiti Malaysia Sabah (UMS) for providing the research facilities.

Highly sensitive determination of sunset yellow FCF (E110) in food products based on Chitosan/Nanoparticles/MWCNTs with modified gold electrode

K Rovina[1], S Siddiquee[1] and S M Shaarani[2]
[1] *Biotechnology Research Institute*, Univeristi Malaysia Sabah, Jln UMS, 88400 Kota Kinabalu, Sabah, Malaysia
[2] *Faculty of Food Science and Nutrition*, Universiti Malaysia Sabah, Jln UMS, 88400 Kota Kinabalu, Sabah, Malaysia

E-mail: [1]rovinaruby@yahoo.com, [1]shafiqpab@ums.edu.my, [2]fatihah@ums.edu.my

Abstract. Sunset Yellow belongs to the family of azo dyes, commonly used in food industry. High consumption of Sunset Yellow can cause health problem to human. Due to arising of the health issues, there are several analytical methods available for determination of Sunset Yellow. However, these methods are required skilled manpower, complicated procedures, time consuming and high cost. Herein, an electrochemical sensor was developed based on the combination of chitosan (CHIT), calcium oxide nanoparticles (CaONPs) and multiwall carbon nanotubes (MWCNTs) sensing film for detection of Sunset Yellow in food products. Electrochemical behavior of the modified gold electrode in the presence of Sunset Yellow was studied by using cyclic voltammetry (CV) and differential pulse voltammetry (DPV). The morphological characteristics of CHIT/CaONPs/MWCNTs were observed under scanning electron microscope and transmission electron microscope. Under optimal conditions, the DPV was detected with different concentrations of Sunset Yellow in the range of 0.9 to 10 ppm, with detection limit of 0.8 ppm. The developed method has successfully applied for monitoring the presence of Sunset Yellow with different food products including candy, royal jelly, ice cream and soft drink with satisfactory results.

1. Introduction

Sunset Yellow FCF (E110) belongs to the azo group with formula of $C_{16}H_{10}N_2Na_2O_7S_2$. This color is commonly used in the production of swiss roll, apricot jam, lemon curd, sweets, beverage mix and packet soups, margarine, custard powders, packaged lemon gelatin desserts, breadcrumbs, packaged instant noodles, cheese sauce mixes and powdered marinades by food industry. Sunset Yellow commonly consisted of 2-hydroxyl-(4-sulphonatophenylazo) naphthalene-6-sulphonate and subsidiary coloring matters together with sodium chloride and sodium sulphate as the principal of non-colorant components. Sunset Yellow is synthesized by diazotizing 4-aminobenzenesulphonic acid using combinations of hydrochloric and sodium nitrite, or sulphuric acid and sodium nitrite. High consumption of Sunset Yellow may cause allergic such as contact urticaria, angioneurotic edema, asthma, contact anaphylaxis and immune suppression [1]. Besides, Sunset Yellow is able to alter the reproductive and neurobehavioral parameters, which lead to allergies, diarrhea and other symptoms [2, 3]. Food and Drug Administration strictly fixed the maximum level of Sunset Yellow in non-alcoholic beverages is less than 100 μg mL^{-1} [4].

Several methods have been developed such as electrochemical sensor, enzyme linked immunosorbent assay (ELISA), high performance liquid chromatography (HPLC), and spectrophotometry. Among these methods, electrochemical sensor has gained much attention and demonstrated the promising tool for monitoring food systems due to high sensitivity and selectivity, portability, low detection and quantification limit, faster analysis, and low cost. Sunset Yellow is electrochemical active compound that undergo reversible reaction during electrochemical process. Up-to-date, different modified electrodes are employed, for example, Ghoreishi et al. [5] have been addressed gold nanoparticles-modified carbon paste electrode (CPE) and Chen et al. [6] have introduced alumina microfibers-modified CPE. Another effort have been developed based on the multiwalled carbon nanotube (MWCNT)-modified glassy carbon electrode (GCE) [7,8], a platinum wire-coated electrode [9], a graphene layer-wrapped phosphotungstic acid (PTA) hybrid film-modified GCE [3] for determination of Sunset Yellow in food products. Carbon nanotubes (CNTs), including single wall carbon nanotubes (SWNTs) and multiwall carbon nanotubes (MWCNTs) are widely applied in electrochemical sensor due to their unique properties such as large specific surface area, good mechanical stability and high electronic conductivity [10, 11]. In this study, a novel electrochemical sensor was developed based on chitosan, calcium nanoparticles and multiwall carbon nanotubes modified gold electrode (CHIT/CaONPs/MWCNTs/AuE) for the determination of Sunset Yellow in food products. The schematic mechanism oxidation of Sunset Yellow took place a one-electron, one-proton reversible reaction on CHIT/CaONPs/MWCNTs/AuE shown in Figure 1.

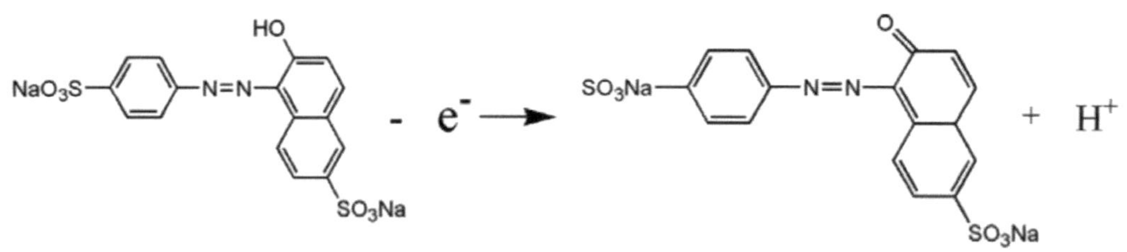

Figure 1. Oxidation mechanism for the electrochemical process of Sunset Yellow [16].

2. Experimental

2.1 Reagents and chemicals

Sunset Yellow FCF (E110), chitosan (CHIT) and multiwall carbon nanotubes (MWCNTs) were purchased from Sigma Aldrich (USA). Calcium oxide nanoparticles (CaONPs) were obtained from Biosensors and Bioelectronics Laboratory, Department of Chemistry, Faculty of Science, Universiti Putra Malaysia. Sunset Yellow (50 ppm) was dissolved into dH_2O as a stock solution and stored at 4°C. Methylene blue (MB) (0.1 mM) solution was prepared in 50 mM Tris-HCl + 20 mM NaCl buffer solutions (pH 7.0) and stored at room temperature. Other chemicals were of analytical reagent grade and used as received. All solutions were prepared using deionized water as the solvents and were conducted the experimental works at room temperature condition of $25 \pm 0.1\ ^0C$.

2.2 Apparatus and equipment

All electrochemical measurements were carried out with a μ-autolab (Ecochemie, The Netherlands) voltammetric analyzer using the software package on NOVA 1.8 for CV and DPV analysis. A three-electrode systems were used in the measurement which composed of a gold electrode (d = 3 mm) as working electrode, Ag|AgCl|KCl as the reference electrode and the platinum wire as counter electrode. The morphological characteristics were observed under Hitachi S-3400N scanning electron microscope (SEM) and transmission electron microscope (TEM) (Tecnai G2 Spirit Bio TWIN, Crezh Republic).

2.3 Preparation of CHIT/CaONPs/MWCNTs

Before fabrication, the gold electrode (AuE) was pre-treatment. The AuE was polished with 3 μM aluminium slurry for 2 min and then sonicated for 2 min. Then, the AuE was rinsed with deionized water about 2 min. For modification of AuE, a 2 % of CHIT solution was prepared by dissolving the CHIT powder in 2 % acetic acid. The solution was stirred for at least 4 h at room temperature until the CHIT was fully dissolved. CaONPs were added into the 2 % CHIT solution with ratio CaONPs:CHIT was 2:5 then sonicated for 20 min. It was stirred for 8 h for highly dispersed colloidal suspension. Again 1 % of MWCNTs was added in the CHIT/CaONPs mixture and sonicated to produce a homogenous suspension. After that, MWCNTs mixtures were allowed the formation of a uniform CHIT/CaONPs/MWCNTs nanocomposite suspension.

2.4 Analysis of real samples

Candy, royal jelly, ice cream and soft drink were selected and purchased from the local super market in Kota Kinabalu, Sabah, Malaysia. The samples preparation procedures are followed previously reported by Sahraei et al. [13]. Firstly, 5.7199 g candy, 10.0742 g royal jelly and 30.0512 g ice cream were respectively dissolved in 100 mL hot pure water (~45°C). Each of the samples was filtered through a 0.45 μm membrane filter to obtain solution without precipitation for subsequent use [14]. The soft drink sample was used directly without any pretreatment. Finally, 0.1 mL sample extract was added into working buffer, and then analyzed according to the optimization electrochemical sensor protocols.

3. Results and discussion

3.1 Morphological characteristic of modified AuE

Firstly, morphology of the modified AuE surface was observed using SEM. Figure 2a, CHIT membrane was showed mesoporous surface that able to immobilize with other nanomaterials. The pure MWCNTs film clearly illustrated in Figure 2b that shows the bundles interlocked together to form netlike and highly mesoporous nanostructure due to the strong π-π electronic interactions and lack of attached oxygenated groups on the surface of MWCNTs [15]. In Figure 2c, CHIT/CaONPs revealed the well dispersed of CaONPs were adhered well to the CHIT nanocomposite membrane. The CaONPs were found well distribution that attached and reacted in the circumference of CHIT membrane, and result the surroundings of the particles were looked brighter against SEM images. Figure 2d shows CHIT/CaONPs/MWCNTs with folding structure of MWNCTs in the middle of CHIT pores. It could be seen that the surface of the CHIT/CaONPs/MWCNTs nanocomposite was relatively rough and possesses distinct interstices, which was good for intensely accumulation of Sunset Yellow and increasing the activity sites. Hence, CHIT/CaONPs/MWCNTs were provided more adsorption sites and exhibited high accumulation efficiency of Sunset Yellow. Additionally, the nanomaterials were analyzed using TEM for providing insight into the origin of enhancement effect for Sunset Yellow oxidation. Figure 2e and Figure 2f show the nanoparticles shape under 30 K magnification of TEM with nanoscale of 50 nm of CHIT/CaONPs/MWCNTs. It was clearly indicated that the CaONPs were well attached onto the nanocomposite membrane with size approximately of less than ~10 nm diameters.

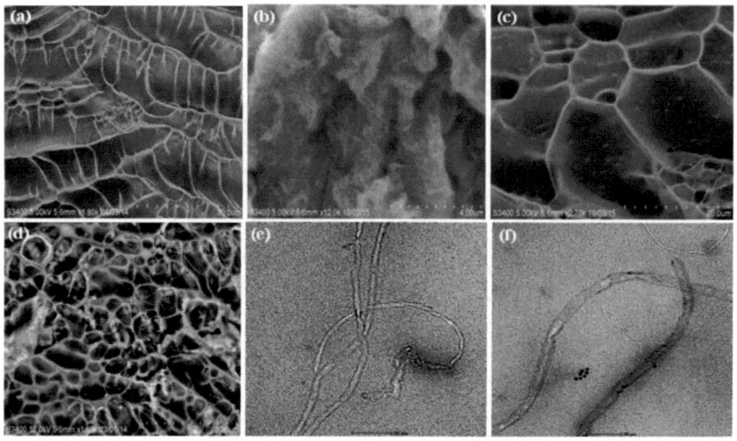

Figure 2. SEM image of the (a) CHIT, (b) MWCNTs, (c) CHIT/CaONPs, (d) CHIT/CaONPs/MWCNTs and, TEM image of (e) MWCNTs, (f) MWCNTs/CaONPs.

3.2 Optimization of the modified AuE

3.2.1 Influence of pH

The electrochemical behaviors of Sunset Yellow with different pH solutions were assayed using CV method. The highest peak current responses on the CHIT/CaONPs/MWCNTs/AuE are shown in Figure 3. The applied supporting electrolyte was measured the pH values in the range of 6.0-8.5. With the pH changing from 6.0 to 7.0, the response of oxidation current increased and reached a maximum at pH 7.0, above pH 7.0, decreasing the current response. Thus, it is indicated that the reversibility of Sunset Yellow oxidation is better at pH 7.0 which suggesting that the number of transferred proton and electron was same. Besides, at pH 7.0, it can be seen that the oxidation signals were more sensitive compare to reduction peak currents. However, the oxidation currents were then decreased obviously in the alkaline condition. The experimental results demonstrate that high acidic or alkaline surroundings have possibility damage the structure of the nanocomposite membrane and make current response decrease [6, 16]. Thus, the oxidation peak current on pH 7.0 was used as analytical signals for Sunset Yellow in subsequent experiments.

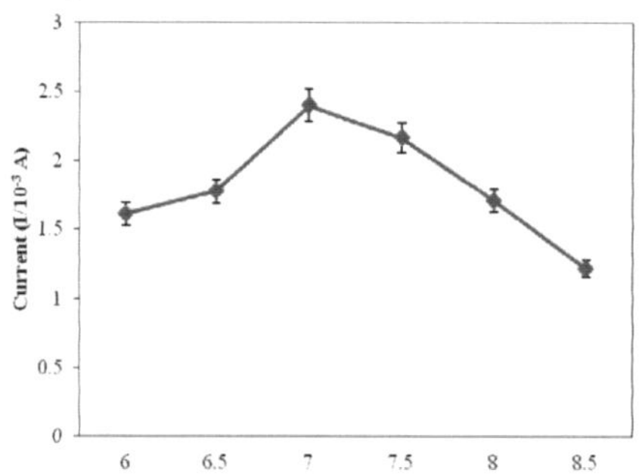

Figure 3. The relationship between the oxidation peak current and pH of Tris-HCl buffer with MB in solution containing 10 ppm of Sunset Yellow.

3.2.2 Influence of accumulation time and scan rate

For handling convenience and further improving the sensitivity, the determination of Sunset Yellow was employed with different scan rate in the range of 0.05 to 0.4 V/s. Cyclic voltammetry was used to investigate the influence of scan rate with the modified CHIT/CaONPs/MWCNTs/AuE surface. From Figure 4A, it is found that the oxidation signal of Sunset Yellow obviously increased from 0.05 to 0.3 V/s, indicating that accumulation efficiency of Sunset Yellow on the AuE surface is always higher.

After 0.3 V/s, the oxidation peak currents of Sunset Yellow were decreased. The high sensitivity and excellent oxidation currents peak was achieved at 0.3 V/s, which was selected as the optimum scan rate.

The interaction time of Sunset Yellow has influenced the performance of the CHIT/CaONPs/MWCNTs/AuE in the presence of MB. In the same solution as above, the interaction time changed from 5 to 40 s when the scan rate fixed of 0.30 V/s. Accumulation time can improve the loading amount of Sunset Yellow on the surface of CHIT/CaONPs/MWCNTs/AuE, and able to amplify the electrochemical signals. Figure 4B represents the influence of different time interaction between oxidation peak currents Sunset Yellow and CHIT/CaONPs/MWCNTs/AuE. The peak current of Sunset Yellow increased gradually and reached a plateau after 30 sec, meaning that 30 sec was sufficient to reach the saturation of Sunset Yellow on the CHIT/CaONPs/MWCNTs/AuE. Due to the electrode surface was completely attached by Sunset Yellow with the increasing accumulation time and lead to increase the peak current [16]. Consequently, 30 sec was chosen as optimum interaction time for detection.

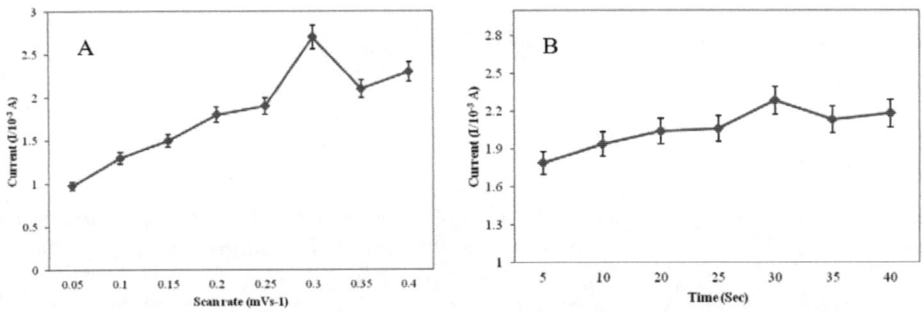

Figure 4. The relationship between the oxidation peak current (A) different scan rate and, (B) different accumulation time in solution containing 10 ppm of Sunset Yellow.

3.3 Electrochemical characterization of the modified AuE

The electrochemical behavior of the bare AuE and modified AuE were characterized by DPV using MB as electroactive indicator (Figure 5). The CHIT/CaONPs/MWCNTs/AuE (curve a) was found highest oxidation peak currents according to CHIT/CaONPs (curve b) and bare AuE (curve c). CHIT/CaONPs/MWCNTs/AuE was attributed large surface area and high accumulation efficiency of Sunset Yellow. Meanwhile, CHIT/CaONPs and bare AuE were exhibited that the oxidation activity of Sunset Yellow was very poor due to less electron and proton transferring. It is clearly indicated that the oxidation signal is more sensitive CHIT/CaONPs/MWCNTs/AuE as compare to the reduction signal.

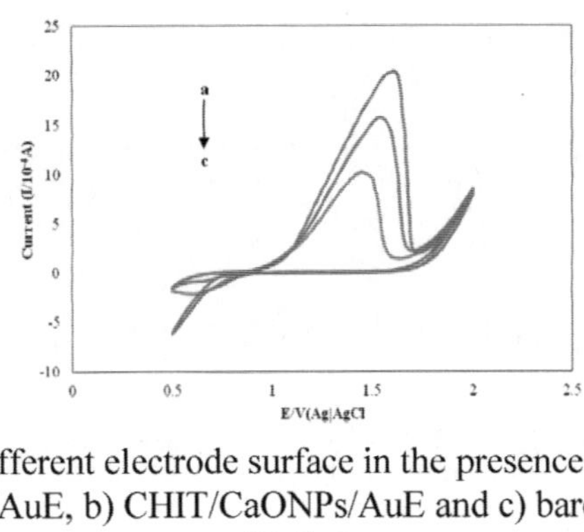

Figure 5. DPV measured different electrode surface in the presence in pH 7.0, Tris-HCl buffer: a) CHIT/CaONPs/MWCNTs/AuE, b) CHIT/CaONPs/AuE and c) bare AuE with 10 ppm of Sunset Yellow.

3.4 Determination of Sunset Yellow

Differential pulse voltammetry (DPV) was performed to investigate the relationship between the peak currents and different concentration of Sunset Yellow. Figure 6 shows the DPV responses with different concentrations of Sunset Yellow using MB as a redox indicator. Under the optimum conditions, DPV was detected with different concentration of Sunset Yellow in the range of 0.9-10 ppm, with detection limit of 0.8 ppm (n = 5). At potential volt -0.18, a linear regression equation was expressed as I (10^{-6} A) = 0.6006 (ppm) + 0.7071, with correlation value of r^2 = 0.9931. These results indicated a very good analytical performance of the developed electrochemical sensor in food products analysis.

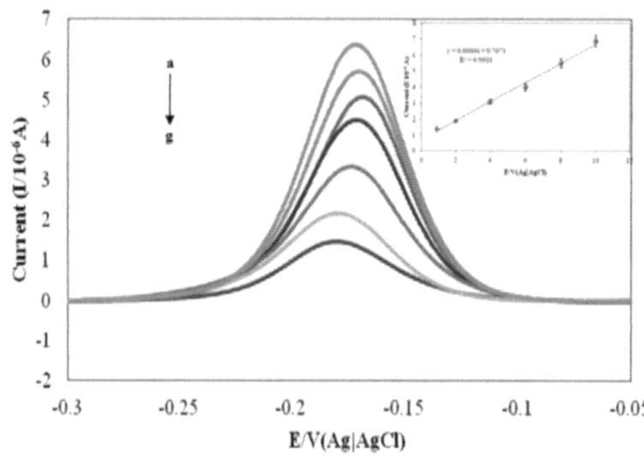

Figure 6. (A) DPV method was measured with different concentrations of Sunset Yellow in analytical buffer (Tris-HCl, pH 7.0) (a-g: 10, 8, 6, 4, 2, 0.9, and 0 ppm). Inset: calibration plots of the oxidation peak currents as a function of Sunset Yellow.

3.5 Analytical application in food products

In order to validate and verify the applicability of the developed method, the CHIT/CaONPs/MWCNTs/AuE was applied for determination of Sunset Yellow at varying food samples purchased from local market, Kota Kinabalu, Sabah. Four different samples were analyzed including candy, royal jelly, ice cream and soft drink. The accuracy of the modified AuE was analyzed by performing a recovery test after spiking the samples. Based on the results obtained, the recovery rates were calculated to be 91.8 to 97.5%, showing the high accuracy and feasible. Three replications were conducted per experiment, and the relative standard deviation was found lower than 1%, revealing that the results obtained by the developed method are acceptable and good precision (Table 1). Thus, the developed method can be successfully applied for detection of Sunset Yellow in food products.

Table 1. Recovery studies of food products.

Samples	**Recovery**	**RSD (%)**
Candy	95.2	0.27
Royal jelly	91.8	0.31
Ice cream	92.3	0.62
Soft drink	97.5	0.38

4. Conclusion

A simple and rapid electrochemical sensor has developed based on the modified gold electrode (CHIT/CaONPs/MWCNTs/AuE) using MB as a redox indicator. The developed method has exhibited high sensitivity and stability for detection of Sunset Yellow. The peak currents have increased with the increasing Sunset Yellow concentrations from 0.9-10 ppm. The detection limit is calculated of 0.8 ppm, which is much lower than traditional methods, with linear coefficiency of 0.9931. The recovery rates of food beverages products are 91.8-97.5 %. The developed method is successfully applied for determination of Sunset Yellow level in food products. The developed electrochemical sensor offers a simple, fast, high selectivity and sensitivity, wide detection range and convenient method for use in food research laboratories.

5. References

[1] Yadav A A, Kumar P D, Dwivedi A and Tripathi M 2012 *Toxicol. Lett.* **208** pp 239–245
[2] Ghoreishi S M, Behpour M and Golestaneh M 2011 *Anal. Meth.* **3** pp 2842–2847
[3] Gan T, Sun J Y, Cao S Q, Gao F X, Zhang Y X and Yang Y Q 2012 *Electrochim. Acta* **74** pp 151–157
[4] Al-Degs Y S 2009 *Food Chem.* **117** pp 485–490
[5] Ghoreishi S M, Behpour M and Golestaneh M 2012 *Food Chem.* **132** pp 637–641
[6] Chen X R, Wu K B, Sun Y Y and Song X J 2013 *Sensors Actuat. B: Chem.* **185** pp 582–586
[7] Zhang W K, Liu T, Zheng X J, Huang W S and Wan C D 2009 *Colloids Surf. B Biointerfaces* **74** pp 28–31
[8] Song Y Z 2010 *Revue Canadienne de Chimie* **88** pp 676–681
[9] Rouhani S 2009 *Anal. Lett.* **42** pp 141–153
[10] Behzad R, Omid R and Ali A E 2014 *Sensors Actuat. B: Chem.* **196** pp 539–545
[11] Virupaxi G, Robert J, Santoshkumar B and Govindarajan T R 2014 *Mater. Sci. Eng. C, Mater. Biol. Appl.* **40** pp 299–307
[12] Ghoneim M M, El-Desoky H S and Zidan N M 2011 *Desalination.* **274** pp. 22–30
[13] Sahraei R, Farmany A and Mortazavi S S 2013 *Food Chem.* **138** pp 1239–1242
[14] Shawish H M, Ghalwa N A, Saadeh S M and Harazeen H E 2013 *Food Chem.* **138** pp 126–132
[15] Zhang K, Lu L, Wen Y, Xu J, Duan X and Zhang L 2013 *Anal. Chim. Acta* **787** pp 50–56
[16] Li J, Wang X, Duan H, Wang Y, Bu Y and Luo C 2016 *Talanta.* **147** pp 169–176

Acknowledgments

This work was supported by grants from the Ministry of Education Malaysia, Fundamental Research Grant Scheme (FRGS) (No. FRGS/1/2014/SG05/UMS/02/4).

IOP Conf. Series: Earth and Environmental Science **36** (2016) 012024 doi:10.1088/1755-1315/36/1/012024

Effect of titania nanotubes on the flux and separation performance of polyethersulfone membranes

M N Subramaniam[1], P S Goh[1], A F Ismail, W J Lau
[1]Advanced Membrane Technology Research Center, *Universiti Teknologi Malaysia*, 81310, Skudai, Johor

Email: [1]mahesannaidu@yahoo.com, [1]peisean@petroleum.utm.my

Abstract. The improvement of membrane performances in terms of separation and permeation is a constant research problem. In this study, polyethersulfone (PES) hollow fiber nanocomposite membranes were fabricated with titania nanotubes (TNT) synthesized via hydrothermal method used as inorganic filler to improve separation and membrane permeation. The membranes were prepared using dry/wet phase inversion process. The concentration of PES was fixed at 18% and TNT was added at concentration of 0.1% wt. The membranes were characterized in terms of pure water permeation rate (PWP), proteins (BSA) rejection, porosity and water contact angle. The results showed that addition of TNT improves the water flux and rejection rate as well as increasing hydrophilicity of the membrane. The flux of membrane is improved more than 20% while the rejection has been improved from 79% to 96%. The addition of TNT into the membrane matrix has shown that photocatalytic property can be introduced to a normal membrane.

1. Introduction

The demand for cheaper and reliable ways to purify products or remove waste generated by industrial processes is ever present. Cost and simplicity are two things that are always taken into consideration when it comes to new technologies. This is where membrane technology can come into the fore to provide a cheap and reliable system that can filter out impurities.

Membrane is universally described as a selective layer of barrier that only allows certain particles to go through it. With the developing technology in the membrane industry, researchers are able to tailor the structure and characteristics of the membrane to fit the profile of whatever feed they intend to use it for. This is where incorporation of non-organic materials comes into play in current study trends. Many different types of inorganic fillers have been added into membranes to improve the performance of membrane in terms of permeance and selectivity [1].

In light of this, titanium dioxide nanotubes (TNT) in particular has been receiving strong interest from researchers as a nanomaterial that not only improve the permeance of membrane, but also incorporate photocatalytic activity it is renowned for into the membrane as a dual functioning membrane [2]. Titanium dioxide is environmentally friendly and has good photocatalysis properties. The synthesis of TNT has many methods, template-assisted method, the electrochemical anodic

[1] To whom any correspondence should be addressed.

oxidation method and hydrothermal treatment are among popular choices [2]. In this study, hydrothermal method is used to synthesize the TNT as they are environmentally friendly, easy to handle, and able to produce nanotubes with minimal issues [3]. Furthermore, they are easily modifiable and produce nanotubes in random alignment [4]. The downside of this method is their relatively long reaction time and the difficulty to obtain nanotubes with same sizes [3]. Hence, in this study, TNT, which is synthesized using the hydrothermal method with the appropriate reaction time and temperature, is incorporated into PES hollow fiber membranes via physical blending in which polyvinylpyrrolidone (PVP) is used as pore former. The effects of TNT on the membrane performance and characteristic such as water permeance, salt and molecular weight cut off are investigated. Furthermore, morphological studies will be done on both nanotubes formed and membranes fabricated to observe the changes brought about by the addition of TNT into membrane.

2. Experimental

2.1 TNT synthesis

The starting material used for the synthesis of TNT were rutile commercial Titanium Dioxide TiO_2 P25 Degussa supplied by Evonik Industries. The PES used was from Radel®, polyvinylpyrrolidone (PVP) from Fluka Analytical. Both hydrochloric acid (HCl) and sodium hydroxide (NaOH) used in the synthesis of TNT was from Merck & Co, while the solvent used for the preparation of polymer dope solution was from Across Chemicals. All the chemicals were used as received without further modification or purification. For the synthesis of TNT, 10 M of NaOH was mixed with TiO_2 and stirred for 12 hours using a magnetic stirrer. The blended solution was then transferred into a Teflon-lined autoclave container and placed into a Autoclave block and sealed tightly. The sealed autoclave was then placed into an oven and was heated at 180°C for 24 hours. The heated autoclave was allowed to cool down to room temperature before the content was removed for further treatment. The sample was washed with 0.1M of HCl, and then washed with tap water until the pH of the residual water reached 7. The sample was then filtered and placed in an oven at 60°C to remove any moisture, and then ground into powder form. The sample was analyzed using transmission electron microscope (TEM), x-ray diffraction (XRD) and brunauer–emmett–teller (BET) to determine its morphology and characteristic.

2.2 Preparation of PES/TNT hollow fiber membranes

The PES/TNT dope solution was prepared by dispersing 0.1wt% of TNT into n-methylpyrrolidone (NMP) solvent and placed into a sonicator for 60 minutes to ensure 100% dispersion of TNT in the solvent. Next, pvp K90 was added into the dope solution. Then, 18 wt% of PES polymer pellets were added into the dope solution and blended until homogeneous using a mechanical stirrer. Finally, 10wt% of ethanol was added into the dope solution drop wise. The hollow fiber membrane was spun using an annular spinneret with an internal diameter of 0.5m and an outer diameter of 1.15mm. Water was used as the bore fluid and external coagulant. The fabricated membrane was then immersed in water for one day to ensure complete removal of solvent and then treated with glycerol to improve integrity of membrane. A neat membrane was also fabricated without the addition of TNT to study the effect of TNT on membrane performance and characteristics.

2.3 Fabrication and performance analysis of fabricated membranes

The membranes were fabricated using the same spinning parameters, with a spinneret dimension of 1.15mm outer diameter and 0.50mm inner diameter, H_2O used as inner and outer coagulation agent, and the membrane were finally treated with glycerol to strengthen its structure. Finally, the membrane was air dried for 48 hours and then underwent performance testing. Pure water flux, bovine serum albumin (BSA) rejection, Contact angle, porosity and pore volume of the membrane fabricated.

2.4 Contact angle
Membrane contact angle test was done via sessile drop method, using distilled water to determine the hydrophilicity and wettability of the membrane. The test was done for 10 times and the results were averaged to improve accuracy.

2.5 Membrane porosity
Membrane porosity can give some information on the performance of the membranes fabricated. Select amount of membranes were cut and the ends were closed using epoxy resin. Then, they were immersed in water for 3 hours and the weights were taken after removing superficial water on its surface using tissue. Then, the membranes were air dried in oven for a night and weighed again. The porosity of the membranes was calculated using equation 1 below:

$$\epsilon = \frac{W0-W1}{V} \quad (1)$$

where ε is the membrane porosity, W0 and WI is the weight of wet and dry membrane in grams, and V is the membrane volume in cm^2.

3. Experimental results

3.1 Characterization of TNT
Three different characterizations were done on the fabricated TNT, namely x-ray diffraction to determine the crystallinity of the nanotubes, TEM to determine the morphology and dimensions of the nanotubes, and finally BET to determine the surface area and porosity of the nanotubes formed. The XRD analysis shows that the nanotubes formed are in anatase form, which is similar to the crystallinity of the starting material used. Figure 1 is the crystallinity graph obtained from XRD analysis.

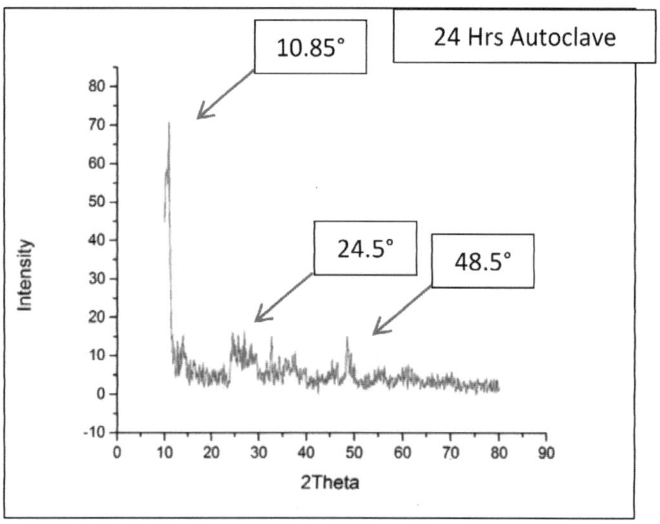

Figure 1. X-Ray Diffraction result on the titania nanotubes (24 hours) formed via hydrothermal method.

As shown in figure 1, few significant peaks appear at angles of 10.85°, 24.5° and 48.5° evidenced the formation of anatase TNT. The XRD patterns are in agreement with recent publications on the formation of TNT [5,6]. Since the crystallinity of the nanotubes formed is similar with the starting material used, TiO_2 P25 Degussa, it can be deduced that the alkaline hydrothermal treatment does not affect the crystallinity of the nanotubes.

However, looking into the surface area of the nanotubes formed, there was an increase of up to 500%, where the BET analysis on the TNT formed is 197.7021 m^2/g, in agreement with literatures published elsewhere [5-7], while the commercial TiO_2 P23 Degussa has a surface area of only 40.0 m^2/g to 50.0 m^2/g. This massive increase is attributed to the tubular form of TNT, which increases the surface area greatly compared to the nanoparticles which is grain shaped.

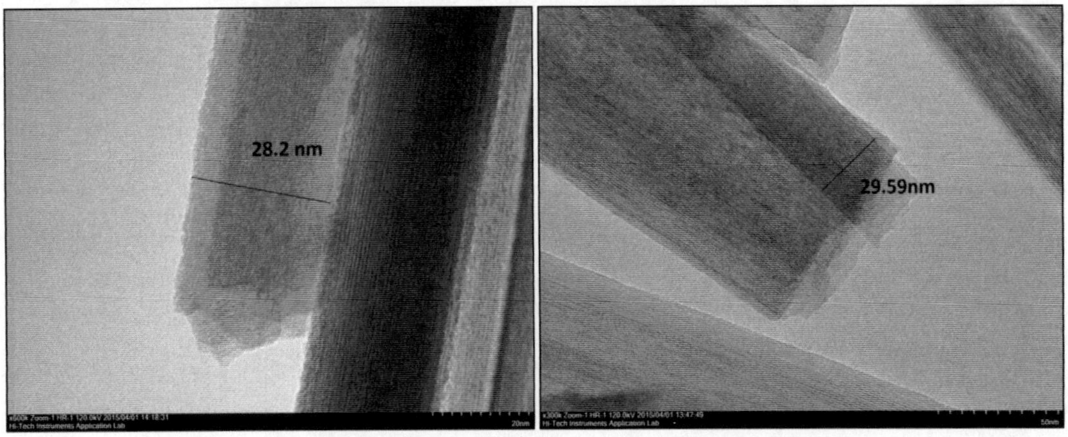

Figure 2. TEM Images of TNT treated vis hydrothermal method for 24 hours

Figure 2 shows the results of TEM analysis done on the morphology and structure of the TNT fabricated. Both images show the diameter of the nanotubes formed is at the range of around 29 nm, while the images also shows the inner wall of the nanotubes formed. Furthermore, both images indicate that the tubes formed are open ended. However, clear images of single images were not taken as the nanotubes faced agglomeration due to their high surface area, with high surface area increases the collision frequency among nanoparticles, which promotes agglomeration among nanoparticles [8].

3.2 Pure water flux and Bovine Serum Albumin (BSA) rejection test

The neat and mixed matrix membrane (MMM) fabricated was tested to determine the changes in pure water flux and the BSA rejection.

Table 1. Performance comparison between neat and MMM membrane

Membrane	Pure water flux	BSA Rejection	Contact Angle	Porosity
Neat	15.70 L/m^2h	79.71%	87.4 ± 2.0	69.77 g/cm^3
TNT 0.1%	19.61 L/m^2h	96.79%	69.0 ± 2.4	67.79 g/cm^3

The first test done was to see the difference in the pure water flux of the membrane. As seen in the table 1, the pure water flux of the membrane added with TNT as an additive is higher compared to the neat membrane fabricated, with an improvement of nearly 20%. The addition of TNT in the PES membrane has improved the flux as the structure of TNT has hydroxyl groups which are polar in nature. The presence of hydroxyl group has improved the pure water flux of the membrane [9]. The molecular weight cut off (MWCO) of the membrane shows that the addition of TNT into the polymer matrix improved the rejection from 79.71% to 96.79%. This indicates that the addition of smaller sized nanoparticles into the polymer matrix enhances its permeability and its separation property [10].

3.4 Contact angle of membrane
The contact angle analysis shows that the contact angle reduced from 87.4° ± 2.0° to 69.0 ± 2.40, supporting the fact that the addition of TNT has increased the hydrophilicity of the membrane. This result is similar to the work done by [11] where contact angle of the membrane was reduced after the introduction of TNT into the polymer matrix. This is due to the fact that the addition of TNT has introduced higher amount of hydroxyl radicals, which enhances the hydrophilicity of the membrane.

3.5 Membrane porosity analysis
Membrane porosity analysis was done to determine how the addition of TNT into the polymer matrix affects the membrane porosity. The porosity of the neat membrane was calculated at 69.77 g/cm^3 while the porosity of the MMM reduced to 67.79 g/cm^3. This can be explained by the fact that the addition of TNT into the polymer matrix hinders the phase inversion process, hence the slight addition of the inorganic TNT has reduced the porosity of the membrane by a small amount [10]. This also explains why the addition of TNT has improved the rejection performance of the membrane compared to neat PES membrane.

4. Conclusion
This study has shown that addition of TNT as a nanocomposite into membranes have the ability to improve the performance of PES based membranes in terms of its pure water flux and also its rejection ability. The unique tubular structure of the TNT also is an upgrade compared to nanoparticles in terms of its higher surface area. Furthermore, the OH hydroxyl group that is present in the TNT also improves the hydrophilicity of the membranes incorporated with it.

5. References

[1] Kustiningsih I and Purwanto W W 2014 *Int. J. Tech.* **5** 133
[2] Yuan Z Y and Su B L 2004 *Colloids Surfaces. A Phys. Eng. As.* **241** 173
[3] Liu N, Chen X, Zhang J and Schwank J W 2014 *Catal. Today* **225** 34
[4] Vu T H T, Au H T, Tran L T, Nguyen T M T, Tran T T T and Pham M T 2014 *J. Mater. Sci.* **49** 5617
[5] Chen X and Mao S S 2007 *Chem. Rev.* **107** 2891
[6] Paulose M, Prakasam H E, Varghese O K, Peng L, Popat K C and Mor G K 2007 *J. Phys. Chem.*14992
[7] Lee K, Mazare A and Schmuki P 2014 *Chem. Rev.* **114** 9385
[8] Suttiponparnit K, Jiang J, Sahu M, Suvachittanont S, Charinpanitkul T and Biswas P 2011 *Nanoscale Res. Lett.* **6** 1
[9] Kim K S, Lee K H, Cho K and Park C E 2002 *J. Membr. Sci.* **199** 135
[10] Han L, Xu Z, Yu L, Wei Y and Cao Y 2010 *Iranian Polym. J.* **19** 553
[11] Shaban M, AbdAllah H, Said L, Hamdy H S and Abdel Khalek A 2014 *Chem. Eng. Res. Des.* **95** 307

Acknowledgement
The authors would like to express their sincere gratitude towards Ministry of Education for the Fundamental Research Grant Scheme (Vot no.: 4F306) and Universiti Teknologi Malaysia for providing financial assistance in this project.

Decolourisation of palm oil mill biogas plant wastewater using Poly-Diallyldimethyl Ammonium Chloride (polyDADMAC) and other chemical coagulants

A Y Zahrim [1, 2, 3, 4*] and Z DDexter[1]

[1] Chemical Engineering Programme, Faculty of Engineering, *Universiti Malaysia Sabah*, Jalan UMS 88400 Kota Kinabalu, Sabah, Malaysia
[2] Water Research Unit, *Universiti Malaysia Sabah*, Jalan UMS 88400 Kota Kinabalu, Sabah, Malaysia
[3]Sustainable Palm Oil Research Unit, *Universiti Malaysia Sabah*, Jalan UMS, 88400 Kota Kinabalu, Sabah, Malaysia
[4] Energy Research Unit, *Universiti Malaysia Sabah*, Jalan UMS 88400 Kota Kinabalu, Sabah, Malaysia

E-mail: [1]zahrim@ums.edu.my

Abstract. Palm oil mill effluent was expected as a future source of renewable biogas. Nevertheless,colours in palm oil mill biogas plant wastewater (POMBPW) causes negative perception among the public and the wastewater is difficult to be treated biologically. In this study, the performance of various chemical coagulants i.e., calcium lactate, magnesium hydroxide, ferric chloride, aluminium chlorohydrate i.e. CK-800, CK-1000, and polyDADMAC, forPOMBPW colour removal were investigated. PolyDADMAC (1,000 mg/L) shows best colour removal (~48%). The main coagulation process with polyDADMACcould be due to charge neutralization-bridging mechanism. The zeta potential analysis supports the finding where the value became positive as the dosage increases. The addition of polyDADMAC has increased the conductivity of the treated wastewater up to 9.22%; however, the final pH is maintained (8.0-8.3). It can be deduced that polyDADMAC has potential to treat POMBPW at low dosage.

1. Introduction

Biogas plant is adapting the anaerobic digestion technology system where it uses anaerobic bacteria to convert the organic carbon i.e. chemical oxygen demand (COD) and biochemical oxygen demand (BOD) to biogas [1]. Nevertheless, the liquid wastewater from this plant i.e. palm oil mill biogas plant wastewater (POMBPW) still has high level of polluting contents, thus cannot be discharged into the watercourse. Nowadays, polymers such as poly-diallyldimethylammonium chloride (PolyDADMAC) and polyacrylamide (PAM) are used widely in coagulation process as a sole coagulant or in combination with inorganic metals. PolyDADMAC is a synthetic, high molecular weight, and highly positively charged (cationic) polymer [2]. In this study, the colour removal of POMBPW was

[1] To whom any correspondence should be addressed.

investigated by a coagulation/flocculation process using a polyDADMAC as a sole coagulant. In addition, the effect of various coagulants was also investigated i.e. calcium lactate, magnesium hydroxide, ferric chloride, and aluminium chlorohydrate.

2. Materials and methods

The POMBPW was collected from the overflow of anaerobic digester tanks in Lahad Datu, Sabah. Six types of coagulant were used in this study and prepared for their stock solution such as calcium lactate (Molecular mass 308.32 g/mol) (Merck, Germany), magnesium hydroxide (Molecular mass 58.32 g/mol) (Sigma-Aldrich, USA), and ferric chloride hexahydrate (Molecular mass 270.30 g/mol) (Techno Pharmchem, India) were prepared by dissolving their solid form in distilled water. Two coagulants in liquid form were supplied by Chemkimia Sdn. Bhd. Both coagulants are aluminium chlorohydrate; Chemchlor CK-800 and Chemchlor CK-1000 are in 45% wt. Polymer i.e. polyDADMAC (Tramfloc® 724, 40% wt) was supplied by Tramfloc, Inc., Houston, Texas. Details on coagulation/flocculation experiments as well as physico-chemical analysis were described in previous publications [3, 4]

3. Results and discussions

3.1 Effect of various coagulants

Figure. 1 shows the effect of coagulant types on the colour removal of POMBPW at the dosage of 1,000 mg/L. PolyDADMAC gives the highest removal i.e. 47.9%, followed by CK-800 (37.4%), magnesium hydroxide (21.9%), and calcium lactate (9.2%). However, both CK-1000 and ferric chloride did not show any removal. PolyDADMAC might make the flocs undergo charge neutralization and bridging mechanisms [5]. Despite the same chemical (ACH), both CK-1000 and CK-800 have different specific gravity. This specific gravity is the measurement of the chemical strength and has a strong relationship with the basicity and molecular weight of the chemical [6]. Higher specific gravity indicates higher molecular weight. Molecular weight of chemical may affect its dissolution in water where the dissolution rate decreases as the molecular weight increased [7]. Therefore in this case, CK-800 is easily soluble in compared to the CK-1000 which is has high molecular weight. Contrary with magnesium hydroxide and calcium lactate, where there was no rapid formation of flocs after been added into the wastewater. Magnesium hydroxide form precipitates, which have large adsorptive surface area and positive electrostatic surface charge, that can attract the negatively charged particle in the solution through charge neutralization and adsorption [8-10]. Same case with calcium lactate, adsorption and charge neutralization mechanisms are occur during the process, however electronegative behaviour of calcium is lesser than the magnesium, thus it affect the attraction of particle of the solution with the metal.

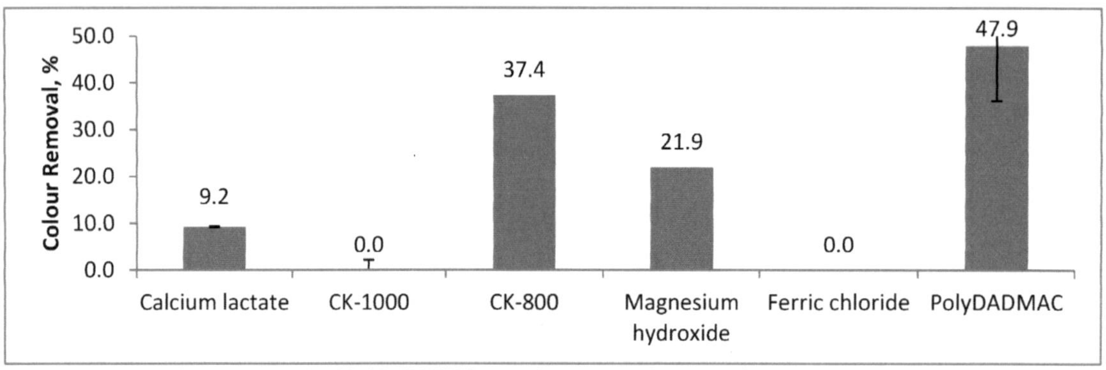

Figure 1. The colour removal of POMBPW by various coagulants at the dosage of 1,000 mg/L, pH = 7.95, Conductivity = 17297 μs/cm

Figure. 2 shows the zeta potential value after the coagulation proces at the dosage of 1,000 mg/L. Initially, the zeta potential value for the wastewater is -13.35 mV, showing the presence of negatively charged particles in the wastewater. The presence of positively charged coagulant adsorbs onto the particle surface thus alter the zeta potential value towards less negative value [11]. However, at low dosage, they did not give a big effect in the wastewater. PolyDADMAC gives a very significant change i.e. -8.01 mV due to high solubility at the given pH. Electrostatic interactions are the main factor for the particles to transfer from the water to the polymer chain during flocculation [12].

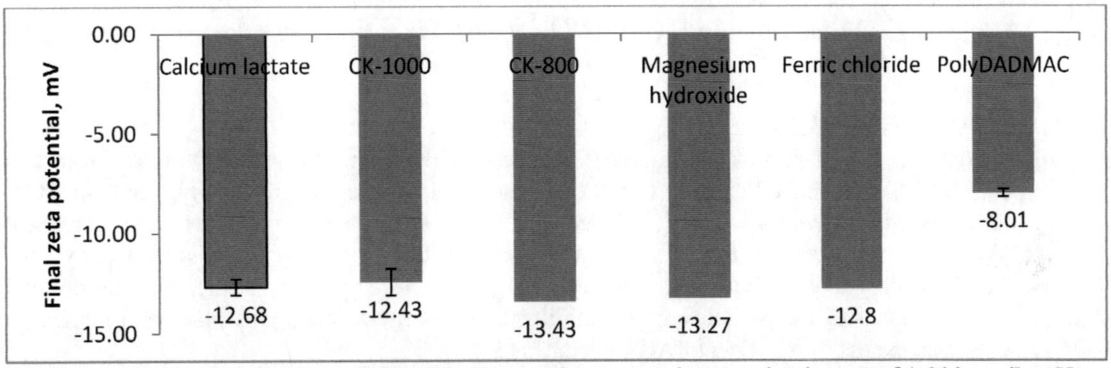

Figure.2. Final zeta potential of POMBPW by various coagulants at the dosage of 1,000 mg/L, pH = 7.95, Conductivity = 17297 µs/cm

Figure. 3 shows the effect of coagulant on treated water pH and conductivity. The pH of the wastewater is increase from the initial pH of 7.95 to 8.45, 8.50, 8.35, and 7.97 in calcium lactate, magnesium hydroxide, polyDADMAC, and CK-800 respectively. The first two coagulants are due to the presence of hydroxide ion (OH^-) when dissociated and dissolved in water and the presence of OH^- ion indicates the properties of the solution to be basic [13]. In polyDADMAC case, it is due to wide range pH that has lift up the value a little. Hydrolysis phenomenon might occur during the process which is at pH higher than 5.5 [3, 14]. CK-800 only shows very small increment. Contrary with the other coagulants, they tend to decrease the final pH to 7.78 and 7.70 for CK-1000 and ferric chloride respectively. This is due to the characteristic of Fe^{3+} and Al^{3+} that behave as Lewis acids, react with OH^- ions (Lewis base) of wastewater precipitate in the form of $Fe(OH)_3$ or $Al(OH)_3$ [15, 16]. Thus, by adding Lewis acid into a solution, it should decrease the pH by changing the balance between H^+ and OH^- ions. Normally, inorganic coagulants such as iron and aluminum salts may lower the alkalinity and pH of the solution into acidic condition [17]. Therefore, the treated wastewater cannot be directly discharge into the watercourse hence it needs to be re-neutralized [3].

The conductivity for raw POMBPW is 17297 µs/cm, yet, this value is still within the range of 6500 - 35900 µs/cm [4, 18]. This parameter indicates of total ionized constituent of water and related to the total sum of inorganic cations or anions that presence in water [19, 20]. Means that, high conductivity indicates the water still contains a large amount of inorganic salt and other ions. The conductivity of treated water using calcium lactate as coagulant has increased by 2.62%.This is due to the excess presence of unreacted calcium lactate [3, 21]. However, the conductivities for treated water using CK-1000, CK-800, magnesium hydroxide, ferric chloride, and polyDADMAC were decreased around 7.50%, 10.39%, 1.72%, 4.61%, and 0.27% respectively. The dissolved organics were removed by adsorption and precipitation along with the settlement of flocs [22]. These findings indicate that the coagulation process might release minimum amount of residual coagulant in the treated water.

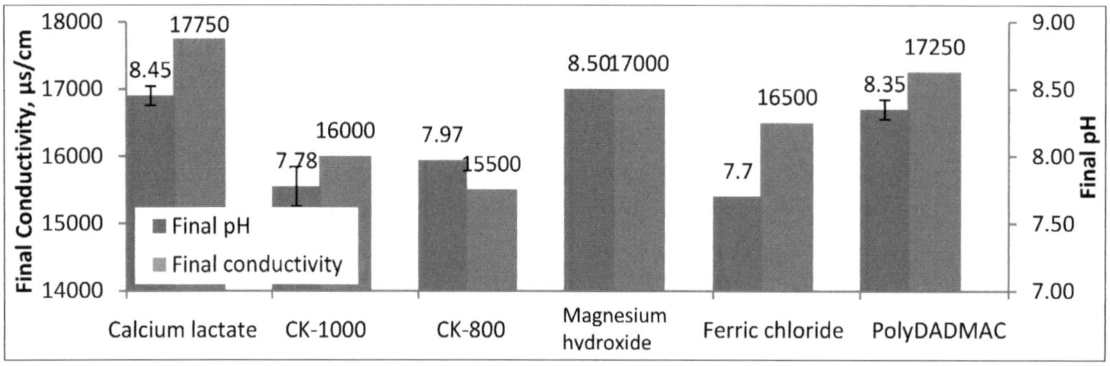

Figure.3. Final pH and conductivity of POMBPW by various coagulants at the dosage of 1,000 mg/L, pH = 7.95, Conductivity = 17297 µs/cm

3.2 Effect of polyDADMAC dosage

Figure. 4 shows the colour removal at different dosage ranging from 500 – 10,000 mg/L. From this figure, polyDADMAC is indeed capable to treat wastewater as a sole coagulant at higher than 1,000 mg/L. The removal is between 40 – 56%; indicates that the characteristic of raw water was varied and polyDADMAC is sensitive to the small change in wastewater characteristics. Similarly, the performance is inconsistent as the dosage increase where it tends to decrease. Ahead of the optimal dosage of polyDADMAC, the particles are re-stabilized and charge reversal take place on the particle surfaces due to an excess of polyDADMAC molecules [3, 23].

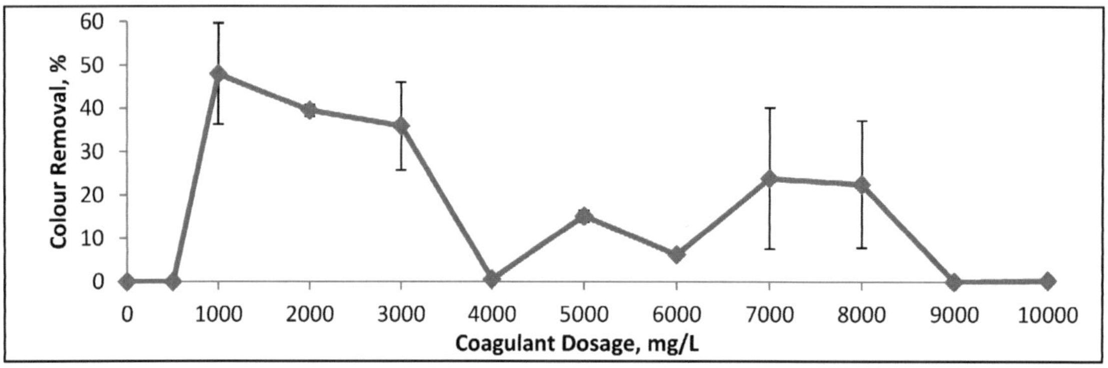

Figure.4. The colour removal of POMBPW by polyDADMAC at different dosages, pH = 7.99, Conductivity = 18,156 µs/cm

Figure. 5 shows the zeta potential at different dosages of polyDADMAC. It clearly shows the zeta potential value is increases as the polyDADMAC dosage increased. Cationic polymer is capable to reduce the negative charge of particles by charge neutralization [3, 24]. Electrical repulsion between particles is reduced, which results the formation of flocs and surface charge reduction [25, 26]. As the polyDADMAC dosage increases, the zeta potential value become greater than zero indicating after complete neutralization mechanism [3] which can be explained by the presence of excess polyDADMAC during the process. The chains of polyDADMAC carry N^+ are attached or adsorbed by the neutralized particles [3, 23].

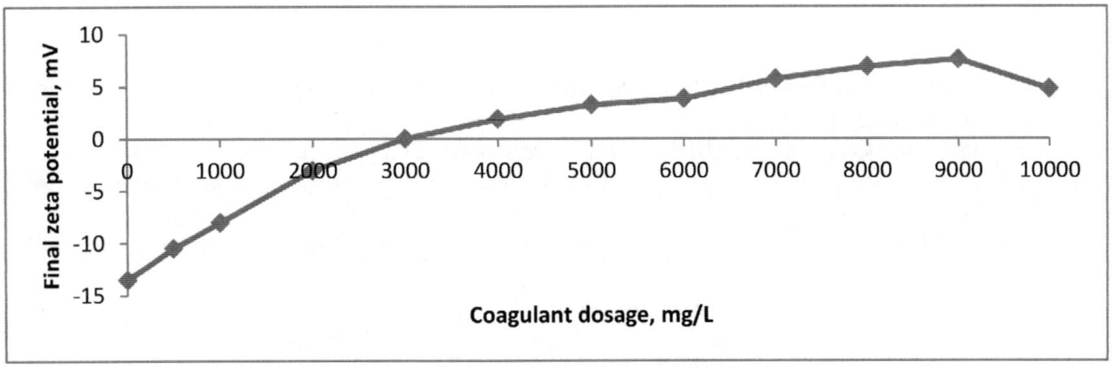

Figure.5. Final zeta potential of POMBPW by polyDADMAC at different dosages,pH = 7.99, Conductivity = 18,156 μs/cm

Figure. 6 shows the pH and conductivity at different dosages by polyDADMAC. The pH shows stable reading between 8.0-8.3 as the dosage increased indicates that it does not consume alkalinity unlike the inorganic coagulant [17]. Inorganic salts in solution form react with alkalinity in the water to form insoluble hydrous oxide that coagulate by sweep flocculation and charge neutralization [27]. The conductivity shows increment of value as the dosage increase. This is due to the presence of unreacted polymer as residues in the wastewater [21]. At optimum dosage of 1,000 mg/L, the conductivity is lesser than the initial value. It was reported that at optimum dosage the conductivity may show zero or minimum increment [3].

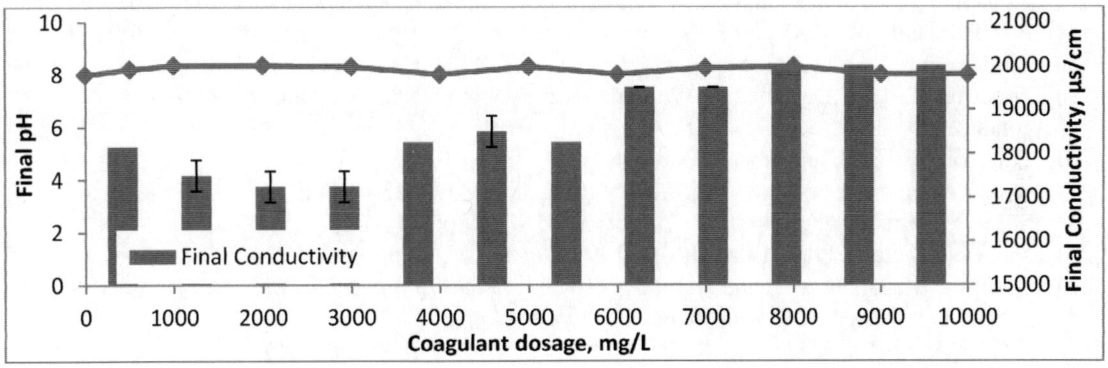

Figure.6. Final pH and conductivity of POMBPW by polyDADMAC at different dosages, pH = 7.99, Conductivity = 18156 μs/cm

4. Conclusions

In this study, at low dosage of 1,000 mg/L, polyDADMAC shows the best removal which is between 40 – 56%. The coagulation process with polyDADMAC may undergo a charge neutralization-bridging mechanism. The zeta potential also shows the value has increase significantly when the dosage was increased. Unreacted polyDADMAC presence in the treated water increase the treated water conductivity. At optimum dosage (1,000 mg/L), the conductivity was lesser than the initial value. The pH was however to be unchanging along the experiment at any dosage. It can be deduced that polyDADMAC has potential to remove the colour of POMBPW at the dosage of 1,000 mg/L. However, the treated water is cannot be discharged into the watercourse yet and a post treatment is highly recommended.

5. References

[1] Chin M J, Poh P E, Tey B T, Chan E S, and Chin K L 2013 *Renewable and Sustainable Energy Reviews*, **26** pp 717-726

[2] Zahrim A Y, Tizaoui C, and Hilal N 2011 *Desalination*, **266** pp 1-16

[3] Zahrim A Y, Nasimah A, and Hilal N 2015 *Journal of Environmental Chemical Engineering*, **3** pp 2145-2154

[4] Zahrim A Y, Nasimah A, and Hilal N 2014 *Journal of Water Process Engineering*, **4** pp 159-165

[5] Pearse M J 2005 *Minerals Engineering*, **18** pp 139-149

[6] Gebbie P 2006 University Central Queensland Campus, Rockhampton

[7] Miller-Chou B A and Koenig J L 2003 *Progress in Polymer Science*, **28** pp 1223-1270

[8] Gao B-Y, Yue Q-Y, Wang Y, and Zhou W-Z 2007 *Journal of Environmental Management*, **82** pp 167-172

[9] Semerjian L and Ayoub G M 2003 *Advances in Environmental Research*, **7** pp 389-403

[10] Zhang G, Li X, Li Y, Wu T, Sun D, and Lu F 2011 *Desalination*, **274** pp 255-261

[11] Tripathy T and De B R 2006 *Journal of Physical Science*, **10** pp 93-127

[12] Zemaitaitiene R J, Zliobaite E, Klimaviciute R, and Zemaitaitis A 2003 *Colloids and Surfaces A: Physicochemical and Engineering Aspects*, **214** pp 37-47

[13] Bruice P Y 2007 Upper Saddle River, NJ: Pearson/Prentice Hall

[14] SNF-FLOERGER (2015, 3 December). Available: snf.com.au/downloads/Preparation_of_Organic_Polymers_E.pdf

[15] Amokrane A, Comel C, and Veron J 1997 *Water Research*, **31** pp 2775-2782

[16] Song Z, Williams C J, and Edyvean R G J 2004 *Desalination*, **164** pp 249-259

[17] Zahrim A Y, Tizaoui C, and Hilal N 2010 *Journal of Hazardous Materials*, **182** pp 624-630

[18] Poh P E, Ong W Y J, Lau E V, and Chong M N 2014 *Journal of Environmental Chemical Engineering*, **2** pp 1174-1181

[19] Fondriest (2015, 7 December). Available: http://www.fondriest.com/news/what-is-conductivity.htm

[20] Emerson (2015, 7 December). Available: www2.emersonprocess.com/siteadmincenter/.../Liq_ADS_43-018.pdf

[21] Solberg D and Wagberg L 2003 *Colloids Surf. A*, **219** pp 161-172

[22] Van Benschoten J E and Edzwald J K 1990 *Water Res.*, **24** pp 1519-1526

[23] Ariffin A, Razali M A A, and Ahmad Z 2012 *Chemical Engineering Journal*, **179** pp 107-111

[24] Lee W and Westerhoff P 2006 *Water Research*, **40** pp 3767-3774

[25] James M, Ebelinga, Kata L, Risbela, Philip L, and Sibrellb 2003 *Aquacultural Engineering*, **29(1)** pp 23-42

[26] Bolto B and Gregory J 2007 *Water Res.*, **41(11)** pp 2301-2324

[27] Zeta-Meter I 1988: Zata-Meter

Acknowledgement

Authors would like to thank the Universiti Malaysia Sabah (SGI0013) and the Ministry of Higher Education, Malaysia (FRGS/2/2013/TK05/UMS/02/1) for funding this work.

Preliminary studies on immobilization of lipase using chicken eggshell

S Salleh, N A Serri[1], S Hena, H A Tajarudin
School of Industrial Technology, 11800 Universiti Sains Malaysia, Malaysia

Email: [1]aziah_serri@usm.my

Abstract. A few advantages of enzyme immobilization are reusability of expensive enzyme, improvement of stability and activity compared to crude enzyme. Various organic components can be used as carrier for enzyme immobilization such as chicken eggshell. It can be used as a carrier for immobilization as its mineral component mostly contains of calcium carbonate. In the present study, Tributyrin method was used to test enzyme activity of *Rhizomucour Miehei, Candida Antarctica* and *Candida Rugosa*. *Rhizomucour Miehei* shows the highest enzyme activity (360.8 mol/min/mL lipase) and was used in further experiment. Experiment was continued to study incubation time for lipase immobilization on eggshell (1-4 hours) and reaction time of esterification of sugar ester (0-72 hours). Two hours incubation time for lipase immobilization was observed and gives the highest yield of sugar ester (78.13%). Fructose and stearic acid as substrate was used for the production of sugar ester. The highest percentage of sugar ester production was shown at 36 hours of reaction time.

1. Introduction
Nowadays, there are many methods by researcher to improve the characteristic of an enzyme and also to ensure the stability of the enzyme. One of the methods is by immobilizing the enzyme. Enzyme immobilization is widely used method in industry [1] . In 1976, it was reported as the first industrial use of enzyme immobilization where they develop the immobilization of *Aspergillus oryzae* amino acylase in order to resolute synthetic racemic D-L amino acid [2]. In order to choose the carrier for immobilization, certain properties have to be followed ; hydrophilicity, ability to resist microbial attack, biocompatibility and availability at low cost [3]. Egg is one of the common foods consumed by humans. According to statistics by Penang Veterinary Services, 360,641.98 metric tonnes of chicken eggs was produced in 2014 [4]. The yolk inside the egg is only the part that can be eaten meanwhile the shell is considered as a solid waste. The cuticle on the outer surface, a spongy calcerous layer and an inner lamellar layer are the three layered structure that made up the hen eggshell [5]. All avian eggshells are made up from the same mineral component, calcium carbonate ($CaCO_3$) also known as calcite [6].

[1] To whom any correspondence should be addressed.

One of the main problems in enzyme immobilization is the cost of the carrier. As the eggshell can be obtained at zero cost, it can be used as the carrier for enzyme immobilization. On top of that, eggshells have a great mechanical strength where it helps in resisting against microbial attack [7]. There are many methods for enzyme immobilization and adsorption is the most common method used. Only weak physical interaction occurred between the carrier and enzyme such as Van Der Waals forces, hydrogen bonding and ionic interactions. This weak binding maintained the native structure of the enzyme [8].

There are a wide range of applications sugar ester in industry such as in food, cosmetics and surfactants in pharmaceuticals [9]. The efficiency of the enzyme lipase for esterification process depends on the choice of substrates [10]. Different lipases favor different selectivity either has high selectivity on short, branched fatty acids or long and medium chain fatty acid [11]. According to Soultani *et al.* [12], 100% of fructose conversion yield obtained when C18 fatty acid was used as acyl donor. In this present study, parameter affecting immobilization of lipase on chicken eggshells was studied together with parameter affecting synthesis of fructose fatty acid ester.

2. Materials and Method

2.1 Materials
Lipase from *Rhizomucour Miehei, Candida Antarctica* and *Candida Rugosa* used in this study were purchased from Sigma. All chemicals used were analytical grade.

2.2 Lipase activity assay
Tributyrin method was used to determine the lipase activity following method suggested by Chattopadhyay and Sen [13]. 1mL of 0.02M NaOH is equivalent to 100 mol fatty acid liberated per minute. The lipase activity was expressed as fatty acid produced per minute per mg lipase. For immobilized lipase on eggshell, 1g of eggshell was used.

2.3 Lipase Immobilization
Before lipase immobilization, the eggshells used were pretreated. Eggshells were boiled in a 0.1% Sodium dodecyl sulfate (SDS) solution for 15 min. The shells were washed thrice with distilled water to remove residual SDS. Next, eggshells were washed with acetone thrice to remove water. Acetone were removed by drying at 60°C for 5 h. Dried eggshells were then crushed and sieved through a mesh to generate uniform size. Lipase was added into conical flask containing eggshells and incubated in incubator shaker at 37°C. After incubation, the conical flasks were stored at 4°C for 16 h. The enzyme-immobilized matrices were then washed with phosphate buffer (50mM, pH7) three times to remove unbound enzyme.

2.4 Esterification
In this study, fructose and stearic acid was used as substrates for the synthesis. Synthesis of sugar ester was done in a 25mL shake flask where 0.5 mol stearic acid, 0.5 mol fructose and 0.2g of immobilized lipase were added into the shake flask. 6mL of ethanol was added as solvent. The mixture was incubated for 72 hours at condition 200 rpm and 37°C. Another set of mixture which acts as a control was prepared simultaneously without lipase. This method was done according to Neta *el at.*, [14]. To determine the content of sugar ester, volumetric method was used by calculating the amount of fatty acid left in the reaction mixture. 0.1g of aliquot was taken from the reaction mixture and added into 20mL of ethanol. Phenolphtalein was used as indicator. The sample was then titrated with NaOH (0.02M). The yield was calculated based on equation 1-3.

$$\frac{V\,NaOH * 0.02}{W\,sample} = X \tag{1}$$

$$\frac{V\ NaOH*0.02}{W\ Control} = Y \quad (2)$$

$$\text{Yield (\%)} = 100 - \frac{X}{Y} * 100 \quad (3)$$

V NaOH indicated the volume of sodium hydroxide, W sample was the weight of the sample and W Control was the weight of the control.

3. Results and Discussion

3.1 Lipase activity

Lipase activity assay was done on three different lipases from *Rhizomucor Miehei, Candida Antarctica* and *Candida Rugosa.* As shown in Table 1, *Rhizomucor Miehei* showed the highest lipase activity that was 360.8 mol/min/mL. Based on this result, *Rhizomucor Miehei* was chosen to be used in further study on parameter affecting immobilization. All the experiments were done in triplicate. Previous study by Palomo *et al.* [15] used all three enzyme in their study but immobilized lipase from *Rhizomucor Miehei* showed 20 times more active than the soluble enzyme.

Table 1. Lipase activity (at pH 7, 37° C)

Enzyme	Activity (mol/min/mL lipase)
Rhizomucour Miehei	360.8 ± 1.6
Candida Antarctica	32.64 ± 1.7
Candida Rugosa	6.83 ± 1.4

3.2 Effect of incubation time on lipase immobilization

Based on Figure 1, 2 hr of incubation time shows the highest lipase activity followed by 1 hr of incubation time. This result was supported by previous research that also acquired 2 hr as the best incubation time [16]. Lipase activity started to drop at 3 and 4 hr. Long incubation time can caused multi-layered adsorption and will decrease the enzyme activity [17].

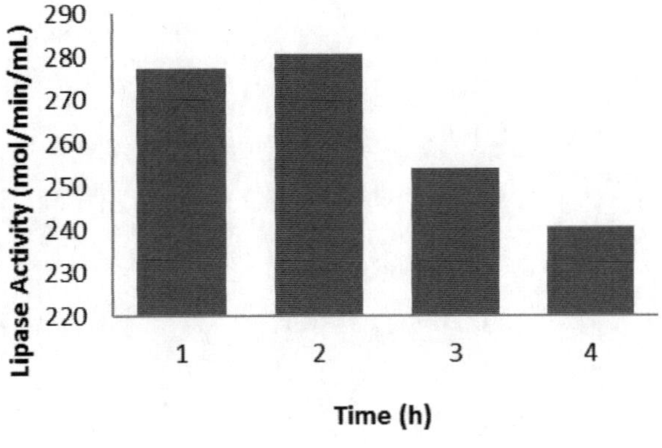

Figure 1. Lipase activity for eggshell immobilized with *R.Miehei*

3.3 Effect of enzyme loading on eggshell immobilized lipase

Enzyme loading was one of the vital parameter should be studied as it will affect the production cost. In this study, 0.2 mL and 0.4 mL was loaded into 1g of eggshell during immobilization. After immobilization, these eggshells were used in synthesis of sugar fatty acid ester (Figure 2) .Both showed the highest yield at 36 hours where 78.13% sugar ester was produced. 0.2 mL enzyme loading showed faster conversions compared to 0.4 mL. Study by Soultani *et al.* [12] which used the same substrates (fructose and stearic acid) achieved 100% conversion yield at condition 60°, 200 rpm using 2-methyl-2 butanol as solvent. Although the maximum yield obtained in this present study was only 78%, lower temperature was used (37°C).

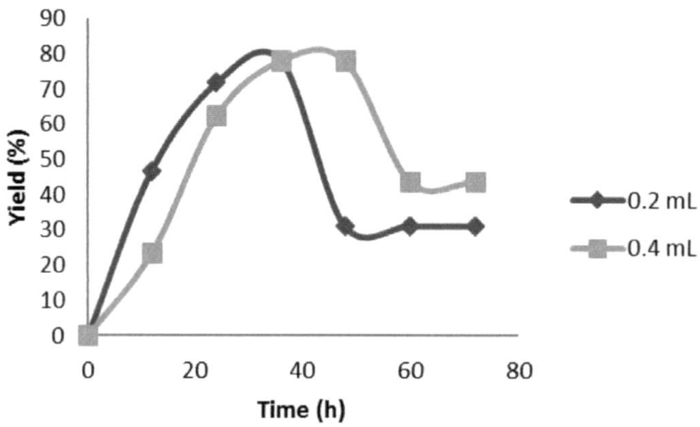

Figure 2. Esterification of fructose and stearic acid with eggshell immobilized lipase using different enzyme loading

3.4 Reaction time of esterification producing sugar ester

Figure 3 showed the reaction time from 0-72 hours of esterification using different incubation time of lipase immobilization. The highest yield of conversion showed at 36 hr (78.13%). After 36 hr, the conversion started to drop and maintained from 48 hr to 72 hr. The same period of time was also obtained by Neta *et al.* [14] when synthesizing sugar fatty acid ester with the highest yield at 37.8 hr. There were quite large differences in yield of conversion between 1and 2 hr and 3 and 4 hr incubation time (Figure 3). This may be due to decrease of lipase activity as shown previously in Figure 1 (Section 3.2)

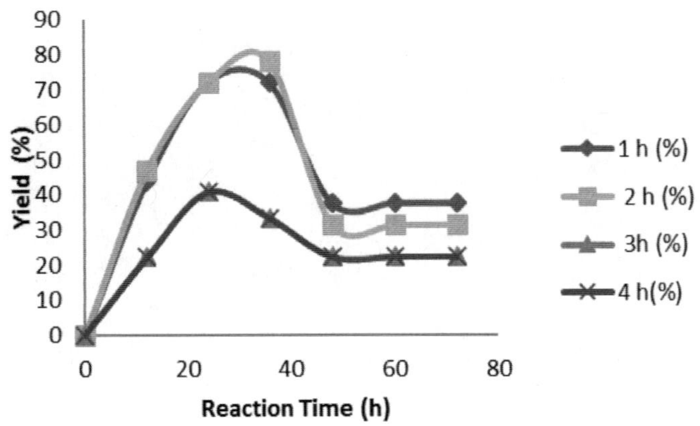

Figure 3. Reaction time for synthesis fructose fatty acid ester using lipase immobilized on eggshells at different incubation time.

4. Conclusion

Based on this study, eggshells can be used as a carrier for lipase immobilization. Lipase from *Rhizomucor Miehei* which obtained the highest lipase activity was used for immobilization onto eggshells with the enzyme loading of 0.2 mL and incubated for 2 hr. 78.13% yield was obtained when the immobilized lipase was used in synthesis of fructose fatty acid ester.

Acknowledgement

This work was supported by Kementerian Pendidikan Malaysia (KPM) FRGS research grant (6711374) and KPM MyMaster.

References

[1] Preeti Joshi H C J and Sunil Kumar Sanghi Subir Kundu 2010 *Microchim Acta* **169** 383-88
[2] Tosa T, Mori T Fuse and and Chibata I 1966 *Enzymologia* 31.
[3] Batista-Viera, B M B A F 2006 Immobilization of Enzymes: A Literature Survey. *In:* GUISAN, J. M. (ed.) *Immobilization of Enzymes and Cells.* Second Edition ed. New Jersey: Humana Press Inc.
[4] Pinang, U. Perancang Dan Penilaian J.P.V.P., 2015 Perangkaan_Ternakan.pdf. , p.17. Available at: http://www.jpvpp.gov.my/index.php/my/peraturan/statistik-jabatan [Accessed January 23, 2016].
[5] Tsai W T, Yang J M, Lai C W, Cheng Y H, Lin C C and Yeh C W 2006 *Bioresource Technology* **97** 488-93.
[6] Nys Y, Gautron J, Garcia-Ruiz J M and Hincke M T 2004 *Comptes Rendus Palevol* **3** 549-62
[7] Makkar H P S and Sharma O P 1983 *Biotechnology and Bioengineering* **25** 595-7
[8] Jesionowski T, Zdarta J and Krajewska B 2014 *Adsorption* **20** 801-21
[9] Flores M V and Halling P J 2002 *Biotechnology and Bioengineering* **78** 795-801
[10] Gumel A M, Annuar M S M, Heidelberg T and Chisti Y 2011 *Process Biochemistry* **46** 2079-90
[11] Adachi S K T 2005 *Bioscience and Bioengineering* **99** 87-94
[12] Soultani S, Engasser J M and Ghoul M 2001 *Journal of Molecular Catalysis - B Enzymatic,* **11** 725–31
[13] Chattopadhyay S and Sen R 2012 *Process Biochemistry* **47** 749–57 Available at: http://dx.doi.org/10.1016/j.procbio.2012.02.003
[14] Neta N D A S, Santos J C S D, Sancho S D O, Rodrigues S, Gonçalves L R B, Rodrigues L R & Teixeira J A 2012 *Food Hydrocolloids* **27** 324-31
[15] Palomo, JM Et Al 2002 *Journal of Molecular Catalysis B: Enzymatic* **19-20** 279–86

[16] Soham Chattopadhyay R S 2012 *Process Biochemistry* **47** 749-57
[17] Li Y, Zhou G, Li C, Qin D, Qiao W and Chu B 2009 *Colloids and Surfaces A: Physicochemical and Engineering Aspects* **341** 79-85
[18] Neta N S Et Al 2011 *New Biotechnology* **28(4)** 349–55

The potential of papaya leaf extract in controlling *Ganoderma boninense*

Z H Tay and K P Chong[1]
Sustainable Palm Oil Research Unit (SPOR), Faculty of Science and Natural Resources.
Universiti Malaysia Sabah, 88400, Jalan UMS, Kota Kinabalu, Sabah, Malaysia

Email: [1]chongkp@ums.edu.my

Abstract. Basal Stem Rot (BSR) disease causes significant losses to the oil palm industry. Numerous controls have been applied in managing the disease but no conclusive result was reported. This study investigated the antifungal potential of papaya leaf extracts against *Ganoderma boninense*, the causal pathogen of BSR. Among the five different solvents tested in extraction of compounds from papaya leaf, methanol and acetone gave the highest yield. *In vitro* antifungal activity of the methanol and acetone extracts were evaluated against *G. boninense* using agar dilution at four concentrations: 5 mg mL^{-1}, 15 mg mL^{-1}, 30 mg $mL^{-1}$and 45 mg mL^{-1}. The results indicated a positive correlation between the concentration of leaf extracts and the inhibition of *G. boninense*. ED50 of methanol and acetone crude extracts were determined to be 32.016 mg $mL^{-1}$and 65.268 mg mL^{-1}, respectively. The extracts were later semi-purified using solid phase extraction (SPE) and the nine bioactive compounds were identified: decanoic acid, 2-methyl-, Z,Z-10-12-Hexadecadien-1-ol acetate, dinonanoin monocaprylin, 2-chloroethyl oleate, phenol,4-(1-phenylethyl)-, phenol,2,4-bis(1-phenylethyl)-, phenol-2-(1-phenylethyl)-, ethyl iso-allocholate and 1-monolinoleoylglycerol trimethylsilyl ether. The findings suggest that papaya leaf extracts have the ability to inhibit the growth of *G. boninense*, where a higher concentration of the extract exhibits better inhibition effects.

1. Introduction

Oil palms in Malaysia are mostly affected by Basal Stem Rot disease (BSR), which is caused by *Ganoderma boninense*. The disease causes a yearly economic loss of up to RM1.5 billion to the oil palm industry [1]. BSR was first discovered in Malaya in 1928 and was reported to be affecting only old oil palm trees but had slowly changed to palms as young as 10-15 years. After 1957, the disease was observed to have spread to younger oil palms even at the nursery stage [2]. Numerous methods have been established in an attempt to control the disease. For instance, cultural control methods such as sanitation, replanting, removing the root masses and digging trenches around the infected plants are adopted to prevent the spread of the disease [3]. Chemical controls such as fungicide are applied but are ineffective. This led to the development of biological control methods using biological agents such as bacteria and fungi to suppress the growth of *G. boninense*. However, no conclusive result has been reported to stop the disease completely, extensive research is required to combat this pathogen.

[1] To whom any correspondence should be addressed.

In the current study, the leaves of papaya were tested for their antifungal potential against *G. boninense*. Papaya leaves have been used in cancer cell line studies for its anti-tumour properties, showing it to be potentially useful in the medical field [5]. Juice made from papaya leaves was shown to increase platelet counts in patients with dengue fever, as compared with the control group [6]. Papaya leaves also exhibit many other antibacterial and antifungal activities. Previous research showed that papaya leaf extract has high inhibition activities towards gram-negative and gram-positive bacteria and fungi *Candida albicans* [7] and *Xanthomonas axonopodis,* which causes Cassava blight disease [8]. Papaya leaf extract was also found to be effective in controlling soft rot disease affecting yam caused by *Mucor circinelloides* and *Rhizopus nigricans* [9]. In this paper we report the antifungal activity of papaya leaf extract against *G. boninense*.

2. Methodology.

2.1 Collection of papaya leaves

Fresh papaya leaves collected from Kota Kinabalu were washed first under running tap water, followed by sterile distilled water. After drying at 40 °C for 72 hours [10], the leaves ground to a fine powder using a mechanical blender (Waring ® Commercial Blender). The dry weight was determined, and the powder was stored in an air-tight container at -20 °C until further use [11].

2.2 Preparation of Ganoderma boninense *culture*

A stock culture of *G. boninense* was obtained from Genetic Laboratory of Faculty of Science and Natural Resources, University of Malaysia Sabah. The mycelia was transferred onto PDA agar plates and incubated at 25 °C. Culture plates fully covered with white mycelia were used within two weeks.

2.3 Solvent Extraction of Papaya Leaves

The compounds of interest in the papaya leaves were extracted using polar and non-polar solvents: methanol, acetone, chloroform, ethyl acetate and n-hexane. One hundred grams of powdered leaves were immersed in conical flasks containing 1000 mL of methanol, acetone, chloroform, ethyl acetate and hexane, respectively. The flasks were then placed in a sonicator (Branson ® 5510) for 10 minutes at 25 °C. The extracts were pump filtered through Whatman No. 1 filter paper. The filtrates were evaporated and condensed using Rota Vapor (BUCHI) at 40 °C, 150 rpm, to obtain the final volume of 1 mL of extract per 10 g of plant sample. The aliquots were then weighed and kept at -20 °C for further use [12]. The extraction yield was determined using the following formula:

$$\text{Yield (\%)} = \frac{\text{Dry weight of extract}}{\text{Dry weight of plant powder}} \times 100$$

2.4 In vitro Antimicrobial Assay

Crude extracts were dissolved in Dimethyl sulfoxide (DMSO), and then a serial concentration of plant solvent extracts at 5, 15, 30, 45 mg mL^{-1} was prepared. Agar with DMSO (no plant extract) was prepared as negative control; agar (no DMSO or extract) was the absolute control.*G. boninense* was taken from the edge of a seven day old culture using a sterile micropipette tip sized 0.8 cm and then placed in the middle of the media. Growth of the pathogen was measured (in cm). Each extract concentration was assayed in triplicates and the mean values were calculated. ED_{50} for both the methanol and acetone crude extracts were determined using SPSS Probit analysis.

2.5 Preparation of Solid Phase Extraction (SPE) papaya leaf extract

The leaf crude extracts were purified using solid phase extraction (SPE) installed with Strata™ C18 cartridges (500 mg/ 6 mL). Absolute methanol (1 mL) was used to activate the sorbent which was equilibrated with sterile deionized distilled water (1 mL) afterwards. Crude extracts were loaded and left in the sorbent matrix for up to a minute, then washed with 1 % methanol (1 mL) to obtain the 'flush fraction'. The 'elute fraction' was obtained by using 5 mL of methanol: acetonitrile (1:1; v:v) [12].

2.6 Identification of Potential Antifungal Compounds using GC-MS
Compounds in the SPE extracts were identified using Gas Chromatography (model: Agilent, 7890A) interfaced with Mass Spectrometer (Model: Agilent, 5975C) which was equipped with Agilent J & W GC capillary column with HP-5MS stationary phase (30 m x 0.25 mm x 0.25 μm) and composed of (5%-phenyl)-methylpolysiloxane. An electron ionization system with ionization energy of 70eV was used. Helium gas (99.999%) was used as the carrier gas at a constant flow rate of 0.8 mL min^{-1}and an injection volume of 1 μL (split ratio of 10:1). With the injector temperature at 250 °C and ion-source temperature at 280 °C, the oven temperature was programmed from 110 °C (isothermal for 2 min) with an increase of 10 °C min^{-1}, to 200 °C, then 5 °C min^{-1} to 280 °C (isothermal for 9 min). Mass spectra were taken at 70 eV (a scan interval of 0.5 secs and fragments from 40 to 550 Da). The total GC running time was 41 min. The relative percentage of each component was calculated by comparing its average peak area to the total area. The software used to handle the mass spectra and chromatograms was GC/MSD ChemStation [2].

2.7 Data Analysis
Each treatment was replicated three times and the results were expressed as mean ± standard deviation. The mean values were subjected to Analysis of Variance (one-way ANOVA), and the Post Hoc tests (SPSS statistical package, version 22) was used to determine the significant differences ($p < 0.05$) between treatments.

3 Results

3.1 Extraction yield of papaya leaves
Powdered papaya leaves were extracted using organic solvents at the ratio of 100 g of leaf samples to 1000 mL of solvents. Different types of solvents resulted in different extraction yields (table 1). Methanol gave the highest yield of extraction (8.24%) while n-hexane gave the lowest (2.06%).

Table 1. Extraction yield of papaya leaves using different solvents

Solvents	Solvent Polarity Index	Yield (%)
Methanol	6.6	8.24
Acetone	5.4	7.20
Ethyl acetate	4.3	6.87
Chloroform	4.1	4.04
n-Hexane	0.1	2.06

3.2 In Vitro inhibitory effect of papaya leaf crude extract against G. boninense
The antimicrobial activity of the papaya leaf crude extracts from the two solvents with the highest extraction yield (methanol and acetone) was studied via *in vitro* antimicrobial assay against *G. boninense* (figure *1* and figure 2). *G. boninense* had grown to the maximum size of the plate in negative control at day 16. There was a significant reduction of growth rates for *G. boninense* in plates with methanol and acetone papaya leaf crude extract with the slowest growth shown in the highest tested concentration of methanol extract (45 mg mL^{-1}). An exception was the case of methanol and acetone extracts (5 mg mL^{-1}), where the inhibitory effect was lower than the negative control containing DMSO during the first week of observation. However, this enhanced growth did not continue until the end of the observation. Statistical analysis indicated that fungal growth treated with both the methanol and acetone crude extracts were significantly lower ($p< 0.05$) as compared to negative control. *G. boninense* were affected less in acetone extracts which were of similar concentrations as the methanol extracts. All the tested concentrations of methanol and acetone papaya leaf crude extracts failed to totally inhibit growth of the pathogen. From Probit analysis, ED_{50} of

methanol and acetone crude extract was determined to be 32.015 mg mL^{-1} and 65.268 mg mL^{-1} respectively.

The morphology of *G. boninense* was observed for plates incorporated with methanol and acetone extracts. In the control and methanol plates, the fungus grew to form a white mat that appeared cotton-like, which grew outwards from the fungal plug [14]. In the acetone extract plates, the mycelia growth zone was unclear, and the structure was not white and cotton-like. The acetone extract at 45 mg mL^{-1} had white cotton-like mycelia but was malformed and stunted (figure *3*).

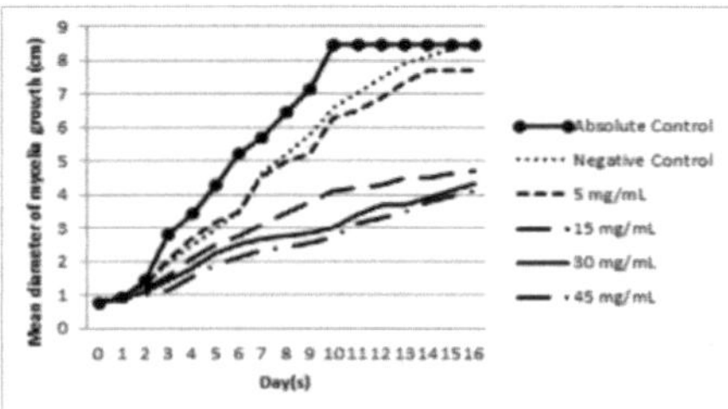

Figure 1. *G. boninense* mycelial growth on PDA incorporated with different concentrations of methanol extract of papaya leaf. Each mean obtained from three replicates. Absolute control: agar without addition of DMSO. Negative control: agar with addition of DMSO.

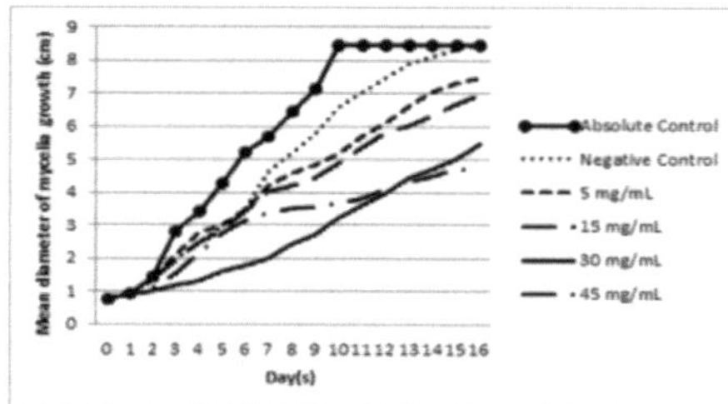

Figure 2. *G. boninense* mycelial growth on PDA incorporated with different concentrations of acetone extract of papaya leaf. Each mean obtained from three replicates. Absolute control: agar without addition of DMSO. Negative control: agar with addition of DMSO.

3.3 GC-MS analysis of antifungal compounds in papaya leaves

The methanol and acetone crude extracts of papaya leaves were fractionated using solid phase extraction into elute and flush fractions. The compounds in the fractions were analysed separately using GC-MS and identified by comparing to the National Institute of Standard and Technology (NIST) database. Different compounds were identified from the eluant and flush fractions of the papaya leaf SPE extracts from methanol and acetone.

Five classes of compounds were identified from the methanol and acetone extracts: carboxylic acid, ester, fatty acid ester, phenol and steroid. Two steroid compounds were discovered in the eluted fraction of methanol SPE extracts: 1-monolinoleoylglycerol trimethylsilyl ether (RT= 35.05 min; 77.32%) and ethyl iso-allocholate (RT= 33.40 min; 22.67 %). In the flush fraction of methanol SPE extracts, the amount of fatty acid ester, dinonanoin monocaprylin (RT=32.63min; 71.78%) was much higher than 1-monolinoleoylglycerol trimethylsilylether (RT= 32.38 min; 20.35%) while octasiloxane, 1,1,3,3,5,5,7,7,9,9,11,11,13,13,15,15-hexadecamethyl- (RT= 34.00 min) was only around 4.84 %. Three compounds were found in eluted fraction of acetone SPE extracts: phenol,4-(1-phenylethyl)- (RT= 13.51 min; 0.09 %), phenol,2,4-bis(1-phenylethyl) (RT= 27.23 min; 0.29 %) and octasiloxane, 1,1,3,3,5,5,7,7,9,9,11,11,13,13,15,15-hexadecamethyl- (RT=28.52min; 0.75%). Flush fraction of acetone SPE extracts contained four compounds with Z,Z-10-12-Hexadecadien-1-ol acetate being the most aundant (RT=20.99min; 46.26%), followed by 2-chloroethyl oleate (RT= 21.11 min; 38.24%), phenol-2-(1-phenylethyl)- (RT= 13.52 min; 15.50 %) and decanoic acid,2-methyl- (RT= 17.21 min; 11.15 %).

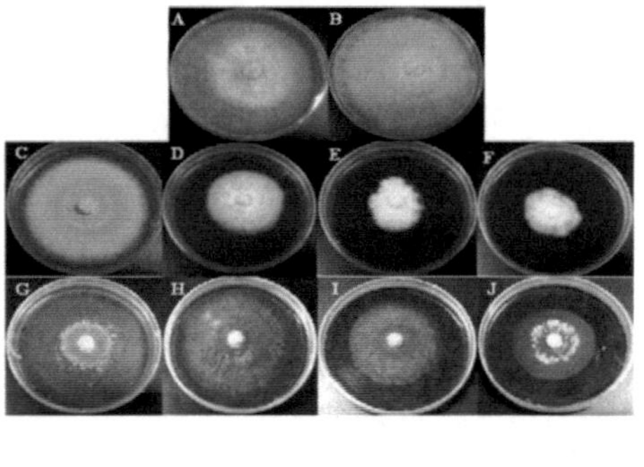

Figure 3. Growth of *G. boninense* on incorporated PDA. (A) Absolute control (B) Negative control with

Methanol extract: (C) 5 mg mL^{-1} (D) 15 mg mL^{-1} (E), 30 mg mL^{-1}, (F), 45 mg mL^{-1}, with Acetone extract: (G) 5 mg mL^{-1} (H) 15 mg mL^{-1} (I) 30 mg mL^{-1}, (J) 45 mg mL^{-1} on day 16.

4 Discussion

4.1 Extraction of papaya leaves using different solvents

The current study showed methanol to produce the highest yield in extracting compounds from papaya leaves as compared to acetone, ethyl acetate, chloroform and n-hexane, consistent with previous work [15, 16]. Higher extraction yield achieved by methanol could be explained by its ability to dissolve more compounds including phenols and endogenous compounds [17]. Besides the polarity of the solvents, the chemical compositions of the sample also determine the yield of extraction. Polar solvents extract polar compounds while less polar solvents extract non-polar compounds [17]. Polar compounds are also easier to extract compared to non-polar compounds [18]. It is likely that papaya leaves contain more polar than non-polar compounds.

4.2 *Inhibitory effects of papaya leaf crude extracts to* G. boninense

Four concentrations of extracts were tested, and the fungicidal effects to *G. boninense* were shown to have a positive correlation, except in the case of methanol and acetone extracts (5 mg mL^{-1}). This was also shown in a previous report on *Microsporum canis* where plant extracts at low concentrations enhanced mycelia growth [19]. It is possible that the low concentration of extract meant that the inhibition only started taking effect after a short while. Methanol extracts showed a higher inhibition than acetone extracts, possibly due to its higher extraction yield resulting in higher concentration of bioactive compounds in the extract. It is also possible that the different compound composition resulted in higher inhibition in the methanol extract.

4.3 Identity of potential inhibitory compounds in papaya leaf SPE extract

The identified compounds were classified to five classes based on their possible activities; carboxylic acid, ester, fatty acid ester, phenol and steroid. Each of these compounds has been shown previously to have antifungal activities.

Phenolic compounds were reported to have antifungal activities against *G. boninense* [20], *Candida* species [21], *Alternaria solani*, *Botrytis cinerea* and *Fusarium culmorum* [22], possibly by diffusing through the fungal membrane and interfering with the synthesis of essential components such as ergosterol, glucan, chitin, proteins and glucosamine [23]. Fatty acids or carboxylic acids are one of the essential components in plants with antimicrobial activities. For instance, the saturated fatty acid, decanoic acid,2-methyl- was shown to inhibit hyphae formation of *C. albicans* [24]. Fatty acid esters extracted from *Phyllanthus amarus* was shown to be inhibit growth of several bacteria species [25]. The identified ester, dinonanoin monocaprylin, in this experiment was shown to have the potential to inhibit pathogens such as *Streptococcus* species and *Escherichia coli* [26].The identified steroids, ethyl iso-allocholate and 1-monolinoleoylglycerol trimethylsilyl ether were also shown previously in various reports to have antimicrobial and antibacterial potential [27, 28], the latter being a major compound in leaf and stem extracts of *Sesuvium portulacastrum* [29].

Generally, steroids were detected in the methanol extract while the phenolic compounds and carboxylic acids were from the acetone extracts. Fatty acid esters were detected in both methanol and acetone extracts. The compounds of different categories could have different inhibitory effect shown in *in vitro* antimicrobial assay.

5 Conclusion

The inhibitory effect to *G. boninense* was greater at higher concentrations of papaya extracts, but the effects were not complete inhibition. ED_{50} of methanol and acetone crude extracts of papaya leaves were determined to be 32.016 mg mL^{-1} and 65.268 mg mL^{-1}, respectively. The eluted and flush fractions of the papaya leaf extracts were carboxylic acid, ester, fatty acid ester, phenol and steroid. These bioactive compounds had the potential in contributing to the inhibitory effect of papaya leaf extracts towards *G. boninense in vitro*. Future research can be conducted to identify the minimum fungicidal concentration of papaya leaf extract in inhibiting growth of *G. boninense*. The extracts can be further purified and re-tested with the pathogen.

6 References

[1] Arif M S, Roslan A, Idris A S 2011 In Proceedings of the Third MPOB-IOPRI International Seminar Integrated OP Pests and Diseases Management 14 November 2011 Kuala Lumpur Convention Centre

[2] Upgade A, Bhaskar A 2013 *Asian J. Pharm. and Clinical* Res. **64** 101-106

[3] Naher L, Yusuf U K, Ismail A, Tan S G, Mondal M M A 2013 *Jacq Australian J. Crop Sci.* **711** 1723-1727

[4] Fauziya S, Krishnamurthy R 2013 *CIBTech J. Pharm Sci.* **21** 25-34

[5] Subenthiran S, Tan C C, Kee C C, Thayan R, Mok B T, Muniady P K, Afzan A, Abdullah N R, Ismail Z 2013 *Evidence-Based Complem. and Alternat. Med.* **2013** 1-7

[6] Peter J K, Kumar Y, Pandey P, Masih H 2014 *J. Pharm. and Biol. Sci.* **92** 29-37

[7] Olahan G S, Amadi J E 2003 *NISEB J.* **34** 117-121

[8] Ngumah C 2012 *Food Sci. and Tech.* **132** 202-209

[9] Nirosha N, Mangalanayaki R 2013 *L Int J. Adv. in Pharm. Bio. and Chem* **23** 473-476

[10] Jayaraman S, Manoharan M S, Illanchezian S 2008 *Trop. J. Pharm* Res **74** 1143-1149

[11] Abdullah S, Gobilik J, Chong K P 2012 *Int J. Phar.m and Pharmaceutical Sci.* **45** 227-230

[12] Abdullah S, Gobilik J, Chong K P 2014 *Int. Conf. on Civil Biol. and Envi. Eng. CBEE*-2014 Istanbul Turkey

[13] Suryanto D, Wibowo R H, Siregar E B M, Munir E 2012 *African J. Microbio.* Res **69** 2053-2059

[14] Kumoro A C, Hasan M, Singh H 2009 *Sci. Asia* **35** 306-309

[15] Joseph B, Priya M R 2010 *Int J. Current Pharm* Res **23** 28-32

[16] Fatiha B, Khodir M, Farid D, Tiziri R, Karima B, Sonia O, Mohamed C 2012 *L Pharmacognosy Communications* **24** 72-86

[17] Razak M F B A, Yong P K, Shah Z M, Abdullah L C, Soh S Y, Choong T S Y 2012 *J. Applied Sci.* **1211** 1207-1210

[18] Abd-Elaah G A, Abo - Amer A E, Soliman S A 2006 *Int. J. Agri. Bio.* **84** 434-439

[19] Chong K P, Atong M, Rossal S 2012 *Rot Plant Path.* **61** 953-963

[20] Barros L, Duenas M, Alves C T, Silva S, Henriques M, Santos-Buelga C, Ferreira I C F R 2012 *Industrial Crops and Products* **41** 41-45

[21] Winkelhausen E, Pospiech R, Laufenberg G 2005 *Bulletin of the Chemists and Technologists of Macedonia* **241** 41-46

[22] Ansari Ma, Anurag A, Fatima Z, Hameed S 2011 *Sci. technology and education Formatex* Res. Center Spain 1189-1195

[23] Murzyn A, Krasowska A, Stefanowicx P, Dziadkowied D, Lukaszewicz M 2010 *PLoS ONE* 58 e12050

[24] Akin-Osanaiye Cb, Gabriel Af, Alebiosu Ra 2011 *Annals of Biol.* Res. **22** 298-305

[25] Mulyono N, Lay B W, Rahayu S, Yaprianti I 2012 *Int. J. Pharm. Biol.* Archives **34** 770-778

[26] Sarayanan P, Chandramohan G, Mariajancyrani J, Shanmugasundaram P 2014 *Int J. Pharm. and Pharmaceutical Sci.* **61** 457-460

[27] Meenakshi V K, Gomathy S, Senthamarai S, Paripooranaselvi M, Chamundeswari K P *2012 J. Current Chem & Pharm Sci.* **24** 271-276

[28] Sheela D, Uthayakumari F 2013 *L BioSci Disc* **41** 47-53

Acknowledgement

The authors wish to acknowledge the support from the Faculty of Science and Natural Resources, Universiti Malaysia Sabah throughout the project.

Fruit waste adsorbent for ammonia nitrogen removal from synthetic solution: Isotherms and kinetics

AY Zahrim[1], Y Lija, L N S Ricky, I Azreen
Faculty of Engineering, Universiti Malaysia Sabah, Jalan UMS, 88400, Kota Kinabalu, Sabah, Malaysia.

E-mail: [1]zahrim@ums.edu.my

Abstract. In this study, four types of watermelon rind (WR) adsorbents; fresh WR, modified WR with sodium hydroxide (NaOH), potassium hydroxide (KOH) and sulphuric acid (H_2SO_4) were used as a potential low-cost adsorbent to remove NH_3-N from solution. The adsorption data were fitted with the adsorption isotherm and kinetic models to predict the mechanisms and kinetic characteristics of the adsorption process. The equilibrium data agreed well with Langmuir isotherm model with highest correlation (R^2=1.00). As for kinetic modelling, the adsorption process follows pseudo-second order for all four types of adsorbents which has R^2 value of 1.0 and calculated adsorption capacity, Q_e of 1.2148mg/g. The calculated Q_e for pseudo-second order has the smallest difference with the experimental Q_e and thus suggest that this adsorption process is mainly governed by chemical process involving cations sharing or exchange between WR adsorbent and NH_3-N in the solution.

1. Introduction

Excess ammonia nitrogen (NH_3-N) discharged into water source can cause environmental problems including eutrophication, depletion of dissolved oxygen and toxicity to fish and other aquatic organisms[1]. Thus, prevention of nitrogen pollution with ammonium removal from waterways is important. Commonly, NH_3-N removal is achieved by ozonation[2], air stripping[3], chemical precipitation[4] and adsorption[5]. Adsorption is one of the best NH_3-N removal methods and it is economically feasible and environmental friendly[6]. Great efforts have been made to identify new low-cost adsorbent and fruit waste is one of promising material as an adsorbent. Fruit crops use around 375,000 hectares (5.4%) of land in Malaysia[7] and has total production of 1.8 million metric tons[8] thus leaving behind a large quantities of wastes resulting from the production and food consumption chain. Fruit wastes are unlimited, abundance and cheap[9] thus makes it convenient for experimental purposes. Disposal of fruit waste adsorbent after adsorption process can be accomplished either by composting or incineration [10].

In this study, watermelon waste was utilized as fruit waste adsorbent in NH_3-N adsorption. Watermelon biomass can be categorised into three main components; the flesh, seed and rind. The flesh constitutes approximately 68% of the total weight, the rind approximately 30%, and the seeds approximately 2%[11]. Watermelon rind (WR) was the only part used for this research. The skin of

[1] To whom any correspondence should be addressed.

fully ripened watermelon contain approximately 20% cellulose, 23% hemicellulose, 10% lignin, 13% pectin, 7 mg/g silica and 12% silica free minerals[12]. Previously, it has been reported that the adsorption capacity of solid adsorbent can be increased by chemical modification or pre-treatment method [13-16]. A comparison of adsorption capacity between unmodified and chemically modified adsorbents with sodium hydroxide (NaOH), potassium hydroxide (KOH) and sulphuric acid (H_2SO_4) was reported in this study.

Previous research proves that WR adsorbent can removed variety of pollutant from wastewater such as heavy metals [17] and dyes [18]. However, available information on the NH_3-N adsorption by watermelon rind still limited. The main purpose of this study is to determine the isotherms and kinetics adsorption in order to obtain an insight into the equilibrium characteristics of ammonia adsorption by watermelon rind. The adsorption isotherm and kinetic modelling were studied by fitting the equilibrium data into different types of isotherm (Langmuir, Freundlich, Temkin) and kinetic (Pseudo-first order, pseudo-second order and Intra-particle) models.

2. Materials and Methods

Watermelon rinds were collected from Kota Kinabalu area. The rinds were washed thoroughly with tap water and rinsed with distilled water to remove any impurities. Next, the WR samples were cut into small pieces (around 1 to 2 mm^2) and dried in an oven at 40°C for 48 hours to a constant weight[19]. For chemically modified adsorbent, the WR was soaked in 1 L NaOH solution (20mmol/L), 1 L KOH solution (20mmol/L) and 1 L H_2SO_4 solution (20mmol/L) respectively for 24 h[13]. The watermelon rind was then dried further in an oven at 40°C for 48 hours.

Synthetic ammonium solutions were used throughout the adsorption tests. First, ammonium chloride salt (NH_4Cl) with calculated weight and distilled water were used for the preparation of a stock NH_3-N solution of 50 mg/L. Then, the experimental solutions at the desired ammonium concentrations were prepared by diluting the stock solution with distilled water. The initial pH was adjusted to 7.0 using dilute solutions of NaOH (0.1 N) throughout the experiments [19].

Batch experiments consisted in analysing the zeta potential and effect of contact time and mass of adsorbent on the NH_3-N adsorption by all four types of WR adsorbent. All the samples were adjusted to the optimum pH prior to the addition of the adsorbent. Sample solutions were withdrawn at certain intervals in order to determine the residual concentration by using UV-vis spectrophotometer with HACH Spectrophotometer DR2010 brand at the maximum wavelength of 425 nm. Hence, the amount of ammonia nitrogen adsorbed at equilibrium, qe (mg/g) was calculated as followed:

$$q_e = \frac{(C_0 - C_e)V}{W} \tag{1}$$

where C_0 and C_e (mg/L) are the liquid-phase concentrations of initial adsorbate and equilibrium, respectively. V is the volume of the solution (L) and W is the weight of the adsorbent (g).

2.1. Adsorption isotherm and kinetics studies

The amount of adsorbed solute onto the solid phase (mg/g) versus the solute concentration in the aqueous phase (mg/L) at equilibrium that called sorption isotherm[20]. This was carried out by fitting the equilibrium data to the Langmuir[21], Freundlich[22] and Temkin[23] isotherms. In order to predict the mechanism involved during the present adsorption process and the potential rate controlling steps such as mass transport, pore diffusion and chemical reaction processes, three kinetic models were used to fit the experimental data, namely, pseudo-first-order[24], pseudo-second-order[25] and intra-particle diffusion[26] models for varied type of adsorbents.

3.0 Result and Discussions

3.1. Adsorption isotherm modelling

Figure 1 shows the Langmuir isotherm for four different types of adsorbent. The Langmuir isotherm model exhibited an excellent fit to the adsorption equilibrium data with highest R^2 value of 1.00. The best fit of adsorption data in Langmuir isotherm model confirms the monolayer adsorption of NH_3-N onto WR adsorbents. Fresh WR adsorbent shows the highest maximum monolayer adsorption capacity, Q_m of 1.2192 mg/g and followed by modified WR with KOH (Q_m=1.2051 mg/g), modified WR with H_2SO_4 (Q_m =1.1892 mg/g) and lastly modified WR with NaOH (Q_m=1.1885 mg/g) as shown in Table 1. Several studies on the adsorption of NH_4^+ ions onto different types of fruit waste adsorbents shows different maximum monolayer adsorption capacity; strawberry leaves (Q_m=6.71 mg/g) and strawberry stems (Q_m= 4.62 mg/g)[1], partly decomposed empty fruit bunch (Q_m=0.821mg/g)[20]. Different adsorption capacities of various adsorbent are due to the properties of each adsorbent (structure, surface area) and different experimental condition (initial concentration, mass of adsorbent).

For Freundlich isotherm model, the linear regression coefficient, R^2 of NH_3-N adsorption on the adsorbents ranges from 0.9985-0.9997. The values for Freundlich isotherm constant, n positioned within -38.6 to -92.6 as shown in Table 1. This values showed unfavourable adsorption process as the values does not satisfies the condition $|1|< n<|10|$ [18]. As for Temkin isotherm graph, the constant B is related to the heat of adsorption. Negative values of B Temkin constant were obtained for all four types of adsorbents ranges from -0.0316 to -0.0258kJ/mol. The R^2 values for all four adsorbents positioned within of 0.9985-0.9997 as shown in Table 1.

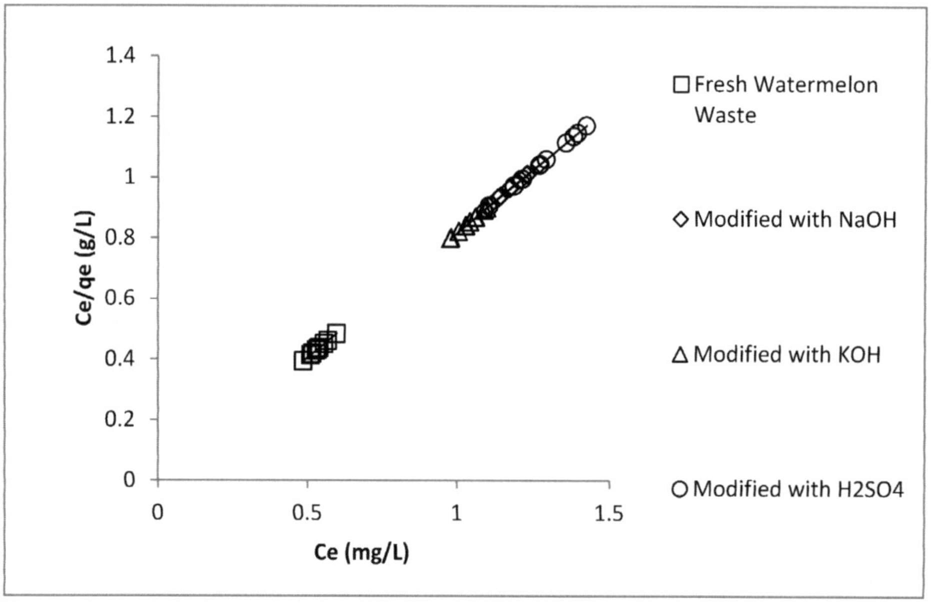

Figure 1. Langmuir Isotherm

Table 1. Isotherm parameters

Isotherm		Adsorbent			
		Fresh watermelon rind	Modified watermelon rind with NaOH	Modified watermelon rind with KOH	Modified watermelon rind with H_2SO_4
Langmuir	Q_m(mg/g)	1.2192	1.1885	1.2051	1.1892
	R^2	1.00	1.00	1.00	1.00
Freundlich	$\|n\|$	92.6	41.7	47.4	38.6
	$K_f((mg/g)(dm^3/mg)^{1/n})$	1.2268	1.2210	1.2238	1.2237
	R^2	0.9988	0.9995	0.9997	0.9985
Temkin	B(kJ/mol)	-0.0134	-0.0295	-0.0258	-0.0316
	R^2	0.9988	0.9995	0.9997	0.9985

3.2. Adsorption Kinetic Modelling

Kinetic modelling is use to investigate the adsorption mechanism and rate controlling step. The concentration of the solution was measured by taking the adsorbent-adsorbate solution at present time intervals as shown in equation 2 below:

$$q_t = \frac{(C_o - C_t)V}{W} \quad (2)$$

where C_o and C_t (mg/L) are the liquid –phase concentrations of adsorbate at initial and at any time, t, respectively. V is the volume of solution and W is the mass of adsorbent used [27]. Figure 2 shows the linearized plot of pseudo-second order for NH_3-N adsorption onto fresh WR, modified WR with NaOH, KOH and H_2SO_4. The correlation coefficients and the other parameters calculated from the all three kinetic models are given in Table 2.

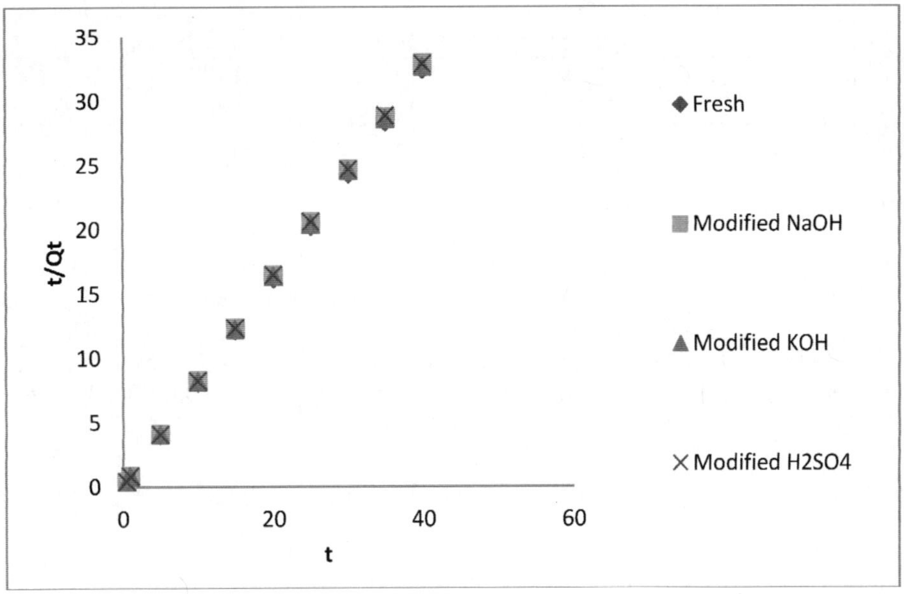

Figure 2. Pseudo-Second order

It was observed that the pseudo first-order kinetic model did not adequately fit the experimental values ($R^2 < 0.60$). Besides that, the difference between experimental and calculated equilibrium adsorption capacity is very high as shown in Table 2. In contrast, the pseudo-second order rate equation for sorption of ammonium onto adsorbents agreed well with the data (R^2=1.00). The calculated equilibrium sorption capacity, $Q_{e,cal}$ for the pseudo second-order was 1.2148 mg/g and it is close to the experimental data values range from 1.2151 to 1.2332 mg/g. This explained that the adsorption process is mainly governed by sharing or exchange of cations between the adsorbent and ammonium ions in the solution [28]. As for intra-particle diffusion model, the experimental data were poorly fitted to intra-particle diffusion model with low R^2 values ($R^2 < 0.60$) compared to pseudo-first and pseudo-second order.

Table 2. Kinetic rate constants

Kinetic Models	Types of adsorbents			
	Fresh watermelon	Modified NaOH	Modified KOH	Modified H_2SO_4
Experimental Qe	1.2332	1.2151	1.2237	1.2165
	Pseudo-first-order model			
$k_1(min^{-1})$	0	-0.0242	0.1740	-0.1019
Calculated $Q_{eI}(mgg^{-1})$	0	0.3022	0.0004	0.2871
R^2	0	0.0096	0.5602	0.1905
	Pseudo-second-order model			
$k_2(min^{-1})$	-616.02	-616.02	-616.02	-616.02
Calculated $Q_{eII}(mgg^{-1})$	1.2148	1.2148	1.2148	1.2148
R^2	1.000	1.000	1.000	1.000
	Intra-particle diffusion			
k_p (mg g^{-1} $min^{1/2}$)	0.1950	0.1921	0.1935	0.1923
R^2	0.1896	0.2545	0.2019	0.5534

4. Conclusion

The adsorption data of NH_3-N onto watermelon rind adsorbents has been fitted with isotherm and kinetic models. The experimental data correlated well with Langmuir isotherm model with highest R^2 value of 1.00 and thus theoretically suggest that the NH_3-N is removed by monolayer adsorption process. The experimental data fitted well with pseudo-second model with R^2 value of 1.00 and has the closest value of experimental and theoretically calculated Q_e of 1.2192 mg/g. Thus, confirm the adsorption process of NH_4^+ ions onto adsorbent is governed by sharing or exchange of cat-ions. The present study shows that the WR have great potential to be used as adsorbent for the NH_3-N removal.

5. References

[1] Liu H, Dong Y, Liu Y and Wang H 2010 *Journal of Hazardous Materials* **178** 1132-6
[2] Yang M, Uesugi K and Myoga H June 1999 *Water Research* **33** 1911-7
[3] Serna-Maza A, Heaven S and Banks C J 2014 *Bioresource Technology* **152** 307-15
[4] Yigit A and Akbal F 2014 *Journal of Selcuk University Natural and Applied Science* 391-6

[5] Ali I, Asim M and Khan T A 2012 *Journal of Environmental Management* **113** 170-83
[6] Ugurlu M and Karaoglu M H *Microporous and Mesoporous Materials* **139** 173-8
[7] Rahim A and Rahman A 2014 *Journal Mechanical* 36-46
[8] Mohammed S M and Rokiah M Y 2006 Fruits and Vegetables for Health Workshop 15-16 August 2006, Seoul, Korea 1-5
[9] Othman N, Mohd-Asharuddin S and Azizul-Rahman M *Applied Mechanics and Materials*: Trans Tech Publ pp 29-35
[10] Krishnie Moodley, Ruella Singh, Evans T Musapatika, Maurice S Onyango and Ochieng A 2011 *Water SA* **37**
[11] Hani M N F, Zahidah W W N, Saniah K and Irwani H M 2014 *J. Trop. Agric. and Fd. Sc* **42** 115-23
[12] Singh R, Kumar J and Nandpuri K 1975 *Indian Journal of Horticulture* **32** 98-101
[13] Feng N, Guo X and Liang S 2010 *Transactions of Nonferrous Metals Society of China* **20** s146-s52
[14] Demirak A, Keskan F, Azahan Y and Kalemca V 2015 *Mugla Journal of Science and Technology* **1**
[15] Chen Y, Liu C, Nie J, Luo X and Wang D 2013 *Clean Technologies and Environmental Policy* **15** 395-9
[16] Ngah W W and Hanafiah M 2008 *Bioresource Technology* **99** 3935-48
[17] Othman N and Asharuddin S M *Advanced Materials Research*: Trans Tech Publ pp 266-71
[18] Ibrahim M B and Sani S 2014 *Open Journal of Physical Chemistry* **04** 139-46
[19] Wahab M A, Boubakri H, Jellali S and Jedidi N 2012 *Journal of Hazardous Materials* **241** 101-9
[20] Zahrim A, Ricky L, Shahril Y, Rosalam S, Nurmin B, Harun A and Azreen I *InCIEC 2014*: Springer pp 989-1001
[21] Langmuir I 1918 *Journal of the American Chemical Society* **40** 1361-403
[22] Freundlich H 1906 *J. Phys. Chem* **57** e470
[23] Temkin M and Pyzhev V 1940 *Acta Physiochim. URSS* **12** 217-22
[24] Lagergren S 1898 *Kungliga Svenska Vetenskapsakedemiens Handlingar* **24 (4)** 1-39
[25] Ho Y-S and McKay G 1999 *Process Biochemistry* **34** 451-65
[26] Weber W J and Morris J C 1963 *Journal of the Sanitary Engineering Division* **89** 31-60
[27] Ahmad M A, Puad N A A and Bello OS 2014 *Water Resources and Industry* **6** 18-35
[28] Jellali S, Wahab M A, Anane M, Riahi K and Jedidi N 2011 *Desalination* **270** 40-9

Acknowledgement
The authors expressed sincere gratitude to PhuiYin Lee, B. Noor Shilah and G. Norizah for technical assistance in this project. This project was financially supported by University Malaysia Sabah under research Project Grant FRG0420-TK-1/2015 and SGI0013.

A review on palm oil mill biogas plant wastewater treatment using coagulation-ozonation

Z D Dexter[1], C G Joseph[2], A Y Zahrim[1*]
[1] Chemical Engineering Programme, Faculty of Engineering, *Universiti Malaysia Sabah*, Jalan UMS 88400 Kota Kinabalu, Sabah, Malaysia
[2] Industrial Chemistry Programme, Faculty of Science and Natural Resources, *Universiti Malaysia Sabah*, Jalan UMS 88400 Kota Kinabalu, Sabah, Malaysia

E-mail: [1]zahrim@ums.edu.my

Abstract. Palm oil mill effluent (POME) generated from the palm oil industry is highly polluted and requires urgent attention for treatment due to its high organic content. Biogas plant containing anaerobic digester is capable to treat the high organic content of the POME while generating valuable biogas at the same time. This green energy from POME is environmental-friendly but the wastewater produced is still highly polluted and blackish in colour. Therefore a novel concept of combining coagulation with ozonation treatment is proposed to treat pollution of this nature. Several parameters should be taken under consideration in order to ensure the effectiveness of the hybrid treatment including ozone dosage, ozone contact time, pH of the water or wastewater, coagulant dosage, and mixing and settling time. This review paper will elucidate the importance of hybrid coagulation-ozonation treatment in producing a clear treated wastewater which is known as the main challenge in palm oil industry

1. Introduction

Palm oil industry gives rise to highly polluting wastewater, known as Palm Oil Mill Effluent (POME). In general appearance, POME is an oily wastewater which is brownish in colour and consists of fine suspended solid at pH of between 4 and 5. The POME wastewater has high biological oxygen demand (BOD) and chemical oxygen demand (COD) [2, 21, 29]. Therefore the discharge of POME without proper treatment will certainly damage the environment including causing aesthetic damaged such as foul smell in the surrounding area. To meet the discharge limit of BOD $<$ 20 ppm as proposed by the Department of Environment (DOE), the POME must be treated effectively before being released into the receiving watercourse. Anaerobic digestion in a closed tank is the most suitable method for the treatment of wastewater containing high concentration of organic carbon such as COD and BOD while releasing biogas [1, 7]. This biogas can be used in the mills to generate electricity and steam for the mill processes. Palm oil industry has adopted the anaerobic digestion system known as biogas plant to support the government effort in preserving the sustainable environment towards zero discharge.

[1] To whom any correspondence should be addressed.

Generally, the biogas plant is unable to produce wastewater that can be discharged directly into the watercourse because the treated water still could not comply with the discharge standard limit set by DOE. Some researchers reported that the wastewater can be reuse in other applications including composting [31] and land application. However, these applications seem impracticable since only less amount of wastewater is required. Thus, large amount of wastewater still need to be discharged in the watercourse without complying with the standard limit. The wastewater colour turned from brownish (POME) to blackish after POME was digested. This extremely coloured wastewater is because of the presence of lignin and its by-product, tannins, humic acids, lipids, fatty acids, melanoidin and phenolic compounds [32, 46, 30]. Consequently, it will create aesthetic problem, reduce the sunlight penetration thus affecting the photosynthesis activity, and the humic substances reaction with chlorine may produces trihalomethanes which is carcinogenic; if large amount of the wastewater are discharged into the watercourse [43, 27, 17, 30]. Figure 1 shows the simplified diagram in palm oil wastewater treatment containing biogas plant.

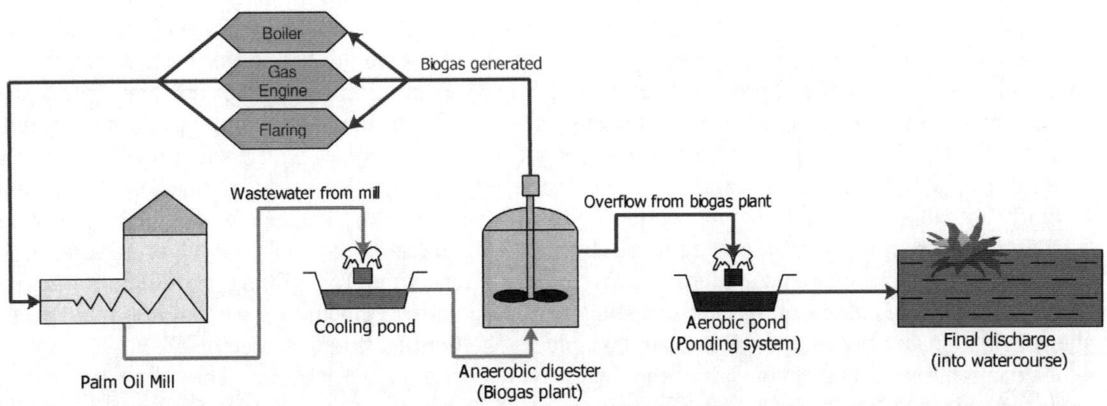

Figure 1. Simplified diagram in palm oil wastewater treatment containing biogas plant in Sabah, Malaysia.

Many studies have used the hybrid technology that combines a conventional physico-chemical treatment and Advanced Oxidation Process (AOP) to treat various kinds of water and wastewater [18, 33, 38]. The biogas plant wastewater characteristics may vary significantly from industry to industry. Due to this variation, the value will be stated in a range to give a brief view regarding the characteristics. Table 1 shows the summary of biogas wastewater characteristics from various places [46, 47, 49].

Table 1. Summary of Biogas Plant Wastewater Characteristic [46]

Characteristic	Value
pH	7.2 – 8.3
Chemical Oxygen Demand (mg/L)	1003 – 13532
Biological Oxygen Demand (mg/L)	440 – 1355
Suspended Solid (mg/L)	290 – 12750
Total Nitrogen (mg/L)	26 – 310
Ammoniacal Nitrogen (mg/L)	45 – 100
Total Phosphate (mg/L)	28 – 85
Colour (PtCo)	1633 – 54200

2. Engineering Parameters for Coagulation-Ozonation Process

Coagulation is one of the most important physico-chemical treatment that commonly used to treat water and wastewater. Every coagulant has their optimum pH, where it is most likely related to the minimum solubility of the coagulant at this point, the coagulation mechanism changes from charge neutralization to sweep coagulation. Metal hydroxides can be produced when the metal coagulants (iron or aluminium salts) are hydrolyzed with water [9]. Hydrolysation and precipitation of ferric salts are less sensitive to pH than aluminium salts, since ferric hydroxide is much less soluble than aluminium salts and precipitates over a wide range of pH [3]. The $Fe(OH)_3$ dominates between a pH of 6 and 10, whereas the $Al(OH)_3$ is present between a pH of 5.5 and 7.5. It means that at a pH above 7.5, some aluminium is resolubilised in the water and no sweep coagulation occurs anymore [9]. Inorganic coagulants such as aluminium salts and iron salts are commonly used in water and wastewater treatment. Both coagulants were found to change the particle size of organic matter in the wastewater towards larger fractions [13]. Usage of inorganic coagulants as a sole treatment is less effective especially for highly concentrated dyes [20]. Therefore, selection of coagulants must be considered carefully by understanding the type of wastewater that has to be dealt with. Increasing coagulant dosage may alter the zeta potential of wastewater towards less negative values. Zeta potential for wastewater is negatively charged and it is also depends on the pH. Aluminium and iron salts which are positively charged will neutralize the wastewater by increasing the particle size, hence increasing the zeta potential to be near zero. However, overdosing will lead to restabilization of particles where the ions will repel each other and produce breakage of flocs. Proper mixing promotes uniform distribution in the coagulation process. When coagulant is added into the water or wastewater, charge neutralization will take place during the first second [9]. Therefore, it is important to distribute the coagulant quickly via rapid mixing so that the coagulant and particles can stick together. Over rapid mixing can cause the restabilization of the colloids. For the slow mixing, it should be sufficient enough to ensure homogenization and not to break the aggregation of flocs [40]. Over mixing might not have effect on the coagulation, however insufficient time for mixing will deteriorate the performance of the coagulant. Recently, a study on single stirring scheme shows that it is possible for the coagulation and flocculation process to take place together in a simpler process[48].

Ozone is a molecule that consists of three negatively charged oxygen atoms. The ozone molecule is very unstable and has a short half-life, causing it to fall back to its original form after a while, according to the following reaction mechanism; $2O_3 \rightarrow 3O_2$ [41].Several researchers have found that pre-ozonation can improve the effectiveness of the coagulation process [12, 4, 28, 44]. Sometimes ozone can act as an oxidant or/and as a coagulant-aid [12]. As a coagulant aid, besides reducing coagulant dosage, it might enhance the surface charge of the complex pollutant to become more negatively charge [14]. Due to its strong oxidizing nature, ozone can oxidize the mostly organic and inorganic compounds [15]. Ozone oxidizes the compounds into insoluble oxides that can precipitate or settle out (normally all are filterable), and occasionally some are foamy (so can be skimmed) [14]. When ozone react with natural organic matter (NOM), the oxidation leads to the formation of aldehydes, organic acids, and aldo- and ketoacids [15]. A study compares the ozone point of application in the several of sources of water samples to remove the disinfection by-product formation potential (DBPFP). Both pre- and post- ozonation processes were better choices to reduce the DBPFP than the conventional methods which by using coagulation process. However, the result showed that pre-ozonation gave better reduction than the post-ozonation process in term of UV_{254}/DOC. 73% and 53% of UV_{254}/DOC percentage removal for pre- and post- ozonation processes respectively [12]. This can be concluded that the overall removal depends on which pre-treatment is used. The post-treatment can act as a "back-up" for the overall process.

Another studies from the same researcher reported that pre-ozonation can enhance the efficiency of coagulation. Aluminium sulphate (alum) is better choice than PAC when reducing the level of TOC and also less total trihalomethane formation potential (THMFP) when combining with ozone. In contrast, total haloacetic acid formation potential (THAAFP) increased after each coagulation, ozonation and their combination. Results also indicated that coagulation after pre-ozonation is not applicable for bromate removal in water [37, 36].Even though many studies have been conducted on the effect of ozone on coagulation, there still some contradiction among different researchers. Some studies have reported that pre-ozonation can reduce the consumption of coagulant in the treatment process [16, 26, 19, 11]. In contrast, the effect of pre-ozonation does not reduce the coagulation dosage [39, 5]. As mentioned earlier, pre-ozonation can act as a coagulant aid in the treatment process. However, several researchers mentioned that when high dosage is applied, the coagulant efficiency will be impaired thus can affect the removal of organic matter [22, 5, 25, 8]. This is due to the ozonation process which breaks down the long chain molecular weight of organic matter into smaller one in which coagulation is less effective [23, 6, 44, 24, 34, 10, 45, 35].Increasing in pH makes the ozone decomposition to occur faster. The OH-radical was produced in alkaline solution as per mentioned in the previous chapter. Thus, increasing the pH will increase the removal efficiency. However, very high pH will cause less removal. This is due to there is an increase of OH-radical scavenging by the carbonate ion which is a predominant species as higher pH [42].

Generally, increasing the ozone concentration causes faster oxidation rate, and increase in removal. However, at greater ozone doses, more extensive degradation of organic matter occurs in which the resulting molecules no longer behave as colloidal particles [16]. Pre-ozonation has shifted the molecular weight distribution of organic matter towards lower average molecular weight values [34, 24] and thus affecting the behavioural of the colloidal. Ozonation contact time may also influence the removal efficiency of organic matter. As the contact time increases, the decolourization rate and COD removal may also increase. Decolourization may occur when the ozone molecule reacts with unsaturated bond of phenolic lignin compound as per mentioned in the previous chapter. However, other substances that are more hydrophilic and polar might be produced during that period. Thus, COD removal may not be significantly increased.

3. Conclusions

Anaerobic digestion is capable in treating POME and at the same time producing biogas which is beneficial to the mills. However, the treated wastewater is highly polluted with blackish in colour and unable to be discharged to surface water. From past experiences, hybrid coagulation-ozonation could treat biogas plant wastewater effectively. In the future, several parameters should be studied in order to ensure the effectiveness of the hybrid treatment such as flocculant dosage, type of metal and polymer flocculants, effect of wastewater pH, settling time as well as sludge generation. The sludge potential to be converted as a fertiliser/soil conditioner would be an interesting topic to be explored. In addition, a study on biological treatment for the chemically treated wastewater also can be carried out.

4. References

[1] Ahmad A, Ghufran R and Abd. Wahid Z 2012 *Journal of Environmental Engineering and Landscape Management* **20** 256-64
[2] Ahmad A L, Chong M F, Bhatia S and Ismail S 2006 *Desalination* **191** 35-44
[3] Bache D H and Gregory R 2007 (United Kingdom)
[4] Barredo-Damas S, Iborra-Clar M I, Bes-Pia A, Alcaina-Miranda M I, Mendoza-Roca J A and Iborra-Clar A 2005 *Desalination* **182** 267-74
[5] Becker W C and O'Melia C R 1996 *Ozone: Science and Engineering* **18** 311-24
[6] Becker W C and O'Melia C R 2001 *Water Science and Technology: Water Supply,* pp 81-8
[7] Borja R, Banks C J and Sánchez E 1996 *Journal of Biotechnology* **45** 125-35
[8] Bose P and Reckhow D A 2007 *Water Research* **41** 1516-24
[9] Boulestreau M and Miehe U 2010 ed K W B gGmbH (Berlin, Germany)
[10] Chen J H, Hsu Y C, Yang H C and Wu K H 2003 *Ozone: Science and Engineering* **25** 323-33
[11] Cheng L, Bi X and Ni Y 2012 *Energy Education Science and Technology Part A: Energy Science and Research* **29** 91-8
[12] Chiang P C, Chang E and Liang C H 2002 *Chemosphere* **46** 929-36
[13] Citulski J, Farahbakhsh K, Kent F and Zhou H 2008 *journal of Membrane Science* **325** 311-8
[14] Eagleton J 1999 ed D Ozone DEL Ozone
[15] EPA 1999 United States Environmental Protection Agency

[16] Farvardin M R and Collins A G 1989 *Water Research* **23** 307-16
[17] Grainger S, Fu G and Hall E 2010 *Water Air Soil Pollut* 1-12
[18] Hai F I, Yamamoto K and Fukushi K 2007 *Critical Reviews in Environmental Science and Technology* **37** 315-77
[19] Jasim S Y, Ndiongue S, Johnson B, Schweitzer L and Borikar D 2008 *Ozone: Science and Engineering* **30** 27-33
[20] Joo D J, Shin W S, Choi J H, Choi S J, Kim M C, Han M H, Ha T W and Kim Y H 2007 *Dyes and Pigments* **73 (1)** 59-64
[21] Lam M K and Lee K T 2011 n *Biotechnology Advances* **29** 124-41
[22] Lefebvre E, Paillard H and Legube B 1990 *Ozone: Science and Engineering* **12** 295-313
[23] Lin S H and Lin C M 1993 *Water Research* **27** 1743-8
[24] Liu H, Cheng F and Wang D 2009 *Desalination* **249** 596-601
[25] Liu H L, Wang D S, Shi B Y, Wang M and Tang H X 2006 *Journal of Environmental Sciences* **18** 453-8
[26] Marchioretto M M and Reali M A P 2001 *Water Science and Technology*, pp 99-106
[27] Mohan S V and Karthikeyan J 1997 *Environmental Pollution* **97** 183-7
[28] Molnar J, Agbaba J, Dalmacija B, Tubic A, Watson M, Krcmar D and Rajic L 2012 *Water Research and Management* **2** 21-8
[29] Najafpour G D, Zinatizadeh A A L, Mohamed A R, Hasnain Isa M and Nasrollahzadeh H 2006 *Process Biochemistry* **41** 370-9
[30] Neoh C H, Yahya A, Adnan R, Abdul Majid Z and Ibrahim Z 2013 *Environ Sci Pollut Res* **20** 2912-23
[31] Norhasmillah A, Puah C, Ibrahim N, Baharuddin A and Choo Y 2013 *Environment, Development and Sustainability* 1-8
[32] Oswal N, Sarma P M, Zinjarde S S and Pant A 2002 *Bioresour Technol* **85(1)** 35-7
[33] Palma C, Carvajal A, Vasquez C and Conteras E 2011 *Chinese Journal of Chemical Engineering* **19** 621-5
[34] Rodriguez F J, Marcos L A, Nunez L A and Garcia M 2012 *Ozone: Science and Engineering* **34** 342-53
[35] Sadrnourmohamadi M and Gorczyca B 2015 *Water Research* **73** 171-80
[36] Selcuk H, Rizzo L, Nikolaou A N, Meric S, Belgiorno V and Bekbolet M 2007 *Desalination* **210** 31-43
[37] Selcuk H, Vitosoglu Y, Ozaydin S and Bekbolet M 2005 *Desalination* **176** 211-7
[38] Shah M 2014 *International Journal of Environmental Bioremediation & Biodegradation* **2** 231-42
[39] Tobiason J E, Reckhow D A and Edzwald J K 1995 *Aqua* **44 (3)** 142-50
[40] Tun L L, Baraoidan W A, Gaspillo P D and Suzuki M 2007 *ASEAN Journal of Chemical Engineering* **7** 49-60
[41] Ullmann F, Gerhartz W, Yamamoto Y S, Campbell F T, Pfefferkorn P and Rounsaville J F 1991 *Ullmann's encyclopedia of industrial chemistry*: VCH)
[42] Umar M, Roddick F, Fan L and Aziz H A 2013 *Chemosphere* **90** 2197-207
[43] VukoviĆ M, Domanovac T and Briki F 2008 *Journal of Environmental Sciences* **20** 1423-8
[44] Yan M, Wang D, Shi B, Wang M and Yan Y 2007 *Chemosphere* **69** 1695-702
[45] Yang J, Zhang L, Lu Z and Nie J 2009 *3rd International Conference on Bioinformatics and Biomedical Engineering, iCBBE 2009,*
[46] Zahrim A Y 2014 *Journal of Applied Sciences* **14** 3149-55
[47] Zahrim A Y, Nasimah A and Hilal N 2014 *Journal of Water Process Engineering* **4** 159-65
[48] Zahrim A Y, Nasimah A and Hilal N 2015 *Journal of Environmental Chemical Engineering* **3** 2145-54
[49] Zahrim A Y, Rachel F M, Menaka S, Su S Y, Melvin F and Chan E S 2009 *World Applied Sciences Journal 5 (Special Issues for Environment)*

Polishing of POME by Chlorella sp. in suspended and immobilized system

F A Lahin, R Sarbatly and E Suali
Faculty of Engineering, *Universiti Malaysia Sabah*, 88400, Kota Kinabalu, Sabah, Malaysia.

E-mail: [1]farhana.abdlahin@ums.edu.my

Abstract. The effect of using suspended and immobilized growth of *Chlorella* sp. to treat POME was studied. Cotton and nylon ropes were used as the immobilization material in a rotating microalgae biofilm reactor. The result showed that POME treated in suspended growth system was able to remove 81.9% and 55.5% of the total nitrogen (TN) and total phosphorus (TP) respectively. Whereas the immobilized system showed lower removal of 77.22% and 53.02% for TN and TP. Lower performance of immobilized microalgae is due to the limited light penetration and supply of CO_2 inside the immobilization materials. The rotating microalgae biofilm reactor was able to reduce the biochemical oxygen demand (BOD) to 90 mg/L and chemical oxygen demand (COD) to 720 mg/L. Higher BOD and COD reading were obtained in suspended growth due to the presence of small number of microalgae cell in the samples. This study shows that suspended growth system is able to remove higher percentages of nitrogen and phosphorus. However, an efficient separation method such as membrane filtration is required to harvest the cultivated microalgae cell to avoid organic matter release into water bodies.

1. Introduction

1.1. Palm Oil Mill Effluent (POME)

Palm oil mill effluent (POME) has been recorded as the largest contributors for water pollutants discharged into water bodies in Malaysia. It is a brownish colored voluminous liquid that contains high turbidity, chemical oxygen demand (COD), biochemical oxygen demand (BOD), oil and grease and suspended solids [1,2]. POME is generated from the oil extraction, washing and cleaning processes in the mill [3]. It is estimated that for each ton of crude palm oil (CPO) produced, 5-7.5 tons of water is required of which 50% will end up as POME [4, 5]. POME is typically high in temperature, 85°C with pH 4.7 and turbidity of 11,000 NTU [6].

[1] To whom any correspondence should be addressed.

The Malaysian government has enforced regulation to limit and monitor the discharged POME effluent through Environmental Quality (Prescribed Premises) (Crude Palm-Oil) Regulations 1977. Under this regulation it is stated that the limit for water discharge from palm oil mill to be BOD3 of 100 mg/L, suspended solids of 400 mg/L, oil and grease of 50 mg/L, total nitrogen of 200 mg/l and pH of 5 – 9 [7].
Ponding system is the most common POME treatment system in Malaysia where more than 85% of palm oil mills are implementing it before discharging into rivers [8]. However, this treatment that is mainly depending on biological treatments anaerobically and aerobically is not entirely efficient to treat POME, which leads to environmental pollution issues [9]. It was identified that the treated POME using ponding system sometimes does not conform to the discharge standard and the removal of nitrogen from POME is usually insignificant because nitrification process is not common in the ponding system [10, 11].

1.2. Treatment of wastewater by Microalgae

The potential of wastewater treatment by using algal remediation was discovered as early as the 1940's where microalgae were proposed to be used for the treatment of a municipal wastewater by Caldwell [12], followed by other researchers [13, 14]. The ability of microalgae to thrive in almost any condition, by only requiring basic necessities such as the supply of CO_2, light illumination and nutrient as well as being able to thrive in has made microalgae a highly versatile organism for wastewater treatment.
Since then, many researchers have explored the potential of microalgae for treatment of various wastewater such as municipal [15-20] Steel plant [21], textile [22], piggery [23], and soybean processing [24, 25] mainly for the assimilation of nitrogen and phosphorus. Development of microalgae based commercial treatment plant has contributed significantly to the reduction of pollutant especially biological oxygen demand (BOD), chemical oxygen demand (COD) as well as nutrients mainly nitrogen and phosphorus.

Other researchers have also looked into the potential of using microalgae as cultivation medium for the purpose of biomass cultivation for biofuel production with a simultaneous mechanism of nutrient uptake [26-30]. The combination of wastewater treatment and microalgae biomass is an economical method for biofuel as wastewater is a free source of abundant nutrients.
Microalgae based wastewater treatment has a number of benefits over the conventional wastewater treatments, namely chemical and biological treatment. Aeration process in biological treatment requires a high cost of mechanical aeration as well as the management of sludge produced at the end of the treatment. As for chemical treatment, the chemical used requires a great amount of cost other than contributing to other environmental impact caused by chemical contamination. On the other hand, the microalgae bioremediation offers a more cost effective approach where the biomass produced at the end of the treatment can be used for various purposes such as biofuel production or as feedstock for many industries. Furthermore, the microalgae treatment can serve a dual purpose of wastewater treatment and CO_2 sequestration at the same time [31].

1.3. Design of Suspended Culture Cultivation

Microalgae are typically cultured in a suspended growth system, which mimics the nature of microalgae natural growth mechanism in the environment. The method of cultivation are usually divided into two main systems; open and closed. Open system may utilize either big shallow ponds, raceway system or circular ponds. The choice of a specific system is influenced by the basic properties of the microalgae, the local climactic conditions and the costs of water and land. These factors means that pond volume and area should be minimized and the cell density in the culture should be maximized to have an economical process [32]. However, the main disadvantages of the open pond are the CO_2 emission to the atmosphere, evaporation losses, risk of contamination and the large area required.

Potential contamination to microalgae culture can be minimized by utilizing closed system for cultivation. The concept of closed system has been explored since a few decades ago [33, 34]. Many advances in the design and operation of closed photo-bioreactors in culturing microalgae such as the flat plate reactor [35, 36] or the tubular reactor [37-40]. The fundamental principle of these designs is to reduce the light path so as to increase the amount of light available to each cell.

These reactors have advantages: 'clean' microalgae culture, high light utilisation efficiency which leads to high productivities, temperature control and the ability to be used outdoors in natural sunlight. The wider range of species can be cultivated since contamination is avoided and the reactors can be operated over a much wider climactic range than open system [32]. However, the closed system may require a higher cost in construction compared to an open system.

1.4. Immobilized culture of microalgae

Immobilized microalgae cell is defined as a cell that is prevented from independently moving from its location to all parts of an aqueous system either naturally or artificially [41]. Advantages available by immobilization of microalgae cell are the high biomass concentrates, avoidance of any further utilization of filtration process, and the ability to entrap more than one microorganism as well as being a simple process that is able to be conducted even by non-professionals [42]. However, the major issues in microalgae immobilization are the efficiency of the system to remove pollutants, cost of materials, and cost of immobilization process itself.

The wastewater industry is already familiar to large scale biofilm processes [43]. A scalable microalgae biofilm system could be incorporated into wastewater treatment process to achieve a dual benefit of inexpensive nutrient supply and treated water. Given enough surface area, microalgae biofilm can potentially achieve higher growth compared to suspended growth system [44]. It can contribute to increased density and lower land and water requirement of matrix-immobilized culture [45] without the associated costs of the matrix.

By using biofilm system, it is expected that little or no separation of microalgae and water is required before effluent discharge [46, 47]. Study by Christenson and Sims [48] involved designing and testing a rotating microalgae biofilm reactor for sewage treatment. They were able to achieve effective nutrient reduction as well as biomass production by using cotton substratum. Comparison between immobilized culture submerged in polystyrene and suspended culture in the same condition was done by Johnson and Wen [49]. The immobilized culture showed a greater yield of biomass and lipid content compared to microalgae cultured in the suspended growth system. However, photoinhibition and diffusion limitation of nutrients or CO_2 may limit the performance of immobilized microalgae system [50, 51]. This research was focused on the comparison between application of immobilized and suspended growth microalgae in nutrient uptake in POME and simultaneous cultivation of microalgae biomass for further use.

2. Materials and methods

2.1. Wastewater samples

POME samples was collected from Lumadan Palm Oil Mill (5°15'19"N and 115°39'24"E). All collected POME was filtered to remove large solid and stored at 4°C after sampling before utilized for experiment.

2.2 Microalgae species

Chlorella sp. were isolated locally by Borneo Marine Research Institute, University Malaysia Sabah from the lake located in Likas Stadium, Kota Kinabalu. A 20 mL sample was acquired from the culturing chamber of Borneo Marine Research Institute and directly sent to the lab for culturing. Chlorella sp. was chosen due to its variety in post–harvest application [52-55]. The microalgae was cultivated in an adjusted

IOP Conf. Series: Earth and Environmental Science **36** (2016) 012030 doi:10.1088/1755-1315/36/1/012030

Jaworski Medium (JM). The medium contained 0.02 gl^{-1} Ca $(NO_3)_2 \cdot 4H_2O$, 0.0124 gl^{-1} KH_2PO_4, 0.05 gl^{-1} $MgSO_4 \cdot H_2O$, 0.00225 gl^{-1} EDTAFeNa, 0.00225 gl^{-1} $EDTANa_2$, 0.00248 gl^{-1} H_3BO_3, 0.00139 gl^{-1} $MnCl_2.4H_2O$, 0.001 gl^{-1} $(NH_4)6Mo7O_{24} \cdot 4H_2O$, 0.00004 gl^{-1} cyanocobalamin, 0.00004 gl^{-1} thiamine HCl, 0.00004 gl^{-1} biotin, 0.08 gl^{-1} $NaNO_3$ and 0.036 gl^{-1} $Na_2HPO_4 \cdot 12H_2O$.

2.3 Suspended culture
Suspended culture experiment was conducted in a photobioreactor built from transparent materials (Perspex). The laboratory scaled reactor was dimensioned 71 cm long, 31 cm wide and 12 cm deep with a working volume of 20 L. Air diffuser made from 0.84" diameter PVC pipes was installed at the bottom of the photo bioreactor which was connected to an air pump for aeration and also acted as a mixer of the wastewater and microalgae to maximize algae-wastewater contact as well contact with the light supply.

2.4 Rotating Microalgae Biofilm for Immobilized System
The laboratory-scaled microalgae biofilm reactor consisted of cubic tank built using Perspex with dimension 71 cm long, 31 cm wide and 12 cm deep. 2 x 1 in aluminum hollow channel was used to frame out tank and motor parts of the reactor. Rotating drum consisted of two cylinders was set up using 8.9 cm PVC pipe wrapped with cotton and nylon as the growth surface for microalgae. The biofilm was rotated by motors and was partially submerged in POME sample with 40% into aqueous phase and 60% of substrate exposed to gaseous phase. 10 rotation per minute (rpm) ensured that enough time was provided for microalgae exposure to light and nutrients.

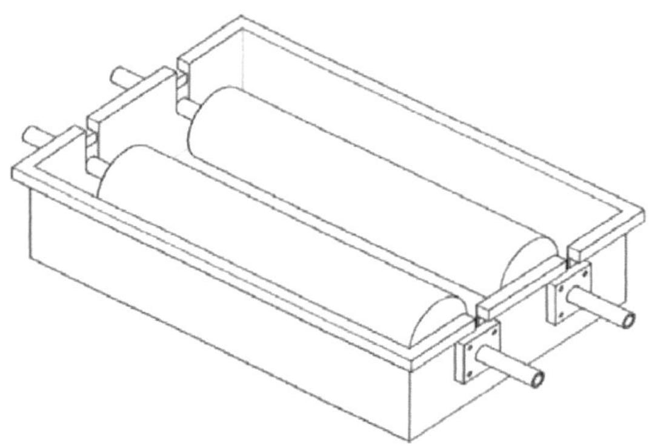

Figure 1. Setup of Rotating Microalgae Biofilm system (not to scale)

2.5 *Sampling*
Experiments were carried out within a 5-day period and sampling was done daily. All samples were filtered out using a syringe filter (0.45 μm nominal pore size with 25 mm diameter, Sartorius filter) in order to remove any suspended solid.

3. Results and discussion

3.1 Nitrogen and phosphorus removals
Figure 2 shows the rapid removal rate of TN in the first 2 days of experiment and become almost constant after the third day. Nevertheless, all three experiments were not able to remove the TN content completely.

Table 1. Removal of Nitrogen and Phosphorus immobilized and suspended system material using *Chlorella* sp.

Parameter	Material	Rpm	Initial Concentration (mg/L)	Final Concentration (mg/L)	Percentage Removal (%)
TN	Cotton	10	54.8	12.	77.22
	Nylon	10	57.7	15.4	73.33
	Suspended Cell		49.7	9	81.9
TP	Cotton	10	30.43	14.3	53.02
	Nylon	10	41.00	18.1	55.85
	Suspended Cells		22.47	10	55.5

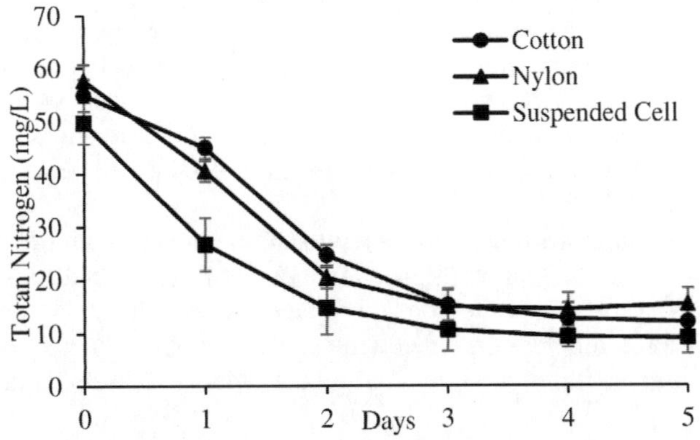

Figure 2. Removal rate of Nitrogen in immobilized and suspended system material using *Chlorella* sp.

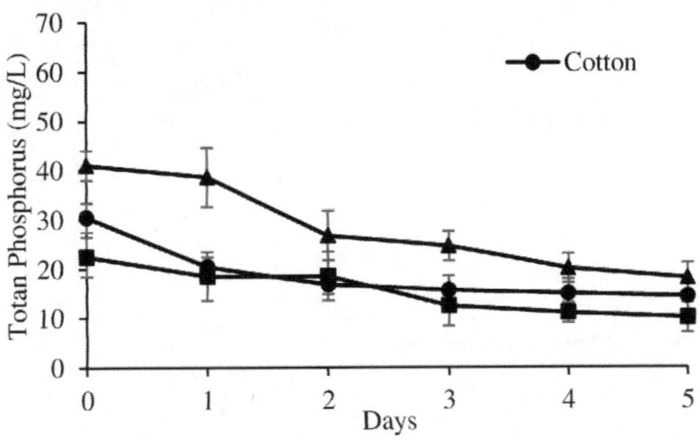

Figure 3. Removal rate of Phosphorus in immobilized and suspended system material using *Chlorella* sp.

From the result, the suspended growth treatment showed the most rapid removal of TN most probably due to the free movement of microalgae cell that provided better contact with the pollutants for nutrient assimilation.

The result obtained also showed that the TP removal was slightly lower than that of TN removal. This may be related to the N: P ratio of the POME tested which was implied by Beuckels *et al.* [56] to have significant effect on the phosphorus uptake by microalgae. Microalgae will accumulate more phosphorus when nitrogen concentrations are high and less when N concentrations are low. This statement suggests that a sufficient concentration of nitrogen is required for an effective phosphorus removal [56]. The same pattern of result was obtained by Gonzalez *et al.* [23] and Yun *et al.* [21] where phosphorus removal was low even with the exhaustion of nitrogen. It is also important to note that chemical precipitation is insignificant for the phosphorus removal since the pH was in the range of 7-8 (result not shown) during the entire experiment. Therefore, it can be concluded that TP removal was by the assimilation of microalgae.

The rotating biofilm removed nitrogen and phosphorus via microalgae propagation through the immobilization material. The propagation of microalgae in cotton rope is faster than nylon due to the higher water affinity of cotton. The suspended growth shows higher removal rate of TN and TP due to greater contact between microalgae cell and POME. This is also due to the higher yield of microalgae biomass which is caused by larger light and CO_2 supply of the suspended microalgae. This is a good vindication of reports in literature that imply the difference in nitrogen and phosphorus removal efficiencies depends on the composition of the culture medium such as the concentration, light intensity N:P ratio and light/dark cycle [57, 58].

However, the result found in this study is contrary to the results reported by Gao *et al.* [59] where biofilm photobioreactor was able to remove higher amounts of nitrogen and phosphorus at >95% and >85% respectively than a conventional photobioreactor. Nevertheless, this result could be affected by the fact that the biofilm photobioreactor was combined with a suspended growth medium. Both of the systems were also assisted by a membrane filtration system before the effluent is discharged.

3.2 BOD and COD removal

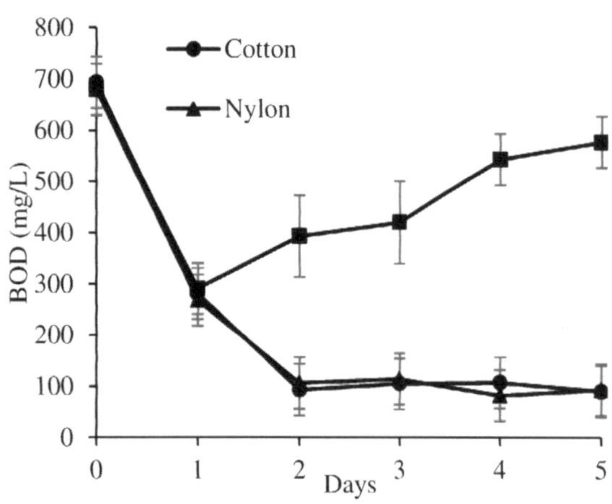

Figure 4. Removal rate of BOD in immobilized and suspended system using *Chlorella* sp.

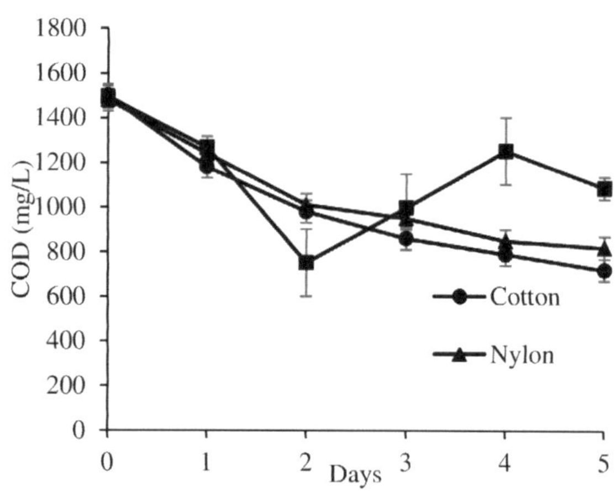

Figure 5. Removal rate of COD in immobilized and suspended system using *Chlorella* sp.

From the obtained data for BOD and COD removal, the immobilized microalgae cell was able to achieve a steady removal throughout the 5-day operation. At the end of experiment, an average of 86% and 51% of BOD and COD respectively was removed from the samples. The BOD was removed well under the specified limit of 100 mg/L of BOD, stated by Malaysian authority [7]. However, comparing both treatment systems, the removal by suspended microalgae cell presented fluctuating and inconsistent readings as indicated by the large range of errors in the plotted data.

The increase in biomass of microalgae with time has resulted in contamination of microalgae cell in the effluent of suspended solid due to the lack of separation between wastewater sample and microalgae cell. This is mainly caused by the motility of microalgae cell itself and has been reflected by the higher turbidity reading in samples collected for suspended cell system (result not shown). However, other literature has reported a contrary result where the immobilized microalgae presented steady but lower nutrient removal due to the limitation of biomass growth inside the immobilization medium of sodium alginate [60], when compared to a suspended cell treatment system. This result might however be affected by the different immobilization methods applied in the research.

The fluctuation of BOD and COD removal in the suspended treatment system shows that microalgae cell separation before effluent discharge is crucial for the application of microalgae to wastewater treatment. Without a solid-liquid separation, a complete treatment of wastewater will not be achieved where the BOD and COD reading will still be high due to the presence of microalgae cell in the discharge. This fugitive cell will cause an eventual eutrophication in the water bodies. With the increasing amount of biomass with time, the separation of biomass and water was becoming more difficult. The microalgae biomass was not able to be completely separated from the sample and had caused the increase in BOD as well as COD concentration in the water sample.

3.3 Biomass recovery

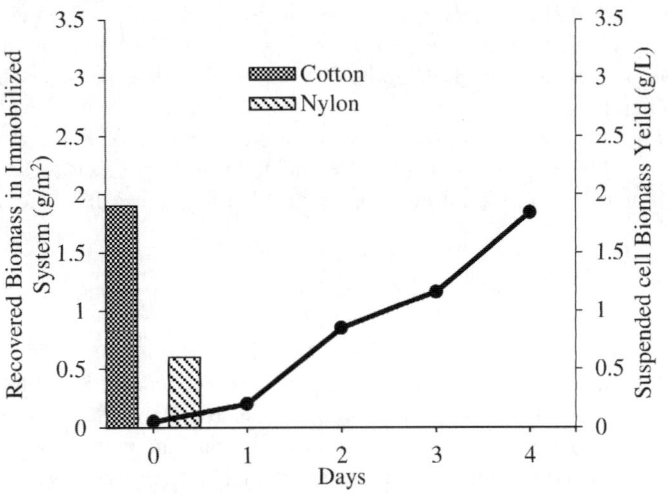

Figure 6. Recovered *Chlorella* sp. biomass from immobilized and suspended cell system

Two different methods were used for the recovery of both immobilized and suspended solids. The biomass yield for suspended microalgae cell was monitored daily whereas the immobilized microalgae was only recovered after the 5-day experiment to avoid any potential interruption to the system.

Comparison between two materials of biofilm shows that cotton rope has a higher yield in microalgae biomass after 5 days of operation compared to nylon rope at 1.9 g/m^2 and 0.6 g/m^2 respectively. This could be contributed by the better hydrophilicity attribute of cotton rope that facilitated the attachment of cell and the distribution of wastewater thoroughly within the biofilm. From previous research, it was evident that different materials of biofilm would result in different growth rates of microalgae. Similar result was obtained by Christenson and Sims [61] and Gross *et al.* [62] found respectively that cotton rope and cotton duct, showed the best microalgae growth among other tested materials. However, the exact reason affecting microalgae growth in an attached growth system is still inconclusive. From the review done by Schnurr and Allen [63], they concluded that many factors affect the growth of microalgae onto biofilm including surface tension, water contact angle/hydrophobicity, surface energy and surface micro patterning.

The effect of biofilm utilization for microalgae growth can be seen based on the comparison between the immobilized and suspended cell system. Figure 6 shows a significant difference between both methods. Microalgae biomass yield in suspended cell is higher when compared to the attached growth in biofilm. Attached growth system within the cotton and nylon rope has limited light penetration that is much needed for microalgae growth. Since the microalgae are deep within the immobilization material, and only in contact with wastewater at certain times due to the rotation of the biofilm, limitation of nutrient occurred which was reflected in the lower yield in biomass. Freely moving cell in suspended solid ensures sufficient light and nutrient supply for photosynthesis and promotes the microalgae growth.

4. Conclusion

The biofilm has shown a high potential in the polishing of POME which showed steady removal for TN, TP, BOD and COD. However, the efficiency of immobilized treatment is limited by insufficient supply of light and nutrient. This study concludes that immobilized microalgae culture has significant limitation for wastewater treatment if compared to suspended cell system. From the suspended cell system operation, it can be concluded that even with the effective removal of TN and TP, the treatment of wastewater will not be completely effective as fluctuation in BOD and COD removal can still be seen which is caused by the existence of residual microalgae biomass in the effluent water. Therefore, it is evident that an efficient biomass separation technique is very important for an effective treatment of wastewater using microalgae. Further study on the combination of suspended treatment system of microalgae with a solid-liquid separation method such as membrane filtration is recommended for higher efficiency of biomass removal from the effluent.

5. Reference

[1] Ahmad A L, Chong M F, Bhatia S and Ismail S. 2006 *Desalination* **191** pp 35 – 44

[2] Ibrahim, M.H., Rupani, P.F., Singh, R.P and Esa, N. 2010 *World Appl. Sci. J.* **11**(1) pp 70-81

[3] Agamuthu, P. 1995. Waste treatment plant. Narosa Publishing House, New Delhi, pp 338-360

[4] Ma, A.N. 1999 Malaysian Oil Palm Growers' Council, Kuala Lumpur, pp 113-123

[5] Ahmad, A.L., Bhatia, S and Ismail, S. 2003 *Desalination* **157** pp 87-95

[6] Ma, A.N., 2000 *Palm Oil Dev.* **30** pp 1-10

[7] EQA, Environmental Quality Act, 1974 Environmental Quality (Prescribed Premises) (Crude Palm-Oil) Regulations 1977

[8] Ma, A.N., Cheah, S.C and Chow, M.C., 1993 Ministry of Science, Technology and the Environment, Malaysia, pp 111 -136

[9] Ahmad, A.L., Sumathi, S and Hameed, B.H. 2005 *Chem. Eng. J* **108** pp 179-185

[10] Chin K.K., Lee S.W and Mohammad H.H. 1996 *Water Sci. Technol.* **34** pp 119 - 123

[11] John, C.K., 1985 *Water Sci. Technol.* **17** pp 781 - 790
[12] Caldwell, D.H., 1946 *Sewage Works J.***18** pp 433-458
[13] Oswald, W.J and Gotaas, H.B. 1957 *Trans Am Soc Civil Eng* **122** pp 73 - 105
[14] Green, F., B., Bernstone, L., S., Lundquist, T., J., and Oswald, W., J. 1996 *Water Sci. Technol.* **33** pp 207–217
[15] Wang, L., Min, M., Li, Y., Chen, P., Chen, Y., Liu, Y., Wang, Y. and Ruan, R. 2009 *Appl. Biochem. Biotechnol.* DOI 10.1007/s12010-009-8866-7
[16] Cho, S., Lee, N., Park, S., Yu, J., Luong, T.T., Oh, Y.K. and Lee, T. 2013 *Bioresource Technol.* **131** pp 515-520
[17] Kim, J., Lingaraju, B.P., Rheaume, R., Lee, J.Y. and Siddiqui, K.F. 2010 Tsinghua Science and Technology ISSNl l1007-0214 04/16, **15** pp 391-396
[18] Hammouda, O., Gaber A. and Abdel-Raouf, N. 1995 *Ecotox. Environ. Safe* **32** pp 205-210.
[19] Singh, G. and Thomas, P.B. 2012 *Bioresource Technol.***117** pp 80-85
[20] Kassim, T.I. 2002 Proceedings of International Symposium on Environmental Pollution Control and Waste Management, 7-10 January 2002, Tunis (EPCOWM'2002). pp 628-632
[21] Yun, Y.S., Lee, S.B., Park, J.M., Lee, C.I. and Yang, J.W. 1997 *J. Chem Tech. Biotechnol.* **69** pp 451-455
[22] Lim, S.L., Chu, W.L. and Phang, S.W. 2010 *Bioresource Technol.* **101** pp 7314-7322
[23] Gonzalez, L.E., Canizares, R.O. and Baena, S. 1997 *Bioresource Technol.* **60** pp 259-265
[24] Hongyang, S., Yalei, Z., Chunmin Z., Xuefei, Z. and Jinpeng, L. 2011 *Bioresource Technol.* **102** pp 9884-9890
[25] Silva-Benavides, A.M. and Torzillo, G. 2012 *J. Appl. Phycol* **24** pp 267-276
[26] Chinnasamy, S., Bhatnagar, A., Hunt, R.W. and Das, K.C., 2010 *Bioresour. Technol.* **101** (9) pp 3097–3105
[27] Caporgno, M.P., Taleb, A., Olkiewicz, Font, J., Pruvost, J., Legrand, J. and Bengoa C., 2015 *Algal Research* **10** pp 232-239
[28] Maity, J.P, Hou, C.P., Majumder, D., Bundschuh, J., Kulp, T.R., Chen, C.Y., Chuang, L.T., Chen C.N.N, Jean, J.S., Yang T.C. and Chen, C.C., 2014 *Energy* **78** pp 94-103
[29] Komolafea, O., Ortab, S.B.V., Monje-Ramirezc, I., Noguezc, I.Y., Harvey, A.P. and Ledesmac, M.T.O., 2014 *Bioresource Technol.* **154** pp 297-304
[30] Hernández, D., Riaño, B., Coca, M., Solanac, M., Bertucco, A. and García-Gonzáleza, M.C. 2016 *Chem. Eng. J.* **285** pp 449–458
[31] Sarbatly, R. and Suali, E. 2014 Algal Research **5** pp 274-282
[32] Borowitzka, M.A., 1999 *J. Biotechnol.* **70** pp 313-321
[33] Juttner F., 1977 *Biotechnol. Bioeng.* **19** pp 1679-1987
[34] Pirt, S.J., Lee, Y.K., Walach, M.R., Pirt, M.W., Balyuzi, H.H.M. and Bazin, M.J., 1983 *J. Chem. Tech. Biotechnol.* **33B** pp 35-58
[35] Pulz, O., 1994 Algal.
[36] Hu, Q., Guterman, H. and Richmond, A., 1996 *Biotechnol Bioeng* **51** pp 51-60
[37] Miyamoto, K., Wable, O. and Benemann, J.R., 1988 *Biotech .Lett.* **10** pp 703-708
[38] Richmond, A., Baussiba, S., Vonshak, A. and Kopel, R., 1993 *J. Appl. Phycol* **5** pp 327-332
[39] Borowitzka, M.A., 1996 *J. Mar. Biotechnol.* **4** pp 185–191
[40] Torzillo, G., 1997 *Taylor and Francis, London*, pp 101-115
[41] Tampion, J. and Tampion, M.D., 1987 *Cambridge University Press*, UK. pp 257
[42] de-Bashan, L.E., Antoun, H. and Bashan, Y. 2005 FEMS Microbiol. *Ecology* **54** pp 197-203.
[43] Wuertz, S., Bishop, P.L. and Wilderer, P.A., 2003 *IWA Publishing*. London

[44] Middlebrooks, E.J., Porcella, D.B., Gearheart, R.A., Marshall, G.R., Reynolds, J.H. and Grenney, W.J., 1974 *J. Water. Pollut. Control. Fed.* **46** pp 2676-2695
[45] US DOE, 1985 Alexandria (VA): Solar Energy Research Institute. Report No: SERI/STR-231-2798. Prepares for U.S Department of Energy
[46] Roeselers, G., Loosdrecht, M. C. M. van and Muyzer, G. 2007 *J. Appl. Phycol.* **20** pp 227-235
[47] Schumacher, G., Blume, T. and Sekoulov, I., 2003 *Water Sci. Technol.* **47** pp 195–202
[48] Christenson, L. and Sims, R. 2011 *Biotechnol. Adv.* **29** pp 686-702
[49] Johnson, M.B. and Wen, Z., 2010 *Appl. Microbiol. Biotechnol.* **85** pp 525-534
[50] Liehr, S, K., Eheart, J. W. and Suidan, M. T., 1988 *Water Res.*, **22**(8) pp 1033–1041
[51] Murata, N, Takahashi, S., Nishiyama, Y. and Allakhverdiev, S. I., 2007 Biochimica et Biophysica Acta (BBA) - *Bioenergetics* **1767**(6) pp 414–421
[52] Spolaore, P., Claire, J.C., Duran, E. and Isambert, A. 2006 *J. Biosci. Bioeng.* **101**(2) pp 87-96
[53] Morais, M.G.D and Costa, J.A.V. 2007 *Biotechnol. Lett.* **29** pp 1349-1352
[54] Borowitzka, M.A. 1997 *J. Appl. Phycology* **9** pp 393-40
[55] Phukan, M.M., Chutia, R.S., Konwar, B.K. and Kataki, R. 2011 *Appl. Energy* **88** pp 3307-3312.
[56] Beuckels A, Smolders, E and Muylaert K, 2015 *Water Res.* **77** pp 98-106
[57] Azov, Y. and Shelef, G., 1987 *J. Water Sci. Technol.* **19**(12) pp 381-383
[58] Aslan, S. and Kapdan, I.K. 2006 *Ecol. Eng.* **28** pp 64-70
[59] Gao F, Yang Z H, Li C, Zeng G M, Ma D H and Zhou L 2015 *Bioresource Technol.* **179** pp 8-12
[60] Jiménez-Pérez, M.V., Sánchez-Castillo, P., Romera, O., Fernández-Moreno, D. and Pérez-Mart´ınez, C., 2004 *Enzyme Microb. Tech.* **34** pp 392–398
[61] Christenson L. and Sims R., 2012 *Biotechnol. Bioeng.* **109**(7) pp 1674–84
[62] Gross M, Henry W, Michael C. and Wen Z. 2013 *Bioresour Technol* **150** pp 195–201
[63] Schnurr, P.J. and Allen, D.G., 2015 *Renew. Sust. Energ. Rev.* **52** pp 418-429

Acknowledgement
Our deep gratitude is extended to Lim Chong Gap and Cheoh Liew Chuan for their effort and assistance in this study. Authors also would like to thank the Ministry of Higher Education Malaysia for financial assistance through grant FRGS0324-TK-1-2013.

A preliminary study for removal of heavy metals from acidic synthetic wastewater by using pressmud-rice husk mixtures

H Ahmad[1], C J Ee[1], N S Baharudin[1]
[1]Environmental Technology Division, School of Industrial Technology, *Universiti Sains Malaysia*, 11800 Penang, Malaysia

E-mail: [1]harlinaa@usm.my

Abstract. The study was carried out to evaluate the effect of combining pressmud and rice husk in the removal efficiencies of heavy metals in acidic synthetic wastewater. The ratios of pressmud to rice husk were varied at different percentages of weight ratio (0%, 20%, 40%, 60% 80% and 100%) and removal of heavy metals concentrations was observed. The result showed that the removal efficiency was increased with the addition of pressmud by up to almost 100 %. Pressmud alone was able to remove 95% to 100% of heavy metals while rice husk alone managed to remove only 10 % to 20% of heavy metals. The study also demonstrated that pressmud behaved as a natural acid neutralizer. Hence, the initial pH of the synthetically prepared acidic wastewater which was below 2 also was increased to pH ranging from 6 to 8.

1. Introduction

Industrial wastewater with a low pH is considered as an acidic wastewater, a common problem for many industries. Increasingly stringent environmental regulations require that pH, among other contaminants, be controlled to certain levels before they are discharged into a natural body of water or municipal sewer system. Other than that, the existence of contaminated-industrial wastewater with hazardous heavy metals such as Cd, Cr, Cu, Ni, As, Pb and Zn is also a big concern among the chemical-intensive industries.

Heavy metals are easily adsorbed by living organisms because of their high solubility in the aquatic environment. Once they enter the food chain, the consumption of large concentrations of heavy metals may start to accumulate in the human body. If the metals are ingested beyond the permitted concentration, they can cause serious health disorders (Barakat, 2010). Therefore, it is necessary to treat metal-contaminated wastewater prior to its discharge to the environment as it will constitute a great risk for the aquatic ecosystem, while the direct discharge into the sewerage system may have a negative effect affecting the subsequent biological wastewater treatment (Balintova et al., 2012).

Generally, the removal of heavy metals from aqueous solutions has commonly been carried out through several processes: chemical precipitation, solvent extraction, ion exchange, reverse osmosis or adsorption. Among these processes, adsorption is considered to be the most facile and effective

[1] To whom any correspondence should be addressed.

method (Wang et al., 2014). Adsorption with a suitable adsorbent can be a simple but effective technique for removing heavy metals from wastewater.

Pressmud is actually compressed waste produced from the filtration of sugarcane juice before crystallization process of sugar. It is usually disposed or sold as immature compost to farmers. Furthermore, pressmud is a source of nutrients for soil and can improve the ability of the soil to hold the water holding capacity. It is a good source of biogas since it contains approximately 5–15% of sugar (Gupta et al., 2010). There is also a study to investigate the potential use of 0.5M H_2SO_4 treated pressmud as cyanide ion sorbent from aqueous solution (Gupta et al., 2012). The potential of pressmud as chelating agent has been studied to be used to reduce the leachability of heavy metals in landfill leachate by mixing with laterite soil (Maheera et al., 2013).

One of the most widely available agricultural waste in Malaysia is rice husk. Rice husk is the agricultural crop residue generated during dehusking process at rice mills. Numerous applications have been proposed for rice husk utilization from which silica production and energy generation are among the most scrutinized (Bazargan et al., 2015). Unmodified rice husk has been reported for their ability to bind metal ions (Kumar & Bandyopadhyay, 2005). The use of rice husk provides a less costly adsorbent than activated carbon as it is cheap and easily available.

This research is a preliminary study to investigate the potential conversion of waste materials that are produced in large quantity into low-cost adsorbent for the removal of several types of heavy metals such as chromium, iron, manganese, nickel, lead and zinc from acidic synthetic wastewater. This research highlights the 'reuse' of the 3R Concept where the elements of the discarded item are used again. The used of rice husk and pressmud, where the pressmud is the compressed sugar industry waste while rice husk is the by-product from rice industry which does not have substantial commercial value can solve the disposal problem of rice husk and pressmud. Both materials have caused environmental problems concerning their disposal. Besides that, the mixture of rice husk-pressmud can remove heavy metals from the wastewater preventing them from entering water stream.

2. Materials and Methods

2.1 Preparation of pressmud
Pressmud was collected from a sugar mill, Malaysian Sugar Manufacturing (MSM) Sdn. Bhd, Seberang Perai, Penang. The pressmud was air-dried at a shady place outdoor for one week. Then, it was sieved through 500μm sieve to remove large and coarse pebbles. The pressmud was later dried in the oven at 60°C for 24 hours to remove moisture until the weight is constant and stored in air-tied bottle.

2.2 Preparation of rice husk
Raw rice husk was obtained from a local rice mill, Leong Guan Sdn. Bhd, Kepala Batas, Penang. The rice husk was washed and soaked in distilled water for 3 days. It was washed with distilled water until it was odorless. The washing process was continued until the water was colorless, air-dried at room temperature and then dried in the oven at 105°C for 24 hours to remove moisture until the weight was constant (Lim et al., 2010). The dried rice husk was grounded and sieved to desired particle sizes in the range of 125 - 250μm (Nhapi, I., et al., 2011)and then stored in air-tied bottles.

2.3 Preparation of acidic synthetic wastewater
Metal solutions (chromium, iron, lead, manganese, nickel and zinc) of 1000mg/L in nitric acid from Fisher Scientific, UK were purchased. Each metal solution was diluted with deionized water to obtain similar initial concentrations of acidic synthetic wastewater containing each type of heavy metal at below 10mg/L and later confirmed with ICP Optical Emission Spectrometer Model Varian 715-ES. The pH of the synthetic wastewater was measured using HACH sensION+ PH3 pH meter before and after the experiment.

2.4 Initial Concentration of Heavy Metals in Acidic Synthetic Wastewater

Several heavy metals were chosen in this study. These heavy metals were usually encountered in leachates from landfills. In this study, the initial concentrations of heavy metals for Cr, Fe, Mn, Ni, Pb and Zn with their concentrations were 2.07 mg/L, 2.28 mg/L, 2.13 mg/L, 2.18 mg/L, 2.12 mg/L and 2.17 mg/L respectively determined by using ICP-OES.

2.5 Initial pH of Acidic Synthetic Wastewater

The synthetic wastewater was prepared with metal solutions of 1000mg/L in nitric acid from Fisher Scientific resulting the prepared synthetic wastewater to be acidic. The initial pH values of the acidic synthetic wastewater were low at between 1.04 and 1.45. The higher the initial concentration of heavy metals, the lower the pH value of the synthetic wastewater. The process of heavy metals removal seemed to occur in acidic condition.

2.6 Batch equilibrium test

A 5±0.02 g adsorbent of pressmud-rice husk mixture at different ratio was weighed and poured into 250mL conical flasks respectively. The adsorbent was designated as 1) pressmud alone (RH 0); 2) pressmud 80% while rice husk 20% (RH 20); 3) pressmud 60% while rice husk 40% (RH 40); 4) pressmud 40% while rice husk 60% (RH 60); 5) pressmud 20% while rice husk 80% (RH 80) and 6) rice husk alone (RH 100). The adsorbent were equilibrated with 100 ml of the prepared acidic synthetic wastewater containing heavy metals (Cr, Fe, Mn, Ni, Pb and Zn) in 250mL conical flasks. The conical flasks were agitated using horizontal shaker at a speed of 180 rpm for 24 hours at room temperature (Annadurai et al., 2002; Gupta et al., 2012).

After reaching the equilibrium, the sorbent-solution was centrifuged at 4000 rpm for 20 minutes to separate the liquid and solid form. The supernatant was filtered with No.601 Whatman filter paper and collected in small plastic bottles. The samples were preserved to a pH < 2 with ultrapure HNO_3 according to U.S. EPA Method 200.7 (USEPA, 1994). The samples were stored at 4°C in refrigerator before analyzed by ICP-OES Model Varian 715-ES. The concentration of heavy metals left in the solution was used to calculate the amounts of heavy metals absorbed by the pressmud-rice husk mixtures.

The percentage removal of heavy metals from the initial solution concentration C_o, was calculated from the following equation.

$$\text{Percentage removal} = (C_o - C_e) / C_o \times 100 \qquad (1)$$

where C_o = initial concentration of heavy metals in the solution (mg/L)
C_e = concentration of heavy metals left in the solution (mg/L)

3. Results and Discussion

3.1 pH Measurement of Rice husk-pressmud Mixtures

Table 1. pH of rice husk-pressmud mixtures

Rice husk-pressmud composition	pH 125-250μm
RH 100	6.40
RH 80	7.29
RH 60	7.41
RH 40	7.85
RH 20	8.21
RH 0	8.04

pH of the rice husk, pressmud and rice husk-pressmud mixtures was measured with pH meter to determine the original pH of the adsorbent. Table 1 showed the pH of rice husk-pressmud mixture. From the result, the pH value of pure pressmud (RH0) was approximately 8.04. Pressmud was a bit alkaline. Unmodified but washed rice husk (RH 100) was a bit acidic which showed pH value of 6.40. As the composition of pressmud increased in the mixtures, the pH of the mixtures also increased but was limited to around pH 8.

3.2 Final pH Measurement of Acidic Synthetic Wastewater

After the batch equilibrium test, the pH value of the acidic synthetic wastewater was measured again. The pH value of acidic synthetic wastewater added with pressmud increased and became neutral. The acidic synthetic wastewater added with 80% pressmud (RH20), 60% pressmud (RH 40), 40% pressmud (RH 60) and 20% pressmud (RH 80) also become neutral with the pH ranged between 7.29 - 7.85. The synthetic water added with only rice husk remained acidic as pH of the solution did not change. These showed that pressmud can act as buffer to neutralize acidic synthetic wastewater as pressmud had high bicarbonate content about 1069mg/L (Maheera et al., 2013).

3.3 Removal of Heavy Metals

The removal efficiency of six heavy metals namely Cr, Cu, Mn, Ni, Pb and Zn were determined with the initial concentrations of acidic synthetic wastewater which were 2.07mg/L, 2.28mg/L, 2.13mg/L, 2.18mg/L, 2.12mg/L and 2.17mg/L to stimulate trace amount of heavy metals with pH value in the range of 1.04 to 1.45 respectively. The removal efficiencies of Cr, Cu, Mn, Ni, Pb and Zn are illustrated in Figure 1-6 as shown below.

3.3.1 *Removal of Cr*

Figure 1 showed that the removal efficiency of rice husk and pressmud alone was between 33.0-39.2%. The removal efficiency of Cr for mixtures of pressmud and rice husk had increased till 44.2-71.8%. Figure 1 clearly showed that rice husk alone could remove less than 40% of chromium at low concentration.

According to analysis of variance (ANOVA), composition of pressmud and rice husk in the mixtures had significant impacts on the removal efficiency of Cr ($p < 0.05$). Pressmud alone had removal efficiency between 30-50%. The combination of pressmud with rice husk clearly improved the removal efficiency of Cr. Pressmud became sticky when it was wet (Maheera et al., 2013). When pressmud was mixed with rice husk, the sticky mixture acted as a pore opener to adsorb Cr. The mixture of rice husk-pressmud had higher removal efficiency than unmodified rice husk and pressmud. The highest removal efficiency achieved at RH 80 which was approximately 72%. Generally, the removal rate of Cr increased from RH 0 to RH 80 proved that higher composition of rice husk was more effective to remove Cr. Rice husk alone, RH 100, showed the lowest removal efficiency because rice husk could not perform well under unfavorable condition as the pH of acidic synthetic wastewater was low at the range of 1.04-1.45. The optimum pH for the adsorption of Cr by rice husk ash was at pH 4 (Anand et al., 2014) while rice husk activated carbon was at pH 2 (Ahmed et al., 2012). When there was no pressmud at RH 100 or only little pressmud at RH 80 to act as buffer, the removal efficiency was low.

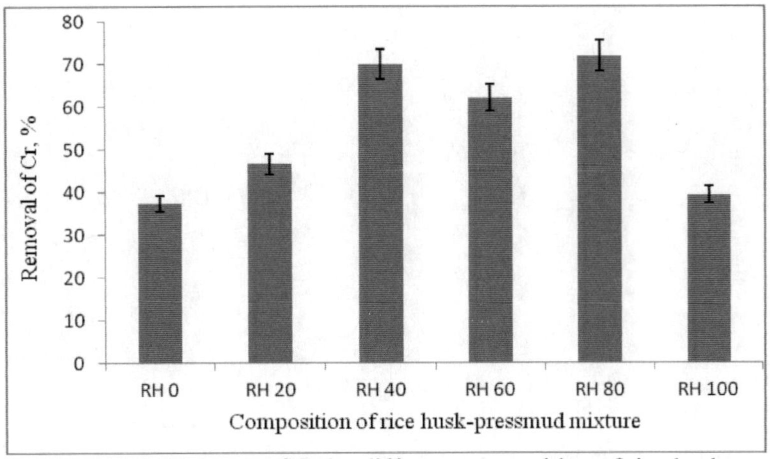

Figure 1. Removal efficiency of Cr by different composition of rice husk-pressmud

3.3.2 *Removal of Fe*

The removal efficiency of Fe was determined with an initial concentration of acidic synthetic wastewater at 2.28mg/L to stimulate a trace amount of Fe with pH value in the range of 1.04 to 1.45 respectively. The removal efficiency of Fe was illustrated in Figure 2 as shown below. According to Figure 2, pressmud alone and pressmud-rice husk mixture could remove 100% of Fe from RH 0 till RH 80, while the removal efficiency dropped at RH 100. Due to the acidic condition, sorption process of rice husk was interrupted since there was no pressmud in RH 100 to neutralize the excessive H^+ or H_30^+ ions in the acidic synthetic wastewater. There was significant correlation between composition ratio of pressmud-rice husk and the removal efficiency since $p < 0.05$ based on ANOVA.

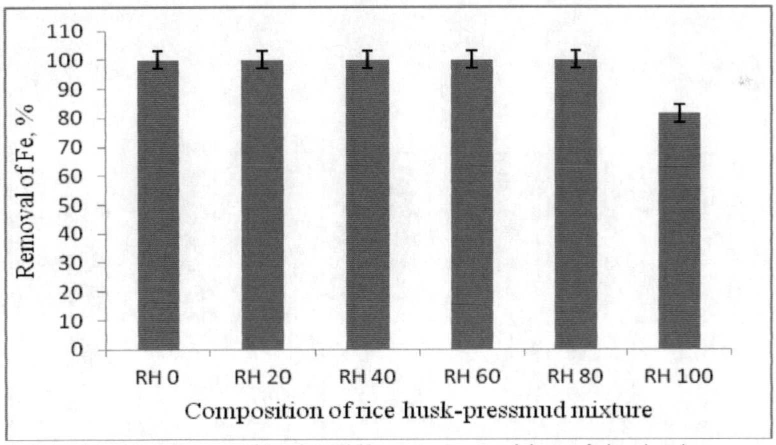

Figure 2. Removal efficiency of Fe by different composition of rice husk-pressmud

3.3.3 *Removal of Mn*

The removal efficiency of Mn was determined with an initial concentration of acidic synthetic wastewater at 2.13mg/L. The removal efficiency of Mn was illustrated in Figure 3. Figure 3 showed that as rice husk ratio was increased, the removal efficiency of Mn decreased. Rice husk alone (RH100) showed the lowest removal efficiency as it was not able to absorb Mn under acidic condition where the pH solution was at 1.04. Composition ratio of pressmud-rice husk had a significant effect on Mn removal since $p < 0.05$ according to ANOVA. For the removal of Mn, pressmud alone seemed to have the highest removal efficiency. Addition of rice husk to pressmud did not improve removal efficiency of Mn. Pressmud alone already removed more Mn than unmodified rice husk.

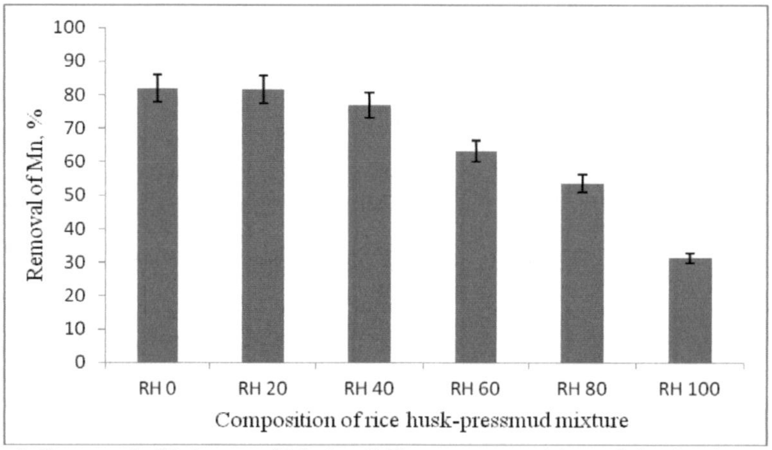

Figure 3. Removal efficiency of Mn by different composition of rice husk-pressmud

3.3.4 *Removal of Ni*

The removal efficiency of Ni was determined with three samples at initial concentrations of synthetic wastewater of 2.18mg/. The removal efficiency of Ni was illustrated in Figure 4 as shown below. Based on Figure 4, the removal efficiency of RH 0, RH 20 and RH 40 were between 69.12-73.73% respectively. The removal trend decreased from RH 60 to RH 100. Thus, Figure 4 showed the pressmud-rice husk mixtures were more effective than pure rice husk and pure pressmud. Combination of pressmud and rice husk increased the performance in removing Ni. Composition ratio of pressmud-rice husk had no significant effect on Ni removal since $p > 0.05$ based on ANOVA. These might be due to the property of pressmud to allow Ni ions to adhere on it and also the ability of pressmud to act as a buffer for the acidic synthetic wastewater so that rice husk could perform well for the adsorption of Ni.

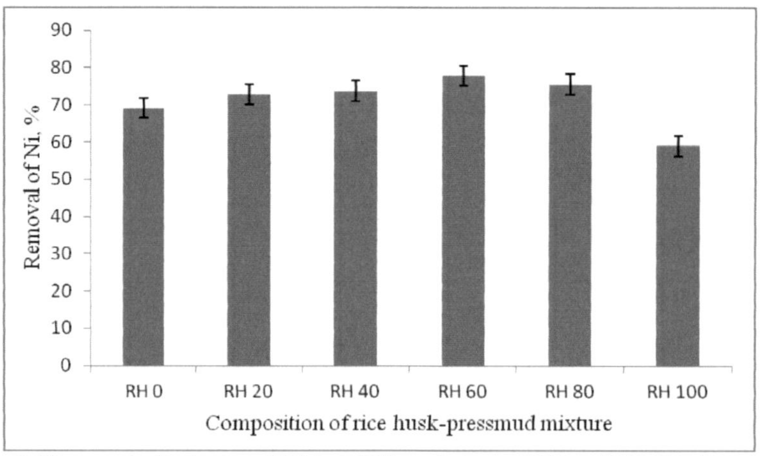

Figure 4. Removal efficiency of Ni by different composition of rice husk-pressmud

3.3.5 *Removal of Pb*

The removal efficiency of Pb was determined with three samples at initial concentrations of synthetic wastewater of 2.12mg/L. The removal efficiency of Pb was illustrated in Figure 5. The figure showed the similar removal efficiency trend from RH 0 to RH 80 except for RH 100. Figure 5 also showed that pressmud alone and pressmud-rice husk mixtures were very effective to remove trace amount of Pb from aqueous solution. The removal efficiency was between 99-100%. Pressmud alone was a good sorbent for Pb After the addition of pressmud to rice husk, the removal efficiency was not much affected and remained high as rice husk alone had been reported as a good adsorbent for Pb ions (Lata & Samadder, 2014).

RH 100, pure rice husk had very low efficiency in Pb removal which was only 27.75% as acidic condition is not suitable for sorption of Pb. According to Wong et al.(2003), the uptake of Pb by tartaric acid of modified rice husk increased when the pH increased from 2 to 3.Since the prepared acidic synthetic wastewater was around pH 2, the removal efficiency of Pb was very low with the absence of pressmud or insufficient amount of pressmud to act as buffer. This was because at low pH, the surface of the sorbent was surrounded by hydronium ions which prevented the metal ions from approaching the binding sites on the sorbent (Wong et al., 2003). Based on ANOVA, there was significant correlation between composition ratio of pressmud-rice husk and the Pb removal since $p < 0.05$.

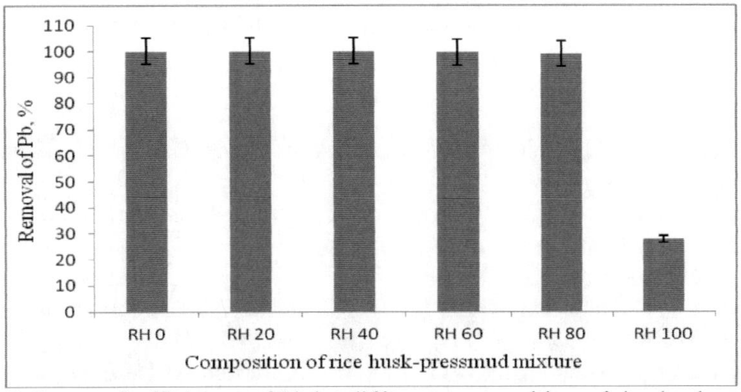

Figure 5. Removal efficiency of Pb by different composition of rice husk-pressmud

3.3.6 *Removal of Zn*

The removal efficiency of Zn was illustrated in Figure 6 as shown below. It was clearly shown in Figure 6 that pressmud alone and pressmud-rice husk mixtures removed 100% of Zn, while rice husk alone had only removed 4.15% of Zn from the acidic synthetic wastewater. Pressmud alone was a good adsorbent in eliminating the trace amount of Zn. Based on ANOVA, composition ratio of pressmud-rice husk had significant effect on the Zn removal since $p < 0.05$.

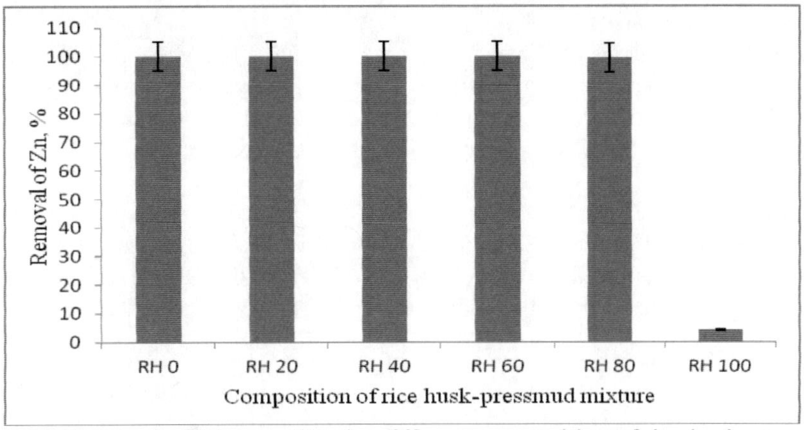

Figure 6. Removal efficiency of Zn by different composition of rice husk-pressmud

4. Conclusion

Pressmud alone had high removal efficiency of up to 90% to remove trace amount of iron, lead and zinc. Rice husk alone had very low efficiency in the removal of heavy metals due to the acidic condition of the acidic synthetic wastewater. By mixing pressmud with rice husk, the removal trend of iron, lead and zinc did not have much difference. However, when pressmud was mixed with rice husk, the removal efficiency of chromium and nickel increased since it acted as a buffer in the acidic synthetic wastewater due to its high bicarbonate content to enhance the sorption of metal ions to the sorbent. There was a decrease in the removal efficiency of manganese as pressmud was added to the rice husk. But, rice husk alone (RH 100) consistently showed low efficiency for removing all of heavy metals as it could not perform well in acidic condition. The composition of rice husk and pressmud mixtures greatly affected the removal efficiency of heavy metals as it improved the chemical characteristics of the mixture in reducing the heavy metals in the solution and also increased the initial pH of synthetically prepared acidic wastewater from below 2 to as high as 8.

5. References

[1] Ahmed I, Attar S J and Parande M J 2012 *International Journal of Advanced Engineering Research and Studies*. **1** pp 92-94
[2] Anand S A, Nagarajappa D P, Sanjeev S and Ramu S 2014 *International Journal of Innovative Research in Science, Engineering and Technology*. **3** pp 2319-8753
[3] Annadurai G, Juang R S and Lee D J 2002 *IWA Publishing*. **47** pp 185–190
[4] Balintova M, Holub M and Singovszka E 2012 *Chemical Engineering Transactions*. **28** pp 2012
[5] Barakat M P 2010 *Arabian Journal of Chemistry*. **4** pp 361–377
[6] Bazargan A, Bazargan M and McKay G 2015 *Journal of Renewable Energy*. pp 512-520
[7] Gupta N, Balomajumder C and Agarwal V K 2012 *Chemical Engineering Journal*. **191** pp 548–556
[8] Gupta N, Tripathi S and Balomajumder C 2010 *Journal of Fuel*. **90** pp 389–394
[9] Kumar U and Bandyopadhyay M 2005 *Bioresource Technology*. **97** 104–109
[10] Lata S and Samadder S R 2014. *International Journal of Environmental Research and Development*. **4** pp 165-170
[11] Maheera M, Syahidah A A, Harlina A, Norli I and Morad N 2013 *International Journal of Chemical and Environmental Engineering*. **4** pp 5
[12] Nhapi I, Banadda I N, Murenzi R, Sekomo C B and Wali U G 2011 *The Open Environmental Engineering Journal*. pp **4** 170-180.
[13] USEPA 1994 The Analysis of Water and Wastes by U.S. EPA Method 2007 Using the Optima 8300 ICP-OES and prep FAST Auto-Dilution/Calibration System, Version 4.4
[14] Wang X M, Pei Y F, Lu M X, Lu X Q and Du X Z 2014 *Springer Science*
[15] Wong K K, Lee C K, Low K S and Haron M J 2003 *Chemosphere*. **50** pp 23–28

Acknowledgment
The authors acknowledged Universiti Sains Malaysia (USM), Malayan Sugar Manufacturing (MSM) Sdn. Bhd., Leong Guan Sdn. Bhd, Kepala Batas as well as to USM Research University Grant (1001/PTEKIND/811227) for their contributions during the study.

Water desalination by air-gap membrane distillation using meltblown polypropylene nanofiber membrane

S Rosalam[1], C K Chiam[2], S Widyaparamitha, Y W Chang, C A Lee
Membrane Technology Research Group, Material and Mineral Research Unit, Faculty of Engineering, *Universiti Malaysia Sabah*, Jalan UMS 88400 Kota Kinabalu, Sabah, Malaysia

E-mail: [1]rslam@ums.edu.my, [2]chiamchelken@ums.edu.my

Abstract. This paper presents a study of air gap membrane distillation (AGMD) using meltblown polypropylene (PP) nanofiber membrane to produce fresh water via desalination process. PP nanofiber membranes with the effective area 0.17 m^2 are tested with NaCl solutions (0.5 – 4.0 wt.%) and seawater as the feed solutions (9400 – 64800 μS/cm) in a tubular membrane module. Results show that the flux decreases with increasing the membrane thickness from 547 to 784 μm. The flux increases with the feed flow rate and temperature difference across the membrane. The feed concentration affects the flux insignificantly. The AGMD system can reject the salts at least 96%. Water vapor permeation rate is relatively higher than solute permeation rate resulting in the conductivity value of permeate decreases when the corresponding flux increases. The AGMD system produces the fresh water (200 – 1520 μS/cm) that is suitable for drinking, fisheries or irrigation.

1. Introduction

The shortage in clean water supply worldwide has driven desalination to become a sustainable process of potable water production. Reverse osmosis and thermal evaporation are the current commercial desalination technologies. However, the high energy consumption in these technologies has retarded their sustainability. The use of membrane distillation (MD) is becoming increasingly recommended to produce fresh water as MD has fulfilled the requirement of process intensification i.e., less mechanical parts, non-pressurized system, low energy consumption and low cost [1-3].

In MD, water vapor transfer across the membrane occurs when a driving force is subjected between the feed and permeate sides. On the feed side, saline water is brought direct contact with a porous and hydrophobic membrane surface; water vapor is thermally driven through the membrane pore. Various designs on the permeate side to maintain the driving force categorize the MD into four types of configurations: direct contact MD (DCMD), air gap MD (AGMD), sweeping gas MD (SGMD) and vacuum MD (VMD).

Nanofiber membranes have been attractive in MD applications in the recent years [4, 5] due to the high porosity, interconnected open pore structure and tailorable membrane thickness. These nanofiber membranes are fabricated by using electrospinning technique whereby polymer pellets (<30 wt.%) are

[1, 2] To whom any correspondence should be addressed.

dissolved in a solvent to form a homogeneous polymer solution. At least 70 wt.% of the solvent is evaporated and released into the air during electrospinning the polymer solution within a high voltage electronic field. As a result, a massive scale of solvent recovery from the dilute air is required which is the major drawback of using electrospinning technique.

In this work, the nanofiber membranes are fabricated by using meltblowing technique where molten polymer instead of polymer solution is used. The polymer pellets are melted at high temperature to form the molten polymer. The molten polymer is then sprayed through a spinneret to form the nanofibers. The applicability of the meltblown nanofiber membranes in AGMD desalination to produce fresh water is investigated in this study.

2. Experimental methods

Meltblown PP nanofiber membranes were fabricated using a melt blowing system (Japan Zetta Co. Ltd.). The polypropylene (PP, Sun-Allomer) was melted at temperature 230°C and sprayed through a spinneret (diameter: 0.8 mm) to a collector at a distance of approximately 240 cm. The nanofibers were formed into tubular membranes with inner diameter 5.5 cm and length 99 cm in that resulted the membrane area as 0.17 m^2. The membrane thickness was varied from 550 to 790 μm.

The nanofibers were sputtering-coated with a thin layer of gold. The morphology of the nanofiber was observed by using scanning electron micrography (SEM) (Carl Zeiss Evo MA10, USA). The diameters of the nanofibers were determined by observing the SEM images.

The membrane thickness was measured with a digital micrometer (MDC-1''M, Mitutoyo, Kawasaki GmbH (Germany). Gravimetric method was employed to determine the membrane porosity. Isopropyl alcohol (IPA) (QRëc) was used as the wetting liquid. The porosity of the membrane was calculated by the equation as follows:

$$\varepsilon = \frac{(w_w - w_d)/\rho_{IPA}}{(w_w - w_d)/\rho_{IPA} + w_d/\rho_{pp}} \quad (1)$$

where w_w is the wet membrane weight, w_d is the dry membrane weight, ρ_{IPA} is the density of IPA (0.786 g/cm^3), and ρ_{pp} is the density of PP (0.946 g/cm^3).

A schematic diagram of the air gap membrane distillation (AGMD) is shown in Figure 1. Two types of feed solutions were tested: NaCl solutions (0.5 – 4.0 wt%) and seawater collected from the beach near Universiti Malaysia Sabah. The volume of the seawater feed used was 2 L. The feed solution was circulated through the middle of the membrane module by a calibrated gear pump implemented with a digital flow meter (Micro pump Gear Pump Drives 75211-35, Cole-Parmer Instrument Cp., Illinios, USA) and passed through a heating water bath (Protech HC-10, Tech-lab MFG, Malaysia). The cooling water was circulated through the copper tube and the temperature was maintained at 20°C. After operating time of 2 hours, permeate was collected using a volumetric flask and the readings of mass were recorded. Each experiment was repeated three times under the same operational conditions, and the average values were reported.

3. Results and discussion

3.1. Characteristics of the meltblown PP nanofiber membrane

3.1.1. Morphology and parameters of the nanofiber membrane

The SEM images of the meltblown nanofibers are shown in Figure 2. The diameter of the fibers ranges from 10^2 to 10^3 nm. These fibers piled up to form membranes with the thickness varies from 470 to 800 μm. The average porosity of the nanofiber membrane with dry weight of 5 g is approximately 90%.

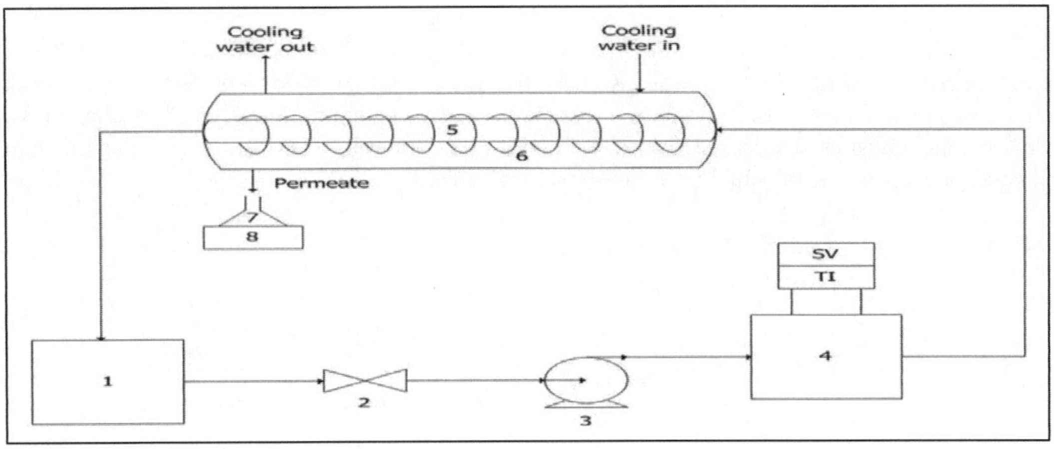

Figure 1. Schematics of AGMD experiment setup: 1, feed reservoir; 2, valve; 3, pump; 4, temperature control bath with heating coil; 5, nanofiber membrane module; 6, cooper tube with cooling water in and out; 7, volumetric flask; 8, electronic weighting balance.

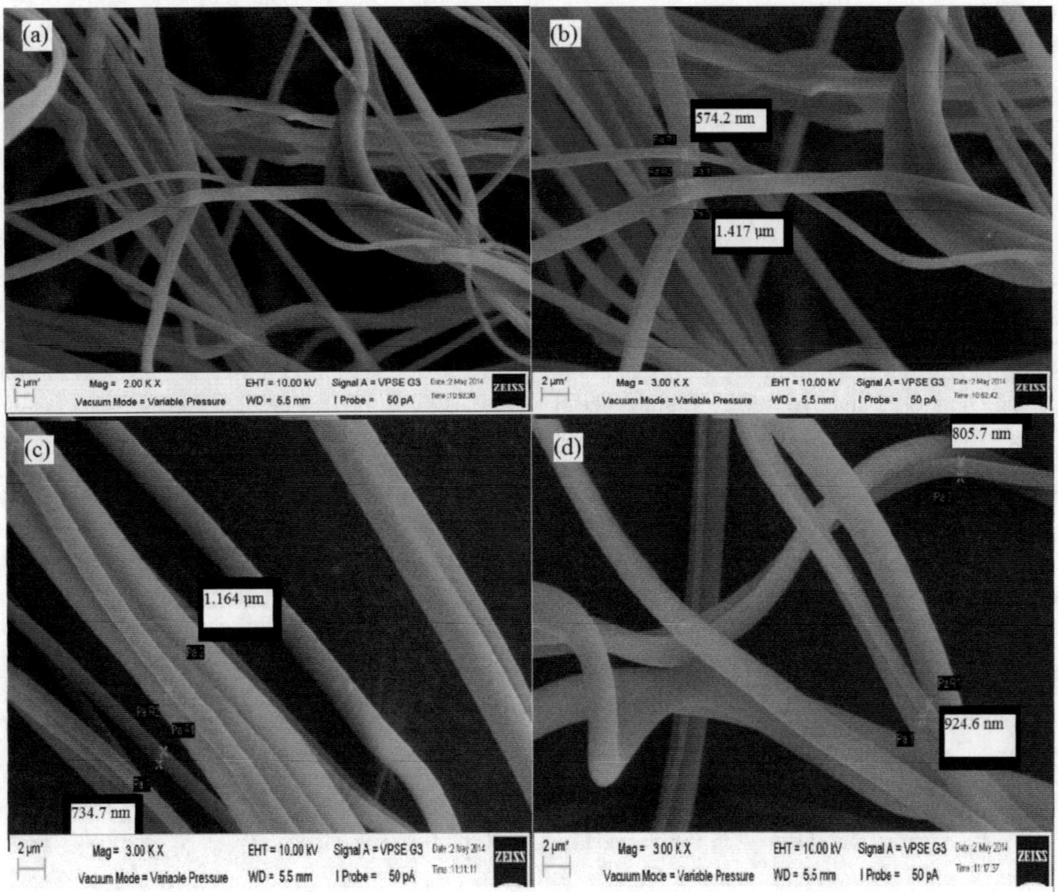

Figure 2. SEM images of the meltblown PP nanofibers at magnification of (a) 2 kX, (b), (c) and (d) 3 kX.

3.1.2. Effect of nanofiber membrane thickness on AGMD flux

Figure 3 shows the variation of flux at the corresponding nanofiber membrane thickness for 2 h AGMD operation when seawater was used as the feed solution. The flux decreases when the membrane thickness increases. The flux reduces significantly i.e., approximately 27% reduction when the membrane thickness increased from 547 to 784 μm. In this AGMD system, the membrane thickness contributes a considerable mass transfer resistance.

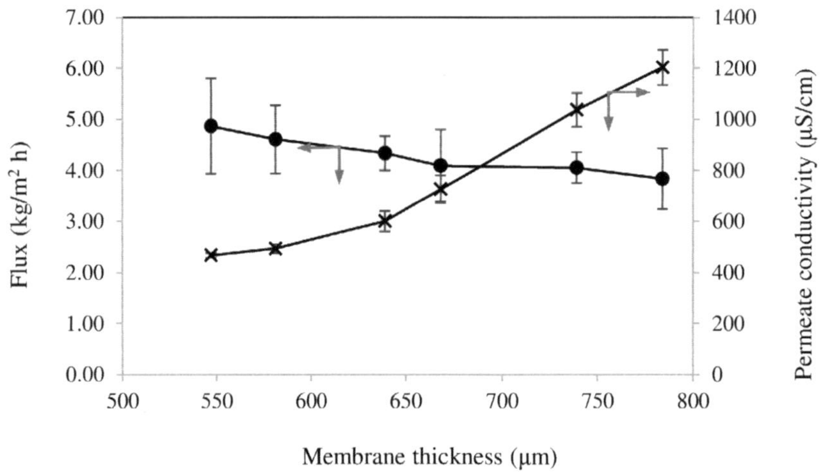

Figure 3. The effect of nanofiber membrane thickness on the flux and permeate conductivity. *(Feed solution: seawater, feed flow rate: 300 mL/min, cooling water temperature: 20°C)*

3.2. Influence of process conditions on AGMD performance

The influence of process conditions on the AGMD flux was investigated. The membrane thickness for these studies was 520 – 525 μm. The experimental results represented the effect of feed flow rates on the fluxes are shown in Figure 4. The flux increases with the feed flow rate for 3.5 wt.% NaCl was tested as the feed solution and the temperature difference across the membrane was 60°C. The dependence of the flux on the feed flow rate indicates the existence of the boundary layer in the feed side. The boundary layer reduces at higher feed flow rate as resulting in a greater flux.

Figure 5 illustrates the experimental values of AGMD flux as a function of the feed concentration. The flux decreases with the feed concentration insignificantly when the feed flow rate was 300 mL/min and the temperature difference across the membrane was 60°C. This indicates that the concentration polarization effect could be negligible in the feed side.

The experimental results to present the effect of temperature differences across the membrane on the fluxes are shown in Figures 6 and 7 for 3.5 wt.% NaCl and seawater as the feed solutions respectively. The flux increases greatly with the increasing temperature differences. This is due to the exponential increase of the vapor pressure with the temperature which can be described by the Antoine equation.

The error bars corresponding to each point reported in Figures 4 – 7 are below 1% where these percentage values were represented by the standard deviations $s(J)$ calculated as follows:

$$s(J) = \sqrt{\frac{\sum_{i=1}^{n}(J_i - \bar{J})^2}{n-1}},\ s(J)_{\text{percen.}}\ (\%) = \frac{s(J)}{\bar{J}} \times 100 \qquad (2)$$

where n is the number of experiments and $\bar{J}$ is the mean flux. The percentages of the deviations are small and the experimental fluxes are acceptable.

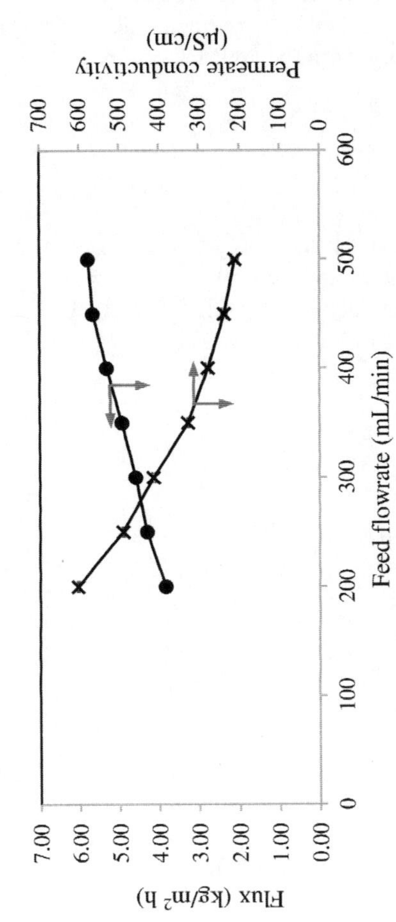

Figure 4. The effect of feed flow rate on the flux and permeate conductivity. *(Feed solution: 3.5 w/w % NaCl, feed temperature: 80°C, cooling water temperature: 20°C)*

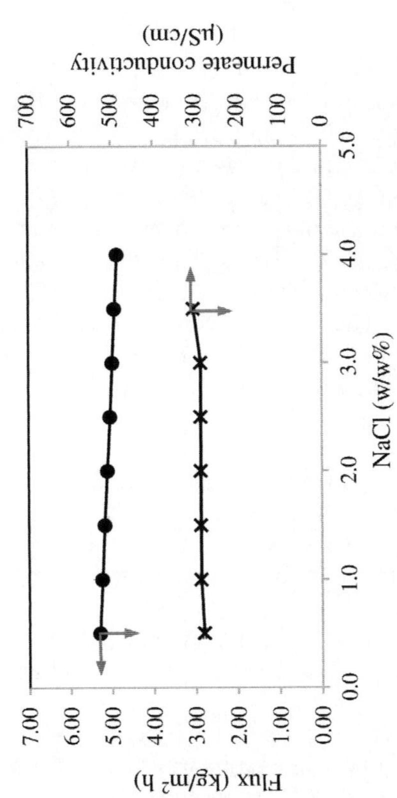

Figure 5. The effect of feed concentration on the flux and permeate conductivity. *(Feed flow rate: 300 mL/min; feed temperature: 80°C; cooling water temperature: 20°C)*

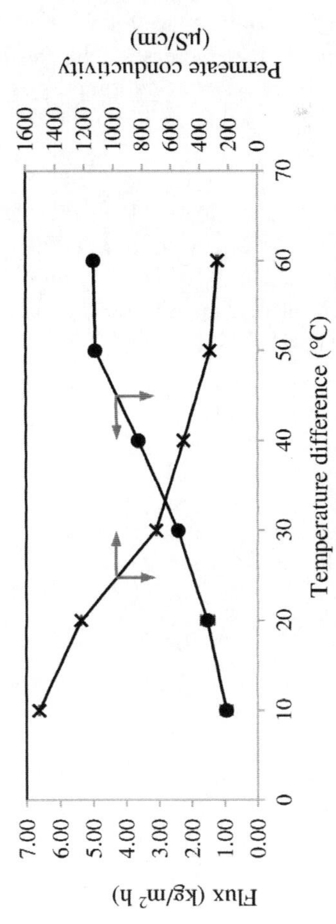

Figure 6. The effect of temperature difference across the nanofiber membrane on the flux and permeate conductivity. *(Feed solution: 3.5 w/w % NaCl, feed flow rate: 350 mL/min, cooling water temperature: 20°C)*

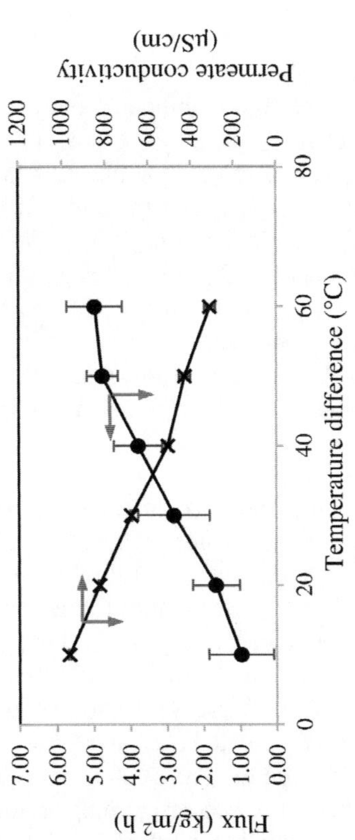

Figure 7. The effect of temperature difference across the nanofiber membrane on the flux and permeate conductivity. *(Feed solution: seawater, feed flow rate: 300 mL/min, cooling water temperature: 20°C)*

3.3. Permeate quality

The effect of the tested conditions on the conductivity values of permeates are reported in Figures 3 – 7. The conductivity values of permeates varied between 200 and 1520 µS/cm, whereas the conductivity values of the feed solutions (the NaCl solutions and the seawater) ranged from 9400 to 64800 µS/cm. As a result, the rejection of salts were above 96% i.e., at least 4% of solute (salts) permeation did take place during the AGMD operation. Interestingly, the conductivity value reduces when the corresponding flux increases as shown in Figures 3 – 7. It is suggested that the water vapor permeation rate is relatively higher than the solute permeation rate. The solute concentration could be diluted by the condensate (liquid water) in permeate. The water produced from the AGMD desalination in this work is fresh water (0 – 1990 µS/cm) which can be used as drinking water (< 500µS/cm), fisheries water (< 500µS/cm), or irrigation (< 2000 µS/cm).

4. Conclusions

This work presented an assessment study of AGMD using meltblown nanofiber membranes in a tubular module for fresh water production via desalination process. The results show that:

(1) The nanofibers made from PP have diameter ranged from 10^2 to 10^3 nm. The average porosity of the nanofibers for 5 g (dry weight) is approximately 90%.
(2) The membrane thickness from 547 to 784 µm contributes a significant mass transfer resistance.
(3) The AGMD flux increases with increasing the feed flow rate. The boundary layer effect on flux reduces when operated at higher flow rate.
(4) The effect of the feed concentration (0.5 – 4.0 wt.% NaCl) on the flux is insignificant.
(5) The AGMD flux increases with temperature across the membrane is attributed to the vapor pressure increase exponentially with temperature based on the Antoine equation.
(6) The rejections of salts are above 96%. The conductivity value of permeate decreases when the corresponding flux increases. The fresh water produced from the AGMD system can be used for drinking, fisheries, or irrigation purposes.

5. References

[1] Lawson K W and Lloyd D R 1997 *J. Membr. Sci.* **124** pp 1
[2] Stankiewicz A I and Moulijn J A 2000 *CEP Mag.* **January** 22
[3] Alkhudhiri A, Darwish N and Hilal N 2012 *Desalination* **287** pp 2
[4] Liao Y, Wang R, Tian M, Qiu C and Fane A G 2013 *J. Membr. Sci.* **425 - 426** pp 30
[5] Dong Z Q, Ma X H, Xu Z L, You W T and Li F B 2014 *Desalination* **347** pp 175

IOP Conf. Series: Earth and Environmental Science **36** (2016) 012033 doi:10.1088/1755-1315/36/1/012033

Characterisation of crude palm oil O/W emulsion produced with Tween 80 and potential in residual oil recovery of palm pressed mesocarp fibre

N H Ramly[1], R Zakaria[1*], M N Naim[1]
[1]Process and Food Engineering Department, *University Putra Malaysia*, Serdang Selangor

Email: [1]rabitah@upm.edu.my

Abstract. Surfactant-assisted aqueous extraction has been proposed as a "green" alternative to hexane extraction for the recovery of oil from plant matters. An efficient aqueous surfactant extraction system usually use an extended type of ionic surfactant with the ability to produce Winsor type III microemulsion, reducing the interfacial tension (IFT) between plant oil and surfactant solution to an ultralow level (10^{-3} mN/m). However, the safe used of this surfactant in food processing is uncertain leading to non-food application of the recovered oil. In the present study, the potential of Tween 80, a commercial food-grade non-ionic surfactant, was evaluated in the recovery of residual oil from palm–pressed mesocarp. The emulsion produced between Tween 80 and crude palm oil (CPO) was characterised in terms of IFT, droplet size, viscosity and phase inversion temperature (PIT). The effect of surfactant concentration, electrolyte (NaCl) and temperature were studied to determine whether a Winsor Type III microemulsion can be produced. Results shows that although these parameters were able to reduce the IFT to very low values, Winsor type III microemulsion was not produced with this single surfactant. Emulsion of CPO and Tween 80 solution did not produce a PIT even after heating to 100°C indicating that middle phase emulsion was not able to be formed with increasing temperature. The highest percentage of oil extraction (38.84%) was obtained at the concentration above the critical micelle concentration (CMC) of Tween 80 and CPO, which was at 0.5 wt% Tween 80 with 6% NaCl, and temperature of 60°C. At this concentration, the IFT value is 0.253 mN/m with a droplet size of 4183.8 nm, and a viscosity of 7.38 cp.

1. Introduction

Crude palm oil (CPO) is usually obtained by mechanical screw press. After this mechanical pressing, there is still 5-11% residual oil per dry weight retained in the palm-pressed fibre (PPF). Subramaniam et. al. (2013) reported that the residual oil from PPF contains higher vitamin E which is 1700-2600 ppm as well as higher carotenes of 1400-1600 ppm compared to common CPO which are 600-1000 ppm and 500-700 ppm respectively. This precious substances left in residual oil from PPF make it significant to be extracted [1].

[1] To whom any correspondence should be addressed.

Surfactant-assisted aqueous extraction process (SAAEP) has been proposed as an environmentally friendly alternative to hexane extraction for the recovery of oil from plant materials such as corn germ and canola seeds [2]. The surfactant reduces the high interfacial tension (IFT), the surface energy exerted by intermolecular interactions, between two immiscible phases of oil and water [2]. An efficient aqueous surfactant extraction system usually has the ability to reduce the IFT value to an ultralow level (10^{-3} mN/m) [3]. Previous researches have shown that ultralow value of IFT contributed to significantly high extraction efficiency and free oil recovery [2,3]. Formulating microemulsion between surfactant and vegetable oils in the form of triglycerides is quite challenging compared to hydrocarbons. The complexity of the large triglyceride molecules as well as its sensitivity to air, light and temperature render the emulsification process difficult compared to hydrocarbon oils. Other than that, when triglycerides undergo decomposition, the products formed i.e. mono-and-di-acylglycerol are also amphiphiles and will change the surface activities such as the critical micelle concentration (CMC) of the formulation [4].

Proper structure and system of surfactant will contribute to the condition whereby hydrophilic-lipophilic balance can be achieved. This is called the middle phase microemulsion. This transition phase can be achieved by electrolyte addition at optimum salinity for anionic surfactants as well as increasing temperature to the phase inversion temperature (PIT) for non-ionic surfactants [5]. Previously, conventional surfactants was not able to reduce the IFT between oil and water to an ultralow level. Extended surfactant, a special surfactant with unique structure of having groups of intermediate polarity in between hydrophilic and lipophilic group of the surfactant, acts like a linker or a bridge that will provide stronger interaction between the hydrophobic group of the surfactant and the hydrophobic oil as well as between hydrophilic group of surfactant and water. This strong interaction will contribute to IFT reduction to the ultralow level [2,3]. However, the health and safety aspects of extended surfactant in food processing application is uncertain.

Tween 80 is a commercial food-grade non-ionic surfactant that has relatively low toxicity, low irritation potential and is widely used in pharmaceutical microemulsion preparation [6]. The structure of Tween 80 comprised of polyoxyethylene (20) and sorbitan monooleate. Compared to Tween 20 and Tween 40, Tween 80 is less viscous and possess a lower CMC, which means that the system will use less surfactant to achieve minimum IFT [7]. In this study, the potential of Tween 80 in the recovery of residual oil from palm–pressed mesocarp was evaluated. The effect of surfactant concentration, electrolyte (NaCl), and temperature was studied to determine whether a Winsor Type III microemulsion can be produced by using this single commercial surfactant. The IFT, droplet size and viscosity was then correlated to the oil extraction efficiency.

2. Materials and Methods

2.1 Materials

Tween 80 (R&M), Sodium Chloride (R&M), and n-hexane were purchased from Evergreen Sdn Bhd. Crude palm oil and palm pressed fibre (PPF) was kindly provided by Palm Oil Mill Sg Tengi under Felda Plantation Sdn. Bhd.

2.2. Determination of Phase Inversion Temperature (PIT)

0.1g of CPO was mixed with Tween 80 based on specified ratios. This CPO-Tween 80 mixture was homogenised using a magnetic stirrer at room temperature. Next, water was added to make 100 %(w/w) mixtures. The mixtures were heated from room temperature until 100°C at an interval of ~ 1.0°C/min while stirring at ~500 rpm on hot plate stirrer. The physical appearance of CPO/Tween 80 emulsion was observed throughout the heating process. The clear temperature and temperature when emulsion turns back to cloudy were reported to determine the presence of PIT and the type of emulsion formed [8].

2.3. Measurement of Interfacial Tension (IFT)

The IFT between CPO and the aqueous Tween 80 solution was measured using a spinning drop tensiometer (SVT 20N). Tween 80 solutions were prepared at specified concentrations and injected into a 10 ml glass capillary tube. Then, one drop of CPO was injected into the capillary. The capillary was rotated until an elongated cylindrical drop was obtained and measurement of IFT was taken at equilibrium. From preliminary observation, the IFT reached equilibrium within 30 minutes to an hour. Therefore, the measurements were taken within this period [3]. All measurements for IFTs involving CPO were maintained at 60°C to prevent crystallisation [9].

2.4. Measurement of Droplet Size

0.1g of CPO was weighed and mixed with specified concentrations of Tween 80 and stirred using a magnetic stirrer until homogenized. Then, distilled water was mixed and homogenized again at 60°C for 10 minutes to form emulsion. After that, 2 ml of emulsion was injected into a glass cuvette and droplet size was measured using Nanoplus. The reading was taken at the region of optimum intensity [10].

2.5. Viscosity Analysis of Extraction Medium

Different concentrations of Tween 80 solution were mixed with 6% NaCl solution. These mixtures were shaken in a water bath shaker at 60°C at 185 rpm for 30 minutes. This condition was considered based on the extraction condition. Next, the viscosity of the solutions was measured using the Brookfield Programmable DV-II Viscometer using spindle no 2.

2.6. Oil Extraction Efficiency using SAAEP

PPF was dried until the moisture content was less than 5wt%. The residual oil content of this PPF after pressing was analysed by Soxhlet extraction and used as the reference total oil to calculate the oil extraction efficiency of SAAEP. The reference total oil of PPF used was 16.94% per dry weight. 8 g of PPF was mixed with 200 g Tween 80 solution in a 500 ml bottle. Then, the mixture was shaken using a horizontal thermo shaker at 60°C at 185 rpm for 30 minutes. The slurry was then centrifuged using a swing bucket rotor at 3500rpm for 1hour. Then, the centrifuge bottle was placed in a freezer for 1 hour and the free oil was scrapped. The middle and solid layers were left to cool at room temperature and the middle layer was pipetted out. The solid was dried and the residual oil after SAAEP was extracted using hexane via Soxhlet apparatus. The residual oil was weighed and subtracted from the reference total oil of PPF to obtain the oil extraction efficiency [2].

3. Results And Discussion

3.1. Effect of Tween 80 concentration on IFT reduction

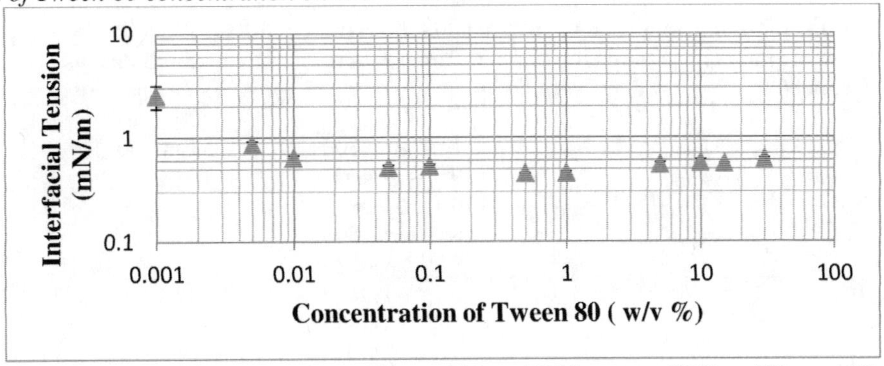

Figure 1. Effect of concentration of Tween 80 on IFT between CPO and Tween 80 solution (60°C, 1 hour).

Referring to Figure 1, Tween 80 concentration was varied in order to observe its effect on the reduction of IFT. The measurements was conducted at 60°C to prevent crystallisation since CPO exists as a semi solid at room temperature [5]. IFT value between pure water and CPO was 4.908 mN/m (data not shown). IFT value between water and oil depends on the types and composition of oil. For example, previous literature has reported that the IFT value between soybean oil and water was 5.3 mN/m [11]. Different types of CPO also give different IFT which was in the range of 6.0-9.6mN/m from the different study [9]. As seen in Figure 1, when very low concentration of Tween 80 was introduced into the system, starting at a concentration of 0.001wt%, the IFT value dropped to 2.518 mN/m. This reduction proves that there was an adsorption of surfactant at the oil-water interfaces due to the interaction between the hydrophobic oil and the hydrophobic molecule of the surfactant and between water molecules and the hydrophilic part of the surfactant [7]. At a concentration of 0.05 wt%, an equilibrium IFT of 0.51 mN/m was obtained and after this concentration, the IFT value reaches equilibrium and cannot be reduced further. The CMC of Tween 80/CPO solution can then be considered as 0.05 wt%. Different surfactant has their own effectiveness level which represents its ability and limit to reduce the IFT to a minimum level [5]. Previous study using different type of surfactant also showed that the IFT between triglycerides and the surfactant solution cannot achieve an ultralow value. IFT of the surfactant with triolein can only reach 2.6×10^{-1} mN/m, whereas IFT with n-octane can reach 4.9×10^{-3} mN/m [4]. If a microemulsion can be produced between the surfactant solution and CPO, a second drop in IFT should be seen when the concentration is increased after the CMC, but this was not the case for Tween 80/CPO emulsion even though the concentration of Tween 80 was increased up to 30 wt%. A very high concentration of Tween 80 however is considered impractical for extraction purposes.

3.2. Effect of temperature on CPO solubilisation

The importance of increasing temperature in non-ionic surfactant is similar to the addition of electrolyte in anionic surfactant which has been proven to cause significant reduction in IFT between some vegetable oils and water in previous literatures [2,3]. Observation of the phase behaviour of CPO/Tween 80 solution at increasing temperature and different concentration of Tween 80 was conducted to investigate the existence of a phase inversion temperature (PIT). PIT occurs between the solubilisation temperature and the cloud point temperature where clear emulsion turns back to a cloudy solutions [8]. From our preliminary experiments, it was found that pure Tween 80 solution in water did not exhibit a cloud point since the solution maintained as a clear phase up to a temperature of 100 °C [4]. The cloud point temperature can be considered as pseudo-phase inversion, in which the inversion from non-ionic detergent-in-water phase (D/W) to water-in-detergent phase (W/D) occurs. Phase inversion temperature (PIT) of emulsion is expected to be near the cloud point of the solution used. The presence of oil in emulsion may lower or raise the cloud point of the surfactant [12].

Table 1. Solubilisation Temperature CPO/Tween 80 at different concentrations

Concentration Of Tween 80	Ratio Of CPO:Tween 80	Composition	Solubilization Temperature
1%	0.1:1	0.1g CPO 1g Tween 80 98.9g Water	No Clear Emulsion
5%	0.1:5	0.1g CPO 5g Tween 80 94.9g Water	78°C
10%	0.1:10	0.1g CPO 10g Tween 80 89.9g Water	75°C
50%	0.1:50	0.1g CPO 50g Tween 80 49.9 Water	80°C

Emulsions of CPO/Tween 80 at room temperature for all concentration of Tween 80 shows a milky appearance until the temperature reached the solubilisation temperature as shown in Table 1. After this temperature the milky solution turn to clear emulsions. This occurs at about 75 to 80 °C for Tween 80 concentration of 5 % and higher. However, until boiling occurs, there was no transition seen from the clear phase back to a cloudy phase. Therefore, there were no cloud points present for emulsions of CPO/Tween 80. It can be deduced that Tween 80/ CPO solution did not exhibit the inversion from Winsor Type I to Winsor Type II microemulsion. As the transition did not occur, the temperature where the middle phase Winsor Type III microemulsion occur also cannot be determined from the phase inversion technique. Nevertheless, it can be inferred that middle phase microemulsion of Tween80/CPO might be produced only at a higher temperature than 75 °C. Temperature close to the water boiling point however was not economically practical for extraction purposes.

3.3. Effect of NaCl on IFT reduction

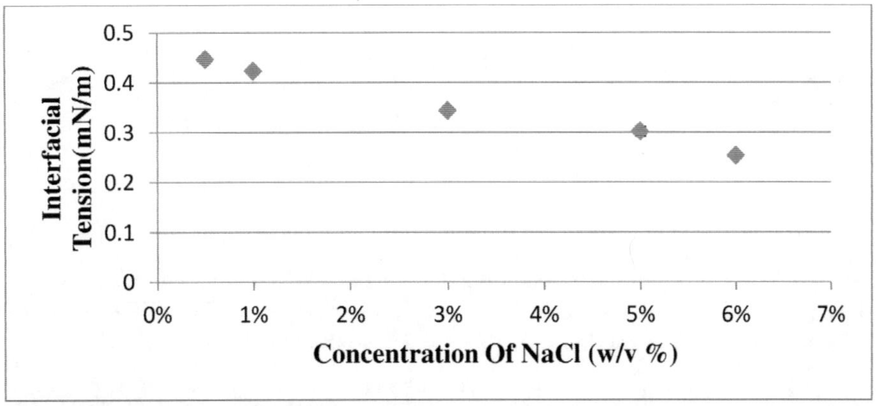

Figure 2. Effect of NaCl addition on IFT reduction. (0.5wt% Tween 80, 60°C, pH 9, 1 hour)

As seen in Figure 2, the addition of NaCl reduces the IFT to a small degree. With up to 6 wt% NaCl, the IFT can be reduced to 0.253 mN/m from 0.449 mN/m with no NaCl. Although the effect of electrolyte was not significant in non-ionic surfactant compared to anionic surfactant, electrolyte still contributes to the mechanism of 'salting out' of the hydrophobic groups in non-ionic surfactant. This process resulted in the hydrophobic groups become more dehydrated and adsorption increases. Increase in adsorption at the interface of oil and water causes the IFT to decrease. In contrast, for anionic surfactant, the repulsion of ionic charges between the hydrophilic/hydrophobic group of ionic surfactant and the electrolyte assists in the IFT reduction of the system. However, in non-ionic surfactant system, as the electrolyte only effects the hydrophobic group, the 'salting out' will occur only when the surfactant molecules are in the monomeric form [5].

3.4 Droplet Size Measurements

Table 2. Tween 80/CPO emulsion droplet size

Concentration Of Tween 80 (w/v %)	Interfacial Tension With CPO (mN/m)	Emulsion Droplet Size (nm)
0% (distilled water only)	4.908	-
0.5 % (without NaCl)	0.449	**1395.8**
0.5 % (with 6% NaCl)	0.253	**4183.8**
30 % (without NaCl)	0.623	**13.5**
30% (with NaCl)	0.238	**19.8**

As the concentration of Tween 80 increases, the size of the emulsion droplet decreases (Table 2). At 30 w/v% Tween 80 concentration, the droplet size became less than 100 nm. This shows that the emulsion size was in microemulsion range although the interfacial tension did not reach an ultralow value. Similarly, previous literature involving the emulsion of palm oil and Tween 40 also showed that increasing concentration of surfactant reduces the size of emulsion droplet to microemulsion range but yet the IFT was not able to reach an ultralow value [13]. The addition of an electrolyte (NaCl) reduces the IFT but did not cause the size of the droplets to reduce.

3.5 Oil Extraction efficiency

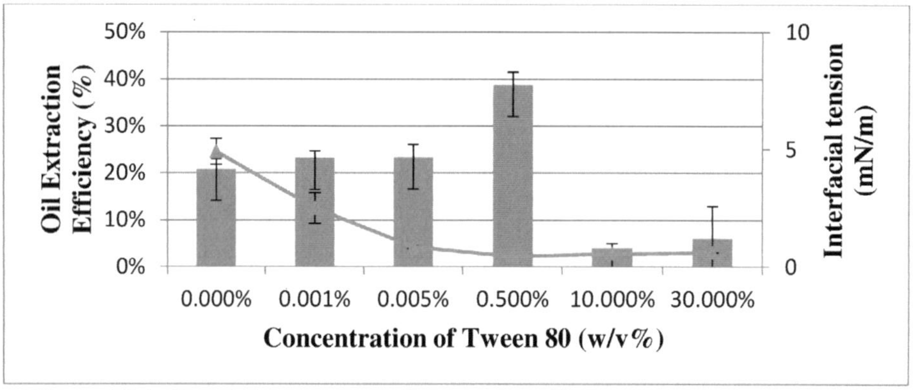

Figure 3(a) Correlation between oil extraction efficiency, concentration of Tween and IFT (6% NaCl, 60°C, 30 minutes extraction, 185 rpm, Fibre size of 0.5mm, Solid to liquid ratio 8g PPF: 200g Tween 80 solution)

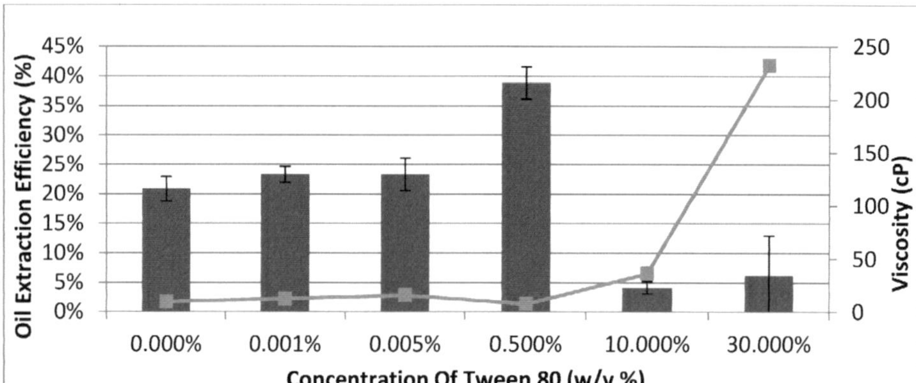

Figure 3(b). Oil extraction efficiency and viscosity of extraction medium vs Tween 80 concentration. (6% NaCl, 60°C Tween 80, 30 minutes extraction, 185rpm, Fibre size of 0.5mm, SLR 8g PPF: 200g Tween 80 solution)

Referring to Figure 3(a), the percentage of oil extraction from PPF using pure water only is about 21%. When Tween 80 was added at 0.001% and 0.005%, the extraction efficiency was almost in the same range although the IFT decreases. These low Tween 80 concentrations may not be enough to reduce the IFT until a value that can detach the oil from mesocarp fibre. However, when the concentration was increases to 0.5%, the oil extraction efficiency increases to about 37%. The highest extraction efficiency correlates with the CMC of Tween 80 in CPO as seen previously in Table 2. At this CMC, the IFT is the lowest, and it is the onset of micelle formation. This contributes to the maximum increase in extraction efficiency. Similarly, previous literature involving extraction of soybean oil from soybean seeds shows that the reduction of IFT increase the extraction efficiency of aqueous extraction process. When the IFT of water/sodium dodecyl sulphate (SDS)/oil reduces from 5.3 mN/m (without surfactant) to 2.1 mN/m with the addition of 3 %(w/w) SDS, soybean oil the extraction efficiency increases from 71.7% to 84.8%. However, further increase in Tween 80 concentration to 10% and 30 % causes the oil extraction efficiency to become much lower than extraction using pure water. Hence, although at 30% concentrations the droplet size is in the microemulsion range which is desirable for extraction, the efficiency is the lowest. Correlation between Tween 80 concentration and viscosity (Figure 3b) shows that the viscosity of 10 and 30 % solution is very high compared to 0.5 wt %. Hence the high viscosity of the solution may reduce the flowability of CPO into the extraction medium. Although high concentration of Tween 80 was able to break up the oil droplets into microemulsion size, the oil is not able to diffuse from the surface of solid mesocarp into the bulk surfactant solution due to the viscous medium. Diffusion of oil is a very important mechanism in extraction [11]. Furthermore, different extraction mechanisms take place at different surfactant concentrations. Usually below and near the CMC, synergism between roll-up and snap-off mechanism resulted in the liberation of free oil phase while at a much higher surfactant concentration than the CMC, solubilisation mechanism takes place which require complicated demulsification process. Therefore, microemulsion system must be optimized at low surfactant concentration to increase the extraction via rollup and snap-off mechanism and to avoid solubilisation mechanism [14].

4. Conclusion

The work shows that the IFT between CPO and Tween 80 solution is greatly affected by Tween 80 concentration up to the critical micelle concentration. The equlibrium IFT achieved is 0.51 at 0.05 wt % Tween 80. The addition NaCl however only has slight effect on the IFT and was not able to reduce the IFT to an ultra-low level. CPO can be solubilised in the Tween 80 solution at a temperature of 75 °C and higher. However, the phase inversion temperature of CPO/Tween 80 emulsion was not observed up to 100 °C temperature limit. Therefore it can be deduced that microemulsion of CPO/Tween 80 solution cannot be achieved at 75 °C or lower. In terms of droplet size, microemulsion size can be achieved using a high concentration of Tween 80 (30% w/v), although the IFT did not drop to an ultra-low range. The highest percentage of oil extraction (38.84%) was obtained at the critical micelle concentration (CMC) of Tween 80 and CPO, which was at 0.5 wt% Tween 80 with 6% NaCl, and temperature of 60°C. At the CMC, the IFT value is 0.253mN/m with a droplet size of 4183.8 nm, and a viscosity of 7.38 cp. As the Winsor Type III microemulsion and ultralow IFT can greatly enhance aqueous extraction process, further works are ongoing to identify method to create the middle phase microemulsion for Tween80/CPO emulsion particularly using a mixed surfactants formulation.

5. References

[1] Subramaniam V, Menon N R, Sin H and Yuen May C 2013 *J. Oil Palm Res.* **25** 116–22
[2] Do L D and Sabatini D A 2010 *J. Am. Oil Chem. Soc.* **87** 1211–20
[3] Kadioglu S I, Phan T T and Sabatini D A 2010 *J. Am. Oil Chem. Soc.* **88** 863–9
[4] Engelskirchen S, Elsner N, Sottmann T and Strey R 2007 *J. Colloid Interface Sci.* **312** 114–21
[5] Rosen M J 2004 (Canada: John Wiley & Sons)
[6] Cho Y-H, Kim S, Bae E K, Mok C K and Park J 2008 *J. Food Sci.* **73** E115–21
[7] Rao A Hameed M-E-R 2012 *Lat. Am. J. Pharm.* **31**
[8] Raman, Ia, Suhaimi H and Tiddy G 2003 *J. Oil Palm Res.* **15** 50–61
[9] Ho C C and Chow M C 2000 *J. Am. Oil Chem. Soc.* **77** 191–9
[10] Bera A, Kumar T, Ojha K and Mandal A 2014 *Fuel* **121** 198–207
[11] Campbell K A and Glatz C E 2009 *J. Agric. Food Chem.* **57** 10904–12
[12] Shinoda K and Arai H 1964 r *J. Phys. Chem.* **68** 3485–90
[13] Chow M C and Ho C C 1996 *J. Am. Oil Chem. Soc.* **73** 47–53

[14] Childs J D, Acosta E, Scamehorn J F and Sabatini D A 2004 *J. Energy Resour. Technol.* **127** 153–62

IOP Conf. Series: Earth and Environmental Science **36** (2016) 012034 doi:10.1088/1755-1315/36/1/012034

A review on the cords & plies reinforcement of elastomeric polymer matrix

S S Mahmood[1], H Husin[1], Mat-Shayuti M.S.[1], Z Hassan[2]
[1]Faculty of Chemical Engineering, *Universiti Teknologi MARA*, 40450 Shah Alam, Selangor, Malaysia
[2]Faculty of Chemical and Natural Resources Engineering *Universiti Malaysia Pahang (UMP)*

E-mail: [1]mia_samihah@yahoo.com

Abstract. Steel, polyester, nylon and rayon are the main materials of cords & plies that have been reinforced in the natural rubber to produce quality tyres but there is few research reported on cord and plies reinforcement in silicone rubber. Taking the innovation of tyres as inspiration, this review's first objective is to compile the comprehensive studies about the cords & plies reinforcement in elastomeric polymer matrix. The second objective is to gather information about silicone rubber that has a high potential as a matrix phase for cords and plies reinforcement. All the tests and findings are gathered and compiled in sections namely processing preparation, curing, physical and mechanical properties, and adhesion between cords-polymer.

1. Introduction

Cord is a structure composed of two or more strands and can formed as plied yarns [1]. Plies are locally represented as unidirectional assembly of cords [2]. Silicone rubbers have unique characteristics as it acts like both organic and inorganic compound properties. They have special properties, including biocompatibility, thermal stability, fine electrical properties, good chemical stability, oxidation resistance, and resistance to temperature [3,4]. Therefore, silicone rubber is chosen in a lot of industries to improve the quality and functionality of product such as electric and electronic equipment, households goods, footwear, and dental application [5-7].

Silicone rubbers were classified into four as shown in table 1, polydimethyl siloxane elastomers (MQ) that have only methyl substituent groups on the polymer chain; second is polydimethyl/vinylmethyl siloxane elastomers (VMQ) which exist with both methyl and vinyl substituent groups on the polymer chain. The third is polymethyl/vinyl/phenol siloxane elastomers (PVMQ) which have methyl, phenyl and vinyl substituent groups on the polymer chain and fourth is poly-Y-trifluoropropylmethyl/vinylmethyl siloxane elastomers (FVMQ) with fluoro, vinyl and methyl substituent groups on the polymer chain. The Q is an abbreviation for silicone rubber that represents quaternary group because silicone atom has four valence electrons [8]. MQ elastomers are rarely used since the methyl groups are difficult to vulcanize. The VMQ elastomer that represents vinyl groups

[1] To whom any correspondence should be addressed.

shortens the curing time of the elastomeric because it reduces the activation energy. The fluorine of FVMQ is distributed in increase the oil and heat resistance while phenyl groups of PVMQ incorporated into the chain to make sure the material remain elastomeric in lower temperature [9].

Table 1. Classification of silicone rubber [10-14]

Class of silicone rubber	Velocity (ms^{-1})
MQ	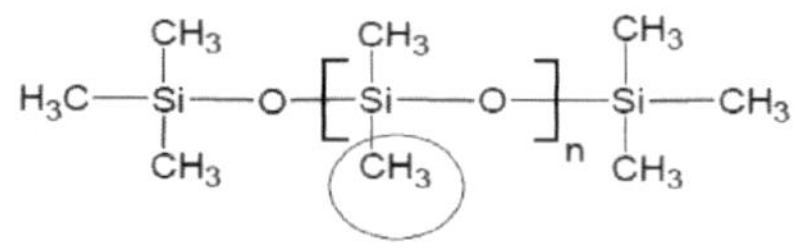
VMQ	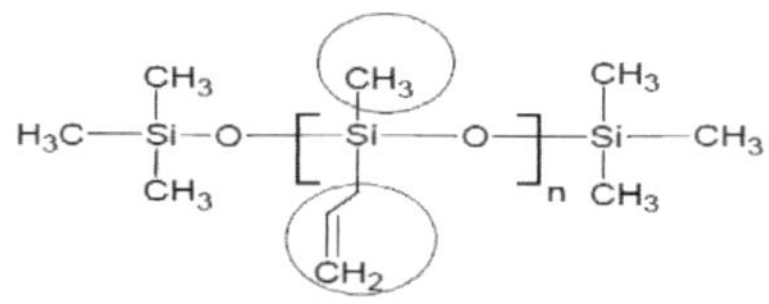
PVMQ	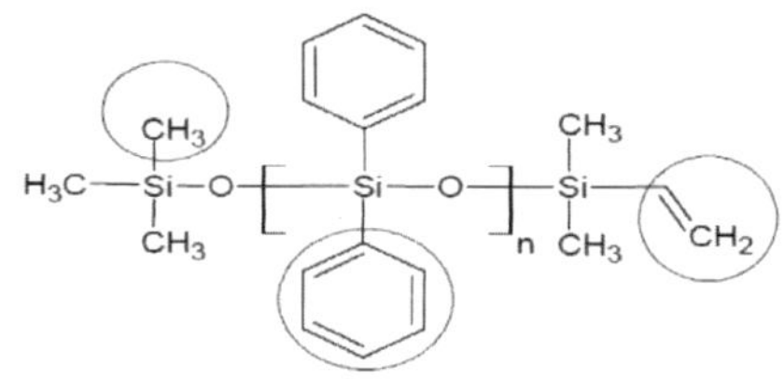
FVMQ	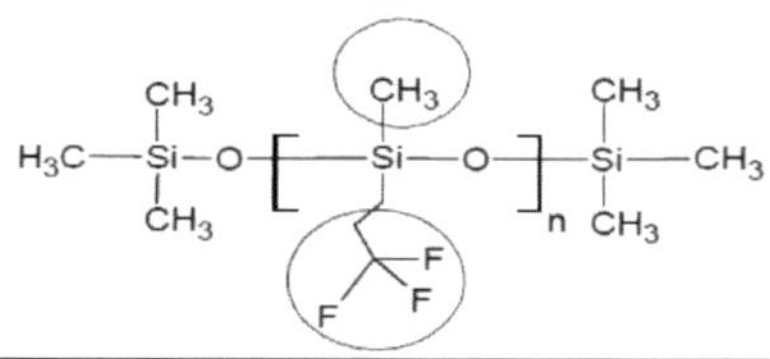

There are two types for silicone rubber which are room temperature vulcanizing (RTV) and high temperature vulcanizing (HTV) [15]. HTV was cured at high temperature that mainly used as a weather shed and RTV used as hydrophobic coating on ceramic surfaces and cured around room temperature [16].

Many research have been written about the upgrading the properties of silicone rubber. Yet, only a few reports were published on used of various cords and plies reinforcements in silicone rubber. Therefore, this paper aims at gathering all information and literature review to design and fabricate silicone rubber with various materials of cords and plies reinforcements. The others study such as the results on mechanical and physical properties, curing and adhesion behaviour of silicone composite also will be summarized.

2. Motivation for studying

In 1960s, the cords materials for rubber were introduced such as polyester[17], steel [18], nylon[19-21]. The fabric cord-inserted rubber compounds was applied in manufacturing of tyres, conveyer belts, and hoses [21,22] and also practiced in materials such as clay [23]. The cross-ply or bias ply tyres design is the first design introduced the cords at angles between +60 until -60 degrees from the direction of travel. After that, the radial tire is approached by laying all of the cord plies at 90 degrees to the direction of travel. This is because the radial design can avoids the plies reinforced in tires to rub against each other as the tire flexes, and reducing the tire's rolling friction that will allowed the vehicles with radial tires design had fuel save economy. In order to have an adequate rigid at the contact with the ground, the entire tyre is surrounded by additional belts that can be made from steel, polyester, and Kevlar [20,24]. Thus, an important aspect that can be taken to produce a better reinforcement polymer product is the specimen directions on recovery behaviours of cords- rubber composites. As reported by Choi & Kim [21] the cutting direction also significant to estimate the aging properties of calendared fabric cord. Then, a new composite for external reinforced polymer had been reported by Figeys et al. [25] material steel plates and carbon fibre reinforced polymers (CFRP) dealt with new material that can be used as external reinforcement which was steel cord reinforced polymer (SCRP) that has a low price and flexible lamination. Other than steel cords, the other types of cords such as nonwoven hemp that replaced glass fibres also being approach by Richardson & Zhongyi [26] in order to provides technological, economical and environmental benefits.

Table 2. Examples types of material reinforcement

	Reinforcement Material	References
Polypropylene	Carbon, Kevlar	[27]
Epoxy resin	Kevlar	[28]
Natural rubber	Steel, Brass plate steel	[20,29-33]
Liquid Phenolic resin	Non-woven hemp	[26,34]
Elastomer matrix	Rayon, nylon	[35]
Polyethylene	Aramid	[36]
Urethane, Silicone rubber	Cotton, fibreglass	[37]

2.1. Cords and plies-rubber composite processing & preparation

Tian et al. [28] explained on the preparation of three-ply Kevlar fiber in epoxy composites. The epoxy resin and curing agent were dried in an oven at 60°C for several hours to get rid of moisture and then soaked in the epoxy resin for some hours until the fibers totally penetrated. After that, Kevlar and epoxy resin were coiled to the mandrel and let to cure at 70°C for 5 hours. Chandra et al. [31] studied on the adhesion of rubber to steel cord by a new testing technique and was further explored by Shi et al. [30]. The base used was natural rubber that being mixed for 8.5 min at 55 rpm and 60°C using an internal mixer. The material of cord used was brass-plated steel cord and was treated by ethanol. The specimens adhere behavior in a carcass was observed. Then, Richardson & Zhang [26] looked out the potency of organic material such as hemp as reinforcement material and liquid phenolic resin as base in their study. The sample without hemp and with two layers nonwoven hemp fabric was fabricated using a resin transfer molding. Sample were kept in the mould at 65°C for four hours as preliminary curing and transferred to an oven at a temperature of 80°C for three hours as post curing. Samples were cut by using diamond saw. Another report published by Carbone et al. [38] introduced the PET cord-rubber composites in order to asses its micromechanical behavior. The PET cords were coated with a Resorcynhol-Formal-dheyde-Latex (RFL) to have a good adhesion with the rubber matrix. The matrix was prepared by mixing all the ingredients using Haake Rheomix 600/610 batch mixer for 12 min at 30 RPM. The mixed compound was processed in a two-roll mill at 80°C and vulcanized by mechanical press at 160°C for 10 min under 1.37 Mpa as to obtain rectangular pads with thickness of c. 1 mm. The two composite types were prepared as discontinuous cord embedded and also as continuous cords completely crossing the specimen over the full length. There also an investigation on high strain rate tensile behaviors of steel cord-rubber composite conducted on the servo-hydraulic machine prepared by Elamnipon et al. [29]. The 45° and 90° cord angles were chosen as respecting the cord axis and 0° cord angle tested with single steel cord. Moghadam et al. [39] prepared HTV Silicone Rubbers by melt mixing in an internal mixer at a rotational speed of 60 r/min and 30 min of mixing times. Then, the blend sample was vulcanized in hot press (Carver hydraulic unit model 3912) at 180°C and pressure of 44482 N/cm2 for 10 min and post-cured in an oven at 200°C for 2 hours under nitrogen.

2.2. Curing

Definition of curing refers to the hardening the polymer material by crosslinking of polymer chains. The crosslinking of polymer does not need a fully cure process for certain cases. [40]. As presented in guidelines [8], there were two types of curing for solid silicone rubber product namely peroxide curing grades and addition curing with platinum catalyzed. For Liquid silicone rubber, it was always addition curing with platinum catalyzed. Curing had been approved to be a better technique for interfaces with silicone rubber compared only using silicone glue such as sealant [41]. Currently, Xue et al. [3] reported that the commercial silicone rubbers were prepared by three curing systems namely platinum-catalyzed hydrosilylation between vinyl groups and Si-H bond, tin- or Titanium-catalyzed condensation reaction of alkoxysilane and thermal-initiated free radical cross linking reaction. However, they had approached a new mechanism of curing namely UV-curing silicone rubber via thiol-ene reaction as to invent a cure mechanism that can unlimited the application of silicone rubber in medicine and biology. Ganesh & Unnikrishnan [42] developed on silicone rubber blend with Ethylene Vinyl Acetate (EVA) and revealed the curing reactions follow first order kinetics. There was also the other research that observed curing characteristics using the same blending of silicone rubber with another polymer [43][44]. Dai et al. [45] had proposed polymeric curing agent (PCA) reinforced silicone rubber composites with low viscosity and volume shrinkage. PDMS/PCA curing system exhibits lower volume shrinkage and longer curing period [45]. The cure characteristics were determined using a moving die rheometer (MDR, Mosanto, USA) at temperature 160°C.

2.3. Physical and mechanical properties

In 1999, the hydrophobicity changes in silicone rubbers based on polydimethylsiloxane (PDMS) already reported by Hillborg & Gedde [46] as their properties of water repellency, high surface resistivity, vandalism resistance, low density and good process ability are suitable for outdoor insulation. Momen & Farzaneh [16] have study three factors namely electrical discharge, adsorption of pollution layers, and UV radiation that cause the loss of hydrophobicity. Tian et al. [28] discovered the three-ply Kevlar/epoxy composites posses low density and have high specific strength and specific modulus. Their low density can be introduced in harsh environment such as a static ring of ship propeller.

The experiment result from Chen et al. [47] shows the mechanical properties of the novel RTV silicone rubbers were better in result compared to reference material such as divinyl-hexa[(trimethoxylosilyl)ethyl]-POSS (DVPS). The influence of the temperature on the mechanical behavior of filled and unfilled silicone rubbers is studied by Rey et al. [5]. The other research made up to approach taguchi method as monitor the significant factor affecting the response and gives the optimum condition to most desirable performance[48]. Heiner et al. [9] discussed on mechanical analysis of two different silicone rubber namely peroxide-cured silicone elastomer and platinum-cured silicone elastomer. The hardness result of peroxide-cured showed 50% higher than platinum-cured. This is because of the side product produce by peroxide-cure in formed of radicals. The radicals tend to crosslinked with the elastomer and improved the hardness. Park [49] investigate the effect of fillers such as glass fiber, wollastonite and fluoro rubber towards silicone rubber composite. The tensile strength of silicone rubber with glass fiber filler increased by 26% compared to silicone without filler [50]. The silicone rubber with glass fiber hardness strength also shows slightly larger than the wollastonite filler. Recently, Brian Swanton from Dow Corning Corporation had introduced a new company product of RTV silicone rubber that was designed to assist in providing more reliability, added cost-efficiency, and reduced performance risks in subsea environments. The new RTV silicone rubber designed as insulation pipeline coating in saltwater production shows an excellent physical properties after curing. [51][52]. In aerospace, silicone rubbers also have contributed as addition-type silicone rubber (ASR) that prepared by vinyl-terminated polydimethylsiloxane (VPDMS) and polymethylhydrosiloxane (PMHS) [6].

2.4. Adhesion between cords or plies -polymer

Adhesive was used to bond two different structures together to form a composite and the forces or mechanism created by chemical bond results on good properties between adhesive and substrate was called as adhesion [53]. The destructive testing to identify the adhesion can be determined by shear, torsion, and peel test. The adhesion to rubber compound also can be determine by pulled out the cords from the rubber using tensile testing machine with the speed of 100mm/min [22]. Picard et al. [54] have compiled information regarding the chemical adhesion of silicone elastomers on primed metal surfaces. The review paper compiles the three mechanism on promoting a strong adhesion namely wetting, priming of the metal substrate and elastomer, and mechanical behavior of the final composite part[54]. Wang et al. [55] had determined the peel strength of different substrate surface to determine the adhesion with liquid silicone rubber. The different substrates used in their research were aluminum, glass, epoxy resin, polypropylene that being treated before coated with liquid silicone rubber and Pt catalyst. Jamshidi et al. [56] reported about the temperature can affect the adhesion between cords and rubber matrix. So, they had measured a wide range of temperatures between 25-170°C of cord-RFL-rubber system and compared with the other adhesion of different cords (N6, N66, and PET) to determine the best vulcanization temperature. Adhesion properties were measured by using three types of aging experiments was performed in Nah et al. [57]. Firstly, thermal aging that prepared in an air oven at 100°C, the second was humidity aging that setup in a humidity chamber (Weiss Technik, model 305B) at temperature 75 °C and 85% relative humidity. Thirdly, salt solution aging, the adhesion specimen were completely immersed in 20wt% NaCl solution at room temperature. After that, the specimens were tested by using a tensile Instron tester at a crosshead speed of 100 mm/min at room temperature in order to determine the maximum force exerted. The result analyzed by ranked a bare steel cord as 0%, and a fully covered cord as 100% [57] [58]. The effectiveness of silane as a primer towards silicone rubber is characterized accordingly to the type of inorganic substrate consideration in Fig.1 and their significance on mechanical and interfacial adhesion was discussed by George et al. [59].

Inorganic Substrates	
Silica Quartz Glass Aluminium (AIO(OH)) Alumino-silicates (e.g. clays) Silicon Copper	Excellent adhesion
Tin (SnO) Talc Inorganic Oxides (e.g. TiO2) Steel, Iron Asbestos Nickel Zinc	Good Adhesion
Lead Marble, Chalk (CaCO3) Gypsum Barytes	Slight Adhesion
Graphite Carbon Black	Poor Adhesion

Figure 1. Effectiveness of silane adhesion according to the type of inorganic substrate [54].

3. Summary

The processing preparation for cords and plies reinforcement depends on type of rubber base whether in solid or liquid form. The substrate treatment preparation also can be varied in order to have a suitable primer formulation. In order to determine the suitable process preparation, the results will be shown in characterization of mechanical and physical properties. Not all cords and plies materials give a good adhesion to its rubber base as they also can results in poor adhesion because of some functional group that can inhibit the curing. Primer such as silane or cords and plies treatment can improve the crosslinking between silicone rubber and reinforcement material. The high crosslinking will in turn contribute to better adhesion.

4. References

[1] Brendan R Reinforcing materials in Rubber Products pp 1–114
[2] Lomov S V., Ivanov D S, Verpoest I, Zako M, Kurashiki T, Nakai H and Hirosawa S 2007 *Compos. Sci. Technol.* **67** 1870–1891
[3] Xue L, Zhang Y, Zuo Y, Diao S, Zhang J and Feng S 2013 *Mater. Lett.* **106** 425–427
[4] Shin-Etsu Silicones 2010
[5] Rey T, Chagnon G, Le Cam J B and Favier D 2013 *Polym. Test.* **32** 492–501
[6] Hao D, Li D and Liao Y 2015 **42036** 1–7
[7] Anon Bonding, Sealing, Potting/Encapsulation and coating with RTV Silicone Rubber Compounds
[8] Anon Solid and Liquid Silicone Rubber Material and Processing guidelines
[9] Heiner J, Stenberg B and Persson M 2003 **22** 253–257
[10] Anon 2004 *Polydimethylsiloxane* **4** 1–4
[11] Bernstorf R A and Ryan D 2007 *Hubbel Power Syst. Inc* 1–8
[12] Gelest Inc. 2013 64
[13] Zhang H and Cloud A 2007 *Sampe*
[14] Abdslam S S 2013 (Universiti Tun Hussein Onn Malaysia)
[15] Barekat M, Razavi R S and Bastani S 2013 *Prog. Org. Coat.* **76** 277–285
[16] Momen G and Farzaneh M 2011 *Rev. Adv. Mater. Sci.* **27** 1–13
[17] Amari T, Themelis N J and Wernick I K 1999 *Resour. Policy* **25** pp 179–88
[18] Radtke A 2012 29–32
[19] Nkiwane L and Mukhopadhyay S K 1999 pp 1045–1053
[20] Savage A W and Diego S 1952
[21] Choi S-S and Kim O-B 2014 *J. Ind. Eng. Chem.* **20** pp 202–207
[22] Rodgers B Reinforcing Material in Rubber Products pp 1–12
[23] M.Joshi, D. Biawas, A. Sarvanan, Roli Purwar R M 2012 *Polym. Compos.* **125** 224–34
[24] Yang X and Olatunbosun O A 2012 *Mater. Des.* **35** 798–809
[25] Figeys W, Schueremans L, Van Gemert D and Brosens K 2008 *Constr. Build. Mater.* **22** 1929–1938
[26] Richardson M and Zhang Z 2001 Nonwoven hemp reinforced composites *Reinf. Plast.* **45** 40–44
[27] Jang B Z 1992 *Compos. Sci. Technol.* **44** 333–349
[28] Tian Y, Guo M J, Zhang J S, Li Y C, Zhang D X, Li D H, Xiao H Y and Jia J 2011 *Adv. Mater. Res.* **311-313** 331–4
[29] Eiamnipon N, Nimdum P, Renard J and Kolitawong C 2013 *Compos. Struct.* **99** 1–7
[30] Shi X, Ma M, Lian C and Zhu D 2013 *Polym. Test.* **32** 1145–1153
[31] Chandra A K, Mukhopadhyay R and Bhowmick A K 1996 *Polym. Test.* **15** 13–34
[32] Tabaddor F 1985 *Compos. Struct.* **3** 33–53
[33] Abou-Kandil A I, Awad A, Darwish N, Shehata A B and Saleh B K 2013 *Int. J. Adhes. Adhes.* **44** 26–35
[34] Dhakal H N, Zhang Z Y and Richardson M O W 2007 *Compos. Sci. Technol.* **67** 1674–1683
[35] Banerjee S and Das C 1987 *Compos. Sci. Technol.* **28** 129–135

[36] Kruijer M P, Warnet L L and Akkerman R 2005 *Compos. Part Appl. Sci. Manuf.* **36** 291–300
[37] Peel L D 1999 (Brigham Young University)
[38] Pastore Carbone M G, Parthenios J, Tsoukleri G, Cotugno S, Mensitieri G and Galiotis C 2013 *Compos. Sci. Technol.* **85** 104–110
[39] Moghadam M, Morshedian J, Ehsani M, Laine R, Kuang L and Umehara N 2014 *IEEE Trans. Dielectr. Electr. Insul.* **21** 244–252
[40] Woehrmann M and Toepper M 2012
[41] Andersson J and Gubanski S M 2007 **14** 137–145
[42] Ganesh B and Unnikrishnan G 2005
[43] Nakason C, Jarnthong M, Kaesaman A and Kiatkamjornwong S 2009
[44] Liu Y, Liu H, Zhang R, Zhou C and Feng S 2013
[45] Dai L, Zhang Z, Zhao Y and Xie Z 2008
[46] Hillborgl H and Gedde U W 1999 **6**
[47] Chen D, Liu Y and Huang C 2012 *Polym. Degrad. Stab.* **97** 308–315
[48] Anon 2004 *Design of Experiments (DOE) Using the Taguchi Approach*
[49] Park E 2007
[50] Anon 2006 D412-06 *ASTM Int. West Conshohoken PA USA* **552** 1–14
[51] Swanton B
[52] Corning D O W XTI-1003
[53] Sperling L H 2006 vol 78
[54] Picard L, Phalip P, Fleury E and Ganachaud F 2015 *Prog. Org. Coat.* **80** 120–141
[55] Wang F, Li Y and Wang D A N 2013 **36** 1013–1017
[56] Jamshidi M, Taromi F A and Mohammadi N 2005 *Iran. Polym. J.* **14** 229–234
[57] Nah C, Sohn B Y and Park S 2002 **16** 653–67
[58] Wire B, Wire H R and Precision W S 2010 *Annu. Book ASTM Stand.* **i** 1–7
[59] George J, Sreekala M S and Thomas S 2001 *Polym. Eng. Sci.* **41** 1471–1485

Acknowledgement
The authors gratefully acknowledge the financial support received from Education Ministry of Malaysia through Research Acculturation Grant Scheme (RAGS) (RAGS/1/2014/TK04/UITM//8), Research Management Centre (RMC), Faculty of Chemical Engineering and Universiti Teknologi MARA, Selangor, Malaysia.

Numerical prediction of kinetic model for enzymatic hydrolysis of cellulose using DAE-QMOM approach

N M Jamil[1] and Q Wang[1]

[1]Faculty of Industrial Sciences & Technology, *Universiti Malaysia Pahang*, UMP, Lebuhraya Tun Razak, 26300 Gambang, Kuantan, Pahang, Malaysia.
[2]Department of Mathematics, *University of South Carolina*, USC, Columbia, South Carolina 29208, USA.

E-mail: [1]norazaliza@ump.edu.my, [2]qwang@math.sc.edu

Abstract. Bioethanol production from lignocellulosic biomass consists of three fundamental processes; pre-treatment, enzymatic hydrolysis, and fermentation. In enzymatic hydrolysis phase, the enzymes break the cellulose chains into sugar in the form of cellobiose or glucose. A currently proposed kinetic model for enzymatic hydrolysis of cellulose that uses population balance equation (PBE) mechanism was studied. The complexity of the model due to integrodifferential equations makes it difficult to find the analytical solution. Therefore, we solved the full model of PBE numerically by using DAE-QMOM approach. The computation was carried out using MATLAB software. The numerical results were compared to the asymptotic solution developed in the author's previous paper and the results of Griggs et al. Besides confirming the findings were consistent with those references, some significant characteristics were also captured. The PBE model for enzymatic hydrolysis process can be solved using DAE-QMOM method. Also, an improved understanding of the physical insights of the model was achieved.

1. Introduction

In [1], Mohd Jamil introduced a mathematical analysis for two separate models for enzymatic hydrolysis of cellulose. It was observed that the number of cellulose chains in both models increased over time due to the effect of enzymes in breaking the cellulose chains into smaller chains; hence more cellulose particles were generated. The limitation of mathematical analysis in measuring one of the enzyme activities is that it would miss the effect of other types of enzymes. Studies by Srisodsuk et al. [2] have shown the importance of synergism between both enzymes, i.e. the simultaneous action of endoglucanase and exoglucanase cellulose during cellulosic hydrolysis. Therefore, an aggregate solution of kinetic model that incorporates the synergism between both enzymes is required. Due to the complexity of the kinetic model proposed by [3] and determining its analytical solution is almost impossible, we have to resort to numerical methods.

Therefore, in this paper, a full PBE model as in [3] is solved numerically by using DAE-QMOM approach. This choice is supported by making a brief comparison between DAE-QMOM approach and asymptotic solution of Model I developed in [1]. In order to evaluate the physical insights of the model, a comparison with the results of Griggs et al. [3] is carried out. Besides confirming the findings are consistent with Griggs, our aim is to further shape the dimensions for developing concepts and relevant implications to integrate hydrodynamic effect within the system.

Structurally, this paper consists of four thematic sections with relevant subsections. In the first section, the paper is introduced. Next, the issues of numerical methods for population balance modelling are addressed. Also, the verification of DAE-QMOM result with the solution of Model I in [1] is presented. In the third section, the solution of the full model using DAE-QMOM is solved and discussed. Finally, a conclusion of this paper is provided.

2. Numerical methods for PBE

The kinetic model for enzymatic hydrolysis of cellulose proposed by [3] is in the form of population balance equation (PBE). To date, various methods have been developed and introduced to solve the PBE problems [4] [5] [6] [7]. Generally, they can be classified into four main categories; moment method, class method (CM), weighted residuals method, and stochastic method. Recently, attention has been given to moment-based method to avoid the computational cost inherent in the integrodifferential equations. The moment-based method mainly includes method of moments (MOM), quadrature method of moments (QMOM) [5], and direct quadrature method of moments (DQMOM) [6].

Method of moments (MOM) was originally derived by [8]. The major issue of MOM is that it suffers from closure problem. An alternative way to overcome the closure problem is by using QMOM, which was first proposed by [5] to study size-dependent growth in aerosols. QMOM is based on the approximation of unclosed terms by using a quadrature formula. Rather than an exact PBE, QMOM gives a series of *N* weights and *N* abscissas that can be converted into *2N* moments. A major advantage of QMOM is, it has fewer variables involved in terms of number of moments.

The quadrature weights, ω_k and abscissas, ξ_k which are required in QMOM could be determined through a product-difference (PD) algorithm developed by Gordon [9]. The algorithm needs 2*N* first moments $\left(m_0, m_1, ..., m_{2N-1}\right)$ of the distribution as an input to produce *N* weights and abscissas. In the algorithm, a tridiagonal matrix is constructed recursively from the moments; and the weights and abscissas are obtained from the eigenvalues and eigenvectors of this matrix. This technique has been validated in many cases of PBE including simultaneous breakage and aggregation, growth, and nucleation [10]. However, PD-QMOM might lead to an ill-conditioned problem for certain cases [11]. An alternative solution for QMOM method without using PD algorithm in solving the PBE has been proposed by Gimbun et al. [12] which is known as DAE-QMOM. In this research, the DAE-QMOM is implemented to solve the kinetic model for enzymatic hydrolysis of cellulose.

2.1. Verification of DAE-QMOM for Model I

An asymptotic solution for Model I as presented in [1] was developed. Model I is a reduced model from the full model of enzymatic hydrolysis of cellulose that only considers EG which breaks cellulose chains randomly. In this section, the DAE-QMOM was applied to solve the PBE in Model I. The results of these simulations from DAE-QMOM were compared to the asymptotic solution for Model I. The DAE-QMOM steps can be summarized as follows:

(a) Simplify the PBE using a moment transformation which leads to a closure problem.
(b) Transform the PBE into a QMOM formulation to eliminate the closure problem.
(c) Solve the DAE system.

For the case *N*=3, the evolutions of the moments for $p(x,t)$ using DAE-QMOM method are shown in Figure 1. The numerical results were compared to the asymptotic solutions that are represented by circle markers on the line in the figure.

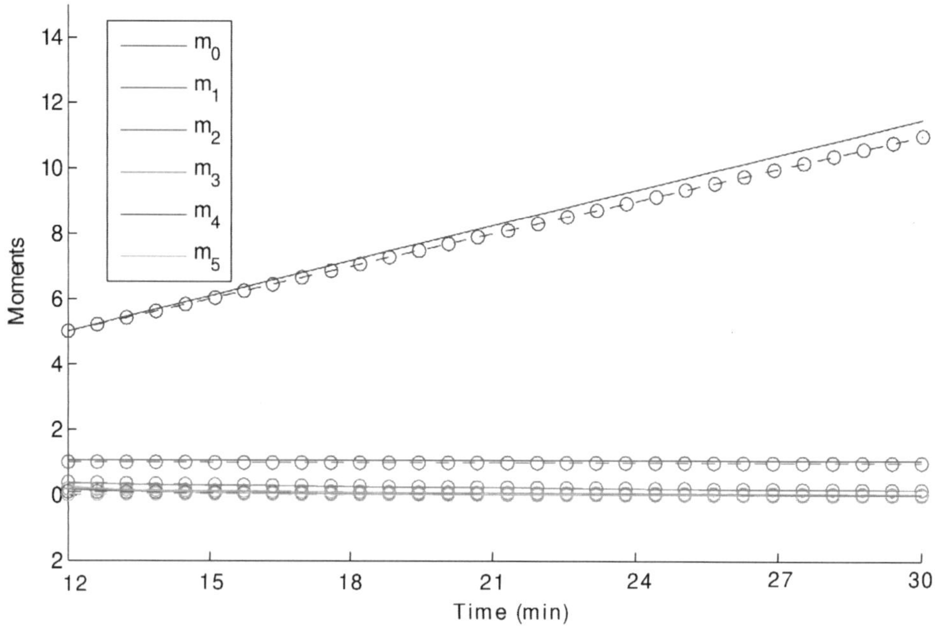

Figure 1. Comparison between DAE-QMOM method (continuous lines) with N=3 and asymptotic solution (circles) for Model I.

The results clearly showed that the zeroth moment of the particle size distribution increased, since during the breakage, the total number of particles increased. This result indicates the generation of shorter chains from longer cellulose chains during hydrolysis which is consistent with the findings reported in [3]. The behaviour of the first moment indicates that the total number of glucan units was conserved until the end of the hydrolysis process.

Excellent agreements were found between DAE-QMOM approach and asymptotic solution for Model I on moment evolution. To conclude, the DAE-QMOM has been validated against the asymptotic solution of Model I. Moreover, these results indicate that the DAE-QMOM is capable of predicting the moment evolutions tested in this study accurately. Next, DAE-QMOM was applied to solve the full kinetic model for cellulose hydrolysis.

2.2. Solving the full model using DAE-QMOM

Next, this study considered both CBH_1 and EG_1 acting simultaneously to depolymerize cellulose, which is also known as the full model. Since the DAE-QMOM was able to predict the moment evolution qualitatively for Model I, the usage of DAE-QMOM was extended to study the behaviour of the full system. The only difference when dealing with the full model compared to Model I was that the behaviour of two different types of population needed to be studied simultaneously, which were free cellulose chains, p, and CBH-threaded cellulose chains, p_B.

The moment operator was applied to p and p_B separately. Then, the dimensionless variables which involved characteristic time, initial total number of enzyme accessible cellulose particles, and thickness of single cellulose chains were introduced. Several parameters were grouped together to form dimensionless parameters. The parameters were chosen to be: $K_d = 1\,mol/m^3$, $p^{(0)}(0) = 100\,ml/m^3$, $R_0 = 1\,nm$, $k_h^{CBH} = 1\,hr^{-1}$, $k_h^{EG} = 1\,m^3mol^{-1}hr^{-1}$, $k_f^{CBH} = 2\,m^3mol^{-1}hr^{-1}$,

$\rho = 0.8\, mol/m^6$, $n = 1000$, and $L = 1000\, m$. For the moment equations, by applying QMOM, it was claimed that:

$$p^{(k)}(t) = \sum_{i=1}^{N} \omega_i(t)\xi_i(t)^k \text{ and } p_B{}^{(k)}(t) = \sum_{i=1}^{N} e_i(t)L_i(t)^k .$$

For $N = 2$, the DAE-QMOM system $M\dot{y} = F$ was developed with:

$$y = \left[p^{(0)}, p^{(1)}, p^{(2)}, p^{(3)}, p_B{}^{(0)}, p_B{}^{(1)}, p_B{}^{(2)}, p_B{}^{(3)}, \omega_1, \omega_2, \xi_1, \xi_2, e_1, e_2, L_1, L_2, R \right].$$

3. Results and Discussion

To study the effect of different enzyme loadings on the moment evolution, we tested various ratios of CBH_1 to EG_1 loading. Here, the total enzyme loading was fixed at $65\, mol/m^3$ and the ratio of each enzyme was varied. Figure 2 shows the zeroth and first moment evolution for p and p_B when the enzyme loading for CBH_1 and EG_1 was similar (CBH_1:EG_1 ratio is 1:1). From the graph, we can see that, $p^{(0)}$ (the total number of cellulose chains) increased due to the breakage event of cellulose chains by EG_1 and CBH_1 in the hydrolysis process. This result indicates the generation of shorter chains from longer cellulose chains during hydrolysis, which is consistent with the findings reported in [3]. Accordingly, the total number of monomeric glucans, $p^{(1)}$ also increased. On the other hand, $p_B{}^{(0)}$ (the total number of CBH_1-threaded cellulose chains) decreased due to the chain fragments generated by CBH_1, which were eventually solubilized by EG_1. This phenomenon implies that the free cellulose chains were generated from the CBH_1-threaded cellulose chains. Subsequently, the total number of CBH_1-threaded monomeric glucans, $p_B{}^{(1)}$ also decreased.

Furthermore, all curves in Figure 2 eventually stopped increasing or decreasing at a certain time and became a straight line. This behaviour indicates that at a certain time, all cellulose chains reached its minimum length and all cellulose cylindrical particles reached its minimum radius owing to the depletion of insoluble substrate during the hydrolysis process. In this situation, both enzymes (CBH_1 and EG_1) stopped their activity of cutting the cellulose chains.

Now, if the rate coefficients of CBH_1 are considered to be very small in the full model, the same result as in Model I will be obtained. In other words, if the effect of CBH_1 from the full model is simply neglected, the solution for Model I can be recovered. Therefore, the solution of the full model using DAE-QMOM is reliable and it can predict the enzymatic hydrolysis process accurately.

For CBH_1 case in Figure 2, the moment evolution was slower than the combined (EG_1+CBH_1) enzyme hydrolysis. However, it can be seen that for this distribution, CBH_1 was capable of solubilizing the substrate efficiently without the aid of EG_1 which agrees with the result of [3]. With a constant total enzyme loading, increasing the fraction of EG_1 also increase the moment. The addition of EG_1 resulted in cooperative hydrolysis of the substrate.

In addition, this study also tested higher than two-order truncation in Taylor series in solving the full model by DAE-QMOM, and it agrees with the second truncation error qualitatively. Therefore, it is sufficient to use the second truncation Taylor series in this study.

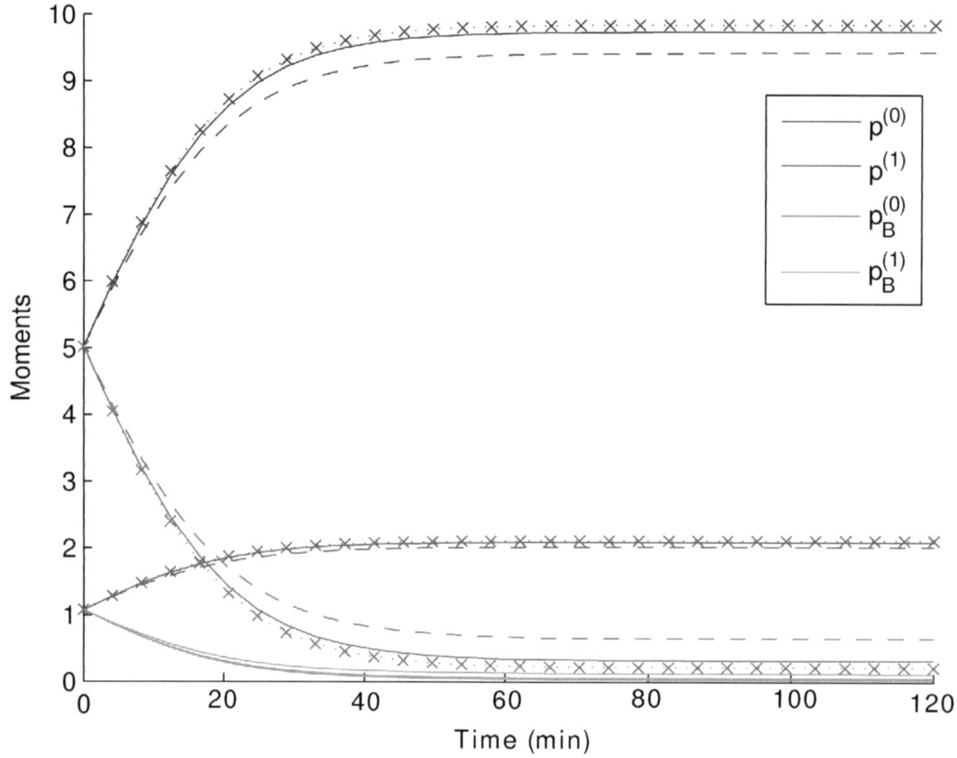

Figure 2. The zeroth and the first moment evolutions during the course of hydrolysis for combined CBH_I and EG_I actions with CBH_I:EG_I ratio of 1:0 (continuous line) compared to ratio 1:1 (dashed lines) and 1:2 (cross line).

4. Conclusion

The DAE-QMOM method was applied to solve Model I which consists of differential equations for the moments and system of nonlinear equations resulted from quadrature approximation as a differential-algebraic equation system. The moments of PBE transformed the integrodifferential equation into a set of simultaneous, ordinary differential equations. The solution from the DAE-QMOM method for Model I was compared to the approximate analytical solution. The results indicate that the method is capable of predicting the moment evolution accurately. Furthermore, the solution for simultaneous action of CBH_I and EG_I in the full model was obtained using DAE-QMOM approach. By examining the evolution of moments during the enzymatic hydrolysis, an improved understanding of cellulose depolymerization can be achieved.

5. References

[1] Mohd Jamil N 2015 *In: Energy and Sustainability V*. WIT Press, pp. 499-510
[2] Srisodsuk, Malee, et al. 1998 *European journal of biochemistry* **251.3** 885-892
[3] Griggs, Andrew J, Jonathan J Stickel, and James J Lischeske 2012 *Biotechnology and bioengineering* **109.3** 665-675
[4] Ramkrishna, Doraiswami 2000 *Academic press*
[5] McGraw, Robert 1997 *Aerosol Science and Technology* **27.2** 255-265.826
[6] Marchisio, Daniele L and Rodney O Fox 2005 *Journal of Aerosol Science* **36.1** 43-73

[7] Marchisio, Daniele L, R Dennis Vigil, and Rodney O Fox 2003 *Chemical Engineering Science* **58.15** 3337-3351
[8] Hulburt, Ho M, and Stanley Katz 1964 *Chemical Engineering Science* **19.8** 555-574
[9] Gordon, Roy G 1968 *Journal of Mathematical Physics* **9.5** 655-663
[10] Marchisio, Daniele L, et al. 2003 *AIChE Journal* **49.5** 1266-1276
[11] John, Volker, and Ferdinand Thein 2012 *Chemical Engineering Science* **75** 327-333
[12] Gimbun, Jolius, Zoltan K Nagy, and Chris D Rielly 2009 *Industrial & Engineering Chemistry Research* **48.16** 7798-7812

The study of solid circulation rate in a compartmented fluidized bed gasifier (CFBG)

S K Wee[1], Y W Pok[2], M C Law, V C C Lee
[1] Faculty of Engineering & Science, Curtin University, CDT 250 98009 Miri Sarawak, Malaysia
[2] Faculty of Engineering and Science, Curtin University, Miri Sarawak, Malaysia

E-mail: [11]wee.siaw.khur@curtin.edu.my, [2]m.c.law@curtin.edu.my

Abstract. Biomass waste has been abundantly available in Malaysia since the booming of palm oil industry. In order to tackle this issue, gasification is seen a promising technology to convert waste into energy. In view of the heat requirement for endothermic gasification reaction as well as the complex design and operation of multiple fluidized beds, compartmented fluidized bed gasifier (CFBG) with the combustor and the gasifier as separate compartments is proposed. As such, solid circulation rate (SCR) is one of the essential parameters for steady gasification and combustion to be realized in their respective compartments. Experimental and numerical studies (CFD) on the effect of static bed height, main bed aeration, riser aeration and v-valve aeration on SCR have been conducted in a cold-flow CFBG model with only river sand as the fluidizing medium. At lower operating range, the numerical simulations under-predict the SCR as compared to that of the experimental results. Also, it predicts slightly different trends over the range. On the other hand, at higher operating range, the numerical simulations are able to capture those trends as observed in the experimental results at the lower operating range. Overall, the numerical results compare reasonably well with that of the experimental works.

1. Introduction

In 2012, Malaysia, the world's second largest producer [1], produced 18.79 million tonnes of crude palm oil on roughly 5,000,000 hectares (19,000 m^2) of land [2, 3]. An annual production of over 4 million tonnes of oil palm shell solid wastes [4, 5] is reported. Gasification is considered to be promising technologies for converting biomass waste into energy. The gasification of solid wastes with steam can efficiently convert 97% of the feedstock into syngas, however it is highly endothermic. To meet its energy demand, a possible solution is to have a separate combustor which burns biomass and interconnects it with the gasifier by transfer lines for heat exchange purpose. However, this type of configuration often complicates the design and operation of the system.

As such, a single compartmented vessel is introduced to improve the conventional design. Compartmented fluidized bed gasifier (CFBG) is a unique compartmented reactor which consists of two reactors, namely the combustor and the gasifier, partitioned at a ratio of 65:35, based on the heating requirement during thermochemical operation. Solids are circulated internally between

[1-2] To whom any correspondence should be addressed.

compartments via a set of non-mechanical valves and risers. Solid circulation rate (SCR) is hence one of the essential parameters for steady gasification and combustion to be realized in their respective compartments.

Several lab-scale cold-flow experimental works had been carried out to study the SCR. Sathiyamoorthy and Rudolph [6] proposed an empirical correlation for SCR based on the different gas flow rates at main bed, V-valve and riser, with a constant depending on the V-valve inlet diameter. Later, He [7] agreed that any increase in the aerations leads to an increase in the gas leakage from the main bed to the V-valve orifice, resulting in the increase in the SCR. Yan and Rudolph [8] developed a new CFBG reactor design which demonstrated that the SCR depends on the gas flow rates to the V-valve and riser. Bhattacharya et al. [9] showed that the SCR is a function of fluidization conditions in upstream bed, V-valve and riser, solid size and upstream bed height. In the latest work done by Chok [10] and Wee [11] in a pilot-scale cold-flow CFBG, their findings are consistent with results as mentioned by previous authors. To date, the first and only numerical modeling of SCR in CFBG was done by Wee [11] using a commercial CFD package. The numerical results also agreed with the experiment findings.

The objective of this paper is to study the extended operating range for main beds aeration (Q_b), v-valves aeration (Q_v), risers aeration (Q_r) and static bed height (H) for higher SCR range by employing a CFD package at cold-flow model with only river sand as fluidizing medium. The results are then compared with Wee (2012) of smaller operating range.

2. Computational Setup and Methods

The schematic of the model is shown in Figure 1. The CFBG of diameter of 0.66m and a height of 1.8m is partitioned into two compartments, the combustor and the gasifier. The combustor or gasifier compartment is neither full cylindrical nor semi-cylindrical but rather a segment of a cylindrical. More details on working principles can be obtained in Wee [11].

The simulation of CFBG is performed by using a commercial CFD package. A multi-fluid Eulerian-Eulerian model, which considers the conservation of mass and momentum for the gas and fluid phases, is applied. The kinetic theory of granular flow is used to describe the solids phase stress. Further details on the modeling strategy can be found in Wee [11]. Table 1 summarizes the flow parameters to be used in the simulation of 3D CFBG. Mesh independence study is carried out to ensure that the numerical results are not significantly affected by number of mesh elements. As a result, fine mesh of 175498 elements is recommended for the SCR simulation:

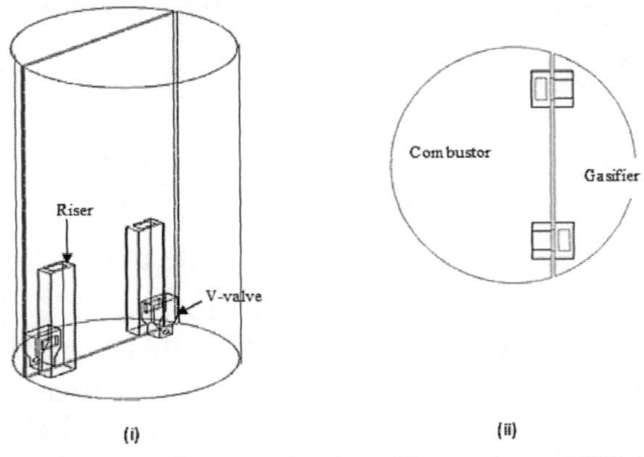

Figure 1. (i) Isometric view (ii) top view of CFBG.

Table 1. Simulation model parameters/ conditions.

Properties of Sand Particles	Description
Mean particle diameter D_p	272 μm
Density, ρ_p	2620 kg/m^3
Minimum fluidization velocity, U_{mf}	0.06 m/s (determined experimentally)
Boundary Condition	**Description**
Combustor/ gasifier inlet condition	$1.3 - 1.7\,U/U_{mf}$ (low) $1.5 - 2.25\,U/U_{mf}$ (high)
Combustor/ gasifier riser inlet condition	$8.0 - 12.0\,U/U_{mf}$ (low) $13.0 - 19.0\,U/U_{mf}$ (high)
Combustor/ gasifier v-valve inlet condition	$5.0 - 9.0\,U/U_{mf}$ (low) $10.0 - 16.0\,U/U_{mf}$ (high)
Initial bed height	0.2-0.45m

3. Results and Discussions

3.1. Effect of Static Bed Height on SCR

Figure 2(a) & (b) shows the effect of static bed height on the SCR. The SCR increases with the increase in bed height. The SCR is dependent on the hydraulic head of the bed.

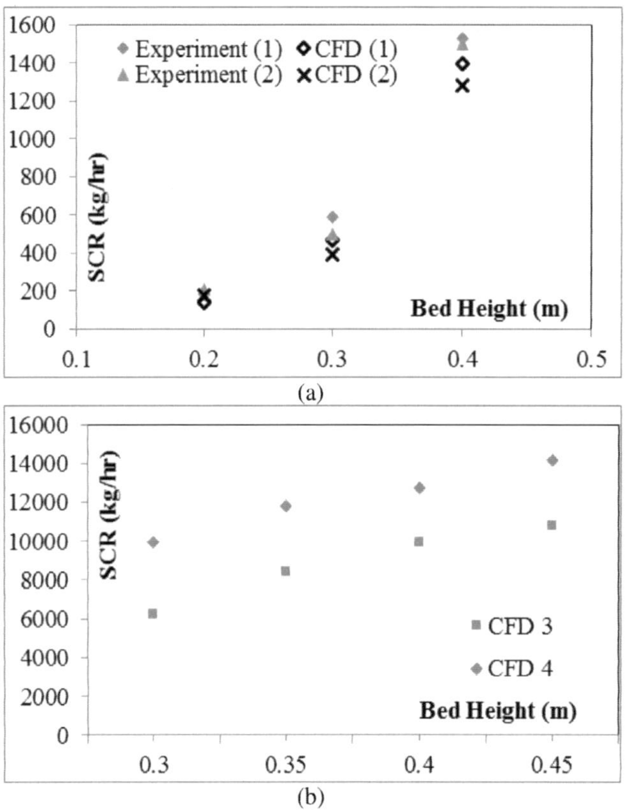

Figure 2(a) & (b). Effect of the static bed height on the SCR.
(1) $Q_b = 1.3U_{mf}$, $Q_r = 12U_{mf}$, $Q_v = 5U_{mf}$; (2) $Q_b = 1.3U_{mf}$, $Q_r = 8U_{mf}$, $Q_v = 9U_{mf}$
(3) $Q_b = 1.5U_{mf}$, $Q_r = 13U_{mf}$, $Q_v = 10U_{mf}$; (4) $Q_b = 2.25U_{mf}$, $Q_r = 19U_{mf}$, $Q_v = 16U_{mf}$

The higher the static bed height, the larger the hydraulic force exerted, thus induces higher SCR. These trends are similar to those observed by other researchers in this type of reactor [7-9]. Numerical simulation under-predicts the SCR values, however the tendency of the profile is still following the experimental ones as shown in Figure 2(a). This could be due to higher U_{mf} as predicted by the numerical standard drag law for the specified sand size. Figure 2(b) is produced by using the same assumption with operating condition dissimilar from experimental setting. Figure 2(b) shows a consistent increment in SCR for higher operating range. The operating static bed height of CFBG is limited by the height of riser. At higher bed height, the expanded bed hinders the outlet flow from the riser hence disrupts the solid circulation between compartments.

3.2. Effect of Main Bed Aeration on SCR

The effect of the main bed aeration on the SCR is illustrated in Figure 3(a). The different main bed aeration does not have any noticeable impacts on the experimental SCR. When the bed is fluidized, the bed pressure drop remains unchanged with further in-crease in the main bed aeration. As a result, one can expect the experimental SCR to remain constant when the compartmented bed is fluidized within this range of velocity. The numerical simulation predicts the similar behavior like those observed in experimental results. The SCR remains nearly constant with the variation in the main bed aeration from 1.3Umf to 1.7Umf. The trend is well captured by the numerical solution. However, the SCR values are under-predicted by around 20%. In order to obtain improved results, numerical simulation needs to be fine-tuned so that wide spectrum of particles size can be set to cater for the actual experimental vicinity.

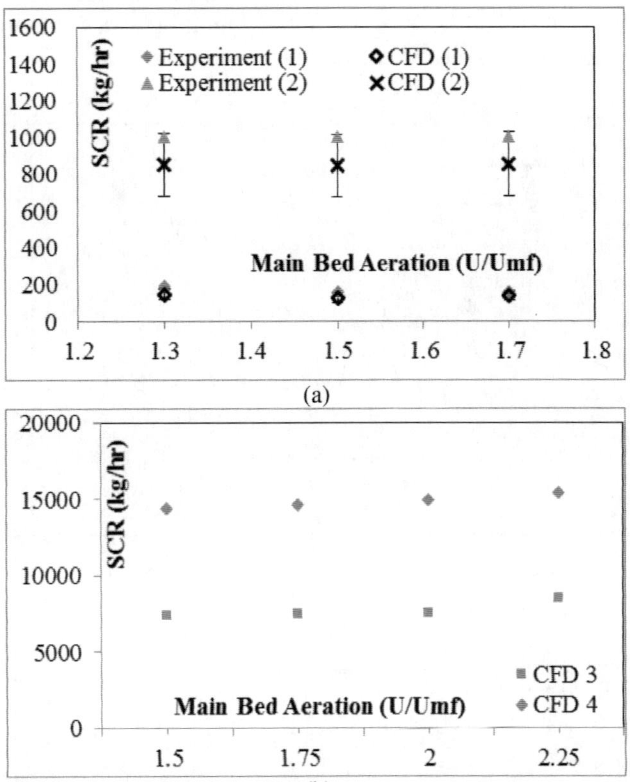

(b)

Figure 3(a) & (b). Effect of the main bed aeration on the SCR.
(1)$H_b = 0.2$m, $Q_r = 12U_{mf}$, $Q_v = 5U_{mf}$; (2) $H_b = 0.4$m, $Q_r = 8U_{mf}$, $Q_v = 5U_{mf}$
(3)$H_b = 0.3$m, $Q_r = 13U_{mf}$, $Q_v = 10U_{mf}$; (4) $H_b = 0.3$m, $Q_r = 19U_{mf}$, $Q_v = 16U_{mf}$

Figure 3(b) is plotted simulation result that have not been confirmed experimentally. Figure 3(b) shows the effect of main bed aeration on the SCR at the higher operating range. SCR increases gradually with the increase in main bed aeration. This observation is opposite to the trends in Figure

3(a). However, these results are consistent with the experimental findings by Bhattarcharya et al. [9] and He [7]. According to Bhattarcharya et al. [9], there are two opposing forces affecting SCR when the bed aeration increases. On one hand, increasing bed aeration leads to the reduction of the driving force due to an increase in bed porosity. This results in a reduction of SCR. On the other hand, gas cross-flow from v-valve aperture to riser increases with the increase of bed aeration. The latter reduces the pressure drop across the v-valve and riser thus increasing SCR. This is also supported by He [7].

3.3. Effect of Riser Aeration on SCR

Figure 4(a) & (b) demonstrates the effect of the riser aeration on the SCR. It is observed that the SCR initially increases then decreases with the increasing riser aeration. It is expected that the increase in riser aeration would increase the SCR due to the higher entrainment rate. However, further increase of the riser aeration may increase the frictional and acceleration pressure drop in the riser. This may eventually result in a reduction in the SCR. This trend is consistent with the results reported by Bhattacharya et al. [9]. He [7] explained this phenomenon as the presence of frictional and acceleration pressure drops. The pressure drop components increases with riser aeration, which results in reduction of solid circulation rate. This is however predicted differently by the numerical simulation at lower operating range. The SCR is linearly increased with the increment in riser aeration for the range of studies. The behaviour at higher riser aeration as discussed before is not apprehended by the numerical solution. This could be attributed to the higher Umf from numerical simulation than actual Umf as discussed earlier. On the other hand, at higher operating range, the numerical simulation is able to capture the reducing trend as observed from the experimental results.

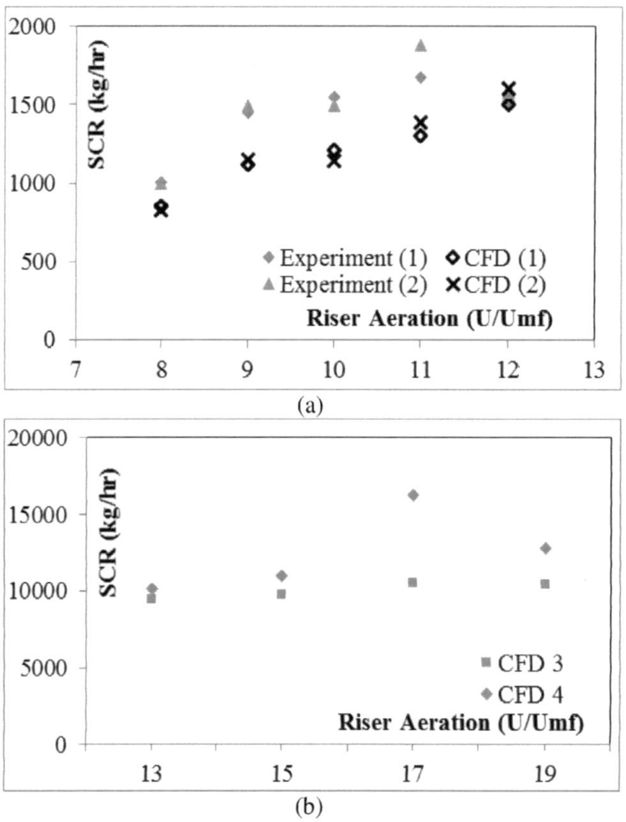

Figure 4(a) & (b). Effect of the riser aeration on the SCR.
(1)$H_b = 0.4$m, $Q_b = 1.3U_{mf}$, $Q_v = 5U_{mf}$; (2) $H_b = 0.4$m, $Q_b = 1.7U_{mf}$, $Q_v = 9U_{mf}$
(3)$H_b = 0.3$m, $Q_b = 1.5U_{mf}$, $Q_v = 10U_{mf}$; (4) $H_b = 0.3$m, $Q_b = 2.25U_{mf}$, $Q_v = 16U_{mf}$

3.4. Effect of V-valve Aeration on SCR

Figure 5(a) & (b) depicts the effect of the v-valve aeration on the SCR. For experiment (1), the SCR in-creases then decreases with increase in the v-valve aeration. Conversely, for experiment (2), the v-valve aeration has no obvious effect on the SCR. Ideally, aeration through the v-valve creates a low pressure region that induces a pumping effect. Greater aeration through the v-valve will therefore lead to a greater pumping effect, where more solids are being drawn which results in higher SCR. Nevertheless, it also in-creases the resistance across the v-valve-to-riser orifice. Thus, the two opposing effects explain the different SCR observed at experimental data.

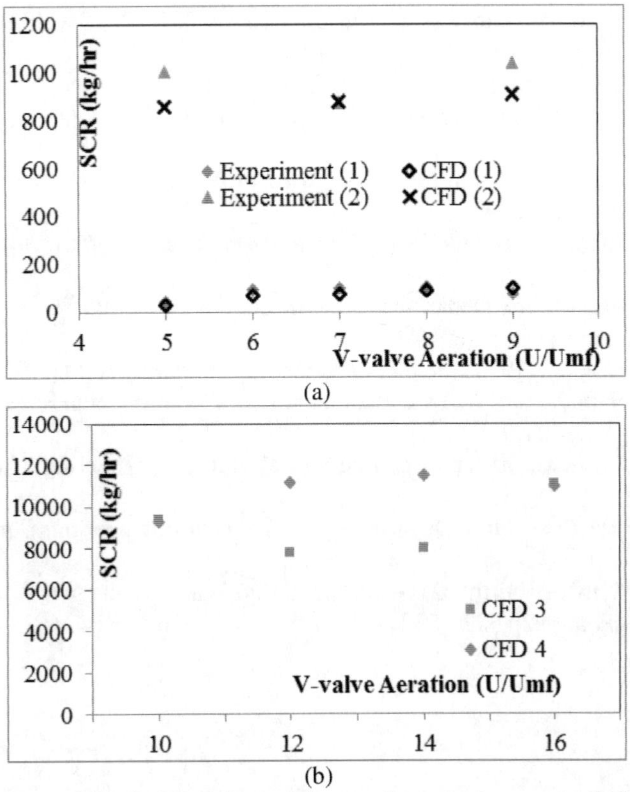

Figure 5(a) & (b). Effect of the v-valve aeration on the SCR.
(1)$H_b = 0.2$m, $Q_b = 1.3U_{mf}$, $Q_r = 8U_{mf}$; (2) $H_b = 0.4$m, $Q_b = 1.7U_{mf}$, $Q_r = 8U_{mf}$
(3)$H_b = 0.3$m, $Q_b = 1.5U_{mf}$, $Q_r = 10U_{mf}$; (4) $H_b = 0.3$m, $Q_b = 2.25U_{mf}$, $Q_r = 16U_{mf}$

The numerical simulation predicts dissimilar tendency of SCR at the lower operating range. The SCR increases linearly with the increase in the v-valve aeration. However, at the higher operating range, the numerical results are somehow agreed with the experimental trends at lower operating range. The pumping effect is well predicted by the numerical simulation but the resistance across the v-valve-to-riser as discussed before is not captured.

4. Conclusion

The solid circulation rate in CFBG with only river sand in cold-flow condition has been studied experimentally and numerically. The SCR is predominantly affected by static bed height, main bed aeration, riser aeration and v-valve aeration. At a lower operating range of the four parameters mentioned, the numerical simulations under-predict the SCR as compared to that of the experimental results. Also, it predicts slightly different trends over the range. On the other hand, at a higher operating range, the numerical simulations are able to capture those trends as observed in the experimental results at the lower operating range. Further experimental work is needed to validate the numerical results at the extended range.

5. References

[1] Ranjeetha P 2013 Kuala Lumpur:*Bloomberg*
[2] MPOP Expects CPO Production to Increase to 19 Million Tonnes This Year 2013 *The Star Online* January 15 http://www.thestar.com.my/story.aspx/?file=%2F2013%2F1%2F15%2Fbusiness%2F12576771&sec=business
[3] USDA Foreign Agricultural Service 2012 *USDA Government*
[4] Teo D C L, Mannan M A and Kurian J V 2006 *Journal of Advanced Concrete Technology* **4 (3)**1-10
[5] Teo D C L, Mannan M A and Kurian J V 2006 *Turkish Journal of Engineering and Environmental Services* **30** 1-7
[6] Sathiyamoorthy D and Rudolph V 1990 *Proc. in Circulating Fluidized Bed Technology III* 505 - 510
[7] He Y 1993 *PhD dissertation* Department of Chemical Engineering, The University of Queensland, Australia
[8] Yan H M and Rudolph V 1996 *Chemical Engineering Community* **147** 133 – 144
[9] Bhattacharya B, Sathiyamoorthy D, Govardhana Rao V and Mahajan S P 1999 *Powder Technology* **101** 191 - 204
[10] Chok V S 2011 *PhD dissertation* Department of Chemical Engineering, Curtin University, Australia
[11] Wee S K 2012 *PhD dissertation* Department of Chemical Engineering, Curtin University, Australia
[12] Bhattacharya B, Sathiyamoorthy D, Govardhana Rao V and Mahajan S P 1997 *Chemical Engineering Technology* **20** 522 - 532

Characteristics of post-impregnated SBA-15 with 12-Tungstophosphoric acid and its correlation with catalytic activity in selective esterification of glycerol to monolaurate

P Y Hoo[1], A Z Abdullah[2]
School of Chemical Engineering, Universiti Sains Malaysia, Engineering Campus, Seri Ampangan, Nibong Tebal, Pulau Pinang, Malaysia.

E-mail: [1]arthur.py.hoo@gmail.com, [2]chzuhairi@usm.my

Abstract. Selective esterification of glycerol and lauric acid to monolaurin was conducted using 12-tungstophosphoric acid (HPW) incorporated SBA-15 as catalyst. They were synthesized with HPW loadings of 10–40 wt. % via post impregnation and characterized in terms of surficial and structural characteristic, acidity and morphology. Relatively high lauric acid conversion (up to 95 %) and monolaurin yield (53 %) were observed while the activity was successfully correlated to the material behaviours, i.e. highly acidic active acid sites within highly uniformed mesopores. The effects of different reaction parameters including reactant ratio (1:1-5:1), catalyst loading (1-5 wt. %) and length of fatty acid were also elucidated. Reduced fatty acid conversion was observed when longer fatty acids were used, thus further strengthen the idea of size selectivity effect provided by the synthesized catalysts.

1. Introduction

The global blooming biodiesel production in recent years has resulted in the corresponding increase in glycerol production [1]. In fact, glycerol is formed at a rate of 10wt% of the total biodiesel synthesized [2]. The situation is made worse with further glycerol contribution from oleochemical and soap making industries [3], ultimately resulting in global glycerol price slump due to oversupply [4]. Thus, conversions of glycerol into other value-added chemicals such as monoglycerides are highly encouraged. Certainly, that was never the sole solution to the matter, however, we see the potential of converting waste (glycerol) to gold (monoglyceride as food addictive). As such, the development of heterogeneous catalysis system for such reaction is worth investigation with its own research merit. Traditionally, the monoglyceride was commercially produced via esterification of glycerol with fatty acids at low temperature (90–120 ℃) using strong homogeneous acids [5]. Though considerably high conversion was achieved, no control whatsoever towards the monoglyceride selectivity was identified. Consequently, extensive purification processes and thus additional costs are needed to enrich the final product and the separation and recycle of homogeneous catalysts.

With such motivation, the development of high-performance heterogeneous catalyst focus on selective monolaurate synthesis was attempted in this study. The exploitation of Keggin-type 12-tungstophosphoric acid (HPW) superacid supported on mesoporous SBA-15 could potentially improve glycerol conversion rate, at the same time provide high selectivity solely towards monolaurate. Our

[1]To whom any correspondence should be addressed

previous study suggested directly synthesized HPW supported SBA-15 could potentially be beneficial to such reaction [6]. However, trapped (and hence inactive) HPW acid sites within the hydrolysed SBA-15 walls, the agglomeration of HPW crystals on the external surfaces and pore deformation at high HPW loading have compromised the catalyst performance. Thus, in this study, we investigated HPW supported SBA-15 via wetness incipient method to minimize the structural modification of parent SBA-15. As such, we would be able to investigate if any improvement on the catalysts could be achieved from both characterization and catalytic performance point of view.

2. Experimental methods

2.1. Synthesis of SBA-15 support and HPW post impregnated SBA-15

The typical SBA-15 support was synthesized as reported in the literature [7]. HPW was incorporated into SBA-15 at 10 to 40wt% loading via incipient wetness impregnation method. The appropriate amount of post synthesized SBA-15 was wetted with deionized water (2.5 ml/g SBA-15). On the other hand, an appropriate amount of HPW was dissolved in deionized water (1 ml/g SBA-15). Under rapid stirring, HPW solution was added dropwise into the wetted SBA-15. The mixture was then dried at 60℃ for 12 hours and followed by another 12 hours at 100℃. The catalysts were then named X wt. %-HPW/IM, with X = 10, 20, 30, 40.

2.2. Characterization of synthesized support and catalysts

Morphology characterizations were done via liquid Nitrogen (N_2) desorption using Quanta-Chrome Autosorb 1C surface analyser and Philips CM12 transmission electron microscope (TEM). The structural properties including available functional groups were investigated with the help of Shimadzu IRPrestige-21 Fourier Transformed Infrared (FTIR) (KBr pellet method). All catalysts were pre-treated in Helium (He) gas and exposed to ammonia (NH_3) mixture (15% NH_3, 85% He) before subject to desorption from 100℃ to 750℃ with ramping of 10℃/min to obtain the acidity profiles via ammonia temperature programmed desorption (NH_3-TPD).

2.3. Catalytic performance assessment

All synthesized catalysts were subjected to the measurement of their respective catalytic performance. A batch reactor consists of a three-necked flask was connected to a temperature probe with temperature controller, magnetic stirrer and vacuum pump. The reactor was used for selective esterification of glycerol with lauric acid with the presence of the catalyst at designated reaction temperature, reactant ratio and catalysts loadings. Samples were collected at appropriate period of times and were analysed using Agilent Technologies 7890A GC equipped with a CP-Sil 5CB (15 m x 0.32 mm x 0.1 mm) column. Lauric acid conversion and the selectivity for monolaurate, dilaurate and trilaurate were calculated accordingly [8].

3. Results and discussions

3.1. Characterizations of HPW post impregnated SBA-15

In order to understand the surficial properties of the synthesized catalyst, N_2 adsorption-desorption test was performed on all the synthesized materials. Table 1 showed the result from the surface analyses. It was evidenced that even after impregnation of HPW (up to 40 wt%), mesoporosity could be observed (55 – 59 Å) with high surface areas (323 – 600 m^2/g). The decrease in surface areas with increasing HPW loading provided indirect evidence of successful impregnation of HPW onto the SBA-15 support due to deposition of HPW anions on the inner pore walls [9]. From Figure 1, Type IV N_2 absorption-desorption isotherms were shown by all catalysts, proving the retained mesoporosity of these catalysts [10]. While on Figure 2, it was obvious that uniformed pores were formed in the virgin SBA-15, though 2 major peaks could be found (around 40 Å and 60 Å). Interestingly, with the impregnation of HPW, both peaks intensities decreased, suggesting successful impregnation of HPW on the inner pores surfaces. Such finding was in agreement with the observations found in Table 1.

Table 1. Summary of surface characteristics of synthesized catalysts.

Catalysts	Total surface area (m^2/g)[a]	Micropores area (m^2/g)[b]	External surface area (m^2/g)[b]	Micropores volume (cm^3/g)[b]	Total pore volume (cm^3/g)[c]	Average pore size (Å)[c]
SBA-15	640	169	471	0.078	0.65	61
10wt%-HPW/IM	600	137	464	0.070	0.62	59
20wt%-HPW/IM	552	119	433	0.061	0.61	58
30wt%-HPW/IM	368	102	267	0.048	0.43	55
40wt%-HPW/IM	323	89	235	0.045	0.40	58

[a]BET desorption method; [b]t-plot method; [c]BJH desorption method

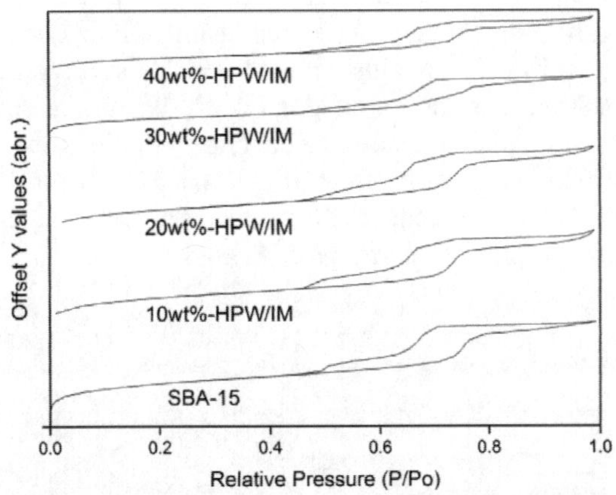

Figure 1. Nitrogen adsorption-desorption isotherms of SBA-15 and HPW post impregnated catalysts of 10wt% to 40wt% loadings.

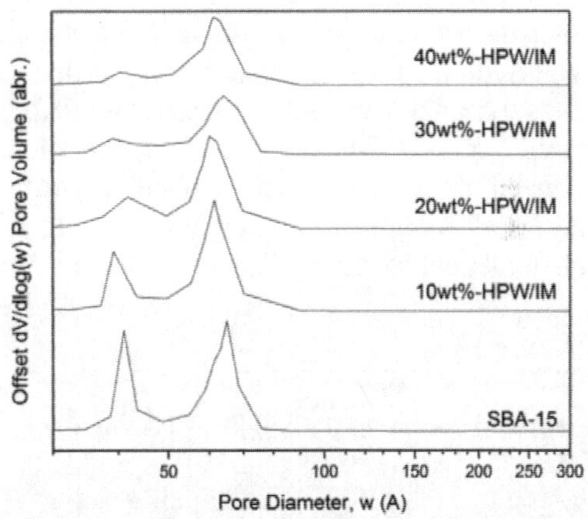

Figure 2. BJH pore size distribution of SBA-15 and HPW post impregnated catalysts of 10wt% to 40wt% loadings.

TEM images of all synthesized catalysts were shown in Figure 3. These TEM images clearly showed long, highly ordered mesoporous hexagonal channels of the famous SBA-15 remained unchanged even up to 40wt% HPW loading. It was also clear that the HPW anions were successfully incorporated into the inner pores of SBA-15 (narrowing of SBA-15 channels), yet without pore blockage (high surface areas remained) unlike the direct synthesized HPW support SBA-15 in our previous study [6]. Via post impregnation method, SBA-15 mesopores networks were stable prior to the addition of HPW anions. As such, any possible deposition of HPW anions could only occurred on the external surfaces or inner pore surfaces of SBA-15 supports, unlike those direct synthesized ones. Furthermore, high solubility of Keggin HPW anions and high mobility of water into the SBA-15 inner pores reduced the agglomeration of HPW that would cause pore deformation. As a conclusion, highly uniformed and long hexagonal mesopores have been successfully synthesized and retained in all catalysts even after impregnation of HPW.

IOP Conf. Series: Earth and Environmental Science **36** (2016) 012037 doi:10.1088/1755-1315/36/1/012037

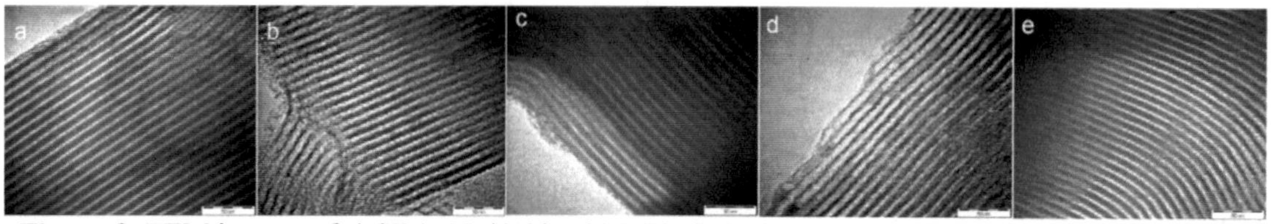

Figure 3. TEM images of a) SBA-15, b) 10wt%-HPW/IM, c) 20wt%-HPW/IM, d) 30wt%-HPW/IM and e) 40wt%-HPW/IM.

The characteristic bands of Si-O-Si condensed silica network of SBA-15 including symmetric (802 cm^{-1}), asymmetric (1220 – 1076 cm^{-1}) and bending vibration of Si-O-Si (459 cm^{-1}) were identified in Figure 4 [9]. Such observation confirmed the successful synthesis of the hydrolysed SBA-15 silica walls. Characteristic bands of Keggin type HPW are also found at the 800 – 1200 cm^{-1} wavenumber region. Interestingly, the increased intensities of Si-OH attributed bands at 3453 cm^{-1} suggested possible interaction between Si-OH with other polar molecules [11]. However, pre-treatment was performed on all the samples to remove physisorbed water, thus the effect of physisorbed water molecules should be of minimal. Such band also showed changes with increasing HPW content supported on SBA-15. It is understood that polyoxoanions (PW^{3-}) could interact with its surrounding via its terminal asymmetric oxygen ($W=O_d$), corner shared asymmetric oxygen ($W\text{-}O_b\text{-}W$) and/or edge shared oxygen ($W\text{-}O_c\text{-}W$) [12]. Based on the understanding that no exchange or sharing of electron has occurred, no ionic or covalent bonding would be formed between the support and acid. However, the interfacial Keggin structured PW^{3-} is electron rich [13]. We deduce that such characteristic can potentially act as electron donors to form hydrogen bond with silanol groups of SBA-15.

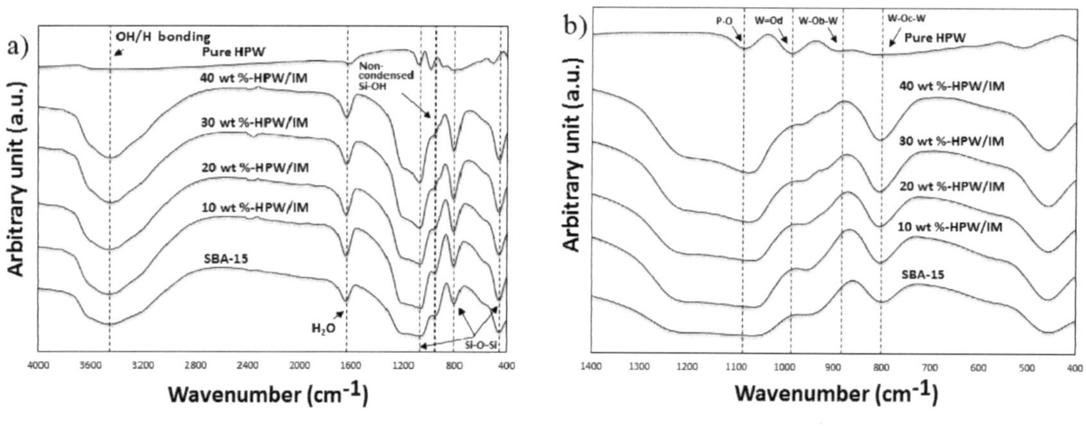

Figure 4. FTIR of all synthesized catalysts at a) 400 – 4000 cm^{-1} and b) 400 – 1400 cm^{-1} wavenumber

The characterization of an acidic catalyst would not be completed without elucidating its acidity profile. Acid strengths classification of these developed catalysts were tabulated in Table 2 [14-16]. As contrast to other reports, the total acidity profiles showed "reverse" volcano shape distribution with increasing acid loadings [17,18]. Thought similar trend was reported [19], we concluded that partial agglomeration of HPW crystal on external surfaces at high HPW loading might have occurred. Nevertheless, comparatively strong acidity profiles (medium and strong acid sites) possessed by 40wt%-HPW/IM was found to be the most beneficial to our study as it was reported that strong Brønsted acid sites were of utmost importance in esterification process [16].

Table 2. Acidity characterization of post impregnated catalysts at different HPW loadings.

Catalysts	Area Under Graph								Relative Acidity
	Weak Acid[a]	%	Medium Acid[b]	%	Strong Acid[c]	%	Total Area	Specific Area	
10wt%-HPW/IM	3.35	66	1.40	28	0.30	6	5	101	2.3
20wt%-HPW/IM	0.81	39	0.94	46	0.31	15	2	40	1.0
30wt%-HPW/IM	2.43	64	1.18	31	0.17	5	4	75	1.4
40wt%-HPW/IM	1.77	8	14.6	70	4.62	22	21	397	8.8

[a] Within temperature range of 100 – 250 ^{0}C [14,15]; [b] Within temperature range of 250 – 450 ^{0}C [15,16];
[c] Within temperature range of 450 – 500 ^{0}C [15,16]

3.2. Catalytic performances of HPW post impregnated SBA-15

The catalytic performances of post impregnated HPW/SBA-15 using wetness incipient method were studied with selective esterification of glycerol with lauric acid producing monolaurate. Parameters included HPW loadings (Figure 5), type of long chain fatty acid (Figure 6), catalyst weight (Figure 7) and glycerol to lauric acid ratio (Figure 8) were studied. From Figure 6, it was observed that while all post impregnated HPW/SBA-15 showed lower conversion compared to pure homogeneous HPW, monolaurate yield of post impregnated HPW/SBA-15 at the end of reaction rivalled to that of pure homogeneous HPW. It was understood that homogeneous HPW could not provide any shape or size control to the reaction. Thus, mono, di and trilaurate might be formed at any given time [7]. Due to active acid sites situated within mesopores, the formation of products could only occur within these pores. Thus, the formation of di and trilaurate are discouraged due to steric hindrance effect within these pores with acid sites. Ultimately, the objective of limiting reaction to form monolaurate was achieved. The effect of steric hindrance was magnified in Figure 6 whereby shorter (capric acid) and longer (myristic and palmitic) fatty acid were used as reactant. Shorter fatty acids were found to have much higher conversion as compared to longer fatty acids. As such, the size constraint effect contributed by the synthesized catalyst was concluded to be effective and significance towards the final products formation.

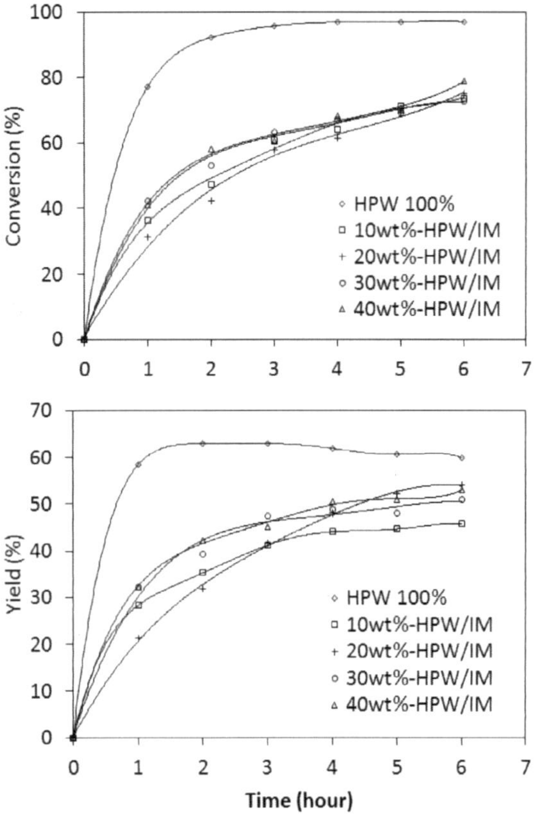

Figure 5. (Top) Lauric acid conversion profiles and (Bottom) monolaurate yield profiles with different HPW loadings.

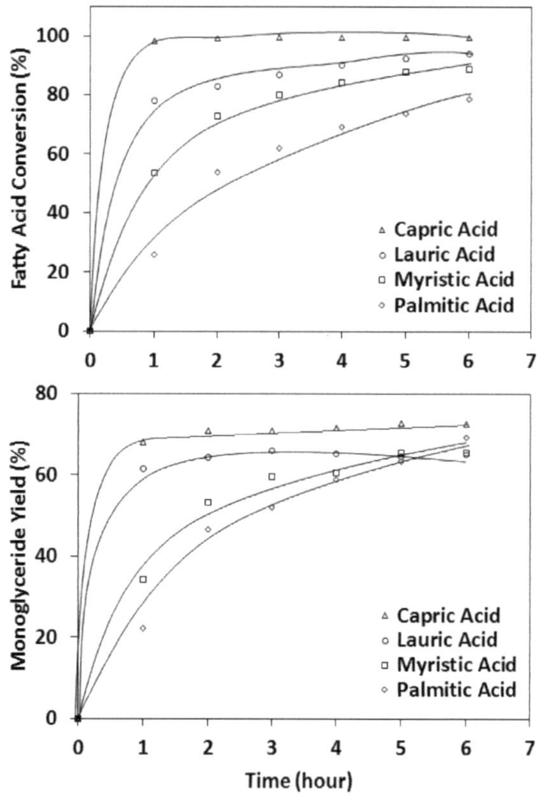

Figure 6. (Top) Long chain fatty acid conversion profiles and (Bottom) their respective monoglycerides yield profiles.

As the most potential catalyst, 40wt%-HPW/IM was chosen to further elucidate the effect of varying reactants ratio and catalyst weight as shown in Figure 7 and 8. When more catalysts were used, higher conversions were achieved. However, the selectivity of monolaurate suffered due to the increased available acid sites encouraged further esterification of monolaurate to di or trilaurate. Although the similar conversions were found when different reactant ratio used, the monolaurate yield was found to be highest at higher ratio. It was understood that esterification of glycerol with fatty acid is reversible, continuous reaction [20]. Dilution of monolaurate by glycerol and Le Chatelier's Principle dictates the formation of monolaurate rather than di or trilaurate.

IOP Conf. Series: Earth and Environmental Science **36** (2016) 012037 doi:10.1088/1755-1315/36/1/012037

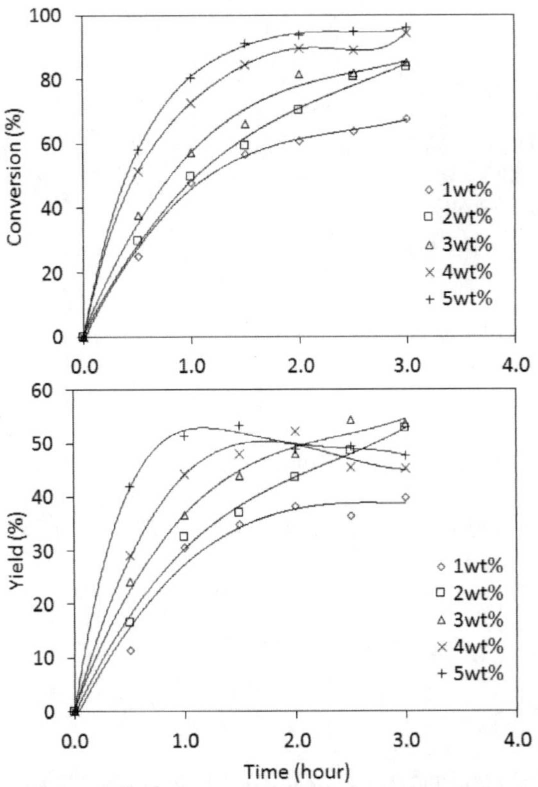

Figure 7. (Top) Lauric acid conversion profiles and (Bottom) monolaurate yield profiles at different catalysts weights used.

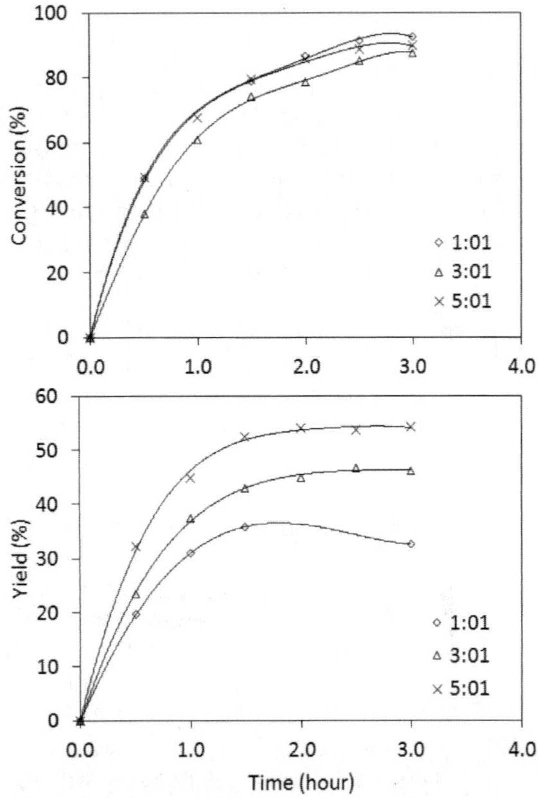

Figure 8. (Top) Lauric acid conversion profiles and (Bottom) monolaurate yield profiles at different glycerol to lauric acid ratio.

4. Conclusion

Post impregnated HPW on SBA-15 with different loadings were synthesized and the surface morphologies, pore structures and elemental analysis were characterized. When compared to directly synthesized HPW on SBA-15 catalysts in our previous study, only low degree of disruption towards the morphologies and pore structures of the support were found, as suggested by both surface characterization and TEM images. At the same time, strong acid sites could be identified from all post impregnated catalysts, especially 40wt%-HPW/IM which possessed the highest acidity from NH_3-TPD analysis. The NH_3-TPD also showed that highly acidic acid sites were accessible situated within the mesopores that would be beneficial to the selectiveness of the catalysts. These developed catalysts showed further improvement on the conversion and monolaurate yield in the selective esterification as tested in this study. Perhaps the most important finding would be the elucidation of the size selectivity of the developed catalysts, as demonstrated by the higher catalytic performance when shorter fatty acid were used. Such result further strengthen the claim whereby the highly active acid sites of these developed catalysts were situated within the mesopores, thus allowing size selectivity to be possible in the tested reaction.

5. References

[1] U.S. Energy Information Administration, E. International Energy Statistics: World Biodiesel Production. 2014 Available online: http://goo.gl/d3NP1F
[2] Rahmat N., Abdullah A Z and Mohamed A R 2010 *Renewable Sustainable Energy Rev.* **14** 987
[3] Bagheri S, Julkapli N M and Yehye W A 2015 *Renewable Sustainable Energy Rev.* **41** 113
[4] Quispe C A, Coronado C J and Carvalho Jr J 2013 *Renewable Sustainable Energy Rev.* **27** 475

[5] Pagliaro M, Ciriminna R, Kimura H, Rossi M and Della Pina C 2007 *Angew. Chem., Int. Ed.* **46** 4434
[6] Hoo P Y and Abdullah A Z 2014 *Chem. Eng. J.* **250** 274
[7] Hermida L, Abdullah A Z and Mohamed A R 2010 *J. Appl. Sci.* **10** 3199
[8] Pouilloux Y, Abro S, Vanhove C and Barrault J 1999 *J. Mol. Catal. A: Chem.* **149** 243
[9] Brahmkhatri V and Patel A 2011 *Appl. Catal., A.* **403** 161
[10] Pierotti R and Rouquerol J 1985 *Pure Appl. Chem.* **57** 603
[11] Wang X, Lin K S, Chan J C and Cheng S 2005 *J. Phys. Chem. B,* **109** 1763
[12] Uskoković-Marković S, Colomban P, Mioč U and Todorović M 2007 *Mater. Sci. Forum* **555** 201
[13] Rao G R and Rajkumar T 2008 *J. Colloid Interface Sci.* **324** 134
[14] Molnár Á, Keresszegi C and Török B 1999 *Appl. Catal. A.* **189** 217
[15] Yang L, Qi Y, Yuan X, Shen J and Kim J 2005 *J. Mol. Catal. A: Chem.* **229** 199
[16] Sakthivel R and Kemnitz E 2008 *Indian J. Chem. Technol.* **15** 36
[17] Balaraju M, Nikhitha P, Jagadeeswaraiah K, Srilatha K, Sai P P S and Lingaiah N 2010 *Fuel Process. Technol.* **91** 249
[18] Zhang J, Kanno M, Wang Y, Nishii H, Miura Y and Kamiya Y 2011 *J. Phys. Chem. C,* **115** 14762
[19] Dong B B, Zhang B B, Wu H Y, Chen X, Zhang K and Zheng X C 2013 *Mater. Res. Bull.* **48** 2491
[20] Macierzanka A and Szeląg H 2004 *Ind. Eng. Chem. Res.* **43** 7744

Acknowledgments
The authors thank the Universiti Sains Malaysia (USM) and School of Chemical Engineering for all the financial and technical supports. A Research University grant (814181) and Transdisciplinary Research Grant Scheme (6762001) from Universiti Sains Malaysia are also gratefully acknowledged.

An overview of oil palm biomass torrefaction: Effects of temperature and residence time

N Yaacob[1], ***N A Rahman**[123], **S Matali**[13], **S S Idris**[13], **A B Alias**[13]
[1]Faculty of Chemical Engineering, *Universiti Teknologi MARA,* 40450 Shah Alam, Selangor Darul Ehsan, Malaysia
[2]Core of Frontier Materials & Industry Applications, *Universiti Teknologi MARA*, 40450 Shah Alam, Selangor Darul Ehsan, Malaysia
[3]Core of Green Technology & Sustainable Development, *Universiti Teknologi MARA*, 40450 Shah Alam, Selangor Darul Ehsan, Malaysia

E-mail: noraz695@salam.uitm.edu.my

Abstract. Biomass is characterized as high moisture content, low bulk and energy density, possesses hygroscopic behaviour and poor grindability material as compared to the superior coal. A thermal treatment called torrefaction is a heating of biomass in a temperature range between 200^{0}C to 300^{0}C under inert atmosphere in order to upgrade biomass properties. Torrefied biomass has many similar characteristics to coal such as low moisture content, high bulk and energy density, hydrophobic and good grindability. This paper reviews the effects of oil palm biomass torrefaction in terms of temperature and residence time. This is because comprehensive studies on torrefaction parameters need to be carried out since different parameters might affect the chemical and physical characteristic of the torrefied product. Hence, this paper aims to discuss the effects of different torrefaction temperature and residence time towards physicochemical characteristic, mass and energy yield as well as calorific value of torrefied oil palm biomass.

1. Introduction

Waste-to-energy conversion of biomass is an important process to meet the demand of energy savings and subsequently reduces the emission of pollutants to the atmosphere [1]. Sustainable Energy Development Authority Malaysia (SEDA) stated that energy demand in Malaysia is predicted to grow 3.6% per annum until 2016 [2]. Renewable energy from biomass has been given priority as an alternative energy source to fossil fuels and this could possibly because biomass is plentifully available throughout the year. Moreover, biomass has carbon-neutral properties because carbon contained in biomass is cycled in the atmosphere when biomass is burnt as fuel [3]. Apart from that, biomass has low sulfur and nitrogen content, therefore biomass is possible to reduce the emission of pollutants such as SO2 and NOx which are commonly released during pretreatment.

Umar et al. [4] mentioned that biomass is expected to be the most prominent renewable energy source of total world primary energy by the year 2050. Interestingly, biomass from agricultural residue is considered as one of the most viable renewable energy [5]. Palm oil industry has contributed to 85.5% of biomass waste in Malaysia [6]. Oil palm or its scientific name *Elaies guineensis* had been introduced in Malaysia for many decades and its oil palm industry has contributes a huge amount of

* To whom any correspondence should be addressed.

biomass residues mainly from its plantation and milling activity [7]. Empty fruit bunches (EFBs), palm kernel shell (PKS), fiber, fronds and trunks are among biomass generated from palm oil industry which has a great potential to be utilized as renewable energy sources.

However, due to high moisture content, low bulk and energy density, hygroscopic behavior and poor grindability, biomass is preferred to undergo thermal pre-treatment process such as torrefaction in order to upgrade its properties. Torrefaction is a heating treatment of biomass which is carried out in the temperature less than 300^{0}C under inert atmosphere. There are various factors need to be taken into account during torrefaction so that a superior torrefied biomass can be produced. This paper highlighted on several operating conditions of torrefaction such as torrefaction temperature and residence time which affects the torrefied products especially in terms of physicochemical properties of biomass. Hence, this paper aims to discuss the effects of different torrefaction temperature and residence time towards physicochemical characteristic, mass and energy yield as well as calorific value of torrefied biomass.

2. Torrefaction

Historically, research on torrefaction has been started in France in the 1930s, but there are limited publications related to this research area [8]. In fact, torrefaction has become an interesting research area in the last two decades and later the process is applied commercially [9]. Torrefaction or mild pyrolysis is defined as thermal treatment of biomass which is carried out in the temperature range between 200^{0}C to 300^{0}C in absence of oxygen at typically less than 30 minutes reaction time [10–12]. The process has been found to lower the moisture content, thereby effectively improving energy density and shelf life of the biomass [13,14]. The process can be divided into three distinct stages namely; light (200^{0}C-235^{0}C), mild (235^{0}C-275^{0}C) and severe (275^{0}C-300^{0}C) [10].

3. Effects of torrefaction parameters

There are a few studies done by previous researchers regarding effects of different heating temperature, residence time as well as heating rate on torrefied biomass. Bergman et al. [15] mentioned that heating period is necessarily important to be considered upon torrefaction process since the biomass is thermally unstable and will rapidly start to decompose during the heating period However, when it comes to cooling period, the torrefied product is much more stable in terms of thermal condition. On the other hand, Williams et al. [16] stated that higher heating rate resulted in greater amount of volatile released thereby resulting different physical features of the char. Abnisa et al. [17] studied pyrolysis of several types of oil palm wastes namely trunk, frond, palm leaf and palm leaf rib and concluded that most of the biochar sample have comparable high heating value (HHV) to certain ranks of coals thus making pretreated biomass as a promising option to convert raw biomass into higher quality solid fuel. Table 1 shows high heating values (HHVs) of raw and torrefied oil palm biomass residues namely empty fruit bunch (EFB), palm kernel shell (PKS), palm mesocarp fibre (PMF) and oil palm frond (OPF).

Based on Table 1, Uemura et al. [13] studied torrefaction behaviour of EFB, PKS and PMF using three different temperatures; 220^{0}C, 250^{0}C and 300^{0}C with a constant heating rate 10^{0}C/min and residence time 30 minutes. Residence time is defined as total period when the feed biomass was inside the torrefaction reactor [23]. They revealed that torrefaction temperature increase the calorific value of all torrefied biomass studied range between 17-23 MJ/kg which higher than that of untreated biomass. Chin et al. [18] conducted torrefaction of EFB at three different temperature, 200°C, 250°C and 300°C for 15, 30, and 45 minutes and reported that the increase of heating value was influenced by temperature and time. However, HHV was greatly affected by temperature as compared to time. This is similar to Sulaiman and Anas [14] studies regarding torrefaction of OPF at three different temperature 230°C, 250°C and 270°C for 50, 30 and 15 minutes respectively. They found out that torrefaction increased the calorific value of the OPF sample by close to 50%. They concluded that higher torrefaction temperatures resulted in higher calorific values of the torrefied biomass. They also predicted that effect of residence time is comparably small for the range of conditions tested. Chew et al. [24] identified that increasing torrefaction temperature and residence time will improve the HHV of biomass whereby HHV increment was in the range of 1-58% in various types of woody and non-woody biomass. In terms of elemental composition, Aziz et al. [5] studied torrefaction behaviour of EFB, PKS and PMF in temperature ranges between 200^{0}C-300^{0}C using thermogavimetry analyser coupled with mass spectrometry (TGA-MS) and concluded that carbon content was increased while hydrogen and oxygen content decreased as torrefaction temperature increased.

Table 1. Mass yield and calorific value of raw and torrefied oil palm biomass residues

Oil palm biomass	HHV of raw biomass (MJ/kg)	Ref.	T (°C)	t (min)	Mass yield (wt%)	HHV of torrefied product (MJ/kg)	Ref.
EFB	17.02	[13]	220	30	43.16	17.17	
			250	30	36.98	17.67	[13]
			300	30	24.18	20.41	
	18.07	[18]	200	15		21.01	
			200	30		23.02	
			200	45		23.80	
			250	15		22.24	
			250	30	-	24.17	[18]
			250	45		24.54	
			300	15		22.63	
			300	30		24.34	
			300	45		25.03	
PKS	16.14	[19]	220	30	77.44	18.85	
	18.81	[20]	250	30	73.83	19.07	[13]
	19.78	[21]	300	30	71.27	21.68	
PMF	19.61	[21]	220	30	63.08	19.03	
			250	30	60.04	19.24	[13]
			300	30	52.45	22.17	
OPF	16.3	[14]	230	50		20.6	
	17.28	[22]	250	30	-	21.5	[14]
			270	15		23.8	

Other than that, mass and energy yield play an important role to determine the effects of temperature or residence time towards torrefied products. Study conducted by Na et al. [25] reported that a high calorific value of the torrefied biomass is indicative of such weight loss during torrefaction. Therefore, torrefied biomass should be evaluated in terms of energy yield so that a high energy density of torrefied biomass can be produced. Chen and Kuo [26] reported that biomass torrefied in less than one hour with light torrefaction are able to produce solid fuel with high energy density. Mass yield is determined as percentage of resulted mass after torrefaction while energy yield is defined as percentage of calorific value of feed materials [27]. The mass and energy yield is closely related to each other since value of mass yield will be used to obtain an energy yield value. Mass yield of torrefied biomass can vary from 24% to 95% of its original weight depending on its hemicellulose content [24]. In terms of fiber compositional analysis, cellulose, hemicellulose and lignin show different thermal decomposition characteristic. Previous study reported that hemicellulose has the thermal decomposition temperature between 220°C-315°C while that of cellulose (315°C-400°C) and finally lignin which undergo gradual decomposition in temperature range from 160°C to 900°C [28]. Aziz et al. [5] reported that based on TGA curves obtained from his study on torrefaction behaviour of EFB, PKS and PMF, hemicellulose degradation was more significant as compared to that of cellulose and lignin. Uemura et al. [13] studied the effects of three different temperatures on torrefaction of oil palm waste and concluded that PMF and PKS exhibited higher energy yield values of more than 95 wt.% compared to EFB energy yield value of 56 wt.%. Strandberg et al. [23] reported that as torrefaction severity increases, the mass and energy yield decreases and their studies also focused on the importance of temperature and temperature process control. They also concluded that effects of torrefaction temperature were more significant than that of residence time.

4. Conclusion
Torrefaction temperature and residence time are the two important parameters that should be considered in torrefaction process. Even though some studies stated that residence time only has a slight effect in torrefaction, further study is required to investigate the effects of residence time on solid char and energy yield and also determination of their optimum values. This is because shorter residence time will make the torrefaction process cost-effective and yet able to produce a superior solid char to be further used in power generation sector. It is also worth mentioning that investigation of residence time varies between individual biomass from different origin either for lignocellulosic or non-lignocellulosic biomass.

References
[1] Liu Z, Zhang Y, Zhong L, Orndroff W, Zhao H, Cao Y, Zhang K and Pan W P 2013 *J. Therm. Anal. Calorim.* **113** 489–496
[2] SEDA 2013 Putrajaya, Malaysia
[3] Chen W-H, Peng J and Bi X T 2015 *Renew. Sustain. Energy Rev.* **44** 847–866
[4] Umar M S, Jennings P and Urmee T 2014 *Biomass Bioenerg.* **62** 37–46
[5] Aziz M A, Sabil K M, Uemura Y and Ismail L 2012 *J. Appl. Sci.* **12** 1130–35
[6] Yusoff S 2006 *J. Clean. Prod.***14** 87–93
[7] Shuit S H, Tan K T, Lee K T and Kamaruddin A H 2009 *Energy* **34** 1225–35
[8] van der Stelt M J C, Gerhauser H, Kiel J H A and Ptasinski K J 2011 *Biomass Bioenerg.* **35** 3748–62
[9] Bridgeman T G, Jones J M, Shield I and Williams P T 2008 *Fuel* **87** 844–856
[10] Chen W H and Kuo P C 2011 *Energy* **36** 803–811
[11] Lu K M, Lee W J, Chen W H, Liu S H and Lin T C 2012 *Bioresour. Technol.* **123** 98–105
[12] Maski D, Darr M J and Anex R P 2010 Pittsburgh, Pennsylvania, 1009376
[13] Uemura Y, Omar W N, Tsutsui T and Yusup S B 2011 *Fuel* **90** 2585–91
[14] Sulaiman S A and Anas M I 2012 *Trends Appl. Sci. Res.* **7** 248–255
[15] Bergman P, Boersma A, Kiel J, Prins M J, Ptasinski K and Janssen F J 2005 *2nd World Conf. Technol. Exhib. Biomass Energy, Ind. Clim. Prot.*78–82
[16] Williams A, Pourkashanian M and Jones J M 2001 *Prog. Energy Combust. Sci.* **27** 587–610
[17] Abnisa F, Arami-Niya A, Wan Daud W M A, Sahu J N and Noor I M 2013 *Energy Convers. Manag.* **76** 1073–1082
[18] Chin K L, H'ng P S, Go W Z, Wong W Z, Lim T W, Maminski M, Paridah M T and Luqman a C 2013 *Ind. Crops Prod.* **49** 768–774
[19] Chin M J, Poh P E, Tey B T, Chan E S and Chin K L 2013 *Renew. Sustain. Energy Rev.* **26** 717–726
[20] Jaafar A and Ahmad M, *World Acad Sci Eng Technol* **60** pp 554–557
[21] Uemura Y, Omar W, Tsutsui T, Subbarao D and Yusup S, *J.Appl. Sci.* **10** 3250–56
[22] Guangul F M, Sulaiman S A and Ramli A 2012 *Bioresour. Technol.* **126** 224–232
[23] Strandberg M, Olofsson I, Pommer L, Wiklund-Lindström S, Åberg K and Nordin A 2015 *Fuel Process. Technol.* **134** 387–398
[24] Chew J J and Doshi V 2011 *Renew. Sustain. Energy Rev.* **15** 4212–22
[25] Na B I, Kim Y H, Lim W S, Lee S M, Lee H W and Lee J W 2013 *Biomass Bioenerg.* **52** 159–165

[26] Chen W H and Kuo P C 2010 *Energy* **35** 2580–2586
[27] Nachenius R W, van de Wardt T a, Ronsse F and Prins W 2015 *Biomass and Bioenergy* pp 1–9
[28] Lu K M, Lee W J, Chen W H and Lin T C 2013 *Appl. Energy* **105** 57–65

Acknowledgement
The authors gratefully acknowledge the financial support received from Education Ministry of Malaysia through Fundamental Research Grant Scheme (FRGS) (600-RMI/FRGS 5/3 (98/2014)), Research Management Centre (RMC), Faculty of Chemical Engineering and Universiti Teknologi MARA, Selangor, Malaysia.

Synthesis and characterization of metal oxide promoted alumina catalyst for biofuel production

S M Anisuzzaman[1], D Krishnaiah, A Bono, S Abang, M Sundang, E Suali, F A Lahin, A Shaik Alawodeen

Chemical Engineering Programme, Faculty of Engineering, Universiti Malaysia Sabah, Jalan UMS, 88400 Kota Kinabalu, Sabah, Malaysia

E-mail: [1]anis_zaman@ums.edu.my, [1]dr.anis.ums@gmail.com

Abstract. Alumina has been widely used as a support in catalysis process which owing to its extremely thermal and mechanical stability, high surface area, large pore size and pore volume. The aim of this study was to synthesize calcium oxide-supported basic alumina catalysts (CaO/Al_2O_3) by impregnation method and to characterize the properties of the catalyst based on its surface area and porosity, functional group, surface morphology and particle size. Impregnation method was chosen for the synthesization of catalyst which involved contacting the support with the impregnating solution for a particular period of time, drying the support to remove the imbibed liquid and calcination process. In the preparation of catalyst, catalytic performance of CaO/Al_2O_3 catalyst was measured at different calcined temperatures (650°C, 750°C and 800°C). Fourier transform infrared spectroscopy (FTIR), scanning electron microscopy (SEM), Mercury intrusion porosimetry (MIP), and particle size analyzer (Zetasizer) was used to characterize the catalyst. The highest total specific area and the total porosity of the catalyst was obtained at 750°C. FTIR analysis basically studied on the functional groups present in each catalyst synthesized, while SEM analysis was observed to have pores on its surface. Moreover, CaO/Al_2O_3 catalysts at 650°C produced the smallest particle size (396.1 nm), while at 750°C produced the largest particle size (712.4 nm). Thus it can be concluded that CaO/Al_2O_3 catalysts has great potential commercialization since CaO has attracted many attentions compared to other alkali earth metal oxides especially on the transesterification reaction.

1. Introduction

The most common preparation of catalyst methods are impregnation, ion-exchange, adsorption and desorption-precipitation. Even though the preparation procedures considerably differ from one catalyst to another, three broad categories can be introduced to classify the catalysts with respect to the preparation procedure which includes bulk catalyst and supports, impregnated catalysts and mixed-agglomerated catalysts [1].

[1] To whom any correspondence should be addressed.

Calcium oxide (CaO) has attracted many attentions compared to others alkali earth metal oxides especially on the transesterification reaction since it possess relatively high basic strength and less environmental impacts due to its low solubility in methanol [2-4]. Besides that, it can be synthesized from cheap sources such as limestone and calcium hydroxide. In addition, CaO has been widely used as a catalyst for reactions involved in organic compounds isomerization such as 5-vinylbicyclo hept-2-ene to 5-ethylidenebicyclo hept-2-ene which showed high activity for organic reactions [5].

The impregnation method involves contacting the support with the impregnating solution for a particular or certain period of time and drying the support to remove the imbibed liquid. Besides that, it helps by activating the catalyst by calcination, reduction or other appropriate treatment. Lithium impregnated CaO is expected to show higher activity towards the transesterification reaction in comparison to pure CaO. Previous studies have indicated that nickel supported alumina (Ni/Al_2O_3) is not completely reduced to the metallic state because of the strong oxide-support interaction between them [6,7]. The major interest in Ni/Al_2O_3 catalyst has centered on interactions between the metal and the support where the metal-support interactions appreciably affect the surface properties of these catalysts and their catalytic properties. Apart from that, Mattisson et al. [8] investigated on the reactivity of some metal oxides supported on alumina with alternating methane and oxygen (application for chemical-looping combustion). Basically in this study, the researchers found that some metal oxides of the transition state metals such as Fe, Cu, Co, Mn, and Ni were feasible candidates to be used as oxygen carriers in a chemical-looping combustion system based on interconnected fluidized beds.

This study was mainly focused on the preparation of calcium oxide-supported basic alumina catalysts (CaO/Al_2O_3) which involved impregnation method. Alumina has been used as a support in catalysis processes due to its extremely thermal and mechanical stability, high specific surface area, large pore size and pore volume. The catalytic performance of CaO/Al_2O_3 catalysts was measured at different calcination temperatures (650, 750 and 800°C). The characterization of the catalyst was done by determining the functional groups by fourier transform infrared spectroscopy (FTIR), observing the surface morphology by scanning electron microscopy (SEM), determination of surface area and porosity by mercury intrusion porosimetry (MIP) and determination of particle size distribution analysis by Zetasizer.

2. Materials and methods

2.1 Preparation of catalysts

CaO/Al_2O_3 were prepared by impregnation method where a calculated amount of calcium acetate ($Ca(CH_3COO)_2 . H_2O$) were dissolved in a certain amount of deionized water, that was stirred at room temperature for 30 min. Then the solution was added drop wise to the basic alumina, and impregnation for 24 h. The solid was filtered and collected by washing thoroughly with deionized water and ethanol, and subsequently dried at temperature of 110°C for 3 h to ensure the residual solvent was removed. The dried catalyst was calcined in the absence of air at required temperature for 2 h.

2.2 Characterization of catalysts

2.2.1 Functional groups determination

FTIR was used to analyse and determine the functional present in the each of the catalysts. In other words, it offers quantitative and qualitative analysis for organic and inorganic samples where it identifies chemical bonds in a molecule by producing an infrared absorption spectrum, where some of the infrared radiation is absorbed by the sample, while some of it is passed through (transmitted).

2.2.2 Morphological characterization
SEM analysis was done to examine the surface morphology of the catalysts. The setting was chosen as follows; working distance (WD) was 6.0mm, 6.5mm and 7.0mm and Electron High Tension (EHT) was 10 kV.

2.2.3 Surface area and porosity
The analysis in determining the surface area and porosity of the catalyst was analysed by using MIP of Thermo Electron Corporation (Pascal Series 440). The mercury porosimetry technique is one of the most highly or recommended methods in terms of investigating the porous structure of solid samples in a quantitative way. Basically, it provides reliable information about pore size/volume distribution, bulk or envelope density, particle size distribution and specific surface for most porous solids.

2.2.4 Particle size distribution
In this study, to determining the particle size of the catalysts used Malvern Zetasizer Nano Series Instrument with laser diffraction technique in the operation. Laser diffraction measures particle size distribution and the particle was reported as a volume equivalent sphere diameter at refractive index of 1.330.

3. Results and discussions

3.1 FTIR analysis
Functional groups of the synthesized catalysts were analysed using FTIR. The results are shown in Figure 1 (a,b and c).

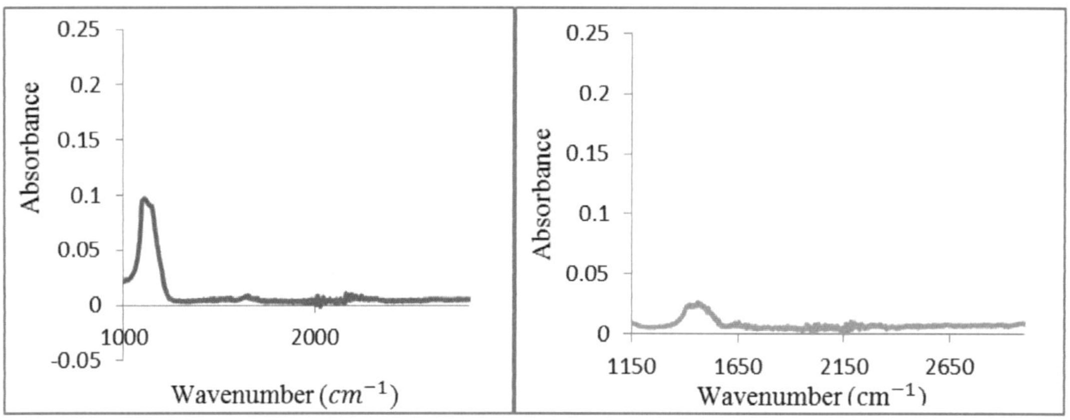

(a) (b)

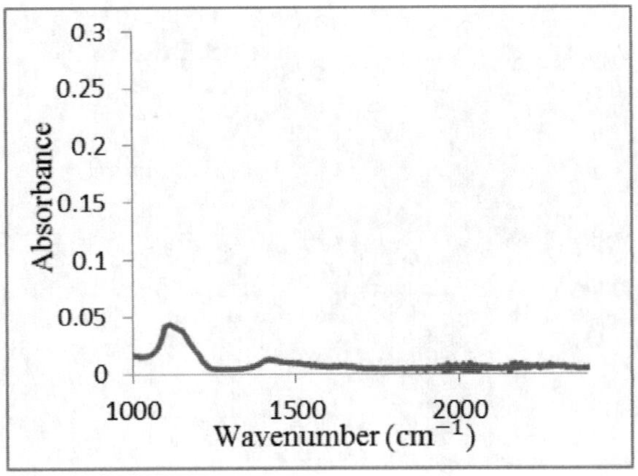

(c)

Figure 1. FTIR Spectra for CaO/Al_2O_3 (a) 650°C, (b) 750°C (c) 800°C

Figure 1 (a) shows that there was presence of broad band at 1236 cm^{-1} can be assigned to C-O bond and carboxylic acid stretching, while the weak strength of peak at 1680 cm^{-1} indicates the functional group of alkenes C=C. In the range between 2220 to 2260 cm^{-1}, those peaks which are small but exposed, indicates the presence of CN triple bond (nitriles), absorptions. As in figure 1 (b), bands at around ~1600 cm^{-1} which can be assigned as C=C stretching. Besides that, it is observed that it is weak band at around ~1700 cm^{-1}, where this can be related to the presence of C=O bonds in carboxylic acids. The CC triple bond (alkynes) and CN triple bond (nitriles) absorptions at around ~2100cm^{-1} are small but exposed. The structure of aromatic compounds may also be confirmed from the pattern of the weak overtone and combination tone bands found from 1650 to 2000 cm^{-1}. While as in Figure 1 (c), which is under calcined temperature of 800°C, it showed C-O absorption between 1080 and 1200 cm^{-1} where these peaks are normally rounded like the O-H and N-H peak where carboxylic acids, esters, ethers, alcohols and anhydrides all containing this peak. Furthermore, it showed the CC triple bond (alkynes) and CN triple bond (nitriles) absorptions that was around 2100-2260 cm^{-1} are small but exposed.

3.2 SEM analysis

The analysis of the surface morphology of CaO/Al_2O_3catalyst with different calcination temperatures were done by SEM S-3400N Hitachi. In this surface morphology analysis, the SEM used a focused beam of high-energy electrons to generate variety of signals at the surface of solid specimens where the signals that derive from electron-sample interactions reveal information about the particular sample such as its crystalline structure, external morphology (texture), chemical composition and orientation of materials making up the sample [9,10]. The SEM images of the CaO/Al_2O_3 with different calcination temperatures are shown in Figure 2 (a, b and c).

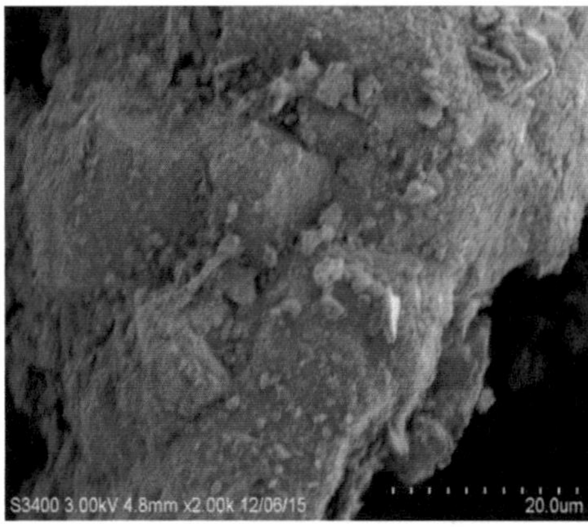

(a)

(b)

(c)

Figure 2. SEM images of CaO/Al_2O_3 (a) 650 °C, (b) 750 °C (c) 800 °C

3.3 MIP analysis

Table 1 shows the results of porosimetry data for CaO/Al_2O_3 with different calcination temperatures

Table 1. Porosimetry data for CaO/Al_2O_3 with calcination temperature of 650°C, 750°C, 800°C

Samples	650°C	750°C	800°C
Total specific area (m^2/g)	8.12	22.97	21.12
Total porosity (%)	20.50	69.31	67.97
Average pore diameter (nm)	3587.80	2640.67	3427.17
Total cumulative volume (mm^3/g)	165.30	400.00	407.63

The catalyst of CaO/Al_2O_3 with temperature of 750°C has surface area of 22.97 m^2/g and decreases at the temperature of 800°C which reduced to 21.12 m^2/g. As for total porosity, it represents 20.50% for temperature of 650°C, and increased dramatically to 69.31 % at 750°C and slightly decreases at 800°C with 67.97%. The reduction of pores and surface area at higher temperature, perhaps, due to the penetration of metal oxide onto the surface of alumina [11]. Basically, this indicated that impregnation process under calcined temperature of 750°C caused the internal pores to collapse, thus, reduced the porosity of 800°C calcined temperature. Apart from that, the total cumulative volume seems increased from calcined temperature of 650°C, 750°C and 800°C with 165.30 mm^3/g, 400 mm^3/g and 407.63 mm^3/g respectively. Most probably, the increase in the cumulative volume is due to the strong bond and structure of the catalysts (calcium oxide promoted alumina). As for calcined temperature of 650°C, it has the largest average pore diameter that is 3587.80 nm compared to calcined temperature of 750 °C and 800 °C, that is 2640.673 nm and 3427.168 respectively.

3.4 Zetasizer analysis

Particle size distribution by intensity (%) of CaO/Al_2O_3 for different calcination temperature is shown in Figure 3.

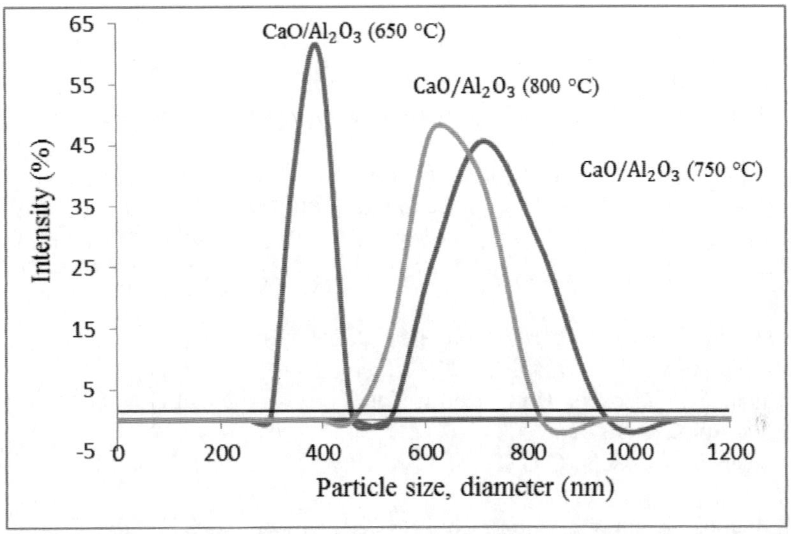

Figure 3. Particle size distribution based on intensity for CaO/Al_2O_3 with calcination temperature of 650°C, 750°C and 800°C.

The particle size distribution was analysed at different calcination temperatures. Based on the Figure 3, the result shows that the particle size distribution by intensity of CaO/Al_2O_3 (650°C), CaO/Al_2O_3 (750°C) and CaO/Al_2O_3 (800°C) are 396.1 nm, 712.4 nm and 615.1 nm respectively. In this analysis, three measurement data were taken for each calcination temperature, and the highest percentage of intensity were taken since it represent the peak of the graph. As for calcined temperature of 650 °C, the intensity results produced were 60.1%, 39.9% and 46.5%. While for calcined temperature of 750 °C, the percentage of intensity obtained were 45.6%, 25.9% and 28.4 %. Intensity percentage achieved by calcined temperature of 800°C were 47.5%, 13.5% and 39.0%. Hence, it can be observed that the peak, which represents particle size, of each calcined temperature was based on the highest percentage of its intensity. Most probably, the difference of particle size of the catalyst is due or affected by the way of grinding the powder form of catalyst [9]. For instance, if the size of a catalyst metal particle is reduced into the nanoscale, its properties initially remain the same as a larger particle. But when the size is smaller than about 10 nm, the movements of electrons in the metal are confined, so their inherent energies are increased. Hence, a finely powdered solid has a larger surface area than an equal mass of a coarsely powdered solid where a larger surface area is a much easier target for colliding molecules. The bigger the target, the easier it is to hit and this shows that fine powders therefore react faster than coarse powders.

4. Conclusion

The CaO/Al_2O_3 was prepared by impregnation method with calcined temperature of 650°C, 750°C and 800°C. As in FTIR analysis, it showed the functional groups present in each catalyst produced, CaO/Al_2O_3. Furthermore, all the catalysts were observed to have pores on the surface by using SEM analysis. Besides that, based on the analysis of MIP method, it can be concluded that CaO/Al_2O_3 at 750 °C produced highest total specific area with the total porosity of the catalyst. In terms of particle size, each of the catalyst produced different particle based on its intensity where CaO/Al_2O_3 (650°C) produce the smallest particle size, while CaO/Al_2O_3 (750°C) produce the largest particle size. Hence, CaO/Al_2O_3 has great potential commercialization since CaO has attracted many attentions compared to others alkali earth metal oxides especially on the transesterification reaction.

References

[1] Perego C and Villa P 1997 *Catal Today* **34** 281-305
[2] Kouzu M, Kasuno T, Tajika M, Sugimoto Y, Yamanaka S and Hidaka J 2008 *Fuel* **87** 2798-806
[3] Kiss A A, Dimian A C and Rothenberg G 2006 *Adv. Synth. Catal.* **348** 75-81
[4] Chai F, Cao F, Zhai F, Chen Y, Wang X and Su Z 2007 *Adv. Synth. Catal.* **349** 1057-65
[5] Baba T, Endou T, Handa H and Ono Y 1993 *Appl. Catal. A-Gen.* **97** L19-L23
[6] Derouane E G, Simoens A, ,Colin C, Martin G A , Dalmon J A , Védrine J C 1973 *J.Catal.* **70** 50–8
[7] Li C and Chen Y W 1995 *Thermochim. Acta,* **256** 457-65
[8] Mattisson T, Jardnas A., Lyngfelt A. 2003 *Energ. Fuel,* **17** 643-51
[9] Anisuzzaman S. M., Joseph C G., Taufiq-Yap Y H, Krishnaiah D and Tay V V 2015 J. King Saud Univ.- Sci. **27** 318-30
[10] Anisuzzaman S M, Joseph C G, Daud W M A W, Krishnaiah D and Ho S , 2015 *Int. J. Ind. Chem.* **6** 9-21
[11] Pua F L, Fang Z, Zakaria S, Guo F, Chia C H 2011 *Biotechnol. Biofuels* **4** 56

Acknowledgements

Authors would like to thank Universiti Malaysia Sabah (UMS) for providing the research facilities.

IOP Conf. Series: Earth and Environmental Science **36** (2016) 012040 doi:10.1088/1755-1315/36/1/012040

Polyhydroxyalkanoates production from waste biomass

A K H Nor Aslan[1], M D Mohd Ali[1], N A Morad[1] and P Tamunaidu[1]*
[1]Shizen Conversion & Separation Technology (SHIZEN) iKohza, Malaysia-Japan International Institute of Technology, *Universiti Teknologi Malaysia (UTM)* Kuala Lumpur, Malaysia

E-mail: [1]pramila@utm.my

Abstract. Polyhydroxyalkanoates (PHAs) is a group of biopolymers that are extensively researched for such purpose due to the biocompatibility with mammal tissue and similar properties with conventional plastic. However, commercialization of PHA is impended by its high total production cost, which half of it are from the cost of pure carbon source feedstock. Thus, cheap and sustainable feedstocks are preferred where waste materials from various industries are looked into. This paper will highlight recent studies done on PHA production by utilizing crop and agro waste material and review its potential as alternative feedstock.

1. Introduction

Polyhydroxyalkanoate (PHA) is one of the most researched bioplastic in hope to replace conventional plastic of today. This biopolymer is a group of naturally-occuring biopolyesters produced by microorganisms under imbalanced growth condition due to carbon source surplus and/or limitation of essential nutrient such as phosphorus and nitrogen [1]. Classification of PHAs depends on the number of carbon atoms present in their monomers as short-chain-length PHAs (scl-PHA; three to five C-atoms), medium-chain-length PHAs (mcl-PHA; with six or fourteen C-atoms) and long-chain-length PHAs (lcl-PHA; more than 14 carbon atoms) [1].

Currently, PHA and its co-polymers have been researched for its potential applications in various industry and fields, for example, poly-(3-hydroxybutyrate) (P3HB) is a homopolymer of PHAs group that are toxicologically safe and exhibit high biocompatibility with mammalian cells, making them suitable for medical applications and food industry [1,2,3]. Additionally, the thermoplastic properties of P3HB and its biodegradability, without generation of toxic by-products, makes it a potential sustainable alternative to conventional plastics [2,3]. Unfortunately, PHA production cost remains substantially expensive (about USD8–11 per kg) compared to other polymers, and has been one of the main obstacle in exploitation of PHA as commercial commodity [2].

Most of the carbon sources used for conventional PHA production are of pure raw material, comprising of pure carbohydrates (glucose, sucrose, maltose, starch), fatty acids and its derivatives, methanol and alkanes [2]. In terms of the production cost, carbon source for microbial growth attributed to 50% of the total costs [3], thus it is desirable to find cheap alternative carbon feedstocks in order to improve sustainability of PHA production while obtaining commercial viability through

[1] To whom any correspondence should be addressed.

fermentation. As such, inexpensive carbon sources such as agro-industrial by-products and agricultural residues have been reported as potential alternative substrates.

Therefore, this paper will review on the current advances involving utilization of biomass from various waste from agriculture, agro-industrial practice and domestic waste for PHA production as well as the challenges and opportunities for commercialisation.

2. PHA Production from Biomass

2.1. Sugar Rich Feedstock

Sugar producing plant such as maple tree, which is known for its sweet sugar sap, has been studied as carbon feedstock for PHA production. Yezza *et al.* employed maple sap a carbon source to *Alcaligenes latus* (now known as *Azohydromonas lata)* for the production of poly-beta-hydroxybutyrate (PHB) and was able to produce up to 78% of this bioplastic monomer [4]. However when feedstock is varied, for example conversion of sugarbeet juice using the same microbe *A. lata*, Wang *et al.* reported lower PHA yield of 66% compared to pure sucrose which gave 80.0% of PHA [5]. This suggests that PHA yield depends on the total chemical composition of the sap/juices used. In another study, Suwannasing *et al.* utilized juice from pineapple for the production of PHB using isolated *Bacillus* sp. SV13 from agriculture soil. The study showed that 56% of PHB could be produced from pineapple juice compared to pure glucose and sucrose (18.5% and 29%) [6].

In a relatively new sugar rich feedstock, Zahari *et al.* showed that oil palm frond can be simply compressed to obtain their plant juice or sap, which contains high glucose of 53.95 gl^{-1} that constitute about 70% of its total sugar content [7]. The study reported a PHA yield of 30.5 g/l, comprising of 75 wt% of the biomass dry weight.

Besides virgin feedstocks, sugar rich by-products are also readily available as raw material for bioconversion into high value products. Molasses is a by-product of sugar manufacturing industries and is currently being utilized as a carbon source for industrial scale fermentations due to its abundance and low price. It is estimated that 12 million ton of beet molasses and 39 million ton of cane molasses are readily available for further conversion [4]. In a study by Albuquerque *et al.,* a three-step fermentation strategy was proposed for cane molasses fermentation to produce PHA. The molasses were fermented to organic acids in the first step followed by PHA accumulation, and then finally, batch fermentation for the production of PHA, where about 30% cell dry weight (CDW) of PHA is produced. [8]. Similarly, soy molasses has also shown positive conversion to PHA resulting in 5-17% of mcl-PHA production using *Pseudomonas corrugata* [9]. While prior studies shows low yield of PHA in molasses feedstock, a recent study done by Akaraonye *et al.* for example has claimed able to produce 61.07% yield of PHA in *Bacillus cereus* SPV. They claimed that by limiting the oxygen supply in their media, this lead to anaerobic condition for growth which may have led to enhanced yield of PHA [10].

2.2. Waste Lipids and Industrial Wastewaters

Cooking oils, specifically waste cooking oils have been considered as potential alternative carbon feedstock for PHA production. Triacylglycerides (TAGS) from the fatty acids of the waste oil is seen as valuable alternative feedstock for the production of mcl-PHAs and lcl-PHAs [11]. A study done by Obruca *et al.* using waste rapeseed oil with precursor propanol is able to produce up to 76% PHA and 8% 3-hydroxyvalerate (3-HV) in co-polymer [12]. Another study using same carbon source showed that the high nitrogen content in heated oil due to thermal degradation was able to increase cell growth and eventually increase PHA yield [13].

Crude glycerol gained more interest as feedstock due to its large quantities as by-product in various industrial processes, particularly in biodiesel production. Crude glycerol from soy-based biodiesel has been studied by utilizing *Pseudomonas corrugata* 388 and *Pseudomonas oleovorans* NRRL B-14682 resulting in accumulation of mcl-PHAs [14]. A study by Mozumder *et al.* yielded the highest PHA content up to 63% by using crude glycerol via a three step fermentation process [15]. In an attempt to reduce production cost further, a study utilizing crude glycerol combined with meat and bone meals as a nitrogen source was done and it resulted in 5.9 gl^{-1} PHAs and a co-polymer PHBV without addition of precursor in the media [16].

Another interesting alternative carbon source is wastewater from industrial practices such as biodiesel wastewater, food processing waste effluent, municipal wastewater and brewery waste effluent [17]. In the first step, organic carbon sources from wastewater are converted into volatile fatty acid in aerobic activated sludge, followed by PHAs conversion by mixing cell cultures in the second step. Generally, PHA concentrations from wastewater are low in current studied settings, with only around 50% production of PHA [17, 18]. Martinez *et al.* managed to accumulate 55% PHA content with 11% of HV when pure *Cupriavidus necator* is fed with pre-treated olive mill wastewater, without synthetic precursor under acidogenic condition [19]. However, another study by Khardenavis *et al.* reported high PHA yield of 67% from mixed microbial culture (MMC) by using rice grain distillery wastewater as carbon source [20]. The high yields were only achieved after adding di-ammonium hydrogen phosphate (DAHP) as extra nitrogen source.

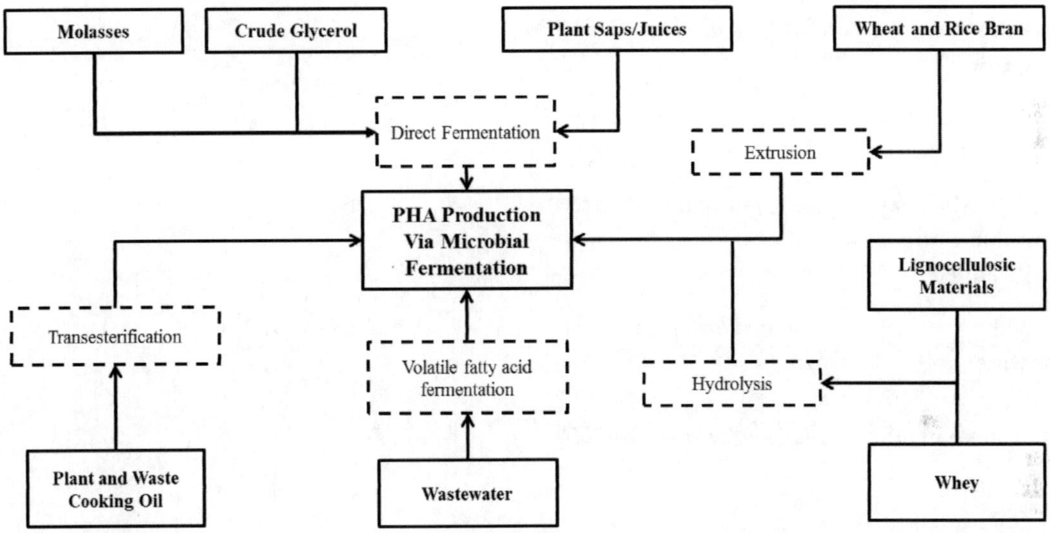

Figure 1. Various feedstock and their respective pre-processing technology prior to microbial PHA

2.3. Lignocellulosic Biomass

Agricultural practices often produces abundant lignocellulosic biomass. Utilization of non-food lignocellulosic materials for PHA production is preferable over food crops. Usually, lignocellulosic materials need to be pre-treated and undergo hydrolysis to convert to sugar monomers. In a study by Cesario *et al.,* PHA production from wheat straw hydrolysate is able to accumulate up to 72% CDW through fed-batch fermentation using *Burkholderia sacchari* DSM17165 [21]. It was also stated that poly(3-hydroxybutyrate-co-3-hydroxyvalerate), PHBV production by utilizing the same carbon source can also be achieved using propionic acid as a co-substrate. Most research on utilizing lignocellulose as carbon feedstock centre around the conversion of the monomer sugars from hemicelluloses.

Other agro industrial residues such as inexpensive mixture of extruded rice bran and wheat bran also has been studied by using haloarchaeon *Haloferax mediterranei,* which resulted in 56% of PHA content [22]. While the resulting PHA is still relatively low, it is important to note that extraction using archaea is inexpensive and safe due to the lack of chemicals used for the separation and purification of the PHA accumulated in the haloarchaea species.

Therefore, the production of PHA depends on many factors, which is the feedstock quality, microbes used, nutrients and pre-processing technology. The PHA production via microbial fermentation of various feedstocks and their pre-processing technology is summarized in figure 1 and table 1 below.

Table 1. PHA production in term of cell dry weight percentage (%CDW) from microbial fermentation of various waste biomass

Biomass	Microorganism	PHA Content, Cell Dry Weight (%CDW)	Ref.
Soy molasses	*Pseudomonas corrugata.*	5.00-17.00	[9]
Molasses (sugarcane)	*Bacillus cereus* SPV	61.07	[10]
Crude glycerol	*Cupriavidus necator DSM545*	62.70	[15]
Wheat straw hydrolysate	*Burkholderia sacchari* DSM 17165	72.00	[21]
Extruded rice bran and wheat bran	*Haloferax mediterranei*	56.00	[22]
Rice grain distillery wastewater with nutrient added	*Activated sludge (MMC)*	67.00	[20]
Olive Mill Wastewater	*Cupriavidus necator DSM 545*	55.00 (11% 3HV co-polymer)	[19]
Waste rapeseed oil	*Cupriavidus necator H16*	76.00 (8% 3-HV co-polymer)	[12]
Pineapple juice	*Bacillus* sp.	48.89	[6]
Sugarbeet Juice	*Azohydromonas lata*	65.60	[5]
Maple Sap	*Azohydromonas lata*	77.60	[4]
Oil Palm Frond Juice	*Cupriavidus necator NCIMB 11599*	75.00	[7]

3. Challenges and Opportunities

Cellulose and hemicellulose from lignocellulosic materials are both excellent non-food carbon source to be used in different biological processes after hydrolysis to monomeric sugars. However, the process is cost inefficient due to natural lignin component that is recalcitrant to degradation. Various pre-treatments are required to loosen up the lignocellulose structure followed by utilization of enzymes or special microbial strains to convert the holocellulose into sugar feedstock which contributed to additional cost [18]. Furthermore, hydrolysis of lignocellulose waste by chemical and/or enzymatic treatments unfortunately releases several inhibitory compounds, such as phenolic, furans and organic acids which can have negative effect to microbial fermentation. To solve this, an adapted tolerant strain of *C.s necator* is engineered, which can grow in diluted bagasse hydrolysates and efficiently reduce acetic acid, formic acid, furfural and acid soluble lignin while producing PHAs [23]. The polyesters produced were mainly PHB (57% on CDW), but also included PHBV. *C. necator* strain mainly utilizes glucose as carbon source leaving potential down streaming of the unutilised sugar components in the treated hydrolysates in the production of biofuels and other chemical.

As for the usage of wastewater for PHA production, it inevitably causes the formation of non-storing PHA population which may adversely affect production of the biopolymer. In a study by Marang *et al.* using a mixed culture dominated by *Methylobacillus flagellatus,* an obligate methylotroph, that cannot store PHA, and *Plasticicumulans acidivorans*, a known PHA producer, has lowered the PHA content from 80 to 66% wt, due to the presence of the non-storing population [25]. The fermentation was enriched with acetate and methanol, as they reduced the effect of methylotroph *M. flagellatus* on the PHA production [25].

Usage of virgin oils such as soybean produces higher PHA compared to waste oils, which may have been due to the presence of impurities in the waste oils [18]. However, the usage of edible oils for PHAs production is no longer considered cost effective due to increase in price. Thus the focus is on non-edible oils and waste cooking oils. Crude glycerol from biodiesel is one of the feedstock that cannot be used as raw material for drug, food and pharmaceutical industries due to the presence of major impurities [15]. Refining crude glycerol is costly, thus the biological conversion of crude glycerol to higher value chemicals, such as PHA, is more desirable. Unfortunately, it is important to note that utilization of glycerol can affect the quality of the polymer by reducing its molecular mass, thus changing the properties of the PHA [15].

Conversely, PHA production using sugar based crops generate higher yield, depending on the sap and strain type used as shown in Table 1. Plant sap or juice consists of high sugar content and as well as minerals such as potassium, nitrogen and other which are essential for PHA accumulation and microbial growth [5, 4, 7]. The surplus of sugar and presence of mineral make the plant juice or sap as potential natural media for microbial fermentation. This is supported by a study by Zahari *et al.* which showed that the PHA extracted from oil palm frond juice produced relatively high molecular mass; an important factor in determining PHA's physicochemical and mechanical properties [7, 25]. The only drawback from using plant saps and juices as carbon feedstock is the negative view on using potential sugar crops for food.

4. Conclusion

PHA differs from conventional plastics as it is non-toxic and biodegradable where the end degradation products are mostly carbon dioxide and water. Current PHA production is still hindered by raw material as carbon feedstock and downstream-processing cost. Various waste biomass from agro-industry and municipal wastes have been looked into as cheap alternative carbon source. However, it has been highlighted in this paper that each alternative feedstock has its challenges towards commercialization of PHA. However, sugar based carbon source produced notably higher PHA in term of CDW, compared to other feedstock and not requiring pre-treatment prior to bioconversion of PHA. The production of PHA may turn to be cost efficient when alternative or underutilized waste or crops are used to make non-edible sugars for PHA fermentation. More study on alternative sugar resources are needed to investigate this notion.

5. References

[1] Tan G Y A, Chen C L, Li L, Ge L, Wang L, Razaad I M N, Li Y, Zhao L, Mo Y, Wang J Y 2014 Polymers. 6(3) 706-754

[2] Shen L, Haufe J and Patel M K 2009 *PRO-BIP 2009.* 111-2 Retrieved from http://www.chem.uu.nl/nws/www/research/e&e/PROBIP2009%20Final%20June%202009.pdf

[3] Bugnicourt E, Cinelli P, Lazzeri A and Alvarez V 2014 *Express. Polym. Lett.* **8(11)** 791–808

[4] Yezza A, Halasz A, Levadoux W, Hawari J 2007 *Appl. Microbiol. Biotechnol.* **77(2)** 269–274

[5] Wang B Q, Sharma-Shivappa R R, Olson J W and Khan S A 2013 *Ind. Crops. Prod.* **43** 802–811
[6] Suwannasing W, Imai T and Kaewkannetra P 2015 *Bioresour. Technol.* **194** 67–74
[7] Zahari M A K M, Abdullah S S S, Roslan A M, Ariffin H, Shirai Y and Hassan M A 2014 *J. Clean. Prod.* **65** 252-260
[8] Albuquerque M G E, Eiroa M, Torres C, Nunes B R, Reis M A M 2007 *Biotechnol. J.* **130** 411-421
[9] Solaiman D K Y, Ashby R D, Hotchkiss Jr A T and Foglia T A 2006 *Biotechnol. Lett.* **28** 157-162
[10] Akaraonye E, Moreno C, Knowles JC, Keshavarz T and Roy I 2012 *Biotechnol. J.* **7** 293–303
[11] Solaiman D K Y, Ashby R D, Foglia T A, Marmer W N 2006 *Appl. Microbiol. Biotechnol.* 2006 **71** 783-9.
[12] Obruca S, Marova I, Snajdar O, Mravcova L and Svoboda Z 2010 *Biotechnol. Lett.* **32(12)** 1925-1932
[13] Verlinden R A, Hill D J, Kenward M A, Williams C D, Piotrowska-Seget Z and Radecka I K 2011 *AMB Express.* **1** 11
[14] Ashby R D, Solaiman D K Y and Foglia TA 2004 *J. Polym. Environ.* **12** 105-112
[15] Mozumder M S I, De Wever H, Volcke E I P and Garcia-Gonzalez L 2014 *Process. Biochem.* **49** 365–373.
[16] Koller M, Bona R, Braunegg G, Hermann C, Horvat P, Kroutil M, Martinz J, Neto J, Pereira L, Varila P.2005 *Biomacromolecules.* **6** 561-5
[17] Liu H Y, VanderGheynst J S, Darby J L, Thompson D E, Green P G and Loge F J 2011 *Water. Environ. Res.* **83** 36-43
[18] Chenyu D, Sabirova, J, Soetaert W and Lin CSK 2012 *Curr. Chem. Biol.* **6 (1)** 14-25
[19] Martinez G A , Bertin L, Scoma A, Rebecchi S, Braunegg G and Fava F 2015 *Biochem. Eng. J.* **97** 92–100
[20] Khardenavis A A, Suresh Kumar M, Mudliar S N, Chakrabarti T. 2007 *Bioresour. Technol.* **98(18)** 3579-3584
[21] Cesario M T, Raposo R S, de Almeida M C M D, van Keulen F, Ferreira B S and da Fonseca M M R 2014 *N. Biotechnol.* **31** 104-113
[22] Huang T Y, Duan K J, Huang S Y and Chen C W 2006 *J. Ind. Microbiol. Biotechnol.* **33(8)** 701-6
[23] Yu J and Stahl H 2008 *Bioresour. Technol.* **99** 8042-8
[24] Marang L, Jiang Y, van Loosdrecht MC and Kleerebezem R 2014 *Int. J. Biol. Macromol.* **71** 74-80
[25] Zahari M A K M, Ariffin H, Mokhtar M N, Salihon J, Shirai Y and Hassan M A 2015 *J. Clean. Prod.* **87** 284-290

Acknowledgement
The authors would like to acknowledge Universiti Teknologi Malaysia (UTM) for providing Potential Academic Staff grant (Vote No. 01K52). Additionally, MyBRAIN15 from Ministry of Higher Education Malaysia as well as Malaysia-Japan International Institute of Technology (MJIIT) for the provision of research support, scholarship and financial assistance to the student. Supports of all sources are greatly appreciated.

IOP Conf. Series: Earth and Environmental Science **36** (2016) 012041 doi:10.1088/1755-1315/36/1/012041

Comparative studies on the alcohol types presence in *Gracilaria sp.* and rice fermentation using *Sasad*

R Mansa[1], H Mansuit[1], C S Sipaut[1], C F Yee[3], S M Yasir[2]
[1] Energy and Materials Research Group, Faculty of Engineering, *Universiti Malaysia Sabah*, Jalan UMS 88400, Kota Kinabalu Sabah, Malaysia
[2] Seaweed Research Unit, Faculty of Science and Natural Resources, *Universiti Malaysia Sabah*, Jalan UMS 88400, Kota Kinabalu Sabah, Malaysia
[3] Faculty of Food Science and Nutrition, *Universiti Malaysia Sabah*, Jalan UMS 88400, Kota Kinabalu Sabah, Malaysia

E-mail: [1]rfmansa@ums.edu.my , [2]elvyeva@yahoo.com , [3]css@ums.edu.my, [4]fychye@gmail.com , [5]suhaimi.yasir@gmail.com

Abstract. Alternative fuel sources such as biofuels are needed in order to overcome environmental problem caused by fossil fuel consumption. Currently, most biofuel are produced from land based crops and there is a possibility that marine biomass such as macroalgae can be an alternative source for biofuel production. The carbohydrate in macroalgae can be broken down into simple sugar through thermo-chemical hydrolysis and enzymatic hydrolysis. Dilute-acid hydrolysis was believed to be the most available and affordable method. However, the process may release inhibitors which would affect alcohol yield from fermentation. Thus, this work was aimed at investigating if it is possible to avoid this critical pre-treatment step in macroalgae fermentation process by using *Sasad,* a local Sabahan fermentation agent and to compare the yield with rice wine fermentation. This work hoped to determine and compare the alcohol content from *Gracilaria sp.* and rice fermentation with *Sasad*. Rice fermentation was found containing ethanol and 2 – methyl – 1 – propanol. Fermentation of *Gracilaria sp.* had shown the positive presence of 3 – methyl – 1 – butanol. It was found that *Sasad* can be used as a fermentation agent for bioalcohol production from *Gracilaria sp.* without the need for a pretreatment step. However further investigations are needed to determine if pre-treatment would increase the yield of alcohol.

1. Introduction

Alternative fuel is needed to reduce the dependency on the energy based on fossil fuel. Biofuels, as an alternative fuel are known as "carbon neutral fuels" because carbon dioxide emitted from vehicles is adsorbed by means of photosynthesis during the growth of plant material [10]. Currently, biofuel is mostly produced from land based crops but the near future could see the alternative usage of marine biomass as a source for biofuel production [1]. Macroalgae biomass has been identified as a potential fermentation substrate for third generation biofuel processes due to its high carbohydrate content. The

[1-5] To whom any correspondence should be addressed.

macroalgae criteria of containing no lignin, also another advantage for it to be suitable biofuel candidates as it can be converted into alcohol without any expensive lignin removal process [12].

Current researcher mostly utilized single culture fermentation for macroalgae fermentation. *Saccharomyces cerevisiae* or yeast is the common fermenting agent used in the work, but there are also other attempt was done by using other types of microbe. Researchers have obtained 2.33% of ethanol from the fermentation of *Gelidium elegans* by using yeast strain, *Saccharomyces cerevisiae* IAM 4178 [14]. In addition, researchers also have obtained 86% theoretical yield of ethanol from the fermentation of *Gracilaria verrucosa* using *Saccharomyces cerevisiae* [7]. As for other type of microbes used in fermentation, a research demonstrated the used of recombinant *Escherichia coli* KO11 in *Laminaria japonica* fermentation with ethanol production of 0.40 g ethanol/g of carbohydrate [6]. Other than that, modified *B.custersii* (KCTC18154P) has also been used and achieved higher yield of ethanol in a continuous reactor of 30°C and 39 hours which was 38.2% from 72.2 g/L of glucose and galactose content [12].

Single culture macroalgae fermentation is happening when the microbes consume fermentable sugars in macroalgae to produce bioethanol anaerobically [9]. Bioethanol is basically produced from biomass by fermentation process of available carbohydrates into bioethanol and carbon dioxide as described in equation below [5].

$$C_nH_{2n}O_n(sugar) \xrightarrow{yields} \frac{n}{3}C_2H_5OH + \frac{n}{3}CO_2 + Heat \qquad (1)$$

However, the idea of macroalgae fermentation using traditional fermentation method was still undiscovered. Traditional fermentation method promotes the significance of microbial coexistence in any substrates. A study shows that in traditional fermentation using mixed culture as agent; molds, yeasts, Lactic Acid Bacteria (LAB) and Acetic Acid Bacteria (AAB) can cooperate with each other in any substrates such as rice and dropped ripe fruit hence producing fermented foods [4]. Yeasts and LAB are also noticeable in brewed rice with *Sasad*, a local Sabahan brewing agent [2]. Hence, this unique criterion of *Sasad* may help in macroalgae fermentation for biofuel as it may eliminate the hydrolysis stage which was initially conducted as a saccharification step for macroalgae. Thus, it is a critical step in the fermentation step as it is not a normal substrate for common ethanol producing agent such as *Saccharomyces cerevisiae* to undergo a successful alcohol metabolic pathway [13].

Thus, this research is a work to determine and compare the alcohol types presence produced in *Gracilaria sp.* and rice fermentation using *Sasad* through a traditional fermentation method.

2. Materials and Methods

2.1. Raw materials preparation

Raw materials used in this research are *Gracilaria sp.*, *Sasad* and rice. *Gracilaria sp.* was purchased from *Sg. Petani*, Kedah Malaysia. *Gracilaria sp.* samples were washed and soaked in tap water to remove sand, salts and other foreign matters. Samples were then oven dried at 50°C until constant weight. Dried *Gracilaria sp.* was then size reduced using a kitchen blender, producing smaller particle size of dried seaweed. Dried seaweed was then kept in an airtight container for further usage.

2.2. Starter culture preparation

Sasad, a local Borneo fermenting agent which was originated from *Kg.Bulu Silou,* Keningau Sabah, Malaysia was used as fermenting agent in this research. It was sized reduced using a manual mortar and then sieve to obtain powder. Then it was stored in a dry container and kept in a fridge for further usage.

2.3. Fermentation

Using modified method of *tapai* or rice wine making production as in [2], fermentation of *Gracilaria sp.* and rice were conducted. 300 g of dry *Gracilaria sp.* and rice were cooked using distilled water.

After cooking, the samples were then allowed to cool down to room temperature before they were inoculated with 75 g of *Sasad*. After cooling, *Sasad* was added into the samples by sprinkling all over them and mixed thoroughly using a wooden spoon. Inoculated samples were then transferred into containers. The containers were left open for 1 day before they were sealed. Fermentation was done for 7 days for all samples at room temperature.

2.4. Distillation

The fermented extract was filtered using vacuum filter before distilled so as to separate solid and liquid content. Then, the product was distilled with simple distillation unit as in figure below. Distillate is collected and filtered using syringe filtration unit and kept in the fridge for further analysis.

2.5. Alcohol analysis

Alcohol presence analysis was conducted using GCMS and FTIR analysis. After distillation, samples were added with sodium sulphate anhydrous, $Na_2SO_4.H_2O$ as drying agent to remove water from samples prior for alcohol GC-MS analysis. Samples were syringed through a 0.45 µm Durapore (PVDF) syringe-driven filter unit (Millex-HV, Millipore, Tokyo, Japan) into 0.5 mL glass vials, sealed with a crimp cap and stored at 5°C to 8°C prior to gas chromatography mass spectrophotometry (GCMS) and FTIR analysis. Alcohol analysis was done by using gas chromatography (Agilent Model 6890 N, USA) with a capillary column (30 m L x 0.25 ID x 0.25 µm film thickness). The automatic injection volume was 1 µL with an inlet split ratio of 200:1. The initial and maximum oven temperatures were 40°C and 100°C, respectively.

3. Results and Discussion

3.1. FT-IR analysis

Figure 1 and 2 shows the FT-IR spectrum of the fermentation product. Rice fermentation, T_1 shows a very interesting spectrum as it differs from *Gracilaria sp.* sample. There are four major peaks can be found from rice fermentation FT-IR spectrum as in figure 2 below. The peaks are OH peak (3389.02 cm^{-1}), C – H peak (2979.36 cm^{-1}), C = C peak (1647.26 cm^{-1}) and C – O peak (1045.87 cm^{-1}). Major peaks from FT-IR analysis shows that several functional groups can be found from the sample. In rice fermentation, T_1 peak results, it shows that alcohol group is most probably presence in the sample. This is due to the significant presence of OH and C – H peak from the analyzed sample.

As for *Gracilaria sp.* fermentation, S_1 shows that three major peaks can be found. The peaks are OH peak (3389.0 cm^{-1}), C = C peak (1640.43 cm^{-1}) and C – O – C peak (1095.06 cm^{-1}). Those peaks represent the possibility of hydroxyl groups, alkene group and ether group presence in the sample.

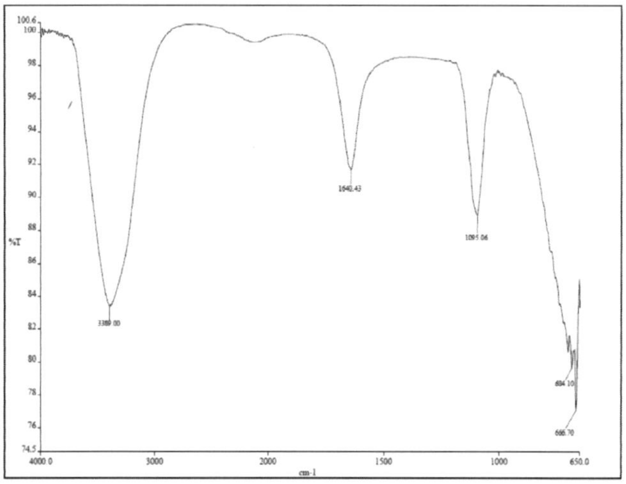

Figure 1. FT-IR Spectrum for *Gracilaria sp.* direct fermentation, S_1.

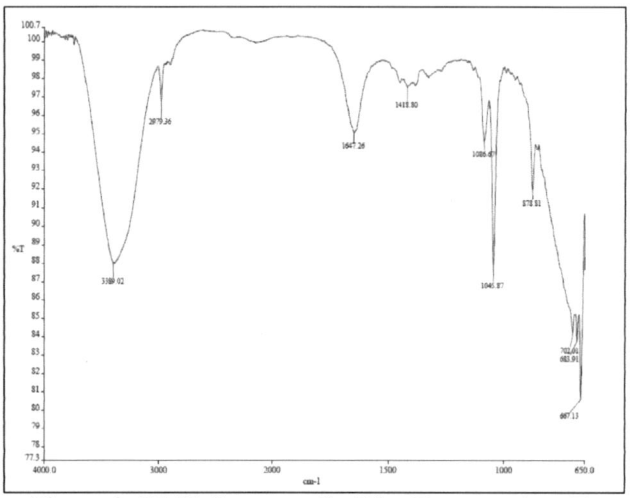

Figure 2. FT-IR Spectrum for Rice Fermentation, T_1

3.2. GC-MS analysis

Figure 3 and 4 shows the GC-MS analysis for all samples. Table 1 shows the compound composition for major component detected from the samples at certain retention time and peak. Based on the result analysis, rice fermentation sample, T_1 shows that there is possibility of ethyl alcohol or ethanol existed in the compound. Other than that, there is high percentage of 1 – propanol 2 – methyl detected which is 87% and 91% at 3.45 minutes and 3.49 minutes respectively. Another component of 1 – butanol 3 – methyl formates might also exist as it was 53% detected at 4.876 minutes. This result can be related to the FT-IR result, as the peak of OH and C–H might exist due to the presence of ethyl alcohol component in the sample. The result is comparable with the finding whereby ethanol and 2 – methyl propanol were produced from the rice fermentation using *Sasad* [2].

Interestingly, *Gracilaria sp.* samples show different compound in the GC – MS analysis result. Sample S_1 represent that there is possibility of 1 – Butanol, 3 – methyl presence in the samples by 86% of detection. The hydroxyl peak presence in the FT-IR spectrum for *Gracilaria sp.* sample might due to the 1 – Butanol, 3 – methyl component which is most possibly presence in the samples. Moreover, there is other component such as 1 – Pentene detected from T_1 samples however it was in a smaller percentage of only 38%. This result however can be related to the C=C peak as detected in the FT-IR result. As for *Gracilaria sp.* sample, 2 – Pentene was 37% detected from S_1 sample which can be related to the FT-IR result.

This research work might be comparable to other research work that utilized mixed culture as bio agent for macroalgae fermentation [8]. Brown seaweed was fermented using mixed culture known as *Nuruk*, which contain yeast, bacteria and fungi and yield ethanol as the major product produced. *Sasad*, on the other hand which holds same characteristic as *Nuruk* however yield 1 – butanol, 3 - methyl as major product. This component usually identified in small quantities in yeast fermentation however research works demonstrated that 1 – butanol, 3 – methyl can also be produced by other microbes such as *E. coli* through conserved and native amino acid biosynthesis pathways [3]. Hence, there is possibility for microorganisms in *Sasad* may help in the production of bioalcohol through macroalgae fermentation.

Table 1. Compound composition from GC-MS analysis

Sample	Retention Time (min)	Main Compound	Kovats Index (KI)	Percentage (%)
T_1	3.144	Ethyl Alcohol	46	50
	3. 453	1 – Propanol, 2 – Methyl	74	87
	3.494	1 – Propanol, 2 – Methyl	74	91
	4.876	1 – Butanol, 3 – methyl	116	53
S_1	3.868	1 – Butanol, 3 – methyl	88	86

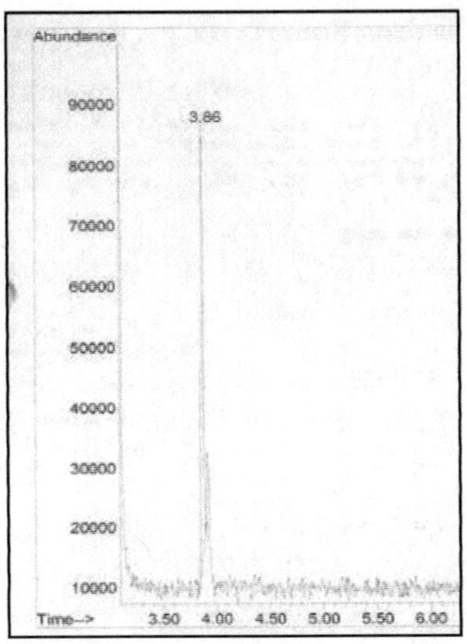

Figure 3. GC-MS Spectrum *Gracilaria sp.* direct fermentation, S_1.

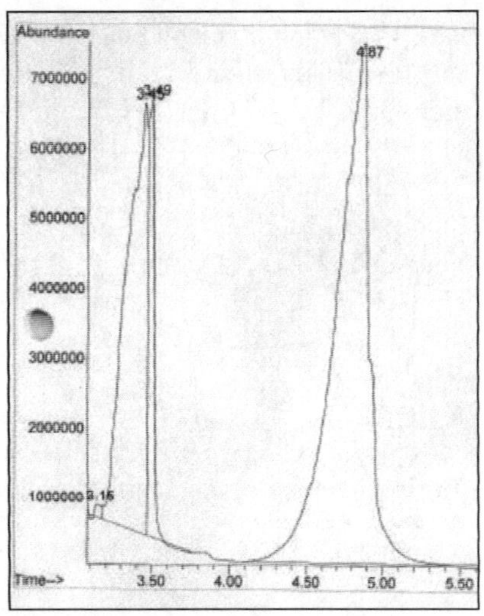

Figure 4. GC-MS Spectrum for Rice Fermentation, T_1.

4. Conclusion

Results demonstrated that the direct fermentation of rice and seaweed shows different type of alcohol yield. Rice fermentation product analyzed shows two types of alcohol which are ethanol and propanol. While, traditional fermentation of *Gracilaria sp.* with *Sasad* had shown the presence of butanol. On the basis of results, *Sasad* is feasible to be utilized as fermentation agent of *Gracilaria sp.* for bioalcohol production. In conclusion, this research work has demonstrated that there is possibility in fermenting macroalgae biomass using other fermentation agent instead of using the common fermentation agent, *Saccharomyces cerevisiae* or baker's yeast. The procedure that was developed and used for the bioalcohol production however requires further development, specifically in the pre-treatment and fermentation part.

5. References

[1] Adams J M, Gallagher, J a, and Donnison I S 2009 *J. App. Phycol.* **21** 5 569–574. doi:10.1007/s10811-008-9384-7
[2] Chiang Y, Chye F and Ismail A M 2006 **2** 1 1–6. Retrieved from http://www.usm.my/mjm/issues/vol2no1/research1.pdf
[3] Connor M R and Liao J C 2008 *Appl. Environ. Microbiol.* **74** 18 5769 - 5775.
[4] Furukawa S, Watanabe T, Toyama Hand Morinaga, Y 2013 *J. Biosci. Bioeng.*, **116** 5 533–539. doi:10.1016/j.jbiosc.2013.05.017
[5] Harun R, Danquah M K. and Forde G M 2010 *J. Chem. Technol. Biotechnol.*, **85**2 199–203. doi:10.1002/jctb.2287
[6] Kim N, Li H, Jung K, Chang H N, Lee P C, Nam H and Cheon P 2011 *Bioresour. Technol.* **102** 16 7466–9. doi:10.1016/j.biortech.2011.04.071
[7] Kumar S, Gupta R, Kumar G, Sahoo D and Kuhad R C 2013 *Bioresour. Technol.* **135** 150–6. doi:10.1016/j.biortech.2012.10.120
[8] Lee S M and Lee J H 2011. *Bioresour. Technol.* **102** 10 5962–7. doi:10.1016/j.biortech.2011.02.006
[9] Mansa R F, Mansuit H, Fong K F, Sipaut C S, Chye F Y and Yasir S M 2013 355–363. doi:10.1007/978-1-4614-6208-8
[10] Oguma M, Lee Y J and Goto S 2012 *Int. J. Automo. Ener.* **13** 1 33–41. doi:10.1007/s12239
[11] Park J H, Hong J, Jang H C, Oh S G, Kim S, Yoon J, Jin Y 2012 *Bioresour. Technol.*, **108** 83–8. doi:10.1016/j.biortech.2011.12.065
[12] Park J H, Yoon J, Park H D, Jin Y, Jung D, Kim S, Lim D J 2011 *Int. J. Hydrogen Ener.* **36** 21 13997–14003. doi:10.1016/j.ijhydene.2011.04.003
[13] Setyaningsih D, Windarwati S, Khayati I, Muna N, and Hernowo, P 2012 *Int. J. Environ. Bioener* **3** 2 98–110.
[14] Yanagisawa M, Nakamura K, Ariga, O, and Nakasaki K 2011 *Process Biochemistry*, **46** 11 2111–2116. doi:10.1016/j.procbio.2011.08.001

Acknowledgments

Authors would like to extend their appreciation and gratitude to Ministry of Agriculture Malaysia for Seaweed Research Unit Grant GPRL013 and Universiti Malaysia Sabah (Grant No: SBK0235 – TK – 2015) for the financial support of the research work.

Hydrothermal pre-treatment of oil palm empty fruit bunch into fermentable sugars

M D Muhd Ali[1], P Tamunaidu[1*], A K H Nor Aslan[1], N A Morad[1], N Sugiura[2], M Goto[2] and Z Zhang[3]
[1]Shizen Conversion & Separation Technology (SHIZEN) iKohza, Malaysia-Japan International Institute of Technology (MJIIT), Universiti Teknologi Malaysia (UTM), Malaysia.
[2]Ecological Engineering iKohza, Malaysia-Japan International Institute of Technology (MJIIT), Universiti Teknologi Malaysia (UTM), Malaysia.
[3]Graduate School of Life and Environmental Sciences, University of Tsukuba, Japan.

E-mail: [1]pramila@utm.my

Abstract. Presently oil palm empty fruit bunch (OPEFB) is one of the solid waste which is produced daily whereby it is usually left at plantation site to act as organic fertilizer for the plants to ensure the sustainability of fresh fruit bunch. The major drawback in biomass conversion technology is the difficulty of degrading the material in a short period of time. A pre-treatment step is required to break the lignocellulosic biomass to easily accessible carbon sources for further use in the production of fuels and fine chemicals. Therefore, this study investigated the effect of hydrothermal pre-treatment under different reaction temperatures (100 - 250°C), reaction time (10 - 40 min), solid to solvent ratio of (1:10 - 1:20 w/v) and particle size (0.15 – 1.00 mm) on the solubilization of OPEFB to produce soluble fermentable sugars. The maximum soluble sugars of 68.18 mg glucose per gram of OPEFB were achieved at 175°C of reaction temperature, 20 min of reaction time, 1:15 w/v of solid to solvent ratio for 0.30 mm of particle size. Results suggest that reaction temperature, reaction time, the amount of solid to solvent ratio and size of the particle are crucial parameters for hydrothermal pre-treatment, in achieving a high yield of soluble fermentable sugars.

1. Introduction

The palm oil industry has brought excessive economic benefits and act as biggest exchange earner to Malaysia till today. Now, Malaysia accounts for 39% of world palm oil production and 44% of world exports [1]. Consequently, the waste biomass generated from this industry is also high. There are six types of biomass produced from this industry include oil palm trunks (OPT), oil palm fronds (OPF), empty fruit bunches (OPEFB) and palm pressed fibers (PPF), palm kernel shells (PKS) and palm oil mill effluent palm (POME).

Hence, oil palm wastes have enormous potential as sustainable biomass resource for the production of alternative fuel and valuable chemicals. Out of these waste biomass, OPEFB is the most

[1] To whom any correspondence should be addressed.

underutilized biomass that is generated almost every day in palm oil mills as it is obtained after stripping off the fresh fruit bunch. The amount of OPEFB was estimated about 22.43 MnT in 2014 [2]. Compared to shells and fibers, OPEFB is not incinerated in mills to produce energy. Several plants using OPEFB have experienced low electricity production, high wear and tear (fouling, sintering, and corrosion) and incomplete combustion [3]. Therefore, it is generally left aside to naturally decompose or used minimally for mulching. At times they are burnt to avoid storage. Therefore, large amounts of OPEFB are still available in the mills for further conversion.

The major drawback in biomass technology is the difficulty of degrading the material in a short period of time. A pre-treatment step is required to break the lignocellulosic compound to easily accessible carbon sources for further use in the production of fuels and fine chemicals. Pre-treatment is a significant stage of fuels and chemical production from lignocellulosic biomass. The purpose of pre-treatment is to alter or remove the hemicelluloses and lignin in order to increase the pore volume and internal surface area as well as decrease the degree of polymerization and crystallinity of the cellulose [4]. Pre-treatment must meet the following requirements: (1) to improve the formation of sugars or the ability to subsequently produce sugars by enzymatic hydrolysis; (2) to avoid the degradation or loss of carbohydrates; (3) to prevent the formation of by-products which are inhibitory to the subsequent enzymatic hydrolysis and fermentation processes; and (4) to be cost-effective.

One renowned green pre-treatment method applied for conversion of lignocellulosic biomass is hydrothermal pre-treatment. The method of hydrothermal pre-treatment is popular due to the usage of water as a solvent and low operation cost. This method has the following advantages, non-chemicals addition and able to open up the plant cell wall structure, and hence, efficient solubilization and degradation of lignocellulosic biomass [5,6]. This method can be a prospective alternative for the pre-treatment of OPEFB resulting in the efficient hydrolysis of the organic matter contained to obtain valuable products for future energy needs.

In this study, the solubilization of fermentable sugars was carried out at reaction temperatures from 100°C to 250°C and reaction time from 10 to 40 min for several of solid to solvent ratio from 1:10 w/v to 1:20 w/v. At optimum conditions, the effect of particle size from 0.15 mm to 1.00 mm was studied. The effects of these parameters towards the solubilization of fermentable sugars from OPEFB were mainly discussed.

2. Material and methods

2.1. Preparation of Sample

The raw materials for this study, which is mill-processed OPEFB was collected from Palm Oil Mill Seri Ulu Langat, Dengkil, Selangor, Malaysia (GPS Coordinates: 02.8507°N, 101.5501°E). It was obtained after oil palm fruits are removed from the fresh fruit bunch during the milling process (threshing). After the threshing process, the OPEFB was shredded into loose fibers using a shredder. The shredded OPEFB was further milled using Crusher (SY-50, China) at Wood Composite Laboratory, Forest Research Institute of Malaysia (FRIM), Kepong, Selangor, Malaysia (GPS Coordinates: 3.2419°N, 101.6370°E) and sieved between 0.15 and 0.50 mm using Laboratory Test Sieve (Endecotts Ltd., UK). Then, it was dried at 60°C overnight and kept in an air-tight container till their prior use. All chemicals and standards used in this study (D(+)-Glucose and acetic acid) was purchased from Wako, Japan.

2.2. Hydrothermal Pre-treatment

The study on the pre-treatment process of OPEFB was conducted using a hot-compressed water system. The shredded OPEFB were placed inside an enclosed stainless steel reactor (OM LabTech Co., Ltd, Japan) with a working volume of 200 ml. The steam pressure inside the reactor is dependent on the temperatures being set. The reactor was loaded with a mixture of 2.5 g of dried sample and distilled water to obtain 1:10, 1:15 and 1:20 w/v of solid to solvent ratio. The reaction temperatures were valued from 100 to 250°C for 10 to 40 min of reaction time to determine optimum soluble sugar content. For these experiments, 0.30 mm of particle size was used. After achieving optimum conditions, the optimum particle size was determined.

2.2. Analytical Methods
Total sugar in soluble products (hydrolysate) was determined by phenol-sulphuric method with some modification [7]. The hydrolysate (0.1 ml) was mixed with 1.9 ml of distilled water, 1 ml of 5 % w/v phenol and 5 ml of concentrated sulphuric acid and was vortex for 30 s. After incubation at room temperature for 10 min, the absorbance of the reaction mixture was measured at 490 nm against distilled water as blank using UV-1800 Spectrophotometer (Shimadzu, Japan). Glucose was selected as a standard considering the major component of biomass. The pH of the liquid portion was determined using a digital pH meter (Mettler Toledo, USA). The concentration of acetic acid was also quantified using a gas chromatograph (GC-8A, Shimadzu, Japan) coupled with a flame ionization detector (FID) and a Unisole F-200 30/60 column (3.0 mm in diameter and 2.0 m in length). 1 µl of sample was injected with nitrogen gas, N_2 as carrier. The injector, detector and column temperatures were kept at 180°C, 180°C and 150°C, respectively.

3. Results and discussion
The preliminary study was done to determine the optimum reaction temperature. For this purpose, the reaction time of 20 minutes, 1:15 solid to solvent ratio and 0.30 mm particles were used. The profile of total soluble sugar obtained is shown in figure 1. It can be seen that amount of total sugar compounds increased gradually with increasing temperature from 100°C to 175°C at 20 minutes reaction time achieving the optimum value of 68.18 mg glucose per gram of OPEFB. The subcritical reactions are recognized to increase the ionization constant of water at high temperature, in which the concentrations of H+ and OH− are increased, thereby assisting the hydrolysis of cellulose and hemicellulose into monosaccharides as mentioned by Lin et al. [8]. The increase in concentration of ions, not only helps in destroying the glycosidic bonds in the cellulose but also provides a shorter time for the conversion of oligosaccharides to monomers [9]. However, further increase in temperature to 250°C reduced the total sugar yield to less than 10 mg glucose per gram of OPEFB.

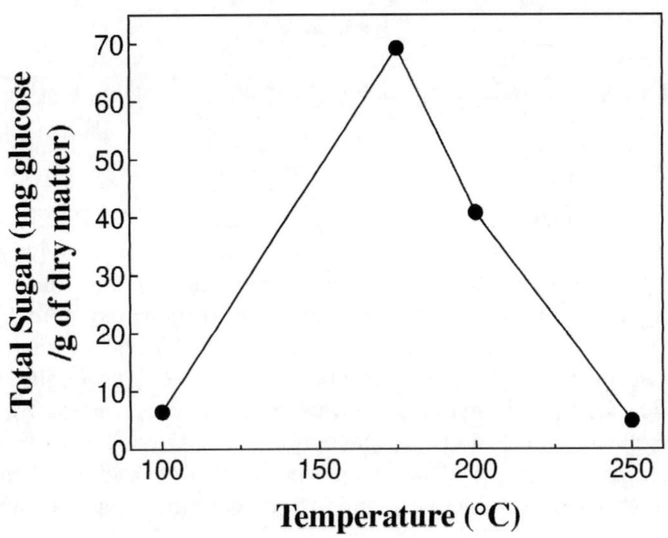

Figure 1. Total sugars at various reaction temperatures for 20 min of reaction time and 1:15 w/v of solid to solvent ratio

Increase in reaction temperature showed a decrease in total soluble sugar yield. This suggests that the sugars are further converted or decomposed at higher temperatures to organic acids, 5-HMF, furfural and other decomposed compounds. The samples were tested and acetic acid was found as a dominant component as shown in figure 2. The decreased total sugar yield after 175°C was attributed to the enhanced degradation of the monosaccharides into by-products and proved by the pH of the liquid as seen in figure 2. The hydrolysate turned from neutral into acidic toward higher reaction temperature treatment due to the accumulation of organic acids. The pH value at optimum reaction temperature (175°C) was 4 and it decreased significantly as temperature increased. This phenomenon can be correlated with the increase in acetic acid concentration. The highest amount of acetic acid was 72.78 mg/ g of dry matter at 250°C. Acetic acid is a well-known valuable by-product from the degradation of complex hemicellulose compound. At elevated temperature will bring about lower pH in solvent, enabling the release of O-acetyl, acetic and uronic acids from hemicellulose [10]. However, this compound is known as a toxic compound that could inhibit the microbial growth in subsequent biological conversion process as stated by Chong *et al.* [11].

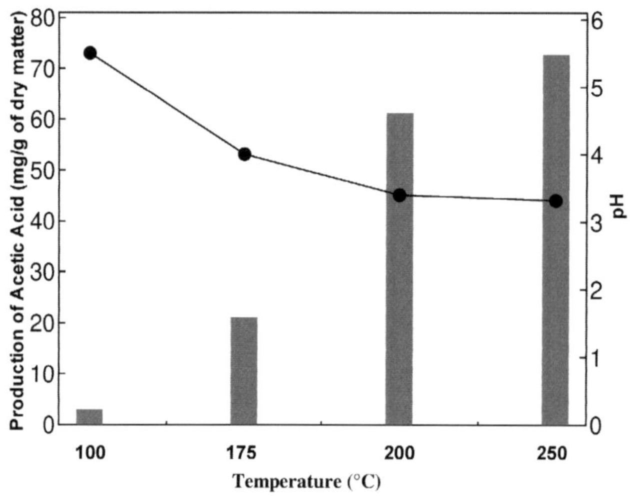

Figure 2. Acetic acid production and corresponding pH at various reaction temperatures for 20 min of reaction time

To further verify the optimum reaction temperature, a ±5°C from 175°C was studied. For this study, the hydrolysis time was varied between 10 to 40 min and the solid to solvent ratio of 1:10, 1:15 and 1:20 was used to simultaneously study the effect of reaction time and solid to solvent ratio. The results as shown in figure 3 showed that the total sugar yields increased from 170°C to 175°C and dropped at 180°C for all reaction time and solid to solvent ratio. The highest total sugar obtained after the hydrothermal pre-treatment at various parameters remained at 68.18 mg glucose for 1:15 solid to solvent ratio at 20 min, and then decreased somewhat by increasing reaction time. The trends were similar at all solid to solvent ratio, establishing the optimum conditions at 175°C and 20 min reaction time. However, the significant sugar yield was achieved at 1:15 w/v solid to solvent ratio as shown in figure 3 (B). This result proved that amount of solid to solvent ratio is one of important parameter to be evaluated.

However, as a whole, the results showed that the degradation of lignocellulosic material through hydrothermal pre-treatment is fast and relatively efficient to produce fermentable sugars. As to compare with common hydrolysis method (acid hydrolysis), it requires longer reaction time between one to two hours to degrade the lignocellulosic material [12,13]. Zakaria and co-workers [6] also reported finding in hydrolysis of oil palm empty fruit bunch under batch water treatment at sub-critical temperature and pressure. It was found that, the maximum total sugars 60.1 mg per g of OPEFB were obtained at 210°C in a batch treatment at reaction time of 20 min.

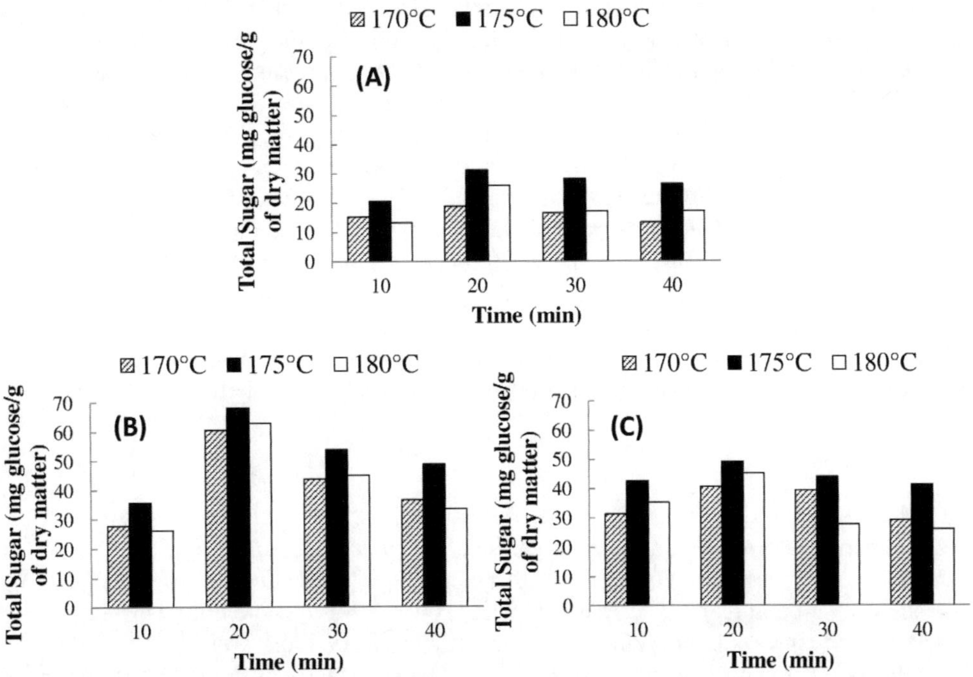

Figure 3. Effect of total sugars at different solid to solvent ratio; (A) 1:10 w/v, (B) 1:15 w/v and (C) 1:20 w/v for reaction temperatures of 170°C, 175°C and 180°C and reaction times from 10 to 40 mins

Generally, a high solid to solvent ratio could promote an increasing concentration gradient, resulting in an increase of diffusion rate that allows greater extraction of solids by solvent [14]. Therefore, at optimum conditions of 175°C of reaction temperature, 20 min reaction time and 1:15 solid to solvent ratio, the size of particle was varied and investigated. The values of solubilized total sugar, ranging from particle size 0.15 mm to 1.00 mm are presented in figure 4.

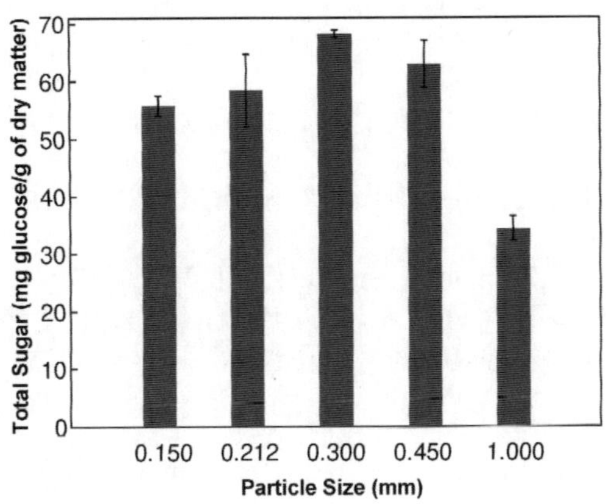

Figure 4. Total sugars obtained at various particle sizes for 1:15 w/v of solid to solvent ratio, 20 min of reaction time and 175°C of reaction temperature

It was found that the highest solubilized total sugar was 68.18 mg glucose/g of dry matter at 0.30 mm of particle size. Meanwhile, at the finest particle size of 0.15 mm, the solubilized total sugar was found as 55.74 mg glucose/g of dry matter. Theoretically, smaller particle size should give more yield because of the higher surface area to volume ratio which enhances the contact between the solvent molecules and plant material during the extraction process [15]. However, this finding was different from the theory as observed, that when the particle size was too fine, the sample in the reactor was compacted and agglomerated when in contact with the solvent at high pressure and temperature.

4. Conclusion

Hydrolysis of OPEFB substrates showed that the hydrothermal pre-treatment method provided an accelerated pre-hydrolysis of OPEFB during the treatment process. The maximum total soluble sugars of 68.18 mg glucose per gram of OPEFB were achieved at 175°C of reaction temperature, 1:15 w/v of solid to solvent ratio and 0.30 mm of particle size at 20 min of reaction time. A significant drop of the soluble sugars in the OPEFB hydrolysate was due to the generation of acetic acid. Results suggest that reaction temperature, time, solid to solvent ratio and particle size are crucial parameters for hydrothermal pre-treatment, in order to achieve a high yield of soluble fermentable sugars. The soluble sugars obtained can be used as feedstock for conversion to biofuels and fine chemicals.

5. References

[1] MPOC (Malaysia Palm Oil Council) 2014 Retrieved from: http://www.mpoc.org.my/The_Oil_Palm_Tree.aspx. Accessed on 2 June 2015
[2] Malaysia Industry-Government Group for High Technology (MIGHT) (2013). Malaysian Biomass Industry Action Plan 2020: Driving SMEs Towards Sustainable Future. Malaysia Industry-Government Group for High Technology, Malaysia Available from: http://www.biomass.org.my/uploads/editor/file/Jan%202015/Msia%20Biomass%20Industry%20Action%20Plan%202020.pdf. Accessed on 18 Jan 2016
[3] Stam A, Livingston W, Cremers M and Brem G 2010 Review article for IEA 1–18
[4] Kumar P, Barrett D M, Delwiche M J and Stroeve P 2009 Ind. Eng. Chem. Res. 48(8) 3713–29
[5] He L, Huang H, Lei Z, Liu C and Zhang Z 2015 Curr. Org. Chem. 19 437–46
[6] Zakaria M R, Hirata S and Hassan M A 2015 Bioresour. Technol. 176 142–8
[7] He L, Huang H, Lei Z, Liu C and Zhang Z 2014 Bioresour. Technol. 171 145–51
[8] Lin R, Cheng J, Ding L, Song W, Qi F, Zhou J and Cen K 2015 Bioresour. Technol. 186 8–14
[9] Rubio M, Tortosa J F, Quesada, J. and Gomez, D. 1998 Biomass Bioenerg. 15 (6) 483–91
[10] Negro M J, Manzanares P, Ballesteros I, Oliva J, Cabaas A and Ballesteros M 2003 Appl. Biochem. Biotechnol. 105-108(1) 87-100
[11] Chong P S, Jahim J M, Harun S, Lim S S, Mutalib S A, Hassan O and Nor M T M 2013 Int. J. Hydrogen Energ. 38 9592–9
[12] Wijaya Y P, Putra R D D, Widyaya V T, Ha J-M, Suh D J and Kim C S 2014 Bioresource Technol. 164 221-31
[13] Rahman S H A, Choudhury J P, Ahmad A L and Kamaruddin A H 2007 Bioresour. Technol. 98(3) 554-9
[14] Tan P W, Tan C P and Ho C W. 2011 Int. Food Res. J. 18(2) 557–62
[15] Yunus M A C and Hasan M 2013 60(1) 21–25 Retrieved from http:// www. jurnalteknologi. utm.my/index.php/jurnalteknologi/article/view/1436 Accessed on 15 May 2015

Acknowledgements

This research was supported by Potential Academic Staff Grant (Vot. No. 01K52) by Universiti Teknologi Malaysia (UTM) and Ministry of Higher Education Malaysia. Furthermore, the financial support for student living and short term exchange program to Japan was provided by Malaysia-Japan

International Institute of Technology (MJIIT) and Japan Student Services Organization (JASSO). Financial support from all these sources are greatly appreciated.

Correlation study of microalgae carbonation in membrane integrated photobioreactor

E Suali[1], R Sarbatly, S R M Shaleh[2], F A Lahin, S M Anisuzzaman
[1]Faculty of Engineering, Universiti Malaysia Sabah, Jalan UMS 88400 Kota Kinabalu, Sabah, Malaysia.
[2]Borneo Marine Research Institute, Universiti Malaysia Sabah, Jalan UMS, 88400 Kota Kinabalu Sabah, Malaysia.

E-mail: [1]emma.suali@gmail.com, [1]emma.suali@ums.edu.my

Abstract. Microalgae ability to utilise CO_2 higher compared to terrestrial plant making it suitable for biomass production and as CO_2 utiliser. This could be one of many ways to preserve a safer and healthier environment with less air pollutant. For study purposes, CO_2 usually transported to microalgae culture broth with the aid of membrane technology to prevent formation of large bubble and to accelerate the carbonation of microalgal media. However, membrane susceptible to accumulation of CO_2, which can cause extreme acidic to microalgal media. This prevents microalgae to assimilate CO_2. Thus, this study proposes correlations to prevent the extreme acidic: which represents the relationship of: (1) CO_2 inlet and accumulation, (2) CO_2 inlet and CO_2 at the membrane-liquid interphase and (3) CO_2 inlet and CO_2 solubility in the media. The correlations were successfully validated with a deviation of less than 20% compared to the theoretical value.

1. Introduction

The membrane technology has received considerable attention as one of the methods to deliver CO_2 into bioreactor for microalgae cultivation [1,2]. The integration of membrane with bioreactor was reported capable of increasing the microalgal biomass production [3,4].The biomass are desired to be processed into biofuel, pharmaceutics product and etc [5].

Another purpose of membrane integration is to increase the CO_2 capture and utilisation by microalgae, which currently under investigation [6]. However, the membrane integration also susceptible to accumulation of CO_2 due to fibre microporous which causes resistance between the gas-membrane, membrane-liquid and gas-liquid interphase in the membrane contactor [7-9]. The accumulation of CO_2 causes extreme acidic to the microalgae culture when pumped out at once off the membrane due to the natural conversion process of CO_2 into carbonate, bicarbonates and acid carbonic [10]. This affects the microalgae growth rate and causes lethal to some microalgae species which have low tolerance of low media pH [11].

The effect of accumulation within the membrane can be avoided by supplying CO_2 at a low concentration but probably will result in insufficient CO_2 for dense microalgae cultured in the bioreactor. Thus, supplying at optimum concentration is the most appropriate way to overcome the problem of accumulated CO_2 in the membrane. However, supplying the optimum CO_2 concentration requires massive experimental works and yet it depends on the membrane type. One of many ways to predict the accumulation is by correlating the CO_2 inlet (feed) with CO_2 solubility (DCO_2). A correlation to predict the relationship of CO_2 inlet, accumulation and DCO_2 with respect to membrane properties is the most suitable method to overcome the accumulation and CO_2 excess. With the correlation, optimum CO_2 concentration can be supplied accordingly. Thus, this work aims to develop the said correlation.

2. Materials and methods

2.1 Membrane photobioreactor system

The setup of the membrane photobioreactor (PBR) is shown in figure 1.

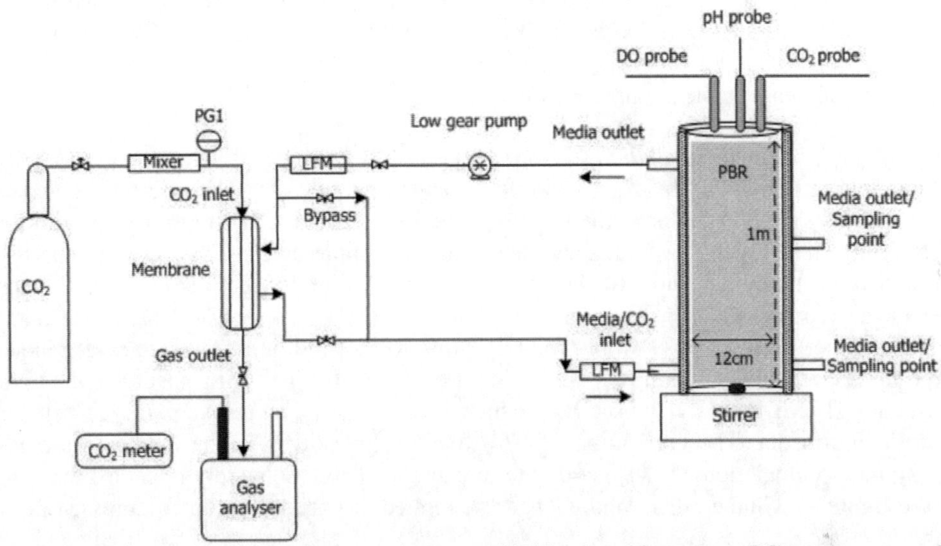

Figure 1. Experimental setup of membrane integrated with photobioreactor, PG: pressure gas, PBR: photobioreactor, LFM: liquid flow rate metre

The membrane that was used to validate the correlation is GFD hydrophobic UFS220 which was made up of 2400 hollow fibre made from polyethylene. The characteristics of the membrane and PBR are listed in table 1.

Table 1. Characteristics of the membrane and PBR.

Features	Membrane	PBR*
Number of fibres	2400	Not applicable
Fibre pore size (μm)	0.5	Not applicable
Length (mm)	495	1000
Total surface area (m^2)	0.80	0.4
Shell internal/PBR diameter (mm)	60	80
Capacity (m^3/s) x 10^{-5}	2.7	Not applicable
Membrane/acrylic thickness (mm)	0.65	3
Shell side/PBR volume (m^3) x 10^{-3}	0.43	23
Volume occupies by hollow fibre (m^3) x 10^{-3}	2.5	Not applicable

The outlet gas of the membrane can be controlled and closed when necessary, according to experimental purpose thus making this membrane physically suitable for the carbonation in microalgal media beside low cost compared to other available commercial membrane in the market. The PBR was operated as a closed system in a controlled temperature at 27°C. The CO_2 and temperature probes and gas analyser were attached in the PBR. The PBR is covered with aluminium foil to avoid any photosynthetic activity that could change the results during the experimental work.

2.2 Microalgal media

This study employs *Chlorella* sp. which was isolated from Tun Fuad Stephens Lake that located at 6°N and 116°E. A modified Jaworski's medium which is lacking in carbon source was used as a medium for the microalgae. The media contained of 0.0124 g/l KH_2PO_4, 0.05 g/l MgS $O_4.H_2O$, 0.00225 g/l EDTAFeNa, 0.00225 g/l $EDTANa_2$, 0.00248 g/l H_3BO_3, 0.00139 g/l $MnCl_2.4H_2O$, 0.001 g/l $(NH_4)6Mo7O_{24}.4H_2O$, 0.00004 g/l Cyanocobalamin, 0.00004 g/l Thiamine HCl, 0.00004 g/l Biotin, 0.08 g/l $NaNO_3$ and 0.036 g/l $Na_2HPO_4.12H_2O$. The media and microalgae were mixed in water which was used as liquid phase in the membrane contactor.

2.3 Evaluation of proposed correlation

The correlation was validated by comparison with experimental data. The microalgae density of the media is 0.0401 kg/m^3. This value is the most suitable for optimum CO_2 solubility and resulted in low risk of fouling. The CO_2 that accumulates inside the membrane and DCO_2 were measured based on CO_2 mass balance that enter and exit the membrane which directly recorded by the CO_2 probe that attached in the bioreactor. The CO_2 probe is connected to CO_2 metre CGP-31. Note that, the CO_2 metre does not measure the total DIC but total CO_2 that soluble in liquid. Thus, to determine carbonate and carbonic acid, the pH solution was taken simultaneously. The pH is a measure of the total ratio of hydrogen ion (H^+) to hydroxyl ions (OH^-) which effective way to measure the carbonic acid and carbonate in the media. The correlation of CO_2 feed and dissolves in the media was measured by subtracting the accumulation of CO_2 inside the membrane. The CO_2 supply is controlled using a CO_2 mixer (see figure 1). On the other hand, CO_2 was supplied in purified grade compressed in 30kg CO_2 cylinder.

3. Theory and calculation

The correlation of CO_2 feed and accumulation with CO_2 that would soluble in a liquid phase in the membrane contactor were developed by first correlating the CO_2 feed with CO_2 concentration that accumulated in the membrane. This follows by correlating CO_2 feed with CO_2 at membrane interphase and maximum solubility in liquid.

3.1 Correlating CO_2 feed with accumulation in the membrane

The inlet and outlet flow rate of CO_2 can be correlated based on mass balance that enter and exit the membrane. The mass flow rate is equivalent with volumetric flow rate multiply with concentration or constant volume multiply with a concentration rate ($\dot{m}_{CO_2} = \dot{V} C_{CO_2}$ or $\dot{m}_{CO_2} = V \dot{C}_{CO_2}$). Thus,

$$\frac{dC_{CO_2(A)}V_m}{dt} = \left(\dot{V}_{(gi)}C_{CO_2(gi)} + \dot{V}_{(Li)}C_{CO_2(Li)}\right) - \frac{dC_{CO_2(Lf)}V_{(Lf)}}{dt} - \frac{dC_{CO_2(gf)}V_{(gf)}}{dt} \tag{1}$$

whereas, $C_{CO_2(A)}$, $C_{CO_2(gi)}$, $C_{CO_2(Li)}$, $C_{CO_2(Lf)}$ and $C_{CO_2(gf)}$ are CO_2 concentration that accumulated in the membrane, at gas inlet, at liquid inlet, at liquid outlet and at gas outlet, respectively. The membrane volume on the other hand is represented as V_m. The V_{AG} and V_{AS} are constant volumetric space that can be occupied by gas phase and liquid phase in the membrane, respectively. The symbol of d/dt and $\dot{V}_{gi}$ shows the non-constant and constant variable (volumetric flow rate) variable that changes with time, respectively. Rearranging,

$$V_{AG}\int\frac{dC_{CO_2(AG)}}{dt}+V_{AS}\int\frac{dC_{CO_2(AS)}}{dt}+V_{(Lf)}\int\frac{dC_{CO_2(Lf)}}{dt}+V_{(gf)}\int\frac{dC_{CO_2(gf)}}{dt}=x1+y1 \tag{2}$$

$\left(\dot{V}_{gi}C_{CO_2(gi)}\right)$ and $\left(\dot{V}_{Li}C_{CO_2(Li)}\right)$ are known constant values and symbolized as $x1$ and $y1$, respectively. In this work, the inlet liquid is microalgal media and was assumed has no dissolved CO_2, thus $\dot{V}_{Li}C_{CO_2(Li)}=y1=0$. The feed CO_2 is forced to exit membrane as dissolved CO_2. Thus, there is no CO_2 outlet in the gas phase ($\frac{dC_{CO_2(gf)}V_{(gf)}}{dt}=0$) and equation (2) becomes equation (3).

$$\frac{dC_{CO_2(A)}V_m}{dt}=\left(x1\right)-\frac{dC_{CO_2(Lf)}V_{Lf}}{dt} \tag{3}$$

The V_{Lf} on the other hand depends on the inlet volumetric flow rate. Since the inlet volumetric flow rate, V_{Li} was constant, the V_{Lf} can be considered as constant as well. Note that, $\dot{V}_{Lf}$ is the volume flow rate of outlet liquid obtained from ($\dot{V}_{Lf}$ × time taken × a). The $\dot{V}_{gi}$ is the volume flow rate of inlet gas respectively. The outlet liquid V_{Lf} lowers compared to V_{Li} about a times due to pressure drop in the liquid side. The a is a constant value found from the comparison of inlet and outlet liquid flow rate in the membrane. The V_{Lf} depends on V_{gi} and membrane conditions. The membrane volumetric space is constant. Thus, integration of equation (3) becomes equation (4).

$$V_m\int dC_{CO_2(A)}+V_{Lf}\int dC_{CO_2(Lf)}=x1\int dt \tag{4}$$

The $\int_{c_0}^{c} dC_{CO_2(A)}=V_m C_{CO_2(A)}-0$ and $\int_0^t dt=t$. Thus, $V_m(C_{CO_2(A)})+V_{Lf}(C_{CO_2(Lf)})=(\dot{V}_{gi}C_{CO_2(gi)})t$. The Henry law stated that the CO_2 that dissolves in liquid can be measured based on: $C_{CO_2(gi)}H=C_{CO_2(Lf)}$. Thus, the equation (1) becomes as in equation (5).

$$C_{CO_2(A)}=\frac{(C_{CO_2(gi)})\sqrt{(\dot{V}_{gi}t-V_{Lf}H)^2}}{V_m} \tag{5}$$

The CO_2 feed into membrane photobioreactor follows the ideal gas theory. Thus, the accumulation also can be predicted based on gas molecular weight (M), flow rate and pressure as follows:

$$C_{CO_2(A)}=\frac{PM_{CO_2}}{RT}\frac{\sqrt{(\dot{V}_{gi}t-V_{Lf}H)^2}}{V_m}\approx\rho_{CO_2}\frac{\sqrt{(\dot{V}_{gi}t-V_{Lf}H)^2}}{V_m} \tag{6}$$

Thus, the equation (5) and equation (6) can be used to predict the accumulation of CO_2 in the membrane.

3.2 Correlating CO_2 concentration at the membrane-liquid interphase

The CO_2 at the membrane–liquid interphase can be defined in term of CO_2 transfer. Figure 2 illustrates the gas flux through the pore of single hollow fibre. The CO_2 flux ($J_{CO_2(m)}$) through the membrane pore is theoretically affected by varying concentration and working pressure as illustrated in figure 2 (a) [12,13]. The increase in pressure drop is one of the main reasons for the flow rate declines [14]. Figure 2 (a) shows the illustration of flows within the membrane. The higher the pressure in gas side, the slower the flow in liquid side of the membrane. This phenomenon can be explained in term of resistance in the gas side of the membrane. In addition, the driving force of the outlet liquid depends on the pressure gradient between the gas and liquid side [15]. The transport phenomena as introduced by Fick which known as Fick's law can be presented as: $J_{CO_2(m)} = -D(dC/dx)$ and ideal gas theory [16,17]. The symbol of D represents as a diffusion coefficient, dC is change in CO_2 concentration and dx represents the distance of the CO_2 travels within the membrane. This equivalent to the mass flow rate of gas in kg/s unit that diffused through the membrane pores $\left(d\dot{m}_{CO_2(m)}\right)$ divide with membrane-pore surface area $\left(A_{pore}\right)$ in m^2 unit and can be presented as follows:

$$J_{CO_2(m)} = -D\left(\frac{dC}{dx}\right) = -\left(\frac{d\dot{m}}{A_{pore}}\right) \quad (7)$$

whereas $d\dot{m} = \dot{m}_{(CO_2)i} - \dot{m}_{(CO_2)m}$. The $\dot{m}_{CO_2(i)}$ and $\dot{m}_{CO_2(m)}$ is the initial mass flow rate of CO_2 in the gas side and CO_2 that diffused or transported throughout the membrane pore that in contact with the liquid side, respectively (see isometric view in figure 2 (b)).

The diffusion flux of gas through hollow fibre membrane has a travel path, dx equivalent to hollow fibre membrane thickness, δ_m (illustrated in figure 2). The rationale of this assumption is CO_2 diffuses out of the hollow fibre through the pore wall which made up the membrane. The concentration gradient of CO_2 on the other hand can be measured as $dC_{CO_2} = C_{CO_2(i)} - C_{CO_2(m)}$, $C_{CO_2(i)}$ is CO_2 initial concentration in kg/m^3 that equivalent to the CO_2 supply from external sources such as $C_{CO_2(gi)}$. The $C_{CO_2(m)}$ on the other hand is the amount of CO_2 that in contact with the liquid boundary but not dissolve. The mass flow rate can be presented as follows: $\dot{m}_{CO_2(i)} - \dot{m}_{CO_2(m)} = D\ ((C_{CO_2(i)} - C_{CO_2(m)})/\delta_m)\ A_{pore}$. Finally, the CO_2 concentration at interphase that diffused through the membrane can be measured as follows.

$$C_{CO_2(m)} = C_{CO_2(i)}\left(\frac{DA_{pore} - \dot{V}_i\delta_m}{DA_{pore} - \dot{V}_m\delta_m}\right) \quad (8)$$

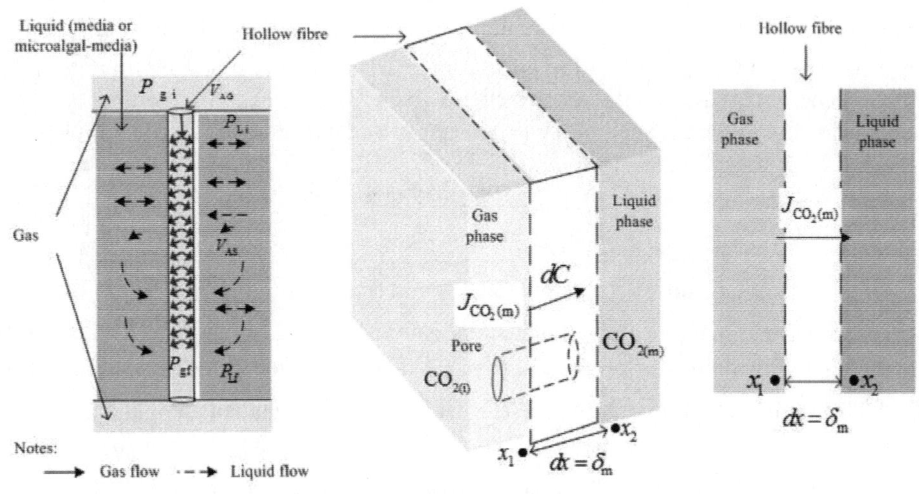

(a) (b) (c)

Figure 2. (a) Illustration of gas flow in a single hollow fibre inside the membrane. (b) Illustration of diffusion flux through a hollow fibre pore and (c) Front view of illustration as shown in (b). V_{Ag} :Volume of gas phase in membrane (m^3), V_{AS} :Volume of liquid phase in membrane (m^3), P_{gi} :Inlet gas pressure (bar), P_{gf} :Inlet gas pressure (bar), P_{Li} :Inlet pressure of liquid side (bar), P_{Lf} :Outlet pressure of liquid side (bar). $dx=\delta_m$: membrane thickness, dC : CO_2 concentration gradient, $J_{CO_2(m)}$: CO_2 diffusion flux through the membrane pore, x_1 :first point, x_2 : second point.

The critical part of this correlation is the diffusion coefficient, D which depends on pore wall, porosity and tortuosity. Thus, the equation proposed by Mason and Malinauskas [18] that relates diffusion coefficient of gas in general condition with diffusion coefficient of gas through a membrane pore, $D=D_e(\varepsilon/\tau)$ should be used. For the selected membrane, the hollow fibre average pore diameter is 0.05 µm. The closed up images of the hollow fibre membrane within the membrane module are shown in figure 3.

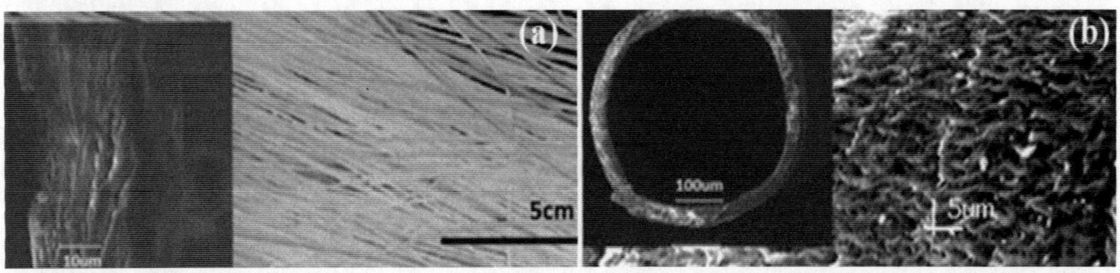

Figure 3. A closed up image of the hollow fibre membrane. (a): Hollow fibres and SEM image of the hollow fibre, (b): SEM image of single hollow fibre and top view of middle-cut hollow fibre

The pore of the tubular shaped-hollow fibre is varied in sphere-like pore (figure 3 (b)). The smaller image in figure 3 (a) is an SEM image of single hollow fibre and in figure 3 (b) is a top view of single hollow fibre. When the pore size of the membrane is smaller compared to mean free path (λ) and the Knudsen number ($K_n=\lambda/d_{pore}$) is larger than one, the CO_2 collides with the pore wall during the diffusion process [19,20].The λ is measured using correlation proposed by Cunningham and Williams

[21].The mean free path λ is 0.06 μm and average K_n is 1.35. Thus, the diffusion coefficient follows the Knudsen theory [22]. The diffusion through the membrane pore is more appropriate in term of effective diffusion coefficient as its diffusion affected by pore size, porosity and tortuosity [23]. Thus, the effective diffusion coefficient in membrane is rearranged and expressed as in equation (9).

$$D_e = \frac{D_{MCO_2} D_{kCO_2}}{D_{MCO_2} + D_{kCO_2}} \tag{9}$$

whereas, D_{kCO_2} represents the Knudsen diffusion and D_{MCO_2} represents the molecular diffusion of CO_2, which in this case was technically described mathematically using Fick's law. The Knudsen diffusion coefficient is related to kinetic theory of gas and expressed as $D_k = u(d_{pore}/3)$ whereas u is molecule velocity. The molecule velocity can be related to an ideal gas theory and expresses as $D_k = (d_{pore}/3)\sqrt{(8RT/\pi M)}$. There are many proposed correlations for Knudsen diffusion coefficient. However, in this work, the correlation that proposed by Cunningham and Williams (1980): $D_{kCO_2} = 4850 d_{pore}\left(\sqrt{T/M_{CO_2}}\right)$ was used. The Knudsen diffusion coefficient for CO_2 is 6.33 x 10^{-2} cm^2/s.

3.3 Overall mass transfer coefficient of CO_2

The CO_2 is in a gas phase within the membrane pore. Thus, the standard molecular diffusion coefficient or simply the diffusion coefficient of CO_2 in the air is 1.39 × 10^{-1} cm^2/s. The overall effective diffusion coefficient is 4.35 x 10^{-2} cm^2/s, porosity is 0.8 and tortuosity is 1.8. The porosity and tortuosity were calculated based on the characteristic of the membrane shown in table 1. The values were used for further calculation of CO_2 at interphase that diffused through the membrane pore, $C_{CO_2(m)}$. The mass transfer resistance within the membrane pore is almost negligible if no media was introduced into the membrane. This resulted in $\dot{V}_{(CO_2)i} \approx \dot{V}_{(CO_2)m}$ and can be proved using equation (8) whereas, the inlet CO_2 supply has a similar concentration with CO_2 at intervals between membrane and liquid.

3.4 Prediction of maximum CO_2 that would dissolve in the media

The CO_2 enters at the top of hollow fibre and forced to exit through the pore of membrane wall which in contact with the liquid phase. Note that, the amount of CO_2 that has transferred into liquid phase is equal or lower compared to gas diffused through the membrane pore $\left(\frac{dC_{CO_2(Lf)}}{dt} \le \frac{dC_{CO_2(m)}}{dt}\right)$. Thus, the CO_2 that would dissolve in liquid phase can be predicted from the amount of CO_2 that diffuses throughout the membrane. In overall, the CO_2 that could dissolve in a liquid is higher compared to solubility without the aid of membrane [1]. The dissolved gas in liquid can be defined in term of diffused CO_2 through the membrane and can be reformulated according to Henry's law: $C_{CO_2(Lf)} = HC_{CO_2(m)}$ whereas $C_{CO_2(m)}$ is concentration of CO_2 diffuses through the membrane pores. The CO_2 that dissolves into media phase can be simply rewritten as in equation (10) or equation (11) regardless of liquid flow rate.

$$C_{CO_2(Lf)} = HC_{CO_2(i)}\left(\frac{DA_{pore} - \dot{V}_{(CO_2)i}\delta_m}{DA_{pore} - \dot{V}_{(CO_2)m}\delta_m}\right) \approx \text{Maximum } C_{CO_2(m)} \tag{10}$$

$$C_{CO_2(Lf)} = H\left(\frac{m_{CO_2(m)}}{t}\left(\frac{\delta_m}{D_m A_{pore}}\right) + C_{CO_2(i)}\right) \approx \text{Maximum } C_{CO_2(m)} \tag{11}$$

The DCO_2 in media is a major factor of CO_2 uptake by microalgae as CO_2 can only be consumed in the media by microalgae. Thus, the prediction of CO_2 that would dissolve in the media is important for CO_2 uptake by microalgae.

4. Results and discussion

4.1 Correlation of CO_2 inlet and accumulation in membrane

The governed correlation (equation (5)) was evaluated by comparison with experimental work. Thus, the effects of operating period and gas flow rate of CO_2 accumulation within the hydrophobic membrane were selected for the comparison. Figure 4 (a) shows the comparison of theory and experimental work on CO_2 accumulates with respect to operating period. The CO_2 inlet ($\dot{V}_{gi}$) is 3.3 x 10^{-6} m^3/s, liquid inlet ($\dot{V}_{Li}$) is 3.3 x 10^{-5} m^3/s, CO_2 concentration ($C_{CO_2(gi)}$) is 1.8 kg/m^3 and the membrane capacity (V_m) is $0.00105m^3$. The experimental work was conducted in multiple times to constant value and three sets of the experiments are represented as points in figure 4 (a). The theoretical correlation on the other hand is represents the line.

The standard deviation between the experimental data was varied from 0.002 kg/m^3 to 0.025 kg/m^3. This resulted in %SD from 5% to 17%. The percentage error between the experiments and theory was less than 16%. The confidence interval (CI) was in the range of -0.018 and +0.196 which, was calculated with 80% of confidence level. This value indicates that there is 80% certainty that the real value of the accumulated CO_2 is not more than 0.196 kg/m^3 and not less than 0.018 kg/m^3 for an operating period of 5s to 55s. For instance, when the carbonation process took about 5s of the operating period, at $\dot{V}_{gi}$: 3.3 x 10^{-6} m^3/s, $\dot{V}_{Li}$: 3.3 x 10^{-5} m^3/s, CO_2 inlet concentration:1.8 kg/m^3 and membrane capacity is 0.00105 m^3, the accumulation will not less than 0.047 kg/m^3 and not more than 0.052 kg/m^3.

The percentage error which defined as the difference between the theory and experiment value in this work is in the range of ±10% to ±16%. The governed correlation is less than 20%. It is agreeable that less than 20% of theory and experiment different is acceptable to validate the correlation [24]. Since, this correlation was governed to predict the accumulation of CO_2 in the membrane so that, right CO_2 feed can be supplied accordingly without causing lethal to microalgae in a bioreactor. Thus, the 20% difference of the real accumulated value within the membrane can be considered as safe to the microalgae. The experimental data shows that suitable working period will result in less than 50% accumulation rate.

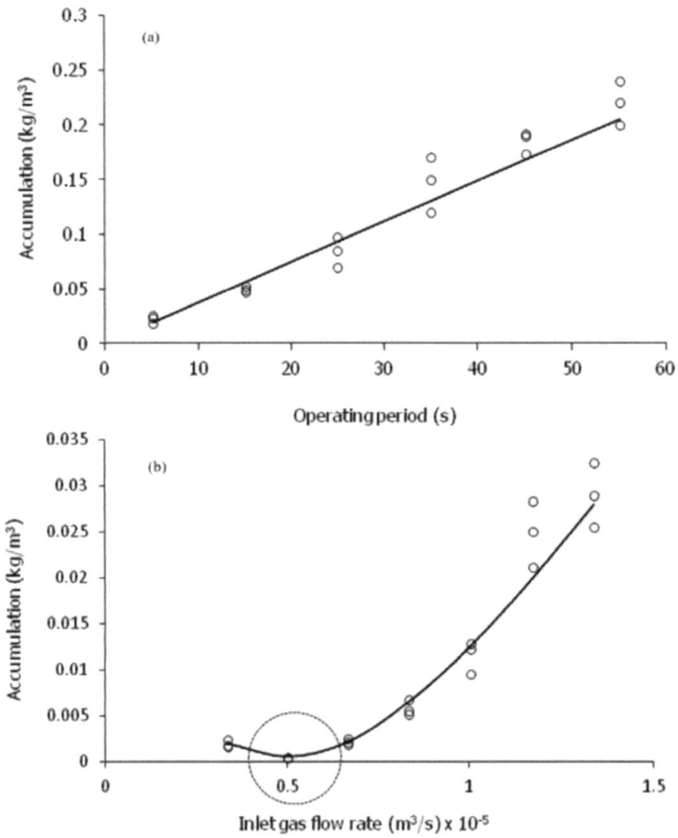

Figure 4. (a) Accumulation of CO_2 with respect to operating period; (b) accumulation of CO_2 with respect to gas flow rate.

The governed correlation also supported by the comparison on the accumulation as a function of gas flow rate when operated at constant liquid inlet and membrane volumetric capacity. The experiment was conducted at $V_m = 1.05 \times 10^{-3} m^3$, $\dot{V}_{gi}$ is in the range of 3.34 x 10^{-5} m^3/s to 1.3 x 10^{-5} m^3/s and $\dot{V}_{Li}$: 3.3 x 10^{-5} m^3/s, with an operating period equivalent to 48s. The result was compared and shown in figure 4 (b). The experimental work was conducted in multiple times to achieve a constant value and three sets of the experiments represented as points in figure 4 (b). The line on the other hand is represented as the theoretical value. The %SD between the data is varied from 12% to 19%. The percentage difference of the experimental data was in the range of 2% to 19% compared to theoretical values. Notice that, the lowest accumulation is at 0.5 x 10^{-5} m^3/s as circled in figure 4 (b). This indicates that at $\dot{V}_{Li}$: 3.3 x 10^{-5} m^3/s, the most suitable CO_2 gas flow rate is 0.5 x 10^{-5} m^3/s.

4.2 Limitation of the correlation of CO_2 inlet and accumulation

The governed correlation was validated based on the operating period and the gas flow rate. This parameter was selected as it has a major impact on the accumulation of CO_2 in the membrane. However, this correlation was governed without taking consideration on fouling that might occur in the membrane due to microalgae cell, in which the fouling reduce the flow rate at the liquid side.

The governed correlation has considered a function of $V_{Lf} = aV_{Li}$ with a as 0.2. This function is limited to $V_{Lf} = aV_{Li}$ that was analysed for V_{Li} lower than 5 x 10^{-5} m^3/s and V_{gi} is lower than 1.67 x 10^{-5} m^3/s as shown in figure 5 (a). Notice that, $a = 0.2$ or $a = 20\%$ is an average of reduction of flow rate as indicates from 15% to 23% in figure 5 (a). Thus, ones have to measure the function of liquid inlet and outlet before using the correlation with regards to microalgae concentration in the media. This correlation can be improved by adding a function of fouling and microalgae density. However, this work requires lots of work and interesting be explored in the future.

The accumulation of CO_2 within the membrane also can be considered as loss for large scale cultivation of microalgae. As discussed in a previous study, 25% of biomass production cost of microalgae is resulted during the carbonation process [25]. In reality, the process to extract pure CO_2 from industry for cultivation of microalgae is costly [26]. Thus, the lower the accumulation rate, the higher the mass transfer of CO_2 into liquid phase. Thus, the requirement of CO_2 to produce an equivalent - amount of biomass is lower. Other factors that affect the solubility of CO_2 include porosity and membrane pores were not included because this variable has more effect on the solubility instead of accumulation.

4.3 Validation of correlation of maximum CO_2 that would dissolve in the media

The correlation in equation (10) and equation (11) were validated as shown in figure 5 (b). The experiment was conducted with CO_2 inlet ($\dot{V}_{gi}$): 3.3 x 10^{-6} m^3/s, liquid inlet ($\dot{V}_{Li}$): 3.3 x 10^{-5} m^3/s, CO_2 concentration ($C_{CO_2(gi)}$): 1.8 kg/m^3 and membrane capacity is equivalent to 0.00105m^3. The theoretical value in figure 5 (b) is indicated as a linear line and the experimental data are indicated as points. With 80% of confidence level, the probability of the true experimental value lies within 0.017 kg/m^3 to 0.057 kg/m^3. For instance, at 0.0203 kg/m^3 of CO_2 inlet, the true value of CO_2 that would dissolve into the liquid phase is not less than 0.017 kg/m^3 and not more than 0.019 kg/m^3. The percentage standard deviation between the final three sets data is 2% to 11%. The different of theory and experiment which is defined as percentage error is 4% to 20%. This value shows that the experimental result and theoretical are within the acceptable range for validation of the governed correlation. Thus, the proposed correlation in equation (10) and equation (11) are acceptable to predict the relationship of CO_2 that would dissolve in the liquid phase when using a hollow fibre membrane contactor. The volumetric flow rate of inlet gas and volumetric flow rate of CO_2 in membrane pore affects the CO_2 solubility. The result can be explained in term of resistance at interphase between the membrane and liquid phase. Based on comparison between inlet gas and inlet media, the inlet media flow rate affected by the volumetric flow rate of gas inlet and vice versa. This creates instability in DCO_2 thus causes the experimental value shown a slight exponential trend at high CO_2 gas inlet.

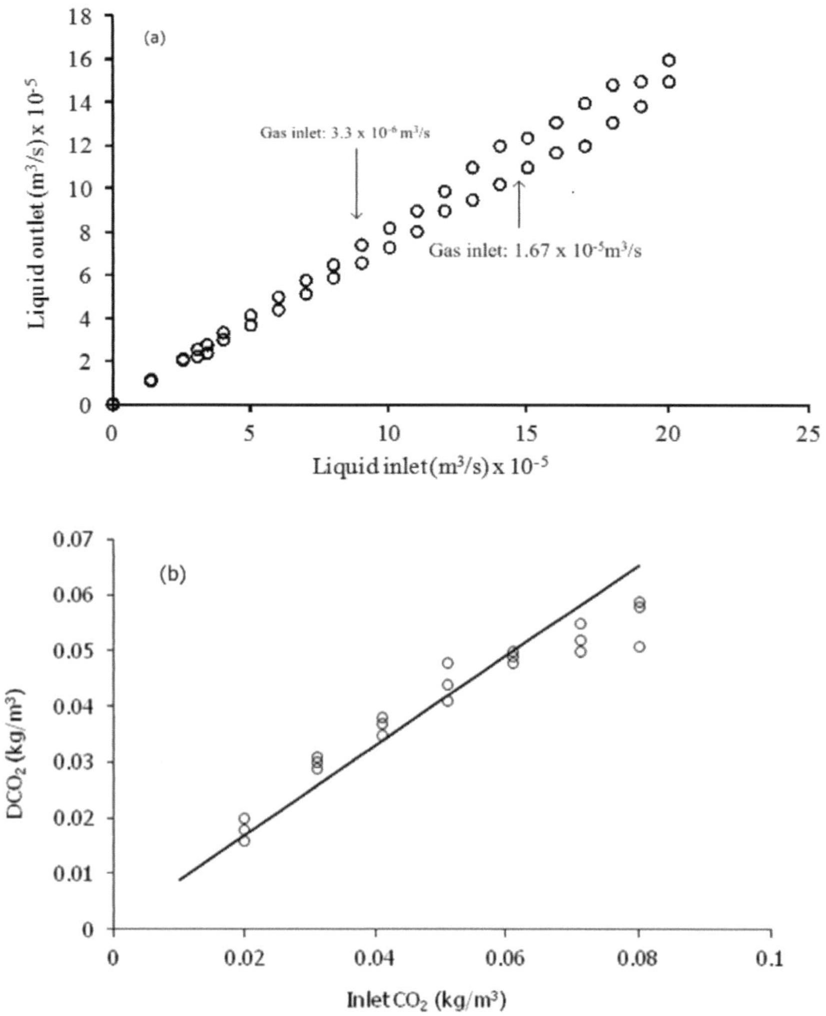

Figure 5. (a) Comparison of liquid inlet and outlet flow rate, (b) Comparison value of DCO_2 between experiment and theory.

4.4 Limitation of the governed correlation

The correlation was validated solely based on Henry constant of CO_2 gas, CO_2 inlet concentration and membrane properties such as thickness and tortuosity. This correlation is used to predict the CO_2 that will pass through the membrane wall from the gas side of the boundary that in contact with liquid side in membrane contactor. The rational for the governed correlation is the CO_2 concentration that would dissolve into media ($C_{CO_2(Lf)}$) must be equal or smaller compared to CO_2 amount that disused through the membrane ($C_{CO_2(m)}$). However, during the validation work, the amount of CO_2 that would dissolve was assumed in its maximum value ($C_{CO_2(Lf)} = C_{CO_2(m)}$) without taking consideration on the resistance on liquid phase. The liquid phase of the membrane is microalgal media with microalgae concentration approximately 0.0401kg/m^3. Thus, the correlation in equation (10) and equation (11) still can be improved with the addition of function of liquid resistance at different microalgae density. However, the proposed correlation is useful to note the maximum CO_2 that would dissolve in the media regarding on CO_2 inlet concentration. The study on the function of liquid resistance is interesting for future work.

The correlation that was proposed and validated can be summarised and presented as in figure 6. Notice that, the correlations takes consideration mainly on the hollow fibre thickness. The reason for this consideration is most of the membrane resistance in transferring gas to liquid phase is from the interaction of the gas molecule with the membrane wall [27]. The media properties also has become one of the major considerations during the governing process of the correlation which in this study, the properties of microalgal media. This includes the concentration of microalgae in the liquid phase and the flow rate of the liquid phase that passing through the membrane to the bioreactor.

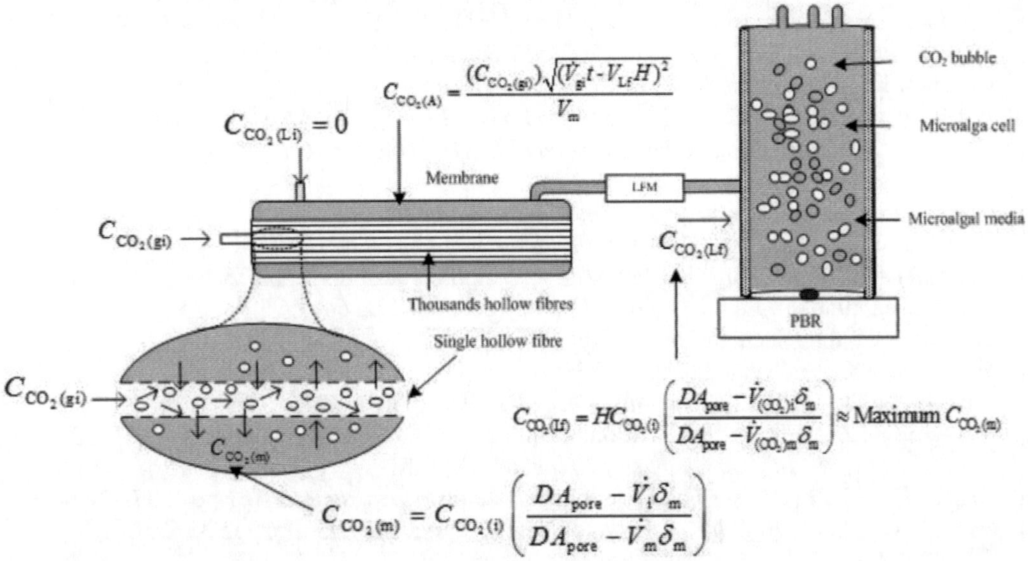

Figure 6. Compilation of proposed correlation of CO_2 feed, accumulation and solubility in microalgal media from the membrane contactor to bioreactor system.

The correlation is important to feed CO_2 into microalgal cultivation system. As discussed in previous studies, microalgae have high ability to utilise CO_2. Thus, from this correlation, optimum utilisation can be achieved and at the same time produces biomass for various purposes [28].

5. Conclusions

It was found that the diffusion coefficient of the CO_2 into the membrane pore has high impact on the gas transfer from the gas to liquid phase in the membrane. The standard deviation and the percentage error of the governed correlation compared to the experimental work is less than 20%. This value is acceptable to predict the DCO_2 by taking consideration on circumstances and equipments that were used to validate the correlation. However, the correlation was only evaluated based on microalgal media at 0.0401 kg/m^3. Other experimental series is required before the correlation can be applied to other media besides microalgal media.

6. References

[1] Carvalho AP, Malcata FX 2001 *Biotechnol Prog.* **17** pp 265–272.
[2] Cheng L, Zhang L, Chen H, Gao C 2006 *Sep Purif Technol.* **50** pp 324–329.
[3] Illman A M, Scragg A H, Shales S W 2001 *Enzyme Microb Tech.* **27** pp 631-635.
[4] Chiu S Y, Kao C Y, Tsai MT, Ong S C, Chen C H, Lin C S 2009 *Bioresour Technol.* **100** pp 833-838.
[5] Raja R, Hemaiswarya S, Rengasamy R *2007Appl Microb Biotechnol.* **74** pp 517-523.
[6] Bayless D, Stuart B 2009 *Energeia* **20** pp 1-6.

[7] Lashkari S, Kruczek B 2010 *J Membr Sci.* **360** pp 442-453.
[8] Kreulen H, Smolders C A, Versteeg G F, Van Swaaij W P M 1993 *J Membr Sci.* **78** pp 197-216.
[9] Mavroudi M, Kaldis S P, Sakellaropoulos G P A 2006 *J Membr Sci.* **272** pp 103-115.
[10] Marshall W J, Bangert S K 2008 5th ed (Edinburg UK: Elsevier Health Sciences) pp 416.
[11] Goswami R C D, Kalita N, Kalita M C 2012 *Ann Biol Res.* **3** pp 499-510.
[12] Metz S J 2003 Thesis (The Netherlands: University of Twente) pp 133.
[13] Abedini R, Nezhadmoghadam A 2010 *Pet Coal.* **52** pp 69-80.
[14] Dindore V Y, Brilman D W F, Feron P H M, Versteeg G F 2004 *J Membr Sci.* **235** pp 99-109.
[15] Stanojevic M, Lazarevic B, Radic D 2003 *FME Transactions.* **31** pp 91-98.
[16] Lebowitz J L, Spohn H 1982 *J Stat Phys.* **28** pp 539-556.
[17] Paradisi P, Cesari R, Mainardi F, Tampieri F 2001 *Phys A: Stat Mech App.* **293** pp 130-142.
[18] Mason E A, Malinauskas A P 1983 *Chemical Engineering Monograph* Vol **17** (New York: Elsevier) pp 194.
[19] Chen X, Pfender E 1983 *Plasma Chem Plasma* P **3** pp 97-113.
[20] Barber R W, Emerson D R 2002 *Adv Fluid Mech.* **32** pp 207-216.
[21] Cunningham R E, Williams R J 1980 (New York: Plenum Press) pp 275.
[22] Knudsen M 1950 3rd Ed (New York: J. Wiley and Sons) pp 64.
[23] Grunwald P 1989 Biochem Educ. **17** pp 99-102.
[24] Ewins D J 2000 Sadhana. 25 pp 221-234.
[25] Suali E, Sarbatly R 2012 *Renew Sust Energ Rev.* **16** pp 4316-4342.
[26] Dubey M K, Ziock H, Rueff G, Elliott S, Smith W S, Lackner K S and Johnston N A 2012 *ACS Division of Fuel Chemistry Preprints.* **47** pp 81-84.
[27] Keshavarz P, Fathikalajahi J and Ayatollahi S 2008 *J Hazard mater* **152** pp 1237-1247.
[28] Sarbatly R, Suali E 2012 *Int J Of Global Environ Issues.* **12(2/3/4)** pp 150-160.

Acknowledgement
This study was supported by FRG0416-TK-1/2015.

Investigation on the quality of bio-oil produced through fast pyrolysis of biomass-polymer waste mixture

S A Jourabchi[1*], H K Ng[2], S Gan[2], Z Y Yap[1]
[1] Faculty of Science, Technology, Engineering and Mathematics, INTI International University, Persiaran Perdana BBN, Putra Nilai, 71800 Nilai, Negeri Sembilan, Malaysia
[2] Faculty of Engineering, University of Nottingham Malaysia Campus, Jalan Broga, 43500 Semenyih, Selangor, Malaysia

E-mail: *amir.jourabchi@newinti.edu.my

Abstract. A high-impact poly-styrene (HIPS) was mixed with dried and ground coconut shell (CS) at equal weight percentage. Fast pyrolysis was carried out on the mixture in a fixed bed reactor over a temperature range of 573 K to 1073 K, and a nitrogen (N_2) linear velocity range of $7.8x10^{-5}$ m/s to $6.7x10^{-2}$ m/s to produce bio-oil. Heat transfer and fluid dynamics of the pyrolysis process inside the reactor was visualised by using Computational Fluid Dynamics (CFD). The CFD modelling was validated by experimental results and they both indicated that at temperature of 923 K and N_2 linear velocity of $7.8x10^{-5}$ m/s, the maximum bio-oil yield of 52.02 wt% is achieved.

1. Introduction

Although the major source of global energy today is fossil fuels, based on its finite amounts, fluctuating prices, associated emission of carbon dioxide (CO_2) to the atmosphere, new energy sources especially renewable ones are being investigated by scientists [1]. The emphasis on renewable sources is ever more important considering the growing worldwide population and hence energy demand. Biomass is one of the important energy sources as it can be directly converted into liquid biofuel through pyrolysis, and can reduce CO_2 footprint. The liquid fuel out of pyrolysis is called bio-oil, and it has not only the highest energy concentration compared to other pyrolysis by-products, but also it is more user friendly in terms of transportation, storage and applications [2–3].

Increasing usage of plastic materials in all industries has becomes a concern since plastic waste mainly stay in the environment forever. Reusing it as a fuel can help in reducing waste as well as producing new energy source. Thus, the feedstock of interest in this project is a mixture of biomass (coconut shell, CS) and plastic (high-impact poly-styrene, HIPS).

The effect of pyrolysis parameters such as reaction temperature and N_2 velocity on final products can be investigated through experiments, but it is important to visualise the dynamics and heat transfer processes so that further improvement can be implemented to the process. The heat transfer and fluid

[1] To whom any correspondence should be addressed.

dynamics play important roles in pyrolysis as the parameters will be affected directly when the biomass-plastic waste is gaining heat inside the reactor. One of the tools is numerical approach such as computational fluid dynamics (CFD) modelling that can simulate the dynamics of the flow inside of the reactor while similar experimental input parameters are being used. Till now, to the best knowledge of the authors, by using CFD modelling, the effect of wall conduction of a fixed bed reactor tube [4], the convective heat transfer at low and high pressure in a fixed bed reactor [5] and the fluid dynamics as well as the heat and momentum transfer during pyrolysis in a fluidised bed [6] have been investigated. In contrast, this study focuses on CFD modelling for a rapid heating fast pyrolysis fixed bed reactor to investigate the effect of reaction temperature and N_2 velocity on the yield of bio-oil produced from a mixture of coconut shell and HIPS. The results were validated by experimental data taken from the existing bio-oil production rig that was modelled.

2. Methodology

2.1. Experimental

Coconut Shell (CS) powder was produced by using the method detailed in [7]. CS powder and recycled HIPS that was supplied by Wespack Waste Management Sdn Bhd in Malaysia and was ground into granules (<2 mm) using a Retsch SM100 Comfort grinder machine were mixed with equal weight. This mixture was pyrolysed in the rapid heating fast pyrolysis rig described in [7], over the same range of reaction temperature and N_2 velocity matrix that was used in [8].

2.2. Computational Fluid Dynamics

Simulations were run by Academic CFD software package, ANSYS version 15.0 on a desktop computer that is configured with Intel® CoreTM i7-3770 @ 3.40 GHz Processor with 8 GB RAM. The pressure was assumed as atmospheric pressure and the pyrolysed materials were assumed to be spherical wood and polystyrene from the material list available in the software. The heating process was assumed to be in steady state condition, and radiation was neglected due to stainless steel wall of the reactor. The geometry for the present work was created by Design-Modeller with a three-dimensional domain for CFD simulation. The fixed bed reactor was modelled as a cylindrical hollow tube and the spherical pyrolysed material positioned in the middle section of the reactor according to the parameters listed in Table 1. The mesh cells for this geometry domain were generated by using ANSYS-Meshing in tetrahedral format. Three boundary surfaces which are inlet, reactor wall and outlet were defined and the details of all the boundary conditions are tabulated in Table 2.

Table 1. Parameters of the geometry.

Parameter	Unit	Quantity
Inner diameter of the reactor	mm	48.7
Outer diameter of the reactor	mm	50.7
Total length of the reactor	mm	200.0
Radius of each biomass powder	mm	2.0

Table 2. Details of boundary conditions.

Boundary zone	Pressure (kPa)	Velocity (m/s)	Temperature (K)
Inlet	101.33	5 combinations[a]	298
Outlet	101.33	Not fixed	Not fixed
Wall	101.33	0	5 combinations[a]

[a] Same points as mentioned in [8]

After confirming by manual calculation of the Reynolds number, laminar flow modelling was considered. For the solver, pressure-based solver was selected by the SIMPLE algorithm for pressure-velocity coupling and the Second Order Upwind for momentum and energy equations. The solutions were implemented 25 times for 5 different N_2 linear velocities and 5 different temperatures based on experiments that were conducted in [8].

3. Results and discussion

From the CFD results, it was found out that the surface temperature of the pyrolysed grains (which is visualised over the 10 cm heating zone of axial length of the reactor and shown in Figure 1 for wall temperature of 1073 K and N_2 velocity of 6.7×10^{-2} m/s) decreases linearly with increase in the distance of the grains from the reactor wall. This characteristic of the temperature distribution is due to the one direction laminar flow of N_2 that mentioned in [9], which is dominated by direct conduction of the heat. This phenomenon can be described based on the existence of conduction heat transfer from inner wall of the reactor to the grains and from each grain to the next neighbour grain toward the centre. As maximum temperature is at the wall, the surface temperature of the grains, which are closer to the wall are higher than grains closer to the centre line of the cylinder as can be seen in Figure 1. With the existence of temperature gradient between materials, the heat transfer will then occur as the higher temperature from the inner reactor wall transferred heat to the lower temperature of grains until equilibrium is reached and steady state condition of heat transfer is reached.

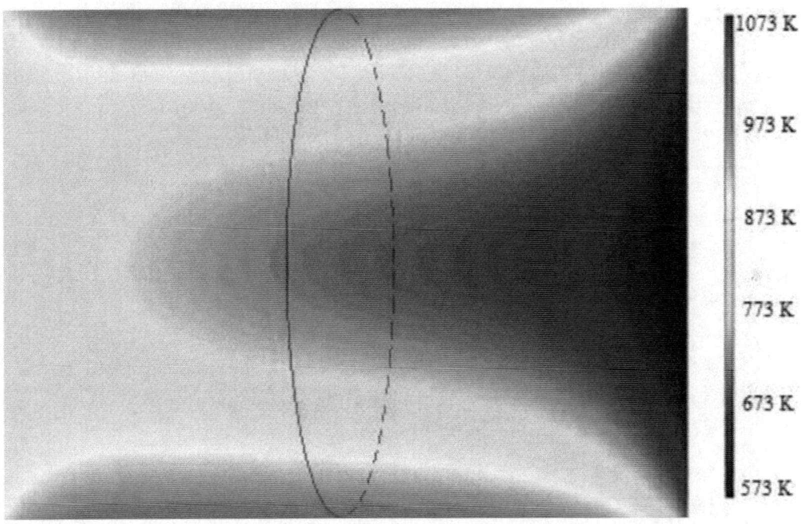

Figure 1. Heat distribution inside the reactor over heating zone for the wall temperature of 1073 K and N_2 velocity of 6.7×10^{-2} m/s

In addition, the above metioned grain temperature distribution is lower when the velocity of the nitrogen gas is lower. This phenomenon is due to the effect of N_2 velocity on the convection that is developing inside the reactor. It means the centrifugal body forces are stronger and the flow was more affected by the natural convection at low Reynolds number [10]. Furthermore, heat loss through the N_2 will be occurred more in higher Reynold numbers [11]. Also it has been observed that the velocity of the N_2 was affected by reaction temperature. The increasing of the velocity is due to the heat that was transferred to the N_2 from the heating sources, hence the kinetic energy of the N_2 increases [4]. The lower the N_2 velocity is, the higher it is affected by reaction temperature and this characteristic can have significant effect on pyrolysis process by having more concentrate high temperature zone in the reactor. The results of the above observation are tabulated in Table 3.

Table 3. Variation of the N_2 velocity

Initial N_2 velocity(m/s)	Maximum N_2 velocity (m/s)	Increment (m/s)	Increment (%)
0.000078	0.00016	0.000082	105.13
0.007000	0.01098	0.003980	56.86
0.013922	0.01940	0.005478	39.35
0.037000	0.04400	0.007000	18.92
0.067000	0.07755	0.010550	15.75

In order to validate the simulated results, the bio-oil production weight percentages were estimated and compared to the experiment results. The estimated bio-oil yield in constant N_2 velocity was calculated according to the developed empirical correlation that generalised version is shown as Equation (1), where T is the reaction temperature.

$$\text{Yield} = \sum (\text{fraction of T} \times \text{estimated bio-oil yield fraction from experiment}) + \sum \text{previous yield} \quad (1)$$

From the simulated results, the yield of bio-oil was measured and the results of both experimental (three experiments were run for repeatability test and shown by using error bars) and CFD simulation are plotted in Figure 2. As can be observed maximum yield of 52.02 wt% with the percentage error of less than 5% can be achieved at a temperature of 923 K and N_2 velocity of 7.8 × 10^{-5} m/s. As the results have a maximum percentage error of below 5%, the validated simulation can be used for further investigation of the pyrolysis process in other input parameters of other materials.

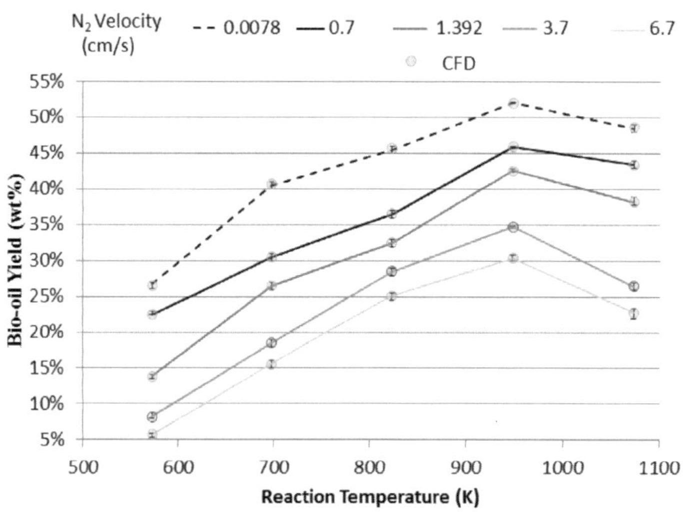

Figure 2. Bio-oil yield in all testing points

4. Conclusions

The optimum conditions to obtain maximum bio-oil yield of 52.02% wt% from the equal weight mixture of ground CS and HIPS are at reaction temperature of 923 K and N_2 linear velocity of 7.8 × 10^{-5} m/s. The results from CFD simulations and experiments have less than 5% difference at all 25 testing points. As the CFD model has been validated, it can be used in further investigations for other testing points or even other pyrolysed materials such as different plastics.

5. References

[1] Bertero M, Puente G and Sedran U 2012 *Fuel* **95** 263
[2] Demirbas A 2007 *Energy Res.* **29 (vol. 4)** 329
[3] Lede J, Broust F, Ndiaye F and Ferre M 2007 *Fuel* **86 (vols. 12 – 13)** 1800
[4] Dixon A G, Nijemeisland M and Stitt E H *Ind. Eng. Chem. Res.* **44** 6342
[5] Guardo A, Coussirat M, Recasens F, Larrayoz M A and Escaler X 2006 *Chem. Eng. Sci.* **61** 4341
[6] Mellin P, Kantarelis E and Yang W 2013 *Fuel* **117** 704
[7] Gui M S Z, Jourabchi S A, Ng H K and Gan S 2014 *Appl. Mech. Mater.* **625** 626
[8] Jourabchi S A, Gan S and Ng H K 2014 *Energy Convers. Manage.* **78** 518
[9] Ofuchi K and Kunii D 1965 *Int. J. Heat Mass Transfer* **8** 749
[10] Hasan W K 2013 *J. Mech. Civ. Eng.* **9** 47
[11] Ferreira L M, Castro J A M and Rodrigues A E 2002 *Int. J. Heat Mass Transfer* ***45*** 951

Acknowledgments

The authors would like to acknowledge INTI International University for funding this work through the internal seed grant INT-FOSTEM-05-01-2014. Additionally, the Faculty of Engineering of the University of Nottingham Malaysia Campus is acknowledged for its support towards this work.

IOP Conf. Series: Earth and Environmental Science **36** (2016) 012045 doi:10.1088/1755-1315/36/1/012045

Hot compressed water extraction curve for palm oil and beta carotene concentration

M S M Sharizan[1,2], M N Azian[1,2*], Y Yoshiyuki[3], A A M Kamal[4], M A Che Yunus[5]

[1]Shizen Conversion and Separation Technology (SHIZEN) ikohza,
Universiti Teknologi Malaysia, Jalan Semarak, 54100 Kuala Lumpur, Malaysia
[2]Malaysia – Japan International Institute of Technology (MJIIT),
Universiti Teknologi Malaysia, Jalan Semarak, 54100 Kuala Lumpur, Malaysia
[3]Department of Chemical Engineering,
Tokyo University of Agriculture and Technology (TUAT), Koganei, Tokyo 184-8588, Japan
[4] Department of Chemical & Environmental Engineering
University of Nottingham Malaysia, Jalan Broga, 43500 Semenyih, Selangor, Malaysia
[4]Centre of Lipid Engineering Applied Research (CLEAR),
Universiti Teknologi Malaysia, Skudai, 81310, Johor, Malaysia

E-mail: [1]mohd_sharizan@yahoo.com, [2]azian.morad@gmail.com,[3] yama_pse@cc.tuat.ac.jp, [4]clear.utm@gmail.com, [5]azizi@cheme.utm.my

Abstract. Hot compressed water extraction (HCWE) is a promising green alternative for palm oil milling. The kinetic characteristic of HCWE for palm oil and it β-carotene concentration was experimentally investigated in this study at the different temperature and pressure. Semi-batch HCW extractor from 120 to 180 °C and 30 to 50 bar was used to evaluated the process for 60 mins of extraction in 10 mins interval. The results obtain using the HCWE process was compared with other extraction method. The oil extraction achieved the maximum extraction rate within 20 mins of extraction in most of the condition and starting to decrease until 60 mins of extraction time. The extraction rate for β-carotene was achieved the maximum rate in 10 mins and starting to decrease until 30 mins. None of β-carotene concentration had been extracted out from the palm oil mesocarp after 30 mins of extraction in all condition. The oil recovery of using HCWE was relatively low compare with the mechanical screw press, subcritical R134b, supercritical carbon dioxide and hexane extraction due to the oil loses in the oil-water emulsion. However, the β-carotene concentration in extracted oil using HCWE was improved compare with commercial crude palm oil (CPO) and subcritical R134a extraction.

1. Introduction

Palm oil industry has contributed significantly to the Malaysia economic since 1950s. In 2014, Malaysia has produced about 20 million tons of crude palm oil (CPO) with the trade value of RM 50 billion. Oil palm can produce two main products that are CPO which being extracted from the palm oil mesocarp, and crude palm kernel oil (CPKO) which extracted from the palm oil kernel. Currently, CPO is extracted using screw press technology with the single or double system preceded by the

* To whom any correspondence should be addressed.

sterilization process in the palm oil mill. The process contribute to the overall oil loses of 8% dry basis (w/w) which about 2% (w/w) and 6% (w/w) in the sterilizer and screw press respectively [1]. The oil loses in screw press system is due to the press pressure limitation where the process are expose to the nut breakage at about 0.0625 % broken nut/bar as demonstrated in Mongana Report (1955) [2]. As a result, the pressed palm oil fiber extraction shows that about 5-6% dry basis (w/w) oil remains in the fiber after pressing process [3].

Thus, hot compressed water extraction (HCWE) process is proposed in this study for CPO extraction. This process used water in subcritical region between the normal boiling point and the critical point; from the temperature of 100 to 374°C and pressure up to 22.08 MPa. Since 1990s, this green process is proven to improve the extractions of food [4], plant material [5], and microalgae [6]. Not limited to that, this process also proven to reduce the extraction time to 20 min from 180 mins using hydro distillation [4]. This improvement would decrease the energy consumption of the process and reduce the operating cost. Although HCWE is never being uses for palm oil, this system seems to be an alternative that provides green and effective processes for CPO extraction.

In this study, the HCWE curve was experimentally examined in semi batch process for about 60 mins of extraction in 10 mins interval at different temperature and pressure. Two process responses were evaluated including overall oil yield (g-oil/ g-dried mesocarp) and β-carotene concentration (ppm). The efficiency of the process are also being compare with the mechanical screw press, subcritical R134a, supercritical carbon dioxide and hexane extraction.

2. Material and method

2.1. Raw material and chemical

Fresh oil palm fruits bunch were collected at Felda Mempaga Satu, Bentong, Pahang, Malaysia. The bunch was treated within 120 mins after the harvesting time using autoclave at 130°C and 40 mins. After that, the mesocarps were shredded out from loose fuitlet as shown in figure 1 prior to the extraction process. The high performance liquid chromatography (HPLC) grade of Hexane (Friendemann Schmidt, Australia) was used for β-carotene analysis and soxhlet extraction. Distilled water generated from water distiller (Favorit, Thailand) was used in the HCWE process.

2.2. *Hot compressed water extraction*

HCWE was done semi-batch wisely using the fabricated equipment in this laboratory as shown in Fig. 2. This equipment is consist of two vessels with one litre volume each; the extraction and cooling vessel which connected by 3/8 inch stainless steel pipeline.

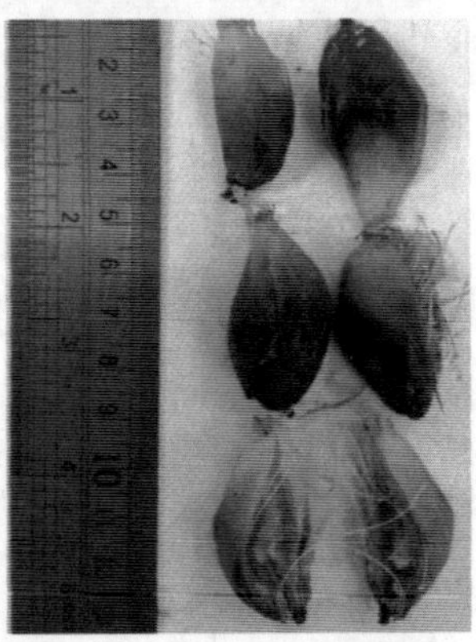

Figure 1. Treated palm oil mesocarp

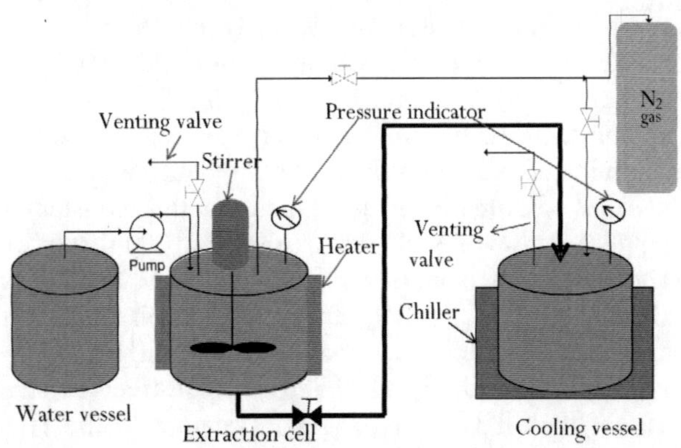

Figure 2. Schematic diagram for hot compressed water extractor

The treated palm oil mesocarp was weighed and loaded into a covered stainless steel mesh cylinder before being placed into the extraction cell. The distilled water was added into the cell. The cell was then securely covered with a stainless steel lid. N_2 gas was then passed through the cell for 2 mins to purge out air and dissolved oxygen. Excess pressure was relieved through the release valve. The temperature was set according to the required experiment. The electrically jacketed extraction cell took 10–15 mins to achieve the desired temperature. The 60 mins extraction time with 10 mins interval started once the set temperature was achieved as indicated by the temperature indicator in the extraction cell. Every 10 mins interval, the water extract was simultaneously transferred into cooling vessel and the new distilled water was adding up into the extraction vessel. The water extract was subjected to the post treatment process prior to further analysis.

2.3. Analysis of oil

In order to recover the palm oil from the water extract, the extract was going through the post treatment process for the oil and water separation by clarifying and centrifuging process. The water extract was clarified using separating funnel for 60 mins at temperature of 60oC to form three distinguish layer of oil, oil-water emulsion and water. After that, the layer of oil and oil-water emulsion was centrifuged at 7000 rpm and 10 mins. The accumulated palm oil was measured and collected prior to the analysis. The overall oil yield was calculated based on the measured palm oil volume using (1).

$$\text{Overall oil yield,} \quad \frac{g_{\text{oil}}}{g_{\text{dried mesocarp}}} = \frac{V_{\text{oil}} \text{x}\, \rho_{\text{oil}}}{W_s} \qquad (1)$$

Where V_{oil} is volume of palm oil recovered from extraction (ml), ρ_{oil} is density of the palm oil (g/ml), W_s is dry mesocarp weight, g. The concentration of β-carotene was measured based on MPOB test method using a spectrophotometer (UV–Vis, Shimadzu, Japan) [7].

3. Result and discussion

3.1. The extraction curve for overall oil yield at different temperature and pressure.

The cumulative overall oil yield extracted by unit time or extraction curve at different temperature and pressure is tabulated in figure 3a, 3b, 3c. At 30 bar and 120°C, the HCW was extracted 35.8 % (w/w) of the oil in 20 mins and it rate was slowly decrease with increasing of only 9 % (w/w) of yield at 50 mins of extraction. After that, the cumulative of extracted oil are constant in 50 to 60 mins of extraction which showing none of the oil being extracted in the last 10 mins of extraction. Similar trend can be observed for temperature of 140 and 180°C at constant pressure of 40 and 50 bar at different magnitude of oil yield. However, the different observation can be seen for temperature of 160°C and constant pressure of 30, 40 and 50 bar. The oil was effectively being extracted at 43 % (w/w) in 10 mins of extraction time. Then, the rate was decreased with extracting only 10 %(w/w) of oil after 40 mins of extraction. The extraction rate is constant in 40 to 60 mins of extraction. At 160°C and 40 bar, 43 % (w/w) oil yield was extracted in 20 mins of extraction.

The maximum cumulative yield of 51 % (w/w) was achieved at 60 mins of extraction. Meanwhile, at 160°C and 50 bar, 50 % (w/w) of oil was extracted in 30 mins of extraction and keep steady until the end of the extraction process. In HCWE process, the extraction starts with the breaking of the cell wall by the hydrolysis of fiber to sugar [8]. This mechanism promotes the entrainment of the oil from its original structure to the HCW and increased the oil extraction as observed by higher extraction in the first 20 mins of the extraction. Not limited to that, the increasing of temperature is also promoting the higher cell wall breaking thus increase the overall oil yield at higher temperature. However, the decreasing of extraction rate at 180°C may be due degradation of the oil at the higher temperature.

The cumulative β-carotene extracted by unit time at different temperature and pressure is tabulated in figure 3(d), 3(e) and 3(f). Similar trend can be observed in β-carotene concentration comparable with oil extraction where the optimum temperature of 160 °C was obtained. However, the effect of pressure on β-carotene extraction was more significant compared with temperature. At 30 bar and 160°C, the concentration of β-carotene extracted is 258.4 ± 0.6 ppm which is very low compare with pressure of 40 bar at 834.5 ± 0.1 ppm. Then, it is slightly increases to 857.7 ± 0.4 ppm at 50 bar. As the time increase, the rate of β-carotene extraction was decrease where approximately 85 % of the compound was extracted out after 20 mins extraction and everything will be extracted out in 30 mins. Thus, the optimum condition 160 °C, 50 bar and 30 mins was identified as the optimum condition of β-carotene concentration.

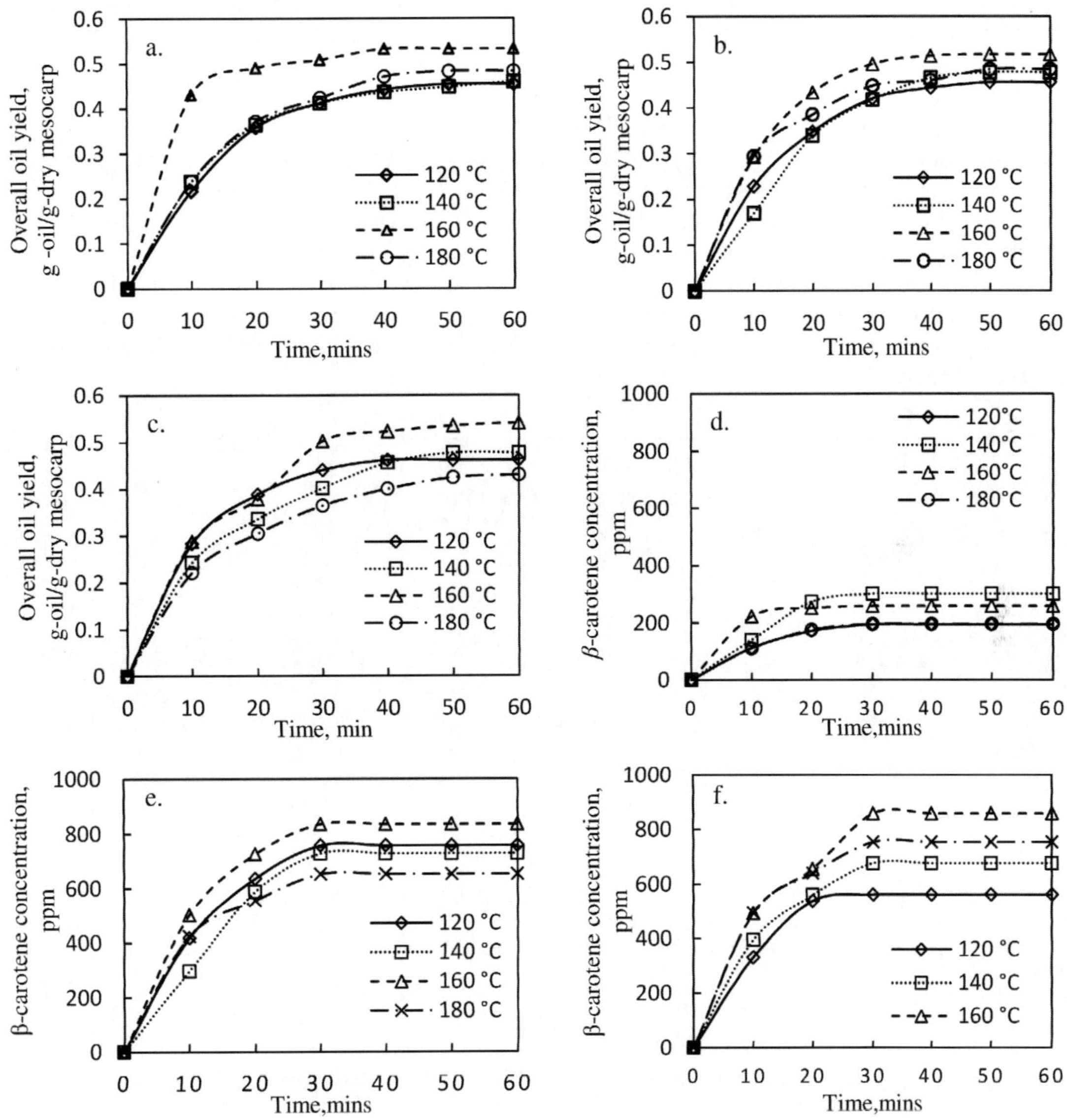

Figure 3. The extraction curve for overall oil yield and β-carotene at different temperature and constant pressure; (a.) Overall oil yield at 30 bar (b.) Overall oil yield at 40 bar (c.) Overall oil yield at 50 bar (d.) β-carotene concentration at 30 bar (e.) β-carotene concentration at 40 bar (f.) β-carotene concentration at 50 bar

As the temperature increased, the solubility of *β*-carotene in HCW was also increased by the decreasing of dielectric constant, ε of water. The ε of water decreases from 75 at 20°C to 60 at 120°C and decreased to 40 at 160°C [9] which make it behaves like the organic solvent. This situation is providing the suitable condition for the compound to soluble in HCW thus increased the extraction efficiency. After that, the rapid increased of extraction rate which observed at temperature of 160°C was possibly caused by the significant reduce of the hydrogen bonding between water molecule at this temperature and above as discussed by Shinoda (1977) [10]. The reducing of the hydrogen bonding promotes the interaction between *β*-carotene and water. After that, *β*-carotene extraction was decreased at 180°C is possibly caused by the thermal hydrolysis and polymerization. Furthermore, the increased of pressure from 30 to 50 bar has decreased the oil extraction rate and shift the effective extraction time from 10 mins at 30 bar to 30 mins at 50 bar. As the conclusion, the develop extraction curve for both components later will be used to characterise the kinetic behaviour of the process [11].

3.2. Comparison with other extraction method

The comparison between HCWE and others extraction is shown in figure 4. As shown in figure 4, the optimum HCWE condition give the lowers overall oil yield of 0.54 g-oil/g-dried mesocarp compare with supercritical CO_2 at 0.77 g-oil/g-dried ginger [12], hexane extraction at 0.76 g-oil/g-dried mesocarp and subcritical R134a at 0.66 g-oil/g-dried mesocarp [13] and lab scale screw press with microwave pretreatment process at 0.62 g-oil/g-dried mesocarp [14]. This probably caused by the oil loses in the oil-water emulsion after extraction process. Thus, effective oil-water emulsions separation need to be establish to improve the oil recovery. The yield for the commercial CPO is not available from the literature.

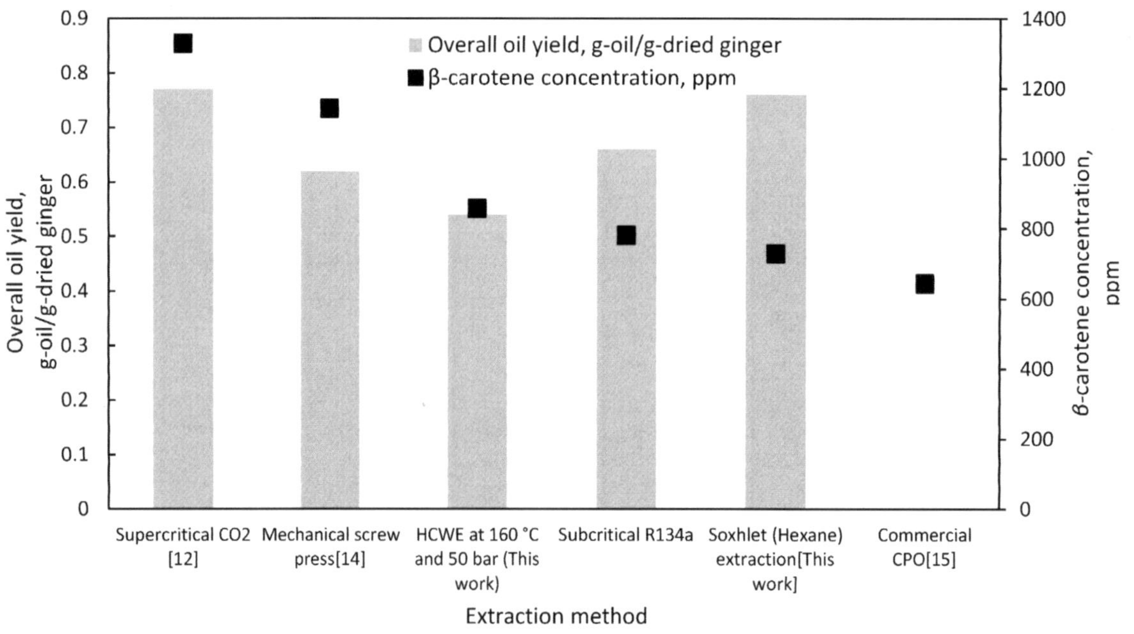

Figure 4. Comparison between HCWE and other extraction method

However, it is important to note that HCWE was able to produce the palm oil with high concentration of *β*-carotene compare with hexane extraction, subcritical R134a and commercial CPO. This would create an added value for the palm oil product with the higher *β*-carotene concentration.

4. Conclusion

As a conclusion, the HCWE curve for overall oil yield and β-carotene was successfully developed at various temperatures of 120 to 180°C and pressure of 30 to 50 bar. These curves can be utilized to characterise the kinetic behaviour of the process with mass transfer mechanism and solubility correlation. Besides that, further research on oil-water emulsion needs to be explore to improve the overall oil yield.

References

[1] Nik N N A, Ahmad A, Omar F M, Banana A A S, Md. Zaidul I S and Ab. Kadir M O 2008 *Sep. Pur. Technol* **60** 272-277
[2] Research on production and storage of palm oil, in: Mongana report, Institute for the Advancement of Scientific Research in Industry and Agriculture (IRSIA) Congo 1955.
[3] Choo Y M, Yap S C, Ooi C K, Ma A N, Goh S H and Ong A H 1996 *J. Am. Oil Chem. Soc* **73** 599-602
[4] Wiboonsirikul J and S Adachi 2008 *Food Science and Technology Research* **14** 319-328
[5] Md S M S, Morad N A, Mohamad A N A, Mohd Y Y A and Che Y M A 2014 *Sep Pur Technol* **124** 141-147
[6] Herrero M, Cifuentes A and Ibanez E 2006 *Food Chem* **98** 136-148
[7] A compendium of test on palm oil products, palm kernel products, fatty acids, food related products and others, in: Palm oil, Malaysian Palm Oil Board (MPOB) Kuala Lumpur, Malaysia, 2005.
[8] H Silvamany and J M. Jahim 2015 *Malaysian Journal of Analytical Sciences* **19** 77 - 87
[9] Uematsu M and Frank E U 1980 *J. Phys. Chem. Ref. Data* **9** 1291-1306
[10] Shinoda K 1977 *J. Phys. Chem* **81** 1300-1302
[11] Sovová 2005 *J. Supercritic Fluid* **33** 35-52
[12] Lau H L N, Choo Y M, Ma A N and Chuah C H 2006 *J. Food Lipids* **13** 210-221
[13] Mustapa A N, Manan Z A, Mohd A C Y, Nik N N A and Ab K M O 2009 *J. Food Eng.* **95** 606-616
[14] Tan C H, Ghazali H M, Kuntom A, Tan C P and Ariffin A A 2009 *Food Chem* **113** 645-650
[15] Nagendran B, Unnithan U R, Choo Y M and Sundram K 2000 *Food Nutr Bull* **21** 189-194

Acknowledgments

The research is funded by Universiti Teknologi Malaysia (UTM) under University Grant (GUP) grant no. Q.K130000.2543.06H91. The financial assistant from Universiti Malaysia Perlis (UniMAP), Malaysian Ministry of Education (MOE) and Malaysian-Japan International Institute of Technology (MJIIT) is greatly appreciated. The help and assistant from En Nasir, the Plantation Manager from Felda Mempaga 1, Bentong,Pahang for fresh fruit bunch supply is greatly acknowledged.

IOP Conf. Series: Earth and Environmental Science **36** (2016) 012046 doi:10.1088/1755-1315/36/1/012046

Metabolomics-based prediction models of yeast strains for screening of metabolites contributing to ethanol stress tolerance

Z Hashim[1], **E Fukusaki**[2]
[1] Department of Bioprocess Engineering, Faculty of Chemical Engineering, Universiti Teknologi Malaysia, 81310 Skudai, Johor, Malaysia
[2] Department of Biotechnology, Graduate School of Engineering, Osaka University, 2-1 Yamadaoka, Suita, Osaka 565-0871, Japan

E-mail: [1]zanariahhashim@utm.my, [2]fukusaki@bio.eng.osaka-u.ac.jp

Abstract. The increased demand for clean, sustainable and renewable energy resources has driven the development of various microbial systems to produce biofuels. One of such systems is the ethanol-producing yeast. Although yeast produces ethanol naturally using its native pathways, production yield is low and requires improvement for commercial biofuel production. Moreover, ethanol is toxic to yeast and thus ethanol tolerance should be improved to further enhance ethanol production. In this study, we employed metabolomics-based strategy using 30 single-gene deleted yeast strains to construct multivariate models for ethanol tolerance and screen metabolites that relate to ethanol sensitivity/tolerance. The information obtained from this study can be used as an input for strain improvement via metabolic engineering.

1. Introduction

Increasing energy usage, high cost and environmental concerns require for the development of sustainable and renewable fuels and chemicals. Fuels and chemicals have long been produced from plant and animal-based resources, but this has raised issues such as competition with food, high prices and unethical land-use practices. A sustainable alternative is to produce directly from abundant and cost-effective renewable resources by microbial conversions.

To solve the problems faced by current biofuel microbial productions, several strategies have been implemented. Mainly there are three approaches; strain development using metabolic engineering, raw material development which includes searching for new and cost effective resources, and process optimization to obtain high yield and low cost. Metabolic engineering of microbial strain has been considered to be a powerful approach to improve biofuel manufacturing. In particular, the potential of metabolic pathways to produce biofuel molecules from microbes has been described [1]. Currently ethanol dominates the biofuel market. Besides ethanol, other candidate compounds for biofuels such as higher alcohols or fatty acid-derived compounds [2] have also been identified. Although yeast has a native pathway to produce ethanol, production yield is low for commercial biofuel production. In

[2] [2] To whom any correspondence should be addressed.

addition, ethanol is toxic to yeast, and thus one strategy to improve ethanol production is via increased ethanol tolerance.

Metabolomics, the exhaustive profiling of metabolites [3,4] contained in a cell or an organism is a powerful tool to study phenotypic changes. Metabolomics offers an unbiased view on cellular functions and links between genotype and phenotype [3]. Previously, Yoshida *et al.* [5] demonstrated that metabolite profile under a standard growth condition is able to predict the lifespan of yeast mutant strains. Using the same concept, we now apply metabolite profiling to construct multivariate models to predict ethanol tolerance and screen for responsible metabolites. The candidate metabolites obtained from this study can be linked back to their putative pathways and considered for metabolic engineering, for example using gene deletion or overexpression strategies, to improve ethanol tolerance in yeast. It is known that the screening of candidate genes/pathways to be modified in metabolic engineering has been difficult due to the complexity of metabolic landscapes [6,7], and the candidates are often picked from known pathways. Thus, a direct measurement of metabolites may reveal better information about metabolite alteration and provide a semi-rational approach for metabolic engineering.

2. Materials and method

2.1. Yeast strains

Yeast BY4742 (*MATα leu2Δ0 lys2Δ0 ura3Δ0 his3Δ1*) and its isogenic derivatives, *yfgΔ* (*MATα leu2Δ0 lys2Δ0 ura3Δ0 his3Δ1 yfg::kanMX*) were purchased from Open Biosystems (Huntsville, AL, USA). The 30 yeast strains used were:
std1Δ, ace2Δ, sut1Δ, hcm1Δ, mig1Δ, tec1Δ, fkh2Δ, skn7Δ, kar4Δ, mss11Δ, bas1Δ, swi5Δ, pho2Δ, rgt1Δ, crz1Δ, ino4Δ, mga1Δ, cup2Δ, rtg1Δ, rlm1Δ, tos4Δ, mbp1Δ, sok2Δ, hms2Δ, hot1Δ, stp2Δ, mot3Δ, rpn4Δ, srb2Δ, aft1Δ
The gene function based on gene ontology (GO) annotation can be obtained from SGD (Saccharomyces Genome Database). These strains were chosen randomly to represent a wide variety of ethanol tolerance values.

2.2. Yeast culture and metabolite extraction

Frozen stocked (-80 °C) yeast strains were plated onto YPD agar plate (yeast extract 10 g/L, peptone 20 g/L, dextrose 20 g/L, agarose 20 g/L) with added geneticin G418 200 μg/mL (except for wild-type strain) and grown at 30 °C for two days. The cultivation and sampling were performed as described previously [8]. Briefly, pre-cultured yeast cells were grown in YPD medium at 30 °C in a rotary shaker (200 rpm) to OD_{600} = 1.0 (optical density at 600 nm, equivalent to 1 mg cells-dry weight) and harvested using a rapid filtration system. After washing with 5 mL of water, the filter membrane containing yeast cells was immediately folded, inserted into a 2.0 mL Eppendorf tube and plunged into a liquid nitrogen bath to quench metabolism. Time from start pouring culture broth until quenching was kept within 30 seconds. Samples were then stored at -80 °C until extraction.

Prior to extraction, the filter-bound cells were disrupted with MM 301 mixer mills (Retsch, Haan, Germany) at 20 Hz for 1 min. 500 μL of extraction solvent (acetonitrile / methanol / water = 2 / 2 / 1 with added 1.2 μg/mL of 1,4-piperazinediethanesulfonic acid (Dojindo, Kumamoto, Japan) as an internal standard) precooled to -30 °C was added, and extraction was carried out on ice for 15 min. After centrifugation at 15000 rcf for 5 min, the supernatant was transferred to a new tube. A second round of extraction was done with another 500 μL of extraction solvent. Supernatants from both extractions were combined, and 500 μL of chloroform was added. After vortex and centrifugation, the upper polar phase was transferred to a new tube (about 450 μL). 400 μL water was added to the remaining mixture (containing chloroform), vortexed and centrifuged. Another 450 μL of upper polar phase was collected and combined with the previous polar extract. Finally, after centrifugation, 700 μL of polar extract was collected and filtered (0.2 μm PTFE, Millipore, MA, USA), and applied to a centrifugal concentrator VC-96R (Taitec, Osaka, Japan) for 2 h. The remaining extract was then transferred to a glass vial and ready for LC analysis.

2.3. Ethanol tolerance measurement
Frozen cells were plated onto YPD agar plates and let to grow for two days. A single colony from each strain was transferred into 1 mL of YPD medium placed in a deep 96-well plate. This plate was set on a plate shaker (1050 rpm) at 30 °C, and cells were let to grow overnight (about 16-18 h). The overnight pre-culture broth from each well was diluted 100 times in fresh YPD medium and distributed to a 96-well microplate, 100 μL per well. The plate was incubated in a 30 °C oven with no shaking for 4 h. After 4 h, 25 μL of 35% ethanol (for 7% ethanol stress condition) or sterilized water (for stress-free condition) was added, and incubation was continued for 8 h. OD_{600} was measured every 1 h. Growth rate for mutant and parental strains were compared in stress and stress-free conditions.

2.4. Metabolite profiling
The LC-MS platform consists of ACQUITY Ultra Performance LC and LCT Premier XE (Waters, Milford, MA, USA). Separation was achieved on Acquity UPLC HSS T3 column (2.1 x 150 mm, 1.8 μm particle size, Waters), using reversed phase chromatography with ion pairing agent tributylamine in the aqueous phase. Solvent A is 97/3 water/methanol with 10 mM tributylamine and 15 mM acetic acid; solvent B is methanol. Solvent gradient profile is as follows; 0 min, 0% B; 2.5 min, 0% B; 5 min, 20% B; 7.5 min, 20% B; 13 min, 55% B; 15.5 min, 95% B; 19 min, 0% B; 25 min, 0% B. The solvent flow rate is kept at 200 μL/min. Other LC parameters are autosampler temperature 4 °C, injection volume 10 μL (full loop), and column temperature 25 °C. Ionization method was ESI in negative mode. The mass spectrometer was tuned and calibrated before use, and final analysis parameters are: capillary voltage 2200 V, sample cone voltage 25 V, desolvation temperature 300 °C, source temperature 120 °C, cone gas flow 50 L/h, desolvation gas flow 500 L/h. Leucine enkaphalin 2.0 ng/μL was used as a reference solution, with 5.0 μL/min flow rate. The scan speed was 0.1 s (10 scans/s) and mass range was 70-900 Da.

2.5. Data processing
Raw data from LC-MS was read using Masslnyx (proprietary MS software from Waters). The mass window was set at 0.05 Da. Based on the retention time values extracted, peak identification method was done using Quanlynx (Waters) and verified by visual inspection. The peak list table was constructed, for each semi-quantified metabolite, composing of metabolite type and abundance in each yeast mutant strain. Identification is based on 48 validated metabolites, and 63 putative metabolites from literature [9]. Peak intensities were expressed as the peak area after alignment and baseline correction where necessary, and normalized to internal standard and total raw signal.

2.6. Multivariate data analysis and data visualization
Multivariate data analysis was performed using SIMCA-P+ ver. 12 software (Umetrics, Umeå, Sweden). Peak list table in Excel format was imported to SIMCA-P+ and multivariate plots (PCA, OPLS and OPLS-DA) were computed automatically. Data was scaled to unit variance.

3. Results and discussion

3.1. Yeast growth
The gross specific growth rate μ (h^{-1}) was determined by the relation between cell mass concentration and time, based on the kinetic model of cell growth. The result is shown in Table 1. The specific growth rate is a key control parameter in industrial application of yeast and serves as a "fitness" indicator of yeast strains [10], therefore monitoring it is of prime importance. Out of the 30 strains,

only *srb2*Δ demonstrated slow growth (approximately 12% lower gross specific growth rate compared to wild-type BY4742), whereas all other strains exhibited comparable growth rates.

Table 1. Growth kinetics under YPD medium for 30 yeast mutant strains used in this study. (RSD: relative standard deviation, measured for n = 3)

Strain no.	Strain name	Gross specific growth rate, μ (h^{-1})	Relative standard deviation, RSD % (n =3)	Doubling time, t_d (h)
1	*std1*Δ	0.46	0.77	1.50
2	*ace2*Δ	0.48	0.79	1.43
3	*sut1*Δ	0.47	0.51	1.48
4	*hcm1*Δ	0.44	0.59	1.57
5	*mig1*Δ	0.46	0.72	1.51
6	*tec1*Δ	0.46	0.47	1.50
7	*fkh2*Δ	0.44	1.33	1.57
8	*skn7*Δ	0.47	0.78	1.48
9	*kar4*Δ	0.46	0.55	1.51
10	*mss11*Δ	0.46	0.45	1.50
11	*bas1*Δ	0.41	0.43	1.71
12	*swi5*Δ	0.44	0.28	1.57
13	*pho2*Δ	0.44	0.19	1.56
14	*rgt1*Δ	0.46	0.29	1.50
15	*crz1*Δ	0.45	0.50	1.55
16	*ino4*Δ	0.47	0.92	1.48
17	*mga1*Δ	0.47	0.88	1.48
18	*cup2*Δ	0.47	0.46	1.49
19	*rtg1*Δ	0.47	0.24	1.47
20	*rlm1*Δ	0.45	0.79	1.55
21	*tos4*Δ	0.46	0.50	1.50
22	*mbp1*Δ	0.45	1.19	1.55
23	*sok2*Δ	0.42	0.53	1.67
24	*hms2*Δ	0.46	0.09	1.50
25	*hot1*Δ	0.47	0.68	1.49
26	*stp2*Δ	0.46	0.28	1.52
27	*mot3*Δ	0.42	0.82	1.66
28	*rpn4*Δ	0.42	0.24	1.67
29	*srb2*Δ	0.36	2.46	1.91
30	*aft1*Δ	0.46	0.54	1.50
Wild-type	BY4742	0.41	0.59	1.67

3.2. Ethanol stress tolerance

Ethanol stress tolerance for 30 yeast mutant strains and wild-type strain was measured by comparing the gross specific growth rates under stress condition (7% ethanol) with non-stress condition. The result is shown in Table 2. Based on the stress tolerance (Y2/Y1) values, there were five strains classified as ethanol sensitive which are *swi5*Δ, *mbp1*Δ, *stp2*Δ, *mot3*Δ, and *srb2*Δ (indicated in **bold** in Table 2). These strains have 10% or more reduction of tolerance value compared to wild-type strain ($p > 0.1$).

To compare current finding with previous works, we searched for the five genes (*SWI5*, *MBP1*, *STP2*, *MOT3*, *SRB2*) in Saccharomyces Genome Database (SGD, http://www.yeastgenome.org/). Table 3 summarizes gene description, competitive fitness and ethanol resistance under null mutation. Indeed, the loss of these genes resulted in decreased competitive fitness (except for *SWI5* which has been reported as both increased and decreased competitive fitness). Moreover, while there was no data for *SWI5* and *MOT3* regarding ethanol sensitivity, *MBP1*, *STP2* and *SRB2* all showed decreased ethanol resistance (Table 3). Thus, our result is consistent with previous reports.

Table 2. Growth rate of yeast strains under stress-free (normal growth), Y1 and stressed (7% ethanol), Y2 conditions and the relative stress tolerance Y2/Y1. **Bold** indicates ethanol sensitive strains, defined as those having 10% or more reduced ethanol tolerance compared to BY4742 (RSD: relative standard deviation, measured for n = 3)

		Stress-free (no ethanol)		7% ethanol		Stress tolerance
Strain no.	Strain name	Y1 (h^{-1})	RSD (%)	Y2 (h^{-1})	RSD (%)	Y2/Y1 (%)
1	*std1Δ*	0.46	0.77	0.26	6.32	55.7
2	*ace2Δ*	0.48	0.79	0.25	8.64	51.4
3	*sut1Δ*	0.47	0.51	0.28	7.99	59.1
4	*hcm1Δ*	0.44	0.59	0.24	8.12	54.8
5	*mig1Δ*	0.46	0.72	0.26	4.46	57.1
6	*tec1Δ*	0.46	0.47	0.26	5.89	56.4
7	*fkh2Δ*	0.44	1.33	0.25	14.16	57.4
8	*skn7Δ*	0.47	0.78	0.30	8.81	63.7
9	*kar4Δ*	0.46	0.55	0.24	3.02	52.3
10	*mss11Δ*	0.46	0.45	0.24	0.83	50.9
11	*bas1Δ*	0.41	0.43	0.28	4.90	68.9
12	***swi5Δ***	**0.44**	**0.28**	**0.18**	**4.34**	**41.0**
13	*pho2Δ*	0.44	0.19	0.25	1.21	56.6
14	*rgt1Δ*	0.46	0.29	0.25	2.04	55.1
15	*crz1Δ*	0.45	0.50	0.26	6.43	58.5
16	*ino4Δ*	0.47	0.92	0.29	2.56	62.4
17	*mga1Δ*	0.47	0.88	0.24	5.55	50.9
18	*cup2Δ*	0.47	0.46	0.24	4.14	50.6
19	*rtg1Δ*	0.47	0.24	0.25	9.47	53.7
20	*rlm1Δ*	0.45	0.79	0.23	7.05	51.0
21	*tos4Δ*	0.46	0.50	0.21	6.55	46.4
22	***mbp1Δ***	**0.45**	**1.19**	**0.17**	**5.59**	**38.7**
23	*sok2Δ*	0.42	0.53	0.25	11.64	59.1
24	*hms2Δ*	0.46	0.09	0.25	5.97	55.1
25	*hot1Δ*	0.47	0.68	0.23	6.99	48.6
26	***stp2Δ***	**0.46**	**0.28**	**0.18**	**3.21**	**40.1**
27	***mot3Δ***	**0.42**	**0.82**	**0.17**	**2.53**	**41.4**
28	*rpn4Δ*	0.42	0.24	0.22	7.16	52.0
29	***srb2Δ***	**0.36**	**2.46**	**0.13**	**4.68**	**37.1**
30	*aft1Δ*	0.46	0.54	0.23	0.09	50.1
Wild-type	BY4742	0.41	0.59	0.21	3.12	49.9

Table 3. Gene description, competitive fitness and ethanol resistance for five genes related to reduced ethanol tolerance, retrieved from Saccharomyces Genome Database (SGD).

Gene	Description	Competitive fitness (null mutation)	Ethanol resistance (null mutation)
SWI5	Transcription factor that recruits Mediator and Swi/Snf complexes; activates transcription of genes expressed at the M/G1 phase boundary and in G1 phase; required for expression of the *HO* gene controlling mating type switching; localization to nucleus occurs during G1 and appears to be regulated by phosphorylation by Cdc28p kinase; *SWI5* has a paralog, *ACE2*, that arose from the whole genome duplication	Decreased/ Increased	No data
MBP1	Transcription factor; involved in regulation of cell cycle progression from G1 to S phase, forms a complex with Swi6p that binds to MluI cell cycle box regulatory element in promoters of DNA synthesis genes	Decreased	Decreased
STP2	Transcription factor; activated by proteolytic processing in response to signals from the SPS sensor system for external amino acids; activates transcription of amino acid permease genes; *STP2* has a paralog, *STP1*, that arose from the whole genome duplication	Decreased	Decreased
MOT3	Transcriptional repressor and activator with two C2-H2 zinc fingers; involved in repression of a subset of hypoxic genes by Rox1p, repression of several DAN/TIR genes during aerobic growth and ergosterol biosynthetic genes in response to hyperosmotic stress; contributes to recruitment of Tup1p-Cyc8p general repressor to promoters; involved in positive transcriptional regulation of *CWP2* and other genes; relocalizes to the cytosol in response to hypoxia; can form [MOT3+] prion	Decreased	No data
SRB2	Subunit of the RNA polymerase II mediator complex; associates with core polymerase subunits to form the RNA polymerase II holoenzyme; general transcription factor involved in telomere maintenance	Decreased	Decreased

3.3. Metabolites detected and peak list

A total of 111 metabolites were detected in yeast extract, as shown in Table 4. Metabolites indicated in **bold** are those validated in this study (48 metabolites), while the rest is putative metabolites based on the retention time and *m/z* values from Lu *et al.*, 2010 [9] (63 metabolites). We selected liquid chromatograph with mass spectrometer (LCMS) for metabolite quantification platform as this method has been reported to cover a broad range of metabolites. Additionally, the use of an ion pairing reagent tributylamine in the mobile phase improved the retention of polar metabolites such as sugar phosphates and nucleotides that are central in primary metabolism. A peak list table was constructed, listing the relative abundance of these 111 metabolites, for each yeast mutant sample. Here, the metabolite intensities were calculated as relative to internal standard and were normalized to total raw signal. This dataset (sample vs. metabolite intensities) was then subjected to multivariate data analysis such as PCA and OPLS-DA as described in the following sections.

Table 4. List of metabolites detected from yeast extracts. (* indicates structural isomers that are non-separable with current method)

Amino acids (35)	
Alanine	Aminoadipic acid
4-aminobutyrate	Aconitate
Serine	Arginine
Proline	Tyrosine
Valine	3-phosphoserine
Threonine/Homoserine*	*N*-acetyl-glutamine
Hydroxyproline	Acetyllysine
N-acetyl-L-alanine	*N*-acetyl-glutamate
Leucine/ Isoleucine*	Tryptophan
Asparagine	Cysthathione
Ornithine	Glutathione
Aspartate	*S*-adenosyl-L-homocystein
Homocystein	Glutathione disulfide
p-aminobenzoate	CDP-ethanolamine
Glutamate	UDP-*N*-acetyl glucosamine
Methionine	D-glucosamine-1/6- phosphate*
Histidine	*N*-acetyl-D-glucosamine-1/6-phosphate*
Orotate	
Organic acids (28)	
Glyoxylate	**Pantothenate**
Glycolate	2-keto-isovalerate
Pyruvate	Citraconate
Lactate	Hydroxycaproate
Glycerate	Phenylpyruvate
Fumarate	Phenyllactate
Succinate	Phosphoenolpyruvate
Nicotinate	2-isopropylmalate
Malate	Xanthurenate
2-Oxoglutarate	2,3-diphosphoglycerate
β-Glycerate	2-dehydro-D-gluconate
Shikimate	D-gluconate
3-Phosphoglycerate	6-phospho-D-gluconate
Citrate /Isocitrate*	2-hydroxy-2-methylbutanedioate
Sugars/ Sugar phosphates (17)	
Erythrose-4-phosphate	**ADP-ribose**
Ribose-5-phosphate	**UDP-glucose**
Ribulose-5-phosphate	**ADP-glucose**
Fructose-6-phosphate	**Ribulose-1,5- bisphosphate**
Glucose-1-phosphate	5-phosphoribosyl-1-pyrophosphate
Glucose-6-phosphate	Trehalose-6-phosphate
Sedoheptulose-7-phosphate	Trehalose/Sucrose/Cellobiose*
Frcutose-1,6-bisphosphate	**DHAP/GAP***
	sn-glycerol-3-phosphate
Nucleotides / nucleotide derivatives (27)	
TMP CMP UMP AMP GMP	Cytidine
CDP UDP ADP GDP	Uridine
dTMP dTDP dCTP	Inosine
CTP UTP ATP GTP	3',5'-cyclic AMP
NAD NADH	Xanthosine-5'-phosphate
NADP NADPH	
FMN FAD	
Others (4)	
Indole	**Acetyl-coenzyme A**
Thiamine	Coenzyme A

3.4. Principal component analysis (PCA)

As the first step in multivariate data analysis, principal component analysis (PCA) was performed. PCA provides an unbiased overview of the dataset without any prior information. In PCA, each observation is plotted in a new dimensional space so that the dataset will be fitted to new axes known as principal components or PC, and so that the data is represented in low-dimensional planes.

Here, it was found that ethanol sensitive mutants, indicated by red triangles in Figure 1, showed a tendency to cluster at the lower right quadrant of the PCA score plot. When the ethanol sensitive strains were picked up, together with wild-type strain, and plotted in a new PCA plot (Figure 2), some clustering was observed. Although the strains were not completely arranged by their stress response values, *srb2*Δ which has a significantly lower growth rate under 7% ethanol (0.135 h^{-1}) was clearly discriminated from other strains (Figure 2).

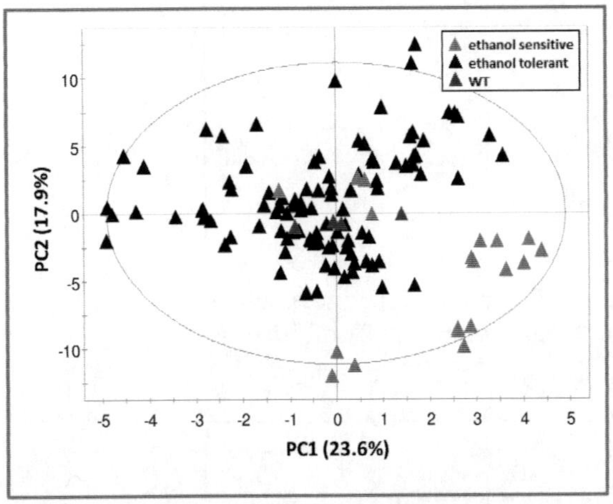

Figure 1. PCA score plot of metabolite profile of 30 yeast mutant strains and wild-type strain WT (Scaling: unit variance).

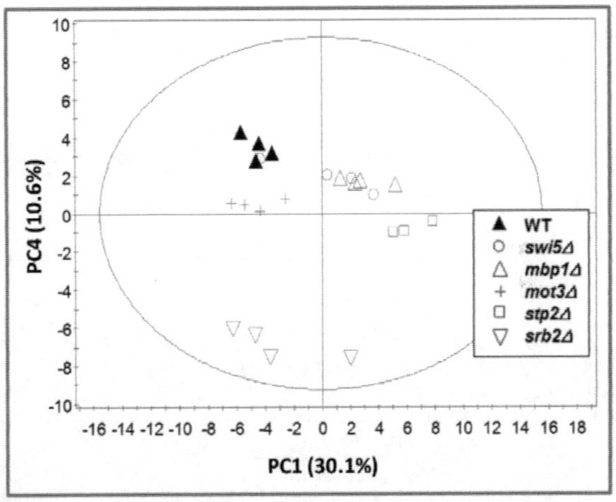

Figure 2. PCA score plot of ethanol sensitive strains and wild-type strain. Strain *srb2*Δ (indicated by reverse open triangles) was clearly separated from other strains.

3.5. Stress tolerance prediction model

To find which metabolites that have a strong contribution to the ethanol stress response, analysis using OPLS-DA (orthogonal projection to latent structure-discriminant analysis) was performed. The yeast strains were each pre-assigned to two classes, class 1 which is ethanol tolerant, and class 2 which is ethanol sensitive, based on the growth rate under 7% ethanol, relative to wild-type strain (Figure 3). From this graph, the corresponding S-plot was obtained (Figure 4), which indicates the contribution of metabolites towards the OPLS-DA model. It was found that citrate/isocitrate and uridine 5'-monophosphate (UMP) were important in ethanol sensitivity and ethanol tolerance, respectively. Tryptophan, which is well known to have effects on ethanol stress in yeast, was also identified.

IOP Conf. Series: Earth and Environmental Science **36** (2016) 012046 doi:10.1088/1755-1315/36/1/012046

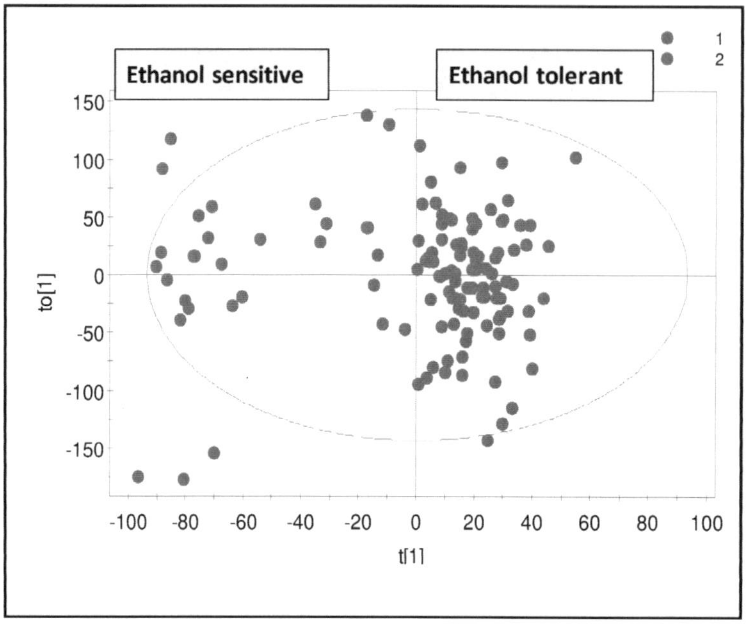

Figure 3. OPLS-DA plot based on the pre-assigned classes of ethanol tolerant (class 1) and ethanol sensitive (class 2) of 30 yeast mutant strains.

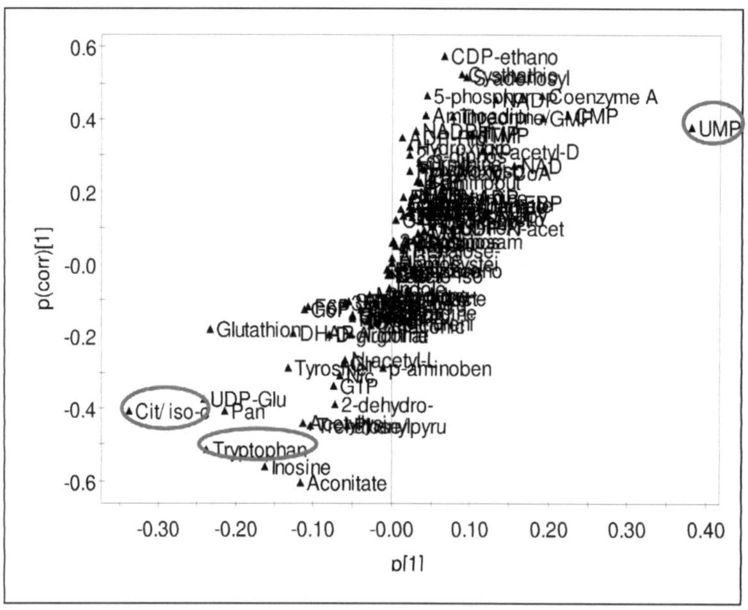

Figure 4. S-plot corresponding to the OPLS-DA model, showing metabolites contribution towards the pre-assigned classes. Citrate/isocitrate and tryptophan contribute to ethanol sensitivity while uridine 5'-monophosphate (UMP) contributes to ethanol tolerance.

Next, prediction model using OPLS (orthogonal projection to latent structures) was performed to yield a linear prediction profile. The horizontal axis shows the predicted ethanol stress from metabolome data, and the vertical axis shows the actual measured stress response (from Section 3.2). Here, a model with good linearity and prediction ability, indicated by R^2 and Q^2 values respectively, was obtained (Figure 5). $Q^2 > 0.8$ is considered as an excellent model, and the difference between R^2 and Q^2 must be within 0.2 - 0.3 (SIMCA-P+ User guide).

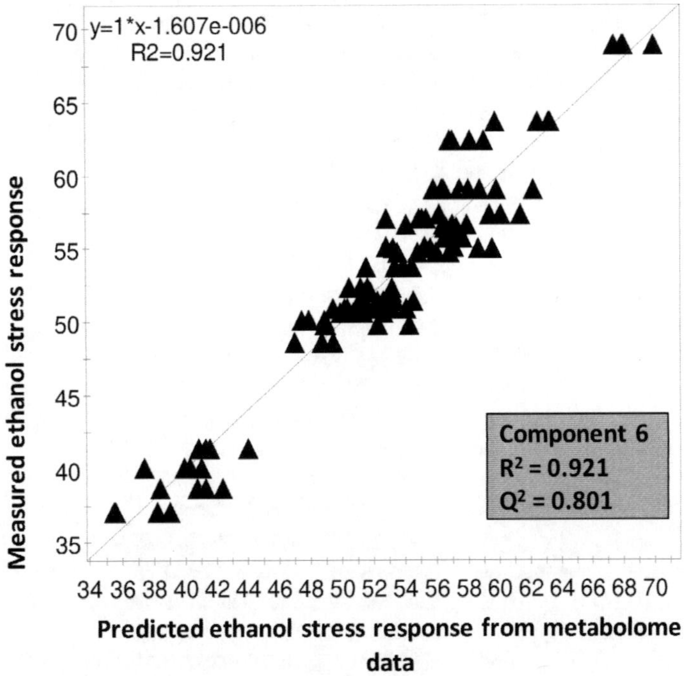

Figure 5. OPLS prediction model of ethanol stress response (ratio of growth rate under 7% ethanol to non-stress growth) based on metabolome data.

This result demonstrates that the metabolite profile of yeast strains under normal growth condition has the potential to be used in phenotype prediction. When the variables influence on OPLS modeling was closely examined using VIP (variables influence on projection) values, we could see that amino acids and many central metabolism compounds contribute greatly to the model. Figure 6 shows the metabolite list with VIP values larger than 1, which are considered important to the model. Aminoadipic acid and hydroxyproline topped the list. Although there is no report about the relation of these two metabolites with ethanol stress response in yeast, increased levels of aminoadipic acid in stressed rats and increased levels of hydroxyproline in stressed humans were previously reported. Aminoadipic acid is an intermediate in the metabolism of lysine and saccharopine. Decker *et al.* [11] investigated the relation of amino acid levels with stress in rats. Levels of aminoadipic acid in the pancreas were significantly increased (more than 2 times) in stressed rats. Similarly, hydroxyproline has been demonstrated to increase its level in urine under stressful condition among humans [12].

Tryptophan and trehalose, two known metabolites that have influence on ethanol stress response in yeast, were also identified, and this result is consistent with previous findings by other researchers. For example, intracellular trehalose accumulation was found to lead to higher ethanol tolerance [13,14] while strains lacking 'metabolism of tryptophan' genes were sensitive to ethanol [15]. Our results demonstrated that VIP analysis gives us a hint of which metabolites that may be influential or has a tendency towards ethanol stress response, and validation experiments under the specific stress condition will be needed for confirmation of these metabolites importance.

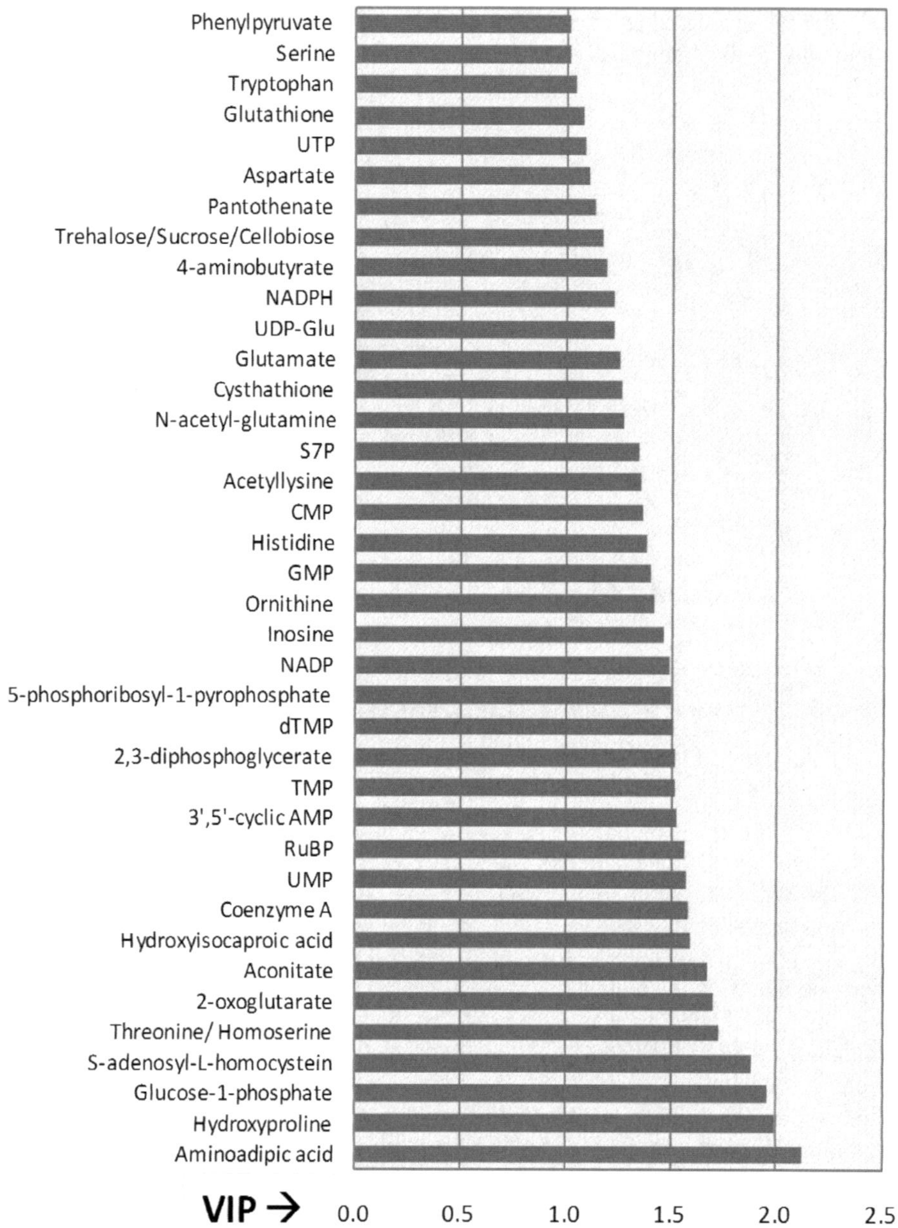

Figure 6. List of metabolites with VIP (variables influence on projection) value > 1.0, considered as influential in the OPLS prediction model of ethanol stress response.

4. Conclusion

In the current study, metabolite profiling of 30 yeast strains each lacking a single gene was performed. Using the metabolite profile, various models for ethanol stress tolerance were built. The prediction models provide information regarding which metabolites that contribute to such a behavior. From OPLS-DA model, citrate/isocitrate and tryptophan contribute to ethanol sensitivity while uridine 5'-monophosphate (UMP) contributes to ethanol tolerance. From OPLS model, aminoadipic acid and hydroxyproline were found to contribute to the tolerance in yeast, while trehalose and tryptophan, which have been reported previously, were also found but with less importance. Taken together, these metabolites can be considered as candidates for metabolic engineering. This study serves as a preliminary screening for semi-rational selection of metabolites and/or metabolic pathways. Moreover, the metabolite profile studied was under a standard growth condition and thus may also be extended for prediction of other phenotypes.

5. References

[1] Fortman J L, Chhabra S, Mukhopadhyay A, Chou H, Lee T S, Steen E J and Keasling J D 2008 *Trends Biotechnol.* **26** 375
[2] Steen E J, Kang Y, Bokinsky G, Hu Z, Schirmer A, McClure A, Del Cardayre S B and Keasling J D 2010 *Nature* **463** 559
[3] Fiehn O 2002 *Plant Mol. Biol.* **48** 155
[4] Fukusaki E and Kobayashi A 2005 *J. Biosci. Bioeng.* **100** 347
[5] Yoshida R, Tamura T, Takaoka C, Harada K, Kobayashi A, Mukai Y and Fukusaki E 2010 *Aging Cell* **9** 616
[6] Alper H, Moxley J, Nevoigt E, Fink G R and Stephanopoulos G 2006 *Science* **314** 1565
[7] Alper H and Stephanopoulos G 2007 *Metab. Eng.* **9** 258
[8] Hashim Z, Teoh S T, Bamba T and Fukusaki E 2014 *J. Chromatogr. B. Analyt. Technol. Biomed. Life Sci.* **966** 83
[9] Lu W, Clasquin M F, Melamud E, Amador-Noguez D, a Caudy A and Rabinowitz J D 2010 *Anal. Chem.* **82** 3212
[10] Van Hoek P, Van Dijken J P and Pronk J T 1998 *Appl. Envir. Microbiol.* **64** 4226
[11] Decker R, Stuckrad-Barre S v, Milakofsky L, Hofford J M, Harris N and Vogel W V 1995 *Life Sci.* **57** 1781
[12] Lee H M, Kim Y M, Lee C H, Shin J H, Kim M K and Choi B Y 2011 *J. Prev. Med. Public Health* **44** 74
[13] Kim J, Alizadeh P, Harding T, Hefner-Gravink A and Klionsky D 1996 *Appl. Envir. Microbiol.* **62** 1563
[14] Jung Y J and Park H D 2005 *Biotechnol. Lett.* **27** 1855
[15] Yoshikawa K, Tanaka T, Furusawa C, Nagahisa K, Hirasawa T and Shimizu H 2009 *FEMS Yeast Res.* **9** 32

Acknowledgements
The authors would like to thank Fukusaki Lab, Osaka University Japan for the strains and technical support. Z.H. acknowledges the Ministry of Higher Education Malaysia/ University Teknologi Malaysia for financial support (Q.J130000.2744.01K94).

IOP Conf. Series: Earth and Environmental Science **36** (2016) 012047 doi:10.1088/1755-1315/36/1/012047

Simulation of air velocity in a vertical perforated air distributor

T N W Ngu[1], C M Chu[1*] and J A Janaun[1]
Chemical Engineering Programme, Faculty of Engineering, Universiti Malaysia Sabah (UMS), 88400, Sabah, Malaysia

E-mail: chrischu@ums.edu.my

Abstract. Perforated pipes are utilized to divide a fluid flow into several smaller streams. Uniform flow distribution requirement is of great concern in engineering applications because it has significant influence on the performance of fluidic devices. For industrial applications, it is crucial to provide a uniform velocity distribution through orifices. In this research, flow distribution patterns of a closed-end multiple outlet pipe standing vertically for air delivery in the horizontal direction was simulated. Computational Fluid Dynamics (CFD), a tool of research for enhancing and understanding design was used as the simulator and the drawing software SolidWorks was used for geometry setup. The main purpose of this work is to establish the influence of size of orifices, intervals between outlets, and the length of tube in order to attain uniformity of exit flows through a multi outlet perforated tube. However, due to the gravitational effect, the compactness of paddy increases gradually from top to bottom of dryer, uniform flow pattern was aimed for top orifices and larger flow for bottom orifices.

1. Introduction

Perforated pipes are often used in industry for chemical vapor deposition, ventilation system, and bubble column reactor [1]. In Asia, two major types of dryer are widely used by industries for paddy drying purpose: fluidized bed and inclined bed dryer [2]. The moisture content (MC) of paddy rice at harvest around 28-35% dry basis [3]. Paddy should undergo drying to reduce the moisture content to 12-15% to be suitable for storage purpose to prevent production of mycotoxins . However, air is known to be distributed unevenly in commercial dryers. Therefore, it is necessary to minimize the uneven air flow in dryer. Sangkyoo and team [4] discovered that the flow rate distribution between orifices of perforated pipes could be adjusted by varying aspect ratio of orifices. A new conceptual configurations for paddy drying, Laterally Aerated Moving Bed (LAMB) as proposed in Tan [5] with an inner vertical aerated tube was installed. The aerator is a vertical perforated tube which function as to distribute hot air to surrounding. The designed vertical inner tube reduces pressure drop hence power required is lesser than fluidized bed dryer [5]. In brief, LAMB drying will lead to more uniform drying of grain.

Drying is a common process in industrial field. Air flow or air velocity play important roles in drying rate. However, air flow is difficult to measure while process is operating. Therefore, computational fluid dynamics (CFD) was applied to model fluid flow situations.

2. Methodology

Schematic diagram of a LAMB dryer is shown in Figure 1. Air distribution through orifices of inner perforated tube in LAMB dryer was focused on this study by using Phoenics CFD and SolidWorks 2013. Phoenics CFD and SolidWorks 2013 were used to simulate the air flow and build up the geometry that to be simulated respectively. Phoenics is a CFD software package that able to analyze fluid flow by showing the prediction of fluid flow such as air, water and etc. Predictions are based on Navier-Stokes equation. The simulation can be viewed holistically by using color contour. Phoenics solves finite domain by using mass, momentum, energy conservation equations. In brief, CFD analysis come into the picture to replace the tedious and time consuming manual calculation. Besides, SolidWorks 2013 was used to build up a multi outlet pipe. SolidWorks is a three dimensional solid modelling program used to produce computer design models and has to be converted to .stl before import to Phoenics CFD.

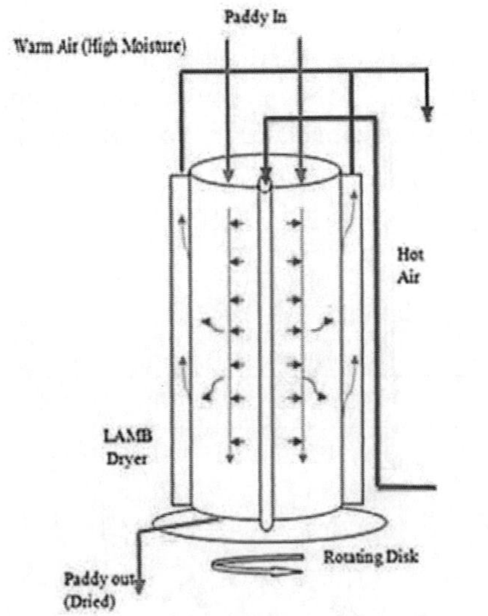

Figure 1. Schematic Diagram of LAMB for Paddy Drying [2]

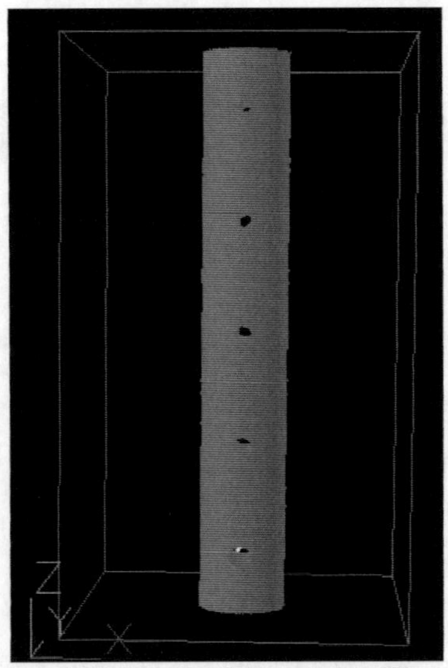

Figure 2. Geometry imported to Phoenics CFD

Table 1. Specification Sheet of Vertical Perforated Pipe

Perforated pipe length, L (mm)	200
Pipe inner diameter, ID (mm)	22.46
Pipe outer diameter, OD (mm)	26.68
Diameter of orifices from top to bottom of vertical perforated pipe, D_i (mm)	
At 0° and 180°	4.62, 4.92, 5.35, 6.07, 8.00
At 90° and 270°	4.76, 5.11, 5.66, 6.72

2.1 Assumptions and Geometry Construction

Assumptions were made in the simulation of the flow in a perforated pipe: 1) it was one-dimensional where the main flows were not considered to be modified by secondary flows. If the flow was along the pipe axis, the radial flow would be neglected; and if the flow was a discharge from a perforated hole, only the flow perpendicular to the cross-sectional flow area of the hole was important; 2) the fluid was considered to behave as an ideal gas as the pressure would not exceed 20 bar, the limit of ideal gas behavior, 3) the system was isothermal since the maximum temperature difference would be

30K, with minimum heat loss, and 4) the flow was at steady state with a pump constantly running as fluid mover. The geometry of thea flow domain was created in Cartesian coordinates form. The geometry consisted of a straight vertical pipe with 22.46 mm internal diameter and 26.68 mm outer diameter with height of 200 mm. In order to achieve uniform fluid distribution at top orifices, a module of vertical multi outlets perforated pipe was specially designed with various orifices diameter and located at uniform intervals. In this study, 4 orifices were located at uniform longitudinal intervals of 40mm along the pipe at both 90° and 270°, with the orifices at both ends measured at 40mm from either end of the pipe; while 5 orifices were located at 4 intervals of 40mm at both 0° and 180°, where the orifices at both ends were located at 20mm from either end of the pipe.

Specification sheet of a perforated tube is shown in Table 1, and the geometry imported from SolidWorks into the CFD simulation software is shown in Figure 2.

2.2 Simulation Set-Up

Simulations with different mesh/grid size are used in Phoenics CFD. Settings include dimension of geometry, grid size, boundary conditions, and number of iterations. Mesh size with in increment order are used in the simulation. In this study, as far as mesh size is concerned, three types of settings were done, namely coarse, fine, and finest.

Boundary conditions of simulations (Case 1) are shown in Table 2.

Table 2. Setting for CFD simulation

Setting	Simulation		
	Coarse	Fine	Finest
Fed air velocity (m/s)	0.10	0.10	0.10
Grid resolution (units)	59 (X) 44 (Y) 60 (Z)	86 (X) 86 (Y) 80 (Z)	80 (X) 80 (Y) 90 (Z)
Operating temperature (^{o}C)		90	
# iterations		1,000	

2.3 Mesh Independence Study

For this investigation, 3 simulations were carried out. Three simulations with lesser numeric iterations (1,000) were carried out to analyze the influence of the mesh grid. Three simulations with different mesh grid; coarse, fine and finest mesh. A domain height of 0.2 m and domain length and width of 0.1 m was chosen. In this study, only inlet been set at bottom of perforated tube which function as to introduce hot air in z direction (upward). Contour plot (a 2D color maps that convey information and data) was applied in simulations to show the direction of fluid flow. The purpose of carry out simulations at different grid size was to discuss the meshing requirements for a valid result. The more accurate/fine the mesh, the more accurate the converged solution will be. Convergence is an important identification for all CFD users to ensure a valid solution. In brief, convergence with mesh independent study must be ensured for every simulation or else extra iterations have to be carried out. In this study, error of less than 1 % was set as targeted convergence percentage. Independent of the mesh solution study helps to avoid inaccurate results in CFD. Since the deviation in average value between fine grid and finest grid was only 0.3 %, therefore, simulation named case 2 with grid size of $80 \times 80 \times 90$ (cells in x direction x cells in y direction x cells in z direction) (250 total cells) was simulated with 10,000 iteration. The simulation set up of a perforated tube for Case 1 is shown in Table 3.

From the result in Table 3, convergence was reached for all simulations. The monitored value on error percentage was used to judge whether convergence achieved through the simulation.

Convergence was fully achieved as far as steady state is concerned. As the data regarding monitor error (%) for pressure and velocity, it can be seen that the error percentage drop as mesh/grid size increases.

Table 3. CFD set-up for case 1

Setting	Case 1		
	Coarse	Fine	Finest
Fed air velocity (m/s)		0.10	
Grid resolution (units)	59 (X) 44 (Y) 60 (Z)	80 (X) 60 (Y) 80 (Z)	118 (X) 116 (Y) 140 (Z)
# iterations		1,000	
		Error (%)	
Pressure	1.98×10^{-1}	2.49×10^{-1}	3.04×10^{-1}
Velocity	2.59×10^{-1}	3.12×10^{-1}	2.60×10^{0}
Temperature	0	1.73×10^{-3}	5.05×10^{-3}
Average velocity at probe location of 0.048884 × 0.048593 × 0.016250 (X location × Y location × Z location)	0.019110	0.020830	0.020891
Time required (min)	145	114	558

2.4 Theoretical Calculations

In this study, the simulated results produced by Phoenics CFD were compared with the theoretical calculated values. The basic principles used by most researchers to model the flow behavior of hot air in the internal pipe were the conservation of mass and the conservation of energy.

Assuming that the density of hot air remains unchanged, the theoretical velocity at the orifices can be calculated by using equation (1) derived from the conservation of mass on a perforated tube as shown schematically in Figure 3 [6]:

$$V_i = \left(1 - \frac{i-1}{n}\right) V_1 \tag{1}$$

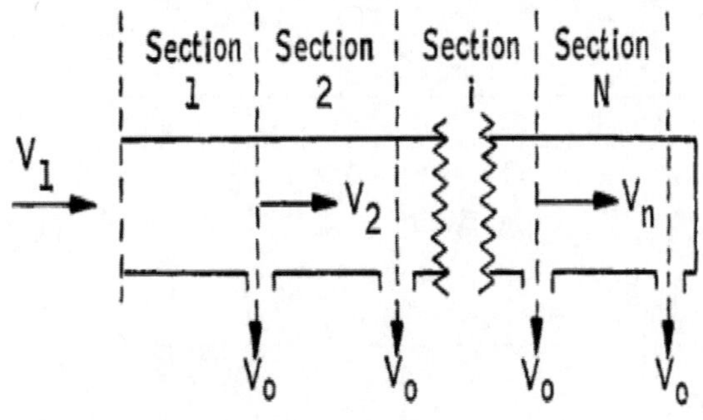

Figure 3. Expanded Schematic of Perforated Tube [6]

Due to the loss of mass along the pipe from the inlet, the mass flow hence the velocity would progressively reduce but the conservation of energy ensured that pressure increased. This would predict that discharge rate at orifices further away from the inlet would increase.

3. Results and Discussions

3.1 Steady State Simulations

Velocity profile of a perforated tube for case 1 is shown in Figure 4. From velocity profiles below, air distribution increases from top to bottom of perforated tube. However, flow at the same level of orifice did not distribute evenly for large grid size.

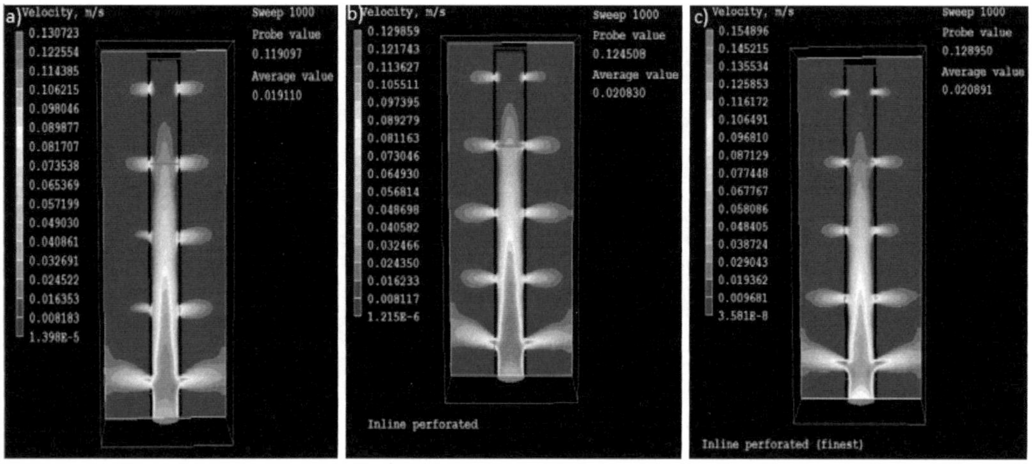

Figure 4. Velocity Profile for Case 1 a) coarse, b) fine, c) finest

Table 4 was the set up for Case 2, where the mesh size was 'fine' from Case 1. The number of iterations was now set at 10,000. Velocity profile of the perforated tube for Case 2 is shown in Figure 5.

Table 4. Simulation setup after mesh independent study (Case 2)

Setting	Case 2
	Exquisite
Fed air velocity (m/s)	0.10
Grid resolution (units)	80 (X) 80 (Y) 90 (Z)
# iterations	10,000
	Error (%)
Pressure	2.23×10^{-1}
Velocity	9.10×10^{-1}
Temperature	4.21×10^{-3}
Time required (min)	1,705

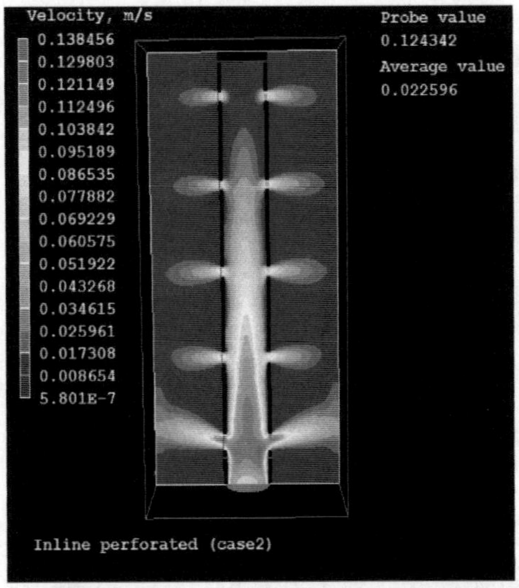

Figure 5. Velocity Profile for Case 2

From velocity profiles above, fluid flows at the highest velocity at middle of the multi outlet perforated pipe which represent hot air has higher flow rate at middle of perforated pipe. Besides, uniform flow was achieved for top 3 orifices and the largest flow at bottom orifice.

3.2 Mean Values From Phoenics CFD and Theoretical Values

While the air flow of top orifices seemed to be uniform as shown in Figure 5, it is insufficient to justify whether uniform flow had been achieved by looking at the velocity plot that was simulated. Therefore, 6 data collection planes were installed at inlet and orifices (outlet) of perforated pipe as dummy objects. At these dummy objects mean velocity and mass flow rate were calculated by Phoenics CFD with a user-defined routine feature known as In-Form 13. The mean values show that uniform flow was achieved for top orifices.

As stated in section 2.4, theoretical values for velocity can be calculated by using the equations listed. Both mean values obtained from Phoenics CFD simulations and theoretical values were compared and converted into graphs. Both graphs show the same trend. To conclude, computational analysis was performed on set geometry and the analysis was similar to the theoretical calculated values. The two methods mutually corroborate each other. Mass flow rate and velocity comparisons are shown in Figures 6 and 7 respectively.

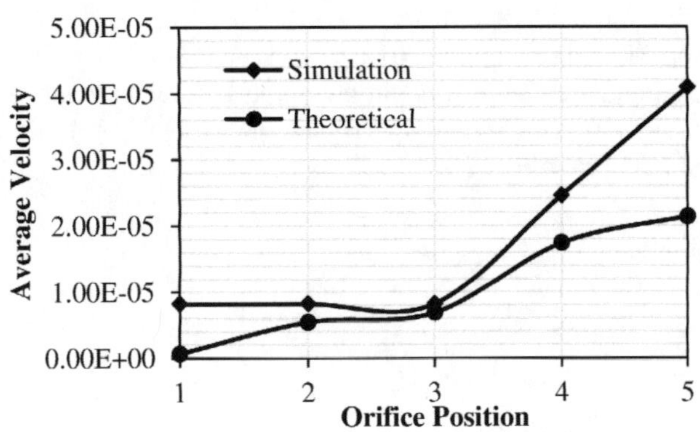

Figure 6. Mass flow rate of simulated result and theoretical value

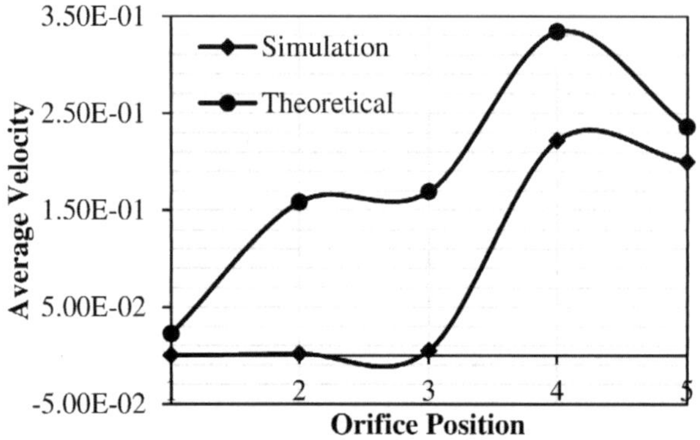

Figure 7. Velocity of simulated result and theoretical value

4. Conclusion

This research study has been carried out to investigate the performances of multi outlet perforated tube. CFD simulation of hydrodynamics of perforated tube has been carried out at different design configuration. Since geometry of multi-outlet perforated tube was not readily unavailable in Phoenics CFD, Solidworks 2013 was used for the geometry assembly before it could be imported into the CFD analysis software. Mesh independent study was carried out to ensure a reliable and converged result. Besides, fluid flow at different velocity in the same perforated pipe configuration was carried out. The purpose was to ensure that the designed perforated pipe actually distribute uniform fluid flow at top orifices within a range of inlet velocities.

From the designed perforated pipe, the relationship between the orifices diameter and the location of orifice with respect to the inlet as datum level H was deduced by a correlation of the ratios between i-th and i+1-th hole with the ratio of the pipeline flow area to the total orifices area, $\frac{A_p}{\sum A_o}$ which was equal to 1.67. The diameters of adjacent orifices were found to follow the ratios relationship $\frac{D_{i+1}}{D_i} = (\frac{H_i}{H_{i+1}})^{0.25}$.

5. References

[1] Melanie P-P 2012 Simulation of Perfusion Flow Dynamics for Contrast Enhanced Imaging, *Thesis*. University of Toronto.

[2] Sarker M S H, Ibrahim M N, Ab. Aziz N, and Mohd Salleh P 2014 *Energy Conversion and Management* **77** 389–395 doi:10.1016/j.enconman.2013.09.038

[3] Wimberly J E 1983 *Technical Handbook for the Paddy Rice Postharvest Industry in Developing Countries*. Int. Rice Res. Inst.

[4] Sangkyoo L, Namsoo M, and Jeekeun L 2012 **26** 2751–2758. doi:10.1007/s12206-012-0721-z

[5] Tan KY 2014 Design and Fabrication of Single-Tube Laterally Aerated Moving Bed (LAMB) Dryer for Agricultural Biomass, *Final Year Project Thesis*, Universiti Malaysia Sabah

[6] Florham P N J 1968 *IEC PROCESS Des Dev* **7**

Peroxidase extraction from jicama skin peels for phenol removal

T Chiong[1,2], S Y Lau[2], E H Khor[2], M K Danquah[2]
[1] Curtin Sarawak Research Institute, Curtin University Sarawak, 98009 Miri, Sarawak, Malaysia
[2] Department of Chemical Engineering, Curtin University Sarawak, 98009 Miri, Sarawak, Malaysia

E-mail: [1,2]chiong.tung@postgrad.curtin.edu.my, [1]johnlsy@curtin.edu.my, [1]khor.ee.huey@curtin.edu.my, [2]mkdanquah@curtin.edu.my

Abstract. Phenol and its derivatives exist in various types of industrial effluents, and are known to be harmful to aquatic lives even at low concentrations. Conventional treatment technologies for phenol removal are challenged with long retention time, high energy consumption and process cost. Enzymatic treatment has emerged as an alternative technology for phenol removal from wastewater. These enzymes interact with aromatic compounds including phenols in the presence of hydrogen peroxide, forming free radicals which polymerize spontaneously to produce insoluble phenolic polymers. This work aims to extract peroxidase from agricultural wastes materials and establish its application for phenol removal. Peroxidase was extracted from jicama skin peels under varying extraction conditions of pH, sample-to-buffer ratio (w/v %) and temperature. Experimental results showed that extraction process conducted at pH 10, 40% w/v and 25°C demonstrated a peroxidase activity of 0.79 U/mL. Elevated temperatures slightly enhanced the peroxidase activities. Jicama peroxidase extracted at optimum extraction conditions demonstrated a phenol removal efficiency of 87.5% at pH 7. Phenol removal efficiency was ~ 97% in the range of 30 – 40°C, and H_2O_2 dosage has to be kept below 100 mM for maximum removal under phenol concentration tested.

1. Introduction

Rapid development and blooming of various industries such as petroleum refineries, wood preservation, resins and plastics, pulp and paper, dyes and other chemicals contribute significantly to the economics of many countries. However, these industries also generate huge amount of industrial effluents, of which if discharged untreated or partially treated, could cause serious ecological problems to aquatic lives and human health. This is due to the presence of xenobiotic and recalcitrant organic pollutants in the wastewaters, which include phenols and its derivatives. Phenolic compounds are mostly toxic and have been classified as hazardous pollutants [1]. Therefore, developing effective and sustainable treatment technologies for industrial wastewaters is of vital importance and continues to be one of the greatest challenges to the environmental engineering and science.

[1] To whom any correspondence should be addressed.

Conventional treatment methods for phenol removal from wastewater include microbial degradation [2, 3], adsorption on activated carbon [4, 5], and chemical oxidation such as ozonation [6, 7] and Fenton reaction [8, 9]. The aforementioned methods, though showing high removal efficiencies for phenol, suffer certain disadvantages such as lengthy start-up for microbial acclimatization [10], microbial growth inhibition due to high concentrations of phenolic compounds [11], intensive cost and energy for carbon regeneration [12], and disposal concern of reagents used during reaction process [12]. Due to the challenges associated with conventional phenol removal methods, research focuses have been geared towards finding an alternative method that is versatile for a wide range of reaction conditions with little or no environmental impacts for phenol removal. Enzymatic approach has been proposed by researchers for phenol bioremediation owing to its potential advantages over conventional treatment methods. These include application to bio-refractory compounds, operation at high and low contaminant concentrations, operation over a wide range of pH, temperature and salinity, absence of delays associated with the acclimatization of biomass, reduction in sludge volume (no biomass generated) and ease of controlling the process [13].

Among the various types of enzymes, peroxidases have been identified as a suitable candidate for the treatment of phenolic contaminants and related compounds. Activated by hydrogen peroxide, peroxidases catalyze the oxidation of aromatic compounds, forming free radicals which undergo spontaneous polymerization. The polymerized product precipitates out from the solution and can be readily separated by sedimentation and filtration techniques [14]. Peroxidases are ubiquitous in nature and can be found in plants, microorganisms and animals. Horseradish peroxidase (HRP) has been explored extensively for its potential in decontamination of phenolic compounds from wastewater [15-20]. However, most of the studies conducted were using purified HRP which imposed high cost to the removal process. This has triggered the search for peroxidases from other less expensive sources with comparable or better stability and efficiency in phenol removal. Peroxidases from soybean seed hulls [1,14,21], turnip [22], bitter gourd [23], radish [24], cauliflower [25] and potato [24] have been evaluated for their performances in phenol removal, and the results obtained were encouraging and promising. Other peroxidases that have been identified, extracted and characterized include oil palm leaves [26], rice leaves [27], sweet potato [28] and tomato [29].

The aim of this study is to explore the possibility of using jicama skin peels as the source of peroxidase for phenol removal. The skin of jicama is not edible and therefore is discarded as waste. Our preliminary study has shown that peroxidase exists in the skin peels of jicama. The use of this agricultural waste for peroxidase extraction is not in competition with human food consumption. This approach offers a chain of green technology as waste from one source is transformed into a useful product to treat waste from a different source. The extraction of peroxidase from jicama skin peels will be studied from the aspect of pH, sample-to-buffer ratio and temperature. Jicama peroxidase extracted at optimum conditions will then be evaluated for its performance efficiency in phenol removal under varying reaction conditions which include pH, temperature and hydrogen peroxide (H_2O_2) concentration.

2. Materials and methods

2.1. Preparation of crude extract

Jicama purchased from local market was washed thoroughly with distilled water, and then the skin peeled off and chopped into small pieces. The chopped jicama was weighed and mixed with buffer solutions based on weight (g)/volume (mL) percentage (w/v %). Homogenization of enzyme mixtures was carried out at 400rpm for 1 h at room temperature (unless stated otherwise). The enzyme extract was then filtered through four layers of cheesecloth before being subjected to centrifugation and sonication. The enzyme collected was considered as crude extract and stored at 4^0C until further use.

2.2. Analytical procedures
Enzyme activity of jicama peroxidases was measured using a colorimetric assay containing phenol, 4-aminoantipyrene (4-AAP) and hydrogen peroxide. This assay is a modification of that developed by Wu *et al.* [20] in which the assay mixture consists of 250µl of 9.6mM 4-AAP, 100µl of 100mM phenol, 100µl of 2mM hydrogen peroxide, 450-500µl of 100mM phosphate buffer (pH 6.0) and 50-100µl of enzyme solution. Prior to significant substrate depletion, activity was proportional to the rate of formation of a coloured product which absorbs light at a peak wavelength of 510nm with an extinction coefficient of 7100 L/mol.cm based on the conversion of H_2O_2.One unit of activity is defined as the number of micromoles of H_2O_2 consumed per minute at pH 6.0 and 25°C.

Phenol concentrations were determined colorimetrically using 4-AAP and potassium ferricyanide in an alkaline buffer medium. Phenolic compounds react with 4-AAP under alkaline conditions to yield an intermediate species which is oxidized in the presence of the potassium ferricyanide reagent. The resulting compound is a quinone-type dye which absorbs light at 510nm. The colour intensity is linear with respect to phenol concentration, provided that this concentration does not exceed 0.1mM in the cuvette [30]. The absorbance was measured at 510nm after 5 minutes.

2.3. Extraction of jicama peroxidase

2.3.1. Effect of pH on peroxidase extraction. The effect of pH on peroxidase extraction from jicama peels was evaluated by mixing buffer solutions (0.1M) of different pH values to the weighed amount of jicama. The range of pH being studied was from pH 3 to 10. One gram of jicama peels was added to 10ml of buffer solution at certain pH, and homogenization was carried out as previously described. Enzyme sample-to-buffer ratio (w/v %) in this aspect was 10%.

2.3.2. Effect of sample-to-buffer ratio on peroxidase extraction. Sample-to-buffer ratio (w/v %) for jicama peroxidase extraction was investigated at 10%, 20% and 40%. The maximum w/v % that can be achieved was based on the minimum amount of buffer solution needed to cover all the pre-weighed quantity jicama peels for proper homogenization.

2.3.3. Effect of temperature on peroxidase extraction. Jicama peroxidase extraction was carried out at 25°C, 35°C and 45°C. For 35°C and 45°C, the extraction process was carried out in water bath with constant stirring. The temperatures of the water bath were monitored to ensure constant temperature throughout the extraction process.

2.4. Batch treatment of aqueous phenol by using jicama peroxidase
The efficiency of jicama peroxidase in removing phenol from aqueous solution was evaluated under varying reaction conditions such as pH, temperature and H_2O_2 concentration. Jicama peroxidase was prepared based on the optimum conditions in extraction process. The reaction mixture consisted of buffer solution, phenol solution and enzyme. The enzymatic reaction was initiated through the addition of H_2O_2 into the mixture, and the process was carried out in incubator with constant shaking for 24 h to ensure maximum phenol removal. Phenol concentration in the reaction mixture before and after reaction process was assayed according to section 2.2, and percentage of phenol removal calculated.

2.4.1. Effect of pH on phenol removal.
Phenol removal processes were carried out in buffer solutions of various pH ranging from pH 4 to 9. The molarity of the buffer solutions was kept constant at 0.1M. The reaction mixtures were incubated at 30°C with constant shaking for 24 h.

2.4.2. Effect of temperature on phenol removal.
Reaction mixtures containing 1 mL of 10mM phenol were treated with 1.5 mL jicama peroxidase in the presence of 0.2 mL of 10mM H_2O_2 at the optimum pH. Reaction mixtures were incubated at temperatures 30, 40, 50 and 60°C respectively for 24 h with constant shaking.

2.4.3. Effect of H_2O_2 concentration on phenol removal.
The effect of H_2O_2 concentration on phenol removal was performed by varying the concentrations of H_2O_2 added into the reaction mixtures while keeping other reaction conditions (pH and temperature) optimum. The range of concentration of H_2O_2 being evaluated was from 0.2mM to 4mM.

3. Results and discussion

3.1. Effect of pH on peroxidase extraction

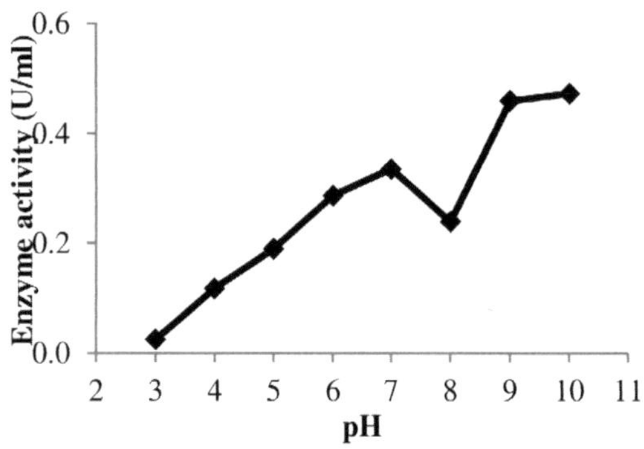

Figure 1. Peroxidase extraction as a function of pH. Extraction conditions: 1g of jicama peels was mixed with 10 mL of buffer solutions of various pH (10% w/v) and homogenized at 400 rpm and 25°C for 1 h.

The pH condition of the medium plays an important role in enzyme extraction process as it affects the ionization state of enzyme amino acids side chain. As shown in figure 1, low pH did not favour the extraction process for jicama peroxidase. This could be due to the detachment of haem prosthetic group from the poly-peptide chain [31]. Crude enzyme of jicama extracted at high acidic regions showed low enzyme activities. Peroxidase activity increased as the pH value approached neutral. However, there was a dip in peroxidase activity at pH 8, which is slightly basic. Peroxidase activity increased again at higher alkalinity regions. The maximum peroxidase activity for jicama was observed at pH 10, with an activity of 0.473 U/mL. This could be due to most of the amino acids side chain in jicama peroxidase molecules are positively charged. In a highly basic medium, the positively charged amino acids are attracted to hydroxide ions, hence releasing the peroxidase molecules into the medium.

3.2. Effect of sample-to-buffer ratio on peroxidase extraction
As can be seen from Figure 2, enzyme activity obtained increased when the amount of jicama used in the homogenization medium was increased. This is because the extracts are more concentrated with peroxidase molecules. The highest sample-to-buffer ratio that can be achieved with jicama was 40%. Beyond this ratio, the volume of buffer solution was not sufficient for proper homogenization. Sample-to-buffer ratio of 40% (w/v) will thus be used for subsequent parameter studies.

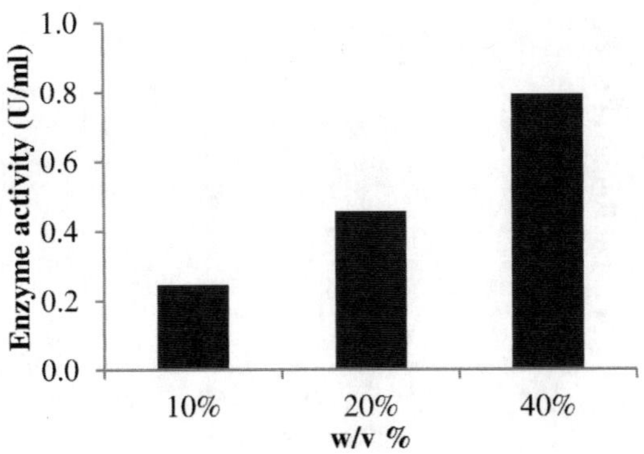

Figure 2. Peroxidase extraction as a function of sample-to-buffer ratio (w/v %). Extraction conditions: jicama peels was mixed with 0.1M buffer pH 10 at various w/v % and homogenized at 400 rpm and 25°C for 1 h.

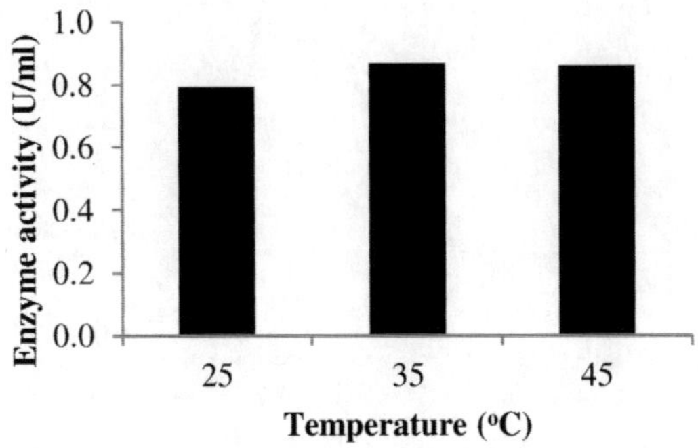

Figure 3. Peroxidase extraction as a function of temperature. Extraction conditions: jicama peels was mixed with 0.1M buffer pH 10 at 40% w/v and homogenized at 400 rpm at various temperatures for 1 h.

3.3. Effect of temperature on peroxidase extraction
From figure 3, it can be seen that extraction of jicama peroxidase at temperatures of 35°C and 45°C enhanced enzyme activity for ~ 9.6 %. However, in view of the extra cost incurred for extraction process at higher temperatures, it is therefore more economical to extract jicama peroxidase at room temperature (~25°C).

3.4. Effect of pH on phenol removal
Figure 4 shows the dependence of phenol removal efficiency on pH for jicama peroxidase. When 1 mM stock phenol solution was treated with jicama peroxidase, phenol removal efficiency of more than 80% was observed at pH values 6 to 7, with the optimum occurring at pH 7. Removal efficiency decreased to less than 40% below pH 5 and above pH 8. This is a result of variations in enzyme protein structure in response to varying pH. The enzyme molecules undergo structural modifications as a result of protonation and hydroxylation effects which could potentially obscure the enzyme active sites before causing denaturation and permanent loss of functionality. Moreover, the decrease in removal efficiency at pH above 8 could be attributed to the formation of phenol conjugated base since the pK_a of phenol at 25°C is 10. This conjugated basic form does not permit the phenolic compounds to act as hydrogen donors, thus hindering binding onto the surface of the enzyme active sites.

Previous studies on phenol removal catalyzed by horseradish peroxidase demonstrated optimal operating pH at pH 8 [17, 18], which was slightly basic. Another work by Wright and Nicell [21] showed that nearly complete removal of phenol was observed over a pH range of 5 to 9, with the maximum removal at pH 6 by using high dose of soybean peroxidase. Other studies using soybean peroxidase exhibited optimum pH for phenol removal at pH 6 [1] and pH 7 [32]. For jicama peroxidase, subsequent enzymatic reactions will be conducted using buffer pH 7.

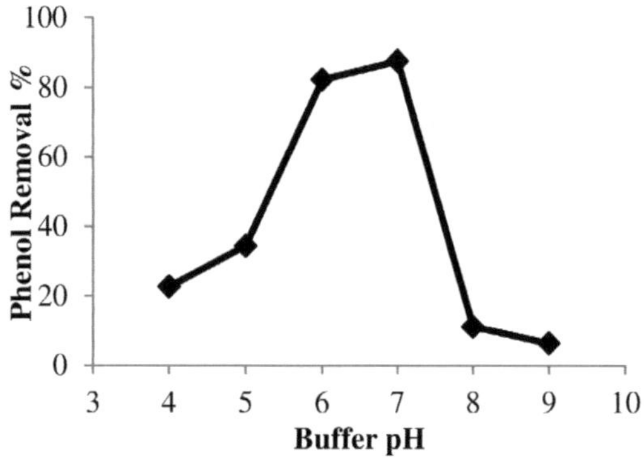

Figure 4. Phenol removal using jicama peroxidase at different pH conditions. Experiments were performed using 1.5 mL jicama peroxidase and 1 mL of 1mM phenol in buffer solutions of various pH at 30°C in the presence of 0.2mM H_2O_2.

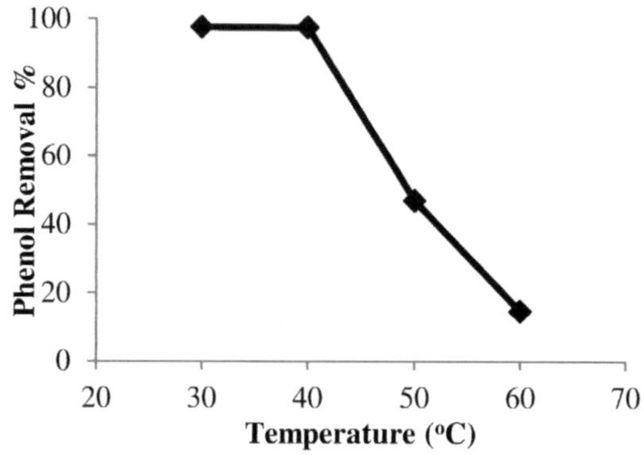

Figure 5. Phenol removal using jicama peroxidase at different operating temperatures. Experiments were performed using 1.5 mL jicama peroxidase and 1 mL of 10mM phenol in buffer pH 7 in the presence of 0.2mM H_2O_2.

3.5. Effect of temperature on phenol removal

The effect of temperature on phenol removal was examined by incubating the reaction mixtures containing 1 mL of 10 mM phenol stock solution at various temperatures, ranging from 30°C to 60°C. As depicted in figure 5, phenol removal efficiency of greater than 90% was observed over the temperature range of 30 – 40°C. Moderate increases in temperature from ambient conditions results in enhanced collision of the enzyme molecules with substrate molecules to form more products. However, further increase in temperature to above 50°C caused a decrease in removal efficiency to less than 50%. This is attributed to thermal denaturation of enzyme molecules, causing loss of its active sites to catalyze substrate molecules. The thermal stability of peroxidases is governed by the haem prosthetic group, which under elevated temperatures is released to form apoenzyme. The transient enzyme formed is less stable and more susceptible to thermal inactivation as compared to the native enzyme [33]. Higher temperature conditions distort the structure of the enzyme, causing a limited binding capacity of its active sites onto substrate molecules.

3.6. Effect of H_2O_2 concentration on phenol removal

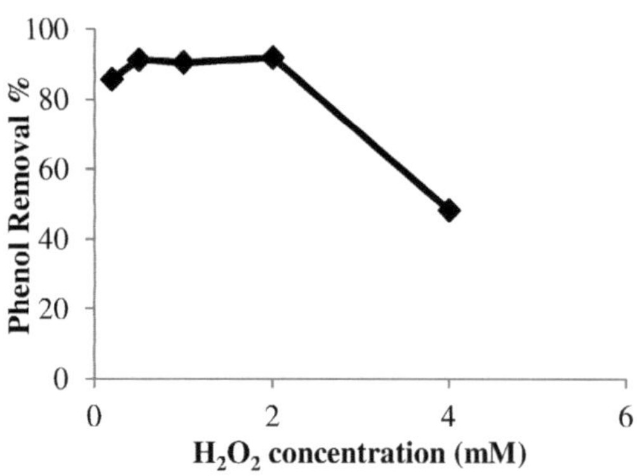

Figure 6. Phenol removal using jicama peroxidase at different H_2O_2 concentrations. Experiments were performed using 1.5 mL jicama peroxidase and 1 mL of 10mM phenol in buffer pH 7 at 30°C.

Hydrogen peroxide (H_2O_2) is required in peroxidase-catalyzed reactions to oxidize the native enzyme molecules into Compound I which then accepts an aromatic compound into its active site and carries out its oxidation [18]. Low concentrations of H_2O_2 limit the enzymatic reaction rate while excessive amounts could result in inhibitory effect on the enzyme activity [34]. According to figure 6, phenol removal efficiencies of more than 85% were recorded over a wide range of H_2O_2 concentrations, from 0.2 mM to 2 mM. However, enzymatic inhibition effect was observed at H_2O_2 concentration beyond 2 mM where the efficiency dropped to less than 50%. High concentrations of H_2O_2 inhibit peroxidase catalytic activity by irreversibly oxidizing the enzyme ferriheme group which is vital for peroxidase catalysis [22]. Therefore, phenol removal process using jicama peroxidase has to be carried out at H_2O_2 concentration less than 2 mM to ensure maximal removal.

4. Conclusions

Present study demonstrated that plant peroxidase can be extracted from agricultural and bio-wastes for phenol removal application. The peroxidase extraction process established in this work is simple and does not require the usage of many chemicals. The crude enzyme from jicama peels was subjected to phenol removal reaction and the results showed that the enzymatic process was dependent on pH, temperature and concentration H_2O_2. The extracted jicama peroxidase was not subjected to any further purification, and its activity in the homogenous crude enzyme solution was uniform. Relatively, enzyme volume was used in place of enzyme concentration as one of variable parameters in phenol removal reaction. However, its effect on phenol removal efficiency was not significant for the range being studied (data not shown).

The phenol removal efficiencies exhibited by jicama peroxidase was encouraging and this revealed the potential of jicama peroxidase in enzymatic wastewater treatment. The use of low cost materials and processes for peroxidase extraction coupled with moderate enzymatic reaction conditions and limited use of harsh chemicals creates a sustainable and environmentally friendly operation with low process economics. Nevertheless, further studies into optimization of reaction conditions such as reaction time and substrate dosages are essential to better understand the factors affecting the performance of this enzyme. The effects and interactions between various extraction conditions such as pH, sample-to-buffer ratio and temperature as well as the effect of H_2O_2 on enzymatic activities should also be addressed.

5. References

[1] Caza N, Bewtra J K, Biswas N and Taylor K E 1999 *Water Res.* **33** 3012
[2] Krastanov A, Alexieva Z and Yemendzhiev H 2013 *Eng. Life Sci.* **13** 76
[3] Thomas S, Sarfaraz S, Mishra L C and Iyengar L 2002 *World J. Microbiol. Biotechnol.* **18** 57
[4] Hameed B H and Rahman A A 2008 *J. Hazard. Mater.* **160** 576
[5] Mukherjee S, Kumar S, Misra A K and Fan M 2007 *Chem. Eng. J.* **129** 133
[6] Manojlovic D, Ostojic D R, Obradovic B M, Kuraica M M, Krsmanovic V D and Puric J 2007 *Desalin.* **213** 116
[7] Turhan K and Uzman S 2008 *Desalin.* **229** 257
[8] Maciel R, Sant'Anna Jr G L and Dezotti M 2004 *Chemosphere* **57** 711
[9] Yavuz Y, Savas Koparal A and Bakir Ögütveren Ü 2007 *Chem. Eng. Technol.* **30** 583
[10] Firozjaee T T, Najafpour G D, Asgari A and Khavarpour M 2012 *Chem. Ind. Chem. Eng. Q.* **19** 173
[11] Gernjak W, Krutzler T, Glaser A, Malato S, Caceres J, Bauer R, *et al.* 2003 *Chemosphere* **50** 71
[12] Busca G, Berardinelli S, Resini C and Arrighi L 2008 *J. Hazard. Mater.* **160** 265
[13] Karam J and Nicell J A 1997 *J. Chem. Technol. Biotechnol.* **69** 141
[14] Wilberg K, Assenhaimer C and Rubio J 2002 *J. Chem. Technol. Biotechnol* **77** 851
[15] Klibanov A M, Alberti B N, Morris E D and Felshin L M 1980 *J. Appl. Biochem.* **2** 414
[16] Klibanov A M and Morris E D 1981 *Enzyme Microb. Technol.* **3** 119

[17] Nicell J A, Bewtra J, Taylor K, Biswas N and StPierre C 1992 *Water Sci. Technol.* **25** 157
[18] Nicell J A, Bewtra J K, Biswas N, St. Pierre C C and Taylor K E 1993 *Can. J. Civ. Eng.* **20** 725
[19] Cooper V A and Nicell J A 1996 *Water Res.* **30** pp 954-964
[20] Wu J, Taylor K E, Bewtra J K and Biswas N 1993 *Water Res.* **27** 1701
[21] Wright H and Nicell J A 1999 *Bioresour. Technol.* **70** 69
[22] Duarte-Vázquez M A, Ortega-Tovar M A, García-Almendarez B E and Regalado C 2003 *J. Chem. Technol. Biotechnol.* **78** 42
[23] Akhtar S and Husain Q 2006 *Chemosphere* **65** 1228
[24] Dec J and Bollag J-M 1994 *Biotechnol. Bioeng.* **44** 1132
[25] Deva A N, Arun C, Arthanareeswaran G and Sivashanmugam P 2014 *J. Environ. Chem. Eng.* **2** 1148
[26] Deepa S S and Arumughan C 2002 *Phytochem.* **61** 503
[27] Ito H, Hiraoka N, Ohbayashi A and Ohashi Y 1991 *Agric. Biol. Chem.* **55** 2445
[28] Leon J C, Alpeeva I S, Chubar T A, Galaev I Y, Csoregi E and Sakharov I Y 2002 *Plant Sci.* **163** 1011
[29] Marangoni A G, Brown E D, Stanley D W and Yada R Y 1989 *J. Food Sci.* **54** 1269
[30] Ghioureliotis M and Nicell J A 1999 *Enzyme Microb. Technol.* **25** 185
[31] Vámos-Vigyázó L and Haard N F 1981 *C R C Crit. Rev. Food Sci. Nutr.* **15** 49
[32] Bódalo A, Gómez J L, Gómez E, Bastida J and Máximo M F 2006 *Chemosphere* **63** 626
[33] McEldoon J P and Dordick J S 1996 *Biotechnol. Progr.* **12** 555
[34] Mohan S V, Prasad K K, Rao N C and Sarma P N 2005 *Chemosphere* **58** 1097

Acknowledgments
The authors would like to thank Curtin Sarawak Research Institute (CSRI) for providing the financial support for this project through the CSRI Academic Grant Scheme.

IOP Conf. Series: Earth and Environmental Science **36** (2016) 012049 doi:10.1088/1755-1315/36/1/012049

2-[(Hydroxymethyl)amino]ethanol in water as a preservative: Study of formaldehyde released by Taguchi's method

W Wisessirikul[1], S Loykulnant[2], S Montha[2], T Fhulua[3] and P Prapainainar[4,5]
[1]Interdisciplinary Graduate Program in Advanced and Sustainable Environmental Engineering, Thailand Advanced Institute of Science and Technology and Tokyo Institute of Technology, *Kasetsart University*, Bangkok 10900 Thailand
[2]Natural Rubber Focus Unit, National Metal and Materials Technology Center, NSTDA, 111 Paholyothin Rd., Klong 1, Klong Luang, Pathumthani 12120 Thailand
[3]Environment Research Unit, National Metal and Materials Technology Center, NSTDA, 111 Paholyothin Rd., Klong 1, Klong Luang, Pathumthani 12120 Thailand
[4]National Center of Excellence for Petroleum, Petrochemicals and Advance Materials, Department of Chemical Engineering, Faculty of Engineering, *Kasetsart University*, Thailand
[5]Center of Excellence of Nanotechnology, *Kasetsart University*, Thailand

E-mail: [4,5]fengpwn@ku.ac.th

Abstract. This research studied the quantity of free formaldehyde released from 2-[(hydroxymethyl)amino]ethanol (HAE) in DI water and natural rubber latex mixture using high-performance liquid chromatography (HPLC) technique. The quantity of formaldehyde retained in the solution was cross-checked by using titration technique. The investigated factors were the concentration of preservative (HAE), pH, and temperature. Taguchi's method was used to design the experiments. The number of experiments was reduced to 16 experiments from all possible experiments by orthogonal arrays (3 factors and 4 levels in each factor). Minitab program was used as a tool for statistical calculation and for finding the suitable condition for the preservative system. HPLC studies showed that higher temperature and higher concentration of the preservative influence the amount of formaldehyde released. It was found that conditions at which formaldehyde was released in the lowest amount were 1.6 %w/v HAE, 4 to 40 °C, and the original pH. Nevertheless, the pH value of NR latex should be more than 10 (the suitable pH value was found to be 13). This preservative can be used to replace current preservative systems and can maintain the quality of latex for long-term storage. Use of the proposed preservative system was also shown to have reduced impact on the toxicity of the environment.

1. Introduction

Thailand is a manufacturer and exporter of natural rubber (NR), making the latter one of the plants important to the country's economy [1]. NR is an elastomer or elastic hydrocarbon polymer that can

[4,5] To whom any correspondence should be addressed.

be used for processing in manufacturing centres, with the associated NR industries being highly competitive in domestic and overseas markets. Most NR is exported as crude rubber and NR latex, which can be transformed into concentrated latex that can itself be prepared for market by rolling slabs of latex for continuous use in many products such as rubber gloves and tires.

NR liquid from rubber trees is called latex. It is a sticky and milky colloid which is drawn off by making a tap and collecting it in vessels. Then the latex is prepared by refining it into rubber ready for processing. Since NR latex putrefies within 6 hours after it is incised from the rubber tree due to natural destabilization by heavy microbial contamination [2], preservation of the latex is necessary. A suitable preservative is used to maintain the quality of latex and to allow long-term storage. The preservative substances can be either a sterilizer or a stabilizer.

Currently, an ammonia system is widely used as a standard preservative for NR latex. Ammonia is used with tetramethylthiuram disulfide (TMTD) and zinc oxide (ZnO) for reducing the harmful effects of ammonia toxicity [3]. However, there are several disadvantages related to environmental pollution and health. Ammonia imparts a strong odor and is easy to evaporate. Its emitted fumes cause corrosion of factory buildings and roofs. The amount of ammonia in latex that affects the qualities of latex and makes unstable the properties of latex is difficult to control [4]. Moreover, TMTD produces carcinogenic nitrosamines and allergens in rubber products which cause contact dermatitis, while ZnO is a source of heavy metal in waste sludge and effluent water.

As a result of the preceding consideration, it has been proposed to used 2-[(hydroxymethyl)amino]ethanol (HAE) as a sterilizing and preserving solution. It is a formaldehyde-releaser with respect to free formaldehyde. This formaldehyde is used as preservative to prevent bacterial and disinfectant [5, 6], making HAE a good choice as a preservative for NR latex, instead of the ammonia system which widely used as the standard. HAE also reduces the environment toxic, thereby opening up potential applications in agriculture. This research focussed on the quantities of free formaldehyde released from HAE in DI water and in NR latex mixture with different ratios of preservative systems in order to investigate the latex quality deterioration and study the effects of various parameters on the amount of formaldehyde released: concentration of formaldehyde releaser, pH, and temperature.

2. Material and Methods

2.1. Design of experiments

Taguchi method was used in the form of orthogonal array of three factors and four levels in each factor to reduce the number of experiments from a full factorial [7] (64 experiments) to sufficient, but lower, number of experiments (16 sampling experiments). The concentration of HAE was varied as 0.1, 0.2, 0.8, and 1.6 %w/v, while the pH was either kept at the original condition or adjusted to 11, 12 or 13. The temperature was chosen as either 4, 28, 40 or 90 degrees Celsius, and the storage time was set as 24 hours. Minitab program was used as a device for statistical calculation and finding the suitable conditions for using NR preservative system. The signal-to-noise ratio (SN ratio) characteristic is a number used to improve quality through variability reduction. The concept of SN ratio for this study is "smaller is better" [8], with the actual ratio given by equation (1):

$$S/N = -10\log_{10}(\frac{1}{n}\sum y^2) \tag{1}$$

where n is the number of the tests in a trial and y is the value of the quality characteristic measured from the test.

2.2. Sample preparation

Samples of 0.2, 0.4, 1.6 and 3.2 mL of HAE were diluted into 200 mL DI water or 200 mL NR latex, and 0.1 mL of EDTA was added. The pH was adjusted with KOH; then the solution was put in a 1000 mL closed bottle. 0.5 g of solution was used in a titration process to determine the quantities of formaldehyde in the solutions at the beginning. After the settling of the solution and allowing formaldehyde release in the air for 24 hours, the outlet of a cartridge sampler was connected to a sampling pump with flexible tubing. The pump was turned on, and 800 mL of air was injected into the bottle through the cartridge sampler at a flow rate of 80 mL/min, and simultaneously, the same volume of gas for removed to a 2,4-dinitrophenylhydrazine (DNPH) cartridge for analysis by high-performance liquid chromatography (HPLC). The amount of formaldehyde in the remaining solution was analysed by titration.

2.3. Formaldehyde titration

The titration method for formaldehyde followed was that in reference [9]. Briefly, 0.5 g of formaldehyde solution was diluted with DI water into 100 mL. 20 mL of the solution was added into 25 mL of 0.05M I_2 and 10 mL of 1M NaOH. The solution was then put in a dark room for 15 minutes before 15 mL of 1M H_2SO_4 solution was added. Any excessive iodine was titrated by 0.1 M $Na_2S_2O_3$, resulting in a solution that was almost clear. After that, 1 mL of 1% w/w starch solution was added for use as an indicator in the solution. Then, 0.1M $Na_2S_2O_3$ was used for titration again, with the solution being added until the color changed from blue to clear.

A blank test was done using 20 mL of water for titration. For it, the calculation of the 0.1M $Na_2S_2O_3$ consumption is shown in equation (2). Each 1 mL of 0.1M $Na_2S_2O_3$ is can be reacted with 1.5 mg of formaldehyde while 1 mL of 0.05M I_2 can be reacted with 1.5 mg of formaldehyde.

$$[HCOH, ppm] = (V_{B1} - V_{Na_2S_2O_3}) \times 15 \times [Na_2S_2O_3] \times \frac{1000}{WS} \tag{2}$$

where $[HCOH]$ is the concentration of formaldehyde (ppm), V_{B1} is the volume of 0.1 M sodium thiosulphate solution (mL), $V_{Na_2S_2O_3}$ is the volume of 0.1 M sodium thiosulphate solution in the blank experiment (mL), $[Na_2S_2O_3]$ is the concentration of sodium thiosulphate solution (M), and WS is the weight of formaldehyde solution (g).

2.4. HPLC analysis of free formaldehyde

To the formaldehyde derivative in the DNPH cartridge sampler was added acetonitrile for the purpose of eluting each sample. Liquid chromatography was set according to the manufacturer's recommended conditions. 10 μL sample solutions were analyzed with an inertsil ODS-80A column (length 159 mm, internal diameter 4.6 mm) with 55% acetonitrile as mobile phase, and the concentrations were detected with photodiode array (PDA) detector (lamp type D2&W) at the wavelength of 360 nm. The flow speed used was 1.5 mL/min. After detection the peak area and peak height of formaldehyde were shown in the chromatogram results. The retention time of the sample derivative analyzed via HPLC was compared with the retention time of the standard formaldehyde, and the peak area of sample solution was interpreted from the standard curve of the calibration equation in order to calculate the concentration of the released formaldehyde.

3. Results and discussion

As mentioned previously, Taguchi's method was used to design the experiments, allowing the number of experiments to be reduced from 64 to 16 by orthogonal arrays (3 factors and 4 levels in each factor). Results and trends from both titration method and HPLC tend to agree, but since titration is less accurate and not accurate enough for the purposes of prediction, HPLC results are focussed on this section. Table 1 shows experimental conditions and corresponding concentrations of free formaldehyde released from HAE solution in water. Normalized concentration of free formaldehyde is calculated as the actual concentration of free formaldehyde multiplied by the ratio of 0.1 %w/v HAE to the actual concentration of HAE used. Taguchi's method can predict the mean value and SN ratio of the concentrations of free formaldehyde in all possible experiments based on statistical calculation. However, the predictions from Taguchi's method could not obtain the exact same value as the results from laboratory experiments due to normal variance in statistical analysis.

IOP Conf. Series: Earth and Environmental Science **36** (2016) 012049 doi:10.1088/1755-1315/36/1/012049

Table 2 is the response table for SN ratio of concentration of free formaldehyde using the concept of "smaller is better". The delta was calculated from the difference between the maximum and the minimum SN ratio of each factor, with a high delta value indicating a high influence of the factor on released formaldehyde. Thus, the order in which parameters affect the released formaldehyde from the most to the least is as follows: concentration of HAE, temperature, pH.

Table 1. Concentrations of formaldehyde released from HAE solution diluted in DI water

Concentration of HAE (%w/v)	pH	Temperature (°C)	Concentration of formaldehyde (μg/ml)	Normalized concentration of formaldehyde	Predicted Taguchi Results (μg/ml)
0.1	None	4	0.183	0.183	0.046
0.1	13	28	0.304	0.304	0.137
0.1	11	40	0.325	0.325	0.503
0.1	12	90	1.008	1.008	1.146
0.2	None	28	0.395	0.198	0.452
0.2	13	4	0.275	0.138	0.530
0.2	11	90	2.397	1.199	1.841
0.2	12	40	0.354	0.177	0.608
0.8	None	40	0.391	0.049	0.154
0.8	13	90	1.066	0.133	1.176
0.8	11	4	0.403	0.050	0.611
0.8	12	28	0.388	0.049	0.322
1.6	None	90	0.767	0.048	1.095
1.6	13	40	0.429	0.027	0.242
1.6	11	28	0.437	0.027	0.621
1.6	12	4	0.631	0.039	0.319

Table 2. Response table for SN ratio for concentration of free formaldehyde

Level	Concentration of HAE (%w/v)	pH	Temperature (°C)
1	8.697	20.361	21.506
2	11.198	16.355	20.497
3	23.991	17.335	20.614
4	29.294	19.129	10.563
Delta	20.597	4.007	10.943

The released formaldehyde is shown to depend on HAE concentration, pH, and temperature, and the SN ratios of the concentration of free formaldehyde released from HAE solution are shown in Figure 1 for each of these three factors. The concept of smaller is better shows here that the high value in the graph indicates a lesser amount of free formaldehyde. It was found that the concentrations of free formaldehyde increased with increasing the concentration of HAE. Also, higher concentration of HAE tends to result in decreased normalized free formaldehyde concentration. Temperature increase has an influence on the increase of formaldehyde released, with the highest release at 90°C (because this temperature is close to the boiling point of formaldehyde solution, 98°C). The graph of SN ratio of concentration of free formaldehyde versus pH is bell-shaped with a maximum of formaldehyde released at a pH of 11 and with decreasing normalized concentration of free formaldehyde with further increasing pH value. On the other end of the spectrum, the pH at the original condition, for which no KOH was used, was, when diluted with DI water, approximately 10, from which KOH affects the amount of formaldehyde released. A higher pH resulted in the ratio of concentration of free formaldehyde decreasing.

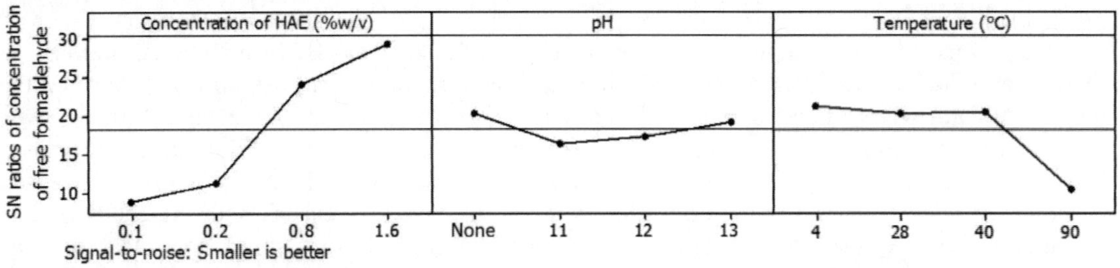

Figure 1. Effect of parameters on SN ratios of concentration of formaldehyde

For comparison purposes, the 16 experiment designed for DI water using Taguchi's method are set beside 4 further experiments with NR latex. HAE was dissolved in the latex at a fixed temperature at 28°C, and the results were compared to those obtained for identical conditions with DI water. However, of these four experiments, two—0.1 and 0.2 %w/v HAE—are not shown due to complications from reactions between HAE and proteins in latex. NR latex can be kept for more than 1 month only when it is preserved by more than 0.4 %w/v HAE [10]. Experimental conditions and results are shown in Table 3. The concentrations of free formaldehyde from NR latex at 0.8 and 1.6 %w/v HAE were close to those released from DI water solutions. The normalized concentration of free formaldehyde in NR latex decreased with increasing concentration of HAE. The pH at the original condition when diluted with NR latex was 7.83, and as for DI water, at higher values of pH, the SN ratio of concentration of free formaldehyde decreases and then rises again for yet higher pH. Table 4 shows the predicted Taguchi results for HAE solution in water at different temperatures.

Table 3. Concentrations of formaldehyde released from HAE solution diluted in NR latex

Concentration of HAE (%w/v)	pH	Temperature (°C)	Concentration of Formaldehyde in NR Latex (μg/ml)	Normalized concentration of formaldehyde in NR Latex	Normalized concentration of formaldehyde in DI water
0.8	12	28	0.441	0.055	0.049
1.6	11	28	0.505	0.032	0.027

Table 4. Suitable conditions from prediction by Taguchi's method

Concentration of HAE (%w/v)	pH	Temperature (°C)	Predicted Taguchi Results (μg/ml)
1.6	13	4	0.238
1.6	13	28	0.246
1.6	13	40	0.240
1.6	13	90	1.175

The results from the prediction by Taguchi's method can be used to specify suitable experimental conditions, those for obtaining the lowest concentration of free formaldehyde being 1.6 %w/v HAE, temperature in the range of 4 to 40°C, and the original pH. Using the original pH, however, is problematic, since EDTA requires a pH of more than 10 (in order to coordinate metal ions such that they form circular structures, i.e., chelate circles) [11] and since this original pH is itself 7.83 (i.e., more than 10). Referring to the SN ratios obtained, it then follows that the best choice, in the range pH >10, is a pH of 13. At this value, one obtains a high SN ratio and complications with EDTA.

4. Conclusion

This study focused on the quantity of free formaldehyde released from HAE in DI water and natural rubber latex solutions. The concentration of HAE, temperature, and pH has significant effects on the released formaldehyde and also on the quality of latex after preservation. The released formaldehyde increased with increasing concentration of HAE. Temperature increases resulted in increased amount of formaldehyde released, with the highest amount being released at 90°C. KOH affected the released formaldehyde. Higher pH led to the decreasing normalized concentration of free formaldehyde. The concentrations of free formaldehyde from NR latex were close to those released from DI water. The experimental conditions which cause the release of the least amount of formaldehyde were as follows: 1.6 %w/v HAE, temperatures in the range of 4 to 40 °C, and the original pH (less than 10). Nevertheless, the pH value in NR latex should be more than 10 due to the use of EDTA in the system, leading to a pH value of 13 giving the best results in the suitable pH range. Based on the findings in this work, HAE could be used to replace the typical ammonia system and maintain the quality of latex for long-term storage. Agricultural applications also remain open, since HAE results in reduced impact on environmental toxicity relative to the existing ammonia systems.

Acknowledgements

The authors acknowledge the Natural Rubber Focus Unit (NRFU), the National Metal and Materials Technology Center (MTEC), and the National Science and Technology Development Agency (NSTDA) for providing lab facilities and financial support. The authors would like to thank Dr. Sekiguchi Hidetoshi, Department of Chemical Engineering, Tokyo Institute of Technology (TIT), Japan, as exclusive consultant and TAIST-Tokyo tech program, NSTDA, for a kind support. We also thank Dr. Aree Thanaboonsombut and Ms. Panida Muangkasem, Environment Research Unit, MTEC, NSTDA, for helpful advice and use of mechanical facilities.

References

[1] Thailand R R I O 2008 Ministry of Agriculture and Cooperative
[2] Attanayake A P, Karunanayake L and Nilmini A H R L 2012 University of Sri Jayewardenepura, Sri Lanka
[3] Blackley D C 1997 Chapman & Hall London
[4] ATSDR A.f.T.S.a.D. Registry Editor 2004 Division of Toxicology U.S. Public Health Service Atlanta GA
[5] Latorre N, Silvestre J F and Monteagudo A F 2011 *Actas Dermo-Sifiliográficas* **102** 86-97
[6] Chaikumpollert O *et al* 2007
[7] Bolboacă S D and Jäntschi L 2007 *Entropy* **9** 198-232
[8] Ghani J A, Choudhury I A and Hassan H H 2004 *J Mater Process Tech* **145** 84-92
[9] Wu P E, Chang C C and Chou S S 2003 *J Food and Drug Anal* **11** 8-15
[10] Jitboon S 2008 Prince of Songkla University
[11] Sanguanthammarong P R 2008 a d c f T r industry Editor

Synthesis and characterization of silver-copper core-shell nanoparticles using polyol method for antimicrobial agent

N Hikmah[1], N F Idrus[2], J Jai[3] and A Hadi[4]
Faculty of Chemical Engineering, *Universiti Teknologi MARA*, Shah Alam 40450, Malaysia

Email: [1]hikmah_nurul3989@yahoo.com, [2]norfaeqah@salam.uitm.edu.my, [3]junejai@salam.uitm.edu.my, [4]hadi9598@salam.uitm.edu.my

Abstract. Silver and copper nanoparticles are well-known as the good antimicrobial agent. The nano-size of particles influences in enhancing the antimicrobial activity. This paper discusses the effect of molarity on the microstructure and morphology of silver-copper core-shell nanoparticles prepared by a polyol method. In this study, silver-copper nanoparticles are synthesized through the green approach of polyol method using ethylene glycol (EG) as green solvent and reductant, and polyoxyethylene-(80)-sorbitan monooleate (Tween 80) as a nontoxic stabilizer. The phase and morphology of silver-copper nanoparticles are characterized by X-ray diffraction (XRD) and Field emission scanning electron microscope (FESEM) and Transmission electron microscope (TEM). The results XRD confirm the pure crystalline of silver and copper nanoparticles with face-centered cubic (FCC) structure. FESEM and TEM analysis confirm the existence of Ag and Cu nanoparticles in core-shell shape.

1. Introduction

Combination of two different metals nanoparticles were known as bimetallic nanoparticles. In recent years the unique properties of bimetallic nanoparticles had lead to a new applications and the possibility of making new products, for example catalyst, sensors, and substrates for surface-enchanced Raman scattering [1]. This area of nanoscience was gaining mounting attention in the field of catalysis also due to the synergistic effects [2].

Bimetallic nanoparticles have four types of mixing patterns which are core-shell nanoparticles, sub-cluster nanoparticles, mixed (alloy) nanoparticles and multi-shell nanoparticles [3]. Bimetallic nanoparticles, either as alloys or as core-shell structures, exhibited unique electronic, optical and catalytic properties compares to pure metallic nanoparticles [4,5]. Silver-copper bimetallic nanoparticles have attracted much attention because of its high electron conductivity and its use for preparation of lead-free solder alloy [6]. There has been wide-ranging research into the synthesis of metal particles such silver and copper by the polyol method [7,8]. Over the past decade the polyol process has been used to prepare elemental Co, Ni, Cu, Ag, Au, Pt, Pd, Cd, and Fe and also bimetallic alloys of Co-Ni, Ag-Pd, Au-Pt, Ag-Cu and Fe-Pt [9]. The polyol involved in the reaction acted as solvent, stabilizer and reducing agent which limits particles growth and avoid agglomeration [10].

[1-4] To whom any correspondence should be addressed.

Ag-Cu nanoparticles were emerging as another important material in this new era due to its great variety applications in electronic devices, dental and biological. Both Ag and Cu nanoparticles have great potential in variety applications due to their outstanding electrical and thermal properties resulting great demand in the market. Ag nanoparticles have received great attention because of their excellent broad-spectrum antimicrobial activity, while silver nanoparticles have proved to be most effective as it has good antimicrobial agent against bacteria, viruses and other eukaryotic micro-organism[1,11]. According to Carroll et al., the metal nanoparticles such as Ag and Cu were found to have antibacterial activity. The bactericidal influence of metal nanoparticles was attributed by their high surface to volume ratio and small size, which allow them to interact closely with microbial membranes and it was not just due to the release of metal ions in solutions. The antibacterial properties of the metal nanoparticles found applications in numerous fields such as devices and medical instrument, food processing and water treatment[9]. Yoon, et al., reported that silver and copper nanoparticles have the properties of microbe inactivation and heavy metal removal. The nano antimicrobial metals can be used effectively by coating them on the sufaces that require antimicrobial functions in water treatment filters and medical devices [12].

However, Ag nanoparticles in huge ratio retarded the antimicrobials life time. Also, Cu nanoparticles were often easily get oxidize and unstable in atmosphere [13]. Bimetallic Ag-Cu nanoparticles were reported to successfully prevent Cu nanoparticles from oxidation and thus enhance the antibacterial activities. Moreover, bimetallic Ag and Cu nanoparticles were also shown higher surface reactivity and therefore being used to improve interfacial and catalytic process [14].

In this study, the Ag-Cu nanoparticle were synthesized through the polyol method using ethylene glycol (EG) as solvent and reductant, and polyoxyethylene-(80)-sorbitan monooleate (Tween 80) as a nontoxic stabilizer. The core-shell nanopaticles was studied because of the less widely researched regarding this type of pattern.The crystal lattice information of the system as well as morphology and size of crystallite were investigated using XRD, while FESEM and TEM analysis were conducted to study the morphology and physical size and shape of the Ag-Cu nanoparticles.

2. Experimental Procedure

2.1. Materials and Synthesis Procedure

Silver nitrate ($AgNO_3$) and copper nitrate ($Cu(NO_3)_2.3H_2O$) were purchased from Sigma Aldrich Malaysia and were used as precursors. Ethylene glycol (EG, Sigma Aldrich) was used as solvent and reducing agent. Tween 80 (Sigma Aldrich) was used as stabilizer. Acetone (Merck Malaysia) was used as a solvent. All chemical were used as purchased without any further purification.

Silver-copper nanoparticles were prepared by using polyol method involving reaction of 0.8M silver nitrate ($AgNO_3$) and 0.2M copper nitrate ($Cu(NO_3)_2.3H_2O$). The preparation of Ag-Cu core-shell nanoparticles has been carried out in two steps. The first step was the synthesis of Cu inner core, and the second step was the preparation of Ag outer shell. At first, 0.2M copper nitrate trihydrate ($Cu(NO_3)_2.3H_2O$) was dissolved in 100mL of EG solution where EG was used both as a solvent and reducing agent. Then, Tween 80 was added to the solution as a stabilizer that prevents the mixture from oxidation or agglomeration. The mixture was heated and vigorously stirring for one hour with temperature of 140°C. 0.8M silver nitrate ($AgNO_3$) was added and the mixture was continuously heated for another 4 hours. Then, the solution was cooled to room temperature. The solution was then centrifuged at 10,000 rpm twice for 30 min each. 20 ml acetone was added to the solution in order to separate the precipitated and the surfactant. The precipitate silver-copper nanoparticles were then rinsed with deionized water and were dried thoroughly at room temperature. The experiment were repeated for other samples by varying the molarity of the two precursors of $AgNO_3$ and $Cu(NO_3)_2.3H_2O$ of 7:3M and 6:4M, respectively.

2.2. Characterization

The phase and crystal structure of Ag-Cu nanoparticles were characterized using X-Ray diffraction (XRD, Rigaku Ultima IV), with Cu Kα radiation (λ= 1.54056 nm). Scans of 2θ were taken in the range between 30° -80° at scanning rate of 1.2°/min. The mean crystallite size of the catalyst was estimated using the Williamson-Hall method by referring to the most prominent peak (111). The surface morphology of Ag-Cu nanoparticles was characterized using a DSM 982 Gemini Supra 40 VP FESEM. Particle morphology and size distribution were determined by TEM. Specimens were prepared by placing a drop of sample onto a 200-mesh holey carbon copper grid. The excess liquid was removed from grids with tissue paper and air-dried before examined with a TEM (JEOL 200CX).

3. Results and Discussion

3.1. XRD analysis

The phase and crystal structure of the synthesized nanoparticles are studied using XRD analysis. XRD patterns of Ag-Cu nanoparticles with different molarity ratios of 6:4M, 7:3M and 8:2M at the temperature of 140°C and reaction time 4 hours are shown in Figure 1.

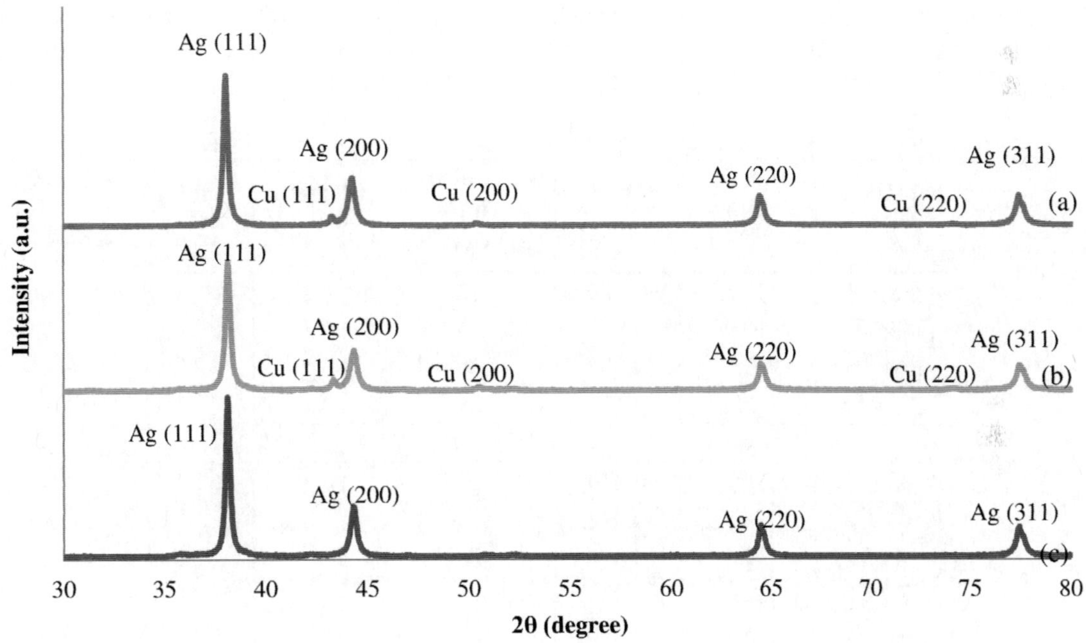

Figure 1. XRD patterns of Ag-Cu nanoparticles with different molarity Ag-Cu at (a) 8:2M, (b) 7:3M and (c) 6:4M.

As can be seen in the pattern, at molarity ratios of 7:3M and 8:2M, the main characteristic diffraction peaks for silver are observed at 2θ = 38.141, 44.34, 64.499 and 77.40 with correspond to crystallographic planes of (111), (200), (220) and (311), respectively for 7:3M and at 2θ = 38.06° (111), 44.278° (200), 64.4° (220) and 77.36° (311) for 8:2M. There are three main characteristic diffraction peaks for copper observed for 7:3M which are at 2θ = 43.338° (111), 50.54° (200), and 74.137° (220). While, for 8:2M sample, the main characteristic diffraction peaks for copper are observed at 2θ = 43.279° (111), 50.356° (200) and 74.157° (220). The results are similar to the results as reported by Min Huang et al., where all seven characteristic diffraction peaks of silver and copper appeared. The Miller indices are specified above the peaks and in the bracket. All the peaks are well indexed to face centered cubic (fcc) phase of crystalline Ag (JCPDS card no.4-783) and fcc phase of crystalline Cu (JCPDS card no. 4-836). The silver nanoparticles contain high atomic density facet as (111) in all samples that are known to be highly reactive [6].

At molarity ratio of 6:4M, four main characteristic diffraction peaks are observed at $2\theta = 38.121^\circ$, 44.301°, 64.46°, and 77.419°, they are assigned to the (111), (200), (220) and (311) crystallographic planes of fcc silver crystals. However, no characteristic diffraction peaks for copper are observed. This is predicted due to the low concentration of copper in the precipitated as the effect of the insufficient reaction. As the concentration of copper increases to be 0.4M, the chemical reaction is predicted to be incomplete. Therefore, the peaks of copper are invisible in the pattern. Moreover, the invisible of copper peaks is also contributed by the peaks intensity of silver is more strength than copper due to the different crystal size. Hence, the low concentration of copper causes the difficulty in observing the copper peaks in the pattern.

From the results it can be concluded that, both Ag and Cu phases are clearly identified on the samples at the molarity ratios of 8:2M and 7:3M. Accordingly, there is no obvious XRD peak arising from possible alloy, mixture or aggregates phases of Ag and Cu are observed. However, only silver peaks are visible in the 6:4M sample due to incomplete reaction. Therefore, it is safe to say that the Cu-Ag core-shell nanoparticles are possible to be synthesized using polyol method with the molarity ratios Ag to Cu between 8:2M and 7:3M.

XRD results also found that the crystallite size of the synthesized Ag-Cu core-shell nanoparticles, calculated using Scherrer's equation for the most intense peak in the pattern, is distributed within the range of 27-32 nm for silver and 24-56 nm for copper, as shown in Table 1.

Table 1. XRD parameters of Ag-Cu core-shell nanoparticles

2-Theta	Intensity (%)	FWHM	Crystal size (nm)	Element	(h k l)
		Sample 6:4 M			
38.121	100	0.279	32	Ag	(1 1 1)
44.301	40.9	0.371	24	Ag	(2 0 0)
64.46	25.3	0.372	26	Ag	(2 2 0)
77.419	27	0.418	25	Ag	(3 1 1)
		Sample 7:3 M			
38.141	100	0.319	27	Ag	(1 1 1)
43.338	3.3	0.181	56	Cu	(1 1 1)
44.34	34.4	0.376	23	Ag	(2 0 0)
50.54	3.3	0.413	21	Cu	(2 0 0)
64.499	25.7	0.416	23	Ag	(2 2 0)
74.137	1.6	0.372	27	Cu	(2 2 0)
77.401	27.7	0.471	22	Ag	(3 1 1)
		Sample 8:2 M			
38.06	100	0.294	30	Ag	(1 1 1)
43.279	5.4	0.286	31	Cu	(1 1 1)
44.278	38.2	0.374	23	Ag	(2 0 0)
50.356	1.7	0.303	30	Cu	(2 0 0)
64.4	26	0.4	24	Ag	(2 2 0)
74.157	1.1	0.346	30	Cu	(2 2 0)
77.36	28.6	0.435	24	Ag	(3 1 1)

3.2. FESEM and EDS analysis

The FESEM micrographs are taken to support the XRD result and also to study the surface morphology of the Ag-Cu core-shell nanoparticles. Figure 2 shows the enlarged image SEM micrograph of the Ag-Cu nanoparticles at different molarity. It can be seen in Figs. 2(a) and (b) that the surface morphology and roughness of both samples are quite similar. The particles are uniformly distributed in nanoscale size with a narrow size distribution. Almost all particles are found in spherical shape.

However, the shape and size distribution in 6:4 M micrograph cannot be clearly identified. As can be seen in Fig. 2(c), the surface is rough and agglomerates. It is also difficult to identify the individual particles. It is due to the incomplete reaction of reduction of reactants to produce Cu particles. As described above, the high concentration of Cu in solutions made the reduction reaction become incomplete. The findings are good agreement with XRD results.

(a) 8:2 M (b) 7:3 M

(c) 6:4 M

Figure 2. FESEM images of Ag-Cu nanoparticles prepared at 140°C for varied molarity

The samples were also examined using EDS analysis to determine the actual elemental concentration. The result is listed in Table 2.

Table 2. EDS data of Ag-Cu core-shell nanoparticles

Sample	Element	Weight (%)	Atomic (%)
8:2 M	Cu L	18.25	27.48
	Ag L	81.75	72.52
7:3 M	Cu L	13.83	21.42
	Ag L	86.17	78.58
6:4 M	Cu L	12.47	19.47
	Ag L	87.53	80.53

3.3. TEM analysis

TEM analysis is conducted to investigate the physical size, size distribution and morphology of the synthesized nanoparticles. Figure 3 shows the images of Ag-Cu core-shell nanoparticles at different molarity. It can be seen that at the concentrations of Ag^+ of 0.6 M and 0.7 M, the synthesized nanoparticles exhibit perfect core-shell structures. Interestingly, it is also found that few of Cu encapsulates more than one Ag nanoparticles in one shell. However, as soon as the Ag^+ concentration increase to 0.8M, the nanoparticles tend to aggregate into chainlike or branchlike structure.

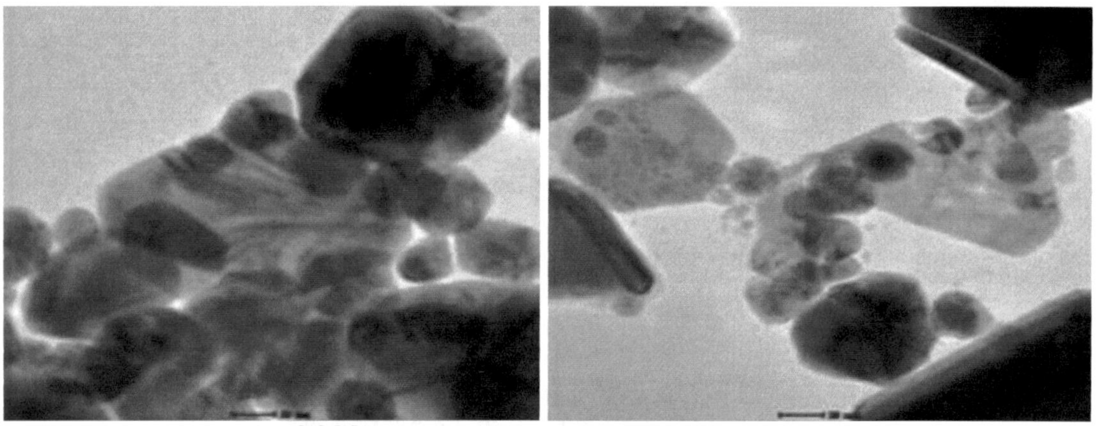

(a) 8:2 M (b) 7:3 M

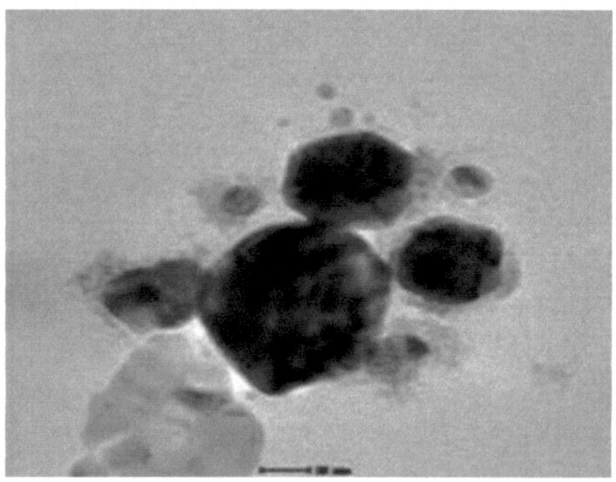

(c) 6:4 M

Figure 3. TEM images of Ag-Cu nanoparticles prepared at 140°C for varied molarity

IOP Conf. Series: Earth and Environmental Science **36** (2016) 012050 doi:10.1088/1755-1315/36/1/012050

4. Conclusions

Silver-copper core-shell nanoparticles were successfully synthesized by a polyol method with wide range of size and various shapes. The synthesized nanoparticles were pure crystalline with face centered cubic structure with the crystallite size distributed in the range of 27 nm to 32 nm for Ag and 24 nm to 56 nm for Cu. Finally, FESEM and TEM techniques confirmed the presence of Ag and Cu nanoparticles.

5. Reference

[1] Rai M, Yadav A and Gade A 2009 *Biotechnol. Adv.* **27** 76–83
[2] Search H, Journals C, Contact A and Iopscience M **025402**
[3] Rahman L, Qureshi R, Yasinzai M M and Shah A 2012 *Comptes Rendus Chim.* **15** 533–538
[4] Chen Z, Mochizuki D, Maitani M M and Wada Y 2013 *Nanotechnology* **24** 265602
[5] Valodkar M, Modi S, Pal A and Thakore S 2011 *Mater. Res. Bull.* **46** 384–389
[6] Huang M, Yang X, Zhao J and Wang Q 2013 *Res. Chem. Intermed.* **40** 1957–1964
[7] Thi My Dung Dang T T T L 2012 *Adv. Nat. Sci. Nanosci. Nanotechnol.* **3**
[8] Zhu H, Zhang C and Yin Y 2005 *Nanotechnology* **16** 3079
[9] Carroll K J, Reveles J U, Shultz M D, Khanna S N and Carpenter E E 2011 *J. Phys. Chem. C* **115** 2656–2664
[10] Zhao T, Sun R, Yu S, Zhang Z, Zhou L, Huang H and Du R 2010 *Colloids Surf. Physicochem. Eng. Asp.* **366** 197–202
[11] Tran Q H, Nguyen V Q and Le A-T 2013 *Adv. Nat. Sci. Nanosci. Nanotechnol.* **4** 033001
[12] Yoon K-Y, Hoon Byeon J, Park J-H and Hwang J 2007 *Sci. Total Environ.* **373** pp 572–575
[13] Lee C, Kim N R, Koo J and Lee Y J *Nanotechnology* **26** 455601
[14] Tan K S and Cheong K Y 2013 *J. Nanoparticle Res.* **15** 1537

Acknowledgments
This work was financially supported by the Research Management Institute (RMI) through FRGS Project No. 600-RMI/FRGS 5/3 (26/2013) and we would like to thank to UiTM Shah Alam for supporting the project.

Preparation of polyacrylnitrile (PAN)/ Manganese oxide based activated carbon nanofibers (ACNFs) for adsorption of Cadmium (II) from aqueous solution

N Abdullah[1], N Yusof[1], J Jaafar[1], AF Ismail[1], F E Che Othman[1], H Hasbullah[1], W N W Salleh[1], N Misdan[2]

[1] Advanced Membrane Technology Centre (AMTEC), Faculty of Petroleum and Renewable Energy Engineering, *Universiti Teknologi Malaysia*, 81310 Skudai, Johor Bahru, Malaysia
[2] Faculty of Engineering Technology, *University Tun Hussien Onn Malaysia*, 86400 Parit Raja, Johor, Malaysia

E-mail: [1]norhaniza@petroleum.utm

Abstract. In this work, activated carbon nanofibers (ACNFs) from precursor polyacrylnitrile (PAN) and manganese oxide (MnO_2) were prepared via electrospinning process. The electrospun PAN/MnO_2-based ACNFs were characterised in term of its morphological structure and specific surface area using SEM and BET analysis respectively. The comparative adsorption study of cadmium (II) ions from aqueous solution between the neat ACNFs, composite ACNFs and commercial granular activated carbon was also conducted. SEM analysis illustrated that composite ACNFs have more compact fibers with presence of MnO_2 beads with smaller fiber diameter of 437.2 nm as compared to the neat ACNFs which is 575.5 nm. BET analysis elucidated specific surface area of ACNFs/MnO_2 to be 67 m^2/g. Under adsorption study, it was found out that Cd (II) removal by ACNFs/MnO_2 was the highest (97%) followed by neat ACNFs (96%) and GAC (74%).

1. Introduction

Water contamination by cadmium has led to numerous studies being conducted for its removal. In general, cadmium (Cd) is categorized as heavy metal that is normally found as ores in other elements and exists in the state of 2+ in aqueous solution. It is primarily used for metal plating and coating operation, including transportation equipment machinery and baking enamels as well as in nickel-cadmium and solar batteries. The metallurgic industries and mining mentioned earlier serve as anthrophogenic factor for its presence in water. Exposure of cadmium has been profound to be associated with renal damage, osteoporosis, negative bone effects and cancer. Due to that, the maximum contaminant level set by World Health Organization and Environmental Protection Agency (EPA) is 0.005 mg/L [1].

[1] To whom any correspondence should be addressed.

Due to hazardous impacts of cadmium exposure, many methods are being employed for its removal such as precipitation, coagulation-flocculation, flotation, ion exchange, membrane filtration and adsorption [2]. Among the aforementioned techniques, adsorption has been found to be the most feasible method to eliminate metallic ions imputed by wide range of adsorbents are available, simpler technique and economic. Conventionally, activated carbons (ACs) in powder or granular forms are used as the common materials in adsorption method. Recently, application of activated carbon nanofibers (ACNFs) as adsorbents has attracted greater attention due to their excellent adsorption capacities attributed by their smaller fibre diameter, more concentrated pore size distribution. There are several methods can be used to fabricate nanofibers such as melt fibrillation, island-in-sea, and gas jet technique [3]. Nevertheless, it seems that electro-spinning provides more advantages in term of high production rate and relatively low in cost as compared to the other techniques mentioned previously.

To prepare electrospun ACNFs, polymer polyacrynitrile serves as a good carbon precursor as it has high carbon yield. Recently, many researches have focused on incorporating various metal oxides such as magnesium oxide, aluminium oxide and manganese oxide into PAN-based polymer solution. It is believed that impregnation of metal oxides as additives could increase the specific surface area of ACNFs as well as the pore volume of electrospun ACNFs. Therefore, this study aimed to prepare PAN/MnO_2-based ACNFs by the process of electrospinning. The electrospun ACNFs were evaluated in term its morphological changes and its performance towards adsorbing cadmium (II) ions from aqueous solution.

2. Experimental procedures

2.1 Materials

Polyacrynitrile (PAN) with molecular weight of 150 000 kDa was purchased from Sigma Aldrich (USA). Similarly, N,N-Dimethylformamide (DMF) and manganese (IV) oxide (MnO_2) powder with 99% purity were also purchased from Sigma Aldrich (USA). Stock solution of cadmium (II) ions was prepared using cadmium nitrate (Sigma Aldrich, USA) while pH of metal solution was adjusted by using 0.01mol of NaOH and HCl solutions.

2.2 Preparation of PAN-MnO nanofiber mats by electrospinning

In brief, dope solution of PAN-MnO_2 was prepared by mixing 2 wt% of MnO_2 (based on total weight solid) in DMF for 5 hours followed by addition of 10 wt% PAN. The dope solution was stirred on mechanical stirring plate to allow it to become homogenous. The solution suspension was subsequently fabricated by electrospinning process, under fixed electric field of 15kV/20cm and suspension rate of 1.5mL/hour. The collected nanofibrous mats were allowed to dry for at least a day prior carbonisation and activation process.

2.3 Carbonisation and chemical activation of nanofibers

The carbonisation and activation of electrospun nanofibers were carried out by following procedures described by Yusof *et.al* [4]. The nanofiber mats were initially stabilised in air at 280°C for 2 hours, with heating rate of 2°C/min followed by carbonisation was carried out in nitrogen gas at 600°C for 1 hour with heating rate of 5°C/min. The CNFs were activated by chemical activation method by soaking into KOH for 3 hours with ratio 3:1 KOH to CNFs prior heated under carbon dioxide gas at 800°C for 3 hours. The activated CNFs were later rinsed by distilled water until pH 7±0.5 and kept dry in oven at 100°C for 12 hours.

2.4 Characterisation

The morphology of the activated nanofibers were analysed by Scanning Electron Microscopy (SEM) while the specific surface area was examined by using Brunauer-Emmett-Teller and functional groups analysis was carried out by Fourier Transform Infrared (FTIR).

2.5 Comparative adsorption study of Cd

Comparative adsorption study of Cd (II) ions was analysed by investigating the adsorption performance of ACNFs/MnO_2, neat ACNFs and commercial GAC. 0.025g of the respective adsorbents was placed into conical flask containing 50 mL Cd(II) solution with initial concentration of 3.1 mg/L and pH of the solution fixed at 5.0±0.5 . The samples were agitated on rotary shaker with speed of 180 rpm at 30°C for 48 hour to reach equilibrium. The sample solutions were later filtered and examined by using Atomic Absorption Spectrometer (AAS).

The adsorption capacity and the percentage of removal of Cd (II) ions of the each type of activated carbons were calculated by following equations [5]:

$$q_t = \frac{(C_o - C_t)V}{M} \quad (1)$$

And the removal efficiency was calculated using :

$$R(\%) = \frac{(C_o - C_t)}{C_o} \times 100 \quad (2)$$

where C_o and C_t is the initial and final concentration of the metallic solution ($mg.L^{-1}$), V is the volume of the solution (L), and M is the weight of the adsorbent used (g).

2.6 Desorption of PAN-MnO_2 activated nanofibers

Desorption of ACNFs used in this study was performed to identify the recovery rate and reusability PAN-MnO_2 based ACNFs. The ACNFs used for batch adsorption study were rinsed with distilled water to eliminate any residual solutions and subsequently dried in oven at room temperature. Desorption of metal ions were carried out by immersing the ACNFs/MnO_2 in 1M HCl and agitated at 30°C with speed of 100 rpm for 1 hour. The concentration of Cd ions in the solution was analysed using atomic absorption spectrometer (AAS) and the desorption ratio (D) was calculated as follows:

$$D(\%) = \frac{mg\ of\ metal\ desorbed}{mg\ of\ metal\ ion\ adsorbed\ into\ ACNFs} \times 100 \quad (3)$$

3. Results and discussion

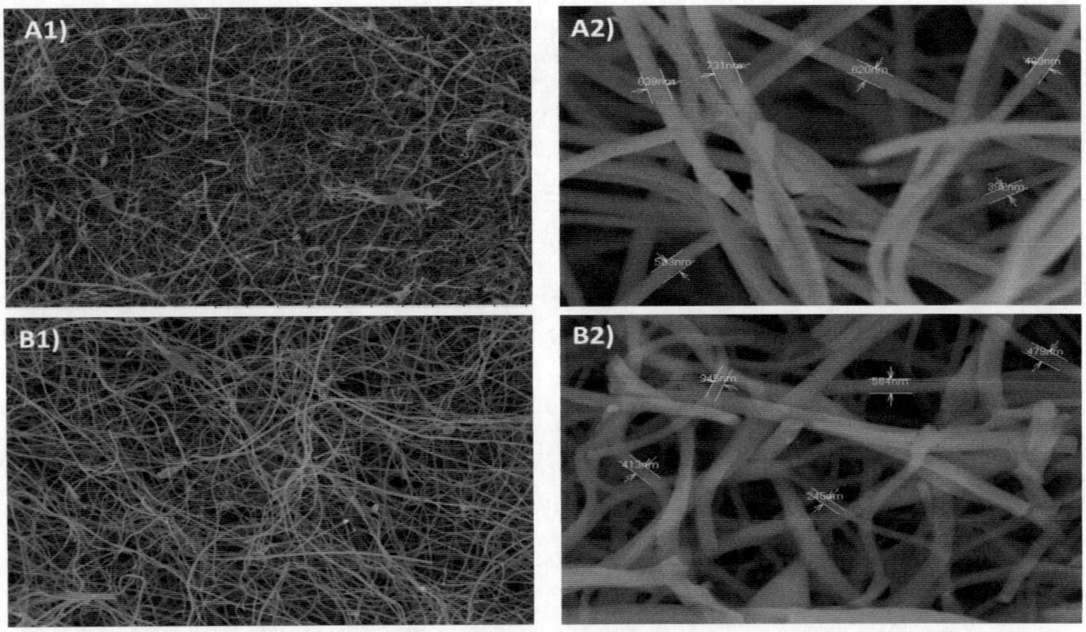

Figure 1. SEM micrograph analysis of A)neat ACNF, B)ACNF/MnO_2 where 1)500x magnification and 2)10000x magnification

Figure 1 exhibits the SEM images of neat ACNFs and ACNFs/MnO_2. From the figure, it can be seen that ACNFs/MnO_2 illustrates more compact nanofibers in comparison to the neat ACNFs which is more aligned and less compact. When compared to the diameter of ACNFs without and with MnO_2, it is revealed that ACNFs embedded with MnO_2 has smaller diameter in range of 473.2 nm while neat ACNF possess fibers with higher average diameter of 575.5nm. In addition, the presence of beads in composite ACNFs may be caused by the MnO_2 particles. The morphology of electrospun ACNFs obtained in this study is concurring to the study conducted by Choi *et.al* [6] in which the ACNF/MnO_2 exhibits as compact nanofibers with dispersion of beads containing MnO_2 particles.

Specific surface area (SSA) analysis of the ACNFs in this study was found to be 16 m^2/g for neat ACNFs and 67 m^2/g for ACNFs/MnO_2, respectively. Theoretically, the SSA obtained for the electrospun ACNFs was low as compared to other composite ACNFs reported elsewhere. Nevertheless, similar result was obtained by study of Wang *et.al* [7] that also demonstrated lower SSA of carbon nanotubes/manganese oxide. It is believed that the carbonisation and activation process of the composite nanofibers in this study might be incomplete, resulting to low SSA when analysed by BET analysis. For complete activation with low loading of manganese oxide, the specific surface area may increase up to 500 m^2/g however any further increment of MnO_2 loading will yield lower SSA as reported by Watanabe *et.al* [8].

Table 1. Comparative study of removal of Cd (II) ions by granular activated carbon (GAC), neat ACNFs and ACNFs/MnO_2. Operating conditions: membrane dose=1g/L' stirring speed = 180 rpm, temperature= room temperature, and contact time = 48 hours.

Cd (II) initial concentration (mg/L)	Adsorbents	Removal efficiency (%)	Adsorption capacity (mg/g)
	GAC	74	5.4
3.1	Neat ACNFs	88	6.2
	ACNFs/MnO_2	97	6.6

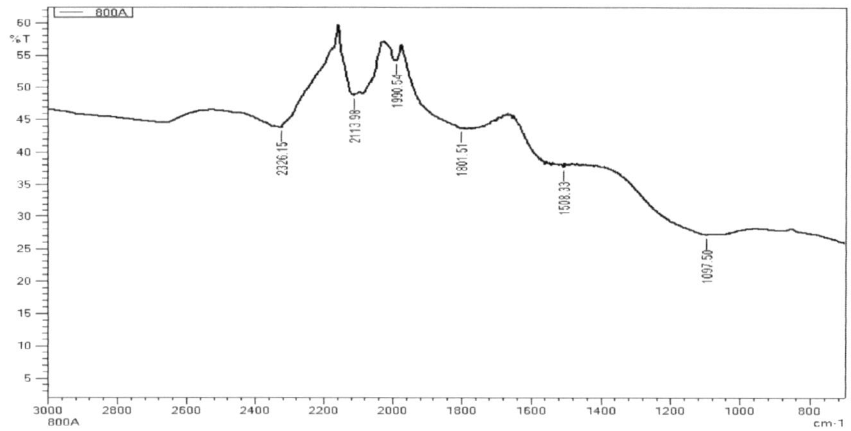

Figure 2. FTIR spectrum of ACNFs/MnO_2

Table 1 tabulates the comparative adsorption study of Cd (II) by granular activated carbon, pure ACNFs and ACNFs/MnO_2. From the table it can be seen that highest removal of Cd (II) was achieved when composite ACNFs/MnO_2 was used as adsorbent in the solution which is 97% followed neat ACNFs (96%) while adsorption by GAC illustrated lower removal efficiency which is 74 %. In theory, the removal capability can be linked to the SSA where more active adsorption sites are available. Nevertheless, in this study the SSA was found to be low. Research by Wang and co-workers [7] stated that despite specific surface area may enhance the adsorption performance, it is however no direct correlation between metal ion adsorption capacity of ACNFs with their specific area, specific pore volume and pore diameter. Thus, the adsorption capacity may also attributable to the amount of surface total acidity present on the surface of ACNFs/MnO_2. Based on this statement, it can be depicted that the adsorption of Cd (II) ions onto ACNFs occur by means of chemisorption where it is mainly controlled by the strong interaction between Cd (II) ions and surface functional groups such as carboxyl and hydroxyl [9,10].

As depicted by figure 2, there are five major peaks observed located at 2326.15 cm^{-1}, 2113.98 cm^{-1}, 1990.54 cm^{-1}, 1801.51 cm^{-1} and 1097.50 cm^{-1}. The peak 2326.15 cm^{-1} is assigned as O-H stretch from strongly hydrogen-bonded –COOH. Peaks 2113.98 and 1990.54 cm^{-1} basically attributed to the strong stretching of C=C groups and peak 1801.51 and 1097.50 indicated as stretching vibrations of C-O groups. These detected functional groups may play role for adsorption of Cd (II). In addition, the mechanism of chemisorption can basically confirmed by testing two common sorption models which are Langmuir model and Freundlich model based on their respective equations, and evaluation of their regression value (not performed in this study) [5]. It is also demonstrated that the adsorption capacity of composite ACNFs used in this study is 6.1 mg/g that is higher as compared to the commercial GAC (4.6 mg/g). The higher adsorption capacity can be explained by the role of manganese oxide itself that functions as an adsorbent and it has higher affinity towards metallic ions [11].

For desorption study, it was found out that ACNFs used in this study has higher regeneration rate which was more than 95%. This high reusability exhibits its potential as another promising adsorbent for heavy metal adsorption.

4. Conclusion

This present study reports the preparation of ACNFs/manganese oxide by electrospinning process for removal of cadmium (II) ions from aqueous solution. In this study, ACNFs shows promising adsorption capacity upon adsorbing Cd (II) as compared to the commercial GAC. The ACNFs incorporated by MnO_2 has found to be more compact and possessed smaller diameter as compared to the neat ACNFs. In this study, the adsorption process is believed to occur by means of chemisorption that involves the interaction of Cd (II) ions with surface functional groups, regardless of their low specific surface area. The sorption/desorption study indicates the excellent reusability of this new type of activated carbon. Further research on exploring more potential of these ACNFs is recommended.

5. References

[1] Dinesh M, Kunwar P S 2002 *Water Research* **36** pp 2304-2318
[2] Barakat, M.A (2011) *Arabian Journal of Chemistry* **4** pp 361-377
[3] Ramakrishna S, Fujihara K, Teo W, Yong T, Ma Z and Ramaseshan R 2006 *Materials Today* **9** pp 40-50
[4] Yusof N, Ismail A F, Rana D and Matsuura T 2012 *Materials Letter* **8** pp 16-18
[5] Kampalanonwat P and Supaphol P 2014 *Energy Procedia* **56** pp 142-151
[6] Choi G, Jung H, Kim C, Yang K S, and Lee W 2005 *Applied Chemistry* **9** pp 49-52
[7] Wang S G, Cong W X, Liu X W, Yao Y W, Gao B Y and Yue Q Y 2007 *Separation and Purification Technology* **58** pp 17-23
[8] Watanabe N, Yeh J, Lee G, Miyawaki J, Mochida I and Yoon S 2010 *The American Carbon Society, Carbon Conference*
[9] Khalkali R A and Modivari R 2005 *Polish Journal of Environmental Studies* **14** pp 185-188
[10] Gadhave A and Waghmare J 2014 *International Journal of Chemical Sciences and Application* **5** pp 56-67
[11] Han R P, Zou W H, Li H K, Li Y H and Hi J 2006 *Journal of Hazardous Materials* **137** pp 134-142

Acknowledgement

The authors would like to acknowledge the financial support from Malaysian Ministry of Education and Universiti Teknologi Malaysia under Grant No. R.J130000.7842.4F279 and Q.J130000.2542.05H46. The authors would also like to acknowledge the technical and management support from Research Management Centre (RMC), Universiti Teknologi Malaysia.

Synergistic effect of Cu-doped into $CesZrO_2$ nanocatalysts

M N A Shah[1], K N Ismail[2], J Jai[3], N Faeqah[4] and A Hadi[5]
Faculty of Chemical Engineering, *Universiti Teknologi MARA*, 40450 Shah Alam, Selangor, Malaysia

E-mail: [1]muhamad_nazri10@yahoo.com.my, [2]knoor@salam.uitm.edu.my, [3]junejai@salam.uitm.edu.my,[4] norfaeqah@salam.uitm.edu.my, [5]hadi9598@salam.uitm.edu.my

Abstract. $Ce_{0.75}Zr_{0.25}O_2$ and $CuO/Ce_{0.75}O_{0.25}O_2$ nanocatalysts have been successfully prepared by combination of microemulsion and deposition precipitation methods. The synthesized nanocatalysts revealed as the cubic structure, homogeneous solid solution and the particles were in nanoscales. The copper particles were well deposited and dispersed on the entire of $Ce_{0.75}Zr_{0.25}O_2$ surface and pores which reflected by the lower the surface area of $CuO/Ce_{0.75}Zr_{0.25}O_2$ than $Ce_{0.75}Zr_{0.25}O_2$. The additional of copper catalysts able to active the catalysts to lower temperature up to < 250°C and stabilize the microstructure of catalysts.

1. Introduction

Nowadays, the total air pollution release to the atmosphere increase rapidly simultaneously with the increasing of the mass production of automobiles worldwide especially in the urban area. As we well known, air pollution contributes to the human health and environment damages. In order to avoid the huge effect of air pollution, many automobile manufactures has introduced the exhaust gas after treatment system or called as three-way catalysts (TWCs) in automobile engines to reduce the level of emission released to the tolerable limits. However, it is still insufficient to reach the desired emissions level, especially during cold-start portion of driving. As reported by Heck and Farrauto [1], the majority of carbon monoxide and unburned hydrocarbon emissions (60%-80% of the total emitted) are produced during the cold-start condition. Therefore, the emission from automobile still becomes the major contributor of air pollution in the atmosphere.

The catalytic components in three-way catalysts mainly consist of the noble metals such as Pt, Pd, and Rh as the active catalysts and cerium oxide (CeO_2) as oxygen storage/release component (OSC). As an OSC, CeO_2 plays the important role in increasing the effectiveness of TWCs during the cold start condition by maintaining the oxygen concentration in catalytic reaction. However, CeO_2 is easily to be sintered leading to the reduction of active surface area and the capabilities of oxygen to storage and release under fluctuating in exhaust system reduced during exposed to high temperature [2,3]. Consequently, it will reduce the effectiveness of TWCs in reducing emissions concentrations.

[1-5] To whom any correspondence should be addressed.

The extensive research and development has been performed to overcome the drawback of CeO_2 at high temperatures, such as the incorporation of zirconium ions into the cerium structure. Li et al., [4] described that the insertion of Zr^{4+} ions into the cerium crystal creates a homogeneous solid solution, which can stabilize CeO_2 against sintering and enhance the thermal stability. In addition, this combination also enhanced the surface area and increased the dispersion of active catalysts, leading to the high catalytic performance of the TWCs [5]. The best ratio of cerium and zirconium for the ideal combination materials was the $Ce_{0.75}Zr_{0.25}O_2$ which promotes a good thermal stability and high surface area as reported by another author [5–7].

The introduction of copper as active catalyst into $CeZrO_2$ has been intensively explored Escribano et al., [8] reported the possibility of copper oxide as an active catalyst to oxidise the carbon monoxide and volatile organic compound. Radlik et al., [9] concluded that inclusion of CuO into $Ce_xZr_{(1-x)}O_2$ catalysts could promoted the oxygen mobility and water-gas shift reaction at lower temperatures in range 100 - 200°C. The replacement of copper oxide as active catalyst in TWCs offered a cheaper metal and a viable solution from both an economical and catalytically point of views.

Numerous combination preparation techniques have been applied to develop the metals loaded $Ce_{0.75}Zr_{0.25}O_2$ catalysts, such as co-precipitation and impregnation [10,11], sol gel and incipient wetness impregnation [12], solid state chemical reaction and incipient wetness impregnation and etc. However, the combination of microemulsion and deposition precipitation methods to synthesis the ceria-zirconia mixed oxides and loaded catalysts still have limited exploration. Even thought, the microemulsion was able to enhance the homogeneous mixing of catalysts components and narrow size distribution with tunable size [13,14]. Besides, the deposition precipitation method aiming to prepare well dispersed active catalyst particles on support, with a narrow particle size distribution [15,16] Therefore, the combination of microemulsion and deposition precipitation method was proposed to enhance the thermal stability and microstructure properties of the TWCs.

In our work, the microemulsion method was employed to synthesize the $Ce_{0.75}Zr_{0.25}O_2$ nanomaterial as support catalysts and the copper catalysts were loaded via deposition precipitation method. The synergistic effects of copper doped into ceria-zirconia structure were explored through x-ray diffraction, N_2-adsorption desorption, field emission scanning electron microscopy, transmission electron microscopy and temperature programmed reduction. We aspired that our work may help to understand the synergistic effects copper doped into ceria-zirconia in term of catalytic properties especially the activation catalysts at lower temperature.

2. Materials and Experimental Procedure

2.1 Catalyst preparation

In this research, the catalysts were synthesized in two steps of preparation procedure. In the first step, $Ce_{0.75}Zr_{0.25}O_2$ catalyst support was produced using water/oil microemulsion synthesis. The method was adopted from Laguna et al., [17]. The microemulsion system consisted of cetyltrimethylammonium bromide (CTAB, Merck Malaysia, 99%) as the surfactant, 1-butanol (Sigma Aldrich Malaysia) as the co-surfactant, n-octane (Merck Malaysia) as the continuous oil phases and cerium nitrate ($Ce(NO_3)_2$, Merck Malaysia, 99.9%) and *zirconium nitrate* ($ZrO(NO_3)_2$, Merck Malaysia, 99.9%) aqueous solutions as the dispersed phase. Typically, two microemulsions with different aqueous phases were prepared. In microemulsion 1, the CTAB, 1-butanol and n-octane were mixed and stirred. The mixture of $Ce(NO_3)_2$ and $ZrO(NO_3)_2$ in a percent weight ratio of 0.75:0.25 as an aqueous phases. Then the aqueous phases of containing Ce^{3+} and Zr^{4+} ions was added under vigorous stirring until the transparent liquid. The same procedures were applied for aqueous phase of NaOH for microemulsion 2. These two microemulsions were mixed under constant stirring, allowing the reacting species to come in contact with each other due to the dynamic nature of the dispersed droplets. This leaded to precipitation of particles within the aqueous micro droplets and the reaction to take place between the hydroxide, OH^- ions and metals ions (Ce^{3+} and Zr^{4+}). The surfactant layer around each droplet acted as a barrier restricting the growth of the particle beyond the water domains. The mixture of microemulsion 1 and 2 was stirred for 3 hours to ensure the colloidal precipitates (oxidation state zero) formed and then be aged for 24 hours. The colloidal precipitates were washed with ethanol (3 times) and di-ionized water (3 times) by centrifuge at 10,000 rpm for 10 minutes. The precipitates were dried at temperature of 110°C for 24 hours and was calcined at 500°C for 6 hours in heating rate at 5°C/min.

In the second step, $CuO/Ce_{0.75}Zr_{0.25}O_2$ nanocatalysts were synthesized by deposition-precipitation method. The preparation procedure was adopted from Hadi & Yaacob [13] with a slightly modification. The synthesized $Ce_{0.75}Zr_{0.25}O_2$ was suspended in a solution of copper nitrate, $Cu(NO_3)_2$ at 4wt% of copper loading. The mixture was stirred for 3 hours. Cu^{2+} ions were expected to be deposited onto the surface and pores of $Ce_{0.75}Zr_{0.25}O_2$ and it was exclusively precipitated by adding an aqueous solution of 0.5M Na_2CO_3 to keep the pH at constant of 10 and the temperature raised-up to 70 °C. The suspension further stirred for 2 hours. The resulting solid was washed with de-ionized water in a centrifuge for 3 times and dried at 100 °C for 24 hours. Finally, the sample was calcined at 500 °C for 5 hours.

2.2 Characterizations
The phase and crystal structure were characterized by X-ray Diffraction (XRD) of Rigaku Ultima IV using Cu Kα radiation (λ = 0.15418 nm). The scans of 2θ were taken in range 20° to 70° at scanning rate of 1.2°/min. The crystallize size were calculated by Scherrer's equation, referring the most prominent peak (111). The crystalline phases were identified according to the reference data from the International Center Diffraction Data (ICDD). The surface area of the samples was determined using the Autosorb-1 Quanta Chrome Instrument TUSA. For morphology studies, the micrographs of samples were obtained by a field emission scanning electron microscopy, DSM 982 Gemini Supra 40 VP and transmission electron microscopy Tecnai G2 20. The catalytic properties of samples were examined by temperature programmed reduction (Chemisorb 2750, micromeritics brand). In this evaluation, a 50 mg sample was pre-treated under N_2 (20 ml/min) at 200°C for 1 hour, followed by cooling to 50°C. The sample was exposed in hydrogen 5% (v/v) H_2/Ar (flow rate: 20 ml/min) from 50 to 700°C at a heating rate of 10°C/min [9]. The consumption of hydrogen was detected by thermal conductivity detector.

3. Result and Discussion
3.1 X-ray Diffraction Analysis
Figure 1 shows the XRD patterns of the samples. All of the peak profiles show the typical main reflections of CeO_2 cubic phase with fluorite structure and it matches with XRD patterns of pure CeO_2 (ICDD No: 21306). The characteristics peaks are located at 2θ = 28.59, 33.03, 47.30, 56.03, and 59.18 which corresponding to (111), (200), (220), (311) and (222) planes, respectively. No segregation peaks are observed, only the symmetrical XRD patterns of synthesized $Ce_{0.75}Zr_{0.25}O_2$ to pure CeO_2 are observed, which suggest that the prepared sample is homogeneous. This finding may be explained through the recent studies by Dobrosz-Gómez et al., [5], which reported that the molar ratio of Ce^{4+} and Zr^{4+} ions have great significant effect to the phase structure. They found that the insertion of zirconia into ceria at molar ratio from $0 \leq x \leq 0.25$ will produce cubic phase solid solution, while at the molar ratio of $0.5 \leq x \leq 0.75$, the tetragonal phase was formed. They also claimed that the ions of Zr^{4+} incorporated into the CeO_2 lattice to form a solid solution. Therefore, in this study we might indicate that the insertion of zirconium (Zr^{4+}) ions have dissolved into the lattice of CeO_2 to form a $CeZrO_2$ solid solution. A similar result also reported by Li et al., [4,18,19].

The diffractogram pattern of the $CuO/Ce_{0.75}Zr_{0.25}O_2$ catalyst as shown in figure 1(c) also presents the symmetrical patterns to the cerium oxide which the phase structure is maintained as the cubic fluorite structure and no CuO phase is detected. It may cause of the low concentration of copper components deposited in the catalyst (< 4 wt%), which under X-ray diffraction limit. Besides, the Radlik et al., [9] have suggested that the insertion of copper catalysts are well dispersed on the entire catalysts surfaces although the CuO phase not detected via XRD analysis. It is may be due to highly dispersion of CuO on the $Ce_{0.75}Zr_{0.25}O_2$.

The low intensities of the peaks for all samples indicate that the crystallite sizes of the samples are in very small particles. Jani et al., [7] reported that, the width of the peaks and intense signifies the hugeness and smallness of crystallite size. As can see in table 1, the crystallite size increases during the additional of copper catalysts up to 8.1 nm and the broadening peak (111) for all samples shows the parallel finding which the broad peak correspond to small crystallite size. Conversely, the crystallite size of cerium oxide has bigger size up to 9.1 nm than cerium zirconium oxides which is 6.6nm. It is due to the insertion of small ionic radius of Zr^{4+} (0.084 nm) into ceria (Ce^{4+} = 0.097 nm and Ce^{3+} = 0.114) structure, make the degrees of crystal lattice shrinking and the crystallite size decreases [4,5].

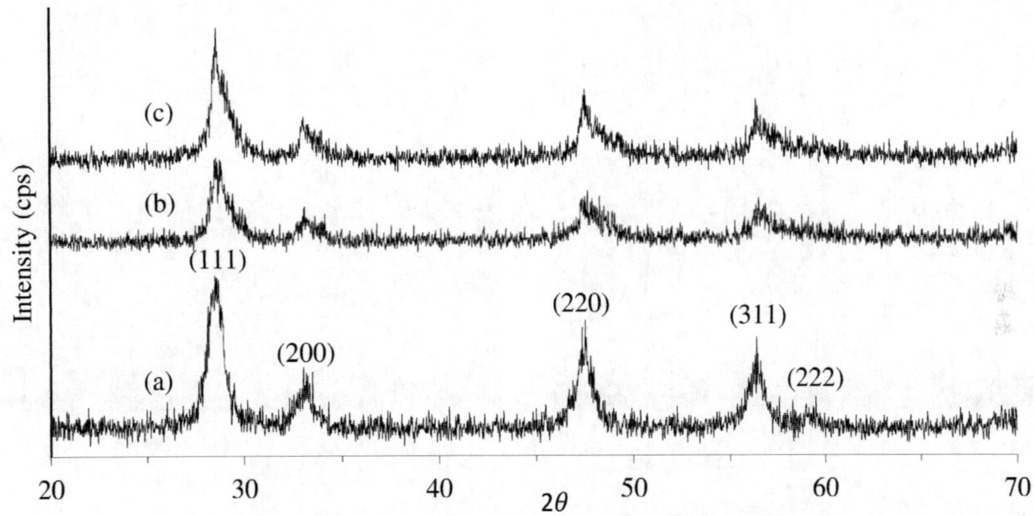

Figure 1. XRD patterns of the synthesized samples
((a); CeO_2, (b); $Ce_{0.75}Zr_{0.25}O_2$ and (c); CuO/ $Ce_{0.75}Zr_{0.25}O_2$)

3.2 Specific Surface Area

The surface area of $Ce_{0.75}Zr_{0.25}O_2$ and $CuO/Ce_{0.75}Zr_{0.25}O_2$ nanocatalysts are listed in Table 1. The surface areas of samples decrease after the deposition of CuO onto $Ce_{0.75}Zr_{0.25}O_2$. This indicates the well dispersion of CuO inside of the pores and on the surface of $Ce_{0.75}Zr_{0.25}O_2$. During deposition process, the Cu^{2+} ions are filled well onto the available pores on the $Ce_{0.75}Zr_{0.25}O_2$ and this phenomenon leads to the formation of homogeneous solid solution $CuO/Ce_{0.75}Zr_{0.25}O_2$. According to the BET surface area, the diameter of particles can be estimated using the following equation (1) with assumption the mixed oxides are spherical.

$$d_{BET} = \frac{6}{S\rho} \qquad (1)$$

Where, the S is specific surface area, m^2/g and ρ is density of the particles g/m^3. By assuming, the density of $CuO/Ce_{0.75}Zr_{0.25}O_2$ is not obviously changing due to very small amount of CuO exist, density of $Ce_{0.75}Zr_{0.25}O_2$ and $CuO/Ce_{0.75}Zr_{0.25}O_2$ is 6.940 g/cm^3. It can be seen also the size of $CuO/Ce_{0.75}Zr_{0.25}O_2$ is 69.2 nm which is bigger than the size of $Ce_{0.75}Zr_{0.25}O_2$, indicating the influence of CuO crystal in $Ce_{0.75}Zr_{0.25}O_2$ structure. This findings support the XRD result; the increment of crystallite size by the addition of CuO.

3.3 Field Emission Scanning Electron Microscopy, FESEM

To observe the surface morphology and roughness, FESEM analysis was performed on the $Ce_{0.75}Zr_{0.25}O_2$ and $CuO/Ce_{0.75}Zr_{0.25}O_2$. FESEM micrographs of samples are shown in Figure 2. The results show that the roughness and morphology for both samples are quite similar, even though the deposited of copper oxide. It supports the XRD and BET results. The high portions of small pores are observed on the surface of samples. However, the presence of CuO reduces the number of available pores, indicating the CuO well deposited on the pores of $Ce_{0.75}Zr_{0.25}O_2$, as shown in Fig, 2(b). Meanwhile, both samples show a uniform particles shape which is spherical and CuO particles are homogenously dispersed on the surface and pores of $Ce_{0.75}Zr_{0.25}O_2$. Based on the results, it is safe to say that the microemulsion method able to produce the uniform shape distribution with high available pores and it leads to the homogeneity of CuO dispersion on $Ce_{0.75}Zr_{0.25}O_2$.

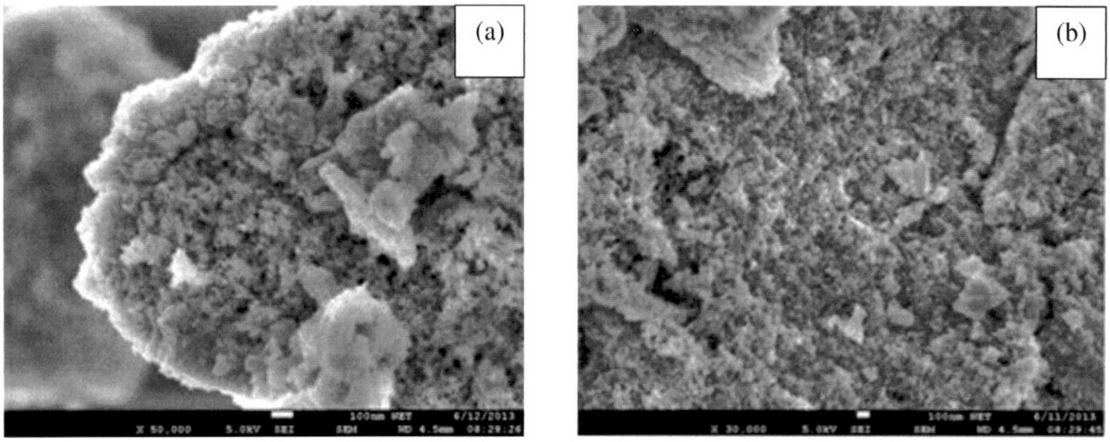

Figure 2. FESEM micrographs of synthesized samples
((a) $Ce_{0.75}Zr_{0.25}O_2$, (b) $CuO/Ce_{0.75}Zr_{0.25}O_2$)

,
3.4 Transmission Electron Microscopy, TEM

In meantime, the structural evolution and physical size distribution as well as the shape of the synthesized catalysts are also studied by transmission electron microscope (TEM). The micrographs of $Ce_{0.75}Zr_{0.25}O_2$ and $CuO/Ce_{0.75}Zr_{0.25}O_2$ are presented in Figure 3. Nano sizes of particles are observed in both samples. The particles are distributed in a narrow size distribution with an average particles size of is around 8 nm for $Ce_{0.75}Zr_{0.25}O_2$ and around 10 nm for $CuO/Ce_{0.75}Zr_{0.25}O_2$. A slightly agglomerate of particles is also observed in both samples. It is due to strong interaction between ultrafine particles $Ce_{0.75}Zr_{0.25}O_2$ and $CuO/Ce_{0.75}Zr_{0.25}O_2$. Besides, almost all of particles are in spherical shape, as found in FESEM analysis. No explicit CuO particles can be visually identified on TEM micrographs of $CuO/Ce_{0.75}Zr_{0.25}O_2$ due to the highly dispersion of CuO, a good agreement to XRD results.

Table 1. Textural properties of samples

Samples	Density, kg/m^3	Surface Area (m^2/g)	Crystallite/Particle Size, d (nm)		
			d_{XRD}	D_{BET}	d_{TEM}
CeO_2	7215	24.5	9.1	33.9	-
$Ce_{0.75}Zr_{0.25}O_2$	6940	45.9	6.6	18.8	8
$CuO/ Ce_{0.75}Zr_{0.25}O_2$	6940	12.5	8.1	69.2	10

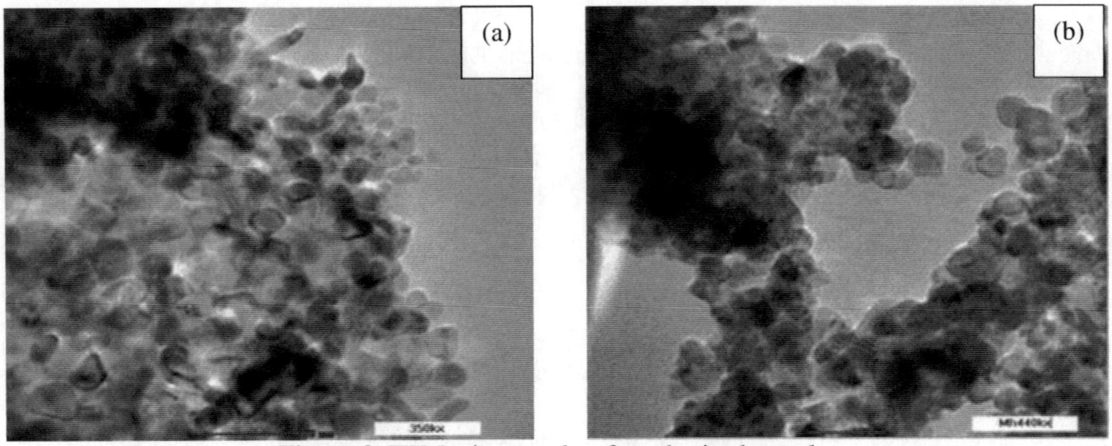

Figure 3. TEM micrographs of synthesized samples
((a) $Ce_{0.75}Zr_{0.25}O_2$, (b) $CuO/Ce_{0.75}Zr_{0.25}O_2$)

3.5 Oxygen mobility

In order to investigate the catalytic properties, $CuO/Ce_{0.75}Zr_{0.25}O_2$ and $Ce_{0.75}Zr_{0.25}O_2$ samples are examined by H_2-temperature programmed reduction. The H_2-TPR profiles, which indicate the reducibility of Ce compounds to release oxygen, are presented in Figure 4. The reduction peak of $Ce_{0.75}Zr_{0.25}O_2$ identify at temperature 466°C, while the reduction peak of $CuO/Ce_{0.75}Zr_{0.25}O_2$ towards to lower temperature up to 250°C are assigned to the reduction of Cerium (4+) to (3+) and the removal of lattice oxygen [4]. It indicates the existence of CuO; the reduction peak progressively shifted towards lower temperature which contributes the enhancement of the catalytic properties to active at lower temperature. A similar finding was reported by Deng et al., [21] & Jung et al., [22]. Therefore the additional of CuO into the $Ce_{0.75}Zr_{0.25}O_2$ structure able to enhance the activity of the catalysts at lower temperature and it also maintain the microstructure stability.

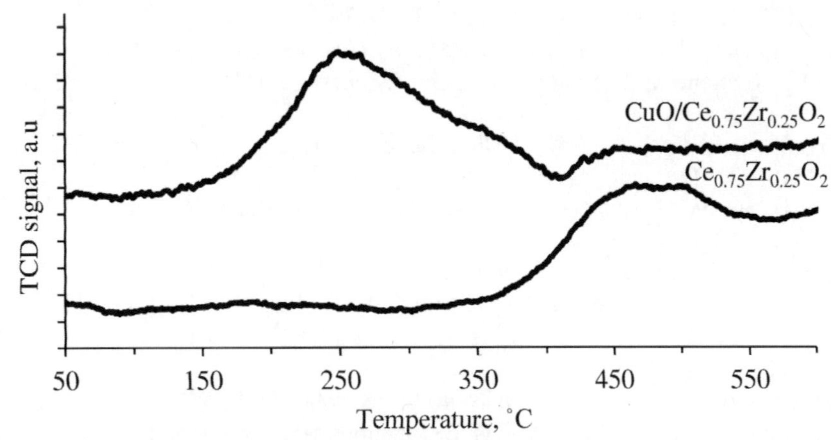

Figure 4. H_2-TPR profile of samples

4. Conclusion

The findings of study could be summarized as follows:- the uniform spherical nanoscales $CuO/Ce_{0.75}O_{0.25}O_2$ and $Ce_{0.75}Zr_{0.25}O_2$ have been successfully synthesized by the combination of microemulsion and deposition precipitation methods. The synthesized $CuO/Ce_{0.75}Zr_{0.25}O_2$ nanocatalysts reveals the highly dispersion of CuO, maintain as a cubic phases and mix as a solid solution. The additional of zirconium into cerium oxide structure able to increase the surface area up to

45.9 m^2/g. The copper catalysts may well deposit on the entire of $Ce_{0.75}Zr_{0.25}O_2$ surface and pores, which is reflected by the decreasing of surface area. According to morphology observation, the roughness of $Ce_{0.75}Zr_{0.25}O_2$ reduces with the presence of CuO. It is also found that the $CuO/Ce_{0.75}Zr_{0.25}O_2$ exists in a homogeneous solid solution and well dispersion of CuO. The particles size of $CuO/Ce_{0.75}Zr_{0.25}O_2$ and $Ce_{0.75}Zr_{0.25}O_2$ is 8 nm and 10 nm, respectively. The addition of copper catalysts able to activate the catalysts at lower temperature as examined by the oxygen mobility in temperature programmed reduction.

5. References

[1] Heck R M and Farrauto R J 2001 *Appl. Catal. Gen.* **221** 443–457
[2] Guillen-Hurtado N, Rico-Perez V, Garcia-Garcia A, Lozano-Castello D and Bueno-Lopez A 2012 *DYNA-Colomb.* **79** 114–121
[3] Shah M N A, Nor S H M, Ismail K N and Hadi A 2014 *Appl. Mech. Mater.* **575** 97–102
[4] Li H, Zhu Q, Li Y, Gong M, Chen Y, Wang J and Chen Y 2010 *J. Rare Earths* **28** 79–83
[5] Dobrosz-Gómez I, García M a. G, Szynkowska M I, Kocemba I and Rynkowski J M 2012 *Catal. Today* **191** 142–145
[6] Laosiripojana N, Chadwick D and Assabumrungrat S 2008 *Chem. Eng. J.* **138** 264–273
[7] Jani N A M, Hadi A and Ismail K N 2013 *Adv. Mater. Res.* **701** 207–211
[8] Escribano V, Delhoyomartinez C, Fernandezlopez E, Gallardoamores J and Busca G 2009 *Catal. Commun.* **10** 861–864
[9] Radlik M, Adamowska M, Łamacz A, Krztoń A, Costa P Da and Turek W 2013 *React. Kinet. Mech. Catal.* **109** 43–56
[10] Kamiuchi N, Haneda M and Ozawa M 2013 *Catal. Today* **201** 79–84
[11] Han W, Zhang P, Tang Z, Pan X and Lu G 2013 *J. Sol-Gel Sci. Technol.* **66** 526–532
[12] Wang J, Shen M, Wang J, Gao J, Ma J and Liu S 2012 *J. Rare Earths* **30** 748–752
[13] Hadi A, Ismail K N and Shah M N A 2015 *Soc. Behav. Sci.* **195** 2051–2060
[14] Boutonnet M, Lögdberg S and Elm Svensson E 2008 *Curr. Opin. Colloid Interface Sci.* **13** 270–286
[15] Hadi A and Yaacob I I 2004 *Catal. Today* **96** 165–170
[16] Mäki-Arvela P and Murzin D Y 2013 *Appl. Catal. Gen.* **451** 251–281
[17] Laguna O H, Centeno M a., Boutonnet M and Odriozola J a. 2011 *Appl. Catal. B Environ.* **106** 621–629
[18] Huang W, Yang J, Wang C, Zou B, Meng X, Wang Y, Cao X and Wang Z 2012 *Mater. Res. Bull.* **47** 2349–2356
[19] Fernández-García M, Martínez-Arias a., Iglesias-Juez a., Belver C, Hungría a. B, Conesa J C and Soria J 2000 *J. Catal.* **194** 385–392
[20] Deng Q-F, Ren T-Z, Agula B, Liu Y and Yuan Z-Y 2014 *J. Ind. Eng. Chem.* **20** 3303–3312
[21] Jung C R, Kundu A, Nam S W and Lee H-I 2007 *Appl. Catal. Gen.* **331** 112–120

Acknowledgement
The authors would like to acknowledge FRGS No project 600-RMI/FRGS 5/3 (26/2013). We would also thank to Universiti Teknologi MARA (UiTM) for supporting the research.

The effect of organic solvent, temperature and mixing time on the production of oil from *Moringa oleifera* seeds

Q Ghazali, N H M Yasin[1]

Faculty of Chemical and Natural Resources Engineering, Universiti Malaysia Pahang, Lebuhraya Tun Razak, 26300 Gambang, Kuantan, Pahang, Malaysia

E-mail: [1]hidayahyassin@ump.edu.my

Abstract. The effect of three different organic solvent, temperature and mixing time on the production of oil from *M.oleifera* seeds were studied to evaluate the effectiveness in obtaining the high oil yield based on the percentage of oil production. The modified version of Soxhlet extraction method was carried out to extract the oil from *M.oleifera* seeds by using hexane, heptane and ethanol as the organic solvent. Among the three solvents, it is found that heptane yield higher oil from *M.oleifera* seeds with maximum oil yield of 36.37% was obtained followed by hexane and ethanol with 33.89% and 18.46%, respectively. By using heptane as a solvent, the temperature (60°C, 70°C, 80°C) and mixing time (6 h, 7 h, and 8 h) were investigated to ensure the high oil yield over the experimental ranges employed and high oil yield was obtained at 60°C for 6 h with percentage oil yield of 36.37%. The fatty acid compositions of *M.oleifera* seeds oil were analyzed using Gas Chromatography/Mass Spectrometry (GC-MS). The main components of fatty acid contained in the oil extracted from *M.oleifera* seeds was oleic acid, followed by palmitic acid and arachidic acid, and small amount of behenic acid and margaric acid.

1. Introduction

Moringa oleifera tree origin is from North West India but nowadays it has been found all around the world and is naturalized in many locales. This might be because it possesses the special ability to endure in harsh environment and it grows in many different soil types as well as it is drought resistance [1]. The tree carries some physical characteristics such as it is filled with long and delicate-looking branches, which are covered with small and oval-shaped dark green leaves [1]. The height of the tree can reach up to 10 m naturally but for cultivation purposes, the tree often being cut back each year into 6 to 12 feet maximum so that the seeds and leaves can be harvested easily [2].

The most utilized part of the *M.oleifera* tree is the seeds which are used in the production of oil. The oil production from *M.oleifera* seeds are important because it have many advantages over other organic sources for example, it can be used to reduce blood pressure and lower glucose level, skin preparation and ointment, perfumes and also for cooking [3,4]. It also possesses the antimicrobial activity to effectively

[1]To whom any correspondence should be addressed

fight against common bacteria such as *Escherichia coli* and *Staphylococcus aureus* [5].

After the *M.oleifera* seeds are extracted for its oil, the seed cake can be used for other applications. According to a study made by Ali *et al.* (2010), *M.oleifera* seed cake can be used in water treatment as it acts as a natural coagulant for use in drinking water treatment [6]. Results showed that the turbidity removal using *M.oleifera* oil as the natural coagulant can be reached up to 96.23%. Emmanuel *et al.* (2011) proposed that *M.oleifera* seed cake can be added to the fertilizer mix as a decent natural, ecofriendly and biocompatible organic fertilizer [7]. The experimental results indicate the mineral content of the soil have increased when the *M.oleifera* seed cake mixed with the fertilizer which in turn improved the production of the harvest.

Oil can be extracted from *M.oleifera* seeds by traditional method. The seeds from mature pods are roasted, crushed and placed in boiling water for 5 minutes. After straining and left for overnight, the *M.oleifera* oil floats to the surface [3]. However carrying out this traditional method is not efficient as it is highly energy intensive and time consuming and only can produce low percentage of oil yield. Therefore, the researchers come up with different methods to maximize the oil recovery from *M.oleifera* seeds. Various methods have been used from the previous researchers to extract the oil from *M.oleifera* seeds such as cold press method [8], supercritical carbon dioxide extraction [9], aqueous enzymatic method [10] and solvent extraction method [1].

In this investigation, oil extraction using the method of solvent extraction was carried out using hexane, heptane and ethanol as the solvent. Solvent extraction has been reported by Mani *et al.* (2007) using hexane, petroleum ether and acetone [11]. Experimental Soxhlet method using hexane as the solvent, and also the effect of heat treatment on yield of oil expressed from *M.oleifera* seeds have been reported by Adejumo *et al.* (2013) [1]. However, there are limited data on the processing factors needed for the ideal extraction of *M.oleifera* oil especially for the type of solvent used. There are many factors influencing the oil extraction yield such as the method of extraction, temperature of extraction, seed particle size, ratio of solvent to the *M.oleifera* seed powder, pre-treatment conditions and residence time between the solvent and the seed [11]. However, this research are only focussing on the organic solvent, temperature of extraction and contact time of the solvent with the *M.oleifera* seeds to increase the oil extraction yield. Therefore, the main objective of this research is to evaluate the suitable set of parameters in obtaining the highest oil recovery within the studied range from *M.oleifera* seeds.

2. Material and methods

2.1 Sample preparation

The dry *Moringa oleifera* seeds were purchased from Mitomasa Sdn. Bhd. The wings and coats of *M.oleifera* seed were removed manually. Only the good quality or undamaged seeds were selected to be used in the experiment. Solid to solvent ratio was 1:10, therefore 10g of the *M.oleifera* seeds were ground to a fine powder using an electrical grinder and hand-operated mortar.

2.2 Organic solvents for oil extraction

Hexane, heptane and ethanol (99% of purity) were used as organic solvents to leach out oils from the insoluble solid structure of the *M.oleifera* seeds. During the selection of organic solvents, the solid to solvent ratio was maintained as 1:10 for the oil extraction [12]. Temperature and mixing time were kept constant at 60^oC and 6 hours respectively. The mixing rate was adjusted at 400 rpm for the whole experiment.

2.3 Oil extraction using solvent extraction method

The modified version of Soxhlet extraction method was used for oil extraction process. Extraction set up mainly consists of a condenser, retort stand, hot plate stirrer, 1 L beaker and 250 ml round bottom flask. The *M.oleifera* powder was placed in the round bottom flask which connected to the condenser. The ratio of solid to solvent that was used in this experiment was 1:10 [12], thus 100 ml of solvent was added round bottom flask containing 10 g of *M.oleifera* powder. A magnetic stirrer in the round bottom flask was used to mix up the powder with the solvent thoroughly. During this process, the round bottom flask was heated in the beaker filled with water and the solvent was evaporated and moved to the condenser where the steam was converted into liquids which then drip back into the round bottom flask. The solvent-oil mixture in the round bottom flask was collected at the end of the extraction process. Figure 1 shows two sets of Soxhlet apparatus used for solvent extraction.

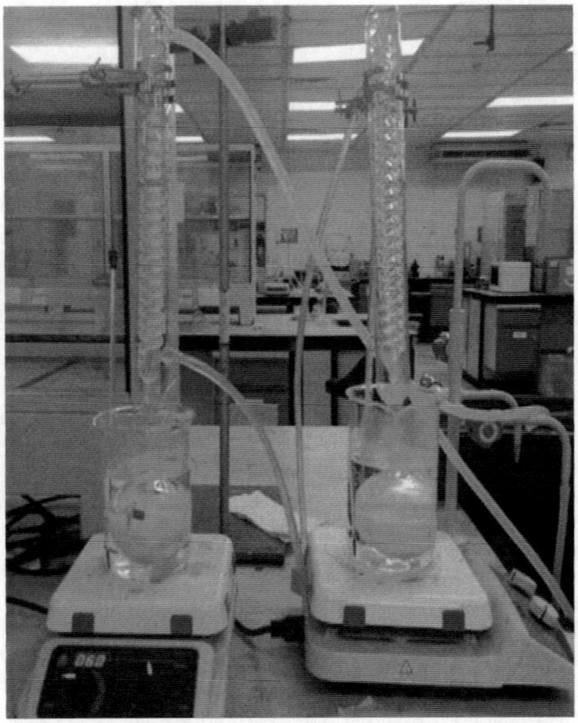

Figure 1. Modified Soxhlet extraction method for solvent extraction process

The flask containing a mixture of *M.oleifera* powder, solvent and oil were separated at the end of the extraction process. A filter paper was used to separate the powder from the solvent-oil mixture before it was dried in the oven overnight at 60°C. Then, the powder was being weighted for the determination of the percentage of oil yield. The oil was recovered by evaporating off the solvent using rotary evaporator where it was heated at the temperature higher than boiling point of the solvent until the solvent finally evaporated and leaving behind the extracted oil. All experiments were conducted in triplicate.

2.4 Extraction parameters to enhance the oil yield

The best condition of extraction parameters such as temperature (60°C, 70°C and 80°C) and mixing time (6, 7 and 8 hours) were evaluated to increase the oil recovery from *M.oleifera* seeds using the best solvent from previous result with the highest percentage of oil yield. The details of the experiment are described in table 1.

Table 1. Variables of the different parameters in the oil extraction

Experiment	Constant variable	Manipulated Variable
1	Temperature and mixing time	Solvents
2	The best solvent and mixing time	Temperature
3	The best solvent and temperature	Mixing time

2.5 Determination of oil yield percentage

The percentage of oil yield was calculated using following formula:

$$X = \frac{\text{Weight of oil extracted (g)}}{\text{Weight of } M.oleifera \text{ powder before extraction (g)}} \text{ X } 100\% \quad (1)$$

where X indicates the percentage of oil yield (%) and weight of oil extracted from *M.oleifera* seed can be calculated by subtracting the weight of *M.oleifera* powder before and after the extraction [1].

2.6 Gas chromatography analysis

The fatty acid profiles of *M.oleifera* oil were analyzed using Gas Chromatography/Mass Spectrometry (GC-MS) after each extraction using Agilent Technologies (G3171A, China). The samples were injected into capillary column HP- 5 (30 m length, 0.25 mm ID). Helium is used as the carrier gas with a flow rate of 1 ml/min. The inlet temperature and the detector temperature were kept constant at 250^{0}C and 280^{0}C respectively with temperature programming of 100^{0}C for 3 minutes with increase at 10^{0}C per minutes up to injector temperature 250^{0}C [13].

3. Results and discussion

3.1 The effect of organic solvent

The *M.oleifera* seeds were extracted with different solvents which are hexane, heptane and ethanol at constant temperature of 60°C for 6 hours to evaluate the effectiveness of the solvents in producing the maximum oil yield from *M.oleifera* seeds. The result of the effect of organic solvent on the oil yield of *M.oleifera* seeds were compared with the other researches [1,9,12] as presented in table 2. As can be noticed in table 2, heptane showed a better solvent compared to hexane and ethanol. According to Conkerton *et al.* (1995), the yield and quality of the oil extracted by heptane were similar to that extracted by hexane [12]. This is might be due to the fact that heptane and hexane are both non-polar solvent while ethanol is a polar solvent. Ethanol yields about 20% lower than heptane and hexane. This result is similar to the research conducted by Nwabueze & Okocha (2008) [14]. They also find that the non-polar solvent is better solvent for extracting oil from African breadfruit (*Treculia africana*) seed oil.

Table 2. The effect of organic solvents on the oil yield of *M.oleifera* seeds

Organic Solvents	Experimental data (%)	Adejumo *et al.* (2013)[a]	Palafox *et al.* (2012)[b]	Conkerton *et al.* (1995)[c]
Heptane	36.26±2.48	-	-	30.4±0.19
Hexane	35.58±10.08	33.7	29	28.4 ± 0.36
Ethanol	16.94±3.06	-	35	-

[a] Heated at 100°C for 6 hours.
[b] Heated at 60°C for 6 hours.
[c] Heated at 88-95°C for 4 hours.

3.2 The effect of temperature

Figure 2 shows the effect of temperature by using heptane as the solvent. The result shows that the *M.oleifera* seeds heated at 60^0C have the highest percentage yield of 36.3% while the lowest oil yield of *M.oleifera* is at 80^0C. The results also showed that there is a reduction in percentage oil yield when the seed is heated above 60^0C. This finding was also found in the study reported by Adejumo *et al.* (2013) where the oil yield from *M.oleifera* seed decreased when the temperature increased from 100^0C until 150^0C [1]. Mani *et al.* (2007) also reported the same finding when hexane is used as the solvent for extracting *M.oleifera* oils at temperature range of 40^0C until 80^0C. The study also revealed that the oil yield was constant up to 60^0C but slowly decreased as temperature increases [11].

3.3 The effect of mixing time

Figure 3 demonstrated the effect of mixing time on the oil yield of *M.oleifera* seeds. The results revealed that 6 hours of extraction resulted in the highest oil yield for *M.oleifera* seeds which is 36.3% of oil yield. This observation shows that the longer the mixing time, the lower the oil yields that can be extracted from *M.oleifera* seeds. This result agrees to the finding reported by Mani *et al.* (2007) that any further increase passes 6 hours in extraction time did not increase the oil yield of *M.oleifera* seeds [11]. This phenomenon is due to low solvent density contained left in the sample after 6 hours [15]. However, the oil yield increased dramatically when the *M.oleifera* seeds were heated from 1 hour until 6 hours because the diffusivities of the oil and solvent increases, which result in high oil yield [11]. Therefore, the maximum oil yield could be achieved even at shorter residence time with an optimal extraction temperature.

Oil yield comparison of the extracted seeds is presented in table 3. It is possible to observe that *Jatropha curcas* has a great ability to extract oil up to 86.1% compared to *M.oleifera*, sunflower and soybean. The ranges of processing parameters investigated by the authors such as applied pressure, moisture contain, preheating time, shell removal and particle size explains the higher yield obtained from *Jatropha curcas*. As can be noticed in table 3, the percentage of oil yield is lower when ethanol used as a solvent even for different temperature and mixing time for sunflower. This observation is supported by the fact that the lower selectivity of ethanol towards oil makes this solvent difficult to extract the oil as much as possible from the seeds [16].

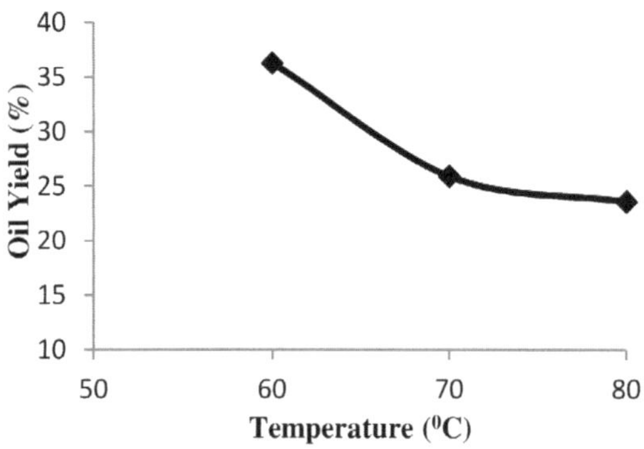

Figure 2. The effect of temperature on the oil yield of *Moringa oleifera*

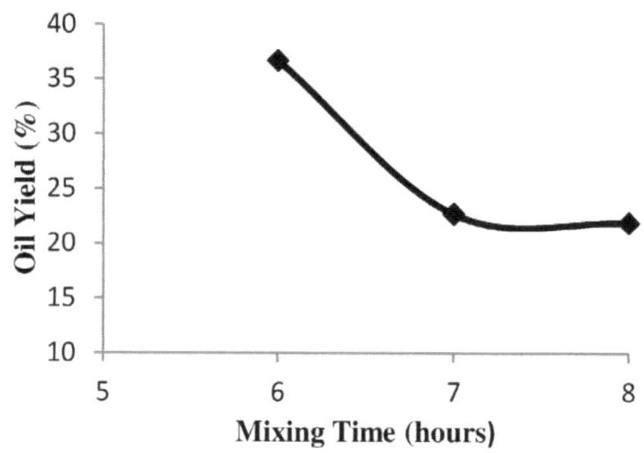

Figure 3. The effect of mixing time on the oil yield of *Moringa oleifera*

Table 3. Comparison of oil yield with the other seeds obtained by Soxhlet.

Seed	Solvents	Temperature (^{0}C)	Mixing time (hour)	Oil yield (%)	Ref.
M.oleifera	Heptane	60	6	36.26±2.48	
M.oleifera	Hexane	60	6	35.58±10.08	
M.oleifera	Ethanol	60	6	16.94±3.06	
Sunflower	Ethanol	60	16	24.5±0.1	[16]
Sunflower	Hexane	130	4	46.2±0.7	[17]
Jatropha curcas	Hexane	60	24	86.1	[18]
Soybean	Ethanol	55	3	25	[19]
Soybean	Ethanol	60	3	25.6±0.2	[20]

3.3 Fatty acid composition of extracted oil

The fatty acid compositions for *M.oleifera* were analyzed by GC-MS using Agilent Technologies Chromatograph. In this present study, the major component of *M.oleifera* oil was oleic acid (C18:1), followed by palmitic acid (C16:0) and arachidic acid (C20:0) and a small amount of behenic acid (C22:0) and margaric acid (C17:0). The result also shows a higher oleic acid composition compared to the values reported by Gaikwad *et al.* (2011) and Abdulkarim *et al.* (2005) [10,13]. According to Adejumo *et al.* (2013) and Ogunsina *et al.* (2014), the high oleic acid content indicates the oil has an excellent quality similar to olive oil which also provides good stability and good thermal stability to *M.oleifera* seed oil [1,21].

4. Conclusion

Extraction of oil from *M.oleifera* seeds by using heptane as the organic solvent showed better results when compared with other solvents studied. The temperature of 60^0C for 6 hours of extraction time gave a maximum oil yield percentage which is 36.37%. In this study, the results also showed the higher percentage of oleic acid oil when compared to previous researches. In addition, based on the fatty acid composition, *M.oleifera* seeds can extract the good quality of oil for further applications.

5. References

[1] Adejumo B A, Alakowe A T and Obi D E 2013 *Int. J. Eng. Sci.* **2(1)** 232
[2] *Moringa Oleifera* 2014 In *I Love Moringa: The Fast-growing, nutritious, delicious gift from god.* http://www.ilovemoringa.com/*MoringaOleifera*.html
[3] Price M L 2007 *The Moringa Tree (*North Fort Myers: ECHO Staff).
[4] Moringasource 2013 *Moringa oleifera - one of nature's best-kept health and nutrition secrets.* http://www.moringasource.com/
[5] Lalas S, Gortzi O, Athanasiadis V, Tsaknis J and Chinou I 2012 *Molecules* **17(3)** 2330
[6] Ali E N, Muyibi S A, Salleh H M, Alam M Z and M. Salleh M 2010 *Water Resour. Prot.* **2(3)** 259
[7] Emmanuel S, Zaku S, Adedirin S, Tafida M and Thomas S 2011 *Agric. Biol. J. North Am.* **2(9)** 1289
[8] Lalas S and Tsaknis J 2002 *J. Food Compos. Anal.* **15(1)** 65
[9] Palafox J O, Navarrete A, Sacramento-Rivero J C, Rubio-Atoche C, Escoffie P A and Rocha-Uribe J A 2012 *Am. J. Anal. Chem.* **3** 946
[10] Abdulkarim S, Long K, Lai O, Muhammad S and Ghazali H 2005 *Food chem.* **93** 253
[11] Mani S, Jaya S and Vadivambal R 2007 *Food Bioprod. Process.* **85** 328
[12] Conkerton E, Wan P and Richard O 1995 *J. Am. Oil Chem. Soc.* **72(8)** 963
[13] Gaikwad M, Kale S, Bhandare S, Urunkar V and Rajmane A 2011 *Int. J. PharmTech Res.* **3(3)** 1567
[14] Nwabueze T and Okocha K S 2008 *Afr. J. Food Sci.* **2(10)** 119
[15] Treybal R E 1980 *Mass Transfer Operations* (New York, USA: McGraw Hill).
[16] Baumler E R, Carrin M E and Carelli A A 2016 *J. Food Eng.* Article in Press.
[17] Ravber M, Knez Z and Skerget M 2015 *Food Chem.* **166** 316
[18] Subroto E, Manurung R, Heeres H J and Broekhuis A A 2015 *Ind. Crops Prod.* **63** 303
[19] Dagostin J L A, Carpine D and Corazza M L 2015 *Ind. Crops Prod.* **74** 69
[20] Toda T A, Sawada M M and Rodrigues C E C 2016 *Food Bioprod. Process* **98** 1
[21] Ogunsina B S, Indira T, Bhatnagar A, Radha C, Debnath S and Gropala Krishna A 2014 *J. Food Sci. Technol.* **51(3)** 503

Acknowledgments

The authors would like to acknowledge the UMP Research Grant (Grant No. RDU 140390) from *Universiti Malaysia Pahang*, which has fully supported this research.

Encapsulation of bioactive compound from extracted jasmine flower using β-Cyclodextrin via electrospray

S Rahmam.[1], M N Naim.[1], E Ng.[1], M Nn Mokhtar[1], N F Abu Bakar[2]
[1] Process and Food Engineering, Universiti Putra Malaysia, 43300 Serdang, Selangor, Malaysia
[2] Faculty of Chemical Engineering, Universiti Teknologi MARA (UiTM), 40450 Shah Alam, Selangor, Malaysia

E-mail: [1]mohdnazli@upm.edu.my

Abstract. The ability of electrospray to encapsulate the bioactive compound extracted from Jasmine flower with β-Cyclodextrion (β-CD) without any thermal-assisted processing was demonstrated in this study. The extraction of Jasmine compound were conducted using sonicator at 70 000 Hz, for 10 minutes and followed by mixing of the filtered compound with β-CD. Then, the mixture was electrosprayed under a stable Taylor cone jet mode at the voltage of 4 – 5 kV, with flow rate of 0.2 ml/hour. The aluminum substrate that used for collecting the deposit was placed at 30 cm from the needle's tip to allow the occurrence of evaporation and droplet fission until the droplet transform to solid particles. Characteristics of solidified bioactive compound from Jasmine flower (non-encapsulated compound) and solidified bioactive compound with β-CD (encapsulated compound) were studied in this work. From SEM images, it can be observed that the particles size distribution of encapsulated compound deposits have better deposition array and did not aggregate with each other compared to the non-encapsulated compound. FE-SEM images of encapsulated compound deposits indicate more solid crystal looks while non-encapsulated compound was obtained in the porous form. The electrospray process in this work has successfully encapsulated the Jasmine compound with β-CD without any thermal-assisted process. The encapsulation occurrence was determined using FTIR analysis. Identical peaks that referred to the β-CD were found on the encapsulated compound demonstrated that most deposits were encapsulated with β-CD.

1. Introduction

Natural fragrance compound extracted from plants have been widely used in many areas such as in pharmaceutical, aromatherapy, cosmetics, as condiment in several food and drinks, and for adding scents to household cleaning products [1]. It can be obtained naturally from fruits, leaves, and flowers such as lavender, mint, citrus, rosemary, lemon [2], Jasmine, and many more. Jasmine flower for example, is a fun-loving fragrance, provides a unique and enthusiastic aroma. It has rejuvenating, boosting and energizing properties, which makes it a natural mind-blowing fragrance. Jasmine is primarily used in the perfume industry and has a very high commercial value due to its therapeutic

[1] To whom any correspondence should be addressed.

properties. The production of perfume from Jasmine fragrance has even started in 1860 [3] and continually expands over the years. The pleasant smell from this flower is coming from the bioactive compounds such as linalool, franesene, and benzyl acetate. However, the extracted Jasmine flower is easy to degrade due to the adversely environmental effect such as undesirable effect of light, heat, pressure, and oxygen [4] that will lead to the decreasing of shelf life.

In order to minimize the compound lost, a proper control after the extraction process is necessary. In this work, the extracted bioactive compound is subjected to the encapsulation process due to the ability of the encapsulated matrix to minimize the permeability of oxygen, water, bacteria and other factor that might change or degrade the compound structure. Encapsulation process has been applied to many areas especially in pharmaceuticals and food industries due to the ability of encapsulation to avoid the adverse factor or condition, increase the stability of bioactive compound [5], enhance bioavailability, and improve the control release of bioactive compounds [6].

Various kind of encapsulation methods have been explored and applied for commercial purpose i.e spray drying and freeze drying. However, due to the heat sensitive issue applies to the extracted bioactive compound, the application of spray drying that involves heat is not preferable. Meanwhile, freeze drying is a simple technique for encapsulating water-soluble essences and natural aromas, as well as drugs, but the major disadvantages of freeze drying are it consumes high energy and take a long processing time [7].

Therefore, an alternative encapsulation method is required to ensure the practicality and the survival of the bioactive compound during and after the encapsulation process. To date, due to the emerging of technology, electrospray has been found to be an alternative method for encapsulation process without the thermal application. The process also provides many advantages such as producing fine monodispersed droplet within nanometer scale, minimum droplet aggregation, controlled charged droplets, and high deposition efficiency due to the charged carries by the produced droplets [8].

Since no thermal application is needed, the drying process that occurs during the electrospray process is due to the evaporation and droplet fission that continuously take place along the process. The evaporation of water from the droplet increases the charge concentration on the droplet surface and lead to the droplet fission. These two processes occurred repeatedly until a solid body is obtained and deposited onto the aluminium collector or substrate.

The application of electrospray is quite well known in the pharmaceutical area to increase the bioavailability of the drug. However, the application of electrospray for the encapsulation of bioactive compound is limited and lack of literature review. Thus, this study is focusing on the encapsulation of bioactive compounds extracted from Jasmine flower using β-CD via the electrospray technology. In general, cyclodextrin is considered as a good encapsulating agent because of its advantages such as extremely low toxicity, excellent biocompatibility and non immunogecity [9]. In addition, β-CD was specifically chosen as encapsulating agent in this study compared to others is due to its' ability to form inclusion complexes with many organic and inorganic compounds within the cavity of their ring structure that composed of D-glucose units and has been demonstrated from our previous work [10]

The drying process that occurs during the electrospray process is due to the high electrostatic charged droplet fusion that produces progeny droplet when exceeding the Rayleigh limit condition. As a result, the repeatable droplet fission has caused the water evaporation until a solid body is obtained as the deposit.

2. Methodology

2.1 Preparation of Jasmine extract

Raw material in this study which is Jasmine flowers were obtained from Universiti Putra Malaysia (UPM) garden, and the β-CD that was used as encapsulating agent was purchased from Sigma Aldrich (M) Sdn. Bhd.

The extracted Jasmine compound was obtained by sonicating 3 g of Jasmine flower in 20 ml of deionized (DI) water. Total sonication process was conducted for 10 minutes with 15 seconds pause-mode after each 1 minute to prevent the heat to develop during the process. The ultrasonic process increases the solution's temperature due to the vibration of ultra sonicator's probe. Therefore, it is important to ensure the temperature of solution is always at the average of 20°C. In order to achieve this target, the solution was put in a 50 ml beaker, and this beaker was put in a bigger beaker that can fit the 50 ml beaker. Wet tissue that was submerged in iced water was placed between these two beakers to control the solution temperature from increasing and it needed to be changed for every 1 minute.

The extracted Jasmine compound consists of inconsistent particle size distribution after the ultrasonic process. To separate the large and fine particles, the extracted Jasmine flower was centrifuged at 10000 rpm, 20°C, for 30 minutes. Figure 1 shows all the steps to obtain the extracted Jasmine compound.

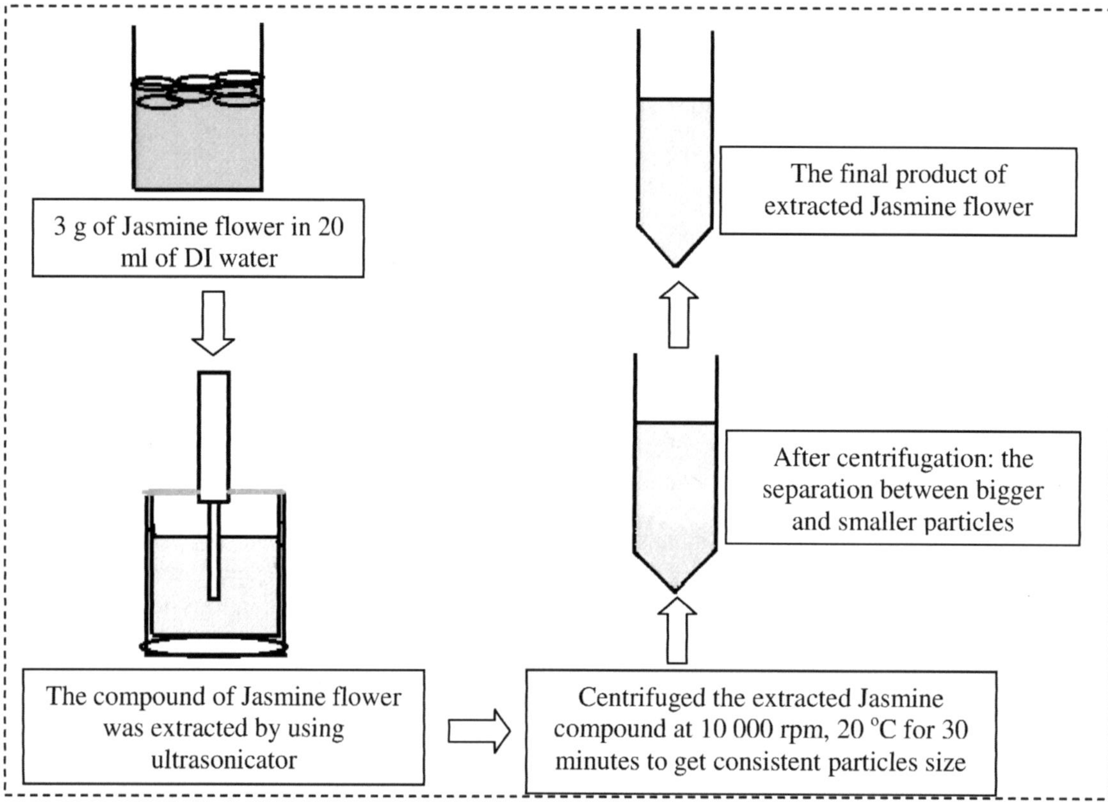

Figure 1. Experimental setup for extraction of bioactive compound from Jasmine flower

2.2 Preparation of Jasmine's extracts with β-CD.

In order to prepare the mixture, 1g of β-CD was mixed with 10 ml of extracted Jasmine compound, and was stirred for 1 day. To ensure the well mixture of β-CD and Jasmine solution, the β-CD was added bit by bit into the solution. Ultimately, two samples were prepared for electrospray, which are pure extracted Jasmine compound (non-encapsulated), and the mixture of extracted Jasmine with β-CD (encapsulated).

2.3 Electrospray

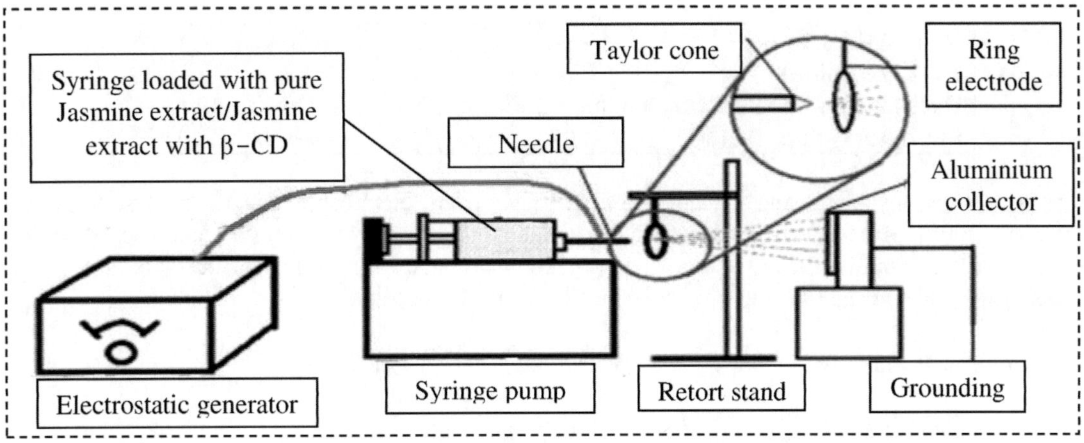

Figure 2. The electrospray experimental setup for producing solidified particles

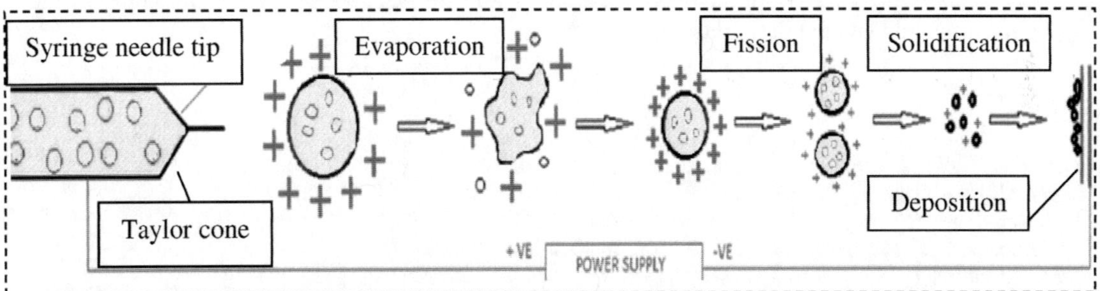

Figure 3. The mechanism of droplet evaporation and fission along the deposition process

As shown in Figure 2 above, electrospray setup consist of electrostatic generator that supply positive charge to the needle, syringe that loaded with jasmine's extract solution, needle, syringe pump, retort stand, substrate, ring electrode, and the negative power supply to the aluminium substrate.

The electrostatic generator supplied high voltage positive charge to the needle and directly to the solution. An intense electrostatic field is formed at the tip of the needle by applying a few kilovolts potential difference between the needle and the surrounding environment. At the tip of the needle, the strong electric field pulls liquid-phase ions toward the liquid-gas interface. The liquid surface at the tip of the needle forms a cusp, called a Taylor cone. The Taylor cone is developed from the combination of hydrodynamic and electrostatic forces, which resulting of fine spray jet and able to produce charged droplets. Evaporating charged droplets quickly reach a point at which they are no longer stable. This condition is known as "Rayleigh instability". At this stage, evaporation process is continuously occurs along the deposition process and reduces the size of droplet. The reduction size of droplet creates the repulsion effect, where the positive charge tends to repel with each other and produce more droplet fission. The evaporation and droplet fission will continuously occur until the droplets are completely dry and transformed into very fine solidified particles before deposited onto the aluminum substrate that was supplied with the contradict charge (negative). The illustration of the mechanism is shown is Figure 3. The conditions of electrospray in this work are tabulated in Table 1.

Table 1. Condition of Electrospray

Parameter	Value
Voltage (kV)	4 – 5
Flow rate (ml/hr)	0.2
Distance from needle tip to aluminium collector (cm)	30
Electrospray duration (hr)	1

The non-encapsulated and encapsulated particle from electrospray were analysed by Gas Chromatography Mass Spectrometry (GCMS) – Head space, Scanning Electron Microscopy (SEM), Field Emission Scanning Electron Microscopy (FESEM), Image J for particle size distribution, Fourier Transform Infra-Red (FTIR), and Transmission Electron Microscopy.

3. Result and Discussion

3.1. GCMS-Headspace

The purpose of conducting GCMS – Head space was to determine the component of bioactive compounds existed in the Jasmine flower. Figure 4 and Table 2 below show the components of bioactive compound existed in the Jasmine flower.

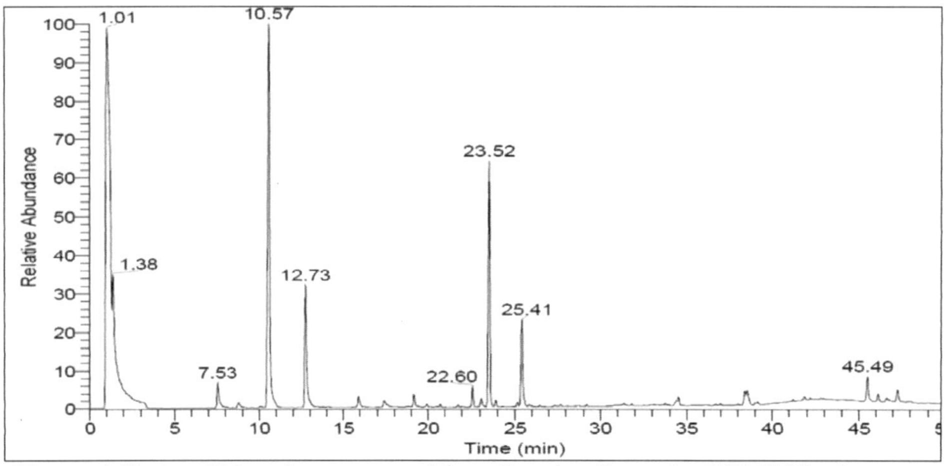

Figure 4. Peaks of bioactive compound from Jasmine flower by GCMS-Headspace

Table 2. Components existed from the extraction of Jasmine flower

Time (min)	Compound
1.01	Ethanolamine
1.38	Alanylglycine
7.53	3-Hexenyl acetate
10.57	Ocimene
12.73	Linalool
22.60	Acetic acid
23.52	Hexanoic acid
25.41	β-Cubebene
45.49	α-Franesene

As observed above, there are many compounds existed in the Jasmine flower. However, based on the research done by [11], the constituent that give essence and has the heat sensitive characteristic (easy to volatile) are lie to the linalool and α -Franesene.

3.2. SEM and FE-SEM

SEM and FE-SEM were used to observe the morphology of the encapsulated and non-encapsulated particle after the electrospray. Figure 5 shows the morphology of particle for non-encapsulated and encapsulated compound.

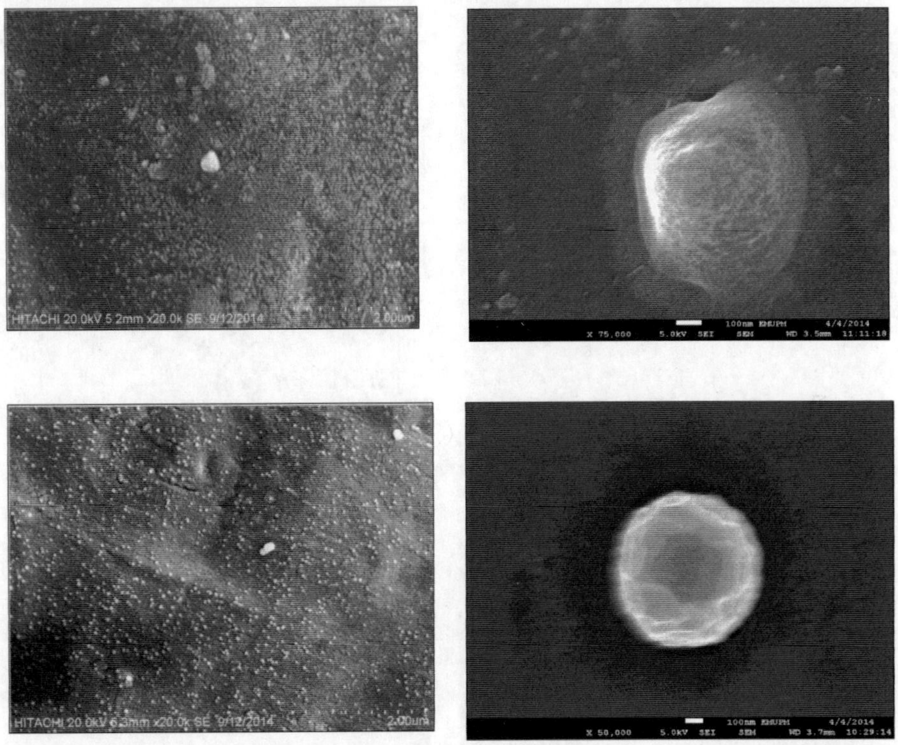

Figure 5. The morphology of solidified particles through SEM and FE-SEM

From SEM, it can be observed that the sphere particle shapes were produced after electrospray process. However, further observation by FE-SEM indicates that the non-encapsulated figure were not in sphere from, more porous, and aggregated compared to the encapsulated ones. In contrast, the particle distributions of encapsulated compounds were observed to be finer, solid-looks and less aggregation between the deposits.

In order to justify the morphology obtained by SEM, high magnification (50,000x) observation was performed using FE-SEM. Based on the FE-SEM images, it can be clearly seen that the non-encapsulated particles indicate a porous-look, poor sphericity, and seems not fully dry when reached the substrate surface. However, the encapsulated particles were deposited in solid form and expected to have the high crystallinity index compared to the non-encapsulated ones. From the FE-SEM images, it can be hypothesized that the non-encapsulated particles were unable to fully dry when undergoes the deposition process even the distance to collect both samples are exactly the same. The encapsulated compound is expected to be fully encapsulated since only crystal look is obtained and no porous form deposits were noticed on the substrate surface. It can be concluded that the Jasmine flower extracts were fully surrounded by β -CD.

3.3. Particles size distribution

The particle size distribution was analysed by using the ImageJ software. By using the software, the distribution of non-encapsulated and encapsulated compound were compared and analyzed.

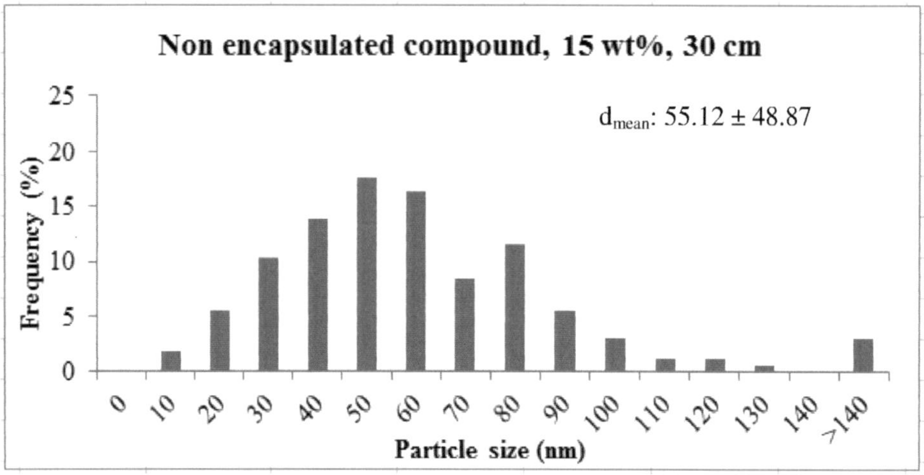

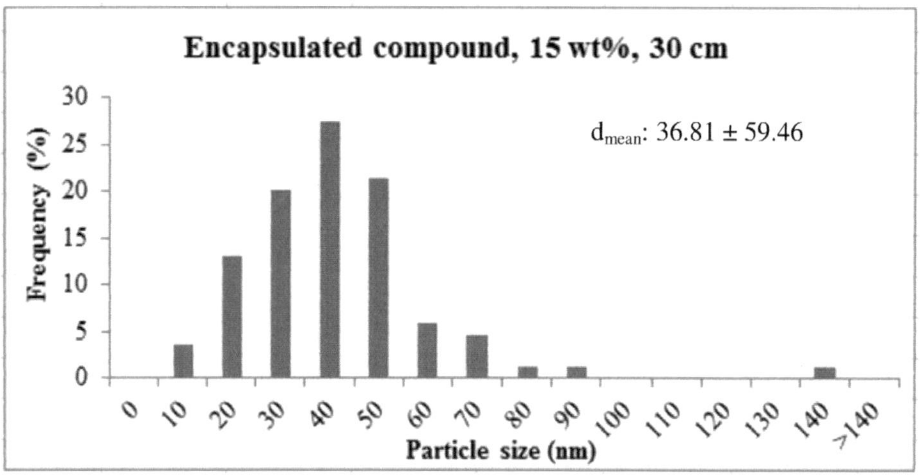

Figure 6. Particle size distribution for non-encapsulated and encapsulated compound collected at 30 cm distance from the needles tip

Based on the graph above, the particle size distribution of encapsulated Jasmine extracts is finer and it has sharp distribution compared to the non-encapsulated compound. This information can be proved through the data of average particle size (d_{mean}). Due to the wider distribution of non-encapsulated compound, the average particle size is bigger than the encapsulated compounds which are 55.12 nm and 36.81 nm respectively.

Since the evaporation is taken place along the electrospray process until the droplets become smaller droplets [12], the different in the obtained results were justified through evaporation rate.

The proposed equation rate to be used is:

$$E_R = \Delta V / t \quad (1)$$

Where,
ΔV : Volume difference
t : Time taken
E_R : Evaporation rate

Therefore;

Table 3. Evaporation rate for non-encapsulated and encapsulated compound

Type of compound	Evaporation rate (nm^3/s)
Non encapsulated	$1.11x10^{11}$
Encapsulated	$1.05x10^{11}$

The evaporation rate of encapsulated compound is slower than non-encapsulated compound is due to the water percentage presence in the initial droplet. Initially, encapsulated compound has 75 wt % of water, while non-encapsulated compound was having 85 wt % of water. Earle (2013) reported that when the water at the surface is gradually removed, the moisture content inside the material is decreased and cause the remaining water inside the droplet to be strongly bounded to its solid compound [13]. Thus, the encapsulated particle took low evaporation rate compared to the non-encapsulated ones. In addition, even though the non-encapsulated compound has higher evaporation rate, but that rates is not high enough to allow the droplet to fully dry before reaching the collector. The distance need to be increased in order to allow more water to be evaporated and transform the evaporating droplet into a solid form.

3.4. FTIR analysis

The FTIR identify the presence of functional group in the compound's structure.

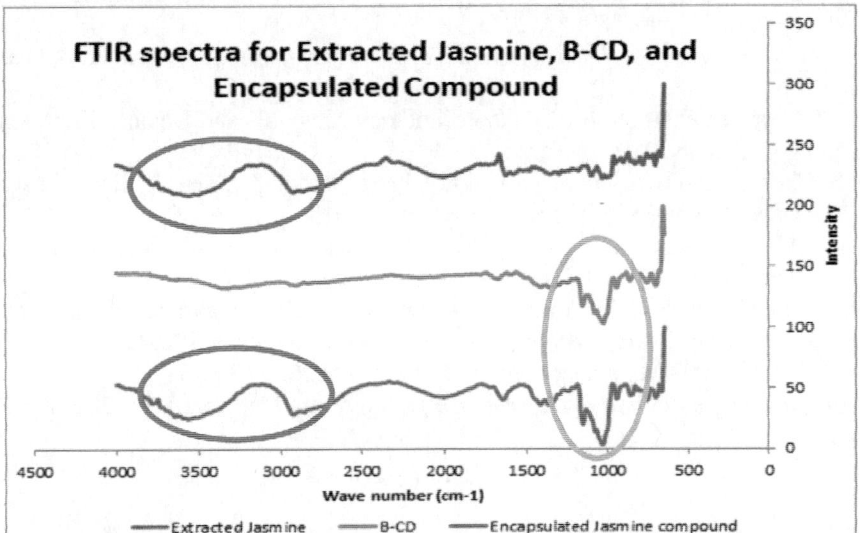

Figure 7. FTIR spectra for the extracted Jasmine, β-CD, and encapsulated of Jasmine compound

As shown in Figure 7, the encapsulated compound consists of identical β-CD between 1100 and 1700 cm^{-1}. The encapsulated compounds also consist of bioactive compound extracts which indicates the broad peaks between 2800 and 3500 cm^{-1}. The β-CD peak that appeared similarly at the encapsulated compound were represented by the wave number of ~1100 cm^{-1} and ~1000 cm^{-1} that corresponding to the C-C and bending vibration of O-H respectively [11]. Meanwhile, the similarities peak between extracted Jasmine and encapsulated compound were represented by the wave number of ~3500 cm^{-1} and ~2800 cm^{-1} that indicates the appearance of OH stretching and C-H stretching. Those functional groups are believed to be signified to the linalool and α-franesene. The similar peak from β-CD and extracted bioactive compound that appear at the encapsulated compound was evidence to the encapsulation of bioactive compound with β-CD.

4. Conclusion

The encapsulation of bioactive compound with β-CD without any thermal-assisted was demonstrated in this research work. The evidences can be seen through SEM, FE-SEM, and FTIR. From SEM result, the morphology of encapsulated compound is found to be finer, solid looks, and less aggregation compared to the non-encapsulated compound. Additionally, the crystal-look is obtained around the spherical shape of encapsulated compound through FESEM that confirms the presence of β-CD. Besides that, the encapsulation of Jasmine extract with β-CD can also be proved through the similarity peak from Jasmine extract and β-CD in the encapsulated peak. The inclusion of β-CD in every single droplet has improves the solidification process compared to the solidification of Jasmine extracts only.

5. References

[1] C. Yuan, Z. Lu, and Z. Jin *Food Chem.*, vol. 152, pp. 140–145, 2014.
[2] C. X. Wang, *J. Ind. Text.*, vol. 34, no. 3, pp. 157–166, 2005.
[3] E. T. Theimer, *Fragrance Chemistry: The Science of the Sense of Smell*. Elsevier, 2012, 2012.
[4] S. F. Hosseini, M. Zandi, M. Rezaei, and F. Farahmandghavi, *Carbohydr. Polym.*, vol. 95, no. 1, pp. 50–56, 2013.
[5] N. Durán and P. D. Marcato, *Int. J. Food Sci. Technol.*, vol. 48, no. 6, pp. 1127–1134, 2013.
[6] P. N. Ezhilarasi, P. Karthik, N. Chhanwal, and C. Anandharamakrishnan, *Food Bioprocess Technol.*, vol. 6, no. 3, pp. 628–647, 2013.
[7] E. S. Nicholas Jan Zuidam, *Encapsulation Technologies for Active Food Ingredients and Food Processing*, pp. 161–185, 2010.
[8] W. Gu, 2013 presented at the IEEE Nanotechnology Materials and Devices Conference National Cheng Kung University, Tainan, Taiwan.
[9] H. He, S. Chen, J. Zhou, Y. Dou, L. Song, L. Che, X. Zhou, X. Chen, Y. Jia, J. Zhang, S. Li, and X. Li, *Biomaterials*, vol. 34, no. 21, pp. 5344–5358, 2013.
[10] S. Saallah, M. N. Naim, M. N. Mokhtar, N. F. Abu Bakar, M. Gen, I. W. Lenggoro, Enzyme and Microbial Technology, vol 64–65, pp 52–59, 2014.
[11] P. Kunhachan, C. Banchonglikitkul, T. Kajsongkram, A. Khayungarnnawee, and W. Leelamanit, 2012 *Evidence-Based Complementary and Alternative Medicine,* vol. 2012
[12] N. Cech, C. Enke, 2002 Mass Spectrometry Review, vol 20, pp 362-387
[13] R. L. Earle, *Unit Operation in Food Processing*, 2, reprint., Elsevier, p. 95. 2013

Ammonia-nitrogen removal from urban drainage using modified fresh empty fruit bunches: A case study in Kota Kinabalu, Sabah

L N S Ricky[1], **Y Shahril**[2], **B Nurmin**[3], **AY Zahrim**[1]

[2]Chemical Engineering Programme, Faculty of Engineering, Universiti Malaysia Sabah, Jalan UMS, 88400 Kota Kinabalu, Sabah, Malaysia

[2]Medical, Faculty of Medicine and Health Science, Universiti Malaysia Sabah, Jalan UMS, 88400 Kota Kinabalu, Sabah, Malaysia

[3]Civil Engineering Programme, Faculty of Engineering, Universiti Malaysia Sabah, Jalan UMS, 88400 Kota Kinabalu, Sabah, Malaysia

E-mail: [1]zahrim@ums.edu.my

Abstract. Highly concentration of ammonia nitrogen in urban drainage could pollute the river and give pungent smell. The strong pungent odours that coming out from the urban drainage may degrade the image a city and could possibly reduce the present of tourist. To minimize the presence of pungent odours, the ammonia nitrogen can be removed from the urban drainage by applying proper adsorbent. In this study, an adsorbent produced through chemical modification of fresh empty fruit bunch (EFB) fibers has been carried out. The maximum adsorption capacity is between 0.01-0.60 mg/g. The finding also shows that the retention time is vital when designing ammonia nitrogen filter.

1. Introduction

The common contaminated substances that discharged from urban residential people to the urban drainage with high content of ammonia-nitrogen which released into rivers would diminish the aesthetic values, dirty and kill aquatic life such as fishes and prawns, and destroy marine ecology. Higher concentration of ammonia-nitrogen discharged from urban drainage into the waterways, also known as nitrogen pollution in hydrosphere [1] which contributed to the eutrophication, corrosion and fouling of water bodies like rivers and lakes, increasing the depletion of dissolved oxygen and toxicity to other aquatic organisms [1, 2]. In addition, ammonia-nitrogen in the rivers produced a strong pungent odour smell [3] also reported in Kota Kinabalu, Sabah [4]. The total tourist that visited Sabah in 2004 was 3.065 million visitors as the third largest tourism industry for Sabah [5]. The strong pungent odour that coming out from the urban drainage may degrade the image of the city and could possibly reduce the present of tourist. In general, the prolonged exposure to this pungent odour smell caused respiratory problems, headaches, vomiting, emotional stresses or depression to physical symptoms such as sensory irritation and nausea [6, 7]. To protect the human and environment as well

[1] To whom any correspondence should be addressed.

to sustain the ecotourism, the drainage water with minimum pollutant is considered to be very important in this regard. From previous studies, there are several techniques to remove ammonia-nitrogen by using chemical precipitation, membrane filtration, ion exchange, air stripping, sand filtration, electrodialysis and biological nitrification.

Empty fruit bunch (EFB) fibers suggested had the potential for the removal of ammonia-nitrogen, nutrient and colour due to its being ecofriendly and abundantly available plant wastes. In the palm oil mill industry, it was reported that EFB-POME compost contained 1.9 % N, 0.6 % P_2O_5, and 2 % K_2O [8]. Malaysia is one of the world largest palm oil exporters, and about 3.0 million tons of oil palm empty fruit bunch (EFB) fibers are produced in every year [9]. after the processing of the fruits in palm oil mill processing, the empty fruit bunch, fibers and shells of oil palm are the primary solid unwanted wastes [10]. All these wastes are called biomass that produced in the oil palm industry in Malaysia [11].

Modification of agricultural wastes may break the lignin structure and consequently increase the ammonia nitrogen removal. Demırak et al.(2015) reported that the modification of the agricultural wastes by using dilute sodium hydroxide solutions were to improve the penetration of modifying agents and increase the sorption capacity [12].

2. Materials and Methods

2.1 Biosorbent preparation

The fresh empty fruit bunch (EFB) fibers were collected from Tawau. The parameters of the empty fruit bunch [13] are summarized in Table 1.

Table 1. Properties of empty fruit bunch (EFB) [13].

Parameters	Shredded EFB
Moisture (%)	24
pH	6.7
C (%)	53.0
N (%)	0.9
C/N	58.9
COD (mg/L)	-
BOD (mg/L)	-
Oil and grease	-
Total solid (mg/L)	-
Cellulose (%)	52.5
Hemicellulose (%)	28.8
Lignin (%)	17.1

These EFB fibers were washed with tap water to remove the unwanted residual salinity and dirtiness such as soil-sand and leftover palm oil and waxy impurities. After that, the half clean EFB fibers were started to shred into small piece and put into a pail and soaked with tap water for repeated 5 times in 30 minutes. Next, these soaked EFB fibers were again washed with distilled water before its dried under the room temperature for 3 days. Modifications of the EFB fibers were carried out by the treatment with 10 mM of NaOH for 12 hours in room temperature. Then, the modified fresh EFB fibers were again washed with distilled water and dried in an oven at 60°C for 48 hours until a constant weight is obtained. For the preparation and modification of the large scale of modified EFB fibers, tap water were used instead of distilled water.

2.2 Retention time study for urban drainage water at Kampung Sembulan Lama, Kota Kinabalu, Sabah

A random urban drainage water sampling was collected at Kampung Sembulan Lama (Point 1 to 4). The distance between each point was 100 cm as shown in figure 1. The collected drainage water samples from four points were took back to the laboratory, Faculty Engineering, UMS. The sample of each point was used to determine the retention time by using 7 grams of modified fresh EFB fibers were added into beaker that contained 500 mL drainage water for 120 minutes. The initial reading of ammonia-nitrogen concentration, pH, conductivity and COD for the drainage water was determined as well.

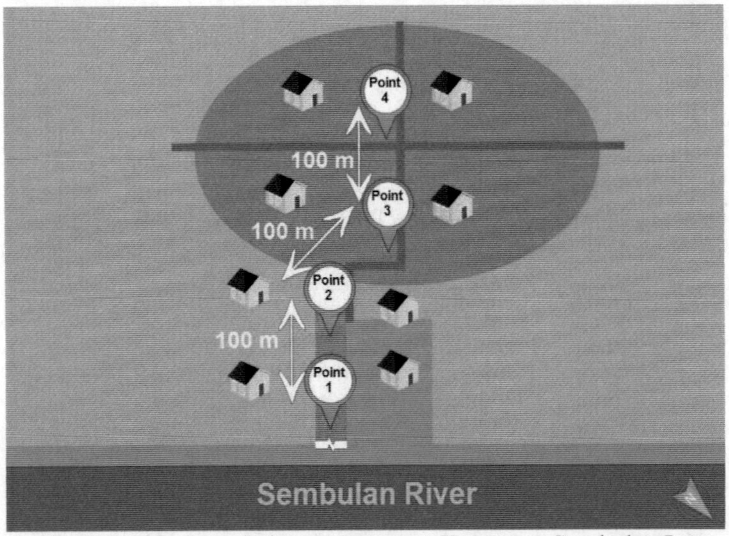

Figure 1. Schematic diagram of urban drainage system at Kampung Sembulan Lama drainage (Point 1 to 4), Kota Kinabalu, Sabah.

2.3 Analysis

The ammonia-nitrogen concentration contents were analyzed according to the volumetric method based on Nessler method [14]. The Chemical Oxygen Demand (COD) was analyzed by using Spectrophotometer HACH DR 6000. The conductivity and pH were measured by using HI 9811-5 portable meter, Hanna instruments.

3. Results and Discussion

It was found that the ammonia nitrogen content and organic matter (measured as Chemical Oxygen Demands) was insignificantly reduced. It is believed that this happen due to insufficient contact time. Therefore an attempt was made to investigate suitable detention time for removing pollutants especially ammonia-nitrogen.

Table 2 showed the characteristics of urban drainage water for the initial concentration on ammonia-nitrogen, Chemical Oxygen Demand (COD), conductivity and pH.

Table 2. The characteristics of the urban drainage water at Kampung Sembulan Lama, Kota Kinabalu.

	Point 1	Point 2	Point 3	Point 4
Initial ammonia-nitrogen concentration (mg/L)	34.71	37.36	31.35	27.77
COD (mg/L)	32	20	41	28
Conductivity (µs/cm)	1930	2320	570	460
pH	7.1	7	6.6	6.5

It appeared that the highest ammonia-nitrogen concentration occurred at point 2 which is 37.36mg/L and the lowest concentration at point 4 is 27.77mg/L. The highest COD was measured at point 3 which is 41 mg/L due to located at the centre of the lagoon area. The COD was diluted when the drainage water flowing out from point 3 to point 1. The conductivity values was observed much higher at points 1 and 2 than points 3 and 4 due to it nearer to the river that connected to the seawater. The conductivity indicates high salinity with present of dissolved ions namely sodium ions and chloride ions [20]. It found that the pH of the drainage water from Kampung Sembulan Lama is in the range from pH 6.5 to 7.1. Ammonia exists in aqueous solution as ammonium ion of the solution which below pH 9.25, the predominant species is NH_4^+ [21]. This proved that there is ammonium content inside the urban drainage water at Kampung Sembulan Lama due to its pH of the drainage water is below pH 9.25.

Based on Table 3, the drainage water collected at Kampung Sembulan Lama at diffrent random points, point 1 to point 4. The investigation of the ammonia-nitrogen removal by EFB fibers was conducted by carry out a series batch of biosorption experiments for adsorption capacity.

Table 3. Performance of ammonia-nitrogen removal using modified EFB.

Adsorption capacity (mg/g)			
Point 1	Point 2	Point 3	Point 4
0.04 (10 min)	-0.09 (10 min)	-0.08 (10 min)	-0.02 (10 min)
0.22 (40 min)	0.03 (40 min)	-0.30 (40 min)	0.01 (40 min)
0.07 (80 min)	0.47 (80 min)	0.08 (80 min)	0.07 (80 min)
0.19 (120 min)	0.60 (120 min)	0.23 (120 min)	0.25 (120 min)

Most of the ammonia-nitrogen solution removal occurs within a 20 minutes contact time. After 20 minutes, the ammonia uptake by the EFB fibers continued in a much slower rate until reached almost 40 minutes when the equilibrium is reached. The adsorption capacity in minutes 40 was showed 0.22 mg/g, 0.03 mg/g, -0.3mg/g and 0.01mg/g at point 1 to 4. From the table 2, the adsorption capacity showed a negative adsorption capacity due to adsorption was occurred by poor adsorption. The negative adsorption indicated that the adsorption behaviour did not fit on *Langmuir* approach which leading to negative slopes [15]. Also, negative values adsorption capacity data was not suitable to plot the graphs as no adsorption occurred on that particular time [16]. The inadequacy for the negative values for ammonia of the isotherm model to explain the adsorption process [17].

While in minutes of 120 minutes, the range of adsorption capacity was 0.19 and 0.6 mg/g at point 1 and point 4. The adsorption is reaching the equilibrium state at time of 40 minutes. The adsorption capacity is 0.22 mg/g at point 1 and 0.01 mg/g at point 4 was adsorbed by EFB fibers in the 40 minutes when all the EFB fibers sites vacant and the ammonia-nitrogen concentration gradient was high because of film diffusion onto the external surface of EFB fibers [1]. This is due to the contact time for the EFB fibers to react with the ammonium ions in the urban drainage water. The cation ion from the ammonium ions will interact with the anions from the surface of the lignocellulosic that is negatively charged [1]. At the time reached 120 minutes, it becomes equilibrium in which there is no interaction happened between EFB fibers and ammonium in the drainage water. Beyond that 120 minutes contact time, the adsorption of the ammonia-nitrogen removal is remain approximately constant. Similarity study also had been done by using different biosorbent such as *P.*ocenica considered a rapid ammonium adsorption process compared to other adsorbents [18]. This can show that the modified EFB fiber for the vacant surfaces were reduced with time increased. The large number of vacant surface sites are available for the sorption is fast at the initial stage and then the

IOP Conf. Series: Earth and Environmental Science **36** (2016) 012055 doi:10.1088/1755-1315/36/1/012055

remaining surface sites are difficult to be occupied because the repulsion between the N-NH_4 on the adsorbent fibers and bulk phase [19].

The retention time for the ammonia-nitrogen removal is at 80 minutes. Davis et al. (2001) suggested that the biological pond containing hardwood mulch and several plant species with retention time of 130 to 250 minutes for removing of several pollutants including ammonia nitrogen. The ammonia nitrogen was reduced to 87% from the initial concentration of 2.4 mg/L [22]. The detention time for ammonia removal was 1.4 to 60.6 days of -86.0 to 96.0 % of long term pollutant for treatment of stormwater runoff by using wetlands [23]. In another study, Ghosh and Gopal (2010) reported that the 77.58 to 99.96 % removal of ammonium nitrogen at retention time of 1-4 days [24]. According to Harmayani and Anwar (2012), the hydraulic retention time was after 60 to 90 minutes for the ammonia (NH_4^+) with the initial ammonia concentration of 0.5 to 5.0 mg/L. The maximum NH_4^+ biosorption showed by using the sawdust was 137.5 mg/g [17]. Dumont et al. (2014) reported that the average ammonium-N adsorption onto wood residues (G30) was 0.154 mg/g during the first 14 days of submersion. The maximum uptake for the ammonium nitrogen (NH_4^+-N) adsorption onto the wood residues (G30) was 0.191 mg/g at constant throughout 60 days [25]. Besides that, Ma et al. (2011) reported that the time needed for the ammonium removal by wheat straw was 240 minutes. The maximum adsorption capacity was 148.7 mg/g with the pH range of 4 to 8 [26].

4. Conclusion

Several past studies have been reviewed and the literatures suggest the potential of modified agricultural wastes for treating the pollutants from urban drainage water. Little knowledge was found for fiber modification for the ammonia-nitrogen removal purpose. Our study found that the filter (modified fresh EFB) retention time is critical during the operation of modified EFB. Besides that, the retention time of minimum eighty minutes is required for effective removal of ammonia-nitrogen from Kota Kinabalu urban drainage water.

5. References

[1] Wahab M A, Jellali S and Jedidi N 2010 *Bioresource Technology* **101** 8606-15
[2] Liu H, Dong Y, Liu Y and Wang H 2010 *Journal of Hazardous Materials* **178** 1132-6
[3] Talaiekhozani A, Bagheri M, Goli A and Talaei Khoozani M R 2016 *Journal of Environmental Management* **170** 186-206
[4] Xiang N J 2013 Water of life from muddy microbes. Insight Sabah. Kota Kinabalu.
[5] Charuruks I B 2015 Sabah tourism board. In: *Ministry of Tourism, Culture and Environment, Kota Kinabalu, Sabah.,*
[6] Nicell J A 2009 *Atmospheric Environment* **43** 196-206
[7] National Research Council Committee on Odors, 1979 Ordors from stationary and Mobile Sources. *Board on Toxicology and Environmental Hazards, Assembly of Life Sciences. National Research Council, National Academy of Sciences,* **Washington, DC.**
[8] Zahrim A Y and Asis T 2010 *Malaysia* **71** 11-7
[9] Sajab M S, Chia C H, Zakaria S and Khiew P S 2013 *Bioresource Technology* **128** 571-7
[10] Omar R, Idris A, Yunus R, Khalid K and Aida Isma M I 2011 *Fuel* **90** 1536-44
[11] Ngadi N and Lani N S 2014 *Science & Engineering* **68(5)** 35-9
[12] Demırak A, KESKİN F, ŞahıN Y and Kalemcı V 2015 *Mugla Journal of Science and Technology* **1** 5-12
[13] Baharuddin A S, Wakisaka M, Shirai Y, Abd-Aziz S, Abdul Rahman N A and Hassan M A 2009 *International Journal of Agricultural Research* **4** 69-78
[14] HACH 2012 *DR/2010 Spectrophotometer Procedures Manual, Spectrophotometer Handbook* (U.S.A

[15] Kiurski J, Adamovic S, Krstic J, Oros I and Vojinovic M M 2011 Adsorption efficiency of low-cost materials in the removal of Zn(II) ions from printing developer *ACTA Technica Corviniensis. Bulletin of Engineering.*

[16] Adediran G O, Nwosu F O and Adekola F A 2000 *Nig. J. Pure & Appl. Sci.* **15**

[17] Harmayani K D and Anwar A H M F 2012 A *International Journal of Environmental Science and Development* **3** 114-7

[18] Jellali S, Wahab M A, Anane M, Riahi K and Jedidi N 2011 *Desalination* **270** 40-9

[19] Mansuri N, Mody K and Basha S 2014 *Internation Journal of Environmental Science and Technology* **11** 1711-22

[20] Fondriest 2015 Conductivity, Salinity & Total Dissolved Solids.

[21] Suneethan M and Ravindhranath K 2012 *Der Pharma Chemica* **4** 214-27

[22] Davis A P, Shokouhian M, Sharma H and Minami C 2001 *Water Environment Research* **73** 5-14

[23] Carleton J N, Grizzard T J, Godrej A N and Post H E 2001 *Water Research* **35** 1552-62

[24] Ghosh D and Gopal B 2010 *Ecological Engineering* **36** 1044-51

[25] Dumont P A, Chadwick D R and Robinson S 2014 *Sustainability, Agri, Food and Environmental Research* **2** 47-67

[26] Ma Z, Li Q, Yue Q, Gao B, Li W, Xu X and Zhong Q 2011 *Chemical Engineering Journal* **171** 1209-17

IOP Conf. Series: Earth and Environmental Science **36** (2016) 012056 doi:10.1088/1755-1315/36/1/012056

Design and simulation of heat exchangers using Aspen HYSYS, and Aspen exchanger design and rating for paddy drying application

J Janaun[1], N H Kamin[1], K H Wong[1], H J Tham, V V Kong, M Farajpourlar
Universiti Malaysia Sabah, Jalan UMS 88400 Kota Kinabalu, Sabah, Malaysia

E-mail: [1]jidon@ums.edu.my

Abstract. Air heating unit is one of the most important parts in paddy drying to ensure the efficiency of a drying process. In addition, an optimized air heating unit does not only promise a good paddy quality, but also save more for the operating cost. This study determined the suitable and best specifications heating unit to heat air for paddy drying in the LAMB dryer. In this study, Aspen HYSYS v7.3 was used to obtain the minimum flow rate of hot water needed. The resulting data obtained from Aspen HYSYS v7.3 were used in Aspen Exchanger Design and Rating (EDR) to generate heat exchanger design and costs. The designs include shell and tubes and plate heat exchanger. The heat exchanger was designed in order to produce various drying temperatures of 40, 50, 60 and 70℃ of air with different flow rate, 300, 2500 and 5000 LPM. The optimum condition for the heat exchanger were found to be plate heat exchanger with 0.6 mm plate thickness, 198.75 mm plate width, 554.8 mm plate length and 11 numbers of plates operating at 5000 LPM air flow rate.

1. Introduction

Drying is a mass and heat transfer process which consisting of the removal of water by evaporation from solid. Drying of agricultural products is one of the preservation methods that used from long time ago. It is an important process for mankind due to the great effect on the quality of the products. The main purpose of drying is to reduce the moisture content in order to prevent the growth of microorganisms. Different drying methods are used to dry different agricultural products (1).

Rice is a staple crops used by half of the earth population as a basic needs (2) including diets, source of income, and is one of a major economic contributors (3), especially in Asia. According to (2), study shows that 586, 787 thousand tonnes of rice are annually consumed. In Malaysia itself, 600000 ha of the total area was covered solely for rice cultivation and a bulk of 2 million tonnes of paddy was produced annually from 400 rice mills of various capacity (4). Drying process for paddy however, has to be carried out right after the paddy is harvested. With the increasing demands, natural drying is no longer favourable as it requires longer time and massive space for drying. The moisture content of paddy typically needs to be reduced to 12-14 wt% (5) before it can be dehusked (3).

[1] To whom any correspondence should be addressed.

IOP Conf. Series: Earth and Environmental Science **36** (2016) 012056 doi:10.1088/1755-1315/36/1/012056

There are many industrial paddy dryers such as inclined bed dryers (IBD), fluidized bed dryers (FBD), re-circulating bed dryers, and continuous-flow dryers. Inclined bed is a common example of fixed bed dryer which possess basic design of paddy drying technology. This type of dryer consists of perforated sheet floor (bed) with a plenum chamber below it that forced air directly or by the assistance of a duct system (6). Paddy that is about to be dried was placed a foot deep and hot air are forced to go through the bulk of paddy from below the bed. Currently, IBD is used by Padiberas National Berhad (BERNAS) as a second dryer after fluidized bed dryer (FBD) with the ability to dry grain with high moisture content of 20-26% w.b. (7). The advantages of this dryer are inexpensive compared to the other dryer and have a faster withdrawal of dried paddy. In addition, the operation of the machine is easy. However, this dryer has been associated with high energy consumption and produces uneven dried product due to an inconsistent air flow distribution. This may affect the head rice yield and rice quality.

Fluidized bed dryer operating in such ways where hot air passes through the bed and then through the suspended grains which is perpendicular to the direction of flow. This type of dryer are able to dry paddy from 24% to 18% moisture content in short period with air temperature 100°C and high air velocity (8) without changing the rice's quality. This dryer produces even drying of grain, rapid heat and mass transfer, and low capital cost (9), however, the power consumption is high, and that it possess a low thermal efficiency.

Re-circulating batch dryer is another dryer that can eliminate moisture gradient problems as the grain/paddy is loaded in a batch and mixed through the entire drying process (10). The dryer consists of drying chamber, a fan and a heater as well as a central auger for transport of paddy from the bottom part to the outlet, which is located at the top of the dryer. Despite its great contribution in solving the problem with moisture gradient, fast drying and ability to produce rice with great quality, this type of dryer used twice the temperature needed for paddy drying compared to the flat-bed dryer, the capital cost is high due to its complicated design.

Continuous-flow dryers can be considered as another version of re-circulating batch dryers, in which it allows the grain to re-circulate in the drying chamber. The difference between these two is, instead of discharging the dried grain from the top, continuous-flow dryer discharges the grain from the bottom part, cooled and transport to the storage. Similar to that of re-circulating batch, it assist in eliminating moisture gradient of paddy drying due to the baffle system exist in the dryer that operates to mix the grain. This dryer offers low operating cost when compared to batch dryer, however, continuous flow dryer is only applicable for large scale process and cannot be used for small scale since it is quite a complicated dryer.

The recent dryer for paddy drying application is Laterally Aerated Moving Bed (LAMB) dryer (11) invented by our group (12). LAMB dryer is also recognized as radial packed bed and consists of a single perforated tube inside the drying vessel. Same as the other conventional dryers, LAMB operates with hot air flow as the drying medium through the perforated tubes and distributes the hot air laterally throughout the drying vessel where the paddy is feed into. The principle of the system is to aerate the paddy bed laterally so that even drying can be achieved more effectively, and lower aeration energy requirement since pressure drop across the bed is smaller due to shorter distance. Typically in large scale paddy drying, hot air is generated by using steam boiler or by burning of rice husk in a furnace.

Nevertheless, understanding on the effective and efficient air heating system for LAMB dryer has not been well established. Choosing the right heating unit is crucial to minimize the operational cost. Various air heating technologies have been developed and each has its advantages and disadvantages. Apart from using electrical, combustible fluids air heater and solar air heater are also commonly used for producing hot air. However, heating air using a furnace powered by combustion of rice husk is very economical but it produces air and solid waste pollutions. Furthermore, the ashes produced from the rice husk combustion often clog the filter of the dryer causing frequent maintenance. In this case, the use of heat exchanger is more favourable. Hence, this research is to determine the suitable heat exchanger to be used to supply hot air for paddy drying in the 2000 kg capacity LAMB dryer, as well as to find the optimum specifications and condition of the said heating unit. In order to find a suitable heat exchanger, this process can be simulated by using ASPEN HYSYS V7.3 and ASPEN Exchanger Design and Rating (EDR).

Heat exchanger is equipment that deals with transfer of heat between two fluids. There are various heat exchangers such as shell and tubes heat exchanger, plate heat, and others. Shell and tubes heat exchanger is broadly used in various industries such as energy generation and preservation system (13).For shell and tubes water to air heat exchanger, usually the fluid with high pressure is located in tubes and fluid with low pressure is located in shell. Through this, it can relieve shell for high pressure and gives a large centreline spacing for tubes (14). Water to air heat exchanger is usually designed as cross counter flow and individual tube passes are connected by bends to form the multi-pass tube. The hot water flow direction are cross to the cold air (15)

Aspen HYSYS and Aspen EDR are comprehensive process modelling tools used for process simulation and process optimization in design. Aspen EDR is a software that delivers the comprehensive range of heat exchanger designs (16). Aspen HYSYS is able to solve the problem to determine the flow rate of hot and cold stream passing through the heat exchanger in different stream conditions. In this software, the heat exchangers are very flexible as it can solve for the problem of temperature, pressures, heat flows and material streams flow. Heat exchanger model can be chosen for analysis purpose in Aspen HYSYS. It is able to operate a heat exchanger and model the heat transfer process (17) occurring insides of the heat exchanger. Aspen Exchanger Design and Rating (EDR) is a software that provides the designs and rating of heat exchanger. It assists the users to reduce the equipment cost up to 10 to 30 % with the effective design of Aspen EDR. The software is able to give an optimum cost arrangement design by providing the process requirement's physical data. The details of heat exchanger will be shown in a specification sheet, setting plan and tubes layout.

2. Experimental procedure

2.1. Aspen HYSYS simulation for heat exchanger

Aspen HYSYS v7.3 was used to simulate a simple model of heat exchanger. The components chosen in Aspen HYSYS were air and water, and the fluid package used in this simulation model was General Non Random two liquid (NRTL). In the component that was labelled 'Inlet cold' was filled with air in component selection and the temperature and pressure inlet cold air was set as ambient condition. The flow rate of air used are 300, 2500 and 5000 litre per minute (LPM). On the other hand, for component labelled 'inlet hot' was filled with water in component selection. The inlet temperature of water was set to 60, 80 and 100°C and the flow rate of hot water was key-in using trial and error method until the minimum value of flow rate was obtained. The Aspen HYSYS will show complete comment if the flow rate of hot water can be used in this system. The outlet air temperatures were set as 40, 50, 60 and 70°C. The other values were left blank and Aspen HYSYS was used to simulate it.

2.2. Aspen exchanger design and rating (EDR)

Aspen Exchanger Design and Rating (EDR) were used to design the heat exchanger after the data of temperature and flow rate of hot water and cold air was obtained. There were two major types of heat exchangers in Aspen Exchanger Design and Rating (EDR), which are shell and tube heat exchanger and plate heat exchanger. A new tab was opened and the types of heat exchanger were chosen and it will re-direct to the navigator. The calculation mode was set as design mode and filled the location of hot fluid. The temperature and flow rate of hot fluid and cold fluid were inserted in the process data and the temperature of hot fluid used was similar to the temperature used in Aspen HYSYS which were 60, 80 and 100°C. The temperature of inlet air was set in ambient condition and the targeted temperature (outlet air) was set as 40, 50, 60 and 70°C. Aspen properties were chosen as the physical properties package for the properties data of both hot and cold stream. The components used in heat exchanger were selected in databank and the percentage of composition of components was filled for both hot stream and cold stream. NRTL was selected as the properties method for the design of heat exchanger. The "run" button was clicked and the design of heat exchanger was simulated by the program. The result of the design can be viewed in the results and discussions section.

3. Results and discussion

3.1. Relationship between heating fluids and cold fluids flow rates and temperatures

Based on this study, the relationship between water's inlet temperature and flow rates with outlet air temperature were determined at different air flow rates, 300LPM, 2500LPM and 5000LPM. The study was performed using Aspen HYSYS and Aspen EDR.

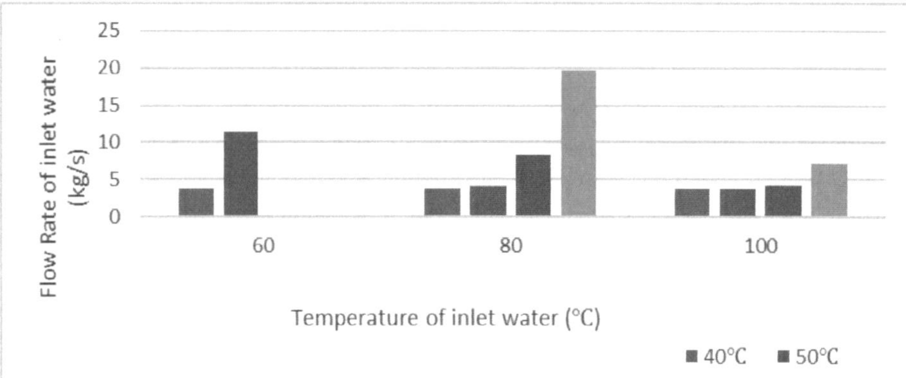

Figure 1. Water inlet Flow Rate against Inlet Temperature with different outlet Air Temperature at 300LPM air flow rate

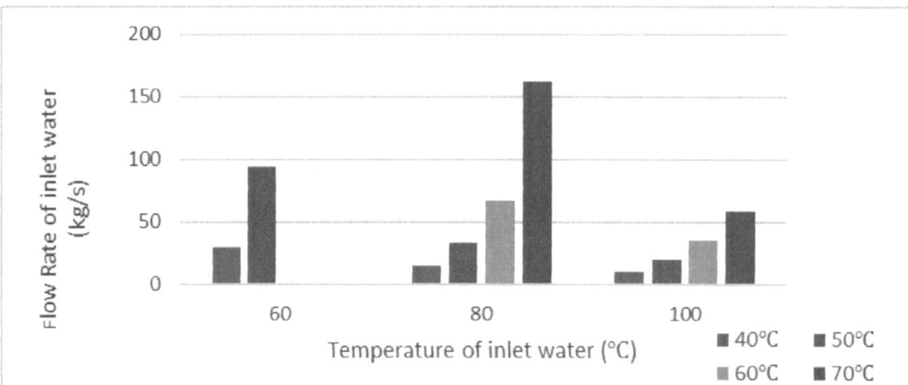

Figure 2. Water inlet Flow Rate against Inlet Temperature with different outlet Air Temperature at 2500LPM air flow rate

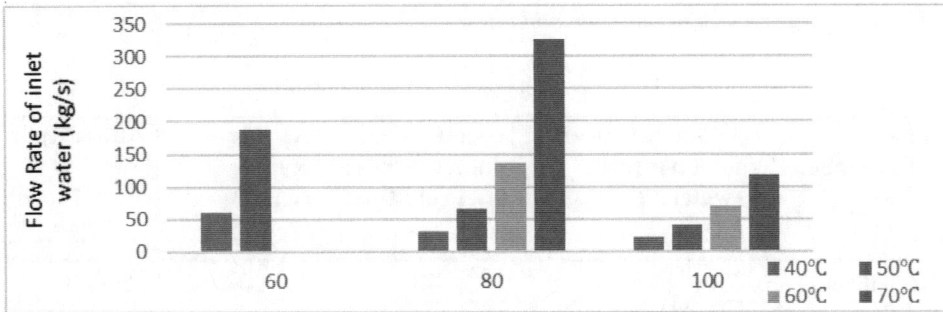

Figure 3. Water inlet Flow Rate against Inlet Temperature with different outlet Air Temperature at 5000LPM air flow rate

Comparing the data obtained by using different air flow rates, the pattern is almost similar for the entire graph as observed in Figures 1, 2 and 3. Referring to the data plotted, it can be seen that at different heating temperature, higher water flow rates are needed to heat up air and reach similar targeted temperature. Figure 1 for instance, represents the heating of air with flow of 300 LPM in a heat exchanger, it needs 3.7 kg/s of water at 60°C to be able to heat up 300 LPM ambient air (25°C) to 40°C. Higher water flow rates at 60°C are needed to heat up the same 300LPM air to 50°C. Whereas, water at 80°C requires lower flow rate to achieve lower targeted temperature and even lower flow rate when the water is at 100°C. This can be implied that, at lower temperature, higher flow rates of heating fluid is required for heating. In contrast, at higher heating fluid temperature, lower flow rates are needed for heating to reach similar targeted temperature.

The heating of 2500 LPM and 5000 LPM of air in a heat exchanger are illustrated by Figure 2 and Figure 3, respectively. The trend of the graph shows that the inlet flow rate of water need to be used was increasing with increasing targeted air temperature. Figure 2 represents inlet flow rate of water at several temperatures required to heat 2500 LPM of ambient air to 40, 50, 60 and 70°C. Based on Figure 2, the trend of graph shows that the inlet flow rate of water required to heat air was increasing with increasing targeted air temperature regardless of the water temperature. From the Figures, it can be observed that higher water flow rates (30.1 kg/s) of water at 60°C was needed to heat 2500 LPM of ambient air to 40°C compared to the water at 80°C (15.4 kg/s) and 100°C (10.4 kg/s) temperature. Similar result can be seen when air was heated to 50°C. Identical trend can be monitored for water with 80 and 100°C temperature where 67.7 kg/s of water at 80°C is needed instead of 35.1 kg/s at 100°C to heat air to 60°C. It shows that water at lower temperature requires higher flow rates to reach the targeted temperature compared to water at higher temperature. Other than that, it can also be monitored that higher air flow rates tend to require higher water flow rates to heat up air. The increase of mass flow rate with increasing flow velocity will lead to increasing heat transfer coefficient and thus have higher heat transfer (18). This trend is also applicable to Figure 3.

Seeing from the perspectives of the whole pattern, it can be said that air flow rate affects the water flow rate required to heat air to certain temperature. By relating the air flow rates and the pattern obtained, it shows that higher air flow rate contributes to higher water flow rates required.

IOP Conf. Series: Earth and Environmental Science **36** (2016) 012056 doi:10.1088/1755-1315/36/1/012056

3.2. Costs of heat exchanger with optimum conditions

Table 1. Costs of Heat Exchangers

Heat Exchanger	Temperature of inlet hot fluid (water, °C)	Targeted Temperature of outlet cold fluid (air,°C)	Air Flow Rate (LPM)	Cost (RM)
Shell and Tubes	60	40	300	34000
Shell and Tubes	80	50	300	34000
Shell and Tubes	100	60	300	34000
Shell and Tubes	100	70	300	34000
Shell and Tubes	80	40	2500	35000
Shell and Tubes	100	50	2500	35000
Shell and Tubes	100	60	2500	35000
Shell and Tubes	100	70	2500	36000
Shell and Tubes	80	40	5000	42000
Shell and Tubes	100	50	5000	42000
Shell and Tubes	80	60	5000	44000
Shell and Tubes	80	70	5000	45000
Plate	-	Low Flow rate	300	-
Plate	80	40	2500	2700
Plate	100	50	2500	2700
Plate	100	60	2500	2800
Plate	100	70	2500	2800
Plate	80	40	5000	3100
Plate	100	50	5000	3100
Plate	80	60	5000	3300
Plate	80	70	5000	3400

The costs of the heat exchangers were determined after the simulation was performed using Aspen EDR and only the heating unit with the highest efficiency was compared. This is to ensure that the heat exchanger is able to meet the requirement needed for paddy drying. Shell and tube as well as plate heat exchanger was chosen for the simulation due to its flexibility as a heating device in a process.

Based on Table 1, it shows that the cost of shell and tubes heat exchanger is higher compared to plate heat exchanger. Shell and tube heat exchanger shows that the cost is increasing with increasing air flow rate and temperature. It can also be observed that to heat up 5000 LPM of air to 70°C, the cost recorded are RM45000. It is significantly higher than those of with lower air flow rates. Plate heat exchanger, on the other hand, cost almost ten times lower than the shell and tube heat exchanger with similar condition (heating 5000LPM of air to 70°C), which is only RM3400. Shell and tube heat exchanger are more flexible compared to plate heat exchanger in term of operating for both small scale and larger scale process and it can stands higher pressure.

4. Conclusions

Two types of heat exchangers, shell and tube, and plate heat exchanger were simulated using Aspen HYSIS to study the temperature profile of each device and to obtain the optimum specifications to be applied on paddy drying and its cost was determined by using Aspen EDR. The desired drying temperature used in this study was 40°C, which was a typical temperature used in industrial paddy drying. For 2000 kg capacity LAMB dryer, the simulation showed that the best heat exchanger was plate heat exchanger with air flow rate of 5000 LPM. The simulations showed that 80°C of water can heat 5000 LPM of air to 40°C, and the cost was RM3100. From the designs, the best specifications for the plate heat exchanger were 0.6mm plate thickness, 198.75mm plate width, 554.8mm plate length and using 11 numbers of plates.

5. References

[1] Wankhade P K, Sapkal R S, and Sapkal V S 2013 *Procedia Eng.* **51** 371–4
[2] Wasim P M. 2002 Pakistan Economic and Social Review. 2nd ed. 153–83.
[3] Akowuah J, Addo A and Bart-Plange A 2012 *J Sci Technol Ghana.* **19**;32(2). Available from: http://www.ajol.info/index.php/just/article/view/81380
[4] Cheong A W. 1998 The rice Industri in Malaysia: A Success Story. APAARI PUBLICATIONS http://www.apaari.org/wp-content/uploads/2009/05/ss_1998_01.pdf
[5] Gummert M, Borlagdan P. IRRI: Dried to Perfection. IRRI Rice Today http://irri.org/rice-today/dried-to-perfection
[6] Román F, Strahl-Schäfer V and Hensel O 2012 *Biosyst Eng.* **112** 359–69
[7] Sarker M S H, Ibrahim M N, Ab Aziz N and Mohd Salleh P 2014 *Energy Convers Manag.* **77** 389–95.
[8] Unknown. Novel Dryers and Recent Developments. FAO CORPORATE DOCUMENT REPOSITORY http://www.fao.org/docrep/t1838e/T1838E0y.htm
[9] Briens L and Bojarra M 2010 *Pharm Sci Tech.* **11** 1612–8
[10] Friesen O H. 1980 Heated-air Grain Dryers Agriculture and Agri-Food Canada
[11] Janaun J, Kong V V and Tan K Y. A Biomass Dryer PI 2015700917
[12] Tan K Y 2014 Design and Fabrication of Single-Tube Laterally Aerated Moving Bed (LAMB) Dryer for Agricultural Biomass. Universiti Malaysia Sabah
[13] Yang J, Oh S R and Liu W 2014 *Int J Heat Mass Transf.* **77** 1144–54.
[14] Fraas A P 1989 Heat Exchanger Design John Wiley & Sons
[15] Yao Y, Huang M, Mo J and Dai S 2013 *Int J Heat Mass Transf.* **64** 173–92.
[16] Aspen Exchanger, Design and Rating http://www.aspentech.com/products/aspen-edr.aspx
[17] Abd Hamid M K 2007 HYSYS: An Introduction to Chemical Engineering Simulation http://eprints.utm.my/3030/2/HYSYS_for_UTM_Degree%2B%2B_Program.pdf
[18] Murugesan M P and Balasubramaniam R 2012 *Research Journal of Engineering Sciences*

Water soluble drug releasing soft contact lens in response to pH of tears

G Kim[1], H Noh[1,2]
[1] Convergence Institute of Biomedical Eng. & Biomaterials
[2] Department of Optometry, Seoul National University of Science and Technology

E-mail: [1]hrnoh@seoultech.ac.kr

Abstract. Human tear characteristics including pH and compositions can vary significantly depending on physical and environmental factors. Contact lenses directly contact with human tears can be swelled or de-swelled depending on the pH of the solution due to the nature of the hydrogel. For examples, anionic hydrogels, when the solution's pH is low, is shrunken due to the electric attraction force within the hydrogel network; the opposite phenomenon appears when the solution is basic. The purpose of this study was to evaluate the extent of water soluble drug, hydroxyl propyl methyl cellulose, released from contact lens according to the pH of the artificial tears. Artificial tears are prepared by mixing lysozyme, albumin, sodium chloride, potassium chloride, and calcium chloride following physiological concentrations. Hydrogel contact lens was thermally polymerized using HEMA, EGDMA, and AIBN. The prepared hydrogel lens was immersed in drug for 3 hours and the eluted drug mass was measured as a function of the time. As a result, the drug was released from the lens for 12 hours in all the pH of artificial tears. At the lower pH of artificial tears (pH 5.8), the total amount of dye emitted from the lens was increased than the total amount of dye emitted at the basic tear (pH 8.4). Also, initial burst at acidic tears was increased within 1 hour. Release pattern of water-soluble drug from hydrogel lens turned out to be different depending on the pH of the artificial tears. When designing drug releasing contact lens, physiological pH of tears should be considered.

1. Introduction

Physicochemical and pharmacokinetic properties of drugs should be considered to properly design a drug delivery system using hydrogels.[1-3] For ocular drug delivery, specificity of tears should also be considered. Tears are made up with water (up to 98%), phosphates, fat, tear proteins, sodium chloride and sodium carbonates.[4] The composition and concentration of tears highly depend on the physical human factors and/or environment. Similar to human blood, tears' pH is typically around 7.5. However, it can vary from 5.8 to 8.3 depending on emotions and/or individual health status.[5] In such, volume of human tears increase or decrease depending on emotional states, such as joy, sadness,

[1] To whom any correspondence should be addressed.

IOP Conf. Series: Earth and Environmental Science **36** (2016) 012057 doi:10.1088/1755-1315/36/1/012057

or depression, and also depending on environmental changes, such as wind or cold. The pH of human tears can be decreased or increased by these changes.

Many research articles regarding pH-sensitive hydrogels for drug release uses have been published in recent years.[1,6-8] pH-sensitive hydrogels are composed of polymeric backbones with ionic pendant groups. In solutions, the pendant groups can ionize and develop positive or negative charges on the hydrogel, generating electrostatic repulsive or attractive forces that lead to increase or decrease water contents in the polymer. These behaviors of cationic copolymers based on poly(dimethylaminoethyl methacrylate) [p(DMAEMA)] and poly(diethylaminoethyl methacrylate) [p(DEAEMA)] have been investigated.[9,10] Peppas studied that the pH-modulated swelling and release characteristics of anionic hydrogels of poly(methacrylic acid) [p(MAA)] and poly(hydroxyethyl methacrylate) [p(HEMA)] for releasing drugs such as salicylic acid.[11]

Similar to pH responsive polymers, the contact lens can also be swelled or deswelled depending on the pH of the solution due to the nature of the hydrogel. In general, when the pH is low, it is shrunken due to the electric repulsive forces within the hydrogel network, the opposite phenomenon appears when the solution is basic.[12] The purpose of this study was to evaluate the extent of water soluble drug release from hydroxyl ethyl methacrylate (HEMA) according to the pH of the artificial tears. We have employed our pH-sensitive hydrogels as a matrix system for the ocular delivery of hydroxyl propyl methyl cellulose (HPMC), non-ionic humectants used in artificial tears or eye drops. We selected HPMC as a model drug, since the drug shows good water solubility and its non-ionic property would not disturb the pH of solutions.

2. Experimental

2.1. Hydrogel polymerization

2-hydroxyethyl methacrylate (HEMA) was used as a main polymer backbone. Ethylene glycol dimethacrylate (EGDMA) and ethylene glycol dimethacrylate (AIBN) were used as a crosslinking agent and an initiator. The reagents (HEMA, EGDMA0.6%, and AIBN 0.4%) were agitated for 30 min. It was molded following thermal polymerization in a cast mold. Polymerized contact lens was rinsed 3 X with deionized water and hydrated in deionized water for 24 hours before drug releasing test to remove unused reagents residues. All chemicals were used as supplied without further purification.

Swelling ratio of hydrogels

Water contents of hydrogels were measured at various pH tear solutions (pH 5.80, 6.45, 7.20, and 8.03). Percentage water (%H_2O) is defined as

$$\frac{H_W - H_d}{H_W} \times 100 \tag{1}$$

Where H_w and H_d are the weights of the wet and dry state of hydrogels, respectively.

2.2. Artificial tears

Artificial tear solutions were formulated in deionized water with the composition of lysozyme (1.2 mg/mL, chicken egg white, ≥90%, Aldrich), albumin (3.88 mg/Ml, bovine serum, ≥96%, Aldrich), sodium chloride (7 mg/mL), potassium chloride (1.7 mg/mL), calcium chloride (3.88 mg/mL). Final pH of the solution was from 5.80 to 8.04.

Elution of dyes or drugs

Water soluble dyes (Shinhan Co.) or hydroxyl propyl methylcellulose (HPMC, Aldrich) were used for release tests. Water soluble dyes were tested to visually observe dye releasing phenomenon. Freshly made hydrogel contact lenses were immersed for 3 hours in dye or drug solutions. Drug containing hydrogels were then soaked in artificial tears (pH 5.80, 6.45, 7.20, and 8.03) for designed time periods. At the time intervals, the solution was analyzed for dye or drug and replaced with fresh artificial tears. UV-Vis spectrophotometer (Mega Array, Scinco, Korea) was used to compare optical density of solutions of dyes or drugs in tears.

3. Results and Discussions

3.1. Physical properties of HEMA hydrogels

Figure.1 shows that the water contents of hydrogels in accordance with the artificial tear solutions' pH. After soaking the lens for 12 hours in artificial tear solutions, the weights of lenses, Hw were measured. Lens was then dried for 24 hours before measured the dried weight, Hd. Following eq.1 the water content of each lens was calculated. Water content of control lens (without dipped in artificial lens) was 32.35±0.66%. Water contents of lenses dipped in artificial tears with different pH (5.8, 6.45, 7.20, and 8.03) were calculated as 41.89±0.74, 43.63±0.86, 46.40±0.59, and 47.12±0.33%, respectively. When the artificial tear is basic, the water content was increased up to 43% in comparison to control lens.

In general, swelling of a gel is affected by many factors including mechanical forces, pH, temperatures, and electrostatic interactions.[1,6] In acidic solution, the network of polymers bearing acidic groups including HEMA and other methacrylc acid grafted gels tend to expand due to electrostatic repulsion. This leads to relative less water contents in lower pH than those in pH 8.03.

In comparison to control hydrogel soaked in PBS solution, water contents of hydrogels soaked in artificial tears appeared to be higher in overall pH ranges. It is considered that the tear proteins adsorbed to hydrogels contributed to weights of gels in wet states.

Figure 2 displays light transmissions (visible, UV-A, UV-B) through hydrogel lenses with various pH of tear solutions. Light transmittance (visible, UV-A, UV-B) of control gel swelled in PBS was 82, 80, 78%, respectively. Regardless of pH of artificial tears, the light transmittance of gels dipped in artificial tears were decreased due to haziness caused by proteins adsorbed to the surface of hydrogels. In specific, UV-A transmittance was decreased to 69.35 ± 3.91, 70.3 ± 1.75, 72.55 ± 5.99, 73.45 ± 2.11% when the gel is soaked in tear solution with pH of 5.80, 6.45, 7.20, 8.03, respectively. As shown in Figure2, transmittance of UV-B light was also decreased to 65.55 ± 4.05, 67.9 ± 1.59, 69.3 ± 5.08, 69.9 ± 2.31%. In accordance with UV light transmittance, visible light transmittance was reduced to 71.45 ± 4.14, 72.15 ± 1.69, 72.9 ± 5.71, 75.7 ± 1.89%. UV and visible light transmittance of the gels were slightly increased as the artificial tears were basic. This was considered to be due to the majority of protein, albumin in artificial tear solution. Apart from synthetic polymers, albumin, a kind of natural polymer, can also swell and deswell in response to pH of solution.[13] Isoelectric point of Albumin is around 5.4 in deionized water and albumin's own charge was reported to zero at the isoelectric pH.[13] It became negative with increasing pH of solution and positive with decreasing pH of solution than isoelectric pH. With increasing basicity of solution to 8.03, the charges of albumin were more negative and tend to be swelled due to electrostatic repulsion forces. As a result, overall light transmittance of gels slightly increased as increasing pH.

3.2. Standard curves of dyes and drugs

Figure 3 A and B show standard curves for dyes and hydroxyl propyl methylcellulose. Linear calibration curves were obtained ($R2 \geq 98\%$) within the concentration range 0.001 mg/ml < dyes <0.35mg/mL and 0.01 mg/mL < HPMC <3.25 mg/mL. Results reported herein were obtained within the linear-working range for dyes or HPMC. With HPMC, calibration curves exhibited two experimentally-useful, linear-like regions at low (0.01 mg/ml <HPMC <0.06 mg/ml) and high (0.1 mg/ml <HPMC <3.25 mg/ml) concentrations as shown in Figure 3B.

3.3. Dye and drug release with varying pH

Figure4 A and B compare cumulative mass of dyes and HPMC released from hydrogels with varying pH of artificial tears. Release of dyes and drugs were continued for 12 hours and after that, it did not released regardless of mass remained in the gels. At the lower pH of the artificial tears such as pH 5.8, the initial bursts (up to 20% of total cumulative release) were maximized in both cases. Drug release from hydrogels was supposed to follow simple diffusion, the amount of released mass at a given time was decreased over time in all pH ranges.

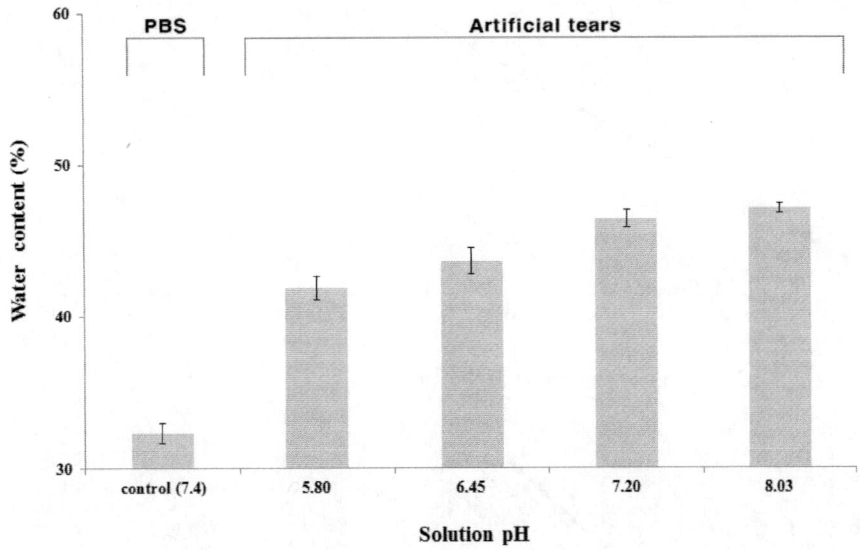

Figure 1. Water contents of hydrogels with varying solution pH

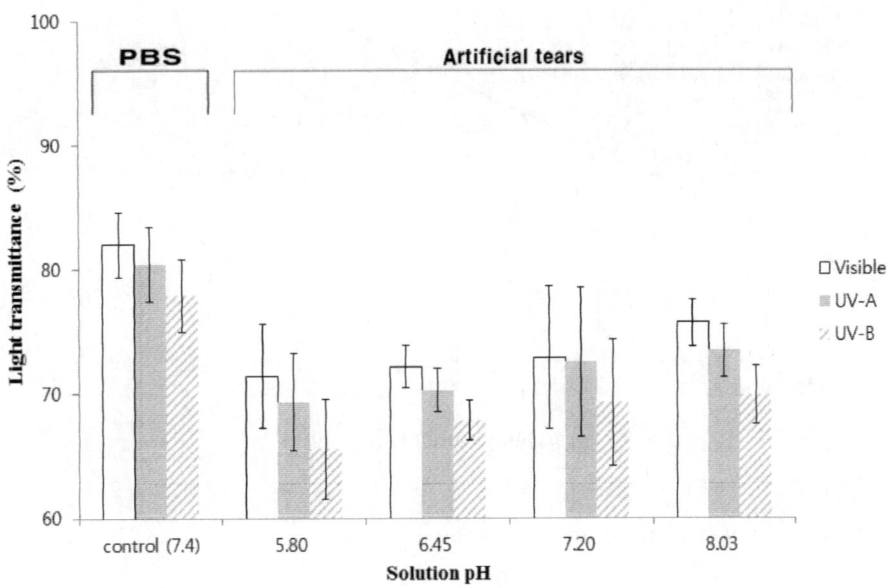

Figure 2. Light transmittance of hydrogel lenses according to various solution Ph.

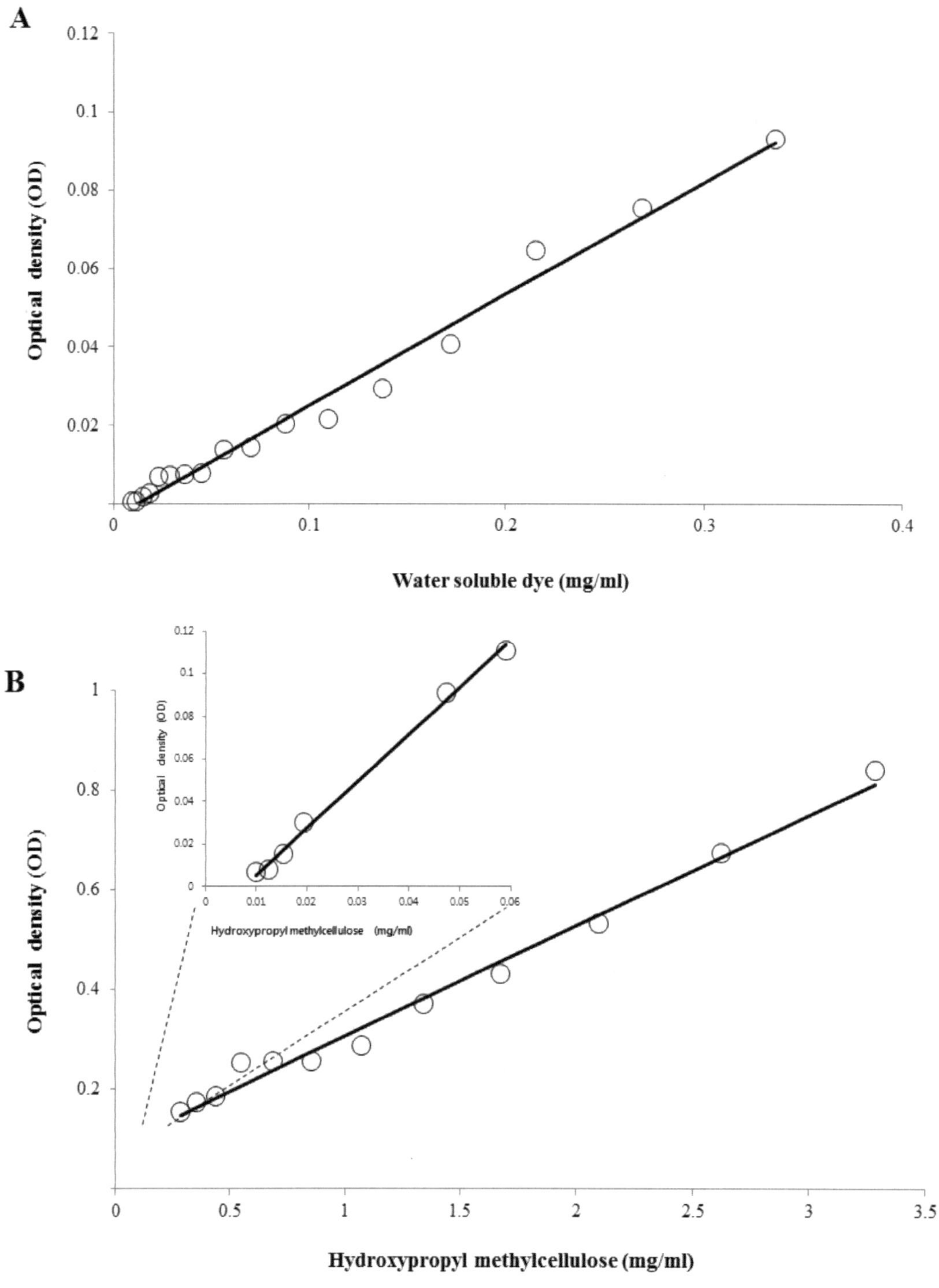

Figure 3. Calibration curves for water soluble dyes (A) and HPMC (B).

IOP Conf. Series: Earth and Environmental Science **36** (2016) 012057 doi:10.1088/1755-1315/36/1/012057

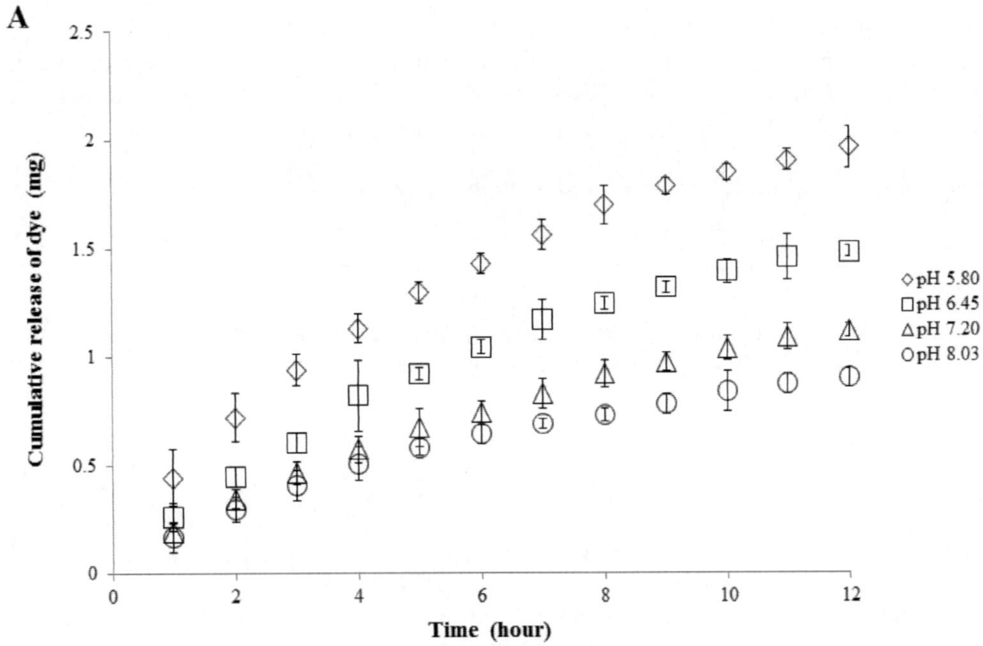

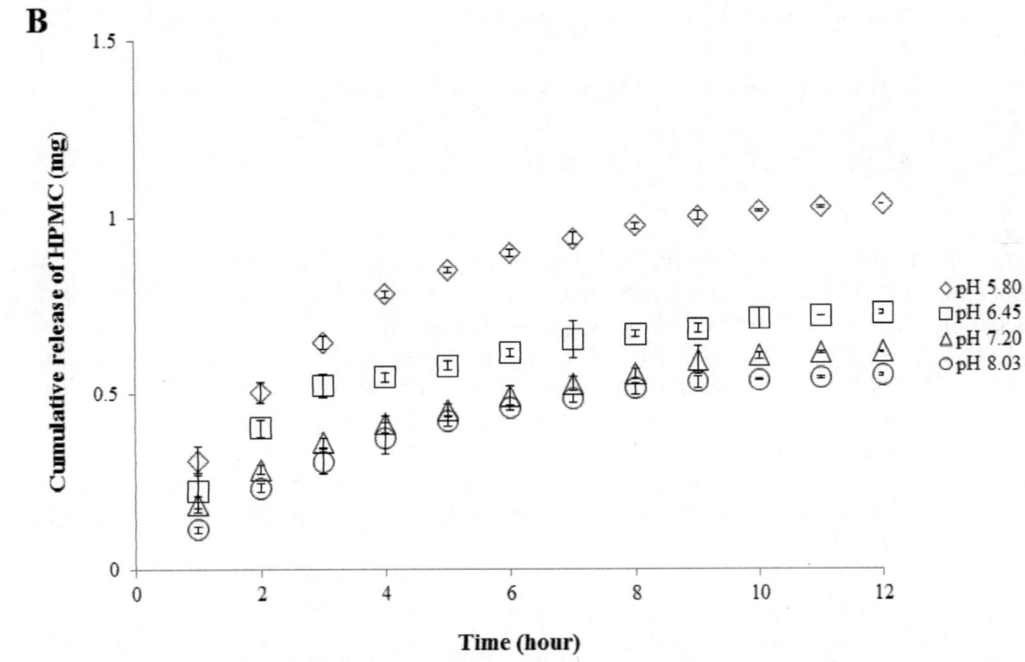

Figure 4. Cumulative release of water soluble dyes (A) and HPMC (B) with varying artificial tears' pH.

We analyzed that in acidic tears (pH 5.80 and 6.45) the amount of released drug was more than twice than that in basic tears. In particular, the cumulative released mass in tear of pH 5.80 was 1.04 mg when that in tear of pH 8.03 was 0.42 mg.

pH sensitive hydrogels such as polyacrylamide and methacrylate gels tends to swell in basic medium attracting more water due to ionization and hydrolysis from acrylic acid.[14] According to Peppas, polymer networks dissociate due to ionization of pendent groups such as acrylic acid with increasing pH. Furthermore, the electrostatic repulsion forces cause the hydrogel to swell, and water passes in. [11]

The total released amount of drug, which diffused through the hydrogel is [7]

$$\text{Total released amount of drug} = \frac{ADC_0}{l}\left(t - \frac{l^2}{6D}\right) \tag{2}$$

Where A is the area of the hydrogel, D is diffusivity, l is membrane thickness, C_0 is the drug concentration within the hydrogel and t is the time. When released amount is plotted against time, the slope is:

$$\text{Slope} = \frac{ADC_0}{l} \tag{3}$$

The diffusivity can be calculated from the slope and the hydrogel area. Applying lens thickness as 90 μm, the diffusivity of gels in tear pH (5.8, 6.45, 7.20, and 8.03) were derived as 5.60 x 10^{-10}, 4.55 x 10^{-10}, 3.13 x 10^{-10}, and 2.82 x 10^{-10} m^2hour^{-1}, respectively. (Table 1) Note that the diffusivity of drugs in hydrogel lens decreased as increasing artificial tears' pH resulting that less drugs delivered at a given time. We reason that the different release rates of dyes and drugs were due to difference in physical properties of dyes and drugs. Typically, it is understood that drug release from hydrogel depends on drugs molecular weights, solubility, and concentration besides from wettability and thickness of gels.

Drug release in short time can be predicted by the following formula.[15]

$$\text{Cumulative drug release} = 4\sqrt{\frac{Dt}{\pi h^2}} \tag{4}$$

Where D is the diffusion coefficient, h is thickness of the gel, t is time. This formula is supposed to be valid only in short time drug release. Following the eq. 3, cumulative drug release found to be increased in proportion to the square root of the release time.
Figure 5 shows that drug exhibited a linear-like release trend over square root of time. With computing eq. 3, drug release was proportional to square root of time with more than 95% agreement. Figure 6 compiles the maximum release amount of drugs with varying pH of artificial tears. In acidic condition (pH 5.80), 0.45mg of HPMC was released and in basic condition (pH 8.03) less than 0.2mg was released from the hydrogels containing same amount of drugs in preparation. Note that as the pH is increased, the maximum release amount of drugs decreased due to increased ionization of hydrogel pendent group.

Table 1. Diffusion coefficient of drug release with varying tear pH.

Tear pH	Slope	Thickness (μm)	Diffusion coefficient, D (x10^{-10} m^2hour^{-1})
5.80	0.20	90	5.60
6.45	0.16	90	4.55
7.20	0.11	90	3.13
8.03	0.10	90	2.82

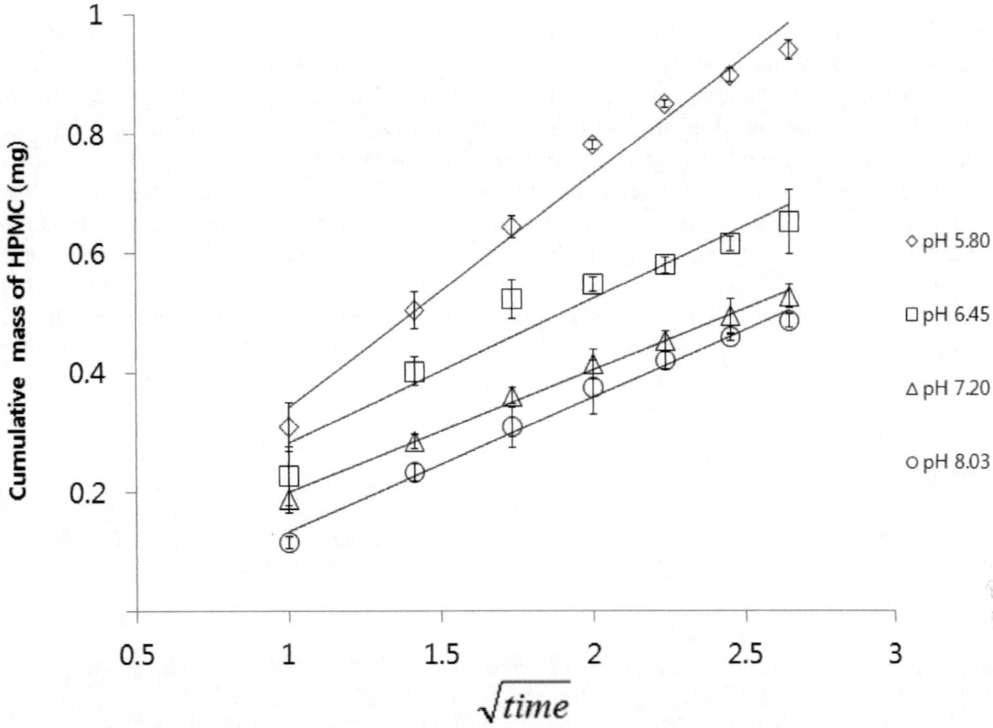

Figure 5. Linear fits for the short time drug release in response to artificial tears'pH.

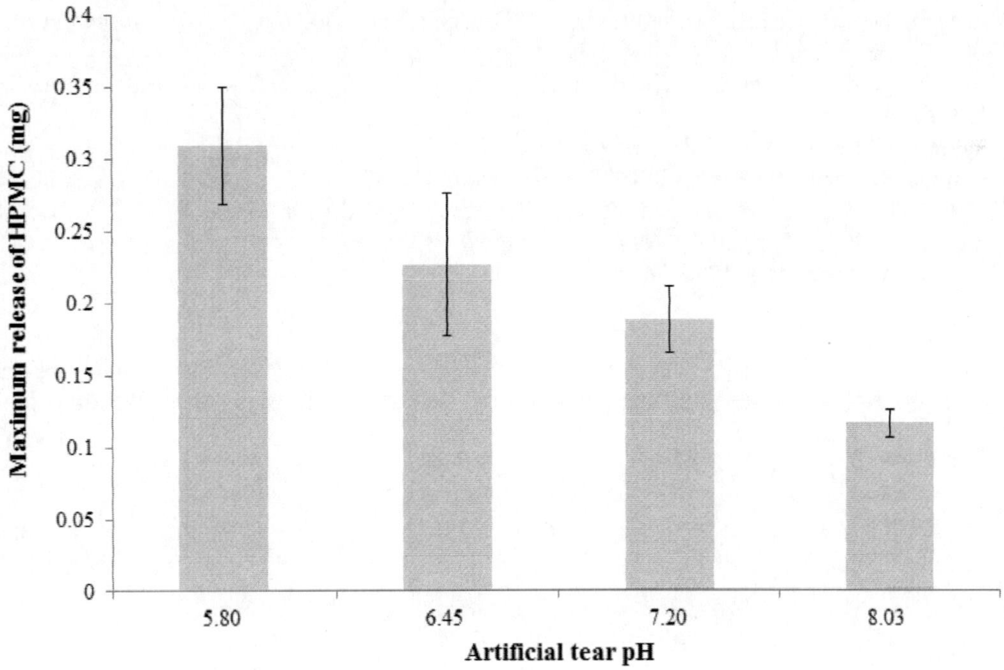

Figure 6. Maximum amount of HPMC with artificial tears' pH released for 12 hours.

4. Conclusions

The ability of pH-sensitive hydrogels to respond to its environment makes them suitable candidates for ophthalmic delivery of water soluble drugs. This hydrogel shows excellent biocompatibility and versatile physical properties. We thermally polymerized a hydrogel contact lens with hydroxyl ethyl methacrylate for water soluble drug delivery. The polymerized gels loaded with drugs were then, dipped in various pH of artificial tears. The drug release from the gel in response to pH was evaluated. As a result, drug was released from the lens for 12 hours in all the pH of artificial tears. At the lower the pH of artificial tears (pH 5.8), the total amount of dye emitted from the lens was increased more than the total amount of drug emitted from the basic tear (pH 8.4) due to the ionization of the carboxylic pendent groups (-COOH). Also, initial burst from acidic tears was increased within 1 hour. Release pattern of both water-soluble dye and drug from hydrogel lens turned out to be different depending on the pH of the artificial tears. We demonstrated that hydrogel contact lens is suitable candidate for ophthalmic water soluble drug career suggesting physiological pH of tears needs to be considered when designing the drug releasing contact lens.

5. References

[1] Shikamura Y, Yamazaki Y, Matsunaga T, Sato T, Ohtori A and Tojo K 2015 *Curr Eye Res* **3** 1
[2] Kojima S, Sugiyama T, Takai S, Jin D, Ueki M, Oku H, Tabata Y and Ikeda T 2015 *Invest Ophthalmol Vis Sci* **56** 2601
[3] Hong J and Noh H 2012 Korean J. Vis. Sci. 14 111
[4] Bohnert J, Horbett T, Ratner B and Royce F 1988 Invest. Ophthalmol. Vis. Sci. 29 362
[5] Abelson M, Udell I and Weston J 1981 *Arch. Ophthalmol.* **99** 301
[6] Marcombe R, Cai S, Hong W, Zhao X, Lapusta Y and Suo Z 2010 *Soft Matter* **6** 784
[7] Hui A, Willcox M and Jones L 2014 *Invest Ophthalmol Vis Sci* **55** 4896
[8] Tieppo A, White CJ, Paine AC, Voyles ML, McBride MK and Byrne ME 2012 *J Control Release* **157** *391*
[9] Manganiello MJ, Cheng C, Convertine AJ, Bryers JD and Stayton PS 2012 *Biomaterials* **33** 2301.
[10] You YC, Dong LY, Dong K, Xu W, Yan Y, Zhang L, Wang K and Xing FJ 2015 *Carbohydr Polym* **130** 243
[11] Ende M and Peppas N 1997 *J. Control. Release* **48** 47.
[12] Brannon-Peppas L and Peppas N 1990 *Biomaterials* **11** 635
[13] Fogh-Andersen N, Bjerrum P and Siggaard-Anderson O 1993 *Clin. Chem.* **39** 48
[14] Tores-Lugo M and Peppas N 1999 *Macromolucules* **32** 6646
[15] Philip P and Nikolaos A 1986 *J Control Release* **5** 23

Acknowledgement
This study was supported by Radiation Technology R&D program through the National Research Foundation of Korea funded by the Ministry of Science, ICT & Future Planning (NRF-2013M2A2A9043274)

IOP Conf. Series: Earth and Environmental Science **36** (2016) 012058 doi:10.1088/1755-1315/36/1/012058

Use of corn steep liquor as an economical nitrogen source for biosuccinic acid production by *Actinobacillus succinogenes*

J P Tan[1], J M Jahim[2], T Y Wu[3], S Harun[2] , T Mumtaz[2]
[1] Chemical and Process Engineering Department, Faculty of Engineering and Built Environment, *Universiti Kebangsaan Malaysia*, 43600 UKM Bangi, Selangor, Malaysia.
[2] Research Centre for Sustainable Process Technology (CESPRO), Faculty of Engineering and Built Environment, *Universiti Kebangsaan Malaysia*, 43600 UKM Bangi, Selangor, Malaysia
[3] School of Science, *Monash University*, Jalan Lagoon Selatan, Bandar Sunway, 46150, Selangor Darul Ehsan, Malaysia

E-mail: [1]jasonjptan2@hotmail.com

Abstract. Expensive raw materials are the driving force that leads to the shifting of the petroleum-based succinic acid production into bio-based succinic acid production by microorganisms. Cost of fermentation medium is among the main factors contributing to the total production cost of bio-succinic acid. After carbon source, nitrogen source is the second largest component of the fermentation medium, the cost of which has been overlooked for the past years. The current study aimed at replacing yeast extract- a costly nitrogen source with corn steep liquor for economical production of bio-succinic acid by *Actinobacillus succinogenes* 130Z. In this study, a final succinic acid concentration of 20.6 g/L was obtained from the use of corn steep liquor as the nitrogen source, which was comparable with the use of yeast extract as the nitrogen source that had a final succinate concentration of 21.4 g/l. In terms of economical wise, corn steep liquor was priced at $ 200 /ton, which was one fifth of the cost of yeast extract at $ 1000 /ton. Therefore, corn steep liquor can be considered as a potential nitrogen source in biochemical industries instead of the costly yeast extract.

1. Introduction

The environmental impact and sustainable issues have drawn the attention of the petrochemical industries to rely on biochemical methods which are environmentally benign and sustainable. In 2004, US Department of Energy (DOE) had identified 12 chemical building blocks which had the potential to be commercially produced via biological approach [1]. Some of the listed chemical building blocks including bioethanol and biosuccinic acid had moved into commercialization [2]. Since 2012, biosuccinic acid has been commercialized with at least 12 production plants around the world are already in operations by 2014 [2]. According to a life cycle analysis by BioAmber [3], bio-succinic

[1] To whom any correspondence should be addressed.

acid production had a net zero carbon footprint. Apart from that, a reduction of 60 % of total energy consumption compared to conventional petrochemical approach was also reported in their study [3]. Recent works in the literature had shown that succinic acid production by fermentation from renewable resources and a greenhouse gas, CO_2 are more cost-effective than the petroleum-based process [4]. Biotechnological production of succinic acid using by-products as a carbon source such as: sugar cane molasses, glycerol and others have been the major focus by many researchers. One of the key factors to be considered in the commercialization of bio-succinic acid is the cost of the fermentation medium. Carbon source in this case, is the major component and it has been the center of focus in bio-succinic acid research. Several cheap carbon sources for example flour, sake lees, cotton straw, rice straw, corn straw, corn core, wheat straw, waste bread, rapeseed meal, sugarcane bagasse, soybean meal, soybean solubles, sugarcane molasses, whey, glycerol and wood hydrolysate had been investigated and reported as substrate for bio-succinic acid production [5]. However, nitrogen source which is the second largest medium components, in this regard, had been given much lesser attention. The present work had an objective to study the potential of corn steep liquor as a nitrogen source in bio-succinic acid to replace the costly yeast extract. *Actinobacillus succinogenes* is used in this study since it is one of the highest yield and intensely researched microbe for bio-succinic acid production [6; 7]. Several aspects are investigated in this study including the yield of succinate, cost of the nitrogen sources as well as the nitrogen content of the corn steep liquor and yeast extract.

2. Materials and methods

A. succinogenes 130Z was purchased from DSMZ - German Collection of Microorganisms and Cell Cultures GmbH, Inhoffenstraße 7B, 38124 Braunschweig, Germany. The stock culture was maintained in Brain heart Infusion (BHI) media. For the fermentation of succinic acid, the mineral media consisted of: 0.2 g magnesium chloride hexahydrate ($MgCl_2.6H_2O$); 0.2 g calcium chloride ($CaCl_2$); 3 g potassium dihydrogen phosphate (KH_2PO_4) and 1 g sodium chloride (NaCl) per liter of medium. Inoculum (10% v/v) was prepared using BHI medium under aerobic condition at 120 rpm and 30 °C for 18 hours. Yeast extract and corn steep liquor with concentration of 15 g/L were employed for the fermentation of succinic acid. The performance of corn steep liquor and yeast extract in succinate fermentation were compared and discussed.

Succinate fermentation was performed in batch mode in 110 ml serum vials using 80ml working volume in sterile condition to study the profile of bacterial growth, succinic acid production, byproduct formation as well as consumption of sugars. Glucose was used as the sole carbon source in the fermentation at a concentration of 40 g/L.

Glucose was analysed using high performance liquid chromatography (HPLC), Agilent 1200 HPLC system (California, USA) equipped with Rezex RPM (Phenomenex, USA) 300 mm x 7.8 mm column and Refractive Index Detector (RID). Water was isocratically eluted at a flow rate of 0.6 ml/min while the column temperature was set at 60 °C. Soluble metabolites such as succinic acid, formic acid and acetic acid were analyzed using Agilent 1100 HPLC system (California, USA) equipped with Rezex ROA column (Phenomenex, USA) and Ultraviolet detector (UVD) at 210 nm. The column was eluted isocratically at a rate of 0.5 ml/min with 0.0025 M H_2SO_4 at 40 °C.

3. Results and discussion

Corn steep liquor is a by-product of corn wet-milling. It is a viscous liquid by-product consists of vitamins, minerals and nitrogen (including amino nitrogen) which is suitable for fermentation. It is an excellent source of organic nitrogen [8]. In the corn processing industry, the corn was steeped by counter current water for approximately two days. The steeped water was concentrated to a 50 % solid content liquid by evaporator. This liquid is called corn steep liquor [8]. On the other hand, yeast extract is the water soluble portion produced by extracting the cell contents from autolyzed yeasts. It is among the best growth stimulant for cell cultures and microbial fermentation. It is normally used with the concentration of 15 g/l in the fermentation.

IOP Conf. Series: Earth and Environmental Science **36** (2016) 012058 doi:10.1088/1755-1315/36/1/012058

In this study, corn steep liquor served well as the nitrogen source for A. succinogenes. As could be seen from Figure 1, Corn steep liquor was not only fermentable; the succinate production was, in fact, comparable with yeast extract. The final concentration of succinic acid using urea in fermentation was 12.61 g/l. Corn steep liquor reported a final succinate concentration of 20.6 g/L which was only 3.7 % lower than the fermentation using yeast extract as nitrogen source (21.4 g/L). Corn steep liquor was able to work as well as the costly yeast extract at the lower cost. This is one of the vital parameters to be considered when moving into industrial scale fermentation.

The nitrogen content of corn steep liquor was comparable to yeast extract. The reason that corn steep liquor has a close performance with yeast extract was because of their similarity in terms of total nitrogen and total amino nitrogen content. Total amino nitrogen is the available of small peptides chain (less than three amino acids). Total amino nitrogen could help to reduce the energy needed for the synthesis of proteins which are essential for the bacterial growth. Moreover, corn steep liquor is rich in minerals and growth factors since it is a byproduct of corn wet milling. In short, corn steep liquor has similar performance as yeast extract while cost approximately one fifth of the cost of yeast extract.

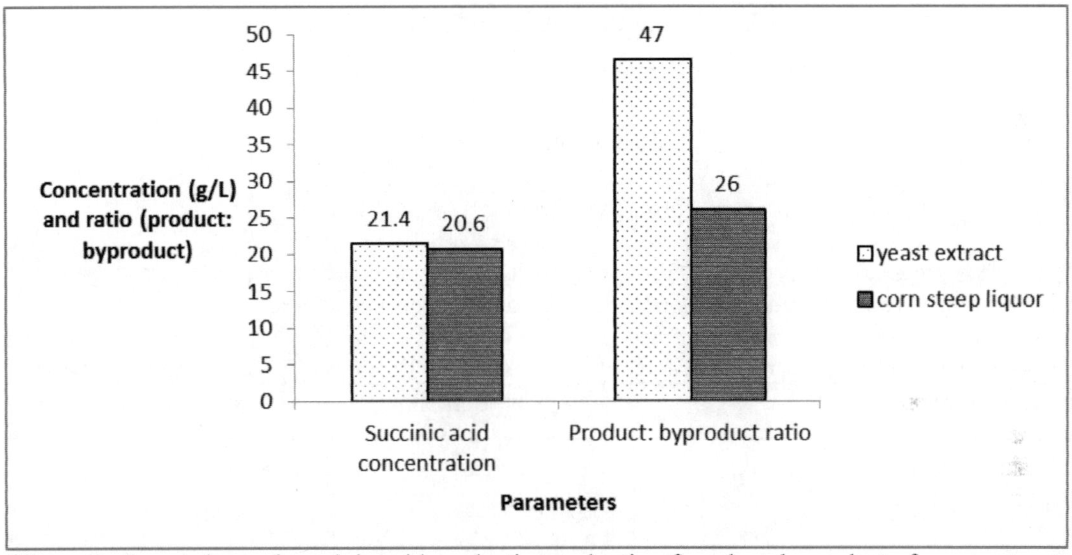

Figure 1. Comparison of succinic acid production and ratio of product: byproduct of yeast extract and corn steep liquor

The byproduct acetate and formate have a closer characteristics with succinate as these are all organic acids which lead to the difficulties during the purification of product succinate. 50-80 % of the cost of organic acid processing in industries is generally imputed to the product purification [9]. The ratio of succinate to by-product in this regard, acts as one of the primary indicator to predict the cost of product (organic acid) recovery, which is the reason it was considered in this study. Corn steep liquor in this case, had a lower product to byproduct ratio as compared to yeast extract.

Byproducts formation from the utilization of yeast extract and corn steep liquor is shown in Figure 2. The fermentation using corn steep liquor has 0.31 g/L of formic acid and 0.48 g/L of acetic acid after 72 hours of fermentation. On the other hand, yeast extract reported comparatively lower titers of byproducts which were 0.16 g/L of formic acid and 0.30 g/L of acetic acid. A lower concentration of byproducts allows an easier and cheaper downstream process. Corn steep liquor was found to be inferior in this regard.

The production of byproducts was also significantly affected by the concentration of total amino nitrogen. Higher amino nitrogen promotes higher flux into product instead of byproducts. Therefore, the byproducts formation was slightly more in corn steep liquor compared to yeast extract. This was due to the higher amino nitrogen content in corn steep liquor as could be inferred from Table 1.

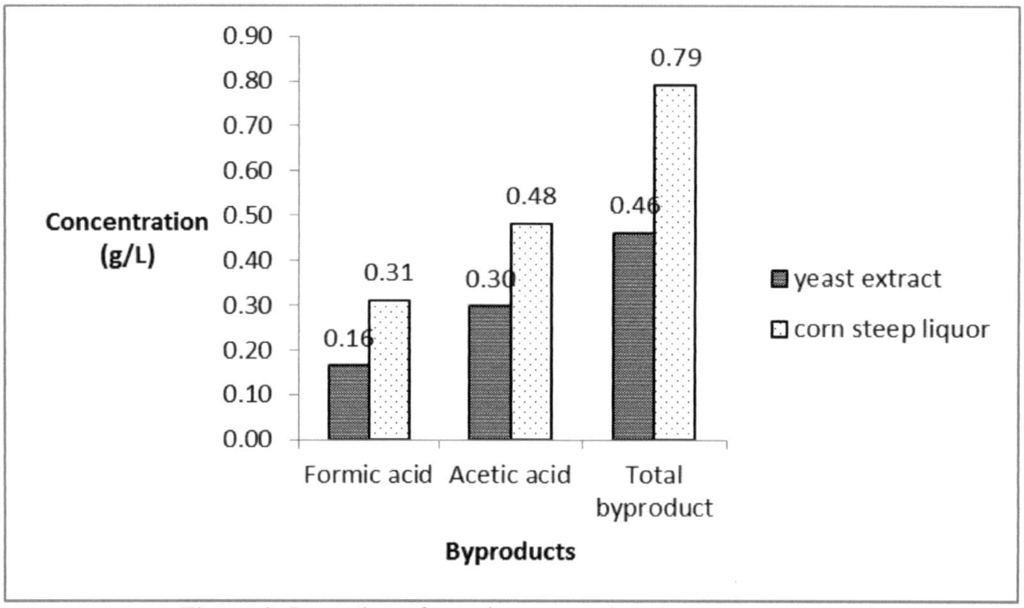

Figure 2. Byproducts formation comparison by yeast extract and urea

In spite of the production of products and byproducts, the nitrogen source content in corn steep liquor and yeast extract was compared in terms of its amount of total nitrogen content.

Table 1. Nitrogen content and cost comparison of corn steep liquor and yeast extract in dry basis

Nitrogen source	Yeast extract	Corn steep liquor
Total nitrogen, %	10-11.8	7.7-8.2
Total amino nitrogen, %	4.5-5.8	2.9-3.3
References	10	8
Cost $/ton	1000	200
Reference	11	12

In order to make bio-succinic acid competitive, the cost of production is among the prime concerns in the stage of commercialization. The cost of raw material in this regard, plays a crucial role in determining the overall production cost of bio-succinic acid. Nitrogen source contributes directly to the cost of raw material. Thefore, the cost of nitrogen source is important to be considered in the selection of nitrogen source. Corn steep liquor in this regard, appeared as a better choice since the price in ton is only $ 200, which was one fifth of the cost of yeast extract that was estimated to be $ 1000 /ton by bbc research [11; 13]. Since nitrogen source is the second largest component in fermentation after carbon source, the cheaper cost of nitrogen will have a pronounced and direct effect on the production cost of succinic acid as a whole. This is especially significant during the commercialization stage where large amount of nitrogen source is needed.

4. Conclusion

In this study, corn steep liquor has been shown to be able to function as a nitrogen source to replace the costly yeast extract. Amino nitrogen content has an impact on succinate yield which were found abundant in corn steep liquor and yeast extract. Apart from having the comparable succinate production, corn steep liquor was estimated to cost at $ 200 /ton, which was one fifth of the cost of yeast extract. Since nitrogen source is the second largest component in the fermentation medium, the utilization of corn steep liquor can significantly reduce the cost of bio-succinic acid production. This can further elevate the competitiveness of commercial bio-succinic acid.

5. References

[1] Werpy T, Petersen G, Top Value Added Chemicals from biomass Volume 1-Results of Screening for Potential Candidates from Sugars and Synthesis gas 2004
[2] ICIS Chemical Business 2012 Retrieved from http://www.icis.com/resources/news/2012/01/30/9527521/chemical-industry-awaits-for-bio-succinic-acid-potential/. February 11, 2014
[3] BioAmber. Retrieved from http://www.bio-amber.com/bioamber/en/products . 11 May 2015
[4] Jiang M., Chen K., Liu Z., Wei P., Ying H., Chang H., 2010 *Appl Biochem Biotechnology* **160** 244-254
[5] Jian Ping Tan, Jamaliah Md Jahim, Ta Yeong Wu, Shuhaida Harun, Byung Hong Kim, Abdul Wahab Mohammad *2014 Industrial and Engineering Chemical Research. American Chemical Society* Volume **53**, issue 42, pp 16123–16134
[6] Guettler MV, Rumler D and Jain MK. 1999 *International Journal of Systematic and Evolutionary Microbiology* Volume **49** Page 207-216
[7] M. J. Van der Werf, Michael V. Guettler, Mahendra K. Jain, J. Gregory Zeikus *1997Archieves of Microbiology* Volume **167** Page 332-342
[8] R. Winston Liggett and H Koffler *Bacteriology Review* Volume **12** Issue 4 Page 297-311
[9] Orjuela, A, A J Yanez, L Peereboom, C T Lira & D J Miller 2011 *Separation and Purification Technology* **83**(0): 31-37.
[10] Acumedia retrieved from http://www.neogen.com/Acumedia/pdf/ProdInfo/7184_PI.pdf
[11] Bbc research. http://www.bccresearch.com/pressroom/chm/global-market-yeast-extracts-autolysates-related-products-worth-$4.7-billion-2015
[12] Poul Ruben Andersen Retrieved from http://www.novozymes.com/en/investor/events-presentations/Documents/10_CMD_CoRE_PORA_FINAL.pdf. April 2015
[13] Hugh G Lawford, Joyce D Rousseau *Applied Biochemistry and Biotechnology* Volume **63-65** Issue 1 Page 287-304

Acknowledgments

This research is funded by the Ministry of Higher Education, Malaysia under LRGS/2013/UKM-UKM/PT/01 on project entitled "Biochemical Platform for Conversion of Diversified Lignocellulosic Biomass to Priceless Precursor and Biobased Fine Chemicals". The authors wish to thank the Ministry of Higher Education, Malaysia for financial support of this work through the grant provided under Long term Research Grant Scheme (LRGS 2013).

Measurement techniques for carbon dioxide sorption capacity on various coal samples: critical review

M Abunowara[1], M A Bustam[1], S Sufian[1], U Eldemerdash[2]
[1]Chemical Engineering Department, *Universiti Teknologi PETRONAS*, 31750 Tronoh, Perak, Malaysia
[2]Chemical Engineering Department, Benha Faculty of Engineering, *Benha University*, Benha, Egypt

E-mail: [1]abunowara1980@gmail.com, [1]azmibustam@petronas.com.my
[1]suriati@petronas.com.my, [2]usama.nour@bhit.bu.edu.eg

Abstract. Underground carbon sequestration is proposed as a geologic disposal technique for the long-term storage of CO_2 emissions to mitigate climate change and air pollution. Coal bed seams have large CO_2 adsorption capacity, long time CO_2 trapping and extra enhanced coal-bed methane recovery (CBM). However, CO_2 sorption capacity is one of significant steps required to be determined accurately in any feasibility evaluation of carbon sequestration. Hence, in lab scale, there are three methods for CO_2 adsorption capacity measurements namely manometric/volumetric, gravimetric and new capsule techniques for gas sorption on variety of sorbents. The manometric and volumetric methods require accurate determination of cell and void volumes and suitable equation of state (EoS). The gravimetric method requires a very accurate sensitive balance and less buoyancy effect and it is the best technique for small amounts (milligrams) of sorbents and the adsorption equilibrium can be mentored. Among all gas adsorption measurement techniques, the newly developed method "capsule method" exhibits the highest CO_2 adsorption capacity on Polish coal by 4.08 mmol/g because capsule method that directly measures CO_2 uptake of solid coal matrix cylinders, without the application of the equation of state (EoS) for CO_2 or volumetric corrections. The main advantage of capsule method is that it is independent of any Equation of State (EoS), and it has no volumetric effects or impurities distort the shape of the gas adsorption isotherm. The disadvantage of capsule method is time-consuming and it is not easy to implement.

1. Introduction

The major excessive emissions of green house gases (GHG_S) are emissions of carbon dioxide (CO_2) and it is one of the main causes of the global warming and air population. However, the major source of anthropogenic CO_2 emissions is the combustion of fossil fuels mainly coal to generate electricity. Therefore, mitigation and controlling CO_2 emissions are essential to address the greenhouse gases consequences [1]. Though at present separating and capturing CO_2 from the flue gas is costly and requires further research [2]. Hence, it is generally agreed that sequestering CO_2 in deep geological formations holds the most viable options for CO_2 large-scale storage of the greenhouse gas.

[1] To whom any correspondence should be addressed.

Carbon sequestration are including storage of CO_2 in deep sandstone formations, deep saline aquifers, gas-rich shales, active or depleted oil and gas fields with enhanced oil recovery (EOR), deep unminable coal bed seams, methane hydrate formations, salt caverns, other geological formations, or the ocean [2-5]. Furthermore, it is a widely accepted that deep unminable coal bed seams can be used as unconventional gas reservoirs for CO_2 sequestration [6]. It is estimated that the CO_2 storage capacity of unmineable coal bed seams in the worldwide is between 140 and 3000 Gt [6], which could simultaneously displace 1.45 × 1013 m^3 of methane reserve [7] and it is reported that one ton of coal can adsorb about 30–35 m^3 of CO_2 at pressures in excess of 5 to 8 MPa [8].

According to G. X. Wang al et [9] predicted that coal bed seams for CO_2 storage with enhanced coal bed methane (CBM) recovery are usually at the depth of 500–1500 m, with the corresponding reservoir pressure of 5–15 MPa and temperature of 30–100°C. In 2007, N. Siemons et al. [10] reported that the target coal seams for CO_2 injection in Europe is usually at great depth, with correspondingly high reservoir pressure of 6-15 MPa and temperature of 35–50°C. In 2010, J. He et al. [11] reported that the temperatures of potential coal seams typically range from 27-52°C and the depths of the coal seams are greater than 756 m, CO_2 sequestration occurs at supercritical condition (T_c = 31°C, P_c= 73.8 bar). Hence, higher pressure maintains more gas to remain adsorbed in place meanwhile higher temperature increases the kinetic energy of the system and enables gas molecules to stay in free-state rather than in the adsorbed state [1].

Economically, some of deep coal bed seams may be infeasible for mining due to inadequate coal seam thickness, poor continuity of the area, or adverse geology. However, under the proper conditions, these coal seams can provide adequate sites for CO_2 storage [11]. Hence, the starting step of evaluating the feasibility and capacity of coal bed seams for CO_2 sequestration is to simulate the coal bed seams conditions in laboratory scale. In laboratory scale, one of the significant parameters is the absolute CO_2 storage potential of the coal matrix and it is the sum of the adsorption and pore-filling capacity.

In the lab scale, there are three techniques to measure the CO_2 sorption capacity on coal samples namely manometric/volumetric, gravimetric methods and there is another newly developed method called capsule method. The manometric/volumetric method is most reliable for CO_2 adsorption capacity measurement at above supercritical pressure and temperature condition. The objectives of this work are to present the current methods for gas sorption on different sorbents and to discuss their advantages and disadvantages and to compare the gas sorption on various coal samples and the experimental results of CO_2 adsorption measurements on variety of coal specimens.

2. Methods of gas adsorption measurement

CO_2 adsorption capacity measurement is one of the major steps in assessing the CO_2 potential storage in any geosequestration reservoirs. The volumetric/ manometric and gravimetric are common techniques where have been utilized to measure gas adsorption isotherms of porous materials such as coal that are flooded with CO_2 under unconfined, isostatic conditions. These techniques are usually applied to determine the Gibbs Surface Excess (GSE), i.e. the amount of adsorbed CO_2 present in the coal in excess of the free, bulk phase present at constant pressure and temperature [22]

The manometric/volumetric method requires accurate determination of cell and void volumes. The amount of gas sorbed is recorded by pressure readings (manometric method) or pressure or volume readings (volumetric method). For high temperatures (27 – 427°C) and for corrosive gases and high pressures (50 – 300 bars) the use of metallic sealings (steel, sliver) always recommended and the amount of the sample depends on the sample cell. The manometric method is most widely used for determining gas sorption capacities on coal [12, 13, 16]. The volumetric method is similar to the manometric method [14-16].

In manometric method as shown in Figure 1, the amount of sorbed gas is measured by reading gas pressure. The excess sorption mass ($m_{excess}^{CO_2}$) is the difference between the mass of gas that has been transferred into the measuring cell up to a given pressure step and the "non-sorption" reference mass [16].

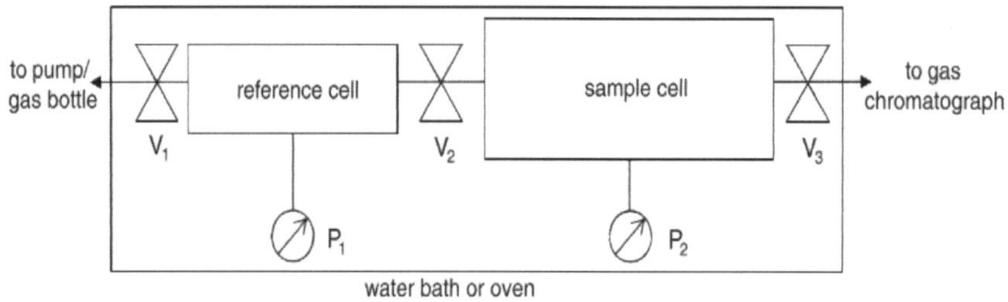

Figure 1. Schematic diagram of manometric sorption setup, V denotes valves and P denotes pressure transducers [16].

$$m^{CO_2}_{excess} = m^{CO_2}_{transfered} - V^0_{void} \cdot \rho^{gas}(T,p) \quad (1)$$

The mass transferred from the reference cell into the measuring cell during *N* successive pressure steps is given by:

$$m^{CO_2}_{transfered} = \sum_{i=1}^{N} V_{ref} \cdot \left(\rho_i^{gas}(p_i,T) - \rho_{i-1}^{gas}(p_{i-1},T)\right) \quad (2)$$

The void volume (V^0_{void})of the sample cell is determined by expansion of a "non-sorbing" gas, which is typically helium. Helium densities are calculated using the equation of state (EoS) by using the Vander Waals equation with *a* and *b* parameters. This procedure also provides the skeletal volume (V^0_{sample}) and the skeletal density (ρ^0_{sample}) of the sample. For gas sorption isotherms, the void volume multiplied by the density of the gas (or supercritical) phase ($V^0_{void} \cdot \rho^{CO_2}(T,p)$, yields the "non-sorption" reference mass, i.e. the amount of gas (supercritical fluid) that would be accommodated in the measuring cell if no sorption takes place.

In volumetric method, the amount of sorbed gas is measured by reading gas pressure or gas volume and the schematic diagram of this method is shown in Figure 2. In order to measure gas sorption amount with volumetric method, volume of measuring system has to be measured precisely at first. Then, adsorption amount is calculated from the change of gas pressure inside the measuring system by means of ideal gas equation. Sample handling is easier in volumetric method than in gravimetric method and the thermal equilibrium reaches rapidly in volumetric method. The amount of gas injected $m^{\text{injected}}_{\text{gas}}$ can be determined from the pump position as it moves forward. Therefore [16]:

$$m^{\text{adsorbed}}_{gas} = m^{\text{injected}}_{gas} - m^{\text{unadsorbed}}_{gas} \quad (3)$$

While

$$m^{\text{injected}}_{gas} = \left(\frac{p\Delta VM}{ZRT}\right)_{pump} \quad (4)$$

$$m^{\text{unadsorbed}}_{gas} = \left(\frac{pV_{void}M}{ZRT}\right)_{sample\ cell} \quad (5)$$

Here *m* denotes the mass of gas, *p* is pressure, *T* is temperature, *M* is the molar mass of the gas species, Z is the compressibility coefficient of the pure gas species and *R* is the universal gas constant. ΔV is the volume change in the pump and V_{void} is the volume of the free gas in the sample cell.

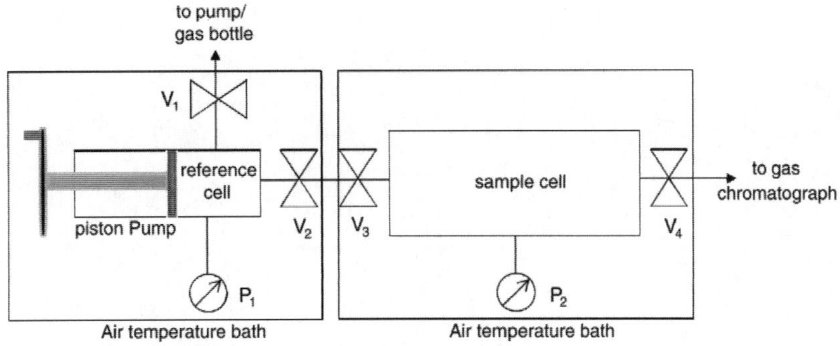

Figure 2. Schematic diagram of volumetric gas sorption setup [16, 17].

In most sorption systems, the excess sorption of the coal sample is the extra amount of gas in the cell than would be present if the coal could not interact with the gas. In volumetric technique, the measurement requires accurate values for the volume of the sample cell, the volume of coal in the sample cell and, in volumetric systems, and the temperature of the system. The volume of the coal is normally calculated from the helium density of the coal; i.e. the density of the coal that is inaccessible to helium [21].

The main advantage of volumetry/manometry is that simple easy to handle. However it needs accurate measurement of the reference and samples. The amount of sorbent material usually needed several grams if only tiny amounts (milligrams) of the sorbent are available then gravimetric measurements strongly recommended. Approach to equilibrium, gas adsorption may last for second, hours or sometime even days and months. The disadvantages of volumetric/manometric adsorption experiments do not give information on the kinetics of the process and small amounts of adsorbent (several milligrams) have shown to be ineffective. They are not practical at extreme low or high pressures as pressure measurements in these regions becomes difficult and accurate of equation of state (EoS) of the sorption gases are not always available [17]. Furthermore, swelling of the coal upon equilibration with CO_2 produces a substantial error in corrections made for the free, bulk phase present in the pores. The reliability of Helium pycnometry as a method to correct for the free, bulk phase (volumetry/manometry) may be questioned, because Helium (He) and CO_2 may not fill the same pore space in coal, and because the probe gas He itself may adsorb to the coal slightly. The thermo-physical properties of pure CO_2 near the critical point, as predicted by the equation of State (EoS), are extremely sensitive to the presence of chemical impurities, in particular to species that are extracted from coal by supercritical CO_2 [22].

In gravimetric method, the amount of sorbed gas is measured by reading sample (sorbent) gravity at constant pressure by means of an accurate balance and less buoyancy effect with the sample either suspended mechanically or by magnetic coupling across the wall of a high-pressure vessel as shown in Figure 3. The maximum amount of amount of adsorbent material is 25g and the temperature ranges 20-250°C and pressures up to 500 bar [18-20].

The excess sorption is determined from the weight change (apparent mass change) of the sample $\left(\Delta m = m_{measured}(T,p) - m^0_{sample}\right)$ recorded during individual sorption steps, where m^0_{sample} is the original sample mass.

The excess sorption is derived from its apparent weight change by a buoyancy correction based on the skeletal volume (V^0_{sample}) of the sample, corresponding to the same reference state as in the manometric procedure. The determination of the skeletal density or volume is performed with helium. The excess sorbed mass is then given by:

$$m^{gas}_{excess} = \Delta m + V^0_{sample} \cdot \rho_{gas}(T,p), \text{ or} \quad (6)$$

$$m^{gas}_{excess} = \Delta m + \frac{m^{0}_{sample}}{\rho^{0}_{sample}} \cdot \rho_{gas}(T,p) \tag{7}$$

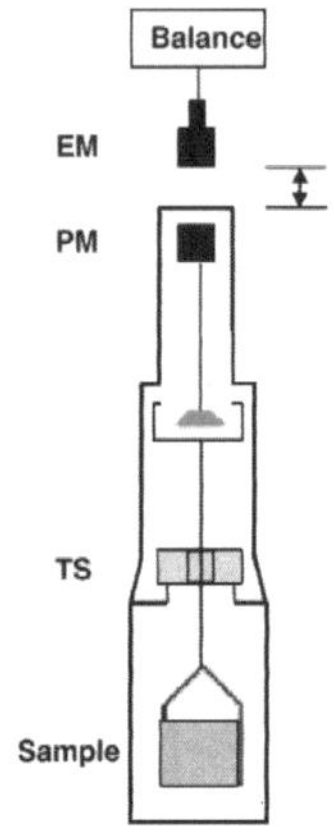

Figure 3. Schematic diagram of magnetic suspension balance (MSB). EM = electro magnet; PM = permanent magnet; TS = titanium sinker [18].

The advantage of gravimetry when compared to volumetry/manometry system is that with gravimetry it is possible to observe the approach to equilibrium during the adsorption process at the balance's data display. Therefore, one can choose arbitrarily a certain small fraction of the pure sorbent sample's mass and a certain time interval and define technical adsorption equilibrium of the system. The main disadvantages of gravimetric are the buoyancy effects, the construction of gravimetric system are costly and high sensitivity to external surrounding environment. Despite attempts made to enhance the accuracy of GSE methods in coal-CO_2 systems, these issues are extremely difficult to resolve. Thus, there is an obvious demand for an independent method to determine CO_2 uptake by coal [22].

Therefore, in 2011, Hol et al. [22] fabricated a new, direct method of measuring the uptake of CO_2 by coal (i.e. by sorption plus pore-filling). The description of the capsule method, the ductile metal capsule is placed in the compaction cell using a capsule holder as shown in Figure 4. A cylindrical coal sample, 4 mm in diameter and 4 mm in length, is jacketed in a tightly fitting, annealed gold capsule and exposed to CO_2 at constant pressure and temperature. Sorption induced swelling of the sample is accommodated by ductile deformation of the capsule. Once the coal is saturated, the capsule is sealed by mechanical loading and the external CO_2 pressure removed. This allows the CO_2 to desorb from the coal and flow into an Al-foil bag attached to the capsule. The volume of the bag, and hence the amount of CO_2 stored in the coal sample, is determined using the Archimedes method.

While the capsule method is time-consuming and not easily executed, the advantage is that it is independent of any Equation of State (EoS), and that no volumetric effects or impurities distort the shape of the isotherm.

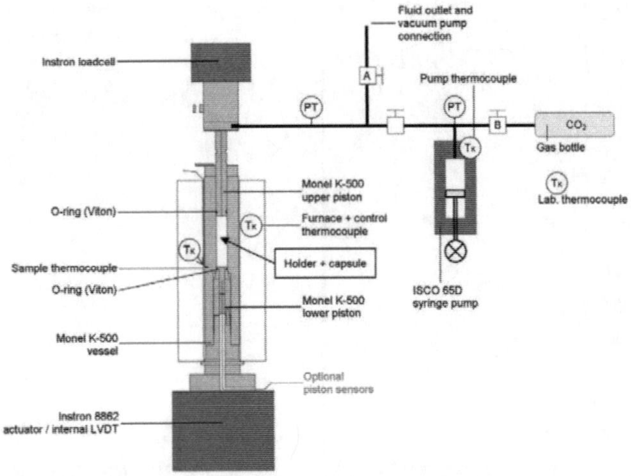

Figure 4. Schematic diagram of capsule method [22, 23].

To calculate the total CO_2 uptake of the samples, they first determined the trapped CO_2 volume by subtracting the empty capsule/ bag volume from that measured in the inflated state. The volume of CO_2 trapped in the bag at ambient conditions (V_{bag}, ml) was then multiplied by the density of pure CO_2 (ρ_{CO_2} in mmol ml^{-1}) at the ambient conditions (*P,T*-ambient) pertaining during weighing, and the result divided by the mass of the sample m_0 in grams, to obtain [23]:

$$N = \frac{V_{bag} \cdot \rho CO_2 (P,T-ambient)}{m_0} \quad (8)$$

Where *N* is the total CO_2 uptake in $mmol g_{coal}^{-1}$ or $mol kg_{coal}^{-1}$

This newly developed technique called capsule method that directly measures CO_2 uptake of solid coal matrix cylinders, without the application of the equation of state (EoS) for CO_2 or volumetric corrections of any kind and it can determine the uptake of any adsorbate by any (swelling) adsorbent, including shale and clay cap rocks.

3. CO_2 adsorption capacity on various coal samples

In lab scale, the manometric, volumetric and gravimetric are common methods and still have being utilized by many researchers for determining gas sorption on different sorbents. In 2010, J. He et al. [24] carried out experiments to measure CO_2 adsorption/desorption isotherms on anthracite coal samples (Kyungdong, Korea) using a static volumetric method at 25°C (298 K) and 45°C (318 K), and up to 15 bar (150 atm).

The adsorption of CO_2 is favorable at low temperatures and dry coal conditions. CO_2 adsorption capacities on wet and dry coals are 1.39 and 1.49 mmol/g respectively as shown in Figure 5. In wet coal, the dissolution of H_2O in the CO_2 phase, absorption of CO_2 in the aqueous phase, and coal structure changes affected the estimation of adsorption amount, especially under supercritical condition.

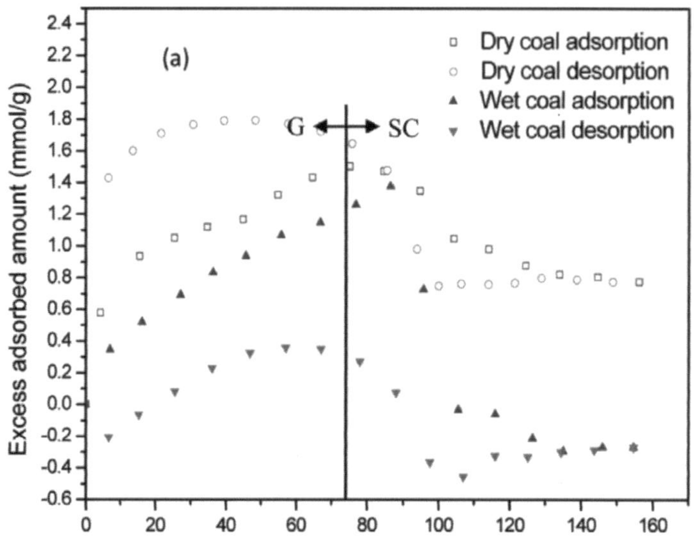

Figure 5. Comparison of adsorption and desorption of CO_2 between dry coal and wet coal (3.64%) at 45°C (318 K) with respect to pressure [24]. G: subcritical and SC: supercritical.

In 2013, H. H. Lee et al. [25] utilized a volumetric gas chromatographic method to measure the excess sorption amount of anthracite Kyungdong coal samples from South Korea and samples size range was 150–500 μm under 45 °C (318K) and 65 °C (338K) and up to 13 MPa. They found out that the total excess adsorbed amount and isotherm shape of mixture on dry and wet coal was similar to the results of pure CO_2. Beyond 4 MPa (40 bar), all curves showed a similar shape due to the increased holdup ratio of the gas phase to the sorbed amount. Therefore the total excess adsorbed amount and isotherm shape of mixture on dry and wet coal were similar to the results of pure CO_2 as shown in Figure 6 and Figure 7. The moisture of the coal reduced the sorption amount compared to dry coal in the mixture system, showing a similar moisture effect on the isotherm of pure CO_2 in the same range of fluid density.

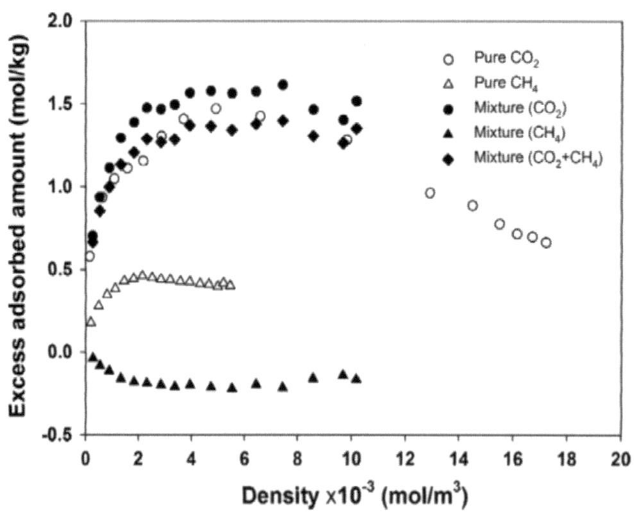

Figure 6. The highest excess adsorbed amount of CO_2 and CH_4 pure and mixture, on dry coal at 45°C (318 K) as a function of density [25].

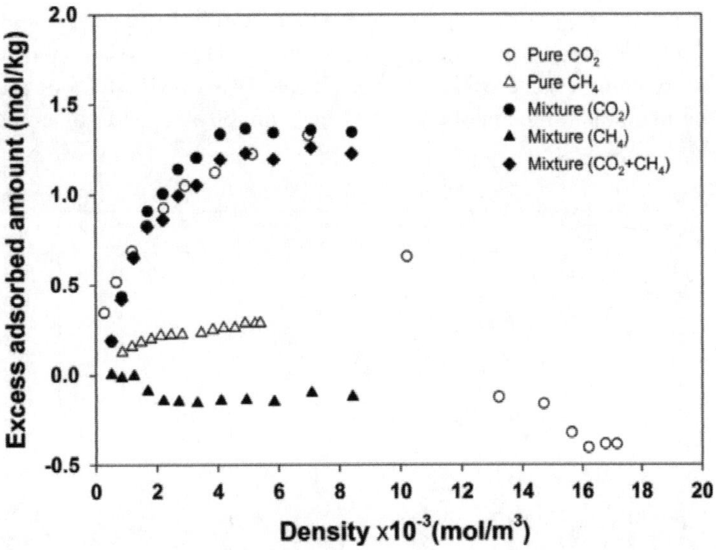

Figure 7. The highest excess adsorbed amount of pure CO_2 and CH_4, and their mixture on wet coal at 45°C (318 K) as a function of density [25].

In 2013, V.N. Romanov et al. [26] utilized manometric technique to measure CO_2 adsorption capacity of coal samples from Central Appalachian Basin, US. The CO_2 adsorption measurements were carried out at 55 °C by introducing CO_2 into the cell to build the pressure step-wise up to 12 MPa. The maximum CO_2 adsorption capacity is 0.97 mmol/g-dry-ash free (daf) as shown in Figure 8. The role of CO_2 dissolution in pore water appears to be marginal.

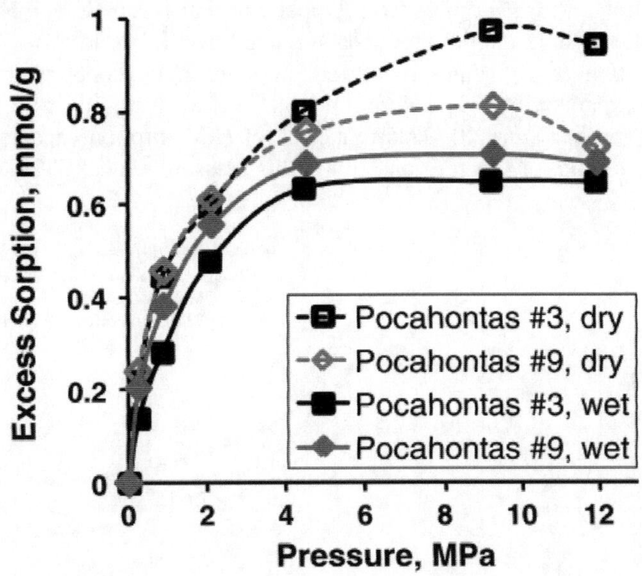

Figure 8. CO_2 excess sorption isotherms (dry basis) on wet and dry Appalachian coals [26].

In 2013, S. Hol et al. [22, 23] newly developed technique called capsule method that directly measures CO_2 uptake of solid coal matrix cylinders. The experiments were performed on high volatile bituminous coal, exposed in the Brzeszcze mine, in the Upper Silesian Basin of Poland. The samples were saturated with CO_2 at a fixed CO_2 pressure between 0 and 16 MPa and a temperature of 40 °C. The total uptake of CO_2 measured using the capsule method continuously increases as a function of increasing CO_2 pressure, up to a value of 4.08 mmol/g coal at 18.10 MPa CO_2 of pressure meanwhile 4.08 mmol/g by using manometric technique meanwhile 3 mmol/g by using manometric technique as shown in Figure 9.

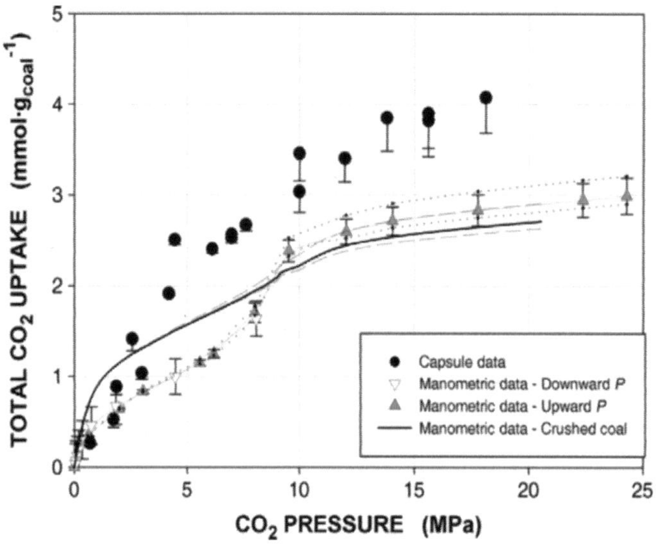

Figure 9. CO_2 comparison between single capsule method and manometric method [23].

In 2012, Weniger et al. [27] conducted the CO_2 adsorption isotherms using manometric technique to measure high-pressure CO_2 adsorption isotherms at 45 °C and at 6 to 10 MPa on as-received coal samples from the Paskov and ČSM mines from Upper Silesian Basin (Czech Republic). The excess sorption capacities of moisture-equilibrated coals ranged from 0.8 to 1.16 mmol/g-daf and the maximum excess sorption (n_{max}) values measured on as-received coals range from 0.96 to 1.41 mmol/g for CO_2. A comparison of CO_2 adsorption isotherms measured on as-received and moisture-equilibrated coal is given in Figure 10. This reduction of CO_2 sorption capacity is due to the steep increase in the specific density of supercritical CO_2 in this pressure range [27-29].

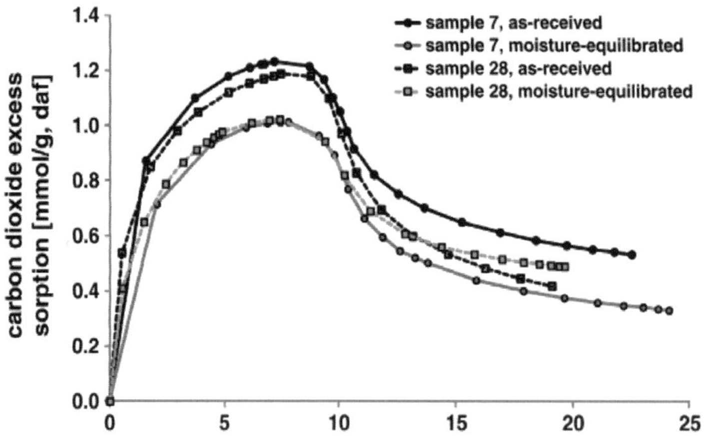

Figure 10. CO_2 sorption isotherms for as-received (open symbols) and moisture equilibrated

(solid symbols) coal from Paskov mine (squares) and ČSM mine (circles) [27].

In 2009, Kelemen and Kwiatek [30] measured CO_2 adsorption on coal samples at 30 °C, dry coal at 75°C and 1.8 MPa using gravimetric gas sorption measurements (Hiden IGA gas sorption system). The rate of gas adsorption for coal varies with the size of coal particles that are smaller than the fissure network of the coal. Gas adsorption isotherm experiments are usually conducted using powdered coal samples to assure that equilibrium is rapidly approached. As shown in Figure 11, the maximum adsorption capacity of CO_2 is 1.40 mmol/g [30, 31].

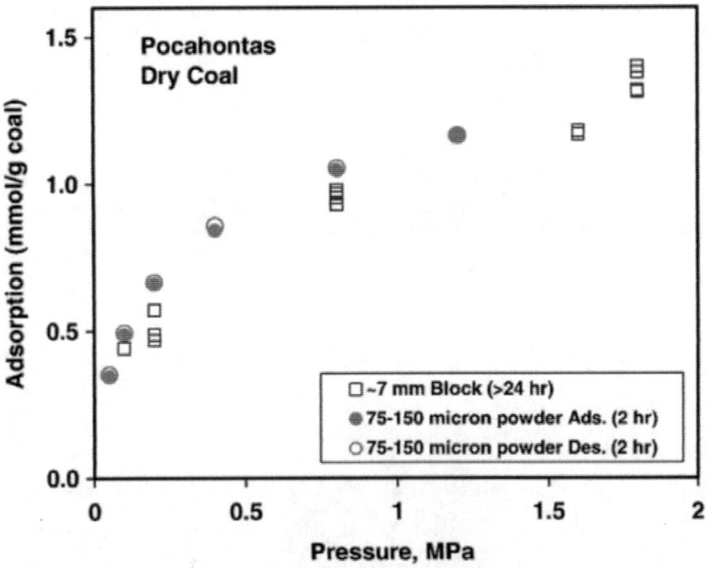

Figure 11. Comparison between CO_2 adsorption results using 75–150 μm powder and ~7 mm block dry Pocahontas Argonne Premium coal [30].

In Table 1 a summary of the maximum of CO_2 adsorption capacities on various coal samples and it is obvious that wet coal samples always have less CO_2 adsorption capacity compared to dried coal samples because the energetic active sorption sites occupied by water [15, 32]. The comparison of all dry and wet coal samples are depicted in Figure 12 and Figure 13. However it noticed that the dry Polish coal sample has the highest CO_2 adsorption on coal by 4.08 mmol/g compared to the other coal samples.

Table 1. Summary of the maximum of CO_2 adsorption capacities on various coal samples.

Coal sample	Technique	Maximum CO_2 adsorption (mmol/g)	Ref.
Dry Polish coal	Capsule	4.08	[23]
Dry Korean coal	Volumetric	1.49	[24]
Wet Korean coal	Volumetric	1.39	
Dry Korean coal	Volumetric	1.48	[25]
Wet Korean coal	Volumetric	1.38	
Dry Appalachian coal	Manometric	0.97	[26]
Dry Czech Republic coal	Manometric	1.41	[29]
Wet Czech Republic coal	Manometric	1.16	
Dry Pocahontas coal	Gravimetric	1.40	[30]

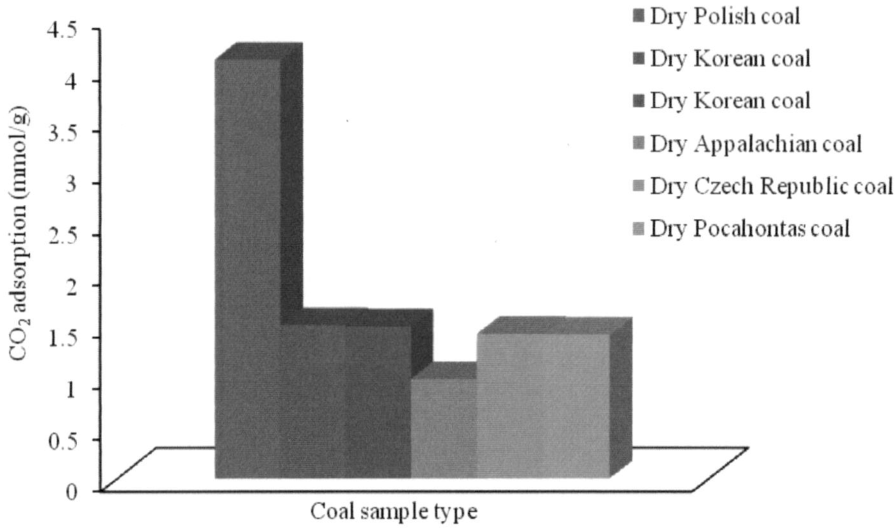

Figure 12. CO_2 adsorption of all dry samples at different conditions.

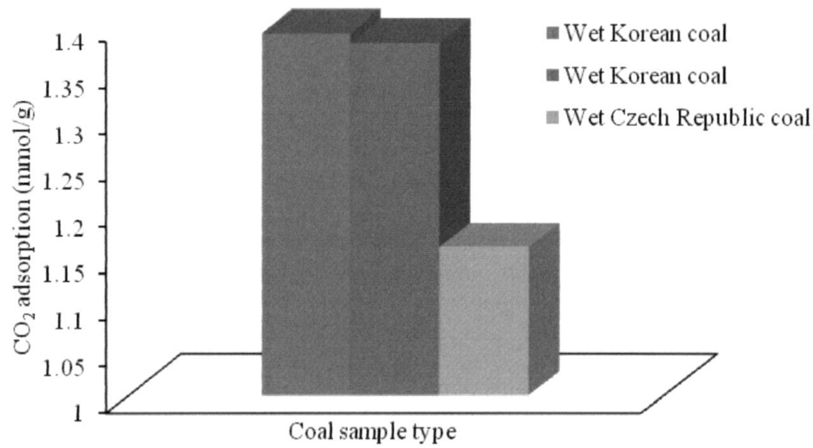

Figure 13. CO_2 adsorption of wet samples at different conditions.

4. Conclusions

The volumetric/manometric is common method for CO_2 adsorption isotherms at high pressures and temperatures on various sorbents because they easy to handle and reach thermal equilibrium rapidly. However, it requires an accurate determination of cell and void volumes and elimination of Equation of State (EoS) errors and small amounts of sorbent are ineffective. Thus, when the amounts of sorbent are small (milligrams) the gravimetric technique is strongly recommended and the approach to equilibrium during the adsorption process can be mentored. The other new capsule method is independent of any equation of State (EOS), and no volumetric effects or impurities distort the shape of the isotherm but it is time-consuming and not easily executed. Therefore, it is noticed that dried Polish coal sample using new capsule method has the highest CO_2 adsorption by 4.08 mmol/g. The presence of moisture in coal reduces the CO_2 adsorption on coal significantly and it makes coal to swell. Hence, wet coals adsorb less CO_2 compared with the dried coals because the water molecules occupy coal's pores. The oxygen-containing functional groups are preferentially occupied by water in moist coal due to its polarity. In coal samples, the low and medium coal ranks usually have high affinity for adsorbing CO_2 than high coal rank (anthracite) due to the low and medium coals have higher surface area and higher oxygen functional groups. All CO_2 adsorption on different coal samples show that there are decline in CO_2 adsorption on coal after exceeding the CO_2 its supercritical condition. This is due to the enormous increase the specific density of CO_2. Furthermore, coal from different coal mine has different physical properties "anisotropic" hence more investigations about the interactions of subcritical and supercritical CO_2 with wet and dry coals should be conducted. In future work, CO_2 adsorption isotherms on coal using new techniques such as capsule method should be explored. Moreover, the effect of impurities such as H_2S, SO_2 and hydrocarbons on CO_2 sorption onto different coal should also be investigated.

5. References

[1] Vishal V, Singh T N and Ranjith P G 2015 *Fuel* **139** 51–58
[2] Harbin S, Twombly N J, West R D and Malhotra V M 2015 *Fuel* **150** 131–138
[3] Liu C J, Wang G X, Sang S X, Gilani W and Rudolph V 2015 *Fuel* **139** 125–132
[4] Day S, Duffy G, Sakurovs R and Weir S 2008 *Int. J. Greenh. Gas Control* **2** 342–352
[5] Li X, Fang Z, 2014 *Int. J. of Coal Sci Technol.* **1** 93–102
[6] Brown T D, Harrison D K., Jones J R and LaSota K A 2007 *Int. J. Environ. Pol.* **29** 474–483
[7] White C M, Smith D H, Jones K L, Goodman A L, Jikich, S A, LaCount R B, DuBose S B, Ozdemir E Morsi, B I and Schroeder K. T 2005 *Energ. Fuels* **19** 659–724
[8] Cook P J, Rigg A and Bradshaw J 2000 *Aust. Petrol. Prod. Explor. Associate J.* **40** 654–666
[9] Wang G X, Wei X R, Wang K., Massarotto P and Rudolph, V 2010 *Int. J. Coal Geology* **83** 46–54
[10] Siemons N, Wolf, Karl-Heinz A A, Bruining J 2007 *Int. J. Coal Geology* **72** 315–324
[11] He J, Shi Y, Ahn S, Kang J W and Lee C-H 2010 *J. Phys. Chem. B* **114** 4854–4861
[12] Harpalani S, Prusty B K, and Dutta P, 2006 *Energy and Fuels* **20** 1591–1599
[13] Li D, Liu Q, Weniger P, Gensterblum Y, Busch A, and Krooss B M 2010 *Fuel* **89** 569–580
[14] Ozdemir E and Schroeder K 2009 *Energy & Fuels* **23** 2821–2831
[15] Reeves S, Gonzalez R, Gasem K A M, Fitzgerald J E, Pan Z, Sudibandriyo M and Robinson Jr, R L 2005 ICBM *Symposium Tuscaloosa* Tuscaloosa Alabama
[16] Busch A and Gensterblum Y 2011 *Int. J. Coal Geology* **87** 49–71
[17] Sudibandriyo M, Pan Z, Fitzgerald J E, Robinson Jr, R L and Gasem K A M 2003 *Langmuir* **19** 5323–5331
[18] Charrière D and Behra P 2010 *J. Colloid Interf. Sci.* **344** 460–467
[19] Ottiger S, Pini R, Storti G, Mazzotti M, Bencini R, Quattrocchi F, Sardu G and Deriue G, 2006 *Environ. Prog.* **25** 355–364
[20] Pini R, Ottiger S, Burlini L, Storti G and Mazzotti M 2010 *Int. J. Greenh. Gas Control* **4** 90–101
[21] Sakurovs R, Day S, Weir S, 2009 *Int. J. Coal Geology* **77** 16–22
[22] Hol S, Peach C J, Spiers C J, 2011*Energy Pro.* **4** 3125–3130
[23] Hol S, Gensterblum Y and Spiers C J 2013 *Fuel* **105** 192–205
[24] He J, Shi Y, Ahn S, Kang J W and Lee C-H 2010 *J. Phys. Chem. B* **114** 4854-4861
[25] Lee H-H, Kim H-J, Shi Y, Keffer D and Lee C-H 2013 *Che. Eng. J.* **230** 93–101
[26] Romanov, V N, Hur T-B, Fazio J J, Howard B H and Irdi G A 2013 *Int. J. of Coal Geology* 118 89–94
[27] Weniger P, Francŭ J, Hemza P and Krooss B M 2012 Int. *J. Coal Geology* 93 23–39
[28] Bae J S, Bhatia S K., 2006 *Energ. Fuel.* 20 2599–2607

[29] Gensterblum Y, van Hemert P, Billemont P, Charrière D, Li D, Busch A., Krooss B M, De Weireld G and Wolf K.-H A A 2009 *Carbon* 47 2958–2969
[30] Kelemen S R and Kwiatek L M 2009 *Int. J. of Coal Geology* 77 2–9
[31] Goodman A L, Busch A, Duffy G.J, Fitzgerald J E, Gassem K A M, Gensterblum Y, Krooss B M, Levy J, Ozdemir E, Pan Z, Robinson Jr R L, Schroeder K., Sudibandriyo M and White C M 2004 *Energ. Fuel.* 18 1175–1182
[32] Liu Y, Wilcox J, 2012 *Environ. Sci. Tech.* 46 1940–1947

Acknowledgments

The authors gratefully acknowledge the financial support provided for this research study through E-science grant (0153AB-B41), Centre of Postgraduate Studies (CGS), Universiti Teknologi PETRONAS (UTP) and Ministry of Science, Technology and Innovation (MOSTI), Malaysia.

IOP Conf. Series: Earth and Environmental Science **36** (2016) 012060 doi:10.1088/1755-1315/36/1/012060

Pollution of PM_{10} in an underground enclosed loading dock in Malaysia

M S Abualqumboz[1], N I Mohammed[2], A Malakahmad[3], A N Nazif[4], A T Albattniji[5]
Department of Civil and Environmental Engineering, Universiti Teknologi PETRONAS, 32610 Bandar Seri Iskandar, Perak Darul Ridzuan, Malaysia.

E-mail: [1]mabualqumboz@hotmail.com, [2]izma.mohammed@petronas.com.my, [3]amirhossein@petronas.com.my, [4]aminanazif@yahoo.co.uk, [5]ayat_tyseer@hotmail.com.

Abstract: The enclosed nature of underground loading docks results in accumulation of motor vehicles emissions. Thus, concentration of numerous harmful air pollutants including PM_{10} particles can increase and reach dangerous levels. This paper aims to study short-term and long-term exposure of PM_{10} particles inside an underground loading dock located in Malaysia. In addition, the correlation with indoor temperature, relative humidity and vehicles flow will be measured. The concentrations of PM_{10} were measured for three consecutive weeks using the real-time air quality monitoring instrument AQM60. Series of statistical tests and multiple linear regression analysis were applied on the data using SPSS software and MATLAB R2013a. The results illustrated that PM_{10} daily average concentration was in compliance with the Malaysian guideline of 150 μg/m^3. Actually, 95 % of instantaneous PM_{10} concentration readings were below 75 μg/m^3. In addition, significant correlation were found between PM_{10} concentration and indoor temperature, relative humidity and the previous concentration. The multiple R and R^2 were 0.91 and 0.83, respectively. PM_{10} concentration was also correlated with motor vehicles flow. In conclusion, health effects of long-term exposure to small repetitive doses of air pollutant inside underground facilities should be studied and appropriate control measures need to be implemented.

1. Introduction

Stable economic development and increase in human population contribute in making surface land scarcer and much expensive especially in highly populated cities [1]. Hence, most governments try to utilize their land resources to maximum extent. Consequently, construction of many facilities including loading docks has been shifted underground [2]. The main source of air pollution inside such underground micro-environments is emissions of motor vehicles [3]. Motor vehicles basically emit particulate matter (PM), carbon monoxide (CO), carbon dioxide (CO_2), oxides of nitrogen (NO_x), sulfur dioxide (SO_2) and hydrocarbons (HC) [1, 4]. The confined environment of underground loading docks and the restricted contact with ambient air, result in quick accumulation of air pollutants that build up and cause poor indoor air quality. Without the installation of proper mitigation measures like

[1] To whom any correspondence should be addressed.

mechanical ventilation systems, underground loading docks can become a dangerous working environment, as both of employees and users will be exposed to high concentrations of harmful gases [2]. PM_{10} particles basically consist of complex suspended particles in the air with different composition and size smaller than 10 µm. These particles enter the respiratory tract, which ultimately causes serious respiratory diseases [5].

Presence of PM_{10} pollution inside enclosed micro-environment has been confirmed by different studies. For example, Li, & Xiang [1] reported that the 14-hour average PM_{10} concentration inside an underground enclosed car park, was 56% higher than the Chinese indoor PM_{10} standard of 150 µg/m^3. Similarly, Johansson and Johansson [6] observed high PM_{10} concentration at an underground station in central Stockholm. In the same context, Raut, et al. [7] stated that PM_{10} average concentration in a confined rail station was 5-30 times higher than those measured on of the busiest streets of Paris.

Previous studies showed that the indoor air quality of underground loading docks has not been explored much yet. Underground loading dock has a unique environment in term of number of people and the time duration they spend inside. Unlike other underground facilities such as car parks and bus station, the workers spent up to 8 hours a day in underground loading docks. Even more, the number of workers is higher than those in other underground spaces. PM_{10} is a very common motor vehicle pollutant and has been linked with respiratory and cardiovascular morbidity [8, 9]. Therefore, this paper is focusing on PM_{10} pollution inside an underground enclosed loading dock.

2. Material and methods

2.1. Sampling area

The study area was selected inside an underground loading dock area located in Kuala Lumpur (KL), which is the capital and the most populous city in Malaysia with more than 1.6 million people [10]. The loading dock is located in the Kuala Lumpur City Center (KLCC) and has the coordinates of 3° 09' 28.3" N 101° 42' 41.7" E. It is heavily utilized because it serves many famous and attractive shopping malls and restaurants, commercial buildings and governmental agencies. The observations showed vehicles rate of one per two minutes during working hours from 8:00 AM to 5:00 PM except break time (1:00 PM to 2:00 PM). Moreover, the loading dock is used for disposing of garbage during night hours. The loading dock is equipped with traditional duct ventilation system that operates only during working hours at a rate of 6 ACH.

2.2. Sampling instrument

Air Quality Monitoring (AQM60) Environmental Station was used for measuring PM_{10} concentration levels inside the loading dock though a particle monitor (Nephelometer) [11]. Nephelometer accuracy was at ± 2 µg/m^3, with range of 0 - 2000 µg/m^3. AQM60 was installed near the administrative office where workers who spend 8 hours every day inside, could be under the risk of exposure to air pollution from motor vehicle. Measurement of PM_{10} concentrations started on Monday December 1, 2014 and lasted for two weeks. AQM60 was adjusted to measure PM_{10} concentration levels after every two minutes. RF modem was used to retrieve the data from the AQM60 unit.

2.3. Data analysis:

Series of descriptive statistical tests have been applied on the collected data using Statistical Package for the Social Sciences (SPSS) software and Microsoft Excel 2013. Descriptive statistical tests include number of valid and missing values, minimum and maximum values, percentiles, average, median and mode. These tests would help organize, summarize and interpret the compiled data. In addition, MATLAB R2013a was used for correlating PM_{10} concentrations with the previous concentration, indoor temperature, relative humidity and traffic flow using multiple liner regression (MLR).

3. Results and discussion

Throughout twenty one days of study, 14,847 valid readings of PM_{10} concentrations were recorded, whereas, damaged and missed vales were approximately 2 % out of the total. The corrupted values irregularly happened and they remain in compliance with the 5 % accuracy of AQM60 [11]. Hence, they were being excluded following the Listwise deletion technique [12].

3.1. Short-Term exposure of PM_{10} concentrations

Table 1 gives the summary of PM_{10} daily average concentrations that have been recorded inside the loading dock during the three weeks of measurements.

Table 1. Daily average PM_{10} concentration inside the loading dock

Parameter	Range	Average
PM_{10} concentration (μg/m^3)	24.82 – 51.05	35.59 ± 7.27

As can be obviously seen in Table 1, the 24-hour average PM_{10} concentration level inside the loading dock fluctuated between 24.82 μg/m^3 and 51.05 μg/m^3, with an average of 35.59 ± 7.27 μg/m^3. Due to the lack of specific air quality guidelines for underground confined micro-environments in Malaysia, the Industry Code of Practice on Indoor Air Quality (ICOP-IAQ) was adapted [13]. PM_{10} concentration levels inside the loading dock was remarkably below the daily averaged PM_{10} guidelines of 150 μg/m^3. Eventhough the World Health Organization (WHO) has a tighter PM_{10} standard of 50 μg/m^3 [14], PM_{10} average concentration is still within the acceptable limit (Figure 1). However, the highest average concentration was slightly higher than the standard with only 1.05 μg/m^3. The WHO guidelines was violated twice during the three weeks of study as the second and the third highest daily averaged PM_{10} concentration were 50.94 μg/m^3 and 46.47 μg/m^3, respectively. The first and the second highest PM_{10} average concentrations were recorded on Sundays, the 7th and the 21th of December due to operating the ventilation system at lower rate than in other days because of the low traffic during weekends. In addition, the cleaning process during weekends cause dust to be resuspended again. The differences in PM_{10} concentrations are deeply linked with traffic flow.

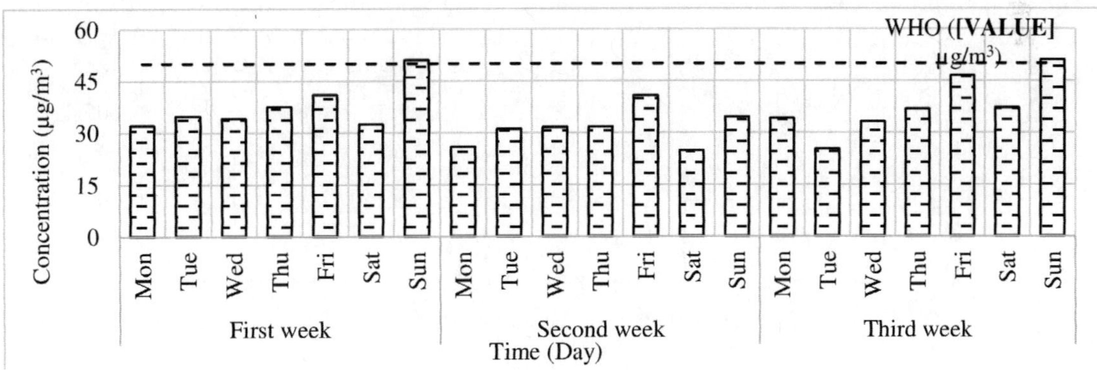

Figure 1: Daily average PM_{10} Concentration level

3.2. Long-Term exposure of PM_{10} concentrations

The instantaneous PM_{10} readings has been studied in order to locate the long-term exposure of PM_{10} concentrations. Valid PM_{10} concentration readings have been summarized using the boxplot diagram. Figure 2 provides information on the instantaneous PM_{10} concentration readings inside the loading dock during the study period of twenty one days. The minimum and maximum values of PM_{10} concentrations were 0.09 μg/m^3 and 656.51 μg/m^3, respectively. This depicts that PM_{10} concentration has a wide range (Difference between the maximum and the minimum values) of 656.42 µg/m^3. Similarly, there are other researchers who have reported extremely high PM_{10} concentrations inside underground facilities [1, 6] .

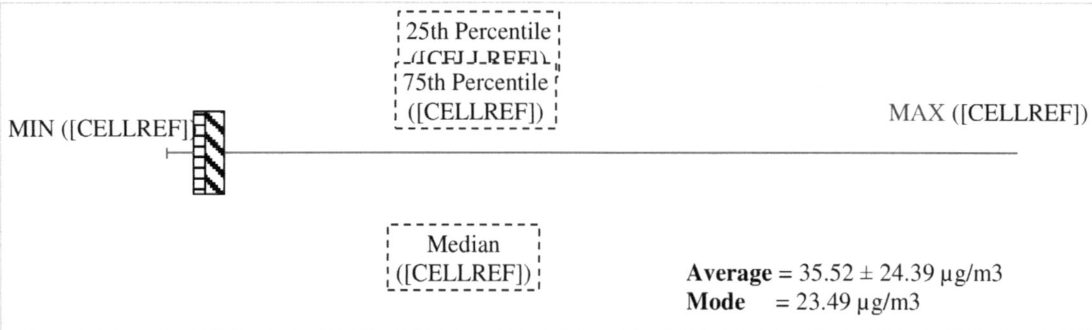

Figure 2: Boxplot for PM_{10} concentrations of the three weeks

The average, median and the mode of PM_{10} concentration levels were 35.52 ± 24.39 µg/m^3, 29.73 µg/m^3 and 23.49 µg/m^3, respectively. These measures reflect the central tendency of the data and where it is centered [15]. Even though a maximum concentration of 656.51 µg/m^3 was recorded, the measures of central tendency of PM_{10} concentration levels were uncommonly low and were between 20 and 40 µg/m^3. This is mainly caused as the mass of PM_{10} readings were low. Specifically, 50 % of PM_{10} concentrations ranged between 20.55 µg/m^3 and 44.13 µg/m^3, while 75 % of the readings were below 44.13 µg/m^3. Moreover, the boxplot in Figure 2 explained that 75 % of PM_{10} concentrations were located in a narrow range of 44.13 µg/m^3, while the remaining 25 % were vacillated in a relatively very wide range of 612.38 µg/m^3. In order to clarify this point, the histogram and cumulative frequency curve were used as shown in Figure 3. The Malaysian PM_{10} limit of 150 µg/m^3 was set as the upper boundary of the last class.

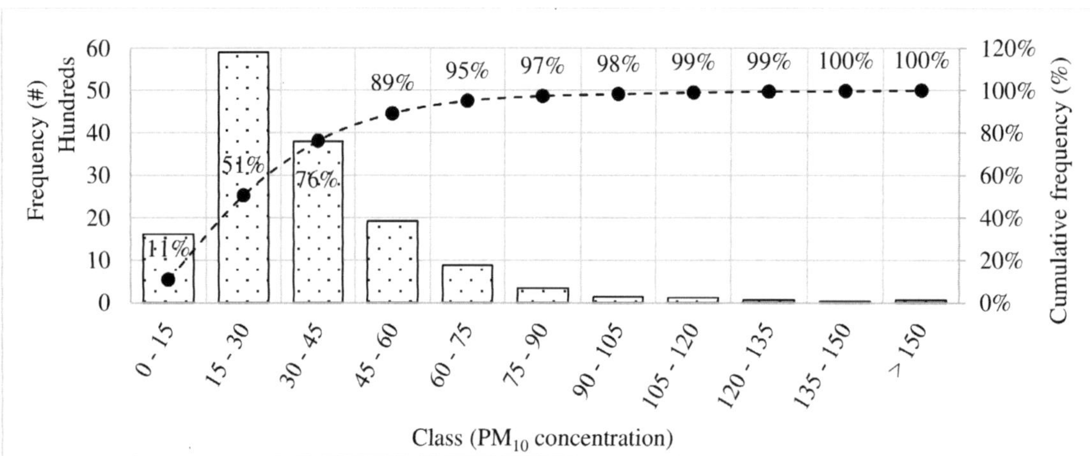

Figure 3: Histogram and cumulative frequency diagram of PM_{10} concentrations

Generally, it is clearly presented in Figure 3 that the mass of the recorded PM_{10} concentrations is concentrated on the left of the figure towards the low concentrations. This case is known as right-skewed distribution [16], in which the most dominant and frequent PM_{10} concentrations would be much closer to zero than the maximum value. It is also illustrated that PM_{10} dominant concentrations ranged from 15 – 30 µg/m^3. In addition, the cumulative frequency curve which is presented in Figure 3 as dashed line, indicates that 95 % of PM_{10} concentration levels were below 75 μg/m^3. In other words, and due to the equal sampling time intervals, PM_{10} concentration level remained below 75 μg/m^3 for 95 % of the time during the three weeks of study. The 95 % of time represents 19.95 days out of the 21 days of sampling.

Figure 2 and 3 clearly state that PM_{10} particles remain low in concentrations most of the time and goes up to alarming levels for very short time span. Therefore, the workers will be exposed to a repetitive, small doses of PM_{10} for long durations. Thus more focus should be paid on health impact of the long term exposure to small and repetitive PM_{10} concentrations. Regardless of PM_{10} guidelines, various studies have proved a positive connection between adverse health effects and long term exposure to repetitive low concentrations of PM_{10}. Schikowski, et al. [17] has found a meaningful and constant detrimental impact of long term exposure to PM_{10} on lung function, while Dominici, et al. [18] and Heinrich, J., et al. [19] linked the long term exposure of small PM_{10} concentrations with increased mortality.

3.3. Temperature levels (° C), relative humidity (%) and vehicles flow

Air temperature and relative humidity (RH) are important factors that influence occupant thermal comfort. Temperature levels inside the loading dock fluctuated between 30.46 ° C and 34.19 ° C with a mean of 32.19 ° ± 0.46 ° C. Matsushita (1993) recorded almost the same range of temperature inside underground car parks in Japan [3]. According to the applicable Malaysian guidelines [20], the recommended range for indoor temperature is 23° C to 26° C. Therefore, indoor climate of the study area is poor. On the other hand, indoor relative humidity recorded a minimum and maximum value of 43.90 % and 67.9 % respectively. The American Society of Heating, Refrigerating, and Air-Conditioning Engineers (ASHRAE) stated that the recommended range for acceptable indoor relative humidity limit is 30 % to 65 % [21]. Indoor relative humidity slightly violated the maximum limit but most of the time it followed the acceptable pattern.

Multiple linear regression (MLR) analysis was used in order to create a prediction model that can estimate PM_{10} concentration based on the previous hour concentration, indoor temperature and relative humidity. The analysis showed a significant correlation between PM_{10} concentration and the parameters with $P < 0.05$ at the 95 % confidence level. The multiple R and R^2 were 0.91 and 0.83, respectively. Looking at details, the previous concentration and indoor temperature showed stronger correlation with PM_{10} concentration than relative humidity with P values almost zero. The P value for relative humidity was > 0.05. The analysis results is shown in equation 1.

$$\boldsymbol{PM_{10}\ concentration} = -13.78 + 0.91 \times \boldsymbol{Previous\ concentration} + 0.58 \times \boldsymbol{Temp} - 0.03 \times \boldsymbol{RH} \quad (1)$$

MLR conducted for the second time and produced an equation that can predict PM_{10} concentration based on the number of vehicles that enter or leave the facility is as shown is equation 2.

$$\boldsymbol{Y} = 93.681 - 0.187 \times \boldsymbol{X} \quad (2)$$

Where Y is PM_{10} concentration and X is the number of vehicles. The model was significant since the P vlaue was < 0.05. The multiple R and R^2 were 0.80 and 0.63, respectively.

4. Conclusion

PM_{10} concentration levels inside the underground enclosed loading dock were in compliance with the applicable air quality guidelines. Even though, the workers will be under long-term exposure to repetitive small concentrations of PM_{10}. The effects on workers should be studied due to long exposure duration and mitigating measures should be taken. Moreover, ambient and indoor air quality guidelines are being used in order to evaluate the indoor air quality of underground confined spaces. However, the underground enclosed micro-environment cannot be considered neither as ambient nor indoor environment. This issue raises the need of enclosed air quality guidelines.

5. References

[1] Li, Y and R Xiang 2013 *Particuology* **11(1)** p 94-98
[2] Maskey, S, T Kang, H J Jung, and C U Ro 2011 *Indoor air* **21(1)** p 12-24
[3] Matsushita, K, S Miura, and T Ojima 1993 *Tunnelling and Underground Space Technology* **8(1)** p 65-73
[4] De Nevers, N 2010 *Waveland Press* Second Edition
[5] Kampa, M and E Castanas 2008 *Environmental Pollution* **151(2)** p 362-367
[6] Johansson, C and P -Å Johansson 2003 *Atmospheric Environment* **37(1)** p 3-9
[7] Raut, J C, P Chazette, and A Fortain 2009 *Atmospheric Environment* **43(4)** p 860-868
[8] Rodopoulou, S, M-C Chalbot, E Samoli, D W DuBois, B D San Filippo and I G Kavouras 2014 *Environmental Research* **129(0)** p 39-46
[9] Kalantzi, E G, D Makris, M N Duquenne, S Kaklamani, H Stapountzis and K I Gourgoulianis 2011 *Atmospheric Environment* **45(39)** p 7121-7126
[10] Saeed, M O, M N Hassan and M A Mujeebu 2009 *Waste management* **29(7)** p 2209-2213
[11] Aeroqual, AQM60 Environmental Station User Guide V 7.3, A. Limited, Editor. 2010, Aeroqual Limited
[12] Enders, C K, 2010 Guilford Publications
[13] Dominick, D, H Juahir, M T Latif, S M Zain and A Z Aris, 2012 *Atmospheric Environment* **60(0)** p 172-181
[14] WHO, WHO Air quality guidelines for particulate matter, ozone, nitrogen dioxide and sulfur dioxide: global update 2005: summary of risk assessment. 2006: World Health Organization.
[15] Hayter, A 2012 *Cengage Learning*
[16] Berenson, M L., D M Levine and T C Krehbiel 2011 *Prentice Hall*
[17] Schikowski, T, D Sugiri, U Ranft, U Gehring, J Heinrich, H -E Wichmann and U Kramer, 2005 *Respir Res* **6(1)** p 152
[18] Dominici, F, M. Daniels, S L Zeger, and J M 2002 *Samet Journal of the American Statistical Association* **97(457)** p 100-111
[19] Heinrich, J, E Thiering, P Rzehak, U Krämer, M Hochadel, K M Rauchfuss, U Gehring and H-E Wichmann 2012 *Occupational and environmental medicine* p oemed 2012-100876.
[20] Yousef, B 2012 *Advances in Mechanical Engineering and its Applications* **3(1)** p 278-283
[21] Ismail, M, M Sofian, N Zafirah and A M Abdullah 2010 *Environment Asia* **3(103)** p e108

Acknowledgment
This work has been supported by Universiti Teknologi PETRONAS (UTP), Malaysia.

IOP Conf. Series: Earth and Environmental Science **36** (2016) 012061 doi:10.1088/1755-1315/36/1/012061

Effect of temperature on Candida antartica lipase B activity in the kinetic resolution of acebutolol

[1]Mariani Rajin, [2]A H Kamaruddin
[1]*Universiti Malaysia Sabah,* Faculty of Engineering, Kota Kinabalu, Sabah
[2]*Universiti Sains Malaysia,* School of Chemical Engineering, Nibong Tebal, Penang

E-mail: [1]mariani@ums.edu.my

Abstract. Thermodynamic studies of free *Candida antartica* lipase B in kinetic resolution of acebutolol have been carried out to characterize the temperature effects towards enzyme stability and activity. A decreased in reaction rate was observed in temperature above 40°C. Thermodynamic studies on lipase deactivation exhibited a first-order kinetic pattern. The activation and deactivation energies were 39.63 kJ/mol and 54.90 kJ/mol, respectively. The enthalpy and entropy of the lipase deactivation were found to be 52.12 kJ/mol and -0.18 kJ/mol, respectively.

1. Introduction

Generally, chiral drugs including beta blockers are available as racemic mixture consists of (S) and (R)-enantiomers. However, the beta blockage properties of most beta blocker drugs reside on the (S)-enantiomer. The (R)-enantiomer is not active and sometimes responsible for the undesired effect. Unfortunately, until recently, most of the beta blocker drugs are still market in racemic form. Among these is acebutolol [N-{3-acetyl-4-[(2-hydroxy-3-isopropylamino)propoxy] phenyl}butanamide, which is effectively used in treating hypertension and cardiac arrhythmias [1,2].

A large number of studies have been done with various strategies and routes to produce pure enantiomer of beta blocker such as asymmetric synthesis, biotransformation and chromatography methods as well as kinetic resolution. Most of the enantiomeric resolution of beta blockers performed by direct and indirect method of high performance liquid chromatography HPLC [3-7]. Although these techniques are capable of producing high enantiomeric excess, they are costly, time consuming and required more rigid condition to be met.

Asymmetric synthesis of acebutolol have been reported earlier. Wang *et al.,* synthesized acebutolol from 4-aminophenol via Fries rearrangement, acetylation of the amino and hydroxyl group, hydrolysis and butyrylation. 51.5% of overall yield was achieved [8]. Acebutolol pure enantiomers have also been synthesized through hydrolytic kinetic resolution (HKR), condensation and followed by reaction with propyl amine, obtained overall yield and optical purity of 47.3 and >98%, respectively [9]. However, both techniques involved various intermediate product couple with multi step and lengthy reaction.

A kinetic resolution (KR) is one of the promising method in preparation of pure enantiomer of chiral drugs including beta blockers. This method offers advantages of less energy consuming and lower production cost [10]. In kinetic resolution, the two enantiomers of a racemate are transformed

[1] To whom any correspondence should be addressed.

into products at different rates [11]. Generally, there are two methods employed in enzymatic kinetic resolution of racemic drugs, which are esterification and hydrolysis. As for beta blockers, enantioselective acylation or transesterification is the common method of enzymatic kinetic resolution that has been applied in the production of its pure enantiomers.

Various types of lipase and solvent was studied for the resolution of propranolol by Chiou *et al.*, [12]. Another example of enzymatic acylation of beta blocker was conducted by Quan *et al.*, which demonstrated a control of the enzymatic acylation position at the hydroxyl or amino group of propranolol [13]. Recently, preparation of pure enantiomer of propranolol [14] and atenolol [15] have been carried out. In both studies, the effects of reaction parameters on the reaction rate and enantiopreference of immobilized *Candida antartica* lipase B (CALB), as well as the effects of different immobilization protocol has been investigated. However, up to now, there is no published data on the thermodynamic study of beta blockers enzymatic kinetic resolution using CALB. Therefore, in this paper, an attempt is made to investigate the thermodynamic properties of acebutolol kinetic resolution catalyzed by CALB. Temperature effect on enzymatic reaction can be determined by understanding the thermodynamic properties such as entropy, free energy change and enthalpy change.

2 .Material and Methods

2.1 Material

The lipase from *Candida antarctica* fraction B (Novozym CALB L) was obtained from Novo Nordisk (Bagsvaerd, Denmark). All other lipases was supplied by Sigma-Aldrich (Michigan). Racemic acebutolol was purchased from Suzhou Yuanfang Chemical Co., Ltd (Shanghai, China). All other chemicals used in this work were of analytical grade and were supplied by Fisher Scientific (Nepean, Ont., Canada) and used without further purification.

2.2 Kinetic resolution of racemic acebutolol

The kinetic resolution of racemic acebutolol through enantioselective transesterification was conducted in a 250 ml Erlenmeyer flask. Racemic acebutolol as substrate was added to 20 ml of selected solvent followed by acyl agent and then by enzyme. The mixture was continuously shaken at 200 rpm and 37 °C in an incubator shaker. Periodically, 1 ml aliquot was withdrawn from the reaction medium for HPLC analysis.

2.3 HPLC analysis

The concentration of the optically pure product from the reaction is determined by HPLC. The HPLC analysis is based on the method explained by Ekelund *et al.*, [16]. Chiralcel OD (25 cm x 4.6 mm I.D, 10 um) type of column from Daicel Chemical Industries is used with a UV detector with a wavelength of 320 nm. The mobile phase used is hexane-ethanol-diethylamine with the ratio of 90:10:0.1. The injection volume and mobile phase flowrate is 1 uL and 1.0 ml/min, respectively.

2.4 Thermodynamic studies on CALB

The thermodynamic studies for kinetic resolution of acebutolol were divided into two parts, mainly on the effect of temperatures on lipase activity and stability. The first part consists of studies on the effect of temperatures on the production rate implemented at various reaction temperatures of 30, 35, 40, 45, 50, 55 and 60 °C. In the second part, the irreversible denaturation rate for CALB under different thermal treatment was studied to determine half-life time as well as the thermal resistance strength. The kinetics of irreversible denaturation was studied by incubating CALB in the substrate without vinyl acetate at temperatures of 45, 50, 55 and 60 °C for variable times of 1, 2, 3 and 4 days. The residual activity of the lipase was then examined by implementing typical kinetic resolution as described in section 2.2.

The thermodynamic properties such as ΔG, free activation enthalpy (kJ/mol) is given by equation (1) and equation (2).

$$\Delta G = -RT \ln K \tag{1}$$

$$\Delta G = \Delta H - T.\Delta S \tag{2}$$

R represents Universal Gas Constant with the value of 8.314 J/mol K. Whereas, ΔH and ΔS represent activation enthalpy (kJ/mol) and activation entropy (J/mol.K), respectively.

Combining equation (1) and equation (2) and solving for ln k gives;

$$\ln K = \frac{\Delta H}{RT} + \frac{\Delta S}{R} \tag{3}$$

The thermal deactivation constant (k_d) used in the estimation of kinetic and thermodynamic parameters is a function of temperature as given by Eyring equation,

$$k_d = T^{\frac{k_B}{h}} e^{\Delta S/R} e^{-\Delta H/R} \tag{4}$$

The linearized form of equation (4) gives

$$\ln\left(\frac{k_d}{T}\right) = \ln\left(\frac{k_B}{h}\right) + \left(\frac{\Delta S}{R}\right) - \left(\frac{\Delta H}{R}\right)\left(\frac{1}{T}\right) \tag{5}$$

Where k_d, T, k_b, h and R represent deactivation constant, absolute temperature, Boltzmann constant, Planck's constant, and gas constant, respectively. From this equation, it can be observed that, by plotting log of k_d as a function of the inverse of absolute temperature, the energy of deactivation E_d can be obtained as the product of angular coefficient of the adjusted straight-line multiplied by the value of universal gas constant R. This approach has been successfully applied by several authors in order to study the thermodynamic properties of various lipases [17-19].

3. Result and discussion

3.1 Effect of temperature on kinetic resolution of racemic acebutolol

The effect of temperature on the kinetic resolution of acebutolol was studied. Figure 1 shows that the conversion increased gradually with the increase of temperature. At low temperature, CALB unable to exhibit its highest activity. The maximum conversion of 42% was obtained when the reaction was carried out at 40°C. Further increase in the temperature brings a significant decrease in the conversion. At high temperature, the enzyme was partially deactivated. Similar behaviours have been reported previously for CALB applied in esterification of lactic acid [20], 1-phenylethanol [21] and ketoprofen [22].

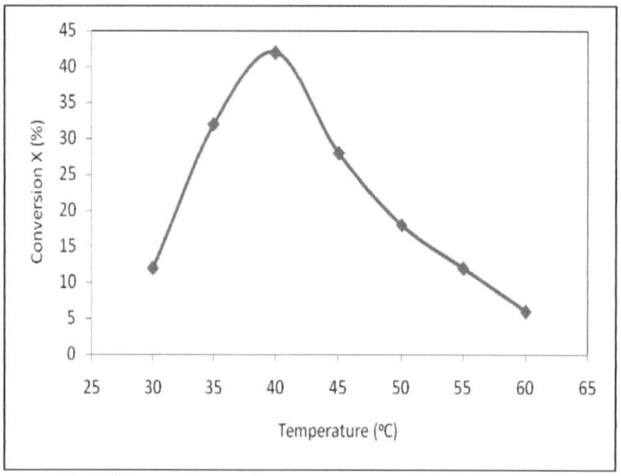

Figure 1. Effect of reaction temperature on the conversion.

3.2 Thermodynamic properties of CALB in kinetic resolution of racemic acebutolol

In this study, the activation energy was determined to be 39.63 kJ/mol. Activation energy is associated with normal enzymatic reaction in formation of product from the substrate [23]. The enzymes with low values of activation energy is favorable for industrial application [24]. Previously, Xin *et al.,* [25] obtained the activation energy of 65.04 kJ/mol based on the Arrhenius law in enzyme mediated transesterification using CALB. By using the same type of lipase, Sun *et al.*, [20] estimated the activation energy for esterification of glyceryl ferulate to be 67.4 kJ/mol.

In the present work, the energy for deactivation was found to be 54.90 kJ/mol. According to Baptista *et al.,* [26], the weakening and breakage of non-covalent bonds that maintain the protein tertiary structure in the more stable folded state may cause reversible thermal unfolding of proteins. In this work, it was found that the deactivation energy was higher as compared to the activation energy. This is similar with the results obtained in other works using different types of lipase [27-29]. This finding shows the predominant effect of deactivation process taking place. In order for the unfolding of the lipase to occur, more energy was needed to overcome the higher energy barrier [28].

It was found that the lipase has an enthalpy of 52.12 kJ/mol.K and entropy of -0.18 kJ/mol.K. Similarly, negative entropy was observed during stability studies of *T. lanuginosus* lipase in transesterification reaction [27].The unique negative entropy attributed to the compaction of enzyme molecule in the system upon denaturation can only be found in biocatalytic systems [30].

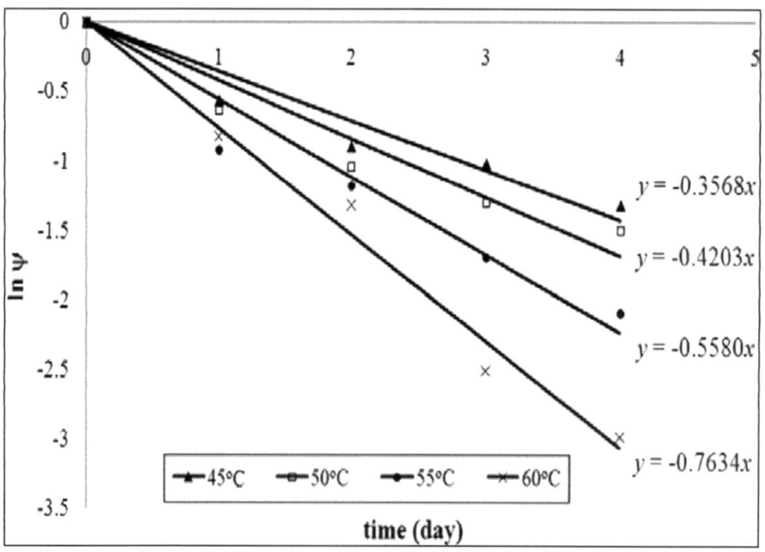

Figure 2. Semi-log plot for determination of denaturation constant

The denaturation constant, k_d at different temperatures can be estimated by plotting ln (v) against time as shown in figure 2, where v represents residual activity. It can be observed that the lipase followed the first-order denaturation pattern. The constant of CALB denaturation progressively increased from 0.012-0.031 h^{-1} with temperature. The deactivation rate was slow at low temperature but it increased rapidly with the increase in temperature. The larger the value of k_d at a particular temperature, the less stable was the enzyme. The same pattern of denaturation constant has been observed for *T.lanuginosus* lipase [27], Aspergillus *oryzae* lipase [28] and *Aspergillus niger* amyloglucosidase [29].

Another important parameter to consider in justifying enzyme stability is the enzyme half-life, $t_{1/2}$. The calculated values for half-life of the lipase were 46.73, 39.60, 29.81 and 21.79 h for 45, 50, 55 and 60 °C , respectively. From the observation, the lipase was considered to have moderate stability. At high temperatures, the irreversible denaturation effect was accelerated and more pronounced. However, the enzyme was strongly deactivated at higher temperatures.

4. Conclusion

Thermal deactivation of *Candida antartica* lipase B occurred due to enzyme distortion disturbing the active conformation of the lipase. Higher value of deactivation energy compared to activation energy implied a predominant effect of deactivation process taking place. The lipase showed good stability at lower range of temperatures but underwent a considerable deactivation at high temperatures.

5. References

[1] De Bono G, Kaye C M, Roland E and Summers A J H 1985 *Am. Heart J* **109** pp 1211-23
[2] Pospisilova M, Kavalirova A and Polasek M *J. Chromatog. A* **1081** pp 72–76
[3] Szymura-Oleksiak J, Walczak M, Bojarski J and Aboul-Enein HY 1999 *Chirality* **11** pp 267–71
[4] Ekelund J, Arkens A V, BrOnnum-Hansen K, Fich K, Olsen L, and Petersen P V 1995 *J. Chromatog. A* **708** 253-61
[5] Ghanem A, Hoenen H and Aboul-Enein H Y 2006 Talanta, **68** pp 602-9
[6] Petersen P V, Ekelund J, Olsen L and Ovesen S V 1997 *J. Chromatog. A* **757** pp 65-71
[7] Kleidernigg O P and Lindner W 2006 *Chromatagrophia* **44** pp 9-10
[8] Wang X Z, Dai L Y, Chen Y Q and Wang J M 2004 *J. Chem. Eng. Chinese U.* **18** pp 729-32
[9] Wang Y, Shen D D, Wang Z Y, Sun L W and Zhu J T 2008 *J. Chem. Eng. Chinese U.* **22** pp 860-5
[10] Long W S, Kamaruddin A and Bhatia S 2005 *J. Membr. Sci.* **247** pp 185–200
[11] Pellissier H 2008 *Tetrahedron* **64** pp 1563–601
[12] Chiou T W, Chang C C, Lai C T and Tai D F 1997 *Bioorg. Med. Chem Lett.* **7** pp 433-6
[13] Quan J, Xu J M, Liu B K, Zheng C Z and Lin X F 2007 *Enzyme Microb. Technol*. **41** pp 756–763
[14] Barbosa O, Ariza C, Ortiz C and Torres R 2010 *Biotechnology* **27** pp 844-50
[15] Barbosa O, Ortiz C, Torres R and Roberto F L 2011 *J. Mol. Catal. B: Enzym.* **71** pp 124–32
[16] Ekelund J, Arkens A V, BrOnnum-Hansen K, Fich K, Olsen L and Petersen P V 1995 *J. Chromatog. A* **708** pp 253-61
[17] Rashid M H and Siddiqui K S 1998 *Process Biochemistry* **33** pp 109-15
[18] Saleem M, Rashid M H, Jabbar A, Perveen R, Khalid A M and Rajoka M I 2005 *Process Biochem.* **40** pp 849-55
[19] Santos A M P, Oliveira M G and Maugeri F 2007 *Bioresour. Technol.* **98 pp** 3142-48
[20] Sun S, Shan L, Liu Y, Song Y and Wang X 2009 *J. Mol. Catal. B: Enzym.* **57** pp 104-8
[21] Suan C L and Sarmidi M R 2004 *J. Mol. Catal. B: Enzym.* **28** pp 111–119
[22] Ong A L, Kamaruddin A H, Bhatia S, Long W S, Lim S T and Kumari R 2006 *Enzyme Microb. Technol*. **39** pp 924–29
[23] Zanin G M and Moraes F F D 1998 *Appl. Biochem. Biotechnol*. **70-72** pp 383-94

[24] Ramos E L, Mata-Gómez M A, Rodríguez-Durán L V, Belmares R E, Rodríguez-Herrera R and Aguilar C N 2011 *Appl. Biochem. Biotechnol* **165** pp 1141-51

[25] Xin J Y, Chen L L, Zhang Y Z, Zhang S and Xia C G 2011 *Food Bioprod Process*, **89** pp 457–62

[26] Baptista R P, Cabral J M S and Melo E P 2000 *Biotechnol. Bioeng.* **70** pp 699–703

[27] Khor GK, Sim JH, Kamaruddin AH and Uzir MH 2010 *Bioresour. Technol.***101** pp 6558–61

[28] Converti A, Del Borghi A, Gandolfi R, Molinari F, Palazzi E, Perego P and Zilli M 2002 *Enzyme Microb. Technol.* **30** pp 216–23

[29] Yu X W and Li Y Q 2006 *J. Mol. Catal. B: Enzym.* **40** pp 44-50

[30] Srinivas R and Panda T 1999 *Bioprocess Engineering* **21 pp** 363-9

IOP Conf. Series: Earth and Environmental Science **36** (2016) 012062 doi:10.1088/1755-1315/36/1/012062

Preliminary studies on the extraction of Glycospanonins in Tongkat Ali extract

S Abirame , K Sivakumar[1], L S Chua[2], M R Sarmidi
[1]Department of Chemical Engineering, Faculty of Engineering, *Universiti Malaysia Sabah, Kota Kinabalu*, Sabah, Malaysia
[2]Institute of Bioproduct Development, Faculty of Chemical Engineering, *Universiti Technology Malaysia*, Skudai, Johor, Malaysia

E-mail: [1]shiva@ums.edu.my, [2]lschua@ibd.utm.my

Abstract. *Eurycoma longifolia,* locally known as Tongkat Ali, is a famous medicinal plant in the family of Simaroubaceae and well known for its aphrodisiac properties from its water extract. The root of *E. longifolia* is used to extract wide range bioactive components of Tongkat Ali. Previous works standardised Tongkat Ali extracts by measuring the concentration of eurycomanone, a quassinoid marker chemical, within the overall extract. There is a newer Malaysian standard that specifies that Tongkat Ali can be standardised to glycosaponin, thus it is desired to determine how extraction parameters such as particle size, extraction temperature, and solvent type affects the glycosaponin content in the extract. The overall study is aimed to determine how the extraction parameters affect the glycosaponin amount in extract. This paper presents the preliminary work where in this study the effect of particle size on overall extract and glycosaponin quantification method development is presented. A reflux extraction method was used to extract Tongkat Ali with a particle size of 0.5 mm, 1.0 mm and 2.0 mm of raw material to study effect of particle size on overall extract. Water and methanol were the two types of solvent used for extraction to study the quantity of glycosaponin.

1. Introduction

Eurycoma longifolia is found in primary and secondary, evergreen and mixed deciduous forest in Burma, Indochina, Thailand, Malaysia, Sumatra, Borneo and the Philippines [1]. *Eurycoma longifolia* Jack (EL) or commercially known as Tongkat Ali in Malaysia, Pasak bumi in Indonesian, Piak and Tung saw in Thailand, and Cay ba binh in Vietnam is a famous medicinal plant in the family of Simaroubaceae. There are three other plant species also known locally as Tongkat Ali which literally means "Stick of a man" and "Ali" of which referring to its aphrodisiac property. The three plant species are *Entomophthora apiculata*, *Polyathia bullata*, and *Goniothalamus sp*. However, EL is the most widely used species for its therapeutic activities [2].

EL is a shrub tree that grows up to 10 metres in height, with long leaves that are green in colour. The leaves are pinnate in shape as the leaflets are arranged in pairs. The flowers of this tree are dioecious, whereas its ovoid-shaped fruits will turn to dark brown colour when they are ripe. A wide

[1,2] To whom any correspondence should be addressed.

IOP Conf. Series: Earth and Environmental Science **36** (2016) 012062 doi:10.1088/1755-1315/36/1/012062

range of chemical compounds have been isolated, especially from the root of EL, which include eurycomanone, eurycomanol, eurycomalactone, canthine-6-one alkaloid, 9-hydroxycanthin-6-one, 14,15!-dihydroxyklaineanone, phenolic components, tannins, quanissoids, and triterpenes [2]. Quassinoids form the major bioactive constituents in this plant and are mainly responsible for its bitter taste [3].

E. longifolia Jack, from the Simaroubaceae family and identified locally as 'Tongkat Ali and Pasakbumi' has been commonly prescribed in traditional medicine as a febrifuge and a remedy for dysentery, glandular swelling, intermittent fever (malaria) [4]. Traditionally, people will boil the Tongkat Ali root chip and drink the decoction [5]. The roots of tongkat ali, often called "Malaysian ginseng", are used as an adaptogen and as a traditional "anti-aging" remedy to help older individuals adapt to the reduced energy, mood, and libido that often comes with age [6].

Studies have provided evidence in support of the claim that a particular compound, eurycomanone from Longjack root can increase testosterone levels in rats. A recent study conducted in humans involved supplementation of 76 of 320 patients suffering from late-onset hypogonadism with a 200 mg serving of a standardized water-soluble extract of Tongkat Ali for 1 month. Data from this study indicate that treatment of late-onset hypogonadism patients with Tongkat Ali extract significantly elevated serum testosterone concentrations. Following treatment, 90.8% of the subjects exhibited normal testosterone values. Another study found that in 14 men who participated in an intensive 8-week weight training program, the seven men who were supplemented with 100 mg/day of aqueous Longjack root extract experienced a 5 % increase in lean mass, a decrease in fat mass, and an average of 1.8 cm increase in arm circumference while those in the placebo group did not. The findings of a later study indicated increased muscle strength and larger quadriceps muscles [7].

Extraction, as the term used in pharmaceutical context, describes the separation of medicinally active components of plant or animal tissues from the inactive or inert components by using selective solvents in standard extraction procedures. Standardization of extraction procedures contributes significantly to the final quality of the herbal drug [8]. The production of standardised extracts, extracts that have a specific measurable amount of bioactive component is an important issue in the production of herbal medicine. Varying processing conditions of extracts can be utilised in order to produce standardised herbal extracts [9]. In this preliminary work, the effects of varying particle size, solvent and temperature on the quantity of glycosaponins and overall extract of Tongkat Ali extracted on a laboratory scale was studied.

A study conducted by Norhidayah et al. [5] use the eurycomanone level to determine the quality of Tongkat Ali product because eurycomanone which is a quassinoid is the major compound in Tongkat Ali. Hence it can be a good indicator on the level of Tongkat Ali in a product. Malaysian Standard (2011) states that eurycomanone level in Tongkat Ali freeze dried extract should be around 0.8-1.5 w/v (%) [5]. However, there is now a Malaysian standard which specifies that Tongkat Ali can be standardised to glycosaponin. According to a study conduct by Talbott et al. [6], hot-water root extracts standardized for known bioactive components such as 1% of eurycomanone, 22% of protein, 30% of polysaccharides, 35% of glycosaponin and verified to be extremely safe at high doses and for long-term consumption [6]. Moreover, there is another study stated in addition to eurycomanone, other active compounds that should be standardized for commercial purposes include the total protein, total polysaccharides and total saponins. Saponins have received much attention recently due to their health benefits such as cholesterol lowering and anticancer properties [10].

Hence, the objective of the overall study is to determine how the extraction parameters affect the glycosaponin extract yield. This paper presents the preliminary work. Therefore, throughout this study we expect to study the effect of particle size on overall extract and also to develop glycosaponin quantification method.

2. Methodology

2.1 Overall Extraction Method

Extraction of Tongkat Ali was studied at a fixed solvent to solid ratio of 10:1 (wt/wt) of Water:Tongkat Ali. 400 ml of deionized water was boiled to 90 °C and maintain at that temperature before adding in the Tongkat Ali by a magnetic stirrer. 40g of 0.5 mm Tongkat Ali particles was added and mixed well. 2ml of the extract was sampled at 5, 10, 15, 30, 60, 90, and 120 minutes. At the end of extraction, the final volume of liquid extract left was recorded. The wet Tongkat Ali particles were weighed. These procedures were repeated by using 1.0 mm and 2.0 mm of Tongkat Ali particles. Each sample of the extract was analysed using UV Vis Spectrophotometry at 350nm wavelength. A calibration curve was plotted.

2.2 Glycosaponins in extracts

2.2.1 Gravimetry method.

The glycosaponin is extracted from sample with methanol and deionized water. Acetone is added drop wisely to completely precipitate the glycosaponin.

Materials

Deionized water, analytical grade of methanol and acetone were used.

Method

1 g of extract was weighed and dissolved in 50 ml of methanol. The mixture was reflux on warm water bath at 50 °C for one hour and filtered. The filtrate was collected. Total methanolic extract was concentrated to 10 ml using rotary evaporator. 10 ml of methanolic extract concentrate was added dropwise to 50 ml of acetone in a pre-weighed beaker in order to precipitate the saponins. The precipitate was dried in oven to constant weight. Amount of glycosaponins was calculated using the formula:

$$\text{Glycosaponins, \%} = \frac{\text{weight of beaker - weight of empty beaker}}{\text{sample weight}} \times 100$$

These procedures were repeated for deionized water extraction.

2.2.2 Spectrophotometric method.

Spectrophotometric method also known as vanillin-sulfuric acid assay is the most commonly selected spectrophotometric method in plant saponin quantification. Extracts were mixed with vanilin (8% w/v) and sulphuric acid (77% w/v) and incubated at 60°C water bath for 10 minutes.

Preparation for spectrophotometric method:

1. Preparation of reagent solution

Vanilin solution with a concentration of 80g/L was prepared by dissolving 0.8g vanilin in 10mL ethanol. 72% sulphuric acid was prepared by dissolving 18.567mL of 97% of sulphuric acid in 6.44mL of water.

2. Preparation of extracts

0.12535 g of desiccated sample is dissolved in 30 mL methanol.

3. Determination of absorption wavelength absorption wavelength was determined based on Cheok *et al.*, 2014 procedure. Blank was prepared by mix 1.0 mL vanillin solution, 1.0 mL of water and 8.0 mL of 72 % sulphuric acid. 1.0 mL of vanilin solution and 1.0 mL extract were mixed, 8.0 mL of 72% sulphuric acid was added. Absorption wavelength was determined using UV visible spectrophotometry.

3. Results and Discussions

3.1 Overall Extract

Figure 1 illustrates the effect of raw material particle sizes on the extraction of eurycomanone. Smaller particles (0.5 mm) shows higher absorbance unit compared to larger particles (2.0 mm) which indicate eurycomanone concentration increases as particle size decreases. This is consistent with mass transfer theory.

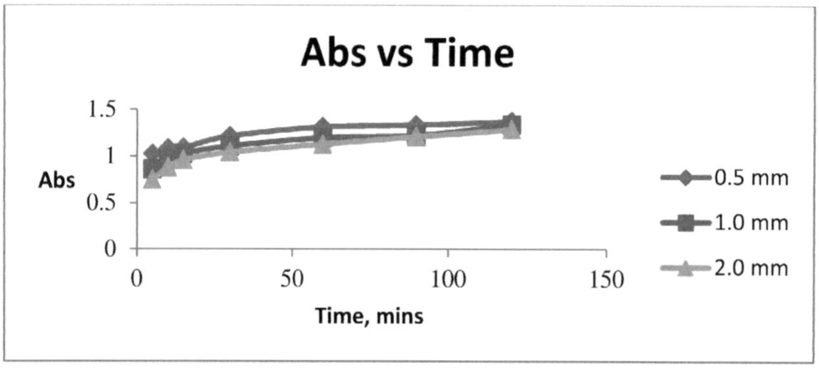

Figure 1. UV-Vis spectrophotometry value for different particle size

Raw material particle size influences the extraction rate by the increase in the total mass transfer area per unit volume when the particle size is reduced. The length of diffusion pathways will also decrease when the particle size decreases, resulting in the increase of mass transfer rate for the process [3]. In this study, three different sizes of raw materials of Tongkat Ali particles were considered. The range of particle size for Tongkat Ali used were from 0.5 mm, 1.0 mm, 2.0 mm.

3.2 Glycosaponin Extract

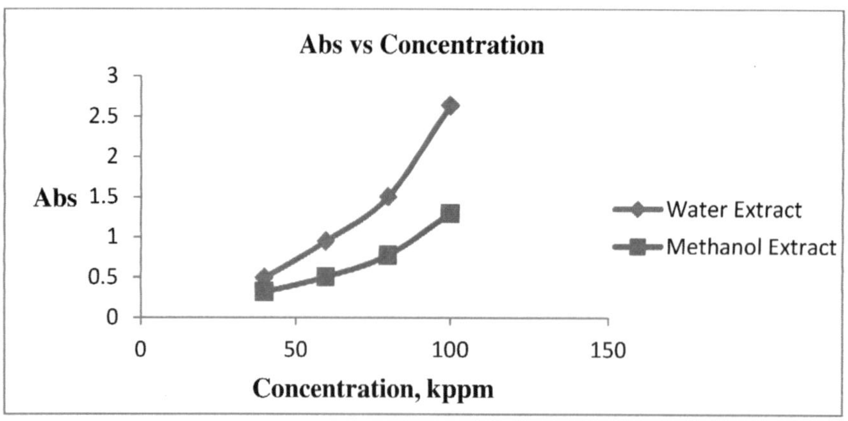

Figure 2. Colorimetric assay for glycosaponins for water and methanol extract.

Figure 2 illustrates the absorbance unit over the concentration of saponins. The spectrophotometric's gives a total saponin value in the quantification of saponins from plant materials. The spectrophotometric technique is simple, fast and inexpensive method to operate. Total saponins also known as vanillin-sulfuric acid assay, is the most commonly selected spectrophotometric method in plant saponin quantification. The basic principle of this method is the reaction of oxidized triterpene saponins with vanillin where sulfuric acid is used as oxidant [11].

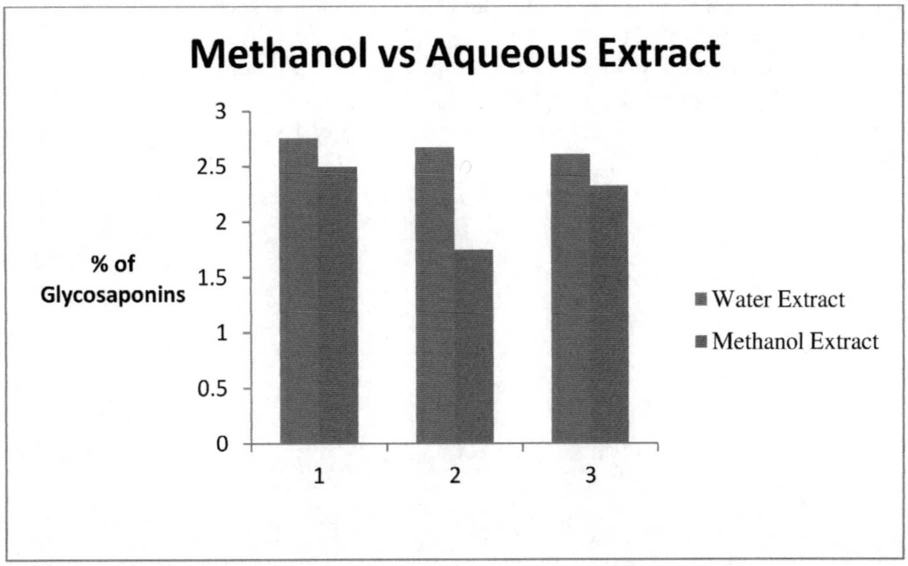

Figure 3. Glycosaponins percentage for water and methanol extract.

Figure 3 illustrates the effect of solvent used on the extraction yields of eurycomanone. From the graph, water extraction yield larger amount of glycosaponins than methanol extraction in every replicate due to their more solubility in water. The presence of water soluble glycone, more soluble in water, plays an important role in the extraction of glycosaponins [12]. Saponins are complex compounds characterized by their structure containing steroidal or triterpernoid aglycone (sapogenin), which linked to one or more oligosaccharide moieties by glycosidic linkage. Saponins are also show strong surface-active properties due to its amphiphilic nature. For example, the presence of a lipid-soluble aglycone and water-soluble sugar chain(s) in their structure [10].

4. Conclusion

The results concluded that smaller particle sizes are better for extraction processes. This is attributed to the larger total surface area presented by smaller particles for extraction. Furthermore, solvent penetration path length decreases when the particle size decreases as the specific surface area of the raw material increases thus influences solubility. There are two solvents for glycosaponins quantification was used, water and methanol. Water is a better solvent for glycosaponin extraction compared to methanol as water extraction produce more glycosaponins than methanolic extraction

5.0 Reference

[1] Faisal G G, Alahmad B E, Mustafa N S, Najmuldeen G F, Althunibat O and Azzubaidi M S 2013 *Journal of Asian Scientific Research,* **3** pp 843-851

[2] Effendy N M, Mohamed N, Muhammad N, Mohamad I N and Shuid A N 2012 *Evidence-Based Complementary and Alternative Medicine,* pp 1-9

[3] Mohamad M, Ali M W, Ripin A and Ahmad A 2012 *Journal Technology (Sciences & Engineering),* **60** pp 51-57
[4] Kavitha N, Noordin R, Chan K L and Sasidharan S 2010 *Journal of Medicinal Plants Research,* **4** pp 2383-2387
[5] Norhidayah A, Vejayan J and Yusoff M M 2015 *Journal of Applied Sciences,* **15** pp 999-1005
[6] Talbott S M, Talbott J A, George A and Pugh M 2013 *Journal of the International Society of Sports Nutrition,* **10**
[7] Gunnels T A and Bloomer R J 2014 *Journal of Plant Biochemistry and Physiology,* **2** pp 1-9
[8] Handa S S, Khanuja S P S, Longo G and Rakesh D D 2008 *International Centre for Science and High Technology,* pp 1-10
[9] Kumaresan S and Sarmidi M R 2003 *Proceedings of International Conference on Chemical and Bioprocess Engineering,* **2** pp 750-754
[10] Harun N H, Aziz A A and Aziz R 2015 *International Conference on Advances in Science, Engineering., Technology & Natural Resources,* pp 27-28
[11] Cheok C Y, Salman H A K and Sulaiman R 2014 *Food Research International,* **59** pp 16-40
[12] Hussain K, Ismail Z, Sadikun A and Ibrahim P 2008 *Natural Product Radiance,* **7** pp 402-408

Comparison of Moringa Oleifera seeds oil characterization produced chemically and mechanically

N A Eman[1,3], and K N S Muhamad[2,3]
Universiti Malaysia Pahang, UMP, Lebuhraya Tun Razak, 26300 Kuantan, Pahang, Malaysia

E-mail: [1,3]eman@ump.edu.my

Abstract. It is established that virtually every part of the *Moringa oleifera* tree (leaves, stem, bark, root, flowers, seeds, and seeds oil) are beneficial in some way with great benefits to human being. The tree is rich in proteins, vitamins, minerals. All *Moringa oleifera* food products have a very high nutritional value. They are eaten directly as food, as supplements, and as seasonings as well as fodder for animals. The purpose of this research is to investigate the effect of seeds particle size on oil extraction using chemical method (solvent extraction). Also, to compare *Moringa oleifera* seeds oil properties which are produced chemically (solvent extraction) and mechanically (mechanical press). The *Moringa oleifera* seeds were grinded, sieved, and the oil was extracted using soxhlet extraction technique with n-Hexane using three different size of sample (2mm, 1mm, and 500μm). The average oil yield was 36.1%, 40.80%, and 41.5% for 2mm, 1mm, and 500μm particle size, respectively. The properties of *Moringa oleifera* seeds oil were: density of 873 kg/m^3, and 880 kg/m^3, kinematic viscosity of 42.2mm^2/s and 9.12mm^2/s for the mechanical and chemical method, respectively. pH, cloud point and pour point were same for oil produced with both methods which is 6, 18°C and 12°C, respectively. For the fatty acids, the oleic acid is present with high percentage of 75.39%, and 73.60% from chemical and mechanical method, respectively. Other fatty acids are present as well in both samples which are (Gadoleic acid, Behenic acid, Palmitic acid) which are with lower percentage of 2.54%, 5.83%, and 5.73%, respectively in chemical method oil, while they present as 2.40% , 6.73%, and 6.04%, respectively in mechanical method oil. In conclusion, the results showed that both methods can produce oil with high quality. *Moringa oleifera* seeds oil appear to be an acceptable good source for oil rich in oleic acid which is equal to olive oil quality, that can be consumed in Malaysia where the olive oil is imported with high prices. In the same time cultivation of *Moringa oleifera* tree is considered to be a new source of income for the country and give more job opportunities.

1. Introduction

Moringa oleifera commonly called drumstick, horseradish or miracle tree is native to southern foothills of the Himalayas and possibly Africa and Middle East and many other places including central and southern America, Mexico and Malaya [1-3]. It belongs to *Moringaceae* family. It is a

[1,3] To whom any correspondence should be addressed.

multi-purpose tree crop with great potentials. The seeds contain 35 - 47% oil. The plant is grown for food and it is an exceptionally nutritious vegetable tree with varieties of potential value [4].
Drumstick grows well in all types of soil, humus rich forest soil being the most ideal. *Moringa oleifera* is one of the nature's gifts to humanity because of its numerous wealth in vitamins and minerals as well as natural anti-oxidants. *Moringa oleifera* is considered as one of the world's most useful trees because almost every part of the tree is useful in one way or another [5].
It was noted that in the tropics, *Moringa oleifera* is used as forage for livestock [6]. As a traditional food plant item in Africa, *Moringa oleifera* has the potential to improve nutrition, boost food security, and foster rural development by enhancing and sustaining rural households as well as supporting sustainable land use and care [1, 2]. All parts of the plant are used in a variety of traditional medicines. The press cake, obtained following oil extraction, is useful as a soil conditioner, and used for water treatment. It is also used as fuel wood source; as an intercrop with other crops and the wood pulp may be used for paper-making. The green pods, fresh and dried leaves are used as vegetable [7].
Due to the high dependence of humans on oil used for both domestic and industrial uses, there is need to look for new source of oil with better method to give a higher yield. This study focuses on the comparison of *Moringa oliefera* seeds oil characterization produced by solvent extraction (chemically) and mechanical press (mechanically). On the other hand, the particle size effect of *Moringa oleifera* seeds on oil yield was studied as well.

2. Material and Method

Dry *Moringa oleifera* seeds were obtained from Kota Bharu, Kelantan, Malaysia. The solvent (n-hexane) was obtained from the Faculty of Chemical and Natural Resources Engineering Laboratory. The mechanical press oil was obtained from MitoMasa Sdn. Bhd., Sabah, Malaysia.

2.1. Pre-treatment

Pre-treatment starts with collection of dry *Moringa oleifera* seeds and continues with the cleaning. In order to get the seeds, the dry pods were removed together with the three papery wings and light wooden shells (figure 1). Cleaning process is crucial because a clean seeds yield clean oil without any impurities. The seeds were crushed using mortar and pestle, and then sieved using sieve shaker and three particle sizes of 2mm, 1mm, and 500μm were collected.

Figure 1. *Moringa oleifera* seeds

2.2. Oil extraction

Oil extraction was carried out using thermal soxhlet apparatus with n-hexane as solvent. Around 20g of the grinded and sieved sample was poured into the thimble. A volume of 200ml of solvent was poured into round bottom flask and heated up to 60-70°C. As the solvent was heated continuously, it starts to evaporate and condenses back into the sample in the thimble. The oil extracted, containing some portion of the solvent was then recycled back to the round bottom flask as it refluxes and the total process of reflux continues until total oil extraction was achieved when the solvent become colourless. A rotary evaporator was used to separate the solvent from the oil, at temperature of 65ºC.

2.3. Oil properties

The oil properties were measured by standard methods of analysis such as: ASTM D445 and Cannon–Fenske viscometers to obtain the kinematic viscosity of oil produced. Density was determined by ASTM D4052. Pour point and Cloud point were measured using K46100 Cloud point and Pour point apparatus by implementing ASTM D2500 standard for Cloud point and ASTM D97 standard for Pour point. pH value was measured using pH meter. Fatty acids composition was determined by gas–liquid chromatography (GLC) [8].

3. Results and Discussion

3.1. Effect of sample size on oil extraction

The percentage of oil yield was calculated using the equation (1):

$$\{(W_o - W_1) / W_o\} \text{ X } 100 \quad (1)$$

Where: W_o and W_1 are the weight of seeds before and after oil extraction, respectively.
The percentage of oil yield was studied by carrying out 3 batches of oil extraction using 200ml of solvent with 20 g of seeds. After each extraction, the residue cake was dried in oven for 1 hour at 50°C and as the drying process ended, the residue cake was weight using electronic weighing machine. From Table 1, it can be concluded that the oil yield was increasing as the particle size decreases which give more exposure of seeds to solvent and more oil was extracted. Figure 2 shows the trend of the oil yield% based on size of samples.

Table 1. Oil yield percentage using different size of samples

Size of sample	Solvent volume (ml)	Weight of seeds (g)	Oil yield %			
			Batch 1	Batch 2	Batch 3	Average
2mm	200	20	36.7	38.1	33.5	36.1
1mm	200	20	39.4	42.6	40.4	40.8
500μm	200	20	42.5	41.3	40.7	41.5

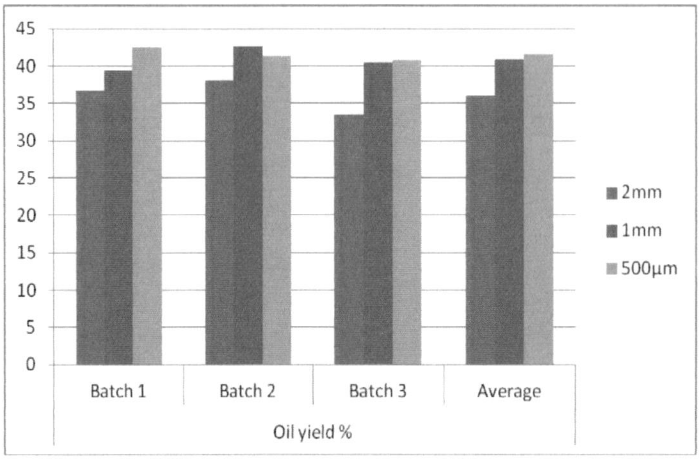

Figure 2. Oil yield percentage using different sample size

3.2. Comparison between oil properties produced by the two methods
Some properties were measured to compare between two methods of oil extraction:

3.2.1. Kinematic viscosity
Kinematic viscosity was measured according to ASTM D445 standard method. The measurement of time of flow for 30ml of oil at 40ºC through calibrated capillary tube number No. 200. The viscosity of oil produced by mechanical press was 12.2 mpas which is more than that from solvent extraction method 9.12 mpas, and this might be due to some suspended particles in the oil produced from mechanical press (Table 2).

3.2.2. Density
The density of oil was measured using Digital Density Analyzer. There was no significant difference in the density of *Moringa oleifera* seeds oil using solvent extraction and mechanical 880 kg/m^3 , and 873 Kg/m3, respectively, as shown in (Table 2).

3.2.3. Cloud point and Pour point
About 50ml of *Moringa oleifera* seeds oil was poured into test tube until it reached the indicated level on the test tube. The sample was placed inside a refrigerator and observed for every one minutes until solid-crystals appeared. Cloud point was measured at the bottom of test jar while pour point was measured at the centre of *Moringa oleifera* seeds oil. The cloud and pour point for *Moringa oleifera* oil was 18ºC, and 12ºC respectively, for both methods. The results are shown in Table 2.

3.2.4. pH
The measurement for pH showed that both methods produced oil of pH value of 6, which means that the extraction method does not affect the pH value of oil produced (Table 2).

Table 2. Oil properties for two methods

Properties	Mechanical Press	Solvent Extraction
Kinematic viscosity (mpas; 40° C)	12.2	9.12
Density (kg/m^3)	873	880
Cloud point (ºC)	18	18
Pour point (ºC)	12	12
pH	6	6

3.2.5Fatty acids

Total fatty acids were more than 90%; the major fatty acid was oleic acid (C18:1) with concentrations of 75.39% (chemically) and 73.60% (mechanically), followed by Gadoleic acid (C20:1) [2.54% (chemically), and 2.40% (mechanically)]. Behenic acid at concentrations of 5.83% (chemically), and 6.73% (mechanically), followed by Palmitic acid (C16:0) with percentage of 5.73% and 6.04% for chemical and mechanical methods, respectively. Traces of C8:0, C16:1, C22:1, and C26:0 were also found in *Moringa oleifera* seeds oil. There was no statistical difference (at the level of 95%) in the fatty acid compositions of the oils produced from the two different ways of extraction. On the basis of the results obtained, the fatty acid composition of *Moringa oleifera* seeds oil showed that it falls in the oleic acid oil category. The *Moringa oleifera* seeds oil had about the same content of C18:1 with olive oil. The fatty acid contents in both methods oil are shown in Table 3. The results obtained here is in agreement with [9-12]. The fatty acid composition of the oil shows that it falls in the category of high-oleic oils and contains a high ratio of monounsaturated to saturated fatty acids. High-oleic oils, although genetically hard to reproduce, recently are gaining importance because of their superior stability and nutritional benefits. A research work showed that high oleic acid *Moringa oleifera* seeds oil to be with high frying quality and stability in comparison with canola oil, soybean, and palm olein oil in frying applications and this will reduce the risk of heart disease and high cholestrol problems [11].

Table 3. Fatty acid percentage in oil produced by two methods.

	Fatty acid %	
Fatty Acid	Solvent extraction	Mechanical press
Oleic acid (C18:1)	75.39	73.6
Gadoleic acid (C20:1)	2.54	2.4
Behenic acid (C22:0)	5.83	6.73
Palmitic acid (C16:0)	5.73	6.04

Fatty acid composition was determined by gas–liquid chromatography (GLC) according to the method used by Tsaknis et al., [8]. Analysis was performed on a Varian 3600 gas chromatograph (Varian, Palo Alto, CA, U.S.A.) equipped with a Supelcowax 10 (Supelco, INC., Supelco Park, Bellefonte, PA) fused silica capillary column 30m X 0.32mm i.d., and 0.25 mm film thickness. The temperature program was 60°C for 10 min and then 2°C/min up to 220°C. Injector and FID temperatures were set at 160 and 280°C, respectively, sample volume was 0.2 mL, the carrier gas was N_2 at a flow rate of 30mL/min, The internal standard used was nonadecanoic acid. Samples were prepared and measured separately in triplicate. The results of GLC are shown in figure 3, and 4, for mechanical press and solvent extraction method, respectively.

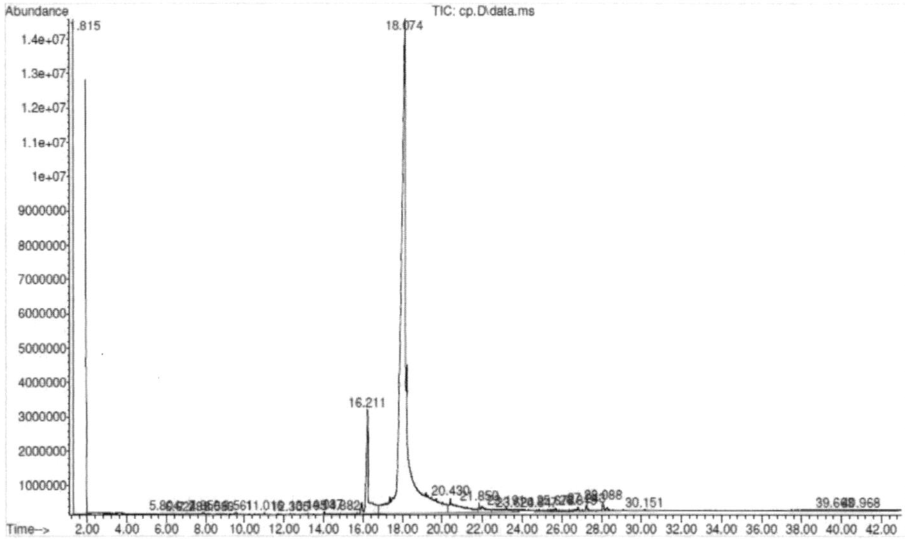

Figure 3. Fatty acids for mechanical method

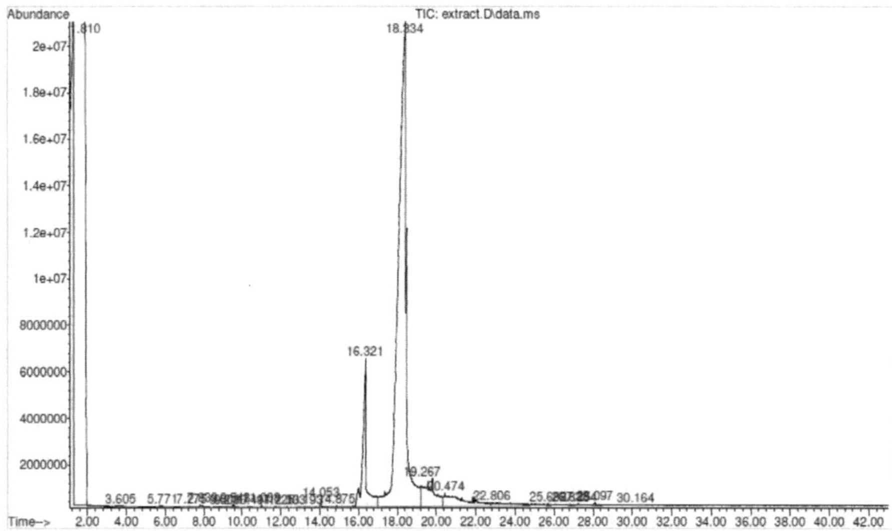

Figure 4. Fatty acids for chemical method

4. Conclusion

Based on results obtained in this research work, it was observed that the particle size of *Moringa oleifera* seeds is affecting the oil yield extracted from seeds. The smaller particles (500 μm) is producing higher oil yield of 41.5% compared to 1mm and 2 mm particle size which produce 40.8%, and 36.1 % , and this might be due to the higher exposure of seeds to the solvent. The oil produced by solvent extraction and mechanical press showed same properties, which means that it is possible to use any available method to get the product with the same quality. The fatty acid analysis showed that *Moringa oleifera* seeds oil has a high Oleic acid percentage which is considered as high quality edible oil with high nutrition value as good as olive oil. The olive oil is imported in Malaysia as a healthy product with high prices. Therefore, it will be of great importance to pay attention to *Moringa oleifera* tree to be cultivated in Malaysia for this purpose and for many other products that can be produced from this tree. In addition, it will generate good income and give more job opportunities in the country.

5. References

[1] Adedokun M O, Oladoye, A O, Olawumi, A T and Laminou, K I 2010 *Obeche Journal* **28** (2) pp 142-146

[2] Odebode, A V, Woes, T O, Oyedeji, O F, Amadi, J O and Sowumi, I L 2010 Proceedings of the 44th Annual Conference of the Agricultural Society of Nigeria (ASN) pp 444-445

[3] Phiri, C 2010. Agriculture and Biology Journal of North America **1** (5) pp 774-777

[4] Ozumba, N A 2011 *Pax Herbal Magazine* **6** pp 7-9

[5] Anwar, F and Rashid, U 2007 *Pakistan Journal of Botany* **39**(5) pp 1443-1453

[6] Dahiru, D, Onubiyi, J A, and Umaru, H A 2006 African Journal of Traditional, Complementary and Alternative Medicines **3**(3) pp 70-75

[7] Folkard G, Sutherland J and Shaw R 2004 Water and Environmental Health at London and Loughborough pp 109 – 112

[8] Tsaknis, J, Spiliotis, V, Lalas, S, Gergis, V, and Dourtoglou, V 1999 *J. Agric. Food Chem.* **47** pp 4495–4499

[9] Lalas S, and Tsaknis J 2002 *Journal of Food Composition and Analysis* **15** pp 65–77

[10] Dalia I, Sánchez M, Jaime L, José A N, Gabriela S Mora, Julia L amd Perfecto P 2015 *Food Chemistry* **187** pp 53–57

[11] Abdulkarim S M, Long K, Lai O M, Muhammad S K S and Ghazali H M 2007 *Food Chemistry* **105** pp 1382-1389

[12] Ricardo A 2011 *Industrial Crops and Products* **33** pp 389-394

Acknowledgment
The authors are grateful to the Research and Innovation Department at Universiti Malaysia Pahang, Pahang, Malaysia, for financial support under grant # RDU 140304.

Effect of moisture content and drying method on the amylose content of rice

J Janaun[1,*], V V Kong[1], C G Toyu[1], N H Kamin[1], P Wolyna[2], J S Lee[2]
[1]Faculty of Engineering, *Universiti Malaysia Sabah*, UMS, Jalan UMS 88400 Kota Kinabalu, Sabah, Malaysia.
[2]Faculty of Food Science and Nutrition, *Universiti Malaysia Sabah*, UMS, Jalan UMS 88400 Kota Kinabalu, Sabah, Malaysia.

E-mail: [1]jidon@ums.edu.my

Abstract. Rice is a primary source to all human beings. It gives energy and supply carbohydrate in human daily life. It is also one of the main sources of employment for people living in the rural area to generate incomes. In this study, determinations of quality of rice dried in an oven and in a Laterally Aerated Moving Bed (LAMB) dryer were performed., tTe drying in an oven was performed with initial moisture content (MC) of 18% reduced to 14% w.b. at temperatures of 35, 45, 55, 65 and 75 0C. The amylose contents of the rice were 20.78, 21.81, 21.53, 21.63 and 22.50 %, respectively. Furthermore, drying of paddy with temperature of 45 0C and initial moisture content of 15 % w.b was performed with different final moisture content of 14, 12, 10 and 8 % w.b using oven drying. The amylose content were 21.79, 20.17, 20.85, 26.35 %, respectively. As for paddy dried in LAMB dryer, it was found that the amylose content were in the intermediate range (20 – 25.50 %). The drying temperature of the LAMB dryer was at 25 0C with 100 to 200 L/min of air flowrate resulted in the intermediate amylose content of the rice despite the final moisture content reduced to 8 % w.b.

1. Introduction

Rice is a primary source to all human beings. It gives energy and supply carbohydrate in human daily life. It has been consumed by larger amount of people on Earth. Besides, it is also one of the main sources of employment for people living in rural area to generate incomes. Rice quality can be divided into two types of characteristics, which are physical and chemical characteristics. Basically, the differences of physical and chemical characteristics might change the consumers' acceptability in terms of few properties such as its colour, shrinkage, nutritional substances, rehydration capability and microstructure [1].

The determination of rice quality is depending on their physical and chemical characteristics. Both of these characteristics may define how well the quality of the rice that has been produced. Based on the physical characteristics, the physical analysis of the rice will be evaluated based on its dimensions, chalkiness, colour and Head Rice Yield (HRY) or known as whole grain percentage. Whereas, in chemical characteristics is to indicate the rice is cooked very well as its texture and

[1] To whom any correspondence should be addressed.

hardness when eaten. The chemical characteristics could be indicated by performing few analysis such as Amylose Content, Gelatinisation Temperature, and Gel Consistency [2].

Harvested paddy usually contains the moisture content in between 20-25% w.b and it must be dried until it reaches the optimum MC around 12-14% for long storage condition[3]. According to the study by Wiset et al. [4], they had stated that the drying conditions of paddy may affect some of the sensory quality and its processing quality that may affects the consumer acceptability. Besides, a study from Cnossen et al. [5] has showed that, the high drying and tempering temperature are correlated to HRY. According to the researcher, the kernels of the rice may fissure during the drying process. If severe drying occurs it may cause the increasing number of kernels that fissured after drying [5].

Therefore, thorough out this study the effectiveness evaluation based on the rice quality of the Laterally Aerated Moving Bed (LAMB) dryer has been performed. This study will focus on the analysis of amylose content for paddy of PR1 variety that will be dried in LAMB Dryer System and using Oven MERMMET in order to understand the eating and cooking quality.

2. Materials and methods

2.1. Materials

Paddy (PR1) from Papar, Sabah was selected as the raw material for this experimental. The harvested paddy with initial moisture content about 18 – 22% w.b was initially determined by using the moisture analyser (Sartorius model MA35) with readability of 0.01%, Commercial Rice (Local Rice), 95 % Ethanol, 1M Sodium Hydroxide, 1M Acetic Acid, Iodine-Potassium Iodide, Amylose from Potato (SIGMA – ALDRICH) and Distilled Water.

2.2. Experimental design

2.2.1. Determination of Amylose Content

The amylose content was determined according to method from Haqim et al. [6]. 0.1 g of grinded rice was added into 100 mL of volumetric flask. 1.0 ml of 95% of Ethanol and 9.0 mL of NaOH was added into the 100 mL volumetric flask. The sample was shaken well and was boiled over the shaking water bath (BS-31) for 10 minutes. After that, the solution was cooled and was made up to 100 mL using distilled water.

Next, 5 mL from the 100 mL sample was pipetted out to another 100 mL of volumetric flask. Then 1 mL of 1 M acetic acid and 2 mL of I-KI was added into the solution and its volume was made up to 100 mL. The solution was shaken and stand for 20 minutes and the absorbance of the solution was determined at wavelength of 620 nm using UV-Vis Spectrophotometer JASCO V-650. The amylose concentration of the sample was determined from the standard curve.

2.2.2. Drying oven method with different temperature (35, 45, 55, 65 and 75 ^{0}C)

25 g of paddy was weighed and dried in an oven (MEMMERT UM 400). The drying process was performed without air ventilation inside the oven. The experimental was carried out with three replications.

The drying of paddy was carried out with three different temperatures at 40^{0}C, 50^{0}C and 60^{0}C in order to achieve 14 % w.b. The moisture loss of paddy was determined by weight loss. Each weight loss of paddy was considered as moisture loss in paddy. The moisture content in w.b was calculated from equation (1).

$$MC = W_0 - W_d \tag{1}$$

where, W_o is the initial weight of the paddy sample (g) and W_d is the weight of the dried sample (g). This method was adopted by Kong and Janaun [7]. Then, the amylose content analysis was determined based on method 2.2.1.

2.2.3. Drying oven method with different moisture content (14%, 12%, 10% and 8% wb)

Drying oven method was performed by repeating method 2.2.2 with the temperature of oven was fixed at 45^{0}C while the moisture content of the paddy was varied from 14, 12, 10 and 8 % w.b. After that, the amylose content analysis was performed by repeating method 2.2.1.

2.2.4. Paddy drying in LAMB Dryer System

Paddy with PR1 variety was dried in the 10 kg LAMB dryer system. Each run of the experiment will have different drying specification as shown in Table 1. The final moisture content of each run of the experiment was different with the different drying conditions.

Table 1: Experimental specification in LAMB dryer

Run	Initial moisture content (% w.b)	Final moisture content (% w.b)	Inner Tube Design Parameters, Z	Temperature (^{0}C)	Flowrate (L/min)
1	23	14.00	22.27	24	100
2	20	8.98	39.60	24	200
3	21	14.00	39.60	24	100
4	15	10.89	39.60	Ambient	200

The quality of the amylose content was determined (method 2.2.1) after the drying process was performed. The sample was mixed vigorously and was taken randomly. On the other hand, run 3 of the experiment, the paddy at 14 % w.b moisture content was taken at three points of the LAMB dryer system. The amylose content of commercial rice which was the local rice from Kota Belud, Sabah with varieties of "TQR, TR8 and TR7" was determined and to be compared with amylose content of rice from LAMB dryer yield.

3. Results and discussion

3.1. Relationship between the Amylose Content in Rice with Different Temperature of 35, 45, 55, 65 and 75 ^{0}C

Based on this study, the relationship between amylose concentration in rice with different drying temperatures of 35, 45, 55, 65 and 75^{0}C were determined at fixed moisture content of 14 % w.b. The drying process was performed in oven drying. The temperatures were selected in order to observe the differences of the amylose concentration in rice at 14 % w.b. Based on Figure 1, the trend of the graph shows that the amylose concentration was increased rapidly as the drying temperature increases.

As been reported reported by Zheng et al. [8], high drying temperature and fastest drying rate may lead to breakage of the paddy kernel and decomposition of the amylose content. Figure 1 shows that, at temperature of 75^{0}C the amylose concentration about 22.50% compare to the other temperatures. This can be explained by relating with Zheng et al. [8] research. The highest amylose concentration was caused by the breakage of the kernel which may cause crack on the surface of the endosperm. This can cause the amylose to leak as the moisture gradient occurs from the endosperm to the surrounding environment. While, at the temperature of 35^{0}C, the amylose concentration was 20.78%. This indicates that, it has less influence on amylose concentration in rice. When relate with the eating quality, which has been reported by Zheng et al. [8], it shows that the drying condition below 45^{0}C will not affect much on the taste quality of the rice.

According to Zheng and Lan [9], paddy drying temperature at below 45 ^{0}C will not appear any significant changes on the taste of the rice, which correlate to the amylose content. Based on Figure 1, the drying temperatures at 45, 55 and 65 ^{0}C does not shows any significant changes on the amylose concentration in rice as amylose concentration was 21.81, 21.53 and 21.63%, respectively. These values demonstrate the amylose concentration were in the intermediate region.

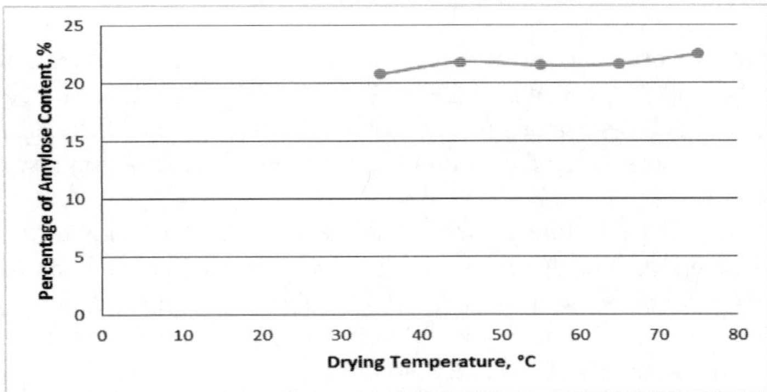

Figure 1. Amylose Concentration, % against Drying Temperature, ^{0}C.

3.2. Relationship between Amylose Concentration in Rice with different Moisture Content at Temperature of 45 ^{0}C in Oven

This study was performed to understand the effect of amylose content at temperature of 45 ^{0}C with different moisture content of 14, 12, 10 and 8 % w.b. The result obtained was tabulated in Figure 2. At 14 to 12 % w.b of moisture content, the amylose concentration was decrease from 21.79 to 20.17% respectively. However, when the drying was continued until the moisture content of the paddy reaches 10 to 8 % w.b, the amylose concentration started to increase gradually with 20.85 to 26.35% respectively.

By relating the concentration of amylose content in rice with the eating quality, it shows that the higher the amylose concentration in rice the lower the eating quality. Based on Zheng and Lan [9], they had stated that once the moisture content of rice has dropped to below 15 % w.b the taste values will decrease as well. Thus, based on Figure 2, 14 to 10 % of moisture content was acceptable to maintain the quality of the dried rice.

At 14 % w.b of moisture content, it was suitable for the milling process at the same time it can preserve the cooking smell and eating quality of the cooked rice. At this moisture content, it shows the condition of the paddy was better so, the weight of the paddy and the milling yield will not decrease. Besides, lower moisture content to 13 % or less was suitable for storage purpose.

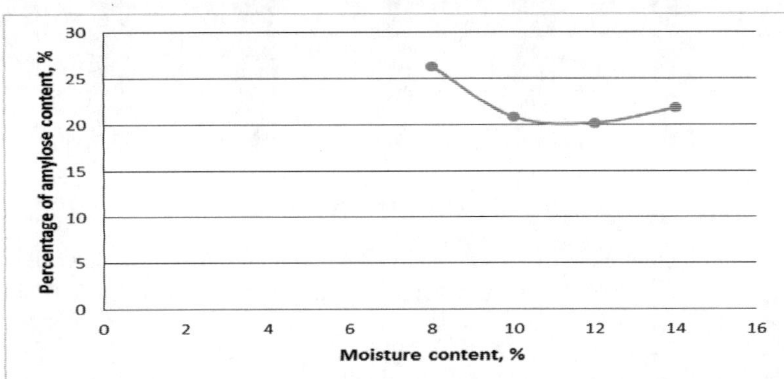

Figure 2. Amylose Concentration, % against Moisture Content Percentage in w.b, %.

3.3. Quality of rice dried in LAMB dryer system

Based on Figure 3, the amylose concentration for Run 1, 2, 3 and 4 was different. The result analysed was 20.67, 24.67, 23.17, and 25.50 %, respectively. The variation of the result obtained was likely due of the inconsistent initial and final moisture content of the paddy with different inner tube design. Therefore, the overall result has stated that Run 4 has the highest value of amylose content, follow by Run 2, 3 and lastly Run 1.

Run 1 with the initial moisture content of 23 % w.b was dried down to 14 % w.b of final moisture content was found that this drying condition of 25 ^{0}C with 100 L/min has the lowest amylose concentration of 20.67% compared to the others. It also similar as Run 3 with initial moisture content of 21 % w.b was dried down to 14 % w.b at 25 ^{0}C with 100 l/min shows that the amylose content was also in the intermediate range of 23.17%. By noticing these two Runs, it shows that the drying conditions were the same, however, the differences was the inner tube design which required for the air distribution during the drying process for the LAMB dryer system. Based on the result obtained, Run 1 has lower amylose concentration compared to Run 3 even though the drying specifications were the same. However, based on the same drying conditions of the LAMB dryer systems, it was found that the inner tube design was also one of the main causes that influence the quality of the rice. It has proved that, the design of the inner tube also may affect the amylose concentration of the rice.

Furthermore, for Run 2 with the initial moisture content of 23 % w.b was dried down to 8.98 % w.b for drying temperature of 25 ^{0}C with 200 L/min and inner tube design of Z = 39.60. The amylose concentration was also in the intermediate range of 24.67%. Thus, the result obtained not only because of the drying temperature, air flowrate and initial moisture content that affect the amylose concentration. But the inner tube design also shows the influence on the amylose concentration. Therefore, the designation of the inner tube may be one of the parameters affecting the amylose concentration. Therefore, the designation of the inner tube may be one of the parameters affecting the amylose concentration.

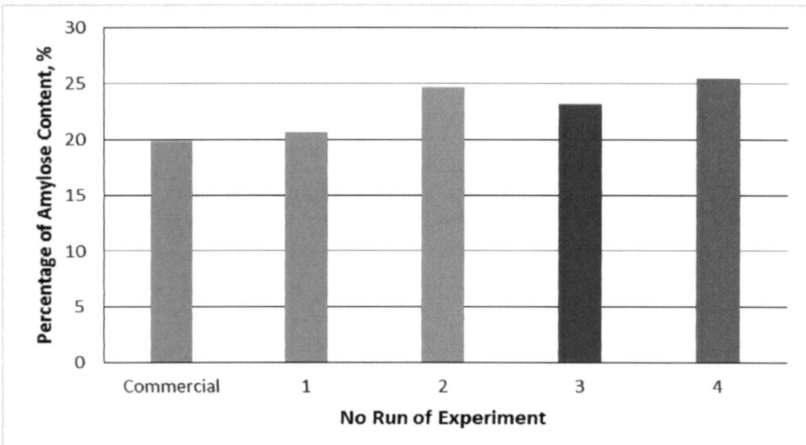

Figure 3. Percentage of Amylose Content for commercial rice and LAMB dryer (experiment run no 1, 2, 3 &4).

3.4. Comparison between Amylose Content of Rice (PR1) with Commercial Rice (Local Rice)

The comparison of the amylose content of rice (PR1) with commercial rice which was the local rice from Kota Belud, Sabah with varieties of "TQR, TR8 and TR7" is summarized in Figure 3.

The amylose concentration has been reported by Jin et al. [10] high amylose content more than 100 mg/L (>25 %) it shows the properties of the rice dry and fluffy after cooked and suddenly it become hard after cooling, while intermediate amylose content in between 80 – 100 mg/L (20 – 25 %)

shows the rice will cooked soft but it is not sticky and lastly, for low amylose content 60 – 80 mg/L (15-20 %) the rice will cooked softly and sticky.

Based on the result obtained, the commercial rice has the lowest amylose content with 19.83%, compared with PR1 Rice samples. The commercial rice sample represents as a control or standard for the PR1 rice variety. According to the rice quality standard, the commercial rice is in the range of low amylose content which will have soft and sticky texture after cooking. As for paddy drying in the LAMB dryer system, the amylose content of PR1 rice samples in experiment no 1, 2, 3 and 4 were in intermediate level with 20.67%, 24.67%, 23.17% and 25.50%, respectively.

The result obtained shows that different rice variety have different amylose content. Therefore, the rice amylose content of rice will have three classifications of low, intermediate and high. Other than that, the commercial rice has three different varieties "TQR, TR8 and TR7" therefore, it has different amylose content. While in this studies, the quality of rice respect to drying process was focused on specific variety of PR1.

4. Conclusion

In this study, the amylose content of the rice dried in an oven and a LAMB dryer were determined. By establish the relationship between amylose content with different moisture content at different drying temperature. This relationship is important to understand the optimal drying temperature for better quality of rice in terms of amylose content.

The amylose content of rice from dried paddy PR1 in LAMB dryer with different inner tube designs were in the intermediate range with the value of 20 – 25.50%, where mostly preferred by the consumers. Hence, it can be concluded that LAMB dryer is a good paddy dryer, and it has the potential to be used in a large scale paddy drying.

References

[1] Luangmalawat P, Prachayawarakorn S, Nathakaranakule A and Soponronnarit S 2008 *LWT - Food Sci. Technol.* **41** pp 716–23

[2] Mujumdar A S 2014 *Handbook of Industrial Drying, Fourth Edition* (CRC Press)

[3] Tirawanichakul Y, Prachayawarakorn S, Varanyanond W and Soponronnarit S 2004 *J. Food Eng.* **64** pp 405–15

[4] Wiset L, Srzednicki G, Driscoll R H, Nimmuntavin C and Siwapornrak P 2001 *Agric. Eng. Int. CIGR J.*

[5] Cnossen A G, Jiménez M J and Siebenmorgen T J 2003 *J. Food Eng.* **59** pp 61–9

[6] Haqim I, Aminah A and Anizan I 2013 *Sains Malays.* **42 pp** 1641–6

[7] Kong V . and Janaun J 2014 Effect-of-temperature-on-drying-of-paddy-in-oven.pdf

[8] Zheng X-Z, Liu C-H, Chen Z-Y, Ding N-Y and Jin C-J 2011 *Dry. Technol.* **29** pp 1297–305

[9] Zheng X and Lan Y 2007 *Agric. Eng. Int. CIGR J.*

[10] Jin L, Lu Y, Shao Y, Zhang G, Xiao P, Shen S, Corke H and Bao J 2010 *J. Cereal Sci.* **51** pp 159–64

A review on intelligent sensory modelling

H J Tham[1*], S Y Tang[1], K T K Teo[2], S P Loh[3]
[1] Chemical Engineering Program, Faculty of Engineering, *Universiti Malaysia Sabah*, Jalan UMS 88400 Kota Kinabalu, Sabah, Malaysia
[2] Electrical & Electronic Engineering Program, Faculty of Engineering, *Universiti Malaysia Sabah*, Jalan UMS 88400 Kota Kinabalu, Sabah, Malaysia
[3] Department of Nutrition and Dietetics, Faculty of Medical and Health Sciences, *Universiti Putra Malaysia*, 43400, Serdang, Selangor, Malaysia

E-mail: [1*]hjtham@ums.edu.my, [1]tangszuyou@gmail.com, [2]kenteo@ums.edu.my, [3]sploh@upm.edu.my

Abstract. Sensory evaluation plays an important role in the quality control of food productions. Sensory data obtained through sensory evaluation are generally subjective, vague and uncertain. Classically, factorial multivariate methods such as Principle Component Analysis (PCA), Partial Least Square (PLS) method, Multiple Regression (MLR) method and Response Surface Method (RSM) are the common tools used to analyse sensory data. These methods can model some of the sensory data but may not be robust enough to analyse non-linear data. In these situations, intelligent modelling techniques such as Fuzzy Logic and Artificial neural network (ANNs) emerged to solve the vagueness and uncertainty of sensory data. This paper outlines literature of intelligent sensory modelling on sensory data analysis.

1. Introduction

Sensory evaluation has been defined as a scientific method used to evoke, measure, analyse, and interpret responses to products as perceived via the senses of sight, smell, touch, taste and hearing [1]. It plays an important role in the quality control of various food productions. Quality of foods encompass many aspects, such as nutritional value, storage stability, safety and sensory quality. The ultimate criterion of sensory evaluation is how customers respond to the food product. Generally, sensory evaluation involved the application of well-established experimental design together with statistical analysis of sensory data.

Sensory attributes on sensory evaluation normally include aroma, appearance, flavour, mouthfeel and taste. These sensory data are uncertain and vague as the response of consumer are relatively subjective. Their responses are greatly influenced by their eating habits, testing ability, testing environment and even their culture [2]. There are three types of sensory testing methods used, which are discrimination test, descriptive testing and affective testing. Discrimination test determines the sensory difference between each of the samples; descriptive test determines the nature or properties of the sensory difference, descriptive test sometimes relates to the sweetness, crispness, hardness of the

[1*] To whom any correspondence should be addressed.

food samples; affective test determines the intensity of preferences of the customers about the food [3,4]. In sensory evaluation, the common scale for acceptance testing is the 9-point hedonic scale that was developed by Peryam and Pilgrim. The sensory panel is required to give opinion in numerical values from 1 to 9 to different types of sensory attributes. The number "1" of the scale normally elucidates 'extremely dislike' and the number "9" is 'extremely like' [1]. However, the hedonic scale is not considered the comparison of the evaluator preference towards the different quality attributes, for examples, in judging the sensory quality of the pumpkin, some of the evaluators may feel that hardness of the pumpkin is more important than color or taste of the pumpkin [5].

For sensory data obtained from different groups of panel, the corresponding mathematical models and the data processing procedures are different. This is because the reliability of the sensory data will influence the final output of the products from the mathematical modelling. The sensory evaluation is based on the knowledge acquired in a sensorial way by the panel of experts that participate in the evaluation process. A suitable mathematical formulation is quite difficult in this type of problems because human perceptions are subjective and not objective, therefore the assessments provided by the individuals are vague and uncertain. Furthermore, the quality of food is decisively determined by its taste which is extremely difficult to model or sense as it is highly influenced by cultural and personal perceptions [6]. Initially, classical computational techniques used in sensory evaluation were based on statistics and factorial analysis (Principal Component Analysis, Correspondence Analysis, Generalized Canonical Analysis and etc). However these methods are not efficient in solving sensory evaluation problems because uncertainties in this type of problems have a non-probabilistic character since they are related to imprecision and vagueness of meanings [7]. These methods as mentioned earlier will cause important data lost because (1) there often exist non-linear relations in sensory evaluation field, (2) they compute with great number of numerical data only, and (3) classical method cannot interpret physical result precisely. Hence, new methods based on intelligent methods such as fuzzy logic, neural networks, data aggregation, classification, clustering and etc are needed for solving the vagueness related to this field. There are several reason to explain why intelligent technique is more efficient: (1) they can compute numerical and linguistic data together while dealing with the uncertain and vague condition, (2) they can consider the nonlinear relationship in the analysis and (3) they can interpret the physical result in more precise way [8].

Although the physical and chemical properties associated with sensory attributes can be measured by some instrument such as textures analyzers, gas chromatography (GC), high-performance liquid chromatography (HPLC) and etc, but these kind of instrument can only measure the intensity of the sensory attributes, but do not provide information about consumer's responses or preferences towards the food product. Nevertheless, they are commonly used in the food industry because their results are more reliable and more objective than those from sensory evaluations. There have only been a few attempts to develop mathematical functions to predict sensory data from instrumental data by correlating them with multivariate regressions and artificial neural networks (ANN). The ANN is a highly simplified mathematical description tool and it can be interpreted as an input output non-linear device [9]. ANN is found to have higher predictive capability than RSM model with limited number of experiments [10]. Another intelligent modeling tools called fuzzy logic is used to solve fuzzy variables with uncertainty and imprecision. The following section will further discuss the introduction and application of intelligent sensory modelling.

2. Fuzzy Logic

Fuzzy logic can be used to convert linguistic data to numerical values. Linguistic data obtained from sensory test are naturally fuzzy and uncertain. The Fuzzy set theory was first proposed in 1965 by Zadeh [11]. Zadeh created this theory that could cover fuzzy variables with uncertainty and imprecision, like linguistic expressions and their relation. Recently, many successful cases of fuzzy set applications can be found in the existing literature, such as pattern recognition, classification, chemical plants, automobiles, fermentation and etc [11]. In the field of food science, several research papers highlighted the application of fuzzy logic on sensory evaluation and food process control [8,11–13].

However, the understanding and applications of fuzzy theory are still scarce in the field of sensory data. The experimental element influenced more than theoretical element in the field of food sensory science. Hence, the application of fuzzy theory has great potential in this field, as sensory data provided are either crisp data or approximate data.

2.1 Theory

Fuzzy logic uses Fuzzy Sets to convert the linguistic data to mathematical data. Fuzzy sets are denoted into several ways. The first is with discrete elements such as the following:

$$A = \{(\mu_{A,i}, X_i)\}, \nabla_X \in X (i = 1,2,3, \dots, n) \quad (1)$$
$$A = \mu_{A,1}/X_1 + \mu_{A,2}/X_2 + \cdots + \mu_{A,n}/X_n \ (i = 1,2,3, \dots, n) \quad (2)$$

where X_i and $\mu_{A,i}$ are the set elements and their membership degrees, respectively, and in the second equation, plus (+) does not mean arithmetic addition. Second, fuzzy sets are denoted with continuous element as follows:

$$A = \{(\mu_A(X), X)\}, \nabla_X \epsilon X \quad (3)$$

where $\mu_A(X)$ is the membership function and X is the element in continuous variables. Membership functions are typically given in three types: triangle, trapezoid, and bell shapes (Figure 1). Membership degrees ranges from 1 to 0.

When designating fuzzy set partitions, the shape and location of the partitions should be determined properly. Fuzzy sets with wide partitions imply more fuzziness, or uncertainty, and those with narrow partitions have a lesser degree of fuzziness. A fuzzy set with a membership degree of 1 at one element only and 0 at the other elements is equivalent to a deterministic value, and that with a membership degree of 1 at all the element becomes a crisp set [11].

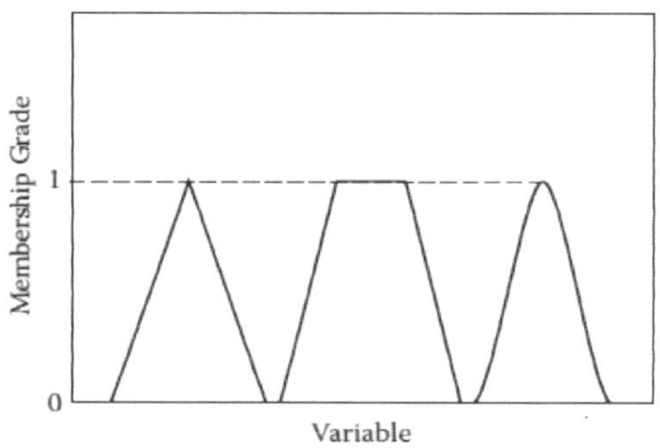

Figure 1. Pattern of membership functions for fuzzy set.

2.2 Application of Fuzzy Sets on Sensory Data

The intensity or preference of attributes obtained from sensory evaluation can be converted to fuzzy sets in two ways. One is to change the sensory data in numerical form to fuzzy values. Another one is converting sensory data in linguistic or hedonic form to fuzzy values.

Fuzzy values can be defined in the following when sensory data are obtained as numerical (intensity scores):

Fuzzy value = {(membership degree)$_1$/(intensity level)$_1$, (membership degree)$_2$/(intensity level)$_2$,...(membership degree)$_n$/(intensity level)$_n$} (4)

IOP Conf. Series: Earth and Environmental Science **36** (2016) 012065 doi:10.1088/1755-1315/36/1/012065

where the intensity level is a sensory score, and its membership degree is the proportion of panelist votes to the total, who rated at the corresponding score.
If the fuzzy values is given in linguistic form, then the panelist vote for the linguistic rating. The vote of the linguistic is correlated as to the membership degree of the linguistic rating. The linguistic levels not just about sensory attributes rating, but also the hedonic or preferences, such as satisfactory, medium, fair and not satisfactory.

2.3 Application of Fuzzy Logic in Sensory Evaluation
Since sensory evaluation are often in linguistic form, fuzzy logic is an important tool whereby vague and imprecise data can be analyzed and concluded according to acceptance, rejection, ranking and the preference of the sensory attributes [5,14–18]. A number of past researches had used Fuzzy logic to analyze sensory data. Perrot et al. [19] provide an overview of the application of fuzzy concepts to the control of the product quality in the food industry over past 10 years. Kavdir and Guyer [20] have also used this approach for apple classification. Martinez [7] discussed the sensory evaluation based on linguistic decision analysis. This analysis shows advantages of using the linguistic 2-tuple decision analysis applied to the sensory evaluation because it provides a guideline for modeling and computing to deal with uncertain and vague information, in a consistent mathematical way. Mukhopadhyay et al. [21] used fuzzy logic to analyze the sensory data responses given by 40 trained panelists regarding the taste, color, aroma and mouthfeel of the chhana podo. The result showed that Fuzzy Logic can be used to validate optimization results obtained from genetic algorithm (GA) (constrained optimization technique). Singh et al. [22] also used this fuzzy approach to analyze the sensory score from 102 panelist that were trained by discussing the definition of quality attributes selected for sensory evaluation, explaining the score sheet and method of scoring. Judges were instructed to tick the respective fuzzy scale factor for each of the quality attributes of the sample after evaluating the samples. The samples were rated as "Not satisfactory," "Fair," "Medium," "Good" and "Excellent." The set of observations were analyzed using Fuzzy analysis of sensory scores. It was also successfully applied for bread prepared from millet-based composite flour [22]. The combination of triangular membership function in Fuzzy logic and five point sensory scale was also implemented in mango drinks [14], dahi powder [23] and instant green tea powder [17]. Uprit and Mishra [18] also calculated normalized fuzzy membership function for the Soy fortified paneer (SFP) samples prepared from blends containing different proportions of buffalo milk of varying fat content and soy milk for their ranking. Normalized fuzzy membership function also has been adopted on the selected coffee products to determine the best quality attributes and rank the products [15]. Debjani et al. [5] also investigated the preference of quality attributes, intensity of quality attributes and overall quality attributes of tea liquor by using Fuzzy logic where the intensity and preferences of sensory scales were represented by triangular membership function. Kaushik et al. [24] applied fuzzy logic techniques to analyze the sensory attributes of mango pulp and litchi juice by using triangular membership distributions function for 5-point hedonic scale.

3. Artificial Neural Network (ANN)
ANN is first introduced in 1943 by Warren McCulloch [11]. ANN is very useful in correlating process variables with non-linear relationships, which is commonly found in experimental data from food processing. ANN has been used in food areas associated with sensory test, sensory quality-based food process control, and control set points of processes [12]. ANN also has been applied in food sciences such as classification of food products [25–27], monitoring authenticity of low-fat yogurt [28], and honey [29], prediction of optimum conditions of frozen cooked rice associated with sensory evaluation [30], evaluation of extra virgin olive oil stability [31], pattern recognition [32], electric nose, machine vision [33], and etc. Most of the application of ANN is not directly correlated sensory attributes with process variables, but using sensory scores with relevant data.

Hybrid type model combining Fuzzy Logic and ANN, which is called Fuzzy Neural Network (FNN) has also been developed. It is reported that FNN is a more powerful approach than ANN itself [34]. FNN has been compared with RSM and MRA (Multiple Regression Analysis) and FNN shows a great accuracy for estimation sensory evaluation scores [34,35].

3.1 Theory

ANN is a network that consists of many neurons that carry non-linear function, and different unit that contains different function are weighted, connected together and combined to produce overall output. ANN consists of input layer, hidden layer and output layer as shown in Figure 2. Before applying ANN, the input and output variables should be established. Next, the type of ANN should be chosen, such as radial basis function, Kohonen self-organizing, recurrent, modular and so on. Input variables are inserted in the first layer, and the calculation steps proceed layer by layer until the last layer. The signals propagate from input to output layer. The hidden layer may affect the accuracy, hence, it is necessary to reasonably determine the hidden layer when they are more correlations between variables [11].

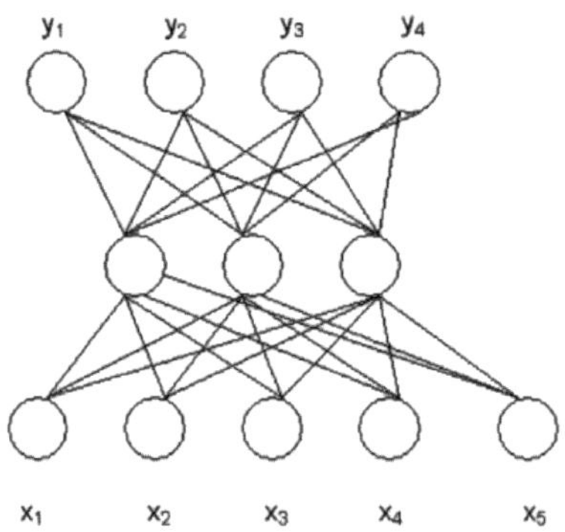

Figure 2. Schematic representation of a multilayer feed-forward ANN.

Each input layer is multiplied with their weight, and the product are summed. The sum deducted by Threshold, which is net, is the input to the transfer function and subsequently produce output value from the neuron. The function can be non-linear function such as sigmoid function, step function and etc. It is necessary to optimize the weight in order to obtain better prediction. Consequently, the training or learning process adjusts the weights by altering them until the errors between outputs given for the training is minimized. This can be done by applying Back-Propagation ANN Algorithm and the weight is changed iteratively [26]. The errors are propagated backwards from final output to input layer.

3.2 Application of ANN on Sensory Data

ANNs have been used for many purposes in food technology especially food classification such as rice [32], cheese [36], coffee [37], eggplant [38], honey [39], tea [40], wine [41,42], fruit beverages[43,44], beer [45] and other alcoholic drinks [46,47]. Bahramparvar et al. [48] used ANN method to predict the total acceptance of ice-cream. The sensory attributes (appearance, flavor, body and texture, coldness, firmness, viscosity, smoothness and liquefying rate) were used as inputs and independent total acceptance was the output of the ANN. Raju Krishnamurthy [49] compared ANN modelling techniques with conventional statistical method such as Multiple Regression Analysis (MRA) to predict consumer liking scores from trained sensory panel scores regarding the 10 market samples of beef bouillon products. Cruz et al. [50] used Levenberg–Marquardt training algorithm with the hyperbolic function as the activation function in the hidden layer and the linear function in the output layer to identify the acceptability and purchase intent of probiotic yoghurt. Tominaga et al. [34]

utilized integration of Fuzzy set theory with ANN to predict the most suitable blending ratio of coffee and its shows a positive result.

4. Conclusions

It is very important to analyze the sensory data obtained from sensory evaluation of food product for marketing purposes, for product control and quality control purposes in food industry. However, there are some difficulties in analyzing these sensory data which are vague and uncertain. Furthermore, human sensory response is relatively subjective and this causes the objective quantification a more difficult task.

Statistical methods for sensory evaluation have been implemented and widely used. However, this method is only suitable to analyse sensory data that have numerical value but not linguistic data. Hence, other methods such as Fuzzy Logic are developed to directly quantify the linguistic data and hence more convenient to analyse the primary data from sensory evaluation. ANN is also one of the intelligent techniques to predict the non-linear relationships between process variables and the sensory data. FNN is suggested to be applied to combine the advantage of Fuzzy Logic and ANN together. FNN is potentially more effective because Fuzzy Logic can solve fuzzy data while ANN can correlate non-linear relationships from sensory test.

5. Reference

[1] Stone H and Sidel J L 2004 *Sensory Evaluation Practices* (San Diego: Academic Press)
[2] Meilgaard M C, Carr B T and Civille G V 2006 *Sensory Evaluation Techniques, Fourth Edition* (CRC Press)
[3] Lawless H T and Heymann H 2010 *Sensory Evaluation of Food - Principles and Practices* pp 473–478
[4] Kemp S, Hollowood T and Hort J 2011 *Sensory Evaluation: A Practical Handbook* (John Wiley & Sons)
[5] Debjani C, Das S and Das H 2013 *J. Food Sci. Technol.* **50** pp 1088–1096
[6] Birle S, Hussein M A and Becker T 2013 *Food Control* **29** pp 254–269
[7] Mart L 2007 *Int. J. Approx. Reason.* **44** pp 148–164
[8] Zeng X, Ruan D and Koehl L 2008 Intelligent sensory evaluation : Concepts , implementations , and applications **77** pp 443–452
[9] Zhang J and Chen Y 1997 *Sens. Rev.* **17** pp 150–158
[10] Mourabet M, Rhilassi A El, Bennani-Ziatni M and Taitai A 2014 *Univers. J. Appl. Math.* **2** pp 84–91
[11] Lee S J 2011 Sensory Evaluation Using Artificial Intelligence *Taylor Fr. Group, LLC* **3** 55–78
[12] Kupongsak S and Tan J 2006 *Fuzzy Sets Syst.* **157** pp 1169–1178
[13] Lee S J, Hong C-G, Han T-S, Kang J-Y and An Kwon Y 2002 *Food Control* **13** pp 301–306
[14] Jaya S and Das H 2003 *J. Sens. Stud.* **18** pp 163–176
[15] Lazim M A and Suriani M 2009 *World Acad. Sci. Eng. Technol.* **3** pp 717–720
[16] Parameswara Rao K and Das H *J. Food Sci. Technol.* **40** pp 576–581
[17] Sinija V R and Mishra H N 2008 *Food Bioprocess Technol.* **4** pp 408–416
[18] Uprit S and Mishra H N 2002 *J. Sens. Stud.* **17** pp 163–176
[19] Perrot N, Ioannou I, Allais I, Curt C, Hossenlopp J and Trystram G 2006 *Fuzzy Sets Syst.* **157** pp 1145–1154
[20] Kavdir I and Guyer D E 2003 *Turkish J. Agric. For.* **27** pp 375–382
[21] Mukhopadhyay S, Majumdar G C, Goswami T K and Mishra H N 2013 F *Food Sci. Technol.* **53** pp 204–210
[22] Singh K P, Mishra A and Mishra H N 2012 *Food Sci. Technol.* **48** pp 276–282
[23] Routray W and Mishra H N 2012 *Food Processing and Preser.* **36** pp 1–10

[24] Kaushik N, Gondi A R, Rana R and Rao P S 2015 *Innov. Food Sci. Emerg. Technol.* **32** pp 70–78
[25] Marini F, Magrì A L and Bucci R 2007 *Chemometrics and Intelligent Lab. Syst.* **88** pp 118–124
[26] Marini F 2009 *Anal. Chim. Acta* **635** pp 121–131
[27] Swider B G- 2011 *Anal. Chim. Acta* **705** pp 283–291
[28] Cruz A G, Walter E H M, Cadena R S, Faria J A F, Bolini H M A and Fileti A M F 2009 *J. Dairy Sci.* **92** pp 4797–4804
[29] Cajka T, Hajslova J, Pudil F and Riddellova K 2009 *J. Chromatogr. A* **1216** pp 1458–1162
[30] Kono S, Kawamura I, Yamagami S, Araki T and Sagara Y 2015 *Int. J. Refrig.* **56** pp 165–172
[31] Silva S F, Anjos C A R, Cavalcanti R N and Celeghini R M dos S 2015 *Food Chem.* **179** pp 35–43
[32] Marini F, Balestrieri F, Bucci R, Magrì A L and Marini D 2003 *Microchem. J.* **74** pp 239–248
[33] Huang Y, Kangas L J and Rasco B A 2007 *Crit. Rev. Food Sci. Nutr.* **47** pp 113–126
[34] Tominaga O, Ito F, Hanai T, Honda H and Kobayashi T 2002 S *J. Food Sci.* **67** pp 363–368
[35] Hanai T, Kakamu A, Honda H, Furuhashi T, Uchikawa Y and Kobayashi T 1996 *Trans. Soc. Instrum. Control Eng.* **32** pp 1113–1120
[36] Cevoli C, Cerretani L, Gori A, Caboni M F, Gallina Toschi T and Fabbri A 2011 *Food Chem.* **129** pp 1315–1319
[37] Bona E, da Silva R S dos S F, Borsato D and Bassoli D G 2011 *Int. J. Food Eng.* **7** pp 1556–3758
[38] Saito Y, Hatanaka T, Uosaki K and Shigeto H 2003 Lecture Notes in Computer Science ed V Palade, R J Howlett and L Jain (Springer Berlin Heidelberg) pp 933–940
[39] Cajka T, Hajslova J, Pudil F and Riddellova K 2009 *J. Chromatogr. A* **1216** pp 1458–1462
[40] Cimpoiu C, Cristea V-M, Hosu A, Sandru M and Seserman L 2011 *Food Chem.* **127** pp 1323–1328
[41] Kruzlicova D, Mocak J, Balla B, Petka J, Farkova M and Havel J 2009 *Food Chem.* **112** pp 1046–1052
[42] Baykal H and Yildirim H K 2013 *Crit. Rev. Food Sci. Nutr.* **53** pp 415–421
[43] Gestal M, Gómez-Carracedo M P, Andrade J M, Dorado J, Fernández E, Prada D and Pazos A 2004 *Anal. Chim. Acta* **524** pp 225–234
[44] Gestal M, Gómez-Carracedo M P, Andrade J M, Dorado J, Fernández E, Prada D and Pazos A 2005 *Appl. Artif. Intell.* **19** pp 181–198
[45] Ceto X, Gutierrez-Capitan M, Calvo D and Valle M Del 2013 *Food Chem.* **141** pp 2533–2540
[46] Ceballos-Magaña S G, de Pablos F, Jurado J M, Martín M J, Alcázar Á, Muñiz-Valencia R, Gonzalo-Lumbreras R and Izquierdo-Hornillos R 2013 *Food Chem.* **136** pp 1309–1315
[47] Jack F R and Steele G M 2002 *Food Qual. Prefer.* **13** pp 163–172
[48] Bahramparvar M, Salehi F and Razavi S M A 2014 *J. Food Process. Preserv.* **38** pp 1080–1088
[49] Krishnamurthy R, Srivastava A K, Paton J E, Bell G A and Levy D C 2007 *Food Qual. Prefer.* **18** pp 275–285
[50] Cruz A G, Cadena R S, Faria J A and Bolini H M A 2011 *Soc. Dairy Techonology, Int. J. Dairy Technol.* **64** pp 549–556

Acknowledgments
The authors would like to acknowledge financial support from Universiti Malaysia Sabah and Ministry of Higher Education Malaysia for RACE0009-TK-2013.

IOP Conf. Series: Earth and Environmental Science **36** (2016) 012066 doi:10.1088/1755-1315/36/1/012066

Sabah snake grass extract pre-processing: Preliminary studies in drying and fermentation

A Solibun[1], K Sivakumar[2]
[1, 2]*Universiti Malaysia Sabah*, Jalan UMS 88400 Kota Kinabalu, Sabah, Malaysia

E-mail: [1]asredpis441@gmail.com, [2]shiva@ums.edu.my

Abstract. *Clinacanthus nutans* (Burm. F.) Lindau which also known as 'Sabah Snake Grass' among Malaysians have been studied in terms of its medicinal and chemical properties in Asian countries which is used to treat various diseases from cancer to viral-related diseases such as varicella-zoster virus lesions. Traditionally, this plant has been used by the locals to treat insect and snake bites, skin rashes, diabetes and dysentery. In Malaysia, the fresh leaves of this plant are usually boiled with water and consumed as herbal tea. The objectives of this study are to determine the key process parameters for Sabah Snake Grass fermentation which affect the chemical and biological constituent concentrations within the tea, extraction kinetics of fermented and unfermented tea and the optimal process parameters for the fermentation of this tea. Experimental methods such as drying, fermenting and extraction of *C.nutans* leaves were conducted before subjecting them to analysis of antioxidant capacity. Conventional oven-dried (40, 45 and 50°C) and fermented (6, 12 and 18 hours) whole *C.nutans* leaves were subjected to tea infusion extraction (water temperature was 80°C, duration was 90 minutes) and the sample liquid was extracted for every 5th, 10th, 15th, 25th, 40th, 60th and 90th minute. Analysis for antioxidant capacity and total phenolic content (TPC) were conducted by using 2, 2-diphenyl-1-pycryl-hydrazyl (DPPH) and Folin-Ciocaltheu reagent, respectively. The 40°C dried leaves sample produced the highest phenolic content at 0.1344 absorbance value in 15 minutes of extraction while 50°C dried leaves sample produced 0.1298 absorbance value in 10 minutes of extraction. The highest antioxidant content was produced by 50°C dried leaves sample with absorbance value of 1.6299 in 5 minutes of extraction. For 40°C dried leaves sample, the highest antioxidant content could be observed in 25 minutes of extraction with the absorbance value of 1.1456. The largest diameter of disc that could be observed at 18 hours of fermentation sample had a pile size of 3 cm that had expanded to 5.9 cm of diameter which indicated the microbe's growth.

1. Introduction

Tea development has been evolving since about four thousand years ago which started the discovery of its leaves, *Camellia sinensis* until advancements in tea processing which enable tea become the most popular drink beverage after water [1,2]. Tea processing involved several steps which starts from the selection of leaves, withering, fermentation for fermented tea-making and sun drying. Green tea,

[1] To whom any correspondence should be addressed.

Oolong and black teas are among the major teas in the market. Green tea is unfermented tea while Oolong and black teas, respectively are semi-fermented and fermented tea.

Currently, the demand of plant-based products in the market as an alternative to modern medicines to treat and prevent oxidative stress-related health diseases such as cancer, diabetes and other chronic diseases have increased significantly due to some claims of efficacy. These claims are possibly due to the phytochemical content which acts as antioxidants such as phenolic, flavonoids, saponins, tannins and lignin presence in plants. These antioxidants have capabilities to neutralize the effect of free radicals and inhibit the cell damaging oxidation reaction to treat and prevent chronic diseases [3,4]. Oxidative stress which is among the contributor to chronic diseases, where reactive oxygen species (ROS) and reactive nitrogen species (RNS) including free radicals were rapidly produced in human cells that led to the occurrence of cell damage [5,3,6].

Clinacanthus nutans (Burm. F.) Lindau which belongs to Acanthaceae family is a small shrub natively grown in tropical Asian countries [7]. In Malaysia, it is known as 'Sabah Snake Grass' or 'Belalai Gajah'(Elephant trunk) among the locals while in Thailand it is known as 'Phaya Yo' or 'Phaya Plong Thong'[3,7]. *C.nutans* has been used as traditional medicines by Asian natives to treat several of health-related diseases. In Malaysia, it is used as a potent alternative medicine to treat cancer [8]. In Thailand, it is used to treat insect and snake bites, skin rashes, virus-related diseases such herpes simplex (HSV) and varicella-zoster lesions (VZV) [3,7]. Indonesians use it to cure dietary-related health disease such as diabetes and unhygienic-related health disease such as dysentery [3,9].

Previous studies reported that the anti-inflammatory [10], analgesic and antiviral properties against varicella-zoster [11] and herpes simplex type -2 [12] virus were found in *C.nutans* leaves extract. According to one study, antioxidant activities towards free radical-influenced hemolysis are present in ethanolic extract of *C.nutans* leaves [13]. Past studies on natural products have developed steps in preparation, preservation and extraction in order to quantify the yield of phenolics compounds [3,14,15,16,17].

However, fresh plants states easily degrade, thus it, preservation methods are needed. The fresh Lamiaceae herbs water content is approximately 75-80% and to preserve its freshness, moisture content must be less than 15% [18]. Drying methods are among the most popular to preserve sample from further deterioration by microorganisms, preserve its biochemical constituents [19], to conserve the desirable qualities and minimize the storage volume [20].Presently, *C. nutans* plants are sold in plants fair and nursery. Some companies are also producing health drinks from *C. nutans* leaves and shoots. There is study about the effect of drying methods on antioxidant activities of *C.nutans* [3] but no literature is available to study the effect of oven-drying at various temperatures on the phytochemical constituents such as phenolic compounds and flavonoids.

Numerous previous studies have reported the ability of fermentation to mobilize the conjugated phenolics that can enhance the content of phenolic and antioxidant properties in fermented samples [21,22,23]. Previous studies also reported that solid state fermentation tends to enhance the antioxidant properties in legumes or cereals. Products such as carbohydrate-degrading enzymes such as xylanases, cellulases, amylase and estrases were produced by microbes during fermentation which are useful to release the bound phenolic compounds in samples during extraction [24,25,26,23,27,28].

Although many studies have been conducted on the effect of drying and fermentation methods on antioxidant capacity of plant products, there are little to no information available about the effect of oven-dried at various temperatures and the effect of pile fermentation in various fermentation duration on microorganism's growth in *C.nutans*. Therefore, this study was conducted to determine the effect of oven-drying at various temperatures and duration on the antioxidant capacity of the sample and the effect of pile-fermentation method at various durations of fermentation on microorganism's growth in *C.nutans*.

2. Materials and Methods

2.1 Plant Materials

Fresh leaves of *C. nutans* with no physical damage were collected from Block B Faculty of Engineering of UMS, Kota Kinabalu, Sabah. Five grams of fresh leaves were plucked from its main tree, washed and rinsed with tap water and distilled water, respectively to remove dirt. Leaves were then pat dry with kitchen towel to blot the excess water and arranged onto aluminium foil.

2.2 Preparation of unfermented C. nutans tea

The preparation of this tea was adapted from the preparation procedure of *C. nutans* tea according to Barek *et al.* [3] and *Ocimum tenuiflorum* tea by Rabeta and Lai [19] with some modifications. The sample (3g) was dried using an oven (Binder, USA) at 50°C for six hours. This step will be repeated for the next sample with different drying durations (12 and 18 hours) and different oven temperatures (40°C, 45°C and 50°C).

2.3 Preparation of fermented C. nutans tea

The preparation of this tea was adapted from the preparation procedure of *C. nutans* tea according to Barek *et al.* [3] and *Ocimum tenuiflorum* tea by Rabeta and Lai [19] with some modifications. The cleaned fresh leaves (3g) went through a withering process inside the incubator (Memmert, USA) at 25°C for 11 hours to reach 60% of moisture content. The withered leaves then underwent a piling fermentation process inside the incubator (Memmert, USA) at 25°C. The withered leaves were piled with thread inside a ceramic cup and the cup opening was closed with aluminium foil before proceeding with the fermentation process. The duration of the fermentation process was varied in this research at 6, 12 and 18 hours. The fermented leaves underwent a drying process inside the oven (Binder, USA) at 30°C to achieve 3% of moisture content in order to stop the fermentation process [29].

2.4 Preparation of leaves infusion

The extraction process was done according to Barek *et al.* [3] which imitated the common tea brewing situations using hot water. The dried unfermented leaves (3 g) were added into 80°C hot distilled water (300 mL) and continuously stirred using a magnetic stirrer. The sample extract (10 ml) was collected at specific durations of 5, 10, 15, 25, 40, 60 and 90 minutes. This process was also used to prepare leaves infusion for dried fermented leaves.

2.5 Determination of total phenolic content

The total phenolic content (TPC) was determined by the Folin Ciocalteu (FC) assay as described by Barek *et al.*, [3] with some modifications. 1 ml of extracted sample was mixed with 0.75 ml of 10-fold diluted Folin-Ciocalteu's reagent. The mixture was left to stand for five minutes at room temperature. 0.75 ml of $NaCO_3$ (60 g/L) solution was added to the solution. The mixture was left for ninety minutes further at room temperature. The absorbance was read at 725 nm using UV-Vis spectrophotometer (V-650 Spectrophotometer, Jasco, Tokyo, Japan). No calibration curve had been done as this was a preliminary study.

2.6 2, 2-diphenyl-1-pycryl-hydrazyl (DPPH) free radical scavenging activity

The DPPH scavenging activity was determined by the method of Barek *et al.*, [3] with some modifications. The DPPH solution was prepared by mixing the DPPH powder with methanol at a concentration of 250 μM. The extracted sample solution (2.5 mL) was mixed with 2.5 mL of DPPH solution. The mixtures were incubated in a dark room at 28°C for 30 minutes. DPPH solution in methanol was used as a blank. The absorbance was read at 517 nm with UV-Vis spectrophotometer (V-650 Spectrophotometer, Jasco, Tokyo, Japan). No calibration curve had been done as this was a preliminary study.

2.7 Microbes Growth Study (Paper Disc Diffusion with modification)

Dried leaves were mixed with 50 ml of distilled water and sterilized paper discs. The paper disc was plated on agar and incubated in room temperature at dark place for seven days. Observation and measurement of the disc diameter was carried out the next day and continued doing so for seven days.

3. Results and Discussion

3.1 Total Moisture Content

The moisture content of *C.nutans* leaves were determined by using A.O.A.C method [30]. Two samples of leaves were weighed at 5 g each and dried inside the oven (Binder, USA) at 105°C for 24 hours. The dried leaves were re-weighed to get the dry weight of the leaves. The moisture content of *C.nutans* leaves was found to be 79.64%.

Table 1: The moisture content of leaves samples

Sample	Moisture Content (%)
A	79.60
B	79.68

3.2 Drying Kinetics for Oven-Drying

The fresh leaves were placed inside the oven and dried at 40°C, 45°C and 50°C. The mass of the leaves were recorded at every hour until the mass reading became constant.

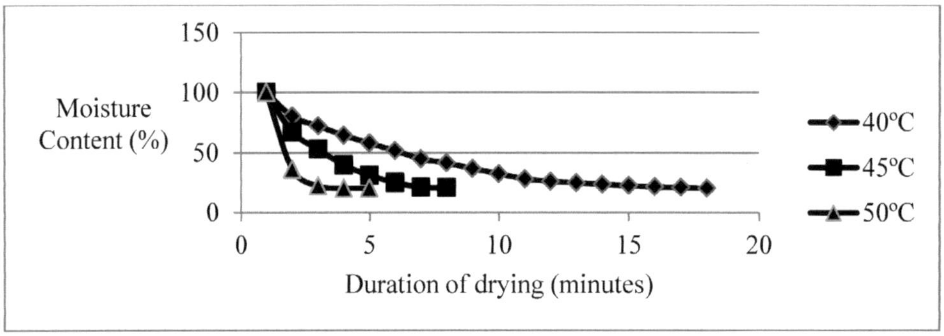

Figure 1: Experimental drying curves of *C.nutans* leaves at different temperatures

Figure 1 illustrates the effect of various oven-temperature and duration of drying on *C.nutans* leaves. At 40°C, the drying rate was minimal and approached equilibrium at 17 hours and the maximum drying rate was at 50°C which approached equilibrium at 4 hours. For all the three samples, the final moisture content was 20%. It also shows that as the temperature increased, the duration for the leaves to reach constant moisture content decreased. This is in line with the drying theory. As the temperature increased, there was an increase in the rate of mass transfer (water) to achieve the similar equilibrium moisture content (approximately 0.98 g water/100g d.m) [31].

3.3 Effect of drying temperatures on phenolic content in unfermented leaves

Figure 2 illustrates the absorbance unit over the duration of extraction on *C.nutans* leaves. The spectrophotometer reading gives the phenolic content value in the quantification of phenolic compounds from plant materials. The 50°C oven-dried leaves had higher absorbance than those dried at 40°C. Processing the samples will affect the content, bioactivity and bioavailability of bioactive compounds [20]. Initial conclusion that can be drawn from this finding was that sample that had been dried at higher temperatures had greater capacity to extract higher phenolic content at shorter duration of extraction.

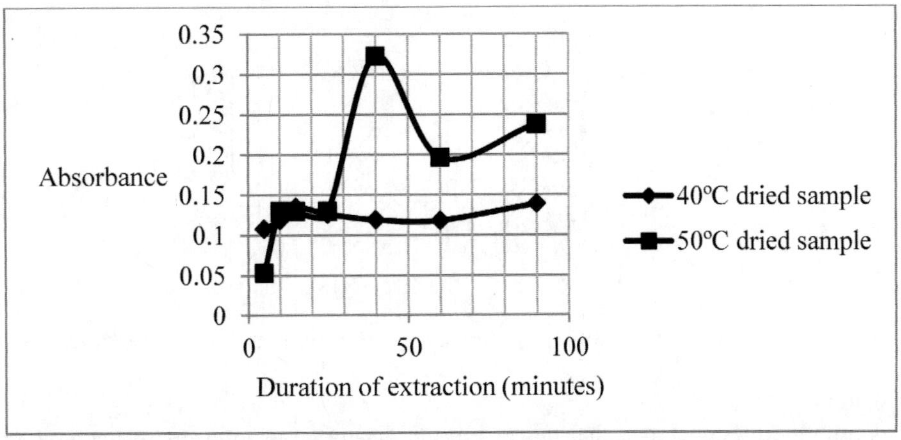

Figure 2. Effect of different dried sample temperatures on the phenolic content

3.4 Effect of drying temperatures on antioxidant content in unfermented leaves

Figure 3 illustrates the absorbance values over the duration of extraction on *C.nutans* leaves. The 50°C oven-dried leaves had the higher absorbance value than those dried at 40°C. Higher drying temperature reduced the times of extraction and costs of processing. Drying at higher temperatures helped increase the rates of deactivation of enzymatic degradation which was responsible for the loss of antioxidant in samples [21]. The same conclusion can be reached as in phenolic content above.

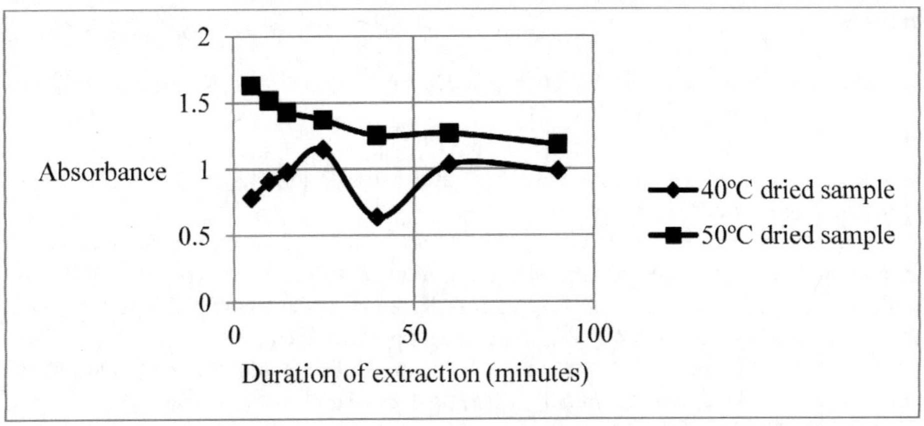

Figure 3. Effect of different dried sample temperatures on the antioxidant activity

3.5 Effect of pile-fermentation on the growth of microorganism's

Figure 4 illustrates the diameter of discs over the pile sizes of fermented *C.nutans* leaves. The 18 hours fermented leaves and pile size (3cm) produces the largest diameter of disc. At present, the species of microbes are unknown, however in pile fermentation of Guangdong Pu-erh tea showed complexity of microbes communities in which fungi were the dominant microbe [32,33]. Initial conclusion can be drawn from this finding, sample with longer duration of fermentation and thicker piles sizes produce higher microorganism's growth.

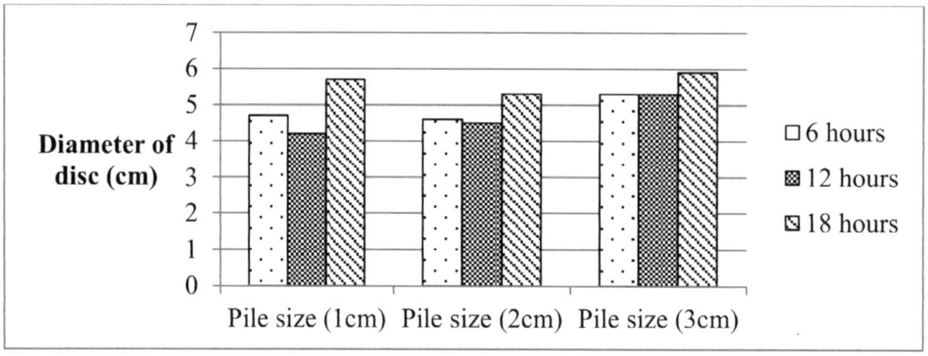

Figure 4. Effect of different fermented sample duration on the microorganism's growth

4. Conclusion

It can be preliminarily concluded that increasing the drying temperatures can reduce the duration of drying and can increase the extraction of phenolic and antioxidant contents from *C.nutans* leaves in shorter durations of extraction. Increasing the duration of fermentation and size of piles can induce the development of microorganism's growth. However, more investigation is needed to find the effect of fermentation on phenolic content, antioxidant activity and microorganism's growth in fermented *C.nutans* leaves. Further work will need to be done to get the standard calibration curve for phenolic content and antioxidant activity for *C.nutans* fermented leaves. Works on rate and yield data for the extraction of *C.nutans* fermented leaves are also conducted. The result of this study will be utilised to determine the critical parameters as well as the range for optimal extraction to ensure the standardization of the leaves extract for overall antioxidant content.

5. References

[1] Erguder, I,B., Namuslu, M., Sozener, U., Devrin, E.,Avci, A., Kocaoglu and Durak, I. 2008 *Alternative Therapies* **14**(3) pp 30-31

[2] Weisburger, J.H. 1997 cancer Lett. **144** pp 315-317

[3] Barek, L. M., Hasmadi, M., Zaleha., A.Z. and Mohd Fadzelly, A.B. 2015 *International Food Research Journal* **22**(2) pp 661-670

[4] Pisoschi, A.M. and Negulescu, G.P. 2011 *Biochemistry and Analytical Biochemistry* **1** pp 106

[5] Bagchi, K. and Puri, S. 1998. *East Mediterranean Health Journal* **4** pp 350–360

[6] Valavanidis, A., Vlachogianni, T., Fiotakis K. and Loridas, S. 2013 *International Journal of Environmental Research and Public Health* **10** pp 3886-3907

[7] Sakdarat, S., Shuyprom, A., Pientong, C., Ekalaksananan, T. and Thongchai, S. 2009 *Bioorganic and Medicinal Chemistry* **17**(5) pp 1857-1860

[8] Roosita, K., Kusharto, C. M., Sekiyama, M., Fachrurozi, Y. and Ohtsuka, R. 2008 *Journal of Ethnopharmacology* **115** pp 72-81

[9] Hariana, H.A. 2008 Swadaya Jakarta pp 56

[10] Satayavivad J, Bunyaoraphatsara N, Kitisiripornkul S, Tanasomwang W. 1996 *Thai J Phytopharm* **3 pp** 7-17

[11] Thawaranantha D, Balachandra K, Jongtrakulsiri S, Chavalittumrong P, Bhumiswasdi J, Janyavasu C. 1992 *Siriraj Hosp Gaz.* **44** pp 285-91
[12] Jayavasu C, Balachandra K, Sangkitjaporn S. 1992 *Com Dis J.* **18** pp 152-61
[13] Pannangpetch, P., Laupattarakasem, P., Kukongviriyapan, V., Kukongviriyapan, U., Kongyingyoes, B. & Aromdee, C. 2007 *Songklanakarin Journal of Science and Technology* **29**(1) pp 1–9
[14] Chan, E.W.C., Lim, Y.Y., Wong, S.K., Lim, K.K., Tan, S.P., Lianto F.S. and Yong, M.Y. 2009 *Food Chemistry* **113** pp 166-172
[15] Chong, K.L. and Lim, Y.Y. 2011 *Journal of Food Quality* **35** pp 51-59.
[16] Ji, H., Du, A., Zhang, L., Xu, C., Yang, M., and Li, F. 2012 *Journal of Medicinal Plant Research* **6**(16) pp 3233-3239
[17] Lin, X., Zhang, L., Lei, H., Zhang, Y. C., Zhu, R., and Ruan, R. 2010 *International Agricultural Engineering Journal* **19**(3) pp 30-37
[18] Diaz-Maroto, M. C., Perez Coello, M. S. and Cabezudo, M. D. 2002 *European Food Research Technology* **215** pp 227- 230
[19] Rabeta , M. S. and Lai, S. Y. 2013 *International Food Research Journal* **20**(4) pp 1601-1608.
[20] Gupta, J.K., Sharma, S.K., Misra, V. and Patel, K. 2011 *International Research Journal of Pharmacy* **12**(2) pp 235-237
[21] Hossain, M.B., Ryan, C.B., Diana, A.B.M and Brunton, N.P. 2010 *Food Chemistry* **123**(10) pp 85-91
[22] Randhir R, Vattem D, and Shetty K 2004 *Innovative Food Sci. Emer. Technol.* **5** pp 235-244
[23] Razak, A.D.L., Rashid, A.N.Y., Jamaluddin, A., Sharifudin, S. A., Kahar, A.A. and Long, K. 2015. *Journal of the Saudi society of Agricultural Sciences* pp 1-7
[24] Bhanja T, Kumari A, Banerijee R 2009 *Bioresour. Technol.* **100**(11) pp 2861-2866
[25] Dey, T.B. and Kuhad, R.C. 2014 *Biotechnology Reports* **4** pp 120-127
[26] Dordevic, T.M., S.S. Siler-Marinkovic and S.I. Dimitrijevic- Brankovic. 2010 *Food Chem.* **119** pp 957-963
[27] Wang CY, Wu SJ, Fang JY, Wang YP, and Shyu YT. 2012 *Food Res Int*; **48** pp 428-434
[28] Yeo, S.K. and Ewe, J.A. 2015 *Advances in Fermented Foods and Beverages* pp 107-122
[29] Yong, Y.K., Tan, J.J., Teh, S.S., Mah, S.H., Ee, G.C.L., Chiong, H.S., and Ahmad, Z. 2013 *Alternat. Med.* pp 1–8
[30] A.O.A.C., 1995 Official Methods of Analysis, 16th ed. The Association of Official Analytical Chemists, Washington.
[31] Miranda, M., Maureira, H., Rodríguez, K., and Vega-Gálvez, A. 2009 *Journal of Food Engineering* **91** pp 297-304
[32] He, G.F., Lin, Y.C. and Xu, F.X. 1987 *Journal of the Tea Science* **7** pp 54-57
[33] Zhao, Z.J., Hu, X.C. and Liu, Q.J. 2015 *International Food Research Journal* **22**(3) pp 1240-1246

Acknowledgement
The author acknowledges the financial assistance provided by the Ministry of Higher Education of Malaysia (MOHE) under Fundamental Research Grant Scheme, FRGS (Project No. FRG0402-TK-2/2014). The authors also would like to acknowledge laboratory assistants of Faculty of Engineering (Chemical Engineering Program), Universiti Malaysia Sabah for the use of laboratory facilities and technical assistance.

Utilizing steel slag in environmental application – An overview

J W Lim[1,2], L H Chew[2], T S Y Choong[2], C Tezara[1], M H Yazdi[1]
[1]Department of Mechanical Engineering, INTI International University, Jalan BBN 12/1, Bandar Baru Nilai, 71800 Nilai, Negeri Sembilan
[2]Department of Chemical and Environmental Engineering, Universiti Putra Malaysia, 43400Selangor, Malaysia

Email: [1]jinwei.lim@newinti.edu.my

Abstract. Steel slags are generated as waste material or byproduct every day from steel making industries.The potential environmental issues which are related with the slag dump or reprocessing for metal recovery are generally being focused in the research. However the chemistry and mineralogy of slag depends on metallurgical process which is able to determine whether the steel slag can be the reusable products or not. Nowadays, steel slag are well characterized by using several methods, such as X-ray Diffraction, ICP-OES, leaching test and many more. About the industrial application, it is mainly reused as aggregate for road construction, as armour stones for hydraulic engineering constructions and as fertilizers for agricultural purposes. To ensure the quality of steel slag for the end usage, several test methods are developed for evaluating the technical properties of steel slag, especially volume stability and environmental behaviour. In order to determine its environmental behaviour, leaching tests have been developed. The focus of this paper however is on those applications that directly affect environmental issues including remediation, and mitigation of activities that negatively impact the environment.

1. Introduction

Various types of slags are generated as waste material or byproduct in metallurgical process. Based on its origins and characteristics, the main slags can be categorised into three major types, such as ferrous slag, non ferrous slag and incineration slag. In terms on its mineral composition, slag usually contains quantities of valuable metals. It's generally possible to recover some values by physical or chemical mineral processing techniques, such as crushing, grinding, classification, hydrocyclone, magnetic seperation, flotation, leaching or roasting [1]. Another feature of steel slag is that the releasing on notable amount of heavy metals which can cause the environmental problem. Furthermore, the steel slag dumping is the most conventional disposal method. The increasing of steel slag dump is not only causing the scarcity of the landfill, but also potentially have an impact to the environment, especially water pollution. Therefore, the metals revovery and utilization of steel slags

[1] To whom any correspondence should be addressed.

are highly desirable. Characterization of the environmental aspects of steel slag helps to evaluate its potential to release contaminants and its suitability as a potential resource [2].

2. Properties of steel slag

2.1 Physical properties

Steel slag aggregates are highly angular in shape and have rough surface texture. They have high bulk specific gravity and moderate water absorption (less than 3%). The grindability index of steel slag is 0.7, in contrast with the value of 0.96 and 1.0 for blast furnace slag and standard sand respectively [3].

2.2 Chemical properties

The chemical composition of slag is usually expressed in terms of simple oxides calculated from elemental analysis determined by X-Ray Fluorescence (XRF). Virtually all steel slags fall within these chemical ranges but not all steel slags are suitable as aggregates. Of more importance is the mineralogical form of the slag, which is highly dependent on the rate of slag cooling in the steel-making process.

2.3 Mechanical and thermal properties

Processed steel slag has favorable mechanical properties for aggregate use, including good abrasion resistance, good soundness characteristics, and high bearing strength. Due to their high heat capacity, steel slag aggregates have been observed to retain heat considerably longer than conventional natural aggregates. The heat retention characteristics of steel slag can be advantageous in its resistance toward high temperature effects.

3. Steel slag characterization

3.1 Scanned Electron Microscopy (SEM)

SEM is a device that utilizes focused beam of high energy electrons to analyze solid material by obtaining topographical, morphological and compositional information of the tested sample. A scanning electron microscope (model S -3400N, HITACHI, Japan) was used to obtain external morphology, crystalline structure and orientation of materials making up the sample at an accelerating voltage of 20.0 kV, beam current of 50 mA and working distance of 9.0 mm – 10.0 mm using backscattered electrons (BSE) objective with aperture of 3. Two images were captured at a low magnification (300×) and high magnification (1000×) respectively. The micrographic appearance of the unreacted EAFS shows distinct mineral phases, presented in Figure 2. The light areas contain a higher proportion of Fe, Mg, Mn and Al while the dark areas contain a higher proportion of CA and Si [4]. From Figure 3, it can be seen that the light areas are larger than the dark areas. This indicates a hig her composition of Fe, Mg, Mn and Al compared to the composition of Ca and Si

Figure 1. SEM micrograph of unreacted EAFS at magnification of (a) 300x and (b) 1000x

3.2 X-Ray Diffraction (XRD)
X-Ray diffractometer (model XRD-600, Shimadzu, Japan) was used for phase identification of unreacted EAFS with Cu Kα radiation at 40 kV and 40 mA and scanned between 2θ = 20°–80° and scanning speed of 0.1°/min. The X-Ray diffraction pattern observation was generated by using powder diffraction method. The XRD pattern was then analyzed using X'Pert High Score Plus software to obtain the mineralogical and chemical composition of the sample slag. Each crystalline solid has its unique characteristic XRD pattern which may be used for its identification. X'Pert High Score Plus software has the database of these XRD patterns so that they could be used to match with the peaks appear in the XRD pattern of unreacted EAFS. Once all the peaks were identified for the chemicals present in the sample slag, the intensity under the peaks was automatically used by the software to determine their composition.

The XRD analysis of unreacted EAFS is shown in Figure 2. The diffraction pattern is heterogeneous, consisting of a mixture of crystalline phases. The analysis identifies calcium oxide (CaO), silicon oxide (SiO_2), hematite (Fe_2O_3), manganese (III) oxide (Mn_2O_3), magnesium oxide (MgO), aluminium oxide (Al_2O_3), phosphorus oxide (P_2O_5), metallic iron (Fe), srebrodolskite ($Ca_2Fe_2O_5$), dicalcium silicate (Ca_2SiO_4) and pyroxene ($CaAl_2SiO_6$).

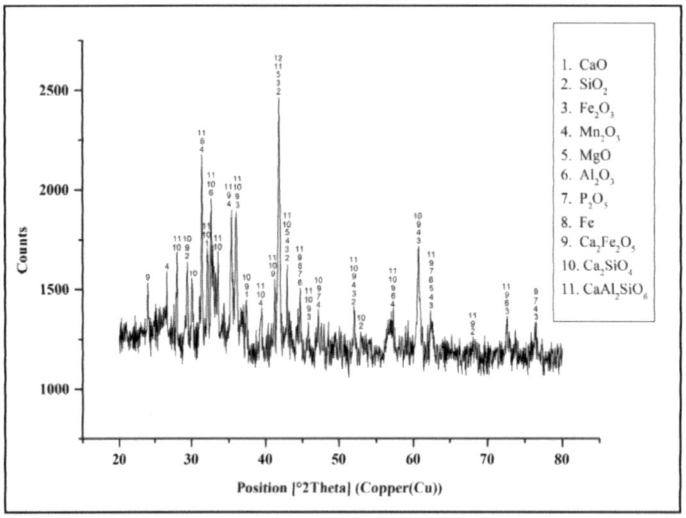

Figure 2. XRD pattern of unreacted EAFS

3.3 Inductively Coupled Plasma Optical Emission Spectrometry

Inductively Coupled Plasma Optical Emission Spectrometry (ICP-OES) is an elemental analytical technique used to determine the composition of a wide variety of materials in liquid form, with excellent sensitivity. In this study, leachates from the experiments manipulating the initial concentration of HCl and EAFS size were diluted and sent for elemental analysis using Inductively Coupled Plasma Optical Emission Spectrometer (model Optima 7300DV, Perkin-Elmer, USA) to determine the concentration of the elements Fe and Ca. Calibration curves for Fe and Ca of known concentration 0 ppm – 9 ppm and 0 ppm – 90 ppm, respectively was prepared for the determination of concentration of Fe and Ca in the samples.

4. Applications of steel slag

According to the Malaysia Environmental Quality (Scheduled Waste) Regulation (2005), Steel slag is classified as non-hazardous waste which can be disposed off to appropriate landfills. The disposal of steel slag requires large landfill area which is rather unfavourable in economical terms. Therefore, several research and development have been conducted to establish the potential use of slag in different applications. However, it might have the risk that different components of slag might elute, especially on the heavy metal. Therefore, it is necessary to examine this metallurgical waste material in appropriate ways. Typical applications of steel slags are sealing aggregate (skid resistant), asphalt aggregate, base, sub-base, construction fills, subsoil drains, grit blasting and waste water treatment. The various applications for steel slag in 2010 (22.3 million tonnes) is illustrated in Figure 3

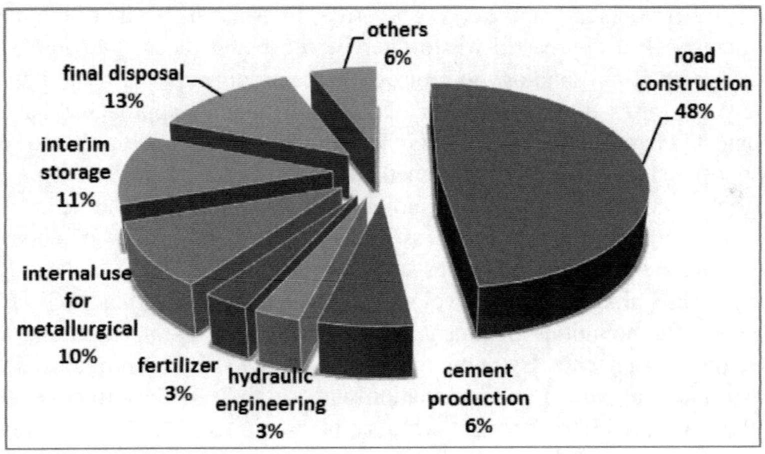

Figure 3. Statistical data of various applications of steel slag (2010) [5]

4.1. Cement production

In cement and concrete industries, slag can be used either as an aggregate or binder in stabilized base courses. In order to conserve natural resources and reduce environmental impact, slags can be used as an aggregate. To diminish the need for cement which is expensive in cost, slag is replaced and used as a binder. Because of these advantages, many researchers have examined adding slag content to cement and concrete [6]. Addition of slag to cement presents some important technical advantages over ordinary Portland cements. These benefits can be enumerated as development of mechanical strengths, low solubility of the hydrates and porosity, lower heat of hydration, excellent durability and stronger aggregate–matrix interface. However, there are some disadvantages such as high shrinkage, formation of micro-cracks and rapid setting. Therefore, the combination of various type of slags (activated, chemical and physical modified) to cement are investigated. For instance, it can boost the hydration ability of slag cement concretes, improving the mechanical performance of cement, enhancing the strengths of cement, autogenous deformation on the cracking, microstructure and durability, resistance.

4.2 Road construction

Industrial by-products such as steel slag can be utilized instead of natural aggregates in order to prevent depletion of resources. Million tons of sand, till and crushed rocks were extracted from mines due to roads construction. It is well known that technical, economical and many environmental benefits are obtained when steel slag has been considered as alternative construction materials [7]. Due to prominent properties of steel slag, it is an ideal aggregate for asphalt surface course materials and road surface treatments. It has been proven that the use of steel slag in road construction mixtures has more advantage compared to conventional asphalt. The investigations of the asphalt mix and road surface have shown very good results in terms of stability, stiffness and durability. The slag asphalt, as expected, provided good friction values, noise reduction levels, enhances resistance to cracking at low temperatures with excellent performance in roughness and good resistance against water permeability[8].

4.3 Phosphurus removal

Phosphorus (P) from agricultural wastewater is one of the major pollutants in natural water that cause the algae growth and eutrophication of lakes. Therefore, Phosphorus is one of the major nutrients that is need to be removed from domestic wastewater before being discharged into water bodies. The application of an appropriate technology to remove the maximum capacity of nutrient is imperative. In order to remove phosphorus from wastewater, various methods or approaches have been attempted, including biological, chemical process (precipitation, metal salt addition) and physical (electro-dialysis, reverse osmosis). Among all the methods, the chemical precipitation is expensive and increasing sludge volume by up to 40%. The biological phosphorus removal needs a lot more volume or space (anaerobic unit) than the other processes. Various research on the potential use of industrial by-products to remove P from wastewater was studied in the late 1980s.

The P removal is along with the development of constructed wetlands (CW). This method is a low cost technology for pollutions treatment, which slag performs an important role in absorbing impurities especially phosphorus. By comparing most of industrial by products, steel slag is a low-cost and abundant material, which its combination with small secondary treatment systems (such as constructed wetlands) is preferred in compared to the others methods [9,10]. The slag contains various metal oxides such as iron oxides and alumina that may be effective in P reduction from domestic, agricultural effluents and municipal [11]. Steel Slag is a light weight porous medium with numerous sites for sorption. Steel slags have different physic-chemical property due to various feedstock ores, fluxes and manufacturing process. These differences caused a board range of P sorption capacity by using slag, ranging from 76.4 to 8390 mg/kg. Many experiments have been carried out in laboratory to evaluate the sorption capacity of slag. These studies have found slag to be a promising substrate for P-removal [12].

4.4 Water and wasteweater treatment

As slag has good sportive characteristics and is low cost, it is widely used in wastewater and water treatment, and may be the alternative of using granular activated carbon. Various research such as adsorptions of dye, heavy metals and organics were carried out by using steel slag. However, the uptake capacity of slag is dependent on the pH solution. The hydration of slag composition in the aqueous solutions provides a high pH. The possible hydration reactions can be occurred with different compositions of slag as followings:

$$CaO + H_2O \rightarrow Ca(OH)_2 \quad (1)$$

$$MgO + H_2O \rightarrow Mg(OH)_2 \quad (2)$$

$$2Ca_2SiO_4 + 4H_2O \rightarrow 3CaO.2SiO_2.3H_2O + Ca(OH)_2 \quad (3)$$

$$2Ca_3SiO_5 + 6H_2O \rightarrow 3CaO.2SiO_2.3H_2O + 3Ca(OH)_2 \quad (4)$$

$$CaO.7Al_2O_3 + 12H_2O \rightarrow CaO.7Al_2O_3.6H_2O + 6Al_2O_3.H_2O \quad (5)$$

With the high pH condition, the slag surface is negatively charged and adsorption metal ions especially cations are preferred. However, heavy metal absorption by using slags can be conducted with a condition of either high temperature or low pH under certain conditions. Yamashita et al. investigated the possible mechanisms of removing dissolved heavy metals from aqueous waste liquids via the converter furnace slag [13]. They suggested that the removal can be attributed to one or more of following effects: adsorption, co-precipitation, hydroxide precipitation as hydroxide, sulphide and ion exchange.

Wastewater from the steel mill contains a high concentration of heavy metals due to the accumulation of filtered particles from the filtration process of wastewater. A direct release of the backwash water into the environment may cause serious effects to both land and aquatic lives. Based on the toxilogical studies, metals are toxic and non-biodegradable and may continue to exist in these water bodies. Furthermore, the heavy metals also have the tendency to accumulate in the food chain. Hence, a strict environmental regulation has been established in an effort to mitigate the heavy metals contamination of the discharge of industrial effluent. Absorption of lead ions by two different types steel slag was carried out in column-type contact process [14]

5. Steel slag leaching

Extraction of metal values from mineral wastes are major steps towards conservation and judicious utilization of our mineral resources which are depleting very fast. Electric Arc Furnace (EAF) Steel Slag is an abundant byproduct in steel industry. The EAF slag contains considerable amount Fe and Ca oxides. By recovering the metal values, these slags would become value added products instead of being considered as pollutants. The EAF slags can be converted to metal chlorides, namely ferric chlorides and calcium chlorides. Ferric chloride is a flocculant used in the wastewater and water industry. Calcium chloride finds application as brine for refrigeration and as adsorbent for fluoride removal.

Extraction of metal values from the EAF slag would convert waste to wealth. The chloride process route has shown a lot of promise. Metal chlorides can be produced from EAF slag by the applications of various chlorinating agents like gaseous chlorine and hydrochloric acid. Ferric Chloride can be used as a flocculant in sewage treatment and drinking water production, and as an etchant for copper-based metals in electronic industries. Calcium chloride applications include brine for refrigeration, dessication and also to remove fluoride in water and wastewater treatment. This EAF slag is free and abundant and is therefore an ideal raw material or low cost production of metal chlorides.

Although it seems reasonable the use of chlorination method for the recovery of refractory metals from EAF slags as their chlorides and oxy-chlorides mixtures, it is the fundamental importance a study that defines the further chemical processing of this mixture. For such, it should be taken into consideration the nature of the desired final products, ferric chloride and calcium chloride as well as their possible uses.

According to the study done by Brocchi and Moura (2008), the conversion of ferric oxide and calcium oxide are relatively much higher than other metal oxides present at a lower operating temperature compared to the reaction with gaseous chlorine [15]. Since the chlorination of ferric oxide and calcium oxide are the only main concern, batch leaching of EAF slag with hydrochloric acid is selected as the chlorination method. The operating parameters are chosen and manipulated by referring to the literature done as well.

6. Conclusion

In this review, widespread usage of slags as a by-product of steel making industries in various applications have been discussed. In terms of steel slag characterization, XRD, SEM and ICP-OES are being used to determine the mineral compostion of steel slag. Based on the literature studies, it is pretty obvious that slags can be the promising replacements for conventional way of application of adsorbent and binder in various fields such as adsorption, construction, road and waste water treatment. The ease of use, low cost, and sources availability are among the few advantages on reusing and utizing the steel slag. On the other hand, the methods of steel slag leaching with hydrochloric acid are developed in order to recover the valuable metals in steel slag. Therefore, by implementing all the remediation, and mitigation of activities as per mentioned above, our environment can be protected.

References

[1] Das B, Prakash S, Reddy P S R, and Misra V N 2007 *Resources, Conservation and Recycling* **50** 40-57.
[2] Nadine M. Piatak, Michael B. Parsons, Robert R. Seal II. Characteristics and environmental aspects of slag: A review. Applied Geochemistry 2014
[3] Hou G H, Wang Z H and Zhu X 2010 *J Yancheng Technol I* **22** 1-4.
[4] Bankole L, Rezan S and Sharif N 2014 *IOP Conf. Ser.: Earth Environ. Sci.*, *19*, 012004. doi:10.1088/1755-1315/19/1/012004
[5] EUROSLAG. Retrieved from http://www.euroslag.com/products/statistics/2010/. 4 March 2015
[6] Frias M, Sanchez de Rojas M. I, and Rodriguez C 2009 *Construction and Building Materials* **23** 1472–5
[7] Suer P, Lindqvist J E, Arm M, and Frogner-Kockum P 2009 *Science of The total Environment* **407** 5110–18.
[8] Ameri M, Hesami S, Goli H, 2013 *Construction and Building Materials*, **49** 611-17
[9] Drizo A, Forget C, Chapuis R P, and Y. 2006 *Water Research,* **40** 1547-54
[10] Chazarenc F, Brisson J, and Comeau Y 2011*Water Science Technology* **56** 109-15
[11] Vohla C, Koiv M, Bavor H J, Chazarenc F, 2011 *Ecological Engineering* **37** 70-89
[12] Pratt C, and Shilton A 2010 *Water Science and Technology* **62** 1713-18
[13] Yamashita K, Ikenata T, Tate K and Nakahara K 1983 Method of Removing Dissolved Heavy Metals from Aqueous Waste Liquids, US Patent, 4377483
[14] Nilforoushan M R, and Otroj S 2008 *Iranian Journal of Chemistry & Chemical Engineering* **27** 69-75
[15] Brocchi E. and Moura, F 2008 *Minerals Engineering* **21** 150-156 doi:10.1016/j.mineng.2007.08.011

Acknowledgement
The authors would like to gratefully acknowledge Ministry of Higher Education (MOHE), and Universiti Putra Malaysia (UPM) for the financial support of this work.

IOP Conf. Series: Earth and Environmental Science **36** (2016) 012068 doi:10.1088/1755-1315/36/1/012068

Assessment of biological Hydrogen production processes: A review

G D Najafpour[1], M H Shahavi, S A Neshat
Biotechnology Research Laboratory, Faculty of Chemical Engineering, Noshirvani University, Shariati Avenue, Babol 47148-71167, Iran

E-mail address: [1]najafpour@nit.ac.ir

Abstract. Energy crisis created a special attention on renewable energy sources. Among these sources; hydrogen through biological processes is well-known as the most suitable and renewable energy sources. In terms of process yield, hydrogen production from various sources was evaluated. A summary of microorganisms as potential hydrogen producers discussed along with advantages and disadvantages of several bioprocesses. The pathway of photo-synthetic and dark fermentative organisms was discussed. In fact, the active enzymes involved in performance of biological processes for hydrogen generation were identified and their special functionalities were discussed. The influential factors affecting on hydrogen production were known as enzymes assisting liberation specific enzymes such as nitrogenase, hydrogenase and uptake hydrogenase. These enzymes were quite effective in reduction of proton and form active molecular hydrogen. Several types of photosynthetic systems were evaluated with intension of maximum hydrogen productivities. In addition dark fermentative and light intensities on hydrogen productions were evaluated. The hydrogen productivities of efficient hydrogen producing strains were evaluated.

1. Introduction

Global warming and accumulation of air pollutions cause by combustion of fossil fuel is one of serious environmental issue of new millennium. Such concern demanded to conduct and search for a new renewable energy source that has no pollution for environment. One of the alternatives for fossil fuels is biohydrogen which can be produced by microorganisms via their photosynthetic systems. The biohydrogen has no pollution and no harm or undesirable by-products produces during formation of hydrogen. Production of hydrogen by combination of water and sunlight in a microbial or green algae cell made this process very attractive[1].

The available processes for production of energy from biomass are divided into two main categories such as thermochemical and biological processes. Thermochemical processes include gasification, liquefication [2, 3], pyrolysis [4] and biomass combustion. Biological processes are involved living organisms include direct and indirect biophotolysis, photo fermentation, dark fermentation and biological water-gas shift reaction. There are known enzymes; hydrogenase and

[1] To whom any correspondence should be addressed.

nitrogenase mostly related to proton reductions which are evolved in all biological hydrogen production processes[5]. A summary of hydrogen production processes can be shown in Figure 1.

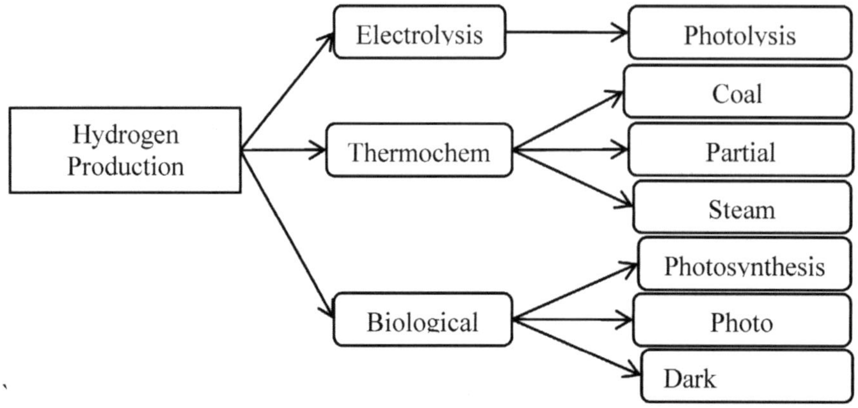

Figure 1. General view of Hydrogen production processes.

2. Biological hydrogen production processes

2.1. Biological water-gas shift reaction

Reaction of carbon monoxide and water resulted in hydrogen and carbon dioxide. For production of pure hydrogen, carbon dioxide should be removed by the means of any carbon dioxide absorbent or any microorganisms can utilize carbon dioxide as substrate.[6] The following reactions are provided the energy for conduction of biological water-gas shift reaction by transferring electron from CO to H_2O:

$$CO + H_2O \rightarrow CO_2 + 2e^- + 2H^+ \quad (1)$$
$$H^+ + 2e^- \rightarrow H_2 \quad (2)$$

Carbon monoxide dehydrogenase and hydrogenase are two enzymes which are responsible for the supply of electron and proton while electron transport and energy supply devoted to cell growth.[7, 8] The overall reaction which is the sum of reactions (1) and (2) is stated as follows:

$$CO + H_2O \rightarrow CO_2 + H_2 \qquad \Delta G = -20.1 \text{ kJ/mol} \quad (3)$$

Oxidation of carbon monoxide can take place in both aerobic and anaerobic conditions; while aerobic oxidation of carbon monoxide is exothermic process; it may generates -283 kJ/mol CO while in anaerobic microbial oxidation process just liberate very limited chemical energy in form of methane and other by-products (-19 kJ/mol CO) [9].

$$CO + \frac{1}{2}O_2 \rightarrow CO_2 \quad (4)$$

Biological water-gas shift reaction basically is oxidation of the carbon monoxide under anaerobic conditions by hydrogenetic microorganisms. There are bacteria can produce hydrogen from carbon monoxide while catalyze the water-gas shift reaction. For instance *Rhodospirillum rubrum*, *Rubrivax gelatinosus* and *Carboxydothermus hydrogenoformans* are gram negative and gram positive actively involved in the above reaction[10]. Pakpour et al.[11] conducted studies on an isolated *Rhodopseudomonas* strain which was locally isolated and they have reported that the isolated strain can generate significant amount of hydrogen (33.5 mmol/l H_2) and the isolated strain was able to utilize 86% of CO uptake and produce hydrogen through water-gas shift reaction.

Najafpour et al.[11] investigated on the effect of various carbon sources for hydrogen production using anaerobic bacteria known as *Rhodospirillum rubrum* which was capable of producing hydrogen via water-gas shift reaction. They used carbohydrate sources such as glucose, fructose and sucrose and organic acids like formate, acetate and malate as carbon sources. They reported that the maximum cell dry weight was obtained from malate and fructose. They also reported that maximum yield of CO conversion and cell growth from acetate and the highest yield of biomass production obtained while the specified organism utilized fructose as carbon source.

Ismail et al.[12] investigated the effect of configuration or type of bioreactor on hydrogen production of *Rhodospirillum rubrum*. Maximum rate of hydrogen production did not lead to the reactor condition; due to instability of process operation obtained at high agitation rate (500 rpm). In contrary, application of micro sparger had positive impact on high hydrogen production yield.

2.2. Direct biophotolysis

Direct biophotolysis is a suitable hydrogen production process which is utilizing microalgae to convert solar energy to hydrogen as chemical energy. Equation (5) shows the reaction of direct biophotolysis.

$$H_2O + \text{Solar energy} \rightarrow H_2 + O_2 \quad (5)$$

Two types of photosystem are involved in production of hydrogen in direct biophotolysis process. One of them is responsible for reducing CO_2 and the second one dividing the water and evolves oxygen. The hydrogenase enzyme in microalgae is responsible to demonstrate the ability of hydrogen production. Due to lack of hydrogenase enzyme in green plants they cannot produce hydrogen even photosynthetic systems are active. The second photosystem generates electron after absorbing the light, and rule of first photosystem is to absorb electrons to ferredoxin by absorbing solar energy, then the electrons which absorbed by ferredoxin transfers to hydrogenase. After transferring the electrons from ferredoxin to hydrogenase, then production of hydrogen is completed[5]. A schematic diagram of direct biophotolysis is shown in Figure 2. Hydrogenase enzyme is very sensitive to presence of oxygen, hydrogen production only can occur under anaerobic conditions (under 0.1% O_2)[13].

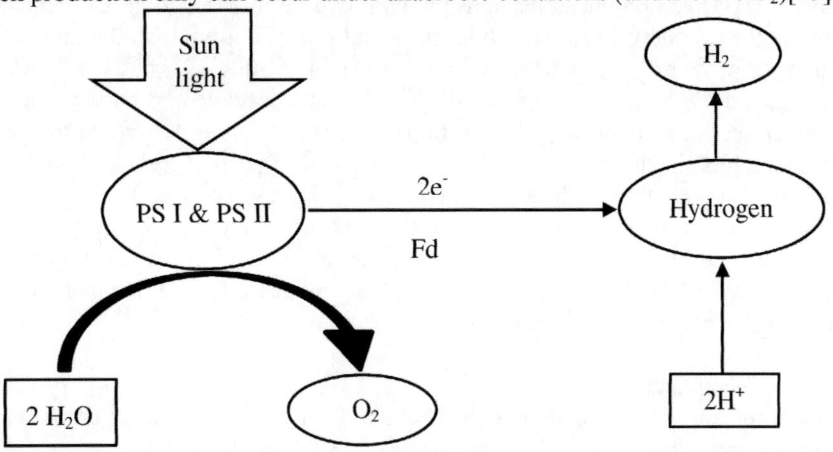

Figure 2. Schematic flow diagram of hydrogen production in direct biophotolysis process.

2.3. Dark fermentation

Dark fermentation is known as active processes operate biologically in absence of light sources. Hydrogen production is conducted in low temperatures (30-80 °C) and in dark condition by anaerobic bacteria and microalgae. These microorganisms consume rich carbohydrate substrates and produce hydrogen under mentioned conditions[14]. A schematic of dark fermentative hydrogen production process showed in Figure 3[15].

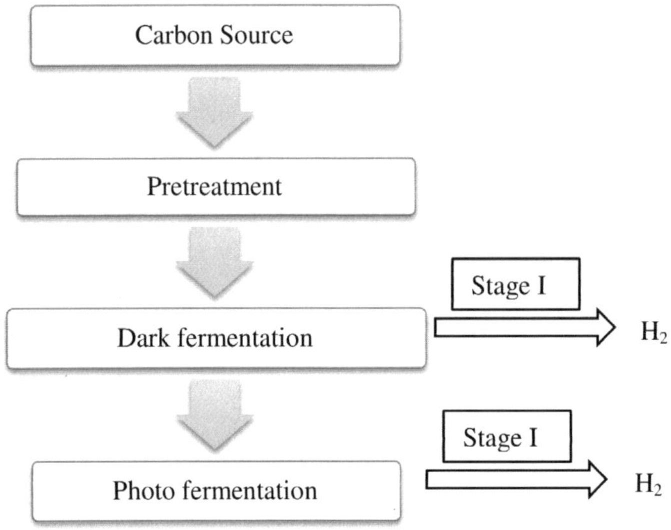

Figure 3. Dark fermentation hydrogen production process.

Products of dark fermentation process are combination of hydrogen, carbon dioxide, methane or hydrogen sulfide. The fraction of methane and hydrogen sulfide depends on the substrate used for hydrogen production. Hawkes et al.[16] conducted studies on optimum hydrogen production in dark fermentation and proposed reactions (6) and (7) for hydrogen production,

$$C_6H_{12}O_6 + 2H_2O \rightarrow 2CH_3COOH + 4H_2 + 2CO_2 \quad (6)$$
$$C_6H_{12}O_6 + 2H_2O \rightarrow CH_2CH_2CH_2OOH + 2H_2 + 2CO_2 \quad (7)$$

The mole of hydrogen produced depends on byproduct, in first reaction (6) acetic acid and 4 moles of hydrogen is the product of process while in second one (7) butyrate is the unwanted byproduct of reaction. In practice no one claimed that achieved to yield of 100%. For the low hydrogen production; it can be explained by the presence of organic acids such as acetate and butyrate in products[16]. An important parameter in amount of hydrogen in product is pH value. According to literature, pH value should be maintained in the range of 5-6[17-19]. Hydrogen partial pressure and HRT are the other important factors in producing hydrogen from the above process.

In a study which was conducted by Van Niel et al.[20] inhibitors of hydrogen production in dark fermentation studied and they reported that after accumulation of hydrogen the pathway of microorganism changed and more reduced substrates produced; lactate, ethanol, acetone and butanol or alanine are some of reduced substrates were reported.

2.4. Indirect biophotolysis

Indirect biophotolysis consists of four stages[21] that sum of them leads a microorganism to produce hydrogen by absorbing light. The four stages are stated as follow:

1. producing biomass by photosynthesis
2. concentrating the produced biomass from first stage
3. aerobic dark fermentation of biomass
4. conversion of acetate which has been produced in third stage

these four steps can takes place inside a microalgae or cyanobacteria cell[22]. The schematic of indirect biophotolysis, illustrated in Figure 4[15].

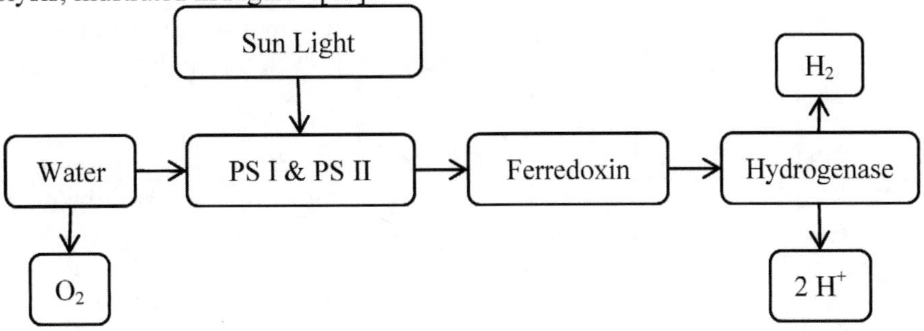

Figure 4. Schematic flow diagram of indirect biophotolysis hydrogen production process

The following reactions are summarized series reactions for production of biohydrogen from indirect biophotolysis for instance in *Cyanobacteria* photosynthesis may generate glucose then the carbohydrate is converted to hydrogen; the reactions are stated as follows:

$$H_2O + CO_2 + \text{light energy} \rightarrow C_6H_{12}O_6 + O_2 \quad (8)$$
$$C_6H_{12}O_6 + H_2O + \text{Light energy} \rightarrow H_2 + CO_2 \quad (9)$$

The required substrates for cyanobacteria for hydrogen production is light as source of energy for producing biomass, air as a source of O_2, water and mineral salts[23]. In this process three enzymes are involved namely: nitrogenases, uptake hydrogenases and bi-directional hydrogenases. These enzymes catalyze the production of H_2 via reduction of nitrogen to ammonia; while catalyzing the H_2 oxidation, the oxidation and synthesis of H_2 [24]. Number of investigations conducted on optimization of hydrogen production by indirect biophotolysis process. Beside the strain of cyanobacteria, the most influential parameters are temperature, pH, gaseous substrate [24, 25], culture age, biomass density and composition of growth medium. For instance increasing the temperature from 30 to 40 °C can enhance (even double) the hydrogen production rate, and optimal pH value in the range of 6.8-8.3 may enhance the hydrogen production [26-28].

2.5. *Photo-fermentation*

Photosynthetic bacteria have the ability to produce hydrogen with the aid of nitrogenase while utilizing biomass or organic acids as carbon source. A schematic diagram of hydrogen production via photo fermentation is shown in Figure 5 [15]. In this process hydrogen production has some draw backs such as high energy demand, low energy conversion of sun light and demanding large areas for the location of anaerobic photobioreactors [29-31]. Various type of biomass such as lactic acid [30], lactate feedstock wastewater [32, 33], sugar refinery wastewater [34] are used for production of hydrogen by photo fermentation. Disadvantage of photo fermentative process is known as an impractical hydrogen production process.

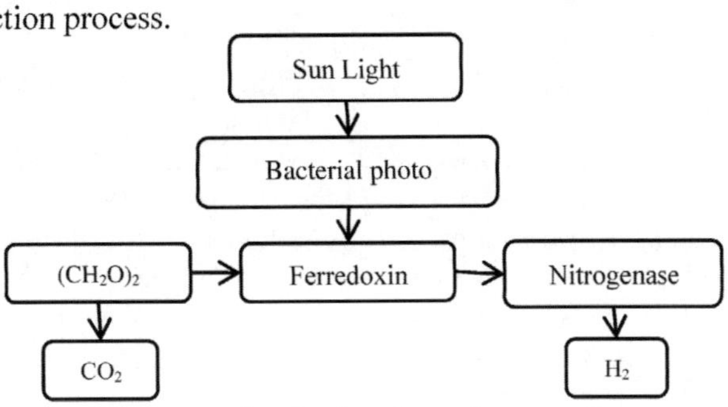

Figure 5. Hydrogen production by photo fermentation process

3. Potential hydrogen producing microorganisms

Four types of microorganisms can produce hydrogen: green algae, cyanobacteria, photosynthetic bacteria and fermentative bacteria. These microorganisms can produce hydrogen in different manner and each organism has its advantage and disadvantages[35]. Table 1 summarized advantages and disadvantages of these potent microorganisms.

Table 1. Summary of variety of potential hydrogen producer microorganisms

Type of microorganism	Name of microorganism	Advantages	Disadvantages	Ref.
Green algae	*Scenedesmus obliquus*	Hydrogen production from water High sun energy conversion	Light requirement for hydrogen production Sensitive to oxygen presence	[36]
	Chlamydomonas moewusii			[37]
Cyanobacteria	*Anabaena azollae*	Hydrogen production from water by nitrogenase enzyme N_2 fixation	Inhibitory of O_2 for nitrogenase Presence of O_2 and CO_2 in the product gas Sun light requirement	[38, 39]
	A. variabilis			[40, 25]
	Nostoc muscorum			[41]
	N. spongiaeforme			[25]
	Westiellopsis prolica			[25]
Photosynthetic bacteria	*Rhodobacter sphaeroides*	Can use different substrate for hydrogen production Uses a wide range of light for hydrogen production	Light requirement for hydrogen production Water pollution problem by fermented broth Presence of CO_2 in product gas	[42, 43]
				[44]
	R. capsulatus			[45]
	R. sulidophilus			[46]
	Rhodopseudomonas sphaeroides			[47]
	R. palustris			[48]
	R. capsulate			[49]
	Rhodospirillum rubnum			[49]
	Chromatium sp.			[50]
	Miami PSB 1071			[50]
	Chlorobium limicola			[50]
	Chloroexu aurantiacus			
	Thiocapsa roseopersicina			[51]
	Halobacterium halobium			
Fermentative bacteria	*Enterobacter aerogenes*	Producing hydrogen without light demand Can use a wide variety of substrates as carbon source No need to oxygen Produces valuable byproducts such as butyrate	Water pollution problem by fermented broth Presence of CO_2 in product gas	[52]
				[19]
	E. cloacae			[53]
	Clostridium butyricum			[54]
	C. pasteurianum			[55]
	Desulfovibrio vulgaris			[55]
	Magashaera elsdenii			[54]
	Citrobacter intermedius			[53]
	Escherichia coli			

Table 2. Hydrogen production from renewable sources

Microorganism	Type	Substrate	maximum H_2 production (mmol H_2/h)	Products of process	Ref.
Oscillatoria sp. Miami BG7	Photosynthetic bacteria	Medium A expect NH_4	0.4	H_2, CO_2, O_2, biomass	[56, 57]
Anabaena cylindrica	Photosynthetic bacteria	Nitrogen-starved medium	1.2	H_2, O_2, biomass	[58]
Rhodopseudomonas capsulate	Photosynthetic bacteria	Lactate and nitrogen sources	5.3	H_2, CO_2, O_2, biomass, fatty acids	[59, 60]
Rhodospiriillium rubrum	Photosynthetic bacteria	Organic compounds	3.0	H_2, CO_2, O_2, biomass, fatty acids	[61, 60]
Citrobacter intermedius	Facultative anaerobic	Cellulose, starch, glucose	11	H_2, CO_2, biomass, fatty acids with high concentration	[54]
Enterobacter aerogens	Facultative anaerobic	Sugar cane	11.36	H_2, CO_2, biomass, fatty acids with high concentration	[60]
Enterobacter cloacae IIT BT-08	Facultative anaerobic	Sucrose containing medium	37.03	H_2, CO_2, biomass, fatty acids with high concentration	[19]

4. Enzymes role in hydrogen production

Fe-nitogenase and Mo-nitrogenase are the most important enzymes involved in pathway of anaerobic organisms liberate hydrogen. In fact protons are reduced to hydrogen while nitrogenase can catalyze nitrogen fixation to generate ammonia. Three types of enzymes are having significant role in hydrogen production in algae and *Cyanobacteria*: the acting enzymes are Hydrogenase [62, 36], uptake hydrogenase and nitrogenase [63].

4.1 Hydrogenase

Hydrogenase is a family of enzymes which catalyze the hydrogen production reaction; the following reaction is catalyzed in presence of hydrogenase enzyme:

$$2H^+ + 2e^- \leftrightarrow H_2 \qquad (10)$$

Most prokaryotic mircroorganisms and some of eukaryotes can liberate Hydrogenase enzyme. These microorganisms can produce or even utilize hydrogen. Equation (10) shows the reaction occurs in presence of hydrogenase enzyme. The reaction is reversible and can goes in both directions. Production or consumption of hydrogen depends on presence of metallic compounds in hydrogenase enzymes. Iron and nickel are two metallic elements which are responsible for leading cells to produce or uptake hydrogen. The hydrogen producer microorganism enzymes usually contains iron and the hydrogen consumer organism contains both iron and nickel [64]. *Chlamydomonas reinhardtii* is a well-known algae species in hydrogen production which is able to liberate hydrogenase [65, 66].

Activity of hydrogenase enzyme was extensively investigated. Also comparison between activities of Hydrogenase in different species was conducted by number of researchers. Winkler et al. [67] reported that the activity of hydrogenase in *C. reinhardtii* was higher than *Scenedesmus sp.*

4.2 Nitrogenase

Nitrogenase in photosynthetic microorganism has significant role in production of hydrogen. High N/C ratio, presence of oxygen and ammonia may inhibit nitrogenase activity [68]. That is due to sensitivity of nitrogenase to photosynthetic microorganisms inhibitors need an oxygen free and ammonia limited media for hydrogen production [69, 70]. Effect of inhibiting materials on hydrogen production of photosynthetic microorganisms has been investigated. Yokoi et al.[71] conducted an investigation on *Rhodobacter* sp. M-19. It has been reported that presence of ammonia completely stopped hydrogen production; that was most probably due to significant changes made in pathway of organism for nitrogen fixation and ammonia generation. In addition, effect of presence of different proteins [72, 69], high nitrogen concentration [72] and high ammonia concentration [69] was investigated.

5. Bioreactors for hydrogen production

Several bioreactor configurations such as: large pounds and tubular photobioreactors for microalgae strains were experimented in hydrogen production. Benemann et al.[63] studied the feasibility of photobiological hydrogen production in large pounds and tubular photobioreactor. They have reported that the production rate of biomass in large pound was higher than tubular photobioreactor. Kumar et al.[73] investigated the effect of bioreactor on biohydrogen production they reported that rhomboid and tapered bioreactors resulted in higher hydrogen production rate than tubular bioreactors. In fact the gas hold up in rhomboid bioreactors was 67% lower than tubular bioreactor.

6. Conclusion

Based on evaluation of various carbon sources; hydrogen is evolved in bioconversion of syngas while utilization of suitable of organic substrate may lead to molecular hydrogen. In evaluation of organic compound through the cell active pathway, it was found that light sources may enhance production of hydrogen in anaerobic processes. Dark fermentation may result in active hydrogen production while biological pigments were synthesized when the light sources available to adsorb light energy for the synthesis of active biological molecules.

7. References

[1] Melis A, Zhang L, Forestier M, Ghirardi M L and Seibert M 2000 *Plant Physiology* **122**(1) pp 127-136
[2] Demirbaş A 2002 *Energy Sources* **24**(1) pp 59-68
[3] Grady J L and Chen, G J 1998 Bioconversion of waste biomass to useful products. Google Patents
[4] Jalan R and Srivastava V 1999 *Energy Conversion and Management* **40**(5) pp 467-494
[5] Ni M Leung, D Y Leung and M K Sumathy K 2006 *Fuel Processing Technology* **87**(5) pp 461-472.
[6] Mohammadi M, Najafpour G D, Younesi H, Lahijani P, Uzir M H and Mohamed A R 2011 *Renewable and Sustainable Energy Reviews* **15**(9) pp 4255-4273
[7] Phillips J R, Clausen E C and Gaddy J L 1994 *Applied Biochemistry and Biotechnology* **45**(1) pp 145-157
[8] Sipma J, Henstra A M, Parshina S N, Lens P N, Lettinga G and Stams A J 2006 *Critical Reviews in Biotechnology* **26**(1) pp 41-65
[9] Antoni D, Zverlov V V, and Schwarz W H 2007 *Applied Microbiology and Biotechnology* **77**(1) pp 23-35

[10] Soboh B, Linder D and Hedderich R 2002 *European Journal of Biochemistry* **269**(22) pp 5712-5721

[11] Najafpour G, Younesi H, and Mohamed A R 2006 *Energy Sources, Part A* **28**(11) pp 1013-1026

[12] Ismail K S, K Najafpour G, Younesi H, Mohamed A R, and Kamaruddin, A H 2008 *Biochemical Engineering Journal* **39**(3) pp 468-477

[13] Hallenbeck P C and Benemann J R 2002 *International Journal of Hydrogen Energy* **27**(11) pp 1185-1193

[14] Lin C Y and Jo C H 2003 *Journal of Chemical Technology and Biotechnology* **78**(6) pp 678-684

[15] Imam J, Singh P K and Shukla P 2013 *Biofuel Technologies*: Springer pp 301-315

[16] Hawkes F, Dinsdale R, Hawkes D and Hussy I 2002 *International Journal of Hydrogen Energy* **27**(11) pp 1339-1347

[17] Fang H H and Liu H 2002 *Bioresource Technology* **82**(1) pp 87-93

[18] Khanal S K, Chen W.-H, Li L and Sung S 2004 *International Journal of Hydrogen Energy* **29**(11) pp 1123-1131

[19] Kumar N and Das D 2000 *Process Biochemistry* **35**(6) pp 589-593

[20] van Niel E W, Claassen P A and Stams A J 2003 *Biotechnology and Bioengineering* **81**(3) pp 255-262

[21] Gaudernack B 1998 *Annual Report National Renewable Energy Laboratory, Golden, Colorado* pp 57-63

[22] Levin D B, Pitt L and Love M 2004 *International Journal of Hydrogen Energy* **29**(2) pp 173-185

[23] Hansel A and Lindblad P 1998 *Applied Microbiology and Biotechnology* **50**(2) pp 153-160

[24] Tamagnini P, Axelsson R, Lindberg, P, Oxelfelt F, Wünschiers and Lindblad P 2002 *Microbiology and Molecular Biology Reviews* **66**(1) pp 1-20

[25] Vyas D and Kumar H 1995 *International Journal of Hydrogen Energy* **20**(2) pp 163-168

[26] Dutta D, De D, Chaudhuri S and Bhattacharya S K 2005 *Microbial Cell Factories* **4**(1) pp 36

[27] Hallenbeck P C 2012 Microbial technologies in advanced biofuels production: Springer pp 15-28

[28] Smith G, Ewart G and Tucker W 1992 *International Journal of Hydrogen Energy* **17**(9) pp 695-698

[29] Argun H and Kargi F 2011 *International Journal of Hydrogen Energy* **36**(13) pp 7443-7459

[30] Fedorov A, Tsygankov A, Rao K and Hall D 1998 *Biotechnology Letters* **20**(11) pp 1007-1009

[31] Tao Y, Chen Y, Wu Y, He Y and Zhou Z 2007 *International Journal of Hydrogen Energy* **32**(2) pp 200-206

32] Tsygankov A, Fedorov A, Laurinavichene T, Gogotov I, Rao K and Hall D 1998 *Applied Microbiology and Biotechnology* **49**(1) pp 102-107

[33] Zhu H, Suzuki T, Tsygankov A A, Asada Y and Miyake J 1999 *International Journal of Hydrogen Energy* **24**(4) pp 305-310

[34] Yetis M, Gündüz U, Eroglu I, Yücel M and Türker L 2000 *International Journal of Hydrogen Energy* **25**(11) pp 1035-1041

[35] Das D, Veziroğlu T N 2001 *International Journal of Hydrogen Energy* **26**(1) pp 13-28

[36] Schnackenberg J, Schulz R and Senger H 1993 *FEBS letters* **327**(1) pp 21-24

[37] Greenbaum E, Lee J and Tevault C 1996 National Renewable Energy Lab., Golden, CO (United States)

[38] Banerjee M, Kumar A and Kumar H 1989 *International Journal of Hydrogen Energy* **14**(12) pp 871-879

[39] Shi D J, Brouers M, Hall D and Robins R 1987 *Planta* **172**(3) pp 298-308

[40] Tsygankov A, Serebryakova L, Sveshnikov D, Rao K, Gogotov I and Hall D 1997 *International Journal of Hydrogen Energy* **22**(9) pp 859-867

[41] Spiller H, Ernst A, Kerfin W and Böger P 1978 *Zeitschrift für Naturforschung C* **33**(7-8) pp 541-547

[42] Fascetti E, D'addario E, Todini O and Robertiello A 1998 *International Journal of Hydrogen Energy* **23**(9) pp 753-760
[43] Sasikala C, Ramana C, and Rao P R 1995 *International Journal of Hydrogen Energy* **20**(2) pp 123-126
[44] Krahn E, Schneider K and Müller A 1996 *Applied Microbiology and Biotechnology* **46**(3) pp 285-290
[45] Peng Y, Stevens P, De Vos P and De Ley J 1987 *Journal of General Microbiology* **133**(5) pp 1243-1247
[46] Singh S and Srivastava S 1991 *International Journal of Hydrogen Energy* **16**(6) pp 403-405
[47] Vincenzini M, Materassi R, Tredici M and Florenzano G 1982 *International Journal of Hydrogen Energy* **7**(3) pp 231-236
[48] Vrati S and Verma J 1983 *Journal of Fermentation Technology* **61**(2) pp 157-162
[49] Zürrer H and Bachofen R 1982 *Biomass* **2**(3) pp 165-174
[50] Gogotov I, Zorin N and Serebriakova L 1991 *International Journal of Hydrogen Energy* **16**(6) pp 393-396
[51] Khan M T and Bhatt J 1991 *International Journal of Hydrogen Energy* **16**(10) pp 683-685
[52] Tanisho S, Wakao N and Kosako Y 1983 *Journal of Chemical Engineering of Japan* **16**(6) pp 529-530
[53] Tanisho S, Suzuki Y and Wakao N 1987 *International Journal of Hydrogen Energy* **12**(9) pp 623-627
[54] Brosseau J and Zajic J 1982 *Journal of Hydrogen Energy* **7**(8) pp 623-628
[55] Bothe H and Kentemich T 1990 Pergamon Press, New York
[56] Kumazawa S and Mitsui A 1981 *International Journal of Hydrogen Energy* **6**(4) pp 339-348
[57] Phlips E and Mitsui, A 1983 *Applied and Environmental Microbiology* **45**(4) pp 1212-1220
[58] Weissman J C and Benemann J R 1977 *Applied and Environmental Microbiology* **33**(1) pp 123-131
[59] Hillmer P and Gest H 1977 *Journal of Bacteriology* **129**(2) pp 724-731
[60] Tanisho S 1996 *Hydrogen Energy Progress* **3** pp 2601-2606
[61] Ormerod J G, Ormerod K S and Gest H 1961 *Archives of Biochemistry and Biophysics* **94**(3) pp 449-463
[62] Greenbaum E 1988 *Biophysical Journal* **54**(2) pp 365
[63] Benemann J 1997 *International Journal of Hydrogen Energy* **22**(10) pp 979-987
[64] Adams M W and Stiefel E I 1998 *Science* **282**(5395) pp 1842
[65] Ghirardi M L, Zhang L, Lee J W, Flynn T, Seibert M, Greenbaum E and Melis A 2000 *Trends in Biotechnology* **18**(12) pp 506-511
[66] Melis A 2002 *International Journal of Hydrogen Energy* **27**(11) pp 1217-1228
[67] Winkler M, Hemschemeier A, Gotor C, Melis A and Happe T 2002 *International Journal of Hydrogen Energy* **27**(11) pp 1431-1439
[68] Koku H, Eroğlu İ, Gündüz U, Yücel M and Türker L 2003 *International Journal of Hydrogen Energy* **28**(4) pp 381-388
[69] Takabatake H, Suzuki K, Ko I.-B and Noike T 2004 *Bioresource Technology* **95**(2) pp 151-158
[70] Zhu H, Wakayama T, Asada Y and Miyake J 2001 *International journal of hydrogen energy* **26**(11) pp 1149-1154
[71] Yokoi H, Tokushige T, Hirose J, Hayashi S and Takasaki Y 1998 *Biotechnology Letters* **20**(2) pp 143-147
[72] Oh Y-K, Seol E-H, Kim M-S and Park S 2004 *International Journal of Hydrogen Energy* **29**(11) pp 1115-1121
[73] Kumar N and Das D 2000 Proceedings of the 13th World Hydrogen Energy Conference: Beijing: Science Press